PERFUMES

ART, SCIENCE AND TECHNOLOGY

PERFUMES

ART, SCIENCE AND TECHNOLOGY

Edited by

P. M. MÜLLER and D. LAMPARSKY

Givaudan Research Company Ltd, CH-8600 Dübendorf, Zürich, Switzerland

Springer-Science+Business Media, B.V.

First edition 1991
Reprinted 1994

Originally published by Blackie Academic & Professional in 1994
Softcover reprint of the hardcover 1st edition 1994

ISBN 978-94-010-5701-1

A catalogue record for this book is available from the British Library

Library of Congress Cataloging-in-Publication Data

Perfumes: art, science, and technology/edited by P. M. Müller and D. Lamparsky
p. cm.
Includes bibliographical references and index
ISBN 978-94-010-5701-1 ISBN 978-94-011-3826-0 (eBook)
DOI 10.1007/978-94-011-3826-0
1. Perfumes. I. Müller, P. M. (Peter M.) II. Lamparsky, D. (Dietmar)
TP983.P365 1991
668′.54—dc20 90-14076
CIP

Preface

The idea of publishing this book on *Perfumes: Art, Science and Technology* grew out of the observation that, on the verge of the 1990s, there was really no state-of-the-art compilation of the relevant know-how on which the fragrance industry is based. It was obvious that such a compilation would be well received, not only by perfumers and fragrance chemists, but also by those involved in related trade and marketing or in the development and distribution of consumer products, by researchers from other fields, by students and, finally, by amateurs of perfumes in general. Therefore, we set out to find competent authors who were willing to contribute to the endeavour, and we did not do this unselfishly; on the contrary, we selected a wish-list of specialists who would provide us with new insight and characterize the trends and research priorities determining the future. Thus, we were counting on learning much ourselves in the course of the project.

We were more than pleasantly surprised by the reactions to our first letter—and so was Elsevier. We certainly had not expected perfumers who are usually much more 'doers' than 'writers' to react in such an enthusiastic way; especially, the spontaneous affirmative answer from the famous E. Roudnitska created a momentum which contributed significantly to the successful completion of this book. But, of course, we should not create the impression that the other authors' chapters are less important, and we thank all of them heartily for their invaluable effort.

The reader who is just glancing quickly at the different chapters might come to the conclusion that this is a rather inhomogeneous book. Chapters which are loaded with literature citations are followed by chapters with none, and even the language changes: one is taken from philosophical perfection to economical bluntness, from artistic non-conformism to scientific heaviness. However, this is not a disadvantage; rather, it underlines the diversity and the fascination of the topic which, after all, is the motivation for those active in the field to dedicate their life to the study of odorants or odours.

The chapters of the book can be read individually, and they all contain a short introduction which sets the stage. It is also possible, however, to read the book as a whole, and the reader who does this will find that its structure has a certain logic. Thus, we hope that the readers who are 'smelling analytically' or reading single chapters enjoy this book as much as those who are 'approaching the composition as a whole'.

We want to conclude this preface by thanking the publishing house, as well as Givaudan. The encouragement and generous support of the latter were important elements for the task.

Peter M. Müller
Dietmar Lamparsky
Dübendorf, June 1990

NOTE ON THE USE OF TRADEMARKS

The use of trademarks in this book does not constitute unreasonable use of these names—even though they may not everywhere be indicated by the symbol ®. Tradenames, especially of perfumes, can be used to designate either the actual product in question or a type of product or family of products not identical with the named original.

Contents

I PERFUMERY AS AN ART

There could be many possible entrées to the topic of 'perfumes: art, science and technology' and it is, therefore, not without significance that the first part of this book is devoted to the art *of perfumery. One might think that the editors just chose the most esoteric and difficult part first with the intention that afterwards the book would be 'downhill work' for the reader, but this was certainly not the case. Aspects of art will come up again and again, and it is actually quite impossible to separate art, science and technology in the field of perfumery. The three categories are too strongly related.*

Chapter 1

The Art of Perfumery

EDMOND ROUDNITSKA

F-06530 Cabris, France

1 THE ROOTS AND THE EVOLUTION OF PERFUMERY

Although history attributes the first perfume container to the period of Darius III (4th century BC), the roots of perfumery reach much deeper into the past.

The first human beings had to rely on their sense of smell to survive, forcing them to recognize the various scents nature provided: animals, of course, but also plants, flowers, trees, fruits, grains, roots and resins. Their awareness of *olfactory abilities* had awakened.

Since odours have a tremendous influence on flavour, people applied their olfactory abilities to the preparation of meals, to 'perfume' their food, even before the term had been created to communicate such impressions.

Condiments, spices and herbs, whose properties they eventually discovered, just like those of grains, fruits and bark, rapidly extended their olfactory know-how. To improve their diet, they extracted oil from oleogenous grains and—fate perhaps lending a hand—they macerated vegetable and herbaceous ingredients into it. These ingredients gave their properties and odours to the oil. It is safe to say that the first rose oil probably dates back somewhat further than we might expect.

Usefulness not being dissociated from pleasure, nutritional or therapeutic values, it was easy to move to adornment perhaps even before the idea of religious offerings was born, because feminine seduction no doubt goes back to Eve. The first representatives of *Homo sapiens* were so observant and sensorially awake, by necessity, that they probably benefited and took pleasure from everything derived from these fragrant materials.

Before he even knew how to make fire, man had already learned from natural fires that heat transformed the odour and flavour of vegetables and flesh. After the art of making fire was discovered, initial applications were no doubt of a culinary nature. However, the fragrant smoke of burning herbs rising to the sky could only trigger the spirit of those first religious practitioners . . . How not to associate smoke and odour? *Per fumar*, *perfumare*, would one day become the term 'perfume'.

Mastering fire and all the progress it engendered—pottery among other skills—led to the construction of the first distillation equipment. The remains of one still were found in Mesopotamia and are said to be 5000 years old. In other words, the Arabs actually did little more than rediscover the alembic.

Used in metaphysics in Syria and Egypt as well as for religious purposes in India, scents and fragrances—except for a few historical setbacks—would become invaluable. Having discovered the antiseptic values of fragrant balms and resins, the Egyptians 'embalmed' their most famous dead, the pharaohs.

In the Middle and Far East, the first application of fragrances in daily life took the form of beauty products rather than perfumes. Ointments, creams and make-up preceded the use of odorants simply for their fragrance.

Thanks to perfumed oils, and later to alcohol, a good extraction solvent and a practical vehicle for the first perfume mixtures, the era of 'smelling good' had begun. But perfumery as it is known today was still far away.

In fact, for several centuries, olfactory practices were limited to simple mixtures, to 'recipes'. The 18th and 19th centuries were years of excess, of terror and conquering wars, followed by the pretentious and ridiculous refinements of the 'Muscadins'. Only the end of the 19th century saw the birth of real perfumery, the result of an evolving industry: the production of odorous substances by chemical reactions. These odorous compounds, added to the old 'recipes', transformed them by setting new fragrant accents, often strong and long-lasting, leading to unprecedented combinations. Later, it would be discovered that some of these new fragrant components were actually hidden in highly complex natural essences from which they could not always be extracted individually. Now available in isolated form, they offer completely different possibilities than when drowned in a natural substrate such as an essential oil or an absolute.

The growing list of chemically defined odorous substances allowed man to enrich simple recipes and to create real 'formulae', which gave way to original olfactory 'forms' invented by formulators.

These formulators had worked with odorous materials of natural or synthetic origin much in the same way that musicians work with tones, or painters with colours. They had combined specific properties of their materials to create a completely new form and not a mere addition to existing forms. With their material, they had *composed* the *olfactory composition*. This stage represents the beginning of perfumery as it is known today; it has existed for just one century.

Of course, perfumery passed through various stages during the past 100 years: craftsmanship, industrialization and commercialization. It had to modernize and evolve under the diverse forces of a changing world. These factors need to be mentioned here in order to understand better the current situation in perfumery, worldwide.

Modern perfumery, therefore, was born at the very end of the 19th century, the initial 'compositions' combining for the first time both natural raw materials and synthetic chemicals. I will only cite the most famous, the still very much alive *Jicky*, created in 1889, whose incredible centennial was smothered by the clamours of a bicentennial. What a pity!

The real champion, the mastermind of this contemporary olfactory composition, François Coty, also gifted in marketing and sales, helped perfumery leave the first stage of rough sketches to enter the world of fine arts. He revolutionized not only compositions, but also bottles, packaging and sales methods. He was the first, and probably the only, real professional in perfume marketing.

Origan (1905, the year of 'fauvism') and *Chypre* (1917, the terrible year), were the glorious masterpieces which Coty revealed to his contemporaries. And *Muguet des Bois*, introduced in 1936, two years after his death but unmistakably one of his creations, was the most beautiful floral evocation ever cultivated on this theme—at a time when resources were limited.

Stimulated by his example, his peers eventually surpassed each other, each through his own ingenuity, each faithful to his own taste. Guerlain, Houbigant (*Parquet*, *Bienaimé*) and Caron (Ernest Daltroff) became renowned for their creations. In each of these perfume houses, the owners were tasteful and perceptive composers, exercising *complete* initiative and responsibility from concept to sale. Success was the sanction of united action. Unfortunately, these qualities were not transmittable and the future of these houses would suffer the consequences.

After World War I, having seen the success of perfumers, renowned fashion designers wanted to exploit this success to stabilize precarious fashion revenues. The first, Coco Chanel, following the failure of Poiret, was exceptionally fortunate to stumble upon Ernest Beaux, who had worked for the Rallet Perfumery in Russia before 1914 and had subsequently moved to La Bocca, Cannes, France, in 1917.

Beaux had created a perfume for Rallet which, under the label of *Rallet no. 1,* had been a failure. Rallet offered it to Coco Chanel and was her supplier during the initial years.† Its launching in 1922 or 1923 as *No. 5* by Coco Chanel (a novice in this field and with limited resources) was far from spectacular. The real introduction was made (probably in 1925) by the powerful Wertheimer Group (Parfums Bourjois) which took over the activities of 'Parfums Chanel'. Beaux, finally a member of the Chanel group, would be the author of several other masterpieces.

Fashion designer Jeanne Lanvin also tried her luck with a modest perfume, *My Sin*, in 1925. She as well would meet a great perfumer who would conjure the most spectacular tetralogy in the history of perfumery: *Arpège* (1927), *Scandal* (1932), *Rumeur* (1932) and *Prétexte* (1937).

Designers Marcel Rochas (in 1944) and Christian Dior (in 1947) followed the footsteps of their two colleagues and also tasted the sweetness of success. In these four cases, when the designer was no longer pulling all the strings, the composition of the perfumes was ensured by true professionals and their distribution was controlled either by the fashion designers or by professionals like Bourjois. The break with tradition was therefore limited and all artistic objectives were maintained.

However, the diversity of new brands introduced by amateurs not familiar with the clever nuances in perfumery, and financed by groups who did not respect the ideals of true perfumery, triggered dramatic changes. Perfume was no longer a respected work of art. It had been reduced to a commercial product sold under the most lucrative of conditions. Its production was given to the highest bidder, without setting any aesthetic priorities, marketing being the choice method to impose it on the public. The entire system was based on this principle, putting a rein on the creative wave of 'parfums à risque', as they were called by the new people in charge. However, the real risk lies in launching any old product; it is an established fact that nine launches out of ten are a failure.

Today, perfumery is uprooted. Only a revolution can redeem it.

2 THE PART OF HEDONISM

Hedonism is attributed to the School of Cyrene which was established by Aristippe (around 435–350 BC), a contemporary of Plato and a follower of Socrates. The School was continued by his daughter Arété and his grandson Aristippe-the-Young, known as the Metrodidacte, who—because of his contacts with the Sophistes and their cynicism—may very well have systematized his grandfathers's principles.

This philosophical doctrine which aimed for an ethic, was founded on the ideal of immediate impressions. It advocated the search for pleasure while

† The year of 1921, which is usually mentioned, must be the year of the perfumer's first creation in La Bocca.

avoiding pain, and preferred intensity to quality. Pleasure was always something good, even if its origin engendered shame.

The philosophy of (immediate) pleasure is the negation of all arduous training, either in sports or the arts, of any intellectual effort, of any creative constraint, of any devotion implying sacrifice, or of any generous altruism.

Those who would place perfume under the sign of hedonism would arbitrarily limit it to sensual and possibly vulgar aspects, completely neglecting the sensory aesthetic quality which has essentially characterized the great perfumes of the 20th century.

For most, hedonism translates into 'pleasant sensations'. In terms of olfactory sensations, a distinction must be made between the perception of the layman, of the perfumer and even of the experienced amateur. The fact that, to ordinary mortals, some odours are considered pleasant and others unpleasant can be attributed to a lack of education and to an innate taste which impedes the layman from making an objective and technical judgement on the value and usage of odours. However, after a short but inevitable emotional reaction, the motivated and experienced perfumer no longer distinguishes between pleasant and unpleasant smells. This is like the music composer who considers notes to be elementary forms which can be combined into intricate music. The composer no longer judges the notes but the rapport he has created between them. The same applies to perfumers. This is one reason why I studied Kant, who so eloquently distinguished between things pleasant and things aesthetic.

The perfumer judges odours in terms of *quality*, *intensity*, and *duration of perceptibility*. The latter must be distinguished from intensity even though duration is defined as intensity in time. *Volatility* is the fourth important component which plays a considerable role in the genesis of a composition, because without volatility there would be no perfume. I have stated the above by convention because these attributes arise not from odours (phenomenon of consciousness) but from the stimulus which determines them, or more precisely, from the reaction between the odorous molecule and the receptive macromolecule to which it will bind. This reaction triggers a change of the normal potential in the living cells which translates into electric impulses, a set of waves which are decoded by the brain. The consciousness of odours is born.

The four main attributes which characterize olfactory perception, namely quality, intensity, duration and volatility, can be grouped under an aesthetic modality which has nothing to do with hedonism. For convenience, quality is classified separately from the other three characteristics because each contributes to and enhances its very meaning.

The progress of physiologists and researchers in determining the intensity of perceived odours, and their mutual commitment to the idea of establishing and using correlation measurements between this intensity and the impact of the corresponding stimulus, are more than welcome.

By analogy, Dr Jacques Sauvan's work shows the future of simulators in perfumery, and the reader will find further details on page 37.

The distinction established above deals with the judgement given to 'odours' as materials. When it comes to appreciating a composition, a perfume, i.e. a complex mixture, the olfactory judgement is no longer linked to the same criteria. Olfactory forms of little and great complexity are now examined. A single chemical substance is a relatively elementary form which can show different discernible sides and shadows, either simultaneously or in the course of its evaporation. For example, methyl ionones can reveal the following facets, depending on the observed isomer: violet, iris, vetiver, wood, leather, liquorice.

Essential oils—even though they are used as elements of composition—are natural, highly complex mixtures containing up to several hundred defined chemical substances. Therefore, they must be judged and used as complex forms. Their use is all the more delicate as they are more complex and not all the elements of their chemical composition are known.

When examining a composition (a perfume), the layman will give a subjective judgement: 'I like it' or 'I don't like it'. The hedonic reaction will be foremost.

The well-informed amateur and the experienced professional artist will also react subjectively. However, they will proceed to a professional evaluation. If their senses of smell and taste have been developed methodically (which does not mean manipulated), they will forget the pleasant (or unpleasant) odour and concentrate only on its aesthetic features, on the construction which develops under their noses in an *objective* olfactory form.

This form must be considered as an entity. Is it incoherent or homogeneous, boring or original, does it emanate an impression of harmony, does it have relief and character, or is it flat? Is it dynamic (without being overwhelming, heady, or heavy)? Does it have volume, is it sufficiently clinging? At the very best, the perfume form will seem both pleasant and *beautiful.*

The true connoisseur of perfumes, like the true art connoisseur, will mentally rebuild any masterpiece with which he is confronted. Aristippe's original hedonism has nothing to do with this, unless it is seen as the search for aesthetic pleasure.

The reason why hedonism was given excessive importance in the judgement of perfumes for such a long time was the double conviction that perfume depended on a *subjective* reaction and that this reaction had to be agreeable. However, to say that the odour of geraniol is similar to that of the rose or that ionone smells like violet—opinions which are shared by most people—is no longer a subjective judgement but an *objective one.* The comparison of a sensation with another which everyone perceives in a more-or-less similar way makes it objective. Geraniol, an objective constituent of the rose smell, substantiates the fact that the 'rose odour' image is formed in a similar way by everyone.

3 HOW PERFUMERY IS LINKED TO THE OTHER FINE ARTS

The thread which links all fine arts arises from one common denominator; they are all born of motivation—a strong need to create a work of art and to do so freely. Artists want to create forms, each using his own specific material and methods—original forms to be recorded in the history of each field of art.

This shared ambition, this identical acceptance of often-unrewarding hardships, creates a link between the members of the world of fine arts, and leads to an understanding and appreciation of one another. Artists also use similar words to designate comparable but different works of art.

They are builders who construct their masterpieces with various materials. We can actually compare the fine arts with a building whose various crafts are managed by an architect. Could not the dream of a chief architect in charge of all the fine arts become reality?

Perfume, when it is a work of art, celebrates the *grand fête* of fine arts and gives quality to life. It is a means of expression like others, but it presents several peculiarities.

Perfume is one of the younger forms of art, less than a century old. Therefore, the general public is less informed on its constitution, forms and specific difficulties than on arts dating back thousands of years. This has led people to believe that it is more difficult to discuss perfume than other arts. We will soon see that this is not the case.

Perfume deals with smell which, just like taste, is a chemical sense. However, sight and hearing are physical and mechanical senses. This chemical nature makes the sense of smell more intimate and more subtle, and therefore amplifies its means of orchestration on human sensitivity. But all sensations, be they physical or chemical, are decoded by the human brain. The correct interpretation of one or the other sense is a question of practice, taste education and experience. Anyone genuinely interested in the evolution of perfume can get involved with no more difficulty than for other arts. It only takes longer because of the double complexity of the materials involved and of their combinations. Just as in music, the composition of perfume involves the most abstract of intellectual acrobatics.

Vocabulary

Sensations arise from external stimuli which excite peripheral cells of the human body. They are almost simultaneously decoded by the brain after a progression of extraordinary but efficient complex processes. This complexity explains the difficulty of creating a specific vocabulary for the various sensory reactions which evolve from such processes.

The problems of language in describing sensations or works of art and the difficulties in communication so readily admitted for odours and perfumes, generally apply to all forms of fine art. This is substantiated by the fact that to express ourselves we all use images which, curiously, are borrowed from our other senses. If a word is lacking in one domain because there is no precedent, the easiest thing to do is to look for an expression in another area which, by association, may also evoke the new form.

What do we say about colours? They are acid, soft, dull, violent, warm, cool, cold, flat, raw, faded, lively, sparkling, flashy, fresh, strident, neutral, sad, happy, aggressive, delicate, soothing, old, sentimental, brilliant, or matt. Blue can be either deep or pale, green is bright, rose is tender; colour X makes colour Y sing, creating special harmonies. There are also screaming colours, dingy greys and immaculate whites, and so on. What about associating colours with nature's products? Apricot, mahogany, amaranth, lemon, orange are examples. How could we not compare the odours of our own substances with those they evoke in nature? One can only speak of the unknown as a function of the known. If we say that geraniol smells like a rose, it is because the rose essence smells of geraniol, which is one of its ingredients.

What clearly shows that sensations of different modalities are related to one another is the fact that the same image words can be used to describe a melody: lively, alert, light, coloured, brilliant, happy, gay, sad, generous, deep, solemn, mournful, throbbing, sleep-inducing, flat, fresh, exciting, enchanting, smooth or suave (like a perfume), soft, savage, wild, choppy, smooth, nostalgic, heavenly, and so on.

Let us not forget that music, thanks to its given laws and mathematical measures, has an advantage over other arts: it has a specific technical vocabulary. Where the music composer has access to specific signs to designate the tone and duration of each note and where he can freely choose his tempo, the perfume composer has but one medium to express simultaneously quality, intensity, duration and volatility which have a mutual influence on the various components of the formula and on its entire olfactory congruence—as if the 'allegro' in a symphony would play an important role in the 'finale' an hour later!

The human voice lends itself to picturesque and 'colourful' imagery because it can be warm, refreshing, icy, drab, white, metallic, leathery, deep, ringing, strong, weak, piercing, whistling, thundering, hard, gentle, neutral, soft, tender, caressing, langorous, captivating, endearing, clear, sombre, cavernous, round, forthright, shrill, acid, sharp, angry, thick, cracked, broken, breaking, piping, thin, oily, altered, dull, tired, trembling, halting, worn-out, frail, smothered, fragile, light, rich, dry, flowing, flexible, mellow, high, low, shrill, vigorous, firm, shaken, tragic, biting, composed, etc. There are probably many other adjectives which 'depict' a voice, but all these are familiar to us and yet they are all foreign to acoustics, which could never provide so rich a vocabulary. This clearly shows that we can only describe things well if we do so with terms of imagery, provided the images are *correct*.

When expressing himself, the perfumer is subject to the same fate. He must be extremely careful in his choice of images and groupings and must avoid all approximations to avoid confusion.

If, in addition to terms of imagery, the colourist possesses over 200 references to the objects of nature, the perfumer has the same references to the odours of nature: herbs, flowers, woods, musky and animal scents, etc. However, when a perfume is said to be fresh, it generally contains products such as bergamot, lemon and mint which give an impression of freshness. A light perfume is rich in volatile products which seem to fly away because they are so light. A green perfume evokes the greenery of prairies, mown lawns, leaves crushed between the hands. A heavy perfume taking time to diffuse itself because it includes several tenacious components like vanillin, musks, or essential oils such as patchouly, sandalwood, labdanum which are related to crystals, gives an impression of compactness and heaviness. Flowery perfumes obviously evoke flowers but it is not always easy to distinguish which ones. And what about woody perfumes: to which types of wood do they correspond? The term powdery applies to perfumes rich in crystallized materials and is often used for cosmetic rice powders. At an advanced stage of evaporation, these perfumes also have an analogous generic form.

To understand what an aldehydic perfume is, i.e. a perfume with a high dosage of aldehydes, the user must know what aldehydes are. The inexperienced amateur who knows nothing about fatty aldehydes can still understand this term by taking a look at *Chanel No. 5*, the first of a series of strong aldehydic perfumes.

This is all a little vague and provides but a blurred image of perfume. To be more explicit about the form of his perfume, the composer should be able to express himself freely, without reservation. He can mention the idea which inspired him but secrecy and prudence towards plagiarists will stop him from divulging his special ingredients—let alone defining their combination.

The musician can describe his composition in detail since its score protects it against imitators. According to the French law of 11 March 1957,[1] great perfumes could also by copyrighted. However, no House which has ever launched a true creation has dared to enforce this law against imitators. Although this would not stop chemical analysts from violating formulae, it would forbid them from profit-making, slow down hasty product launchings and clear the market of imitations. The general public could become connoisseurs again and play a more active role in perfume selection. A jury of qualified experts could expose a perfume which has drawn too much inspiration from another.

To describe olfactory sensations objectively, their corresponding properties must be known chemically, physically and electronically, not forgetting the knowledge of forms. We will then be able to create neologisms (as in music) by fitting a word to an olfactory sensation arising through a chemical substance, a word which will invariably become the *objective* term to describe the odour of this well-defined aroma chemical.

Professor Maurice Chastrette, a chemist-physicist at the University of Lyon, France,[2,3] in cooperation with highly qualified research workers including members of the perfume industry, is looking into possible methodical solutions. By gaining a better understanding of how odorous molecules fix themselves to receptive macromolecules, they can make a rational and accurate selection of descriptive terms to be used in perfumery. These studies have already proved very encouraging.

Meanwhile, we still have ways of communicating our olfactory sensations—our emotions in this field. Communication requires the same effort of imagination needed for creation. A perfumer who has been able to conceive and create a form must find the words to explain it to himself and above all to others. This calls for practice but is beneficial to the composer because such a second effort enables him to understand better what happened during the process of creation.

Though all artistic activities are somehow linked by an invisible thread, the means of expression are peculiar to each art as a function of what it represents. However, each art category has its own technical processes and every artist within a category strives to create his own personal technique.

For this reason, it would be wrong to seek too close a set of analogies from music or from painting for the art of perfumery. Nevertheless, they all share the same aim: harmony. But the perfumer works in a very special universe, one of odours which depend immensely on subtlety.

In the light of these difficulties, the perfumer trainee must develop his taste in conjunction with the other arts. He must be aware of them to be constantly immersed in an aesthetic atmosphere, to see his material in a privileged light. Only then will his materials cease to be enemies and become strong allies.

4 THE ART OF COMPOSITION

Introduction

The composition of perfumes only becomes an art when certain requirements have been met.

A perfume is a 'composition' of odorous products of various origins, diluted in an intelligent proportion (depending on the structure of the formula) with ethyl alcohol, which is necessary to show the composition's value. This is not just a mixture because composing takes place before mixing. To *compose* is to conceive an original and interesting olfactory form. From this conception, a formula must be laboriously defined of the various components and their proportions. By mixing the ingredients of the formula, a concrete basis for the conceived olfactory form is established.

To give this composition an artistic value, its components must be carefully chosen and quantified so that their combination results in a specific, interesting, harmonious and recognizable olfactory form. Only when all these requirements have been met will the mixture turn into a perfume and the perfume into a work of art.

First Prerequisite: an Extended Study of Raw Materials

The perfumer works with several dozen natural essences (essential oils, absolutes, etc.) from various countries and uses thousands (the list is growing) of odorous chemical substances. A considerable number of observations, of well-perceived and correctly interpreted experiments, of efforts of memorizing materials and the multiple and complex 'olfactory' combinations available, are needed to master such an extensive palette. 'Olfactory' is used in the sense of 'aesthetic' because chemistry has nothing to do with it. The number of combinations achieved during a long career is minimal in the light of the infinite number of real possibilities. This is why intuition, as we are about to see, plays a major role in creation.[4]

Since it is impossible to experiment with all possible combinations, intuition will allow the composer to extrapolate from what is known to what is possible and then to what he has projected. Intuition, much like taste, evolves when in touch with the other arts. No education system favours intuition and it is not surprising that there are only a few great composers, perfume composition being neither a science nor a game of chance.

If composition is not a science, the study of its materials must be conducted very methodically to be faster, more efficient and better assimilated. Where there are no methods of composition, there are at least methods of working.

A good approach to memorizing the various odours is to group them in families or series. I initiated my trainees to 15 series: Citrus-like, Rose, Oranger-like, Jasmine, Violet–Orris, Aniseed, Aromatic, Green, Spicy, Woody, Tobacco-like, Fruity, Balsamic, Animal-like, Leathery. This order implies a certain logic: relatively simple series before fairly complex ones, better-known before less-known, light series before heavier and wilder series.

These series are limited to avoid dispersion and confusion. Tuberose can be a sub-group of Jasmine; Narcissus, Lily-of-the-Valley, Lily and Hyacinth can be a flowery sub-group of the 'Green' series; Carnation belongs to the Spicy series, Mimosa to the 'Aniseed', Jonquil to the 'Oranger-like' series, Gardenia to the 'Fruity' series, etc.

Each series consists of a selection of natural products and of their known main chemical constituents. We shall provide a detailed description of the operational approach to these 15 series below (see p. 18).

First, however, it is essential to know the strict disciplines, without which there would be no intelligent, memorizable and profitable olfaction. It is very important to know *how* to smell.

How to Smell

Material

The high number of olfactory analyses that composers conduct on a daily basis, and the essential principle of following the evolution of odorants during long work days, make it impossible for us to implement instruments used by researchers in laboratories of sensory physiology.

Our olfactory testing methods must be simple and practical. The only criticism we hold against our 'smelling-strip' method is not that it lacks in scientific severity but that it has nothing to do with the support materials usually playing a role. This forces us to make a mental effort of transposition, which will become familiar after initial complications, but which does not exclude occasional errors of interpretation or which hides important effects such as *volume*.

The 'strips' used are of white blotting paper, 18 cm long and 1 cm wide. They are folded in two to give them the needed rigidity. One of the extremities is tapered and grooved. Thus, it is easy to insert them into small flasks and only a minimum amount of a given product is adsorbed. We felt that 18 cm was the most acceptable length to ensure that the heat of hands would not impede the olfactory process and that the strips would not bend. The paper weighs 180 g/m^2 (gsm) and must be free of glue.

For delicate tests we use strips of the same paper also 18 cm long but 2·5 cm wide. These strips are neither folded nor cut because we do not dip them into the product being tested. Using a pipette, we place a 20 mg drop on one extremity of the strip so that the drop can spread itself out horizontally on the paper *without extending to the edges of the strip*. The pipette must be held vertically so that the drop is always the same size, and close to the paper (without touching it) to have the shortest fall possible. When the drops spread horizontally, they have a similar surface, ensuring a homogeneous evaporation period and a good comparison basis from test to test. This process, if applied carefully, allows observations of remarkable precision for the experienced nose.

When it comes to comparing two odorants, two persons must work simultaneously so that the strips adsorb the liquid drops coming from the two pipettes at precisely the same time.

It goes without saying that *the strips must be kept away from all odorant products*. Further, it would be most useful for perfumers to agree on the quality and shape of the strips; then the way they judge odorants has a better chance of being comparable.

Method

The composer should practise smelling only *minute* quantities of diluted products.† The olfactive epithelium and its mucous layer thus regenerate more rapidly and can be used more often with increased fidelity. The sense of smell will sharpen, become more refined and be able to discern delicate nuances.

We therefore recommend smelling alcohol bases, diluted according to the intensity of the odorant being tested. This will ensure a more valid judgement. Dip only the tip of the strip into the fragrance and record the time and day of the operation. Breathe evenly during the olfactory process. Smell the odorant with slow deep breaths depending on the intensity of the odorant and on the necessities of your own capacity of concentration: smell only a short time, in a series of one to three inhalations of two to three seconds (sensitivity is soon

† This pertains especially to raw materials. For finished compositions, the ideal concentration of each will be selected for sniffing, first on the blotter and, if possible, by spraying on human skin.

blunted). Quickly write down impressions in the order received. To do well, the atmosphere must be *odourless, temperate, normally humid* and *calm.* Perfect concentration calls for solitude and silence. Smelling is difficult in an atmosphere which is either too cold, too dry or draughty. Take long enough breaks between each series to allow the mucous layer to regenerate. Oxygenation by breathing deeply leads to a rapid regeneration. Before and after each olfactory series, breathe fresh air—take a few steps out-of-doors if possible. The observation of these suggestions has a vital influence on the value of judgements and, by extension, on experience.

Oxygenation is just as much linked to respiration techniques as to the quality of the air. Very few people know how to breathe correctly, even those who have a healthy respiratory system! To breathe correctly means to breathe in deeply through the nose and to breathe out through the mouth to eliminate all the stale air.

Many people have a maldevelopment of the nasal septum, which impedes breathing and considerably affects their olfactory perception. They should not hesitate to consult an otorhinolaryngologist and, if necessary, to undergo surgery. They could only benefit from such an operation, both physically and professionally. It seems quite superfluous here to insist on the harmful effect of stimulants, alcohol and tobacco in particular, on sensory finesse and on memory.

The best olfactory results are achieved in the morning after coming in from the fresh air. Save this moment for precious products whose evolution can be studied conveniently during the first hours of evaporation.

If several series of smelling-strips have to be examined, start with those where the odorants are most evaporated, i.e. the weakest odorants, and work progressively through the more odorant blotters (take breathing breaks between series).

An odorant may have several 'faces', or several nuances. The observer must remain objective and not linger on any one pleasant or unpleasant characteristic as this *could completely prevent him* from discerning the other faces. The observer must make the difficult effort of *reducing his field of observation* to the odorant and *increasing his awareness* of it to capture its every detail. The perfumer's judgement must be objective, i.e. free from the emotional reactions, which mar the amateur's judgement. However, when contemplating his own masterpiece, he becomes a spectator and must make another difficult effort: he must put himself in other people's shoes.

To become very familiar with an odorant, it must not be smelled superficially. All of its faces which arise during the various stages of evaporation must be discovered. Initially, depending on its volatility, the odorant can evolve quite quickly. Therefore, during this first phase, it must be tested several times to ensure that all the various characteristics are recorded. Then the evolution slows down, allowing breaks between observations. In general, a certain number of tests are conducted during the first hour, followed by five or six others during the first day, two or three during the second day and one each morning as of the third day.† Detailed notes are taken at every observation on the odorant's quality and intensity and the test is continued until no smell is recorded at all. Testing should continue for a few days more, as human sensitivity is subject to lapses.

At times, an observer will notice major differences in his evaluation of a given product conducted at different times. These variations can be attributed to changing physiological conditions in relation to health, the air's hygrometric status, temperature, atmospheric phenomena, etc. I mentioned some of these differences in 1943,[5] differences easily explained by the fact that I had had a nasal operation between tests which had noticeably improved my olfactory sensitivity.

The study of an odorant is very delicate, the comparison of two very difficult and the confrontation of several an almost impossible feat. Cunning is called for. The difficulty in comparing is due to the fact that sensations are subject to the law of relativity. Each sensation is relative to those preceding and those accompanying it; each sensation also influences those which follow. Therefore, sensations introduce us not to a definite status but to *the relationship or the ratios between several states*.‡

When testing two smelling-strips, the observer cannot smell them at random. Reflection dictates an intelligent selection in numerous cases: if one of the strips has a weaker odorant it must be tested first. If smelled second it may seem weaker or even scentless because the dulled sense (physiological phenomenon) is joined by the law of relativity (psychological phenomenon). Both factors play an important role in habituation.

If equally intensive, the second strip will also seem weaker than the first. In the observer's mind, these variations in intensity inevitably translate into *apparent* qualitative differences. Practice, reflection and above all self-control help rectify the judgement, but this is not easy.

When comparing two strips, the duration of each olfaction must be very short so that the mucous layer is not saturated by the first olfaction. Take one or two 2-second inspirations before smelling the first strip, and then take a small break—a slow and deep breath to regenerate. The same procedure applies to the second smelling-strip. The break between the two olfactions must be such that the mucous layer can regenerate without losing the mental impression of the first olfaction. This is very subtle. The regeneration period should be as long as possible but—and this is the tragic fact—the olfactory memory loses much of its precision very quickly, thus calling for very short

† These figures are obviously not absolute and may vary. However, it is a fact that we smell each smelling-strip too often and for too long. We thus overburden our mucous layer and become poor judges.

‡ And we shall never have the same state of mind twice since the current minute is related to the memory of all passed minutes and since the future minute will relate to current ones, and so on. Therefore, the same stimulus cannot provoke the same sensation twice. And that is why, Madame, you will find that your perfume is not what it was yesterday. But there is nothing we can do to change that.

intervals between two successive olfactions. To escape this vicious cycle, the interval is limited to a few short seconds during which we breathe in deeply but calmly, all the while concentrating on the sensation. The slightest distraction and everything must be started over again. The interval must be longer if the olfaction was also longer. The speed of our reactions can only be helpful and total concentration is esssential. It is perfectly understandable when perfumers become irritated at being disturbed in their work. It is all the more so, now that we know that an immediate sensation can never be relived or repeated in exactly the same way.

After an initial confrontation, allow several minutes to pass and repeat the process, inverting the order of the strips. The blotters can now be smelled randomly to see if the arising impressions do not repeat themselves chronologically. If this is the case, the respective impression could be attributed to the law of relativity.

During these various confrontations the observer should avoid concentrating on the same 'facet' or peculiarity of each odorant. He must remain objective, go through all the possible facets, and confront them one after the other. He must make an inventory of each odorant within a very limited period of time.

Of course, confrontation series (double olfactions separated by longer breaks) which immediately follow the dipping process will suffer from the composition's too rapid evolution due to its volatility. The tester will not be able to conduct observations as calmly as he can once the evolution has slowed down. It will then be possible to find and verify an identical impression over a period of minutes or hours. This implies that testing conditions must be variable to ensure a diversity of viewpoints.

If there is no specific reason to smell one blotter before another, the tester should ignore the origin of the smelling-strip he is smelling, to avoid external influences. Therefore, two identical blotters should be selected and the reference numbers of the fragrances written down so that they cannot be seen during the olfactory process. The tester can then proceed to a four-step series.

The tester holds a smelling-strip in each hand and first smells the right-hand one (or the left) and observes, for example, that the right-hand strip is more flowery than the left. Then the order is reversed by smelling first the left-hand (or the right) strip. The tester notes whether the first test is confirmed or invalidated. The strips again change hands (still ignoring their references) and the first two actions are repeated as before, making the same mental notes. Thanks to these four repetitions, the beginning of an opinion on the possible infiuence of the order in which the strips are smelled and on the influence of the hand which holds it is formed. The tester looks at the references. If, in all four cases, the same blotter seems the most flowery, the conclusion is confirmed. If, however, there is a divergence of impressions, reasoning is called for. In any case, results must be recorded. Then, the blotters are mixed and a second experiment is started, also in four steps. If, after several experiments, the tester always selects the same strip, he can be certain that the experiment is conclusive. If most tests are in favour of one strip but there are a few olfactory accidents, the result will be less convincing but still valid; if there is a doubt, the odorants are either very similar or the judgement is particularly difficult to make (for example, vanillin).

Comparing several smelling-strips simultaneously can be misleading. To avoid confusion and fatigue, it is best to proceed by a rapid process of elimination until two strips remain. This process of elimination need not be final. It is perfectly possible that a strip might be eliminated during the dipping or first adsorption phase or shortly thereafter and that—if it has developed favourably—it is re-examined during evaporation. The final judgement must be harmonious, encompassing all the various phases of evaporation.

When conducting serious comparisons on several products, it is best to test two at a time, and to extend the testing period over as many days as necessary so that the elimination of the less interesting product of each pair is definite. This is a long procedure but the tester will be certain to have finally selected the two best products which he can then compare with a clear mind.

At the end of this part of the chapter, it is most important to repeat that all olfactory comparisons must be made in comparable conditions: age of concentrates and solutions, simultaneous dipping, quantity dropped on the blotters, and so on.

The Fifteen Most Important Series of Odours

Here now are the 15 series mentioned earlier, giving a comprehensive outline of the principal odorants to be used by beginners, who will of course have to complete each series progressively as they advance in their study of materials.

General Recommendations

1. Try to determine and record the quality and character of the odour (its note, its 'form', what it evokes or suggests); its intensity; its expansion mode (diffusion, volume); its stability or instability; the evolution of the note, its form in time (several days, several weeks); the duration of *perceptibility*. All these traits make up the *attributes* of the odour and give it a *personality*; they are inseparable and will have to be taken into account as a coherent whole. When introduced to a mixture, the odour ceases to be one entity and interacts freely with other odorous bodies.

2. Take note of everything that comes to mind, using the words which arise naturally: if they create an *image*, if they enable a thought to be more precise, if they *surround the contours* of the odours without ambiguity. Avoid 'almosts' at all cost. Try to find the words which unequivocally define the impression so that 20 years later, if confronted with the same impression, the same words come to mind. The odour file—especially in computerized form—can acquire great value by recording all olfactory thoughts and keeping them at hand with the name of the corresponding product.

If the initial olfactory process does not enable the tester to define clearly the impression with words, he must repeat the experiment at different moments and locations. However, do not forget that *only the first impression is pure.* It must therefore be protected and given particular attention. Write all initial, spontaneous reactions even if they may be subject to corrections later. It is particularly instructive to establish the type of judgement errors one has made, to learn to question oneself. This is how a good *critical judgement* is formed—a good portion of it is needed in this job! However, do not confuse spontaneity and impulsiveness.

These are the conditions which need to be created during the study of compositions. Intervals between tests need to be long enough to avoid obsession. Periodically suspend testing over a few months if possible, to try and 'forget' both forms and formulae. Hopefully, this will create a new and impartial judgement.

This is difficult and calls for a strong character, a solid dose of perseverance and honesty . . . what schooling! What an advantage it is to be objective, to have a secure judgement! Going from one topic of study to another, then coming back to the first, and so on, can be a profitable liberation.

3. The products of each series must be observed on the strips as long as an odour is perceptible in the morning after going outside and daily toilet, but before breakfast.

4. The tester must observe his own behaviour attentively, and how his olfactory epithelium responds to the treatments imposed on it. The strips should not be raised closer than 1 cm to the nostrils. The odorous molecules must reach the olfactory receiver and not the surrounding skin, where they would distort subsequent olfactory processes.

N.B. The following operations must be executed with precision, respecting the given timetable; otherwise they cannot be conducted in one morning. However, a general confrontation of all the elements of a series must be attained even though it is already compromised (this is a relative effect and weakened by the next morning) by the time gap between the groups. Full concentration, calm and precise movements are required. Execute brief olfactions limited to a few for each smelling-strip; take notes rapidly. Saturate only the tip of the strip; pay attention to the rapid saturation of the mucous layer. Airing periods should be frequent; they are a must for the regeneration of the mucous layer and for mental rest.

Group 1: Citrus-Like Odorants

8.00 a.m.	*Dilution* (%)
Lemon peel oil (Italy)	10
Bergamot peel oil	10
Bergamot (distilled)	10
Orange peel oil	10
Mandarin peel oil	10

(*continued*)

Group 1: Citrus-Like Odorants—*contd.*

8.45 a.m.	*Dilution (%)*
Bergamot oil	10
Linalool (synthetic)	10
Linalyl acetate (synthetic)	10
Terpineol	10
Terpenyl acetate	10
Limonene	10
9.30 a.m.	
Lemon oil	10
Lime oil	10
Lemongrass oil	10
Lemarome N (Givaudan)	1
Limonene	10
10.15 a.m.	
Orange oil Guinea	10
Orange oil Florida	10
Orange oil Bigarade	10
Decanal	0·1
Methyl anthranilate	1
Limonene	10
11.00 a.m.	
Mandarin oil	10
Decanal	0·1
Methyl anthranilate	1
Methyl *N*-methylanthranilate	1
Limonene	10
Dipentene (racemic)	10

Group 2: Rose Odorants

7.30 a.m.	*Dilution (%)*
Rose oil (Bulgarian type)	10
Rose oil (Moroccan type)	10
Rose de Mai absolute (Grasse)	10
Rose absolute (Morocco)	10
Geranium oil (Bourbon)	10
Geranium oil (Morocco)	10
Palmarosa oil	10

[Five minutes' break out-of-doors.]

Group 2: Rose Odorants—*contd.*

8.30 a.m.	*Dilution (%)*
(First smell the five alcohols just to know their differences in odour, then pass on to the corresponding esters.)	
Geraniol (natural, ex Citronella oil)	10
Geranyl acetate (natural, ex Citronella oil)	10
Geranyl propionate (natural, ex Citronella oil)	10
L-Citronellol	10
D-Citronellol	10
Citronellyl acetate	10
Citronellal	10
Rhodinol (ex Geranium Bourbon oil)	10
Rhodinyl acetate	10
Nerol	10
Neryl acetate	10
[Fifteen minutes' break out-of-doors.]	
10.30 a.m.	
Linalool (synthetic)	10
Linalyl acetate (synthetic)	10
Phenylethyl alcohol	10
Cinnamyl alcohol	10
Ethyl cinnamate	10
[Five minutes' break out-of-doors.]	
11.30 a.m.	
Eugenol	10
n-Nonanol	1
n-Nonanal	0·1
Styrallyl acetate	10

This series—being one of the richest in recognized constituents—will have to be completed with all the esters (acetates, formates, butyrates, valerates, phenylacetates, benzoates, etc.) of all rose alcohols.

For this most important group, 11 smelling-strips must be prepared. Follow the rules in 'How To Smell' carefully. The details there also apply to the other series, because *by not respecting the prescribed duration of each olfaction, it is impossible to make a valid judgement.* This could have serious repercussions.

There will be other opportunities to study each of these products individually: either through methodical studies of related chemical families, or by examining other groups which have common constituents.

Group 3: Oranger-Like Odorants

In this and the following tables, the products marked * contain nitrogen in their molecules and show discoloration when exposed to daylight.

8.00 a.m.	*Dilution (%)*
Orange flower (water phase)	
Neroli oil (Bigarade)	10
Orange flower absolute	10
Orange flower absolute (ex water phase)	10
'Eau de Broûts' absolute	1
Petitgrain oil Bigarade (Italy)	10
Petitgrain oil (Paraguay)	10
Petitgrain oil Citronnier (Italy)	10

[Ten minutes' break out-of-doors.]

9.00 a.m.	
Linalyl acetate (synthetic)	10
Methyl anthranilate	1
Aurantiol (Givaudan)*	1
Indol*	0·1
Indolene (Roure)*	0·1
Phenylacetic acid	1

[Ten minutes' break out-of-doors.]

10.00 a.m.	
Phenylethyl alcohol	10
Linalool (synthetic)	10
Nerol	10
Nerolidol	10
β-Naphthyl methyl ketone	1
β-Naphthyl ethyl ether ('Bromelia')	1
β-Naphthyl methyl ether (Yara-Yara)	1

As in the first two series, the concentrations of the substances vary between 10 and 0·1%. Always bear this factor in mind during the olfactory processes and take notes accordingly. A concentration of 10% should be considered a typical dilution. It applies to the majority of products and has the advantage of being at the right level so as not to traumatize or oversaturate the olfactive epithelium, leaving the objective evaluation unaffected. Further, it is not too different from perfume concentrations, thus enabling us to smell each product under evaporation conditions similar to those of perfume. This fact is extremely important in making a valid assessment.

The image of each product—the image that will stay in the memory of the composer—will be based on its chosen concentration (10%, 1%, 0·1%, depending on the intensity). This image contributes to the practical, regular and efficient judgement as well as the findings, classifications and mental operations of an experienced perfumer.

The composer will find this rule very useful when comparing various ingredients. When he thinks of linalool, it will be a 10% solution of linalool. With regard to its concentration, its form, etc., methyl anthranilate will always evoke the image memorized in the form of its 1% dilution. For example, he will be accustomed to estimating how many grams of a 1% solution of methyl anthranilate can be tolerated by 1000 g of linalool in a 10% solution. This is just the ABC of manipulation. One can indeed estimate the relative intensity of decylic aldehyde at 1‰ or 1%. However, it would be impossible to do so at a 100% concentration because its quality and intensity could not be appreciated at all in its undiluted form.

The empirical establishment of a table of ratios in which mixtures of two or three products would have a defined olfactory stability would be too simplistic and dangerous. Simplistic because the interactions, which arise at the heart of a tincture of 50 or 100 elements (corresponding to a perfume), cannot be measured in the same way as blends which have only two, three or four ingredients. What is the point of knowing that A and B are balanced if the rest of the alphabet makes this balance questionable? Especially since the composition of a perfume has nothing to do with balance itself. In fact, the constituents of a perfume formula are not just 'balanced', they are combined to create a characteristic olfactory form. (It will be more important to observe the chain reaction and correlation between the constituents than to worry about their balance in a formula.)

Such a table would be a dangerous tool because—in addition to the fact that it would demand hard work to establish—it would invariably be an easy way out. The composer who relies fully on such a table would combine his constituents in accordance with prescribed proportions. Not only are these proportions no longer applicable as soon as additional ingredients are added, the composer would never get accustomed to considering each ingredient's correlation to the others. He would be incapable of memorizing these relationships as means of establishing both the physiognomy and the possibilities of an ingredient's complete personality. The composer actually plays with this personality. It is this personality which he tries to assimilate, to integrate into his own nature, to become so familiar with it that he is able to play with it as he wishes. Once he knows an ingredient so implicitly, he can decide whether to introduce it to a new combination. He can foresee—intuitively—if this combination can work out or not, and how. Intuition is not a miracle; it is a flame which is sparked only if the necessary amounts of knowledge, experience, reflection and meditation are available. Intuition is the basis of a slow but steady progress in the study of perfumery. We shall look into intuition in more detail further on.

There are also six crystalline substances in this group. Even more than liquid products, the action (form, behaviour) of crystalline ingredients cannot be appreciated in a 100% concentration because their vapour pressure is lower. In their study, they must be associated with their crystallized form which plays a role in any composition. Crystallized substances often have a high boiling point and evaporate slowly. Thus, they tend to slow down the other constituents in a formula regardless of their individual tenacity.

Nevertheless, the behaviour of crystallized substances cannot easily be systematized; their image must above all reflect their specific action because they all have their own peculiarities. Heliotropine melts at 37°C and boils at 263°C. Coumarin melts at 68°C and boils at 291°C but its persistence is far greater than that of heliotropine, far greater than the difference of boiling points would presume. The tenacity of vanillin, which melts at 83°C and boils at 285°C, is even greater; it is one of the strongest of all odorous substances.

Menthol melts at 51°C and boils at 211°C, whereas camphor melts at 179°C and boils at 208°C. Despite their crystallized form, these odours are intensive and camphor is almost overpowering even though it only melts at 179°C. Another peculiarity of camphor is the fact that its boiling point is quite close to its melting point. Its liquid state is of short duration and it can sublime at normal temperatures. Despite their crystallized form, both these substances can be considered as having a volatility-increasing rather than a retarding effect in a composition.

The three nitro musks, i.e. musk ambrette (m.p. 83°C), musk xylene (m.p. 114°C) and musk ketone (m.p. 137°C) are extremely tenacious, especially the last of these. Their delaying action cannot be denied, although they 'detach' themselves too distinctly at the end of the mixture's evaporation and do not leave the impression of having retained the context of the composition.

Group 4: Jasmine Odorants

8.00 a.m.	*Dilution (%)*
Jasmin concrete (India)	2
Jasmin concrete (Egypt)	2
Jasmin absolute (India)	1·2
Jasmin absolute (Egypt)	1·2
Ylang-Ylang concrete	10
Ylang-Ylang extra oil (Comoro Islands)	10
8.30 a.m.	
Linalool (synthetic)	10
Linalyl acetate (synthetic)	10
Benzyl acetate	10
Benzyl propionate	10
Benzyl formate	10

Group 4: Jasmine Odorants—*contd.*

8.30 a.m.	*Dilution (%)*
Benzyl phenylacetate	10
Benzyl butyrate	10
Benzyl pentanoate	10
9.30 a.m.	
α-Amyl cinnamic aldehyde	10
α-Hexyl cinnamic aldehyde	10
Hedione (Firmenich)	10
cis-Jasmone	1
Benzyl salicylate	10
Nonan-1,3-diol acetate	10
10.00 a.m.	
p-Cresol	1
p-Cresyl acetate	1
Skatole*	0·1
Eugenol	10
Isoeugenol	10
Indole*	1
Indolene*	1
Methyl anthranilate	1
Aurantiol purum*	1

Components which are common to other series—and therefore come up several times—must be studied again individually in the new context and from the new viewpoint. They are all the more important as they are universal. The perfumer must learn as much as possible about them.

Group 5: Violet–Orris Odorants

In this and the following tables, the products marked ** have recently been restricted or forbidden for use in perfumery. Nevertheless, their odour facets are very typical for this group and should not be neglected when studying it. Non-toxic odour substitutes will be available in most of the fragrance-producing houses.

8.00 a.m.	*Dilution (%)*
Orris concrete (Florence)	6·5
Orris absolute	1
Orris resinoid (Florence)	10
α-Irone purum	1

(*continued*)

Group 5: Violet–Orris Odorants—*contd.*

8.45 a.m.	*Dilution (%)*
Ionones	10
α-Ionone extra white	10
β-Ionone	10
9.30 a.m.	
Iralia (Firmenich) = mixture of methyl ionones	10
Raldeine 100% (Givaudan)	10
γ-Methylionone	10
δ-Raldeine (Givaudan)	10
Isoraldeine	10
10.30 a.m.	
Violet leaf absolute	1
Methyl heptinecarbonate** (methyl 2-octynoate)	1
Methyl octinecarbonate** (methyl 2-nonynoate)	1
Methyl 2-nonenoate	10
Cyclamen aldehyde	10
Irival (IFF)	1
2-*trans*,6-*cis*-Nonadienal	0·1
2-*trans*,6-*cis*-Nonadienol	0·1

Ionones and methyl ionones have a rapidly inhibiting action on the olfactory receiver. Each olfaction should be less than a second long to avoid a complete or substantial loss of smell. Bear this fact in mind and proceed by inhaling very briefly, then take a one-minute break; repeat the procedure for as long as possible. To make comparisons, it is best to have a smelling-strip in each hand and always to smell the same blotter from the same hand, holding it horizontally 1 cm away from each nostril. Changing hands—when done in five seconds—should allow the nose to regenerate. However, this regeneration is only relative, making each confrontation difficult. Keep it in mind!

Group 6: Aniseed Odorants

8.00 a.m.	*Dilution (%)*
Star anise oil	10
Anethole	10
p-Methoxyacetophenone	10
Anisyl alcohol	10
Anisyl aldehyde (ex anethole)	10
Anisyl aldehyde (ex *p*-cresol)	10
Anisyl acetate	10

Group 6: Aniseed Odorants—*contd.*

8.00 a.m.	*Dilution (%)*
Methyl anisate (*p*-methoxybenzoate)	10
Dihydro-nordicyclopentadienyl acetate (verdyl acetate)	10
Dihydro-nordicyclopentadienyl propionate (verdyl propionate)	10
9.00 a.m.	
Estragon oil	10
Estragol (=Methylchavicol)	10
Basil oil	10
Caraway oil	10
L-Carvone	10
D-Carvone	10
Coriander oil	10
Eugenol	10
Linalool (synthetic)	10
10.00 a.m.	
Thuja oil (toxic)	10
Tansy oil	10
β-Thuyone (toxic)	10
Spearmint oil	10
Armoise oil (*Artemisia vulgaris* or *A. herba-alba*)	10
Fennel oil (sweet)	10
Fennel oil (bitter)	10
Eucalyptol	10
Camphor	10

Group 7: Aromatic Odorants

(*a*) *First Day*

8.00 a.m.	*Dilution (%)*
Lavender oil	10
Lavandin oil	10
Rosemary oil	10
Sage oil	10
Clary sage oil	10
Linalyl acetate (synthetic)	10
Terpenyl acetate	10
Terpineol	10
Coumarin	10

(*continued*)

Group 7: Aromatic Odorants—*contd.*

(*a*) *First Day—contd.*

9.00 a.m.	*Dilution (%)*
Pine needle oil (Siberian type)	10
Pinus sylvestris oil	10
Turpentine oil	10
Borneol	10
Isoborneol	10
Bornyl acetate	10
Isobornyl acetate	10
Camphor	10
10.00 a.m.	
Eucalyptus globulus oil	10
Eucalyptus citriodora oil	10
Eucalyptol	10
Citronellal	10
Spearmint oil	10
Peppermint oil	10
Pennyroyal oil	10
Menthol crist.	10
Menthone	10

(*b*) *Second Day*

8.30 a.m.	
Rue oil	10
Methyl nonyl ketone	10
Methyl nonyl acetaldehyde	1
9.00 a.m.	
Parsley leaf oil	10
Celery seed oil	10
Carrot seed oil	10
Angelica seed oil	1
Angelica root oil	1
Ambrette seed oil	1
Ambrettolide	0·1
10.00 a.m.	
Cumin oil	10
Cuminaldehyde	10
Valerian root oil	10
Laurel berry oil (toxic)	10
Benzaldehyde	10

Group 8: Green Odorants

8.00 a.m.	*Dilution (%)*
Hyacinth absolute	1
Galbanum oil	1
Phenylethyl alcohol	10
Phenylethyl acetate	10
Benzyl acetate	10
Cinnamyl alcohol	10
Cinnamyl acetate	10
Phenylpropyl alcohol	10
p-Methylphenylethyl alcohol	10
Phenylacetic aldehyde	1
Phenylacetaldehyde dimethylacetal	1
Diphenyl oxide	1
Diphenylmethane	1
Bromostyrene	1
9.30 a.m.	
Violet leaf absolute	1
Methyl heptinecarbonate**	1
Methyl octinecarbonate**	1
2,6-Nonadienal	0·1
2,6-Nonadienol	0·1
Nopyl Acetate	10
Cyclamen aldehyde	10
Cyclamen aldehyde { Dimethylacetal	10
Cyclamen aldehyde { Diethylacetal	10
10.30 a.m.	
Dimethyl phenylethyl carbinol	10
Dimethyl phenylethyl carbinyl acetate	10
Verdyl acetate } chemical names	10
Verdyl propionate } cf. Group 6, first group	10
cis-3-Hexenol	1
cis-3-Hexenyl acetate	1
cis-3-Hexenyl formate	1
Acetaldehyde *cis*-3-hexenyl ethylacetal	1

Many other green notes will be studied later.

N.B. The ingredients already smelled in the violet group have to be reconsidered in terms of green notes, but require a different state of mind.

Group 9: Spicy Odorants

8.00 a.m.	*Dilution (%)*
Clove bud oil	10
Eugenol	10
Isoeugenol**	10
Acetyl eugenol	10
Acetyl isoeugenol	10
Methyl eugenol	10
Methyl isoeugenol	10
Benzyl eugenol	10
Benzyl isoeugenol	10
8.45 a.m.	
Clove stem oil	10
Clove leaf oil (exhausted)	10
Terpene hydrocarbons ex clove oil	10
Allspice leaf oil (Jamaica)	10
Bay oil (St Thomas)	10
Juniper berry oil (Italy)	10
Ginger oil (Sri Lanka)	10
Pepper oil (Madagascar)	10
Elemi oil	10
10.00 a.m.	
Cinnamon bark oil (Sri Lanka)	10
Cinnamon leaf oil	10
Cassia oil	10
Cinnamyl alcohol	10
Cinnamyl aldehyde	10
Eugenol	10
10.30 a.m.	
Cardamon oil	10
Cascarilla oil	10
Terpenyl acetate	10
Eucalyptol	10
Thyme oil	10
Thymol	10
Carvacrol	10
Nutmeg oil	10
Macis oil	10
Marjoram oil	10

Group 10: Woody Odorants

7.30 a.m.	*Dilution (%)*
Cedarwood oil (Virginia)	10
Atlas cedarwood oil	10
Cedrol	10
Cedryl acetate	10
Methyl cedryl ether	10
8.00 a.m.	
Cypress oil	10
Vetiver oil Bourbon	10
Vetiver oil (Indonesia)	10
Vetiver oil (Haiti)	10
Vetiverol	10
Vetiveryl acetate	10
9.30 a.m.	
α-Ionone extra white	10
γ-Methyl ionone	10
Dimethylbenzyl carbinyl acetate	10
Dimethylbenzyl carbinyl propionate	10
Terpenyl acetate	10
Cyclamen aldehyde	10
Terpene hydrocarbons ex clove oil	10
Linalyl acetate (natural ex 'Bois de Rose')	10

[Thirty minutes' break out-of-doors.]

11.00 a.m.	
Coumarin	1
Isobutylquinoline*	1
Ciste labdanum oil dist.	1
Oak moss absolute (Yugoslavia)	
Ether as extr. solvent	1
Benzene as extr. solvent	1
Tree moss absolute	
Benzene as extr. solvent	1

To repeat and memorize

Rosemary oil	10
Petitgrain oil	10
Pepper oil	10
Angelica root oil	10

Group 11: Tobacco-Like Odorants

7.30 a.m.	*Dilution (%)*
Virginia tobacco infusion	20
Maryland tobacco infusion	20
Burley tobacco infusion	20
Ambergris infusion	(Dosage 20 g/l)
8.00 a.m.	
Tobacco absolute	10
Tea absolute	10
Deertongue absolute	10
8.30 a.m.	
Coumarin	10
Tonka beans infusion	10
Hay absolute	10
Flouve absolute	10
Flouve oil	10
Lavender absolute	10
[Fifteen minutes' break out-of-doors.]	
9.30 a.m.	
Clary sage oil	10
Spanish broom absolute	10
Fig leaf absolute	10
Cedarwood oil (Virginia)	10
Cypress oil	10
Vetiver oil (Indonesia)	10
10.15 a.m.	
Ciste labdanum absolute	10
Ciste labdanum oil (dist.)	1
Opoponax resinoid	10
Myrrh resinoid	10
[Fifteen minutes' break out-of-doors.]	
11.00 a.m.	
Orris resinoid	10
Oak moss absolute (ether)	1
Eau de Broûts absolute	1
11.15 a.m.	
Eugenol	10
Honey, white (De Laire)	1
Phenylacetic acid	1
Methylquinoline*	1
Isobutylquinoline*	1

Group 12: Fruity Odorants

	Dilution (%)
8.00 a.m.	
Vanillin	1
Coumarin	10
Heliotropin	10
8.30 a.m.	
Isoamyl acetate	10
Benzyl acetate	10
Benzyl propionate	10
Dimethylbenzylcarbinyl-acetate	10
cis-3-Hexenyl lactate	10
Phenylethyl isobutyrate	10
Benzyl isobutyrate	10
Amyl isovalerate	10
Benzyl isovalerate	10
[Ten minutes' break out-of-doors.]	
9.45 a.m.	
Verdyl acetate } chemical names	10
Verdyl propionate } cf. Group 6, first group	10
2,6-Nonadienal	0·15
Ethyl α-allylacetoacetate	10
Allyl hexanoate**	10
Allyl cyclohexylpropionate**	1
Allyl phenoxyacetate**	1
10.00 a.m.	
Decanal	1
Undecanal	1
Dodecanal	1
[Fifteen minutes' break out-of-doors.]	
10.45 a.m.	
Allyl ionone	1
Ethyl methylphenylglycidate (= Fraise, Aldehyde C-16)	1
Methyl methylphenylglycidate	1
p-Hydroxyphenylbutanone (= Raspberry ketone)	0·1
Ciste labdanum absolute	1
γ-Nonalactone (= Prunolide)	1
γ-Undecalactone (= Peach lactone)	1

Group 13: Balsamic Odorants

8.00 a.m.	*Dilution (%)*
Vanilla beans infusion	15
Vanillin	1
Ethyl vanillin	1
Tonka beans infusion	25
Coumarin	5
Heliotropin	10

[Ten minutes' break out-of-doors.]

9.00 a.m.	
Benzoic acid	10
Benzyl benzoate	10
Linalyl benzoate	10
Methyl benzoate	10
Ethyl benzoate	10

9.30 a.m.	
Styrax oil**	10
Cinnamyl alcohol	10
Benzyl cinnamate	10
Ethyl cinnamate	10
Methyl cinnamate	10
Linalyl cinnamate	10
Phenylethyl cinnamate	10
Phenylpropanol	10

[Ten minutes' break out-of-doors.]

10.30 a.m.	
Benzoe Siam resinoid	10
Peru balsam resinoid	10
Tolu balsam resinoid**	10
Opoponax resinoid	10
Myrrh oil	10
Olibanum oil	10
Elemi oil	10
Galbanum oil	10
Ciste labdanum absolute	10
Labdanum oil (Spain)	1

N.B. Vanillin and ethyl vanillin (and therefore vanilla) have a highly inhibiting effect on the olfactory epithelium. Shorten the inhalation period and lengthen the break between each olfaction. These products are difficult to judge accurately. Try to evaluate the difference in intensity between vanillin and ethyl vanillin. Compare both components in various dilutions—evaluations will vary between three- and five-fold in favour of the ethyl vanillin (in terms of odour strength).

Group 14: Animal-Like Odorants

7.30 a.m.	*Dilution (%)*
Civet infusion	(30 g/litre)
Skatole*	1
Indole*	1
Civetone	1
Phenylacetic acid	1
8.00 a.m.	
Musk natural infusion	(20 g/litre)
Muscone	1
Exaltone (Firmenich)	1
Exaltolide (Firmenich)	1
Musk ketone*	1
[Fifteen minutes' break out-of-doors.]	
9.00 a.m.	
Muscomer (Grau Aromatics)	1
Galaxolide (IFF)	1
Phantolide (PFW)	1
Musk D.T.I. (Firmenich)	1
Musk xylene*	1
10.00 a.m.	
Ambrette seed oil	1
Musk Ambrette**	1
Ambrettolide	0·1
Angelica root oil	1
[Fifteen minutes' break out-of-doors.]	
10.45 a.m.	
Ambergris infusion	(20 g/litre)
Ambrarome absolute	1
Ciste labdanum absolute	10
Ciste labdanum oil (dist.)	1

(continue

Group 14: Animal-Like Odorants—*contd.*

11.15 a.m.	*Dilution (%)*
Cumin oil	10
Cuminaldehyde	1
Castoreum resinoid	1
Costus oil	1
3-Phenylpropanol	10
3-Phenylpropionic aldehyde	10

Group 15: Leathery Odorants

8.00 a.m.	*Dilution (%)*
Raldeine Delta	10
α-Irone	1
Orris resinoid	10
p-tert.-Butylcyclohexyl acetate	10
8.45 a.m.	
Ciste labdanum absolute	10
Ciste labdanum oil (dist.)	1
Ambrarome	1
[Fifteen minutes' break out-of-doors.]	
9.30 a.m.	
'Bois de Teck' infusion	25
Calmus oil (Sweet Flag oil)	10
Birch tar oil (rectified)	1
Angelica root oil	1
Cumin oil	10
Manevoro oil	10
Castoreum resinoid	1
Oak moss absolute (benzene)	1
[Fifteen minutes' break out-of-doors.]	
10.45 a.m.	
Civet infusion	(dosage: 30 g/litre)
Musk seed infusion	(dosage: 20 g/litre)
Costus oil**	1
3-Phenylpropyl alcohol	10
3-Phenylpropionyl aldehyde	10

Reference Marks

It must be said that the study of the aforementioned groups—a determining factor for the initial knowledge of the young perfumer who is learning to classify odours—only has an educational value. Unfortunately, it does not deal directly with the art of composition. The reason is that composition takes place, not with entire families, but with individual constituents whose properties and interactions must be discovered first. Perfumery schooling calls for patience, attention and perseverance.

The attentive and methodical study of raw materials allows the composer to place each piece of olfactory information in a (personal) mental catalogue where he can always find it—if his initial judgement was correct and clear. This mental index is not made up of words (too weak) but of forms, of *olfactory benchmarks* to be precise. Nevertheless, a detailed additional card index should be kept on all odours. It will later serve as a reminder.

During his study of odour families and throughout his career, the perfumer will try to categorize each new odour in a scale of olfactory sensations, such as cool to warm; light to heavy; dry to fatty; bitter to sweet; acid to smooth; sharp to deep; ethereal to balsamic; flowery to fruity; green to candy-like; or transient to tenacious. A single product can be arranged in two or three categories depending on its various facets—for example, lemon in sharp and transient; amyl acetate in ethereal, fruity and transient; labdanum in deep, balsamic, tenacious and fruity; absolute of violet can be simultaneously green, fatty, tenacious and flowery at a high dilution.

These benchmarks have a practical value for the choice of materials the composer will make. By having these categories in mind each time he manipulates an odorous product, the composer can decide: it is dry or it is fatty, it is acid, it is persistent, etc. He will treat and adapt the product in his formula as a function of its olfactory coordinates.

The benchmarks can also be entered in computer databases and in the future (who knows!) be incorporated into thinking memories with sensors—just like the Active Memory (a modest name), an incredible machine which Dr Jacques Sauvan, before the age of microprocessors, had both the imagination and intuition to invent.[6] This would represent a new era for perfumery, one which could only remain creative if man runs the machine and not the other way around. To remain an art, man only has to make essential and deliberate choices and decide on the path of operations. Otherwise, it just becomes an industry.

The incredible amount of knowledge the perfumer progressively acquires consists of a multitude of benchmarks—indicators of all the odours which have struck his attention since early childhood, indicators of each material he has studied, indicators of its reactions observed successively with other materials, indicators of increasingly complex combinations and, finally, indicators of the century's great perfumes (when one has had the chance of memorizing them in their original glory).

The interactions between materials can vary according to each new context.

The more terms there are in a formula, the greater the confusion and the less the formula remains controllable. This is why formulae which run on for several pages must be avoided; existing formulae of this kind are aberrations. It has often been said that great masterpieces are simple, that simplicity keeps the artist on the track to success. This is all the more true because materials have a complex nature.

Every new formula represents a special case which creates a new problem of interpretation. The perfumer who does not want to fall into the boring routine of creating yesterday's perfumes (instead of tomorrow's) is condemned to questioning himself every day, after each newly created formula, even if he has over 60 years' experience.

There would be no art of composition without the first prerequisite we have discussed in such detail. A composition is a construction. However, to construct, the ground must first be prepared carefully. And secondly, before any construction can be started, a plan must be drawn up. This would not be possible without the first and following prerequisites.

Second Prerequisite: Imagination and Formation of Taste

The composer must have aptitudes which go beyond the sense of smell, which can be refined through exercise and limited to a control tool, namely mental and psychological abilities: imagination and taste with a constant willingness to develop them.[7]

Taste is cultivated through an enlightened contact with beauty of all domains; it is enriched by pursuing a critical study of all kinds of harmonies. This results in a personal understanding and synthesis of beauty in general, and in a well-equipped sense of the aestheticism which helps the composer better to confront problems of such a nature. Before being judged by others, it is vital to be capable of judging oneself. Such an attitude includes a solid and severe aesthetic conscience and judgement.

Thanks to repeated exercises of composition and to the acquired aesthetic conscience, an olfactory consciousness will develop which, helped by intuition, can interpret the intimacy of odours. The composer will start *thinking* in odours, will let them penetrate his mind; their universe will become his second nature. It is in this intellectual climate that he will finally start 'composing'. The only other ingredient needed to create a perfume is having *ideas*.

Association of Ideas

Any seed will germinate on a field which has been well ploughed. To have an idea, a pretext, i.e. a motive or already existing challenge, is needed and it seems that every work of art has always been preceded by a pretext. A person who wants to create has a rich catalogue of pretexts. The creative spirit engenders pretexts, in both science and the fine arts. Imagination opens endless opportunities of constructions in a chosen area. This becomes a game—all the more so when experience leads to success which in turn leads to pleasure.

One idea often leads to another. Even the most insignificant detail can trigger an initial association of an idea, which then induces a second and a third idea creating a *chain of association of ideas.* This gives the creative mind the impulse needed to transform his pretext into a masterpiece. Elements of chance may step into this chain either to boost it or set it off in another direction. Sometimes, there is no longer a visible relationship between the first and last points. But there is actually a logical train of thought based on acquired experience and know-how. The more comprehensive the knowledge and cultivated the mind, the more brilliant is the association of ideas and the masterpiece is all the more successful.

An active mind is constantly at work, alert, ready to catch every bit of information that comes along. What is striking about the human mind is the enormous development of the cortex, which is the organ of association of memorized elements in the paleocephalon. The cortex combines all types of information. It is responsible for the faculty of imagination, because to imagine is to associate memorized elements in an original manner.

Memories which are well classified tend to engender *useful* idea associations. However, this is the essential form of activity under which our mind judges, makes decisions, creates inventions. The expression 'to have ideas' is but a mere contraction of 'forming idea associations'. Since these associations depend on our mental disposition, on the degree of tension in our mind and on momentary necessities, the efficient creator needs to have good mental hygiene.

Thus, the creative and sensitive mind will experience no difficulties in having ideas (or emotions). The difficulty will lie in being able to sort them out and choose those which will justify the implementation of the creative process (aesthetic judgement). The prerequisites mentioned before will provide the means as they will have enabled the laborious forging of the technique upon which the entire creation will be based.

Chance

Chances are but a reflection of a lack of attention towards life, an inability to bring oneself to face coincidence. A chance basically represents the unexpected, something that one could neither predict nor know how to make use of. Mathematician Henri Poincaré said, 'Chances are the measure of our ignorance'.[8]

What is gained from a coincidence is no longer a pure chance; it becomes a pretext, which plays but a minor role since nothing has yet been started.

The artist does not base his work on coincidences. He follows an inner voice which prompts him to act and which guides him in the course of his research, based on a technology which thrives on effort and not on fate.

If the user of a specific technique is lucid and sincere, he will accuse his clumsiness, and not fate, if he produces a failure. So, why should success be attributed to luck and chance? If one has developed a taste for aesthetic constructions and if one has made the effort to acquire the necessary knowledge, it is perfectly understandable that we invoke what we call 'chance' for lack of a better term. Basically, reflection has been able to seize the smallest pretext, be it by chance or not, to transform it into an artistic vision.

Is there fate in contemporary music? Yannis Xenakis[9] said that once fate had knocked at his door, it had marked the beginning of a tremendous project. Pierre Boulez, at the beginning of his research, accused the composer who ignored fate of 'running away from his responsibility'.

Technique

A few words, if only a few, must be said about this term, which can be confusing.

It seems that a fairly complete definition of the technique—for instance of a perfumer—would be *the entirety of means used by an artist to create his work of art.* This 'unity' of means includes practical methods and materials, as well as intellectual means. The composition of a perfume, an art of thought *par excellence,* relies considerably on mental faculties to create a technique. This must be understood once and for all because the user 'smells' what the composer has 'thought'.

Another mistake which must be avoided deals with the belief that there are artistic composers and the rest mere technicians. An artist is, by necessity, an experienced technician. He is much more talented if he is an excellent technician. The technique used is not an inferior or minor art; it is very inherent to art and to the artist; it is intimately related to creation, it contributes to the artist's style.

Of course, the mixer of perfume and his collaborator at the weighing scales apply a certain technique to the various manipulations they perform. They can, to a certain extent, be called technicians but they do not actually contribute to the genesis of creation. Their technique is purely mechanical and has nothing to do with the technique of the perfumer except that they are working under his orders.

Formulation

When the composer writes down a formula, his composition is not based on sensation but on the memory of sensations, in other words on abstractions of abstractions. This is a highly intellectual process, regardless of the form aimed at by the composer.

All memorized sensory benchmarks become specific forms in our mind. We will work with these abstract forms by making an effort to evoke and combine them in thought. If necessary, we will also evoke combinations of forms achieved in the past by comparing them with those projected in the present and by intuitively anticipating the results of our project.

Prompted by whatever event, we have had the idea of a new perfume whose rather vague form has developed in our mind. We have allowed the idea to mature and now find ourselves in front of a blank piece of paper.

All the various 'prerequisites' have been mobilized and whisper the first elements which could correspond to our imagined form. We start by writing these names in columns, in a sequence which is dictated, above all, by the *quality* of each product: by the fact that its tonality seems necessary for the envisaged construction. All our ideas land on the paper in a variety of forms which initially add to our general confusion but finally result in our idea of a perfume.

We then put the draft away, somewhere close at hand, and work on other things. The ideas follow their course, become more forceful, more mature, or more precise. The new form becomes so much a part of us that any new odour which seems suitable is immediately added to the list.

Haste makes waste. An incubation period is never a loss of time, but invested time. The suggested materials accumulate on the paper until a selection *must* be made—simply because too much is too much!

In a state of total concentration, we evoke once again the various properties: notes, features, affinities, antagonisms, ease of orientation, risks (instability, discoloration, price). We mix them in thought, we distribute roles which help us define ratios, and the first figures are placed in front of the names. A commitment is made.

The next phase can be called the selection process: names are crossed out, ratios modified, tension becomes palpable as last touches are made, and, after an explosive climax, peace settles in. The formula will take a much-needed rest, to be almost forgotten to prevent the mind from lingering on preconceived ideas, questionable preferences, or ostracisms.

Later, we come back to our embryo of a formula with a clear, objective frame of mind. Renewed efforts of concentration, renewed tension and corrections follow, before we write—in ascending quantitative amounts—the formula of test no. 1, now ready to be weighed out.

This process would be impossible without the aforementioned 'prerequisites'. It would not come into being, it would be a succession of random speculations, less persuasive and more time-consuming. Let us therefore recall them briefly:

1. Profound, methodical, olfactory study on the materials, their inherent features (quality, intensity, volatility, duration), the interactions which enable the various combinations between them (affinities, antagonisms and the resultants).
2. Formation and refinement of taste: contacts with the other fine arts and their creators; general and aesthetic culture.
3. Points 1 and 2 should lead to a safe judgement and a sharp, uncompromising, critical mind, with the olfactory repertoire being improved upon constantly and untiringly.

The amount of time needed for these prerequisites extends over a period of 10 to 15 years. Anyone who thinks he can improvise from the start is sadly mistaken. Such people contribute to the market saturation of failures and mixtures, thereby endangering the future of true composers.

Let us return to our test formula no. 1. Subsequent to dilution in an alcohol solution and a long ageing period, our formula will not only be valuable and well-defined, thanks to our acquired experience, but it will also allow us to discern our original idea to discuss it and then to readjust it mentally.

Our formula no. 1 has now been studied: extensively, at intervals to avoid not only saturation but also possible polarizations and even accepting the risk of tiring of its form. And as from now, the research process takes on a normal rhythm and is guided by the authority of technique, which guarantees a methodical approach in the next phase of development.

We have seen that the perfumer's technique represents all methods and materials used to create a perfume. To a certain extent, it is also linked with our prerequisites. But this does not mean that new associations or constructive ideas cannot arise subsequently to the initial idea.

By proceeding with composure (even in a state of passion) and self-confidence, the imagined form can be achieved after just a few attempts. But the final creation will take months, even years! The creators are not always convincing enough while trying to share their vision with those who will take the risk of confronting the public with the new creation.

When writing his formula, the creator who starts with the smallest amounts of ingredients would like to ensure a maximum amount of sensitivity for the most delicate dosages. However, another main reason is that the general structure of the formula can be understood at a glance. The products are seen for what they are worth; the 'trace components' are immediately discernable in the higher notes of the formula. In short, the complete formula is already smelled in thought.

Further, we have also decided to group crystallized products (in ascending order, according to weight) at the beginning of the formula for two practical reasons: to weigh them easily and to assess their relative importance rapidly. This type of product has a considerable influence on the physiognomy of a perfume. These visual images are an inherent part of composition, they contribute to the formulation and the good comprehension of a perfume.

The Intuition of the Perfumer

The role of intuition in the artistic composition of a perfume is so important and so seldom discussed that the time has come to examine it in more detail. The information provided so far will have given insight into its influence: the framework has been traced, the ambience studiously created, and we are about to see intuition at work. What will be said about the intuition of perfumers will be far more believable now where this invaluable ability has been treated as a 'prerequisite' or 'first resource'.

A superior form of imagination, intuition is also very curious. Little is known about its (not always understood) mechanisms, which have sometimes been described (or denied) by people who undoubtedly have never had any intuition whatsoever.

To immerse oneself completely in the domain of intuition, it is best to surprise it at work. Bergson gives us the opportunity to do so in a page which seems to have been written for us, and which the perfume composer can read and meditate upon without changing a sentence.

Imagine this composer at his desk: to escape from excessive odours, it is located far away from the laboratory where the technician weighs his 'attempts' and where he rarely sets foot to avoid contaminating himself with odours which could impregnate themselves in his hair, his clothes, his fingers. His sense of smell must be treated with deference. It must always preserve or easily find its 'virginity' to have a maximum sensitivity and provide the composer with reliable olfactory information which he then translates into the language of composition. This alone calls for a great deal of imagination.

For months, possibly even years, the composer has multiplied his tests on odorous substances, he has observed his perfumed smelling-strips day after day, he has accumulated observations, reflections, prerequisite trials. Attentive to discovering the particular *form*, the affinities or antagonisms of each essence, each chemical substance, each odour, he considered all these odours as demanding partners, reluctant to part with their favours. Nevertheless, he treats them kindly. Therefore (from Ref. 10):

> Once the subject has been seriously studied, once the documents have been compiled, once all the notes have been taken, a special something is needed to start the work of composition. It often entails a difficult effort to penetrate to the heart of the subject and discover an impulse which will trigger creativity. Once acknowledged, this impulse sets the mind on a known path where he will find the information he had put together and much more: it develops, analyses itself to no end. The farther it goes, the more it discovers; never can everything be said. And yet, if one turns abruptly towards the impulse, a shadow behind us, to catch it, it disappears because it was nothing concrete, it was an incitement of movement, and, although infinitely extendable, it is simplicity in the truest sense of the term.

This incitement is what we call 'intuition'. In the composition of perfumes, intuition can precede the extended study of a given subject. It is then the origin of the study and provokes the selection of materials. An initial intuition may also suggest the idea of a subject to be treated or a path to be trodden. During the course of any research which may result from it, other intuitions may guide the composer.

For an experienced composer, whose documentation and knowledge are rich in years, intuition will suffice to trigger the exploitation of the information he has acquired. This intuition represents a short moment of extreme lucidity during which the composer concentrates on and finds the orientation of a product or the solution to a problem.

With regard to remarkable facts which have been noted, Bergson also knew that[11]

> such a tremendous amount of them were needed to ensure that, in their fusion, they would neutralize all preconceived and premature ideas given by researchers based on their observations. Only in this way does the crude materiality detach itself from known facts . . . It would be impossible for anyone to have distinctive intuitions without having regrouped and confronted a very large number of 'psychological' (or olfactory) analyses.

Every perfume composer who wishes to go beyond simple tasks and to attain the beautiful must take these wise and penetrating suggestions to heart. He will have had to multiply an infinite number of times his attentive manipulations: rigorous, minute, and repeated observations which will enable efficient investigation. He must keep the awareness of the subtle facts of olfaction which escape the normal consciousness but without which there could be no intuition. This is the reason why Bergson glorified effort, disdained facility and recommended 'an intricate way of thinking'.

In the search for art, intuition most often arises under the strength of negation. Like the demon of Socrates, it *shields*, it gives a warning. The composer can limit premature convictions, false manœuvres or unsuccessful attempts, and suppress impulsive gestures. Of course, intuition takes the form of positive ideas, but the composer is often more sensitive to the confused feeling that he may commit an error and wander off his chosen path. A strict 'no' is set against some of the combinations built up in his imagination. Such an exclusion does not result from reasoning, but a reason may be explained and justified afterwards.

'We call intuition the *sympathy* through which we transpose ourselves into the interior of an object to find its uniqueness, in other words, that which cannot be expressed by words.'[12] This truly is how we act when we apprehend odorous bodies in view of organizing them in arrangements of olfactory forms. Whether we find an odour pleasant or unpleasant as such is irrelevant because we almost always feel a certain sympathy for it. This is due to the fact that we do not merely see its agreeable or disagreeable side, its pleasant or unpleasant side which arises from one of our senses (which is a means and not an end). We are attached to the odour by a positive element of association, the material of construction, its potential of harmony, which holds our hope and cannot leave us indifferent. This sympathy increases our attention towards odour tenfold, it makes us more aware and more understanding, and gives us the feeling of penetrating the intimate sphere of the odour, to which we attach a personality. This openness enables us to take a better advantage of the situation.

Our apprehension about the odorous material is not superficial, as is the reaction of the amateur. We explore this form under all its faces and facets and discover psychological benchmarks which will be indispensable in making later identifications or in establishing innovative ideas. However, these benchmarks also enable us to follow with greater ease the fate and the action of each elementary form in the heart of the complex odour which emanates from our blotter. They also make possible our memorizing the value of raw materials, if incorporated in formulae. And what we remember about their pure state is just relative or even misleading. The great wine tasters proceed accordingly. They work with mental benchmarks as we do.

The more we penetrate odours, the more they end up by possessing us. They live within us, become an integral part of us, participating in a new function within us. This is not a *déformation professionnelle*, as the often-misused saying goes, because nothing here has been deformed. On the contrary, we are better informed thanks to a constant awareness worked upon daily throughout a lifetime.

For the perfumer, intuition can manifest itself through ideas which come to mind unexpectedly, even though they have been triggered in some way. A suggestion can impose itself on the composer in the form of a simple association of ideas, i.e. an intuition when smelling an odour for the first time, or when taking a known odour apart and realizing a detail which had previously gone unnoticed. Or maybe a statement will sharpen the mind towards an idea with no apparent link to the thinking process. Sometimes, meditation or contemplation will release an anticipated suggestion made possible because the mind is relaxed. For the intuitive person, a seemingly insignificant fact may be sufficient to provoke an idea.

This idea can assume various forms. However, independent of the initial intuition which may have provoked the beginning of a study, the 'strength of negation' mentioned before may arise in the form of other intuitions which will put the composer on his guard against foolhardy attempts. An interior voice will say 'no' to the questions he may ask himself. The intuitive character of these answers will be seen in their spontaneity by the fact that they have not justified themselves by any preceding argument or technical reasoning.

It would seem that these negative intuitions, like the positive ones, are the result of observations and previous reflections accumulated in our mental computer which bears latent fruit. From this viewpoint, they are original technical manifestations and—we have said so continuously—the fruit of our work. They stand out through the peculiarities we just mentioned, similar manifestations provoked by intelligence alone.

Spontaneity is common to these 'ideas', suggestions and intuitions. They gush forth abruptly as if under the effect of a stimulus and their lifetime is very ephemeral. To be fruitful, they must be quickly captured by the memory, the intelligence and the mind's other faculties, using the complete range of techniques discussed above. An intuition which cannot be exploited immediately must be written down carefully lest it be lost forever. Each composer should have his own 'file of ideas' where he accumulates so-called projects—most of which will probably never see the light of day—to recall and achieve them when the time seems right. If the composer knows how to cultivate his intuition, this file will become the basis of his creativity.

It is a fact that these intuitions arise under the influence of an exterior excitement or interior meditation, but it would seem that they almost always correspond either to current concerns or to dormant worries or desires which the memory may have kept alive.

We said that intuition was a fleeting image, but this mere spark at times can turn into real fireworks! The transient character of intuition is paramount because it relates its influence, its reach, its fragility and its vulnerability to unwelcome or obtrusive matters. It explains and commands the behaviour of the composer. An idea presents itself to the mind, enters the domain of reason and is taken over by intellectual faculties which analyse it to assess its interest. The idea is weighed; its substance is extracted and examined in all its aspects. Taste and intelligence appreciate and make decisions but finally, *judgement*, the synthesis of all faculties, draws the final conclusion.

The complex, ultra-rapid interaction between these mechanisms, which also intervenes during the course of technical operations, constitutes an important part of the creator's technique, making it a superior and highly personal activity.

Not all artists are aware of this interaction. It can happen without their knowledge if they do not have a minimum amount of intellectual curiosity and do not feel the need to make an inventory of their ideas. We have come to the conclusion that many people are capable of intuition, but that out of sheer mental laziness they do not exercise their mind. Without mental exercise, these intuitions escape unnoticed. This clearly indicates the importance of awareness and will in this domain.

The intuition, the vehicle of the idea, brought it without any trimmings; it is up to the mind to do the rest. Without intuition, there was nothing; with it, the tremendous amount of creative work has only just begun. If intuition is the origin of innovative effort, only its symbol is active in the application. According to Bergson, 'intuition, once captured, must find a mode of expression and application which agrees with our thinking process and provides us—in carefully defined concepts—with the solid cornerstones we so much need. This is the perfect condition for terms we call rigour and precision'.[13]

Bergson also demonstrates how fact and idea collaborate (Ref. 14):

> A fact, more or less clearly seen, suggests the idea of an explanation; the scientist asks the experiment to confirm it. During the experiment, he must be ready to abandon his hypothesis or remodel it based on new facts. Scientific research can therefore be defined as a dialogue between the mind and nature (so can artistic research). Mother Nature awakens our curiosity. We ask her questions; her answers may lead to an unexpected development, provoking new questions whose answers suggest yet other ideas. The process can go on indefinitely.

The composer's mind goes through a similar process: an exchange takes place between successive intuitions and intellectual faculties depending on facts. This process has to be accompanied by experimental verification.

During these verifications, his experiments, the composer's mind must remain completely flexible. He must accept all intuitive suggestions; everything is possible, no contradiction should surprise him. Nothing is true or false *a priori.* This is a question of common sense, since, according to the author of *Le Rire*, common sense can be defined as 'the effort of a mind which is constantly adapting itself, changing ideas when objects are changing. This is a mobility of intelligence which bases itself on the mobility of things. It is the moving continuity of our attention to life.'[15]

'Our merit does not consist in being subject to our intuition as much as seizing and discussing it', stated Delacroix.[16]

The intuitive person is an observer: an observer of himself as much as of the universe which surrounds him; an observer who knows how to draw conclusions from his observations. Well, intuition can only thrive on a fertile soil of accumulated observations guarded at the bottom of our consciousness.

In conclusion, the main features of intuition are:

—acute attention,
—disciplined reflection,
—taste for meditation, and
—ability to anticipate.

These four fundamental characteristics, rarely found together in one individual, can be acquired through a patient and persevering education. There is no mystery here; Bergson knew that intuition could be methodically cultivated and developed if the intention to do so was there.[17]

5 CONCLUSION

It has become clear throughout this first chapter that the composition of perfumes, when executed aesthetically—as has been done by the great composers of this century—really is abstract art *par excellence.* This not only concerns the materials, which are now dominated by synthetic products, the forms of which must be interpreted abstractly, but also the compositions, whose contours can only be described by abstract images. It is important to know that the perfume masterpiece is a pure construction of the mind; all legislators should be well informed about this fact.

Formerly, perfume composition was the goal of a cult whose members aspired to universal success, one of high esteem and not of financial reward. This did not stop them from earning money—since that is precisely what attracted financiers. At this time, the general public, familiar with works of art and educated by them, was a connoisseur. Contrary to what has been alleged, perfumes were not reserved for the rich élite.

The first perfumes available on the market were made from very expensive floral essences and were much more costly than today's perfumes, which are largely based on synthetic products. Nevertheless, and despite the fact that diluted (less expensive) perfumes were not yet fashionable, a shopgirl could buy herself a quarter-ounce flask of the real *Arpége* extract and, with just a few drops, enter the illustrious world of royalty.

Today, the same shopgirl can pour a large bottle of eau de parfum or eau de toilette, respectively, over herself for the same price but get only half the satisfaction. The problem now is that society women may have to use the same eau de parfum, thus indicating that the level is once again established from the bottom up.

This essay has been written to give perfumes the chance to become respected again, appreciated and loved; to allow authentic talent to resurge and blossom. This is only possible if their creators have been instructed by authentic, pedagogic and devoted composers. Finally, the formation of a very large public of true connoisseurs should be encouraged, connoisseurs who will not seek an identity in snobbery or fiction, but within the small world of tasteful people, thereby multiplying its number.

REFERENCES

1. Pamoukjian, J. P., *Le Droit du Parfum.* Librairie Générale de Droit et de Jurisprudence, Paris, 1982.
2. Eminet, B., Contribution à l'étude physicochimique des relations structure–activité dans le domaine de l'olfaction. Thesis, University C. Bernard, Lyon, 1983, p. 69, Table 17.
3. Elmouaffek, A., Thesis, University C. Bernard, Lyon, 1988.
4. Roudnitska, E., *L'Esthetique en Question.* Presses Universitaires de France (PUF), Paris, 1977, pp. 100–17 (dedicated to intuition).
5. Roudnitska, E., Durée de perceptibilité des odeurs. *Parfumerie.* Revue, May–June, 1944.
6. Roudnitska, E., *L'Esthetique en Question.* PUF, Paris, 1977, p. 155 (detailed description).
7. *Idem, ibid.*, Genèse de l'oeuvre d'art, pp. 79–128 (detailed description).
8. Poincaré H., *Science et Méthode.* Ernest Flammarion, Paris, 1918, p. 65.
9. Boulez, P., *Penser la Musique d'Aujourd'hui.* Denoël/Gonthier (Editions Gonthier)–C. B. Schott's Söhne Mayence/Mainz, 1963, p. 2.
10. Bergson, H., *La Penseé et le Mouvant*, 12th edn. PUF, Paris, 1940, p. 225.
11. *Idem, ibid.*, p. 226.
12. *Idem, ibid.*, p. 181.
13. *Idem, ibid.*, p. 215.
14. *Idem, ibid.*, p. 230.
15. Bergson H., *Le Rire.* PUF, Paris, 1946, p. 140.
16. Delacroix, H., *Phsychologie de l'art.* Felix Alcan, Paris, 1927, p. 169.
17. Roudnitska, E., *Le Parfum,* 3rd edn. Que Sais-je Collection, No. 1888, PUF, Paris, 1990.

II PERFUMERY AS A TOPIC IN LIFE SCIENCES

Perfumes are perceived through their odour, and the mechanisms of odour perception are, therefore, an important aspect of this part of the book.

An understanding of these mechanisms alone, however, cannot be the goal of intense research in a life science discipline. The elucidation of a sensory system is only of value to humans when completed by an understanding of its role or function in living organisms. It can only then lead to important predictions or innovations. And the study of the genesis of the signals can, of course, be seen in relation to the study of their perception.

Chapter 2

Odours and Perfumes as a System of Signs

J. STEPHAN JELLINEK

DRAGOCO Gerberding & Co., D-3450 Holzminden, Germany

1 INTRODUCTION

In the animal world, odours are a rich language and, for some species, one of the principal systems of orientation and communication. In the context of human civilisation, however, we tend to look upon odours and perfumes primarily as sources of more or less pleasing sensations. In so doing, we deprive ourselves of many opportunities to understand their effects upon human feelings, emotions and behaviour.

This chapter will provide a look at odours in the animal kingdom under the aspect of semiology, the science of signs and their use. Comparisons will be made between animal and human uses of, and reactions to, odours.

2 NATURAL AND ARTIFICIAL SIGNS

At the outset, a clear distinction must be drawn between odours and perfumes. For the purposes of this discussion, odours will be defined as those olfactory properties of people, objects and places which were not brought about purposefully, but which have come about either without human intervention (as in the case of the odour of flowers or of the soil after a rainstorm), or as the result of a human intervention which did not have odour creation as its primary purpose—for example, the odour of wood fire or of perspiration. Odours will be considered to be *natural* signs, in contrast, to perfumes which are part of a purposeful manipulation of olfactory stimuli, typical products of the civilizing effort to control nature and to improve upon it. Perfumes are *artificial* signs.

Odours in the animal world can convey a very wide range of meanings. Among these, the following appear to be of particular importance. Odours can be signs of a presence, either actual or past; they can be signs of a condition or of a quality of things; they can also indicate the individual identity of a living creature or a group to which it belongs. Let us examine these different meanings one by one.

3 ODOURS AS SIGNS OF A PRESENCE

Predators detect the presence of their prey by its odour, and, conversely, the hunted animal may spot the hunter by his. In one case, odour serves as an announcement, a promise; in the other, it is a warning of approaching danger. Odour signs of this type are particularly important in situations where other senses such as sight are limited in their effectiveness, either because of the physiological make-up of the predator or the prey, or because of the terrain (for example, in a dense forest), or because of light conditions as in the case of night-hunting animals.

In the immediate sense of the hunter and the hunted, odour signs play a rather minor role with humans. However, man as the hunter must be aware of the odour signals he emits and move downwind from the hunted animal if he is to be successful. Also, food smells generally have a similarly arousing and stimulating effect upon the hungry human as does the smell of prey upon the predator. Again, in the smell of food, of bread in the baking, coffee in the brewing, steak on the grill, the sense of promise and of anticipation is powerful. Odours such as these are more effective in arousing anticipation than are the most beautiful pictures of food items.

In many animal species, special odour signs are emitted during the mating season by the female or by the male. By helping to bring mates together, sometimes over a considerable distance, such signals are essential to the preservation of the species. It is tempting to see, in the human use of perfumes, a signal of presence similar to the natural mating odour of animals. We should, however, be wary of drawing parallels between the *cherchez la femme* situation and the hunting situation. For, if the female provides herself with an odour in order to be found out, who then is the hunter and who is the hunted?

There is a great deal of social control involved in the way the human female is allowed to use perfume to signal her presence. In Puritan societies, fragrance is permitted only at intensities that are so low that they can be perceived only by the partner who is already in close physical proximity to the woman. Any use of perfume that announces the woman's presence to someone further away, at a 'public' distance, is considered cheap and unladylike. The remarks by a young woman from Kano, Nigeria, a Muslim culture, are typical:[1]

> Of course, in our society here, we don't like very strong perfume, because if you use a strong perfume and you pass by the men, I am sorry to say that this is what they always say: 'She is smelling like a prostitute. She always puts on high perfume like a prostitute so as to draw attention.'

In other cultures there has been less restriction on perfume intensity, either because they were more sexually permissive or because they did not strongly link perfume to sexual invitation.

Perhaps the first man-made perfume ever was the odour of burning meats and grains and fruit in the burnt offering to the gods. It is impossible to reconstruct the meaning which this sign had for the prehistoric worshippers, but the notion of drawing the gods' favourable attention to the particular group or tribe that offered the sacrifice was probably a key element. Through the black fragrant smoke that rose into the skies, the gods were reminded, in a pleasing way, of the presence of the worshippers.

4 ODOURS AS SIGNS OF PAST PRESENCE

All of the attributes of odours are involved in their sign function: odour quality, odour intensity and certainly also tenacity or lasting power. By virtue of the latter, odours can still be present when the creatures or objects to which they belong have long since departed. Sometimes, this property of odours can be a disadvantage to the odour carrier, as in the case of the prey that is captured because it had left olfactory traces behind. In other cases, olfactory traces are left on purpose, for example by ants in order to tell other members of the colony where food is to be found; or by dogs in order to mark their passage and their territory.

In human experience, lingering odours play a major part in creating the atmosphere of a place. The atmosphere of a Catholic church is marked by the lingering traces of incense; the atmosphere of poverty that permeates certain apartment buildings has a great deal to do with the lingering scent of cabbage and other inexpensive foods that have been cooked there.

The fact that the very word 'atmosphere' comes from the Greek *atmos*, meaning vapour, reveals the link between the odour and the mood of places.

Odours as a sign of the past presence of an individual can have powerful emotional impact. This is due to the fact that we tend to experience odours of things and people not just as superficial and arbitrary signs that have somehow been attached to them, but as an expression of the core of their being, of their essence. Here again, there is a strong message in the meaning of a word: the double meaning of 'essence' has certainly not come about by chance. When we smell a thing or a person, we have a sense of its immediate presence. There is a concrete physical basis for this feeling, since every olfactory perception involves the direct contact of molecules that have come from the object or person smelled with the receptor ending of the perceiver's olfactory nerve. In clear distinction to visual and auditory perceptions, we can say that in the act of smelling there is direct contact between the things smelled and an area of our body that is only a few synapses away from the centre of awareness in our brain. Although we are, of course, not explicitly aware of this physical phenomenon, the sense of closeness of the thing smelled is very real. This accounts for the poignancy of the experience where a person whom we love has left us, perhaps forever, but the traces of his or her odour or perfume are still around us. We sense the presence, but we know the absence.

The same phenomenon can also give us a sense of consolation and reassurance. When the master has left the house and the cat stays behind alone, she will lie down to sleep on a piece of clothing that he has recently worn. She knows the absence, but she feels the presence.

5 ODOURS AND PERFUMES AS SIGNS OF A CONDITION

Odour as a sign of condition plays a key role in the mating behaviour of some animals. With many species, the female emits a characteristic odour only during those periods in which she is ready for mating. As any owner of female dogs or cats knows, such signs of condition can be powerful indeed. Again, with humans the natural odour signs of hormonal condition play a far less prominent role. However, recent findings on the synchronisation of the menstrual cycle in females living closely together,[2] and of the emergence, in adolescence, of sex differentiation in body odour,[3] indicate that they do occur.

When a cat that normally loves liver one day refuses a piece after sniffing at it, she is probably responding to an olfactory sign indicating to her that this particular piece is spoiled. Humans also use odour signs to judge the condition of people and objects. They, too, smell at milk and meat to ascertain whether these are still fit for consumption. They judge the cleanliness of laundry and the state of hygiene of a person by their smell, as well. Because this is so, the use of perfumes to suggest, mask or demonstrate a condition is widespread; it accounts, in fact, for a considerable proportion of perfume use. Personal deodorants are perfumed so that people will not have perspiration odours even after physical exertion. Laundry detergents and fabric softeners are provided with substantive and long-lasting perfumes so that the laundry after washing will not only be clean, but also smell clean. Used-car dealers occasionally use a 'new car odour' spray on the interior of the cars they offer for sale because they know that buyers, consciously or unconsciously, use odour as one of the clues for the condition of the vehicle. And today, still, there are in many countries taboos and traditions that regulate not only the amount, but also the kind, of perfume married women and self-respecting ladies are allowed to wear, because strong and rich perfumes are considered to be vulgar.

6 ODOURS AND PERFUMES AS SIGNS OF QUALITIES

Odours can signal not only transitory conditions, but also the permanent qualities of things and people. We judge the quality of leather goods not only by their look and feel, but also by their odour. Consumer tests have shown that creams with a faint odour are considered more likely to be good for sensitive skin than strongly perfumed ones; lotions with a medicinal-type odour are thought more likely to be medicinally effective than lotions with a rich feminine fragrance.[4] A woman's or a man's selection of fragrance for herself or himself is also guided by the feeling that perfumes signal qualities, sporty or elegant, romantic or dynamic.

There is a kind of confusion at work whenever we judge the quality of people, and particularly of products, by their perfume. We respond to the perfume as though it were a natural odour and an expression of the essence of the product or person, rather than recognising it for what it is: an artificially applied signal.

One of the reasons for the Puritans' suspicion and rejection of perfumes is that they sense in them a form of deception. In this, they are not entirely wrong: perfumes certainly do have this potential. The precept, 'Never use perfumes to suggest conditions or qualities of things which they do not actually have, nor to mask harmful qualities', should be one of the two commandments for all makers and users of perfumes. (The other is: 'Never make or use perfumes in such a way that they can be hazardous to health.')

7 ODOURS AND PERFUMES AS A SIGN OF IDENTITY

In our admiration for a dog's ability to recognise individuals by odour, we are being too modest; humans also have this ability, albeit not to the same highly developed degree.

Research on babies 3–12 days old has demonstrated that they react differently to the body odour of their mother than to that of other women or of men—provided the babies are female, and breast-fed.[5] Other studies indicate that mothers can recognise clothing worn by their children by odour, and vice versa;[6] and that young adults can pick T-shirts worn by themselves or by their partner from a set which includes shirts worn by strangers.[7] In these experiments, recognition was never infallible, but it was statistically significant.

Some people, using one perfume exclusively over a long period of time, are thus reinforcing this olfactory dimension of individuality. The use of perfumes in this sense is universal when it comes to branded products such as toilet soaps. Here, fragrance is one of the key signals of the individuality of a brand. This explains the necessity for strict quality control of perfumes. It is not that a deviating perfume is of necessity less pleasing than the established one. The point is that as soon as the fragrance of a product deviates noticeably from the established norm, the consumer draws its identity and all of the other qualities that are associated with this identity into doubt. The manufacturer then receives complaints regarding the product's performance, skin compatability, etc., even if nothing else in the formulation was changed.

Odours can mark identity also in the case of places and environments. Homes and apartments tend to acquire, as a result of their inhabitants' living habits, individual odours. Because of habituation, such odours are noticed more strongly by occasional visitors than by the homes' inhabitants. Children are particularly aware of environmental odours, and many of us carry with us vivid recollections of the odour of grandmother's place. The odour of one's own home, although rarely consciously registered, probably has a great deal to do with the feeling of being at home. In my own case, I feel distinctly not at home in a new apartment as long as it smells of fresh paint and wallpaper glue. It is my apartment only after I have smoked a few pipes in it and impregnated it with my personal odour.

8 ODOURS AND PERFUMES AS SIGNS OF GROUP MEMBERSHIP

The T-shirt tests in which young adults could recognise their own shirt and that worn by their partner[7] also furnished evidence that people can distinguish between female and male body odour—female odour being less intensive and less unpleasant than male. There are many anecdotes regarding the different body odours of different races; however, I am not aware of any controlled studies proving this assumption.

Certainly, in the case of herd animals, there is a herd odour and animals which do not exhibit this odour tend to be rejected and are sometimes subject to physical attack. When cattle ranchers are forced to mix herds, they find that spraying all animals with the same odour makes the animals unable to distinguish brothers from strangers and restores peace.

Group identification probably also plays a major role in the selection of individual perfumes. It explains why many fashion-conscious people like to wear the latest hit fragrance and why career and status-conscious individuals like to wear expensive fragrances in preference to the more common ones sold in drugstores. The woman or the man who consciously selects a highly individual perfume as an expression of their distancing from the common crowd affirms the group message of odours in a negative manner.

9 THE FOUR LEVELS OF SIGNIFICATION

As the examples we have cited have shown, perfumes and odours can communicate at four different levels which we might call the biological, the archetypical, the cultural and the individual levels.

The biological level is supposedly inborn and common to all individuals of a species. All pheromonal odour effects belong to this category, as does probably the distinction between male and female body odour. Odours whose signal character is related to some form of physiological activity, such as the odour of menthol, eucalyptus oil and certain phenols, also belong in this category.

There is probably a group of odour signals the understanding of which is not inborn, but is learned at an early age. They are, moreover, linked to natural phenomena that are common to many parts of the world so that their meaning transcends cultural boundaries. It has, for example, been noted that through the ages, flowers have been associated more closely with women than with men; therefore, flower odours are generally and cross-culturally associated with gentleness and with femininity. Wood fires were a source of mortal danger to our ancient forebears who lived in forests and savannahs—hence the generally arousing nature of smoky odours to man, even today. I think we must be careful in our thinking about archetypical odours. Even though certain phenomena and the odours associated with them may occur and have occurred in many different cultures, the meaning associated with them may differ widely from culture to culture. The odour of soil after the rain is approximately the same everywhere, but it certainly has different meanings in arid countries and in countries with abundant rainfall. The odour of wood burning may indeed spell danger to some of us, but I suspect that to the nomadic inhabitants of the North American plains, before the white man took over, it signalled camp fires and the presence of other human beings rather than a natural danger. To today's city dweller, it signals the tranquillity of an earlier age when wood was used as the fuel for home heating and when a man's home was his castle.

The cultural level of odour meanings is the most pervasive of all; it is also closely linked to the art of perfumery. In each country, for example, a clearly defined notion prevails of how well-cared-for babies smell. This notion is different from country to country and it has demonstrably been shaped by whatever is the leading brand of baby products in the country. In baby products, more than in any other personal products category, brand loyalty tends to be passed on from generation to generation. Another example: to Europeans of the first half of this century, the odour of citronella oil strongly connoted laundry day and clean bed-sheets. This was due to the nearly universal use of citronella as the perfume of laundry soaps. To Americans of the same generation, however, citronella connoted hot summer nights, because this oil was widely used in America as an insect repellent. The degree to which the same odours carry the same messages in different cultures is a sensitive indicator of the degree of homogenisation of the cultures.

A number of individual odour meanings were cited in the examples in the earlier part of this chapter—the perfume of the loved one who has left; the odour of your grandmother's place and of my pipe-smoke-scented apartment. The key event in Marcel Proust's major work *A la Recherche du Temps Perdu* was a highly individual odour experience: dipping a biscuit into a cup of tea, the author experienced an odour sensation that brought back to him the entire world of his childhood. Fortunately for perfumery, such individual experiences make up a limited part of our odour vocabulary and are greatly overshadowed by the culturally shared experiences. Were it not so, perfumery as a cultural phenomenon would not be possible.

Odours as Signs of the Hidden Essence

Odours as a system of signs are different from other important sign systems, such as language and visual signs, in two basic respects. They are perceived as essential rather than superficial and accidental; and they are concrete and material rather than abstract and immaterial. Reference has already been made to our tendency to regard the odours of things and people, more even than their appearance, as signals of their *essence.* This is probably due, somewhat paradoxically, to the greater refinement of our visual perceptions and to our greater reliance upon them. Wherever we have strong visual clues about something, we tend to go by them at the expense of olfactory clues. Only where visual clues are lacking or where we consider them unreliable do we pay great attention to olfactory clues. As a result, olfactory clues acquire the associations of revealing that which is not accessible to the other senses, the hidden core beneath the accessible surface. Other languages are richer than English in their allusions to odour as an indicator of the hidden aspects of things. There is no English equivalent to the German expression: '*Ich kann ihn nicht riechen*' ('I cannot smell him'), in reference to someone who may look perfectly acceptable, but whom for some reason one cannot stand; or: '*Die Sache stinkt mir*' (to me, this is smelly), in reference to a deal or proposition which appears sound on the surface, but in which one expects hidden pitfalls. Nevertheless, there are distinct allusions to smell in such expressions as: 'To be in good/bad odour with someone', 'This is fishy' and in the famous 'There is something rotten in the State of Denmark'.

Although the hidden characteristics revealed by odour are usually negative (why else would one hide them?), there are also occasional hidden positive aspects, as in the French expression '*en odeur de sainteté*'.

Odours as Concrete Signs

The belief in odours as indicators of the essence of things probably rests in part upon the fact that in smelling something, we are touching it, there is the direct contact between the molecules of what we smell and our receptors. In such an intimate encounter, there is no room for deception.

The fact that odour perception is based on material, molecular contact, in contrast to vision and hearing which respond to 'immaterial waves', lends to olfaction, along with taste and touch, a very concrete quality. In somewhat simplified terms, this quality is further emphasised in the processing of the various sensory inputs in the brain: while the neurones of the visual and the auditory system lead to the cortex, the seat of abstract reasoning and analysis, the neurones of the olfactory system first lead to the hypothalamus, to the primitive part of the brain which responds to inputs in their totality rather than analysing them; it is also the part of the brain which directly controls the hormonal system that affects our moods and feelings.

10 LIMITATIONS AND VIRTUES OF THE OLFACTORY SIGN SYSTEM

This concreteness clearly represents a limitation of the olfactory sign system. Odours are always and can only be comments on concrete things: people, things, places, situations. They can never refer to ideas and concepts. From their perspective, the classic philosophers were perfectly justified in calling vision and hearing the higher, and olfaction, taste and touch the lower, senses.

However, the very concreteness of odours also lends specific powers to the olfactory sign system. Because we link the odour of things with their essence, perfumes are powerful in suggesting qualities of things and people. Because olfactory perception always triggers a sense of physical presence, odours can be powerful in creating a sense of anticipation and also a sense of the lingering presence of the past. Pictures remind us of the past, odours make it live again. It might be the direct link of the olfactory system to the hypothalamus which gives odours an immediate, unreasoned emotional impact that is foreign to most sights and sounds.

As we think of the aesthetic impact of odours, the ability of the odour of a flower, of a crushed leaf or of a women's perfume to fill us with a sense of pleasure and well-being, we come to the questions: Can this feeling also be explained in terms of the signal character of odours? Is it learned? Do flowers smell beautiful to us only because we have been taught that they are beautiful? Are women's perfumes beautiful only because we associate them with all the positive qualities of the women that wear them? Or is there something innate, unlearned, a kind of harmony between the chemistry of the odour stimulus and our own physiology? These are questions to which I have no answer.

REFERENCES

1. Bain, H., Measurement of consumer perceptions and evaluation of odors as an aid to perfume selection. In *ESOMAR Seminar on Research for Flavors and Fragrances, Lyon 1989.* Esomar, Amsterdam, 1989, pp. 39–53.
2. Preti, G., Cutler, W. B., Christensen, C. M., Lawley, H., Huggins, G. R. & Garcia, C.-R., Human axillary extracts. *J. Chem. Ecol.*, **13** (1987) 717–31.

3. Sausare-Bazzano, G., Cummings, B., Seller, A. K. & Reisner, R. M., Differences in the lipid constituents of sebum from pre-pubertal and pubertal subjects. *Brit. J. Dermatol.*, **103** (1980) 131–7.
4. Jellinek, J. S., The skin health index as a guide to perfuming cosmetic emulsions. *Dragoco Report*, **30** (1983) 55–65.
5. Schaal, B., Olfaction in infants and children: developmental and functional perspectives. *Chem. Senses*, **13** (1988) 145–90.
6. Schaal, B., Montager, H., Hertling, E., Bolzoni, D., Moyse, A. & Quichon, R., Les stimulations olfactives dans les relations entre l'enfant et la mère. *Reprod. Nutr. Div.*, **20** (1980) 843–58.
7. Schleidt, M., Hold, B. & Attili, J., A cross-cultural study on the attitude towards personal odors. *J. Chem. Ecol.*, **7** (1981) 19–31.

Chapter 3

Semiochemicals: Mevalogenins in Systems of Chemical Communication

WITTKO FRANCKE

Institut für Organische Chemie, Universität Hamburg, D-2000 Hamburg 13, Germany

1 SEMIOCHEMICALS

General Remarks

Besides optical, acoustical and tactile stimuli, living organisms use chemical cues for the transmission of information. Chemotaxis is a general archaic principle, and a wide range of living beings, microorganisms as well as primates, are capable of releasing compounds ('semiochemicals') which are used in intraspecific and interspecific chemical communication.

The importance of the 'chemical channel' and its role in animal communication depends on the development and use of alternative signals. Humans appear to base communication primarily on sound and vision. Strangely, they traditionally try to wash away any personal smell and tend to replace it by an artificial ('individual') odour (perfume) which they use in a somehow synthetic system of odour communication. Nevertheless, there are strong indications that a 'true' chemical communication may also take place: babies are obviously able to identify their own mothers by means of chemical signals, and mothers can distinguish the odour of their own babies. In a recent review[1] the subject is treated more extensively. Apart from parent–child relationship, odours may play a certain role as additional components in human communication; however, there is no compound which will cause a spontaneous unavoidable reaction. In contrast, odour communication seems to be particularly important in insects; these relatively small living beings, who have to take their bearings in a rather spacious environment, developed a multitude of glands for the production and release of volatile signals.

Chemical signalling between individuals obviously evolved several times and for different reasons. Development of cell-aggregating compounds and the establishment of recognition substances appear instrumental to the formation of multicellular organisms; compounds used for communication within such cell aggregations may be regarded as antecedents of hormones and neurotransmitters, etc.

Interspecific and intraspecific chemical defence in host–parasite interactions or in the struggle for living space among animals, plants and microorganisms includes a broad spectrum of compounds which range from behaviour-mediating deterrents or irritants to physiologically active compounds such as growth inhibitors, toxins or antibiotics. Defence compounds mostly act against several species and are often produced in rather large amounts.

Essential oils are believed to be involved in attraction of many insects to their host plants. Co-evolution of plants and herbivores may form a special basis for the origin of systems of chemical communication. Food constituents which attract both sexes will cause the discovery of sexual partners at places where food is found. Mimicking such signals through storage and release of respective compounds by animals with the aim of attracting or deterring others, will lead to the experience that metabolites of food- and/or host-compounds (or appropriate imitations thereof) may be advantageously used to specify a chemical message and thus facilitate its decoding by the receiver. In fact, striking similarities between volatiles from animals and plants point to co-evolutionary processes in the relation between herbivores and their host plants.

While the secretion of compounds which were simply sequestered from the environment or which represent products of primary transformations provides information about the quality of a food source or a potential habitat, endogenously produced semiochemicals merely describe the physiological state of the emitter.[2]

According to their mode of action, semiochemicals which are biosynthesized by the emitter have been classified as intraspecifically active 'pheromones'[3]—sub-groups 'releaser' (causing a change of the behaviour in the receiver) and 'primer' (causing a physiological impact on the receiver)—and as interspecifically active 'allelochemicals'—sub-groups 'kairomone' (beneficial for the receiver) and 'allomone' (beneficial for the emitter); see Fig. 1. A review on the terminology is given by Nordlund.[4]

'Sex pheromones' of moths, produced by the females to attract males, are typical releasers. The 'aggregation pheromones' of bark beetles represent another type of releaser: they are produced by one sex but attract both sexes

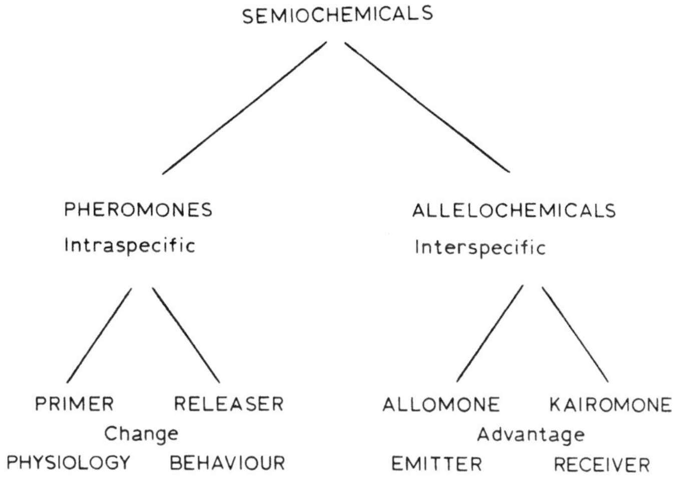

Fig. 1. Terminology of semiochemicals.

and thus induce mass attack to overcome the host tree's resistance during colonization. Trail pheromones, marking compounds and alarm substances are further examples. Typical primers are compounds which are transmitted to suppress development of ovaries in the receiver, to accelerate puberty, block pregnancy or synchronize oestrus. While kairomones are predominantly used during interspecific attacks, allomones represent the whole spectrum of compounds which are used for camouflage and for defence in its widest sense.

Although the above terminology basically represents a useful system, the fact that the classification of semiochemicals is predominantly orientated towards their biological function may sometimes cause confusion. Predators of several bark beetle species locate their hosts by orientating towards their preys' pheromones;[5] with respect to the above definitions, in this example the same compound is produced as an intraspecifically acting pheromone and it is interspecifically 'illegitimately' used as a kairomone. Sometimes an intraspecifically used attractant (pheromone) also acts as a repellent (allomone) for a closely related species. A pheromone of a certain species may also be produced by a predator who uses it as an allomone. Bolas spiders, *Mastophora cornigera,* attract their preys, male moths, by the release of (*Z*)-9-tetradecenyl acetate[6] (Scheme 1, **I-1**), which is one of the most widespread sex pheromone components of female moths.[7] Moreover, a behaviour-mediating releaser may also act at a physiological level as a primer: 9-oxo-(*E*)-2-decenoic acid (**I-2**), which is secreted by the honey bee queen,[8] is a component of her pheromone blend which forms the chemical basis of retinue response and suppresses the development of ovaries in the workers. Last but not least, the terminology for those semiochemicals which are simply sequestered from the environment or which are produced by associated microorganisms and thus not synthesized by the emitter is entirely unclear. The term 'infochemicals', which has been used recently, seems to be diffuse and not very meaningful.

Semiochemicals, however, may not only be ranked in terms of their origin or mode of action, but they may also be classified with respect to their chemical structures. According to the principal processes of primary metabolism, the general variability in chemical structures of compounds used for communication appears not to be unlimited, even with respect to secondary transformations: communication systems seem to be established according to basic concepts which follow distinct rules.

The majority of compounds used for chemical communication are produced from acetate units (acetogenins showing unbranched carbon skeletons), from propionate or succinate units (polyketides with distinct branchings) or from isoprene units (mevalogenins). Figure 2 shows an 'emitter' which under the influence of host compounds and food constituents (with the possible aid of symbionts) secretes a bouquet of metabolites belonging to different chemical classes.

Chemical communication over long distances requires specific signals which can be clearly distinguished as true 'chemical messages', which are different from the 'background noise' caused by odour compounds of the environment. Generally, this is achieved on the basis of unique multicomponent mixtures of less specific compounds which are preferred over the use of species-specific substances, the production of which would require the (costly) development of a species-specific enzyme. In many cases several constituents of the bouquets act synergistically. The aggregation pheromone of the bark beetle, *Pityogenes chalcographus,* is represented by the mixtures of 2-ethyl-1,6-dioxaspiro[4.4]nonane[9] and methyl(*E,Z*)-2,4-decadienoate.[10] The pure spiroacetal (**I-6**) is only slightly active, while the pure ester (**I-5**) is not attractive at all; however, the mixture is highly attractive to both sexes. The spiroacetal is not species-specific as a bark beetle pheromone, and it has been identified, too, as a plant volatile.[11] The above methyl ester is also a component of the secretion of the stink bug, *Euschistus* (Aldrich, J., pers. commun.) and, together with the respective ethyl ester, accounts for the aroma of Bartlett pears.[12]

Enantiomeric composition of chiral compounds plays an important role in semiochemistry. Bark beetles which use chiral oxygenated monoterpenes in

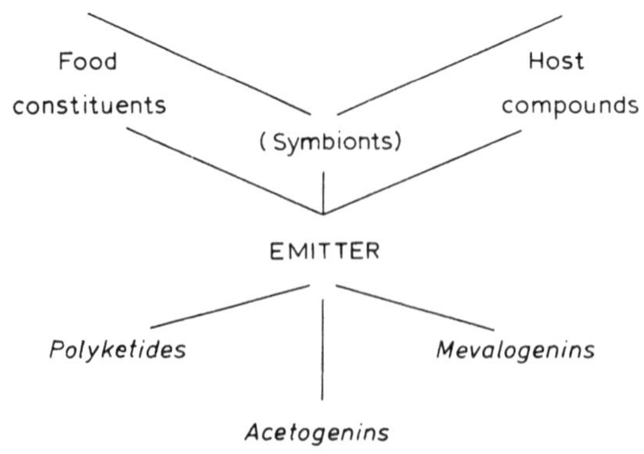

Fig. 2. Biogenetic pathways in the origin of semiochemicals.

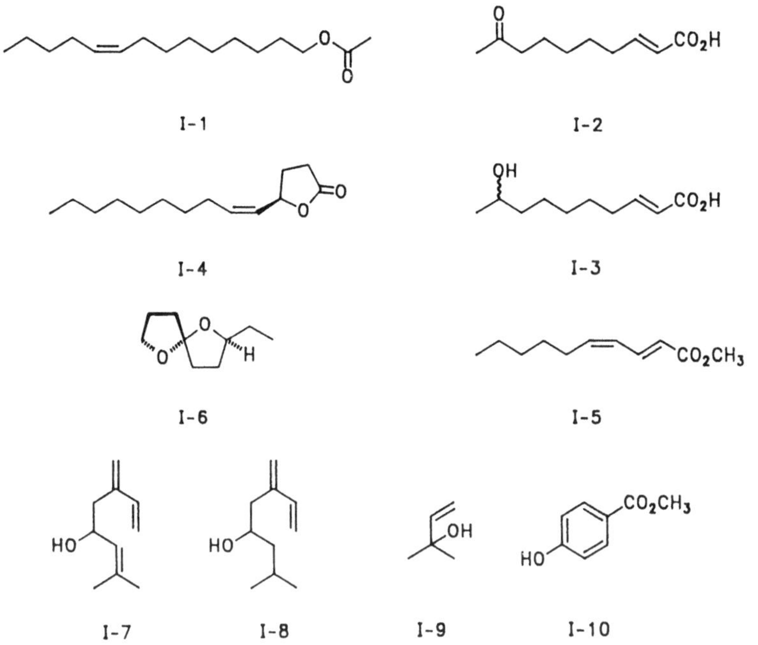

Scheme 1. Various pheromones.

chemical communication may produce species-specific enantiomeric compositions and react highly sensitively to non-natural proportions,[13] the terpene alcohols **I-7** and **I-8** being particularly important in *Ips* spp. The female sex attractant of the Japanese beetle, *Popillia japonica,* is the chiral γ-lactone **I-4**, and male response is significantly reduced by the presence of only 1% of the antipode.[14] In contrast, the bouquet which honey bee queens release to induce retinue response contains both enantiomers of 9-hydroxy-(*E*)-2-decenoic acid (**I-3**), and among other essential compounds the racemate is necessary to elicit this specific behaviour.[15]

Closely related species may produce bouquets which are chemically very similar. Under these circumstances, species specificity of a signal may be based on quantitative differences (including enantiomeric proportions) in the secreted mixture.

Differences in the sensitivity and selectivity at the receptor site as well as differences in the transformation of a perceived signal into a distinct behavioural pattern at the receiver's central nervous system are important elements in the perception of chemical messages. Males of the bark beetle, *Ips typographus,* release the terpenoid 2-methyl-3-buten-2-ol (**I-9**) to attract both sexes of the species,[16] while workers of the hornet, *Vespa crabro,* secrete this compound to induce alarm among conspecific workers.[17] The same alcohol was also found among the volatiles of citrus fruits[18] and hop[19] and has a strong narcotic effect on mice.[19] As another example, besides many other hydrocarbons,

(*Z*)-9-tricosene is particularly widespread among insect volatiles; in most cases it does not seem to show any behaviour-mediating activity. However, houseflies produce and obviously use it as a sex attractant and mating stimulant.[20]

During biosyntheses of active compounds, the action of less specific enzymes may lead to the formation of 'by-products', which cause a 'fine tuning' of the bouquet. If two different species use the same compound as a major attractant, the said minor components need not necessarily be the same and, though intraspecifically inactive, they may well be used as interspecifically active repellents which facilitate species discrimination. Females of the cabbage moth, *Mamestra brassicae,* and the closely related *M. suasa* produce (*Z*)-11-hexadecenyl acetate as the dominant sex pheromone component;[21] however, males of the former species are strongly repelled by small amounts of the respective aldehyde, which is released by the females of the latter species.

Many of the compounds present in a multicomponent secretion may show more than one function: the same compound may act intraspecifically as an attractant and interspecifically as a repellent. For example, the keto-acid **I-2** is intraspecifically active at both the behavioural and the physiological level. Only little is known about antibiotic functions of the secretions: it appears likely that methyl *p*-hydroxybenzoate (**I-10**), which is part of the retinue behaviour-inducing secretion of the honey bee queen,[15] plays such a role, since it is well known for its antibiotic properties. It would be worthwhile to pay more attention to screening semiochemicals and other compounds for these aspects, particularly from insects' secretions.

Differences in vapour pressures and thermodynamic stabilities of the constituents of a multicomponent mixture may reveal information on the age of the signal and/or on the distance of the emitter. However, not all of the secreted compounds must necessarily show any behavioural activity. Many components—and often those which represent major quantities—just act as surfactants, spreading factors or solvents. They may as well represent protecting media or dilutants for thermodynamically unstable semiochemicals or fixatives for highly volatile components. Polar defensive compounds are often embedded in nonpolar media, which after topical application of the mixture carry the active substances through the aggressor's cuticula.[22,23] In short, semiochemicals and accompanying compounds are frequently mixed in a kind of 'cosmetic formulation', including stabilizers and preservation agents. The frass of bark beetles containing their aggregation pheromones acts as a true slow-release system. The connections outlined above are represented schematically in Fig. 3.

Some compounds (sloppily treated as 'trash' in Fig. 3) seem to be 'just there', with no (or unknown) biological function. Such compounds may have had a specific function during earlier evolutionary stages or, as products of less specific enzymes, they may facilitate evolution through introduction of a disposable variant.

Generally, the primary goal of most of the behaviour-mediating odoriferous compounds seems to be the induction of the receiver's attention in a high state

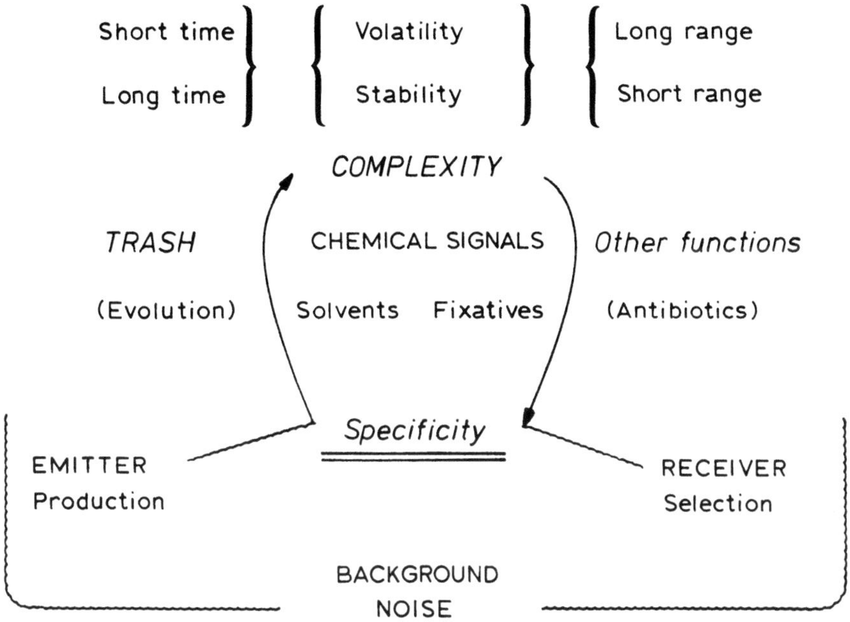

Fig. 3. Specificity and multifunctionality in complex mixtures.

of alertness, which might be one of the reasons for striking structural relationships among semiochemicals. Actually, sex attraction may be regarded in some way as an alarm release. (Thinking in terms of perfumes, humans know this effect very well.)

The present review tries to give a cross-section on semiochemicals and related compounds, aiming at the comparison of mevalogenins which occur in the context of chemical communication. Reviews on structural relationships among semiochemicals showing unbranched carbon skeletons (acetogenins) and on distinctly branched polyketides have been published.[24,25] Excellent reviews on the synthesis of semiochemicals are given by Mori.[26,27]

Isolation and Identification Techniques

Most of the active compounds used as semiochemicals are produced in particularly small amounts, and separation and identification of those components in complex multicomponent mixtures represents a challenge even to laboratories of the highest analytical standard. Isolation and identification techniques are basically the same as in flavour- and aroma-research.

Biologically active material from insects may be obtained by total extraction or from surface washings with organic solvents. The drawback of this rather crude method is the tremendous amount of extracted body lipids which usually cause serious difficulties during further separation steps. Extraction of glands may greatly facilitate the whole procedure. Using on-column GC techniques, solid-sample injection of glands or body tissues gives a true picture of volatile compounds present in the investigated sample. However, these methods represent 'flashlight photos of the state of affairs at a particular moment' and do not necessarily reflect the composition of the 'active principle', which is actually released: some pheromones are generated from precursors, just when they 'leave' the gland. In contrast, head-space collections of volatiles give much better pictures concerning released compounds with regard to both qualitative and quantitative compositions of the blends; an optimal sampling technique is of course most important. For a comprehensive compilation of techniques used in pheromone research see Refs 28 and 29. Since not all of the secreted substances are biologically active, on-line combinations of gas chromatographic separations with electroantennographic methods greatly facilitate the detection of compounds which are actually perceived by the receiver.

Detection of a target compound is usually followed by identification. NMR techniques give the most significant data; however, their application is limited because of the relatively low sensitivity, even of modern instruments (*c.* 1 μg for ^{1}H-NMR and 10–15 μg for a C–H correlation analysis), and since isolation procedures are difficult and time-consuming and often involve large numbers of individuals. In contrast, on-line methods like GC/MS or GC/FT-IR as well as two-dimensional gas chromatography coupled to tandem mass spectrometry (GC–GC/MS–MS) combine high selectivity on the separation side and high sensitivity (picogram to nanogram region) on the detector side; unfortunately the data obtained are often less specific.

The more significant spectroscopic or chromatographic data are *a priori* available on a target compound, the smaller will be the amount needed for structure elucidation. In contrast, the more specific the data must be, which are necessary for structure elucidation (x-ray analysis), the higher will be the required amount of the target compound.

Besides the classical derivatization reactions used in gas chromatography, some special micro-reactions and mass-spectroscopic investigations of the derivatives produced are particularly useful in structure elucidation:[30] catalytic hydrogenation removes double bonds; reaction with various complex hydrides reduces esters and/or carbonyl compounds to alcohols; addition of dimethyl disulphide to unsaturated compounds enables the determination of double-bond positions. In many cases these reactions may be carried out with crude mixtures.

Once the gross structure of an active chiral compound has been identified, determination of the absolute configuration of the natural product is essential because the target organism often reacts differently to non-natural enantiomers or diastereomers or needs defined mixtures for maximum response.

Determination of enantiomeric compositions by gas chromatography may be achieved either by separation of diastereomeric derivatives on achiral stationary phases or by separation on chiral stationary phases. Metal complexes of terpenoids ('complexation gas chromatography'[31]) or amino-acid derivatives of polysiloxanes are used as chiral stationary phases.[32,33] Recently, the use of modified cyclodextrins (Lipodex) opened a new dimension in gas-chromatographic determination of enantiomeric compositions of both natural and synthetic samples of chiral compounds.[34] Considerations of biogenetic relationships of compounds which belong to the same odour bouquet or which occur in a related species may give some ideas on compounds which might be expected in a certain context; based on reasonable assumptions, such 'biosynthetic analysis' may sometimes be particularly useful in identification procedures.

2 MEVALOGENINS

Almost all types of signals, from sex pheromones to highly potent defence substances, are found among the mevalogenins. Many of these compounds seem to be direcly sequestered from plants or represent simple transformation products thereof. It has, however, been shown that *de novo* syntheses may also take place.[35–37] Sometimes associated microorganisms play an important role in the production of terpenoids: they may be involved in *de novo* syntheses as well as in secondary (or tertiary) transformation of plant-derived compounds.[38,39]

Hemiterpenes

'Oxygenated isoprene' has been identified from many species. However, it is not always clear whether such iso-C_5-units represent metabolites of higher terpenes or compounds of *de novo* syntheses via mevalonate. On the other hand, widespread insect volatiles like 3-methylbutanol or 2-methylbutanol and the respective aldehydes, as well as carboxylic acids and respective unsaturated compounds,[23] may certainly also be produced from leucine or isoleucine.

2-Methyl-3-buten-2-ol (Scheme 2, **II-1**) has already been mentioned as an aggregation pheromone of bark beetles[16] and as an alarm pheromone of

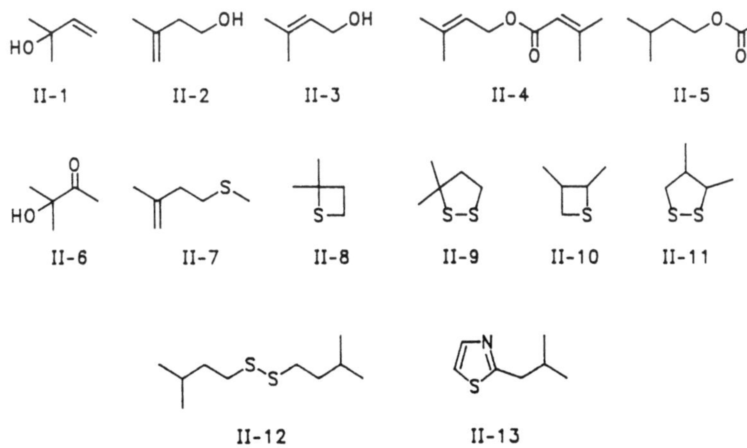

Scheme 2. Hemiterpenes.

hornets.[17] 3-Methyl-3-buten-1-ol (**II-2**) is a pheromone of the larch bark beetle, *Ips cembrae*,[40] and together with its methyl ether and the isomeric 3-methyl-2-buten-1-ol (**II-3**) a component of the complex secretion of the preorbital gland of an antelope species.[41] Esters of these alcohols with a series of long-chain fatty acids are constituents of Dufour's gland secretions of several bee species.[42] Hemiterpenoid esters like 3-methyl-2-buten-1-yl 3-methyl-2-butenoate (**II-4**) were found in the poison glands of workers of the European hornet, *Vespa crabro*;[43] however, nothing has been reported on the biological activity of these esters. In contrast, 3-methyl-1-butyl acetate (**II-5**) is one of the most potent alarm pheromones in honey bees.[44]

A higher oxygenated isoprenoid, 3-hydroxy-3-methylbutan-2-one (**II-6**) is a volatile constituent of ambrosia beetles[45] and induces an extremely high response in electroantennograms (EAG-signals) in *Trypodendron lineatum* (Payne, T., pers. commun.). The same compound was also found in the mandibular gland of the parasitic wasp, *Rhyssa persuasoria.*[46]

Sulphur-containing components account for the characteristic smell of odour marks of many mammals. Several of these compounds showing hemiterpene structures could be identified in the anal gland secretions of mustelids:[47–49] 2,2-dimethylthietane (**II-8**), 3,3-dimethyl-1,2-dithiolane (**II-9**), 2,3-dimethylthietane (**II-10**), 3,4-dimethyl-1,2-dithiolane (**II-11**), bis(3-methylbutyl) disulphide (**II-12**) and 3-methyl-3-butenyl methyl sulphide (**II-7**). The compounds are part of species-specific blends, which are used in odour marking and during defence. The methyl sulphide (**II-7**), the complement of the respective ether,[41] was also found in the urine of the red fox.[50] Cryptic isovaleric acid is represented by 2-isobutyl-1,3-thiazole (**II-13**) and its 4,5-dihydro product which are constituents of the preorbital gland secretion of antelopes;[51] **II-13** was previously known as a natural flavour component of tomatoes.

Monoterpenes

Monoterpenes are particularly important in systems of bark beetle communication,[13,52,53] but they were also identified in many other insect species. As typical plant constituents, monoterpene hydrocarbons play a role in host selection: the pine beauty moth, *Panolis flammea* (Noctuidae) can distinguish between different ratios of α-pinene (Scheme 3, **III-1**) and β-pinene (**III-2**) in selecting its oviposition site.[54] On the other hand, α-pinene and β-pinene are part of the alarm pheromones of certain aphid species.[55] Owing to their habitat, almost all bark beetles attacking coniferous trees contain monoterpene hydrocarbons; however, pronounced 'pheromonal' activity of these compounds could be shown for only a few primitive species;[56] details will be discussed later.

The soldiers of many termite species secrete series of monoterpene hydrocarbons from their frontal glands, α-pinene, β-pinene, limonene (**III-3**), myrcene (**III-4**) and ocimene (**III-5**) being most abundant. Workers of some

III-1

III-2

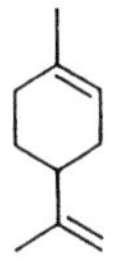
III-3

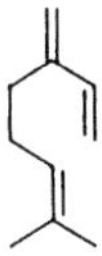
III-4

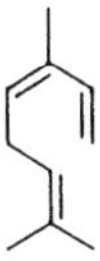
III-5

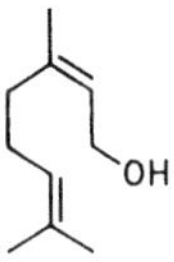

III-6

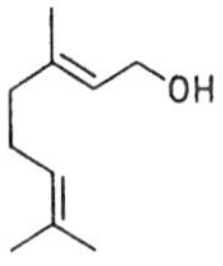

III-7

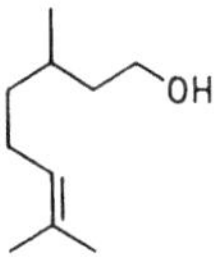

III-8

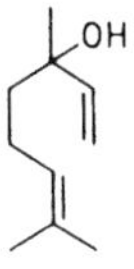

III-9

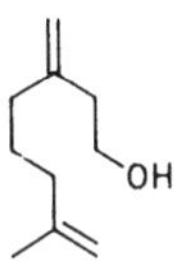

III-10

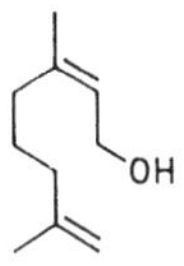

III-11

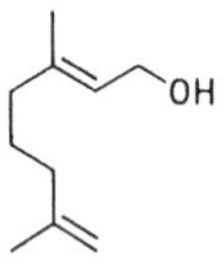

III-12

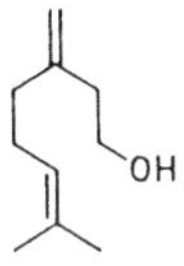

III-13

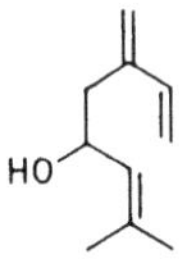

III-14

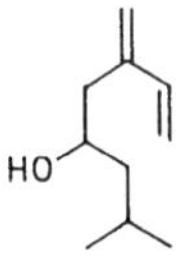

III-15

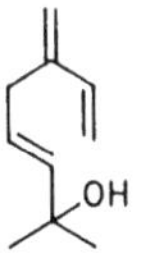

III-16

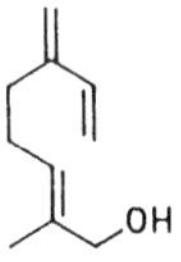

III-17

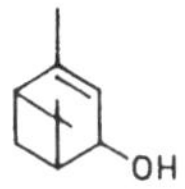

III-18

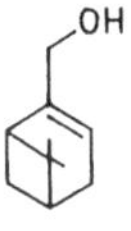

III-19

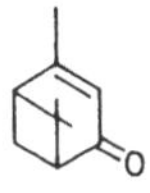

III-20

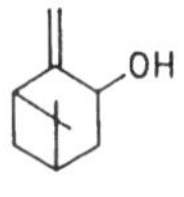

III-21

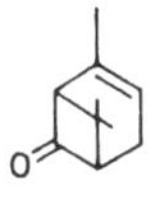

III-22

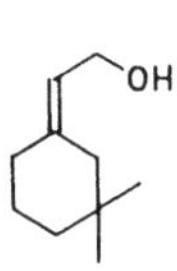

III-23

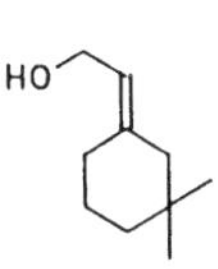

III-24

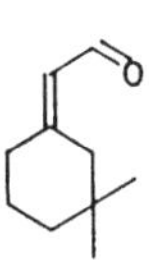

III-25

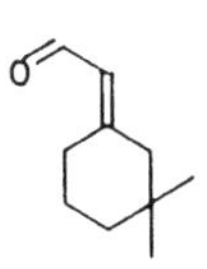

III-26

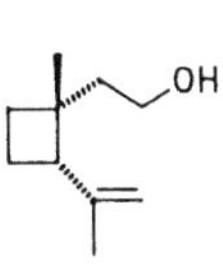

III-27

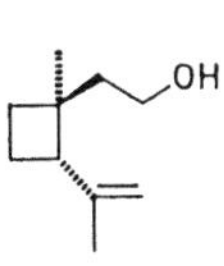

III-28

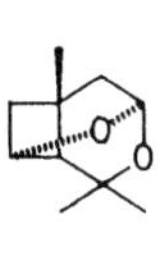

III-29

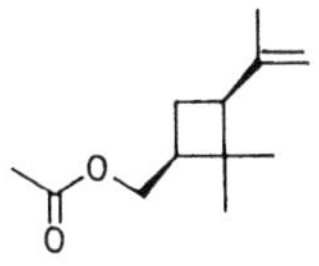

III-30

Scheme 3. Monoterpenes.

ant species secrete the same components from their poison glands,[22,57] and *Tenebrionid* beetles are also capable of producing and using them as part of their defensive secretions.[58] The role of such compounds in male *Pieris* butterflies is unclear.[59] Generally, bees, ants and particularly termites[57,60,61] seem to use monoterpene hydrocarbons for defence or as solvents for more toxic defence compounds. The defensive secretion of the termite *Nasutitermes princeps* contains large amounts of enantiomerically pure (+)-α-pinene,[61a] while the limonene secreted by workers of the ant, *Myrmicaria eumenoides,* proved to be the optically pure (+)-enantiomer.[61b]

Mono-oxygenated open-chain terpenes and terpene esters are abundant in bees,[57] certainly at least in part because of their close contact with flowers during pollination. Typical flower volatiles like nerol (**III-6**), and geraniol (**III-7**) as well as citral (the mixture of the respective aldehydes) and nerolic and geranic acid are particularly widespread. Other common open-chain monoterpenes are citronellol (**III-8**) and linalool (**III-9**). Citronellol is a male sex attractant of the two-spotted spider mite, *Tetranychus urticae.*[62] Neryl formate is an alarm pheromone of the cheese mite, *Tyrophagus putrescentiae,*[63] while a mixture of propanoates of the geraniol isomers, 7-methyl-3-methylene-7-octen-1-ol (**III-10**) as well as (*Z*)- and (*E*)-3,7-dimethyl-2,7-octadien-1-ol (**III-11, 12**) comprise the female sex pheromone of the San José scale insect.[64] A further geraniol isomer, γ-geraniol (**III-13**) is part of the aggregation pheromone of the bark beetle, *Pityogenes quadridens.*[65]

Series of esters of geraniol with acetic, butyric, hexanoic, octanoic and decanoic acids were found in the Dufour's gland secretions of many *Andrenid* bees; also 'true' terpenoid esters like geranyl geranoate and several partially hydrogenated esters of the same type have been identified from bees, wasps and hornets.[57,66] However, no biological function of these compounds has been reported.

Neral, geranial and linalool form part of the aggregation pheromone or mate-seeking cue of *Colletes* bees[67,68] and represent sex pheromone components of a parasitic wasp.[69] Stingless bees, *Lestrimelitta limao,* which do not forage from plants, secrete large amounts of citral during their raids, to disorientate prey bee species while plundering their nests.[69a] Centris bees use a mixture of open-chain monoterpenes to set up territories by marking grass stems.[70] The concentration of citral in the mandibular glands of certain leaf-cutting ants[71] can reach such concentrations (possibly used for defence) that a lemon-tasting drink may be produced from the ethanolic extracts of worker heads. While the biological function of citral in the scent of male *Pierid* butterflies is not entirely clear,[59,65,72] the mixture acts as part of the alarm pheromone of certain mite species.[73] Some ant species[57] and *Anthophorid* bees contain nerolic acid and geranic acid in their cephalic secretions. However, in honey bee workers the acids have been identified along with nerol and citral as components of the Nasonov pheromone,[74] which plays a role in worker attraction, nest finding, foraging and marking of food sources.

Monoterpenes play a particularly important role in host selection and mass aggregation of bark beetles. Insects attacking conifers have to overcome both physical and chemical obstacles during host colonization. Defence through sticky and toxic oleoresin is involved in the most effective mechanisms in tree resistance against bark beetle invasion. However, bark beetles developed a series of strategies to survive and colonize selected host trees successfully. Oxygenated monoterpenes found in bark beetles may be conversion products of host terpenes, they may be simply sequestered from the host or synthesized *de novo.* Under natural conditions all three ways may be realized simultaneously.

Derivatization of host plant monoterpenes by bark beetles may originally have represented a detoxification mechanism, but obviously developed into an important principle in the formation of volatile signals which bark beetles use for intraspecific and interspecific communication. Based on investigations on the effects of the exposure of bark beetles to vapours of monoterpene hydrocarbons,[75–77] it is now generally accepted that the bulk of oxygenated monoterpenes present in bark beetles is generated from resinous compounds of the host trees. Allyloxidation or hydration of unsaturated hydrocarbons followed by secondary reactions such as further oxidation, hydrogenation or rearrangements, etc., seem to be important mechanisms in the production of bark beetle pheromones.[52,78]

While some species oxidize host terpenes more randomly and thus generate an array of rather unspecific volatiles[79] which transmit only little information, others use highly selective enzyme systems for the production of unique olfactory signals. None of the oxygenated monoterpenes identified so far represents a species-specific compound *per se,* and species specificity of pheromonal signals is accomplished by differences in the combination of compounds or their enantiomers and/or different modules in perception. The problem has been extensively reviewed in Refs 13, 52, 53 and 80–83.

The myrcene derivatives, ipsdienol (**III-14**) and ipsenol (**III-15**), which were first identified by Silverstein *et al.,*[84] are typical male-specific aggregation pheromones of many *Ips* spp., but they also play a role in host colonization in other bark beetles such as *Pityokteines*[85] and *Xylocleptes.*[86] The respective ketones, ipsdienone and ipsenone, were recently found to be present in males of *I. sexdentatus* and *I. cribricollis* respectively[82] (redetermination of the species originally regarded as *I. lecontei* showed that it was actually *I. cribricollis*).

The biological function of the ketones in the beetles is unknown; they may be involved in interconversion reactions which lead from ipsdienol to ipsenol.[83] Ipsdienone has already been detected quite early in the essential oil of *Lippia asperifolia* (Rich),[87] and we recently found small amounts of ipsdienol in another Verbenacea (Francke, W. & Meyer, H., unpublished). Ipsdienol also proved to be a major component in the floral fragrance of several species of orchids that are pollinated by male *Euglossine* bees; the racemate was shown to be attractive in field tests.[88] While ipsenol is produced from ipsdienol by a hydrogenation cascade (possibly via ipsdienone and ipsenone), the tertiary alcohol amitinol (**III-16**) represents a product of an allylic rearrangement of ipsdienol. Amitinol, first identified in an American *Ips* bark beetle,[89] is also a constituent of the essential oil of *Ledum palustre.*[90] Recently, another product of an allylic oxidation of myrcene, myrcenol (**III-17**), could be identified among bark beetle volatiles.[91] However, again the compound was already previously known from an essential oil.[92]

Oxygenated monoterpenes, which are found in almost every bark beetle species attacking coniferous trees, include *cis*-verbenol (**III-18**), *trans*-verbenol and myrtenol (**III-19**) which represent primary products of allylic oxidation of the host terpene, α-pinene. Further oxidation leads to the respective carbonyl compounds, verbenone (**III-20**) and myrtenal, which are common bark beetle volatiles, too. Similar transformations of β-pinene give *trans*-pinocarveol (**III-21**) and the respective ketone, pinocarvone. Among these bicyclic terpenes *cis*-verbenol is a particularly important aggregation pheromone in *Ips* spp., while *trans*-verbenol is used by *Dendroctonus* spp. Both sexes of *Ips* species oxidize the chiral α-pinene enantioselectively:[76,93] (*S*)-*cis*-verbenol is produced from (−)-α-pinene, while (+)-α-pinene yields (*S*)-*trans*-verbenol. Verbenone, which in bark beetles appears to be formed largely from verbenols owing to the action of associated microorganisms,[38,39] seems to act as a general inhibitor, which the beetles use to avoid overpopulation and which induces shifting of the attack to another tree.[81,94] In contrast, verbenone in the scent of *Knautia* flowers is attractive to *Zygaenid* moths.[95] Simple rearrangements of verbenone yield the well-known flower constituent, chrysanthenone (**III-22**), which was also found in bark beetles (Francke, W. & Bartels, J., unpublished); however, its biological significance is still unknown.

The cyclohexane derivatives **III-23**, **25** and **26** and (+)-*cis*-1-(2-hydroxyethyl)-1-methyl-2-isopropenylcyclobutane (**III-27**; grandisol), which are probably derived from myrcene (**III-4**) or γ-geraniol (**III-13**), are male sex pheromones of the boll weevil.[96] While **III-24**, another possible cyclization product of γ-geraniol, and grandisol are part of the aggregation pheromones of some *Pityogenes* bark beetles,[65] grandisol and the respective aldehyde, grandisal, are aggregation pheromones of several *Pissodes* weevils.[97] Interestingly, the *trans*-isomer of grandisol was found as a plant constituent,[98] which was called fragranol (**III-28**). The tricyclic acetal **III-29** (lineatin), a higher oxygenated derivative of grandisol, is an aggregation pheromone of several *Trypodendron* sp. ambrosia beetles;[99] the natural product was shown to be the (1*S*,4*R*,5*S*,7*S*)-(+)-enantiomer.[100] Another terpene containing a cyclobutane system is (1*R*-*cis*)-(+)-1-acetoxymethyl-2,2-dimethyl-3-(1-methylvinyl)-cyclobutane (**III-30**), the sex pheromone of the mealy bug, *Planococcus citri*[102] which shows an interesting linkage between two isoprene units. A similar compound prenylated at the isopropenyl moiety and at the geminal dimethyl group (thus representing a diterpene) is known from plants.[102]

Higher oxygenated monoterpenes, showing various biological functions, have been identified in several insect species. The diol (Scheme 4, **IV-1**) is believed to act as some kind of solvent for pheromones of male monarch

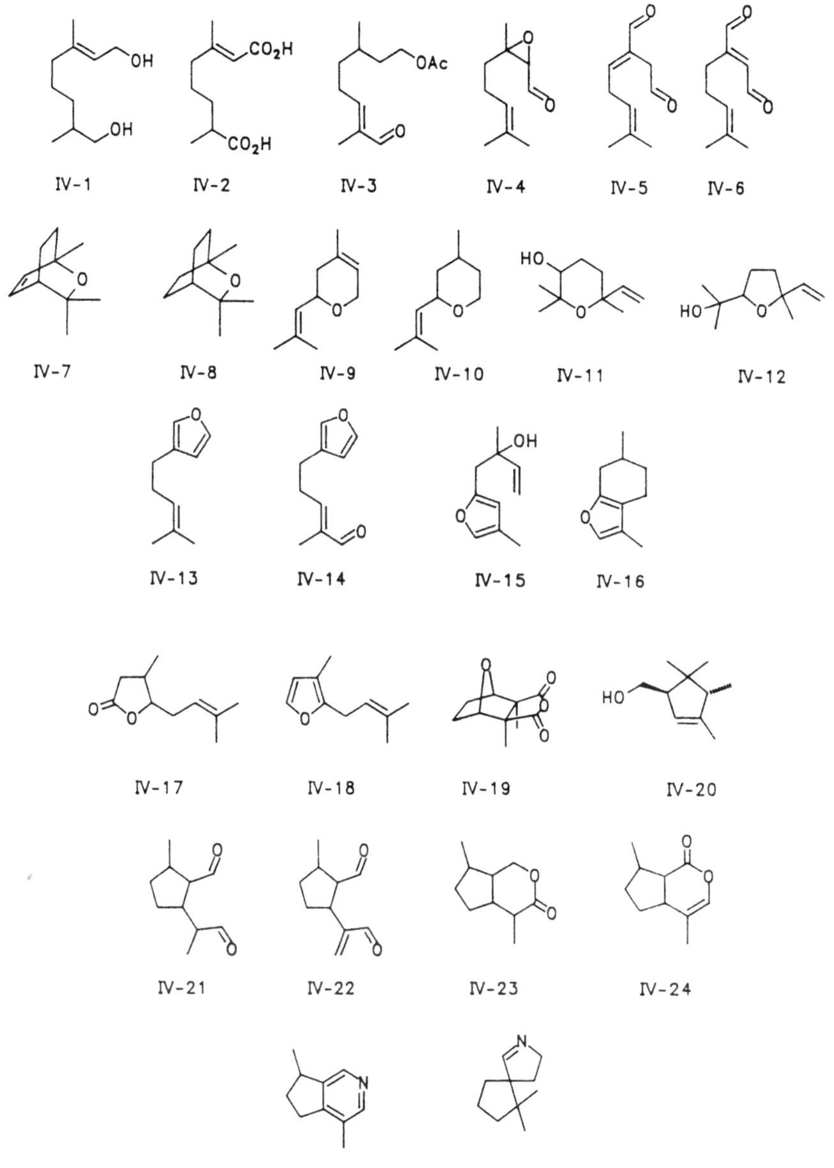

Scheme 4. Higher oxygenated monoterpenes.

butterflies,[103] whereas the respective dicarboxylic acid (**IV-2**) is part of the sex pheromone of the azuki bean weevil.[104] (*E*)-8-Oxocitronellyl acetate (**IV-3**) is a constituent of the defence secretion in rove beetles[105] and is also present in the mandibular gland secretion of *Panurginus* bees.[106] 2,3-Epoxyneral (**IV-4**),[107] the two dialdehydes **IV-5** and **IV-6** (β- and α-acaridial)[108,109] and dehydrocineole (**IV-7**)[110] have been isolated from mites, while 1,8-cineole

(**IV-8**), along with α-pinene, limonene and straight-chain compounds, was found in bugs.[111] Among other components, ant lions use nerol oxide (**IV-9**) as a volatile signal,[112] while digging wasps, *Bembix* spp., emit rose oxide (**IV-10**) from their mandibular glands (Francke, W., Bergström, G. & Tengö, J., unpublished): this had already been previously identified from longhorn beetles.[113] Further oxygen heterocycles which are known as volatile constituents of flowers include the pyranoid- (**IV-11**) and furanoid- (**IV-12**) forms of linalool oxide which are components of the androconial secretion of *Amauris echeria.*[114]

Also, furan derivatives have been described from insects: perillene (**IV-13**) was found in ant species,[115] whereas the aldehyde **IV-14** (perillenal) was identified in a saw fly species.[116] The linalool derivative **IV-15** was isolated from the abdominal secretion of males of *Elymnias thryallis,* a tropical butterfly (Schulz, S., Francke, W. & Roisin, Y., unpublished). It shows an oxygenation pattern similar to that of the well-known plant constituent, menthofuran (**IV-16**). Structural relationships between the γ-lactone **IV-17** (eldanolide), which represents the sex pheromone of the male sugar-cane borer, *Eldana saccharina,*[117] and a plant constituent like rose furan (**IV-18**)[118] are also quite obvious and again point to the close relationship between insect and plant volatiles.

The interesting tricyclic rearranged terpene **IV-19** (cantharidin) is relatively widespread in insects.[22] Males of the meloid beetle, *Lytta vesicatoria* (Spanish fly) are able to produce it from mevalonate and from farnesol and transfer it to the females (who do not produce this defensive compound) during copulation.[119] Cantharidin acts as an intraspecific attractant and appears to play an essential role in female choice.[2]

A variety of cyclopentanoid monoterpenes are known from insects. While they are defence compounds of the carrion beetle, *Necrodes surinamensis,*[120] which represent a unique class of compounds, iridoids occur much more frequently. Various members of this type have been reported from a wide range of insect and plant sources. Isomers of iridodial (**IV-21**) and dolichodial (**IV-22**), as well as lactones like iridomyrmecin (**IV-23**), nepetalactone (**IV-24**) and the respective doubly unsaturated lactones, are widespread defensive compounds among Formicine ants, Staphylinid as well as Cerambycid beetles and larvae of many leaf beetles.[22,23,105,121–123] Dolichodial was also isolated from a plant source, *Teucrium marum.*[124] Nepetalactone, which has long been known from catnip,[125] accounts for the characteristic response of cats to catnip. The compound has also been identified recently as a sex pheromone of the aphid, *Megoura viciae*;[126] the aphid pheromone shows (*4aS,7S,7aR*)-configuration and acts synergistically together with the respective lactol. Actinidin (**IV-25**), an alkaloid derivative of iridoids, is a defence compound of several rove beetle species.[105,127] A second interesting terpene alkaloid is the azomethine **IV-26** (polyzonimine), a defence substance of the millipede, *Polyzonium rosalbum*.[128]

Sesquiterpenes

The majority of sesquiterpenes identified in insects are represented by open-chain compounds. Among other compounds, (E,E)-α-farnesene (Scheme 5, **V-1**) is a component of the sex pheromone of the Mediterranean fruit fly, *Ceratitis capitata.*[129] Acting as an allomone, a mixture of (E,E)- and (Z,E)-α-farnesene, produced by apples during ripening, is an attractant to larvae of the codling moth, *Laspeyresia pomonella,* and acts as an oviposition stimulant for gravid female adults.[130] In the ant, *Solenopsis invicta,* (Z,E)-α-farnesene is a constituent of the Dufour's gland secretion which acts as a trail pheromone.[131] From aphids several isomers of α- and β-farnesene were identified,[55] among which (E)-β-farnesene (**V-2**)[132] shows the strongest effects as an alarm pheromone. Both α- and β-farnesene were found in the tarsal glands of the springbok antelope;[133] (E,E)-α-farnesene was also identified in male Hepialid moths, *Hepialus humuli* (Schulz, S., pers. commun.).

(E,E)-Farnesol (**V-3**) has been frequently reported as a component of the secretions of Dufour's glands of Andrenid bees,[134,135] of the Nasonov glands of honey bee workers, of the labial glands of bumble bees[136,137] and of the mandibular glands of leaf-cutting ants.[71] The same range of carboxylic acids which form geranyl esters in several bee species (see above) are also components of farnesyl esters.[134,135,138,139] Farnesyl hexanoate, the most common farnesyl ester, has also been identified in bumble bees (Francke, W. & Lübke, G., unpublished) and in the tarsal glands of the European roe deer (Schildknecht, H., pers. commun.). Nerolidol (**V-4**) was only seldom found in insects: it is a constituent of the sex-attracting secretion produced by male dung beetles, *Kheper* spp.,[140] it was earlier identified as part of the alarm pheromone of the mite *Tetranychus urticae,*[141] and recently detected in danaid butterflies.[114] (E,S)-Nerolidol has also been reported as an antifeedant for the larvae of the gypsy moth.[142] 2,3-Dihydrofarnesol (terrestrol) is used as a territorial marking substance by several bumble bee species;[136,143,144] the acetate of this alcohol and the respective aldehyde are also present in different ant species.[145] Beside (E)-farnesol, the two hydrated farnesols, 3,7,11-trimethyl-2,10-dodecadien-1,7-diol (**V-5**) and 3,7,11-trimethyl-2-dodecen-1,7,11-triol, are constituents of the temporal gland secretion of the African elephant,[146] and it has been suggested that these compounds are used in scent marking. Respective derivatives of nerolidol such as **V-6** are known from caparrapi oil.[147] (E,E)-Farnesal and the (Z,E)-isomer were identified from the frontal gland of workers of a termite species and are believed to be involved in chemical defence,[148] while males of the rice moth use the same compounds to attract females.[149] The interesting acyclic sesquiterpene **V-7** (nomadone) is a constituent of the mandibular gland secretion of *Nomada* spp.[65] The biological function of this ketone and several very closely related acyclic sesquiterpenes and norsesquiterpene ketones is not yet entirely clear; however, they seem to play a role in intraspecific and interspecific territorial behaviour.

The monocyclic sesquiterpene, (−)-germacrene A (**V-8**) is an alarm pheromone of the alfalfa aphid, *Therioaphis maculata.*[150] This very thermolabile compound was first isolated from a gorgonian coral[151] and, among other terpenes, is also present in the frontal gland of a termite species[152] as well as in the defensive osmeterial secretion of swallowtail butterflies (Papilionids), which contain various additional sesquiterpenes and monoterpenes.[153] The two germacrane derivatives, periplanone A (**V-9**) and periplanone B (**V-10**) are the sex pheromones produced by females of the American cockroach, *Periplaneta americana.* While periplanone B had already been known for some time,[154] and its absolute configuration could be determined ten years ago,[155] structure elucidation of periplanone A still took quite some time.[156] Only recently, additional germacrane derivatives like **V-11** (periplanone D_2) could be identified.[157]

A bicyclic oxygenated sesquiterpene, which shows some structural relationships to the germacrane system, is caryophyllene oxide (**V-12**). It forms a major volatile component of the osmeterial secretion of larvae of the swallowtail, *Papilio memnon,*[153] and it was also identified as a constituent of the essential oil of tropical plants, where it acts as a toxic deterrent to leaf-cutting ants.[158] In bark beetle communication, sesquiterpenes are far less important than monoterpenes: the host-produced hydrocarbon, α-cubebene (**V-13**), which is an attractant for elm bark beetles,[159] appears to represent an exceptional case. However, further examples are likely to be discovered. Other sesquiterpene hydrocarbons such as δ-cadinene (**V-14**) and α-bisabolene (**V-15**) could be identified from the defensive secretions of termites; respective oxygenated compounds were also found.[57] The cadinol **V-16** is a constituent of the frontal gland secretion of *Subulitermes* soldiers,[160] and it is also a component of the wing scents of male *Lycaenid* butterflies, where it is believed to play a role in courtship behaviour.[161] Bisabolene oxide **V-17** is a component of the sex pheromone blends of green stink bugs, *Nezara* spp.[162,163] The occurrence of different mixtures of diastereomers of the epoxide in different species and subspecies is believed to have evolved through evolutionary pressure caused by predators which use bisabolene oxide as a kairomone. The dihydropyran **V-18**, which is the main component secreted by the grasshopper, *Phymateus leprosus,* represents prenylated rose oxide and probably acts as a defensive compound (Francke, W. & Schulz, S., unpublished). Similarly to monoterpenes, appropriate structural elements of sesquiterpenes may form furan subunits: dendrolasin (**V-19**), which plays an important role in the defensive chemistry of *Lasius* ants,[164] is the sesquiterpene complement of perillene (**IV-13**). Further oxygenation of dendrolasin would lead to the structure of ancistrofuran (**V-20**) which is a defensive compound of *Ancistrotermes* termites.[165] Exemplified here by (−)-castoramin (**V-21**), an interesting range of furan-sesquiterpenes occurs in the scent gland of the Canadian beaver, *Castor fiber.*[166] The sesquiterpene analogue of the pyranoid form of linalool oxide (**IV-11**), caparrapioxide (**V-22**), previously known from plant sources, was identified from *Amitermes* termite defensive secretions;[167] its function is as yet unknown.

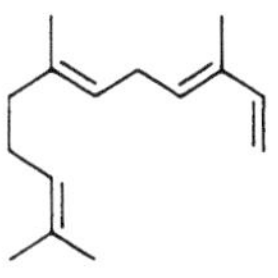
V-1

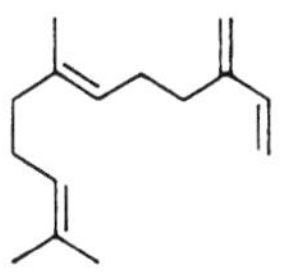
V-2

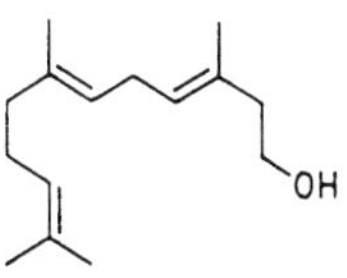
V-3

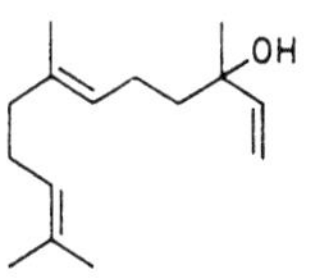
V-4

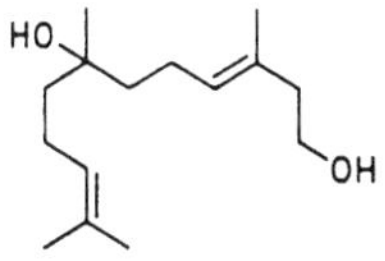
V-5

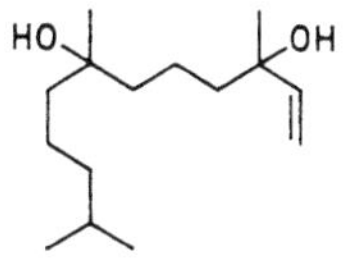
V-6

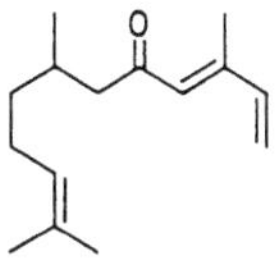
V-7

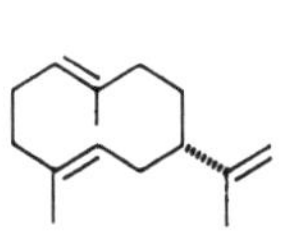
V-8

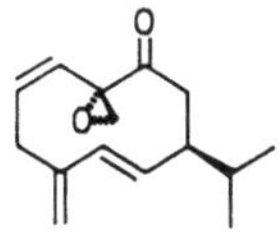
V-9

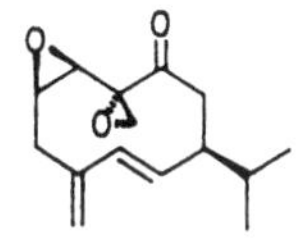
V-10

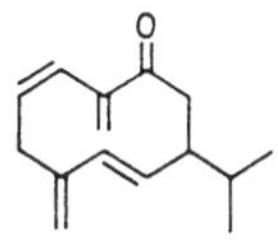
V-11

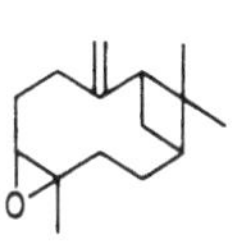
V-12

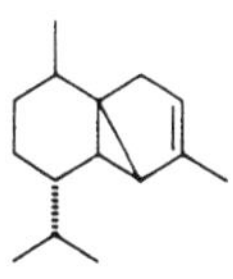
V-13

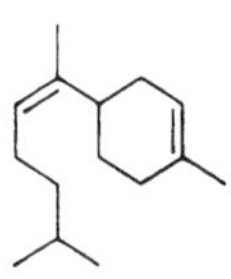
V-14

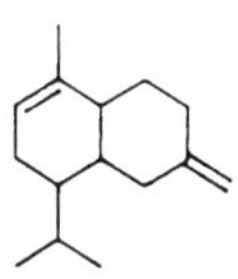
V-15

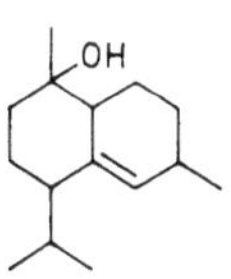
V-16

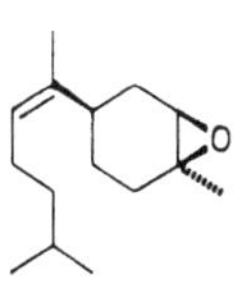
V-17

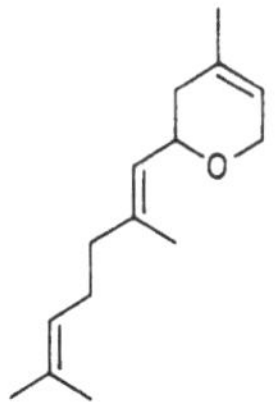
V-18

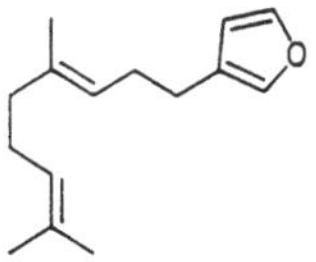
V-19

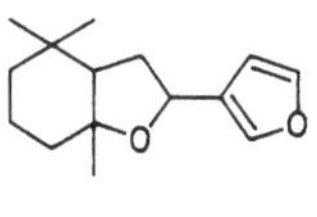
V-20

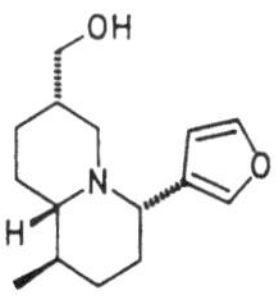
V-21

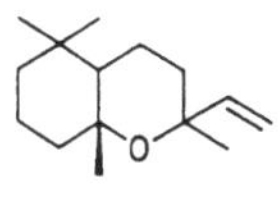
V-22

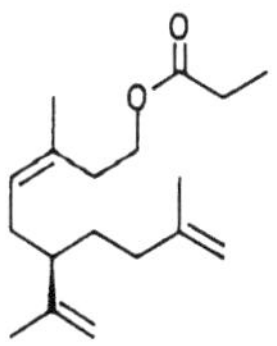
V-23

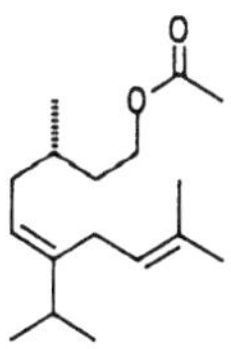
V-24

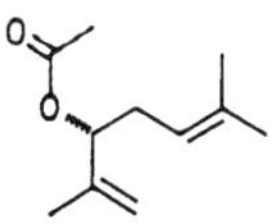
V-25

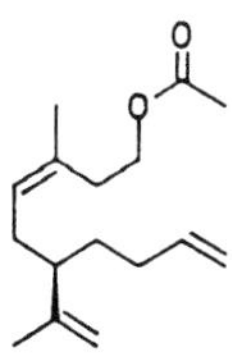
V-26

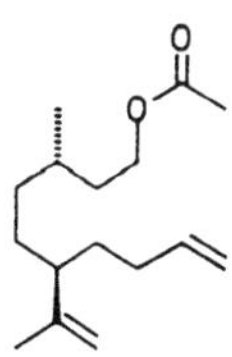
V-27

Scheme 5. Sesquiterpenes.

The 'sesquiterpene' structures of several sex pheromones of female scale insects share an unusual substitution pattern. They represent typical acyclic monoterpene alcohols (see **III-12**, pheromone component of the San José scale insect), which carry an additional side-chain at C-6. The propanoate **V-23** is the pheromone of the white peach scale insect[168] and the acetate **V-24** that of the yellow scale insect.[169] Interestingly, (*R*)-3-acetoxy-2,6-dimethyl-1,5-heptadiene (**V-25**) is the sex pheromone of the Comstock mealy bug.[170] The carbon skeleton of this compound would be generated upon splitting the bonds between C-5 and C-6 in the pheromones **V-23** and **V-24**. The two nor-sesquiterpenes **V-26** and **V-27** make up the sex pheromone of the California red scale insect.[171,172] The close relationships in the chemistry of scale pheromones are striking and point to similar origins. Though nothing is known about the biosyntheses of the compounds, their carbon skeletons indicate that they may be derived from higher terpenes. Actually, cembrene (see Scheme 6, **VI-6**) would yield such degradation products; similar processes are known from plants. Solanone, 5-(1-methylethyl)-8-methyl-(*E*)-6,8-nonadien-2-one, a bis(nor-sesquiterpene) of Burley tobacco, is assumed to be produced through the breakdown of diterpene alcohols, structurally related to cembrene, several representatives of which are well known from tobacco.[173]

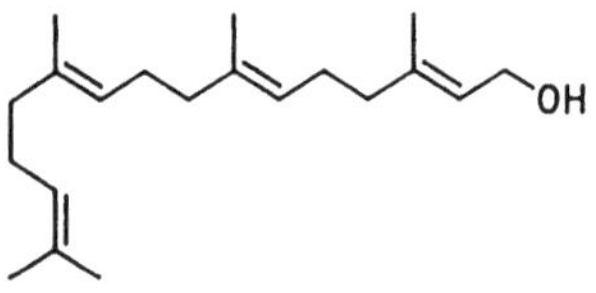

VI-1

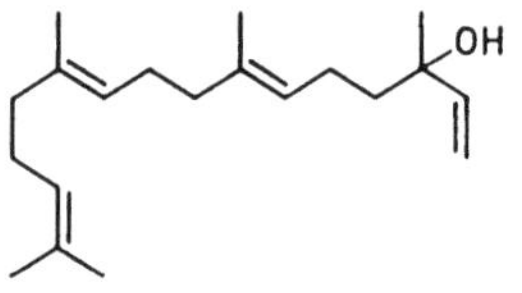

VI-2

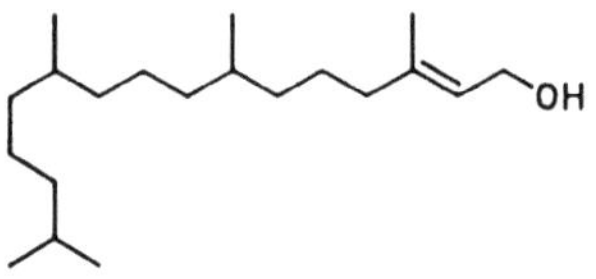

VI-3

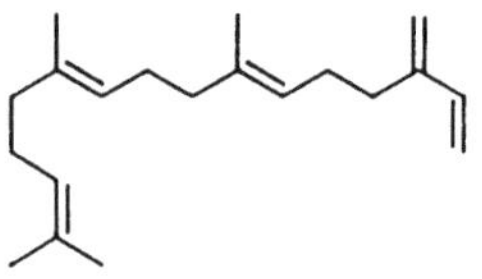

VI-4

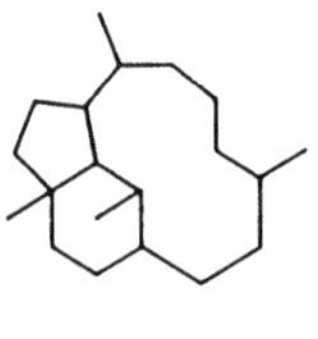

VI-5

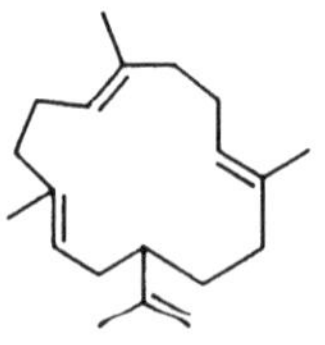

VI-6

Scheme 6. Diterpenes.

Diterpenes

Because of their relatively low volatility, diterpenes seem to be less suitable for odour communication. The acyclic geranyl-linalool (Scheme 6, **VI-1**), possibly the (*R*)-enantiomer, the analogue of nerolidol, is part of the defensive secretion of a termite,[174] while geranylgeraniol (**VI-2**), which is complementary to farnesol, was identified in Dufour's glands of workers of *Formica* ants[175] and in the labial glands of males of several bumble bees, *Bombus* spp.[143] Almost nothing is known about the biological function of these compounds. The 2,3-dihydroalcohol (geranylcitronellol) which is the prenyl homologue of the above-mentioned bumble-bee-marking pheromone, 2,3-dihydrofarnesol (terrestrol), was also found in the labial glands of *Bombus, Psithyrus*[143] and *Pyrobombus* bumble bees.[137] Oxidation products of these alcohols, geranylgeranial and geranylcitronellal, are components of the mandibular gland secretions of *Lasius* ants.[145]

Complementary to the farnesyl esters, esters of geranylgeraniol are also known from *Hymenoptera.*[143,145,175] Geranylgeranyl octanoate is a Dufour's gland product in several North American *Andrena* bees.[139] Respective esters such as geranylgeranyl docosanoate or phytyl dodecanoate were identified from the surface wax of cotton buds and anthers where they act as phagostimulants for adult boll weevils, *Anthonomus grandis.*[176] (*E*)-Phytol (**VI-3**), one of the most typical plant constituents, is a component of the wing gland pheromone of males of the Pyralid moth, *Ephestia elutella* and plays a role in courtship success.[177] Springene (**VI-4**), the prenyl homologue of (*E*)-β-farnesene, is a component of the complex mixture identified from the dorsal gland of the springbok antelope.[133] The vast majority of diterpenes have been identified from the frontal gland secretions of termites,[57] which seem to use them for defence when smearing them on aggressors. Most of these components are derivatives of the tricyclic trinervitane system (**VI-5**). However, monocyclic diterpenes are also known: Cembrene A (**VI-6**) is a trail pheromone of the termite, *Trinervitermes bettonianus.*[178] The compound could be isolated directly from the trail as well as from workers and alate females and proved to be absent in soldiers; however, in another termite species, *Cubitermes umbratus,* the same compound and other monocyclic diterpenes are part of the defence secretion of soldiers.[179] Cembrene was originally identified as a pine constituent.

Norterpenes and Homoterpenes

Many natural products show a typical branching pattern of terpenes without strictly fulfilling the isoprene rule. These compounds may represent either rearrangement or degradation products of higher terpenes, or they may reflect a biosynthesis which involves homomevalonate and/or additional acetate or propionate units. An almost ubiquitous terpenoid is 6-methyl-5-hepten-2-one (Scheme 7, **VII-1**) which represents the product of a formal retro-aldol

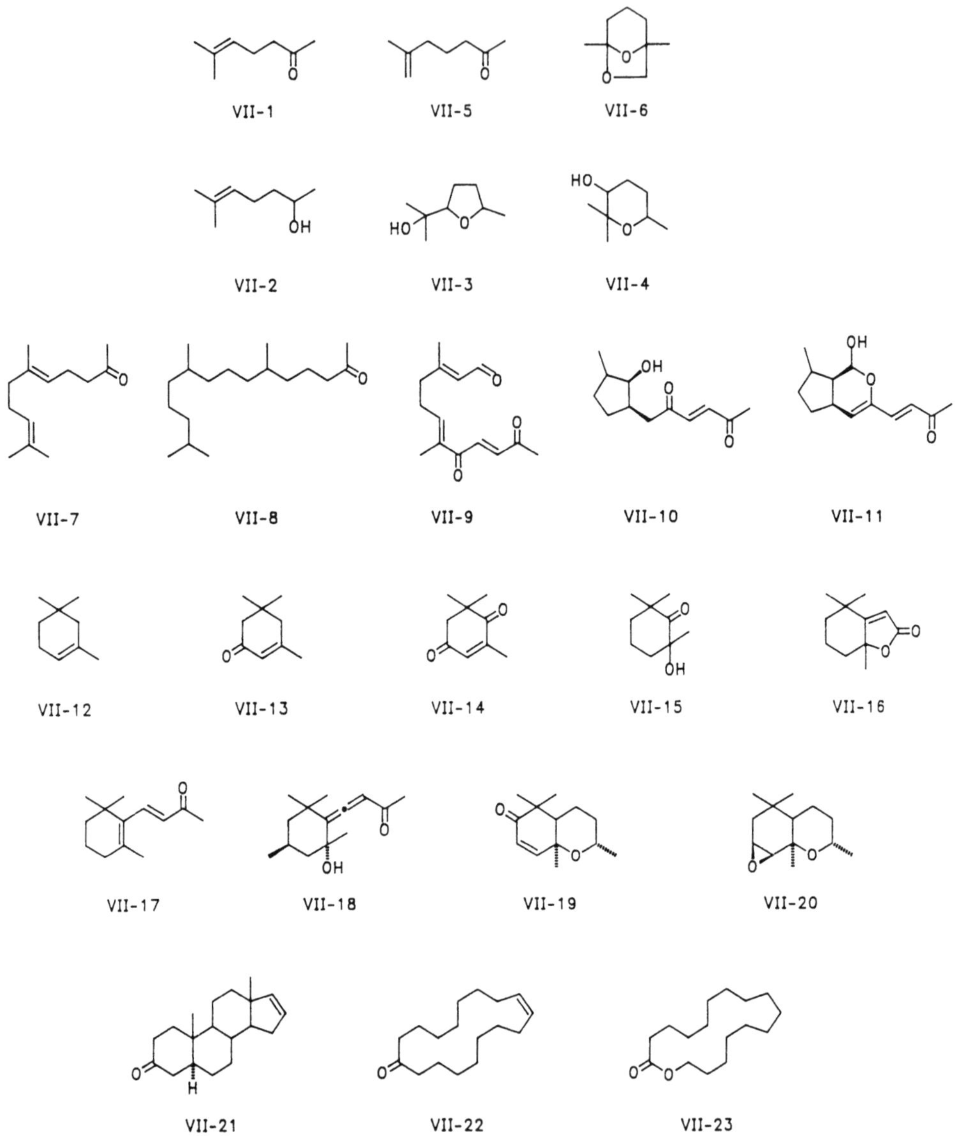

Scheme 7. Norterpenes.

reaction of citral. The terpenoid **VII-1** is a constituent of the essential oils of many plants and has been found in the resin of coniferous trees as well as in the culture media of fungi. It is widespread as a volatile constituent of cephalic and abdominal secretions of ants and bees[57] and was also identified in beetles,[72,180] male butterflies[181] and mammals.[182] Only little is known about the biological activity of this ketone; however, the respective alcohol, sulcatol (**VII-2**), which is frequently associated with the ketone, is an aggregation pheromone of bark beetles, *Gnathotrichus* spp.[183,184] These ambrosia beetles produce species-specific enantiomeric mixtures of the alcohol, and the natural proportions are essential for maximum response. Recently, interesting cyclic derivatives of 6-methyl-5-hepten-2-ol could be identified as bark beetle pheromones: oxidation at the double bond followed by ring closure would yield either 2-(1-hydroxy-1-methylethyl)-5-methyltetrahydrofuran (pityol, **VII-3**) or 2,2,6-trimethyl-3-hydroxytetrahydropyran (vittatol, **VII-4**). While males of *Pityophthorus pityographus* release *trans*-pityol showing (2*R*,5*S*)-configuration,[185] the elm bark beetle, *Pteleobius vittatus,* uses *cis*-pityol and *cis*-vittatol of yet unknown absolute configurations.[180] The structural relations between sulcatol, pityol and vittatol are essentially the same as between linalool (**III-9**) and the pyranoid and furanoid forms of linalool oxides (**IV-11**, **IV-12**). Two other bisnorterpenes, 6-methyl-3,5-heptadien-2-one, which is also present in the tarsal glands of the springbok,[133] and 6-methyl-6-hepten-2-one (**VII-5**) could be identified as volatile constituents of the bark beetle, *Dendroctonus simplex.*[186] The latter ketone might serve as a precursor of the bicyclic acetal, frontalin (**VII-6**), which under physiological conditions would be easily produced from 6,7-epoxy-6-methylheptan-2-one. Frontalin is an important aggregation pheromone in many *Dendroctonus* bark beetles,[13,187] in *Dendroctonus simplex* it occurs as the pure (−)-enantiomer.[186]

Similar degradation processes which yield 6-methyl-5-hepten-2-one (**VII-1**) may furnish geranylacetone (**VII-7**) from a sesquiterpene precursor. Both **VII-1** and **VII-7** are present in the urine of the red fox,[50] in the dorsal gland of the springbok[133] and represent the two main components in the mandibular gland secretion of the cleptoparasitic bee, *Holcopasites calliopsidis.*[188] While no behavioural activity of these compounds is reported for bees, they are believed to possess a communicative function in the fox. Degradation of phytol (**VI-3**) would produce 6,10,14-trimethylpentadecan-2-one (hexahydrofarnesylacetone) (**VII-8**) which is a component of the hairpencils of male *Idea iasonia* butterflies (Schulz, S. & Francke, W., unpublished). The respective alcohol was first identified from the secretion of the male African sugar-cane borer, *Eldana saccharina,*[189] and was also reported as a courtship pheromone of female rice moths, *Corcyra cephalonica.*[190]

The acyclic norsesquiterpene, gyrinal (gyrinidal), **VII-9**, is a defensive compound of water beetles, *Gyrinus* and *Dineutes* spp.[191–193] Cyclic compounds like **VII-10** (gyrinidione) and the bicyclic hemiacetal **VII-11** (gyrinidone)[193,194] were also identified from the secretion of whirligig beetles. Gyrinidone shows some structural relations to the lactol produced from nepetalactone (**IV-24**), which belongs to the set of aphid alarm pheromones discussed above.[126] The three polyoxygenated norsesquiterpenes are highly toxic to fish. By analogy to the formation of iridoids from citral, a biosynthesis starting from farnesal has been suggested;[194] however, higher terpenoids may certainly also serve as precursors.

A second group of degradation products shows a trimethylcyclohexyl moiety and may be derived from carotenoids. β-Cyclogeraniolene (**VII-12**) and isomers represent minor constituents in the frontal gland secretion of

Ancistrotermes cavithorax,[165] while isophorone (**VII-13**), oxoisophorone (**VII-14**) and some other cyclohexanone derivatives were found in the species-specific odour bouquets from male scent organs of Danaine and Ithomiine butterflies.[114] Oxoisophorone and isophorone have also been identified as trace constituents of the defensive secretion of a grasshopper species,[195] while isophorone and the isomeric γ-phorone are volatile constituents of females of the bark beetle, *Ips typographus*.[196] Such terpenoid cyclohexenone derivatives represent widespread plant constituents and are known to be produced by microorganisms also. 2-Hydroxy-2,6,6-trimethylcyclohexanone (**VII-15**) is a component of the supracaudal gland secretion of the red fox.[197] It is part of a mixture which also contains dihydroactinidiolide (**VII-16**) and β-ionone (**VII-17**). Dihydroactinidiolide, a constituent of tobacco and other plant sources, was also identified in queens of the fire ant, *Solenopsis invicta*,[198] and in male scent organs of the butterfly, *Idea iasonia*.[65] The close relationship to the termite sesquiterpene, ancistrofurane (**V-19**), is quite obvious. The particularly widespread β-ionone belongs to a group of bisnorterpenoids which is frequently found in plants. Some ionone derivatives also occur in insects. The allenic ketone **VII-18** was identified from the defensive froth of a grasshopper, *Romalea* sp., and it may originate from allenic carotenoids.[199] The edulane derivative **VII-19** (absolute configuration still unknown) is a male-specific odour component of the danaid butterfly, *Danaus plexippus*,[65] while the epoxide **VII-20** was found in males of other danaids, *Euploea* spp.[200] The compounds show the same structural relationship to caparrapi oxide (**V-22**) as does vittatol (**VII-4**) to the pyranoid form of linalool oxide (loss of a vinyl group). Additionally, the secretion of *Euploea* contained some less oxygenated edulanes, which were also known from plants.[201]

Particularly exciting semiochemicals are the odorous C_{19}-steroids 5α-androst-16-en-3-one, the 'boar-taint steroid' (**VII-21**), and the respective alcohol 5α-androst-16-en-3-ol, which have attracted much attention because of their action as sex pheromones in pigs and their possible importance in human social interactions. The steroidal alcohol was isolated as early as 1944 from pig testes[202] and later in human male axillary sweat,[203] while its glucuronide was known from human urine.[204] The boar-taint pheromone is synthesized in the testes of boars and released to the bloodstream; it is delivered to the salivary glands where it is reduced to the alcohol which represents the actual boar pheromone, stimulating the sow's standing reflex during copulation.

The ketone **VII-21** was also found in man,[205] and there are several contradictory papers concerning the pheromonal effect of the two steroids on human behaviour. Some authors found positive response[206–208] while others reported no influence.[209] Anyway, the compounds seem to contribute to the profiles of axillary odour.[210] Striking similarities between the molecular structures of the musk-smelling androstenone and exciting-smelling macrocyclic compounds such as civetone (**VII-22**) and hexadecanolide (**VII-23**) or exaltolide are quite obvious. Interestingly, **VII-21** was also identified in the occipital gland secretion of the Bactrian camel[211] as well as in vegetables like parsnip and celery.[212]

Some homoterpenes show ethyl branchings, which point to the incorporation of propionate subunits via homomevalonate (Scheme 8, **VIII-1**). The dihydrothiazole **VIII-2** is a volatile from the urine of mice which, together with a second component, is synergistically active in inducing inter-male aggressive behaviour.[213] The compound, a homologue of the antelope heterocyclic **II-13**, shows the carbon skeleton of homomevalonate (and also that of leucine); however, its biosynthesis is still unknown. The homoterpene **VIII-3**, which is

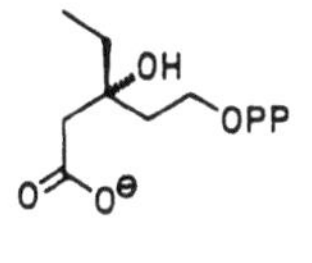
VIII-1

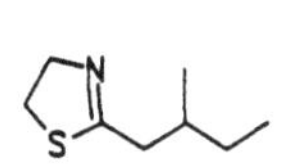
VIII-2

VIII-3

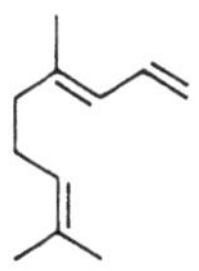
VIII-4

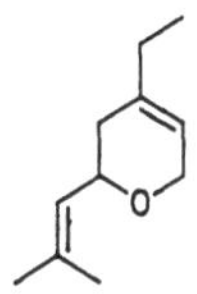
VIII-5

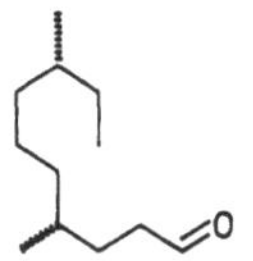
VIII-6

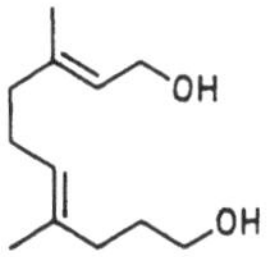
VIII-7

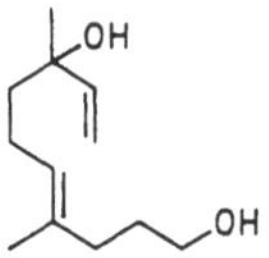
VIII-8

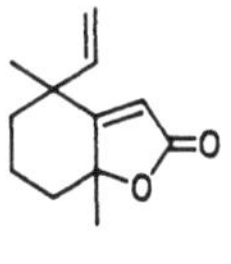
VIII-9

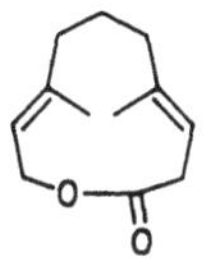
VIII-10

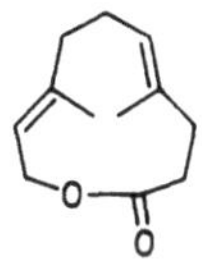
VIII-11

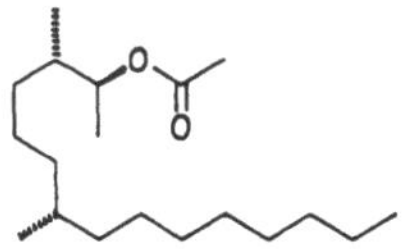
VIII-12

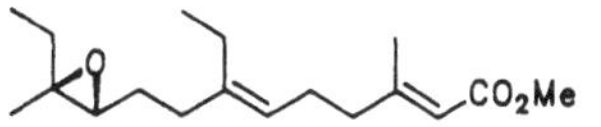
VIII-13

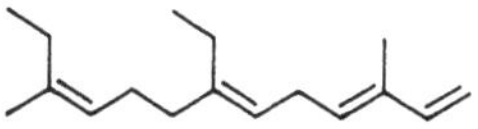
VIII-14

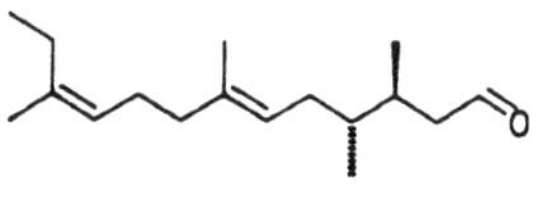
VIII-15

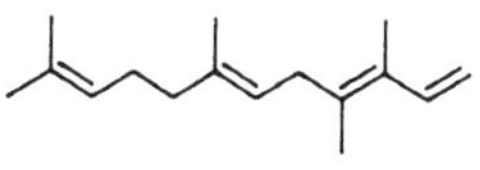
VIII-16

Scheme 8. Homoterpenes.

somehow the complement of the plant-produced **VIII-4**—see Chapter 7 of this book—is produced by the bark beetle, *Ips typographus*, during certain stress situations (Francke, W., Bartels, J. & Kohnle, U., unpublished): its exact biological function is as yet unknown. Another interesting homoterpene is 10-homonerol oxide (**VIII-5**), a main component among the volatile compounds of the thoracic gland of the ant lion *Grocus bore.*[122] (*R,R*)-4,8-Dimethyldecanal (**VIII-6**), the male-produced sex pheromone of the red flour beetle,[214,215] may represent a terpenoid which either shows a bishomologation at the geminal dimethyl group, or a branching caused by the incorporation of homomevalonate as a starter. Formation according to a biogenetic scheme such as alternating acetate–propionate–acetate–propionate–acetate would, however, yield the same carbon skeleton, and thus in fact the compound does not necessarily need to be a terpenoid at all.

The carbon skeleton of **VIII-6** is present in other insect volatiles, too. The 'bishomoterpene' **VIII-7** and the respective dicarboxylic acid, identified from the hair-pencil secretions of *Danaus* spp.,[216] have been interpreted as fixatives for volatiles, particularly important in promoting adherence of pheromone-transfer particles. While **VIII-7** is related to geraniol, complementary linalool derivatives (**VIII-8**) are components of the hair pencils of another danaid butterfly, *Euploea sylvester.*[114]

Homologues of dihydroactinidiolide (**VII-16**) are represented by anastrephin and the bicyclic lactone, 2,6-dimethyl-2-vinyl-7-oxabicyclo[4.3.0]non-9-en-8-one (**VIII-9**), as well as by its 2-epi-isomer, which are components of the sex pheromone blend of male Mexican and Caribbean fruit flies.[217] The compounds seem to be products of an intramolecular cyclization rearrangement of the monocyclic lactone, suspensolide (**VIII-10**), which is another volatile component of the Caribbean fruit fly, *Anastrepha suspensa.*[218] Besides other macrocycles, which clearly represent acetogenins, the lactone **VIII-11**, an isomer of **VIII-10**, was identified as a species-specific component of the aggregation pheromone of the rusty grain beetle, *Cryptolestes ferrugineus.*[219] Similarly to the biogenesis of anastrephin, which was postulated to involve a terpene precursor,[220] compounds **VIII-6–VIII-11** may be produced from a homoterpene precursor or during the 'non-terpenoid' sequence mentioned above.

Several sawfly species use mixtures of stereoisomers of acetates or propanoates of 3,7-dimethylpentadecan-2-ol (diprionol). The acetate **VIII-12** is the most important component in the European pine sawfly, *Neodiprion sertifer,*[221] which actually shows the partial structure of a terpenoid. The elucidation of its biosynthesis as well as the composition of stereoisomers in species-specific blends need further investigation.

The biosynthesis of the so-called insect juvenile hormone (**VIII-13**) of the giant silkworm moth, *Hyalophora cecropia,*[222] includes two homomevalonate units and one mevalonate unit. A corresponding hydrocarbon, bishomofarnesene (**VIII-14**), is a constituent of the Dufour's gland secretion of *Myrmica* ants,[223,224] where it forms a mixture with straight-chain hydrocarbons and other homofarnesenes. Faranal (**VIII-15**), the trail pheromone of the Pharaoh's ant, *Monomorium pharaonis*,[225] shows a structure which points to homomevalonate as the starter in the biosynthesis (like juvenile hormone II), carrying an additional methyl group at C-4. This 'irregular' branching is also present in the homofarnesene **VIII-16**, one of the trail pheromone components of the fire ant, *Solenopsis invicta*.[226]

REFERENCES

1. Alves, L. F., Chemical ecology and the social behaviour of animals, *Prog. Chem. Org. Nat. Prod.*, **53** (1988) 1–85.
2. Meinwald, J., Origins of insect defensive compounds and pheromones—*de novo* biosynthesis versus direct acquisition. Paper presented at the 32nd IUPAC Congress, Stockholm, 2–7 August 1989.
3. Karlson, P. & Lüscher, M., 'Pheromones', a new term for a class of biologically active substances. *Nature*, **183** (1959) 55–6.
4. Nordlund, D. A., Semiochemicals: a review of the terminology. In *Semiochemicals: Their Role in Pest Control*, ed. D. A. Nordlund, R. L. Jones & W. J. Lewis. J. Wiley & Sons, New York, 1981, pp. 13–28.
5. Kohnle, U. & Vité, J. P., Bark beetle predators: strategies in the olfactory perception of prey species by clerid and trogositid beetles. *Z. Ang. Ent.*, **98** (1984) 504–8.
6. Stowe, M. K., Tumlinson, J. H. & Heath, R. R., Chemical mimicry: Bolas spiders emit components of moth prey species sex pheromones. *Science*, **236** (1987) 964–7.
7. Arn, H., Toth, M. & Priesner, E., List of sex pheromones of *Lepidoptera* and related attractants. OILB-SROP/IOBC-WPRS, Paris, 1986. ISBN 92-9067-002-9.
8. Butler, C. G., Callow, R. K. & Johnston, N. C., The isolation and synthesis of queen substance, 9-oxodec-*trans*-2-enoic acid, a honey bee pheromone. *Proc. R. Soc. London Ser. B.*, **155** (1961) 417–28.
9. Francke, W., Heemann, V., Gerken, B., Renwick, J. A. A. & Vité, J. P., 2-Ethyl-1,6-dioxaspiro[4.4]nonane, principal aggregation pheromone of *Pityogenes chalcographus*. *Naturwiss.*, **64** (1977) 590–1.
10. Byers, J. A., Birgersson, G., Löfquist, J. & Bergström, G., Synergistic pheromones and monoterpenes enable aggregation and host recognition by a bark beetle, *Pityogenes chalcographus*. *Naturwiss.*, **75** (1988) 153–5.
11. Kaiser, R., Night-scented flowers, not only attractive to moths. Paper presented at the EUCHEM Conference on Semiochemicals in Plant and Animal Kingdoms, Angers, 1987.
12. Heinz, D. E. & Jennings, W. G., Volatile components of Bartlett pear V. *J. Food Sci.*, **31** (1966) 69–80.
13. Wood, D. L., The role of pheromones, kairomones and allomones in the host selection and colonization behaviour of bark beetles. *Ann. Rev. Entomol.*, **27** (1982) 411–46.
14. Tumlinson, J. H., Klein, M. G., Doolittle, R. E., Ladd, T. L. & Proveaux, A. T., Identification of the female Japanese beetle sex pheromone inhibition of male response by one enantiomer. *Science*, **197** (1977) 789–92.
15. Slessor, K., Kaminsky, L. A., King, G. G. S., Borden, J. H. & Winston, M. L., Semiochemical basis of the retinue response to queen honey bees. *Nature*, **332** (1988) 354–6.

16. Bakke, A., Froyen, P. & Skattebol, L., Field response to a new pheromonal compound isolated from *Ips typographus*. *Naturwiss*, **64** (1977) 98–9.
17. Veith, H., Koeninger, N. & Maschwitz, U., 2-Methyl-3-butene-2-ol, a major component of the alarm pheromone of the hornet *Vespa crabro*. *Naturwiss.*, **71** (1984) 328–9.
18. Moshonas, M. G. & Shaw, P. F., Analysis of flavour constituents from lemon and lime essence. *J. Agric. Food Chem.*, **20** (1972) 1029–30.
19. Hänsel, R., Wohlfahrt, R. & Coper, H., Versuche, sedativ-hypertonische Wirkstoffe im Hopfen nachzuweisen II. *Z. Naturforsch.*, **35c** (1980) 1096–7.
20. Carlson, D. A., Mayer, M. S., Silhacek, D. L., Jones, J. D., Beroza, M. & Bierl, B. A., Sex attractant pheromone of the house fly: isolation, identification and synthesis. *Science*, **174** (1971) 76–8.
21. Toth, M., Szöcz, G., Löfstedt, C., Hansson, B. S. & Subchev, M., Sex pheromone components of *Mamestra suasa*: chemical analysis, electrophysiological activity, wind tunnel activity and field tests in two European countries. *Entomol. Exp. Appl.*, **42** (1986) 291–9.
22. Blum, M. S., *Chemical Defenses in Arthropods*. Academic Press, New York, 1981.
23. Dettner, K., Chemosystematics and evolution of beetle chemical defenses. *Ann. Rev. Entomol.*, **32** (1987) 17–48.
24. Francke, W., Autogenins in systems of chemical communication. *Nova Acta Regiae Soc. Scient. Upsaliensis Serie V*, **C3** (1984) 119–26.
25. Francke, W., Structural concepts in the chemistry of aliphatic pheromones. In *Advances in Invertebrate Reproduction*, Vol. 3, ed. W. Engels *et al.* Elsevier, Amsterdam, 1984.
26. Mori, K., The synthesis of insect pheromones. In *The Total Synthesis of Natural Products*, Vol. 4, ed. J. Apsimon. J. Wiley & Sons, New York, 1981, pp. 1–183.
27. Mori, K., Synthesis of optically active pheromones. Tetrahedron report no. 252. *Tetrahedron*, **45** (1989) 3233–98.
28. Hummel, H. E. & Miller, T. A. (eds), *Techniques in Pheromone Research*, Springer, New York, 1984.
29. Acree, T. E. & Soderlund, D. M. (eds), *Semiochemistry, Flavours and Pheromones*. Walter de Gruyter, Berlin, 1985.
30. Attygalle, A. B. & Morgan, E. D., Strukturbestimmung mit Nanogramm-Mengen durch Kombination mikrochemischer und gaschromatographischer Methoden am Beispiel von Pheromonen. *Angew Chem.*, **100** (1988) 475–94 (Int. Ed. Engl., **27** (1988) 460).
31. Weber, R. & Schurig, V., Complexation gas chromatography, a valuable tool for the stereochemical analysis of pheromones. *Naturwiss.*, **71** (1984) 408–13.
32. Frank, H., Nicholson, G. J. & Bayer, E., Rapid gas chromatographic separation of aminoacid enantiomers with a novel chiral stationary phase. *J. Chromatogr. Sci.*, **15** (1977) 174–6.
33. König, W. A., Benecke, J. & Sievers, S., New results in the gas chromatographic separation of enantiomers of hydroxy acids and carbohydrates. *J. Chromatogr.*, **217** (1981) 71–9.
34. König, W. A., Lutz, S., Mischnik-Lübbeke, P., Brassat, B. & Wenz, G., Cyclodextrins as chiral stationary phases in capillary gas chromatography. *J. Chromatogr.*, **447** (1988) 193–7.
35. Happ, G. M. & Meinwald, J., Biosynthesis of arthropod secretions I. Monoterpene synthesis in ant (*Acanthomyops claviger*). *J. Am. Chem. Soc.*, **87** (1965) 2507–8.
36. Meinwald, J., Happ, G. M., Labows, J. & Eisner, T., Cyclopentanoid terpene biosynthesis in a phasmid insect and in cat mint. *Science*, **151** (1966) 79–80.
37. Mitlin, N. & Hedin, P. H., Biosynthesis of grandlure, the pheromone of the boll

weevil *Anthonomus grandis,* from acetate, mevalonate and glucose. *J. Insect Physiol.,* **20** (1974) 1825–31.
38. Brand, J. M., Bracke, J. W., Britton, L. M., Markowitz, A. J. & Barras, S. J., Bark beetle pheromones: production of verbenone by a mycangial fungus of *Dendroctonus frontalis. J. Chem. Ecol.,* **2** (1976) 195–9.
39. Leufvén, A., Bergström, G. & Falsen, E., Interconversion of verbenols and verbenone by yeasts isolated from the spruce bark beetle *Ips typographus. J. Chem. Ecol.,* **10** (1984) 1349–61.
40. Stoakley, J. T., Bakke, A., Renwick, J. A. A. & Vité, J. P., The aggregation pheromone system of the larch bark beetle, *Ips cembrae* (Heer). *J. Appl. Entomol.,* **86** (1978) 174–6.
41. Burger, B. V. & Pretorius, P. J., Mammalian pheromone studies VI: Compounds from the preorbital gland of the blue duiker *Cephalophus monticola. Z. Naturforsch,* **42c** (1988) 1355–7.
42. Duffield, R. M., Laberge, W. E., Cane, J. H. & Wheeler, J. H., Exocrine secretions of bees IV: macrocyclic lactones and isopentenyl esters in Dufour's gland secretions of *Nomia* bees. *J. Chem. Ecol.,* **8** (1982) 535–43.
43. Wheeler, J. W., Schamin, M. T., Brown, P. & Duffield, R. M., Semiterpenoid esters from the venom of the European hornet, *Vespa crabro* (Hymenoptera Vespidae). *Tetrahedron Lett.,* **24** (1983) 5811–14.
44. Koeniger, N., Weiss, J. & Maschwitz, U., Alarm pheromones of the sting in the genus *Apis. J. Insect Physiol.* **25** (1979) 467–78.
45. Francke, W., Heemann, V. & Heyns, K., Volatile substances from ambrosia beetles. *Z. Naturforsch.,* **29c** (1974) 243–5.
46. Davies, N. W. & Madden, J. L., Mandibular gland secretions of two parasitoid wasps. (Hymenoptera: Ichneumonidae). *J. Chem. Ecol.,* **11** (1985) 1115–27.
47. Schildknecht, H., Wilz, I., Enzmann, F., Grund, N. & Ziegler, M., Über das Mustelan, den Analdrüsenstinkstoff des Nerzes (*Mustela vison*) und Iltisses (*Mustela putorius*). *Angew. Chem.,* **88** (1976) 228 (Int. Ed. Engl. **15** (1976) 242).
48. Crump, D. R., Anal gland secretion of the ferret (*Mustela putorius* forma *furo*). *J. Chem. Ecol.,* **6** (1980) 837–44.
49. Schildknecht, H. & Birkner, C., Analyse der Analbeutelsekrete Mitteleuropäischer Musteliden. *Chem. Ztg.,* **107** (1983) 267–70.
50. Jorgenson, J. W., Novotny, M., Carmack, M., Copland, G. B., Wilson, S. R., Whitten, W. K. & Katona, S., Chemical scent constituents in the urine of the red fox, *Vulpes vulpes,* during the winter season. *Science,* **199** (1978) 796–8.
51. Burger, B. V., Pretorius, P. J., Stander, J. & Grierson, G. R., Mammalian pheromone studies VII: identification of thiazole derivatives in the preorbital gland secretions of the grey duiker *Sylvicarpa grimmia* and the red duiker *Cephalophus natalensis. Z. Naturforsch.,* **43c** (1988) 731–6.
52. Francke, W. & Vité, J. P., Oxygenated terpenes in pheromone systems of bark beetles. *J. Appl. Entomol.,* **96** (1983) 146–56.
53. Byers, J. A., Chemical ecology of bark beetles, *Experientia,* **45** (1989) 271–83.
54. Leather, S. R., Pine monoterpenes stimulate oviposition in the pine beauty moth, *Panolis flammea. Entomol. Exp. Appl.,* **43** (1987) 295–303.
55. Pickett, J. A. & Griffiths, D. C., Composition of aphid alarm pheromones. *J. Chem. Ecol.,* **6** (1980) 349–60.
56. Lanne, B. S., Schlyter, F., Byers, J. A., Löfquist, J., Leufvén, A., Bergström, P. & Norin, T., Differences in attraction to semiochemicals present in sympatric pine shoot beetles *Tomicus minor* and *T. piniperda. J. Chem. Ecol.,* **13** (1987) 1045–67.
57. Wheeler, J. W. & Duffield, R. M., Pheromones of Hymenoptera and Isoptera. In *CRC Handbook of Natural Pesticides,* Part B, ed. E. D. Morgan & N. B. Mandava. Boca Raton, Florida, 1988, pp. 59–206.

58. Gnanasunderam, C., Young, H. & Hutchins, R. F. N., Defensive secretions of New Zealand tenebrionids I: presence of monoterpene hydrocarbons in the genus *Artystona* (Coleoptera, Tenebrionidae). *J. Chem. Ecol.*, **7** (1981) 889–94.
59. Hayashi, N., Kuwahara, Y. & Komae, H., The scent scale substances of male *Pieris* butterflies (*Pieris melete* and *Pieris napi*). *Experientia*, **34** (1978) 684–5.
60. Baker, R. & Walmsley, S., Soldier defense secretions of the South American termite *Cortaritermes silvestri, Nasutitermes* sp.n.d. and *Nasutitermes kemneri*. *Tetrahedron*, **38** (1982) 1899–910.
61. Prestwich, G., From tetracycles to macrocycles: chemical diversity in the defense secretions of Nasute termites. *Tetrahedron*, **38** (1982) 1911–19.
61a. Kaib, M. & Dittebrand, H., The poison gland of the ant *Myrmicaria eumenoides* and its role in recruitment communication. *Chemoecology*, **1** (1990) 3–11.
61b. Everaerts, C., Bonnard, O., Pastrels, J. M., Roisin, Y. & König, W. A., (+)-α-Pinene in the defensive secretion of *Nasutitermes princeps* (Isoptera, Termitidae), *Experientia*, **46** (1990) 228–30.
62. Regev, S. & Cone, W. W., The monoterpene, citronellol, as a male sex attractant of the twospotted spider mite *Tetranychus urticae* (Acarina: Tetranychidae). *Environ. Entomol.*, **9** (1980) 50–2.
63. Kuwahara, Y., Ishii, S. & Fukami, H., Neryl formate: alarm pheromone of the cheese mite *Tyrophagus putrescentiae* (Shrank) (Acarina, Acaridae). *Experientia*, **31** (1975) 1115–16.
64. Anderson, R. J., Gieselman, M. J., Chinn, H. R., Adams, K. G., Henrick, C. A., Rice, R. E. & Roelofs, W. L., Synthesis and identification of a third component of the San José scale sex pheromone. *J. Chem. Ecol.*, **7** (1981) 695–706.
65. Francke, W., Bartels, J., Krohn, S., Schulz, S., Baader, E., Tengö, J. & Schneider, D., Terpenoids from bark beetles, solitary bees and danaine butterflies. *Pure Appl. Chem.*, **61** (1989) 539–42.
66. Francke, W., Schröder, W., Borg-Karlsson, A.-K., Bergström, G. & Tengö, J., Species and sex specificity in the odour composition of two panurgine bees (Hymenoptera, Andrenidae). *Z. Naturforsch.*, **42c** (1987) 169–71.
67. Hefetz, A., Batra, S. W. T. & Blum, M. S., Linalool, neral and geranial in the mandibular glands of *Colletes* bees—an aggregation pheromone. *Experientia*, **35** (1979) 319–20.
68. Cane, J. & Tengö, J. O., Pheromonal cues direct mate-seeking behavior of male *Colletes cunicularius*. *J. Chem. Ecol.*, **7** (1981) 427–36.
69. Robacker, D. C. & Hendry, L. B., Neral and geranial: components of the sex pheromone of the parasitic wasp, *Itoplectis conquisitor*. *J. Chem. Ecol.*, **3** (1977) 563–77.
69a. Blum, M. S., Crewe, R. M., Kerr, W. E., Keith, L. H., Garrison, A. W. & Walker, M. M., Citral in stingless bees: isolation and function in trail laying and robbing. *J. Insect Physiol.*, **16** (1970) 1637–48.
70. Vinson, S. B., Williams, H. J., Frankie, G. W., Wheeler, J. W., Blum, M. S. & Coville, R. E., Mandibular glands of *Centris adani* (Hymenoptera: Anthophoridae): their morphology, chemical constituents, and function in scent marking and territorial behavior. *J. Chem. Ecol.*, **8** (1982) 319–27.
71. Schildknecht, H., Chemische Ökologie — ein Kapitel moderner Naturstoffchemie. *Angew. Chem.*, **88** (1976) 235–72 (Int. Ed. Engl., **15** (1976) 214).
72. Bergström, G. & Lundgren, L., Androconial secretion of three species of butterflies of the genus *Pieris* (Lep. Pieridae). *Zool. Suppl.*, **1** (1973) 67–71.
73. Kuwahara, Y., Matsumoto, K. & Wada, Y., Pheromone study on acarid mites IV. Citral: composition and function as an alarm pheromone and its secretory gland in four species of acarid mites. *Eisei Dobutsu*, **31** (1980) 73–7.
74. Pickett, J. A., Williams, J. H., Martin, A. P. & Smith, M. C., Nasonov pheromone of the honey bee *Apis mellifera* L. (Hymenoptera: Apidae) Part I: chemical characterization. *J. Chem. Ecol.*, **6** (1980) 425–34.

75. Renwick, J. A. A., Hughes, P. R. & Dei, T. T., Oxidation products of pinene in the bark beetle *Dendroctonus*. *J. Insect Physiol.*, **19** (1973) 1735–40.
76. Renwick, J. A. A., Hughes, P. R. & Krull, I. S., Selective production of *cis*- and *trans*-verbenol from (–)- and (+)-α-pinene by a bark beetle species. *Science*, **191** (1976) 199–201.
77. Hughes, P. R., Myrcene: a precursor of pheromones in *Ips* beetles. *J. Insect Physiol.*, **20** (1974) 1271–4.
78. Pierce, H.-D. Jr, Conn, J. E., Oehlschlager, A. C. & Borden, J. H., Monoterpene metabolism in female mountain pine beetles, *Dendroctonus ponderosae* Hopkins, attacking ponderosa pine. *J. Chem. Ecol.*, **13** (1987) 1455–80.
79. Francke, W. & Heemann, V., Das Duftstoff-Bouquet des großen Waldgärtners *Blastophagus piniperda* (Coleoptera: Scolytidae). *Z. Ang. Ent.*, **82** (1976) 117–19.
80. Borden, J. H., Aggregation pheromones. In *Behavior*, Vol. 9 in *Comprehensive Insect Physiology, Biochemistry and Pharmacology*, ed. G. A. Kerkut & L. I. Gilbert. Pergamon Press, Oxford, 1985, pp. 257–8.
81. Kohnle, U., Vité, J. P., Erbacher, C., Bartels, J. & Francke, W., Aggregation pheromones of European engraver beetles of the genus *Ips* mediated by terpenoid pheromones. *Entomol. Exp. Appl.*, **49** (1988) 43–53.
82. Francke, W., Pan, M.-L., Bartels, J., König, W. A., Vité, J. P., Krawielitzki, S. & Kohnle, U., The odour bouquet of three pine engraver beetles (*Ips* spp.). *J. Appl. Entomol.*, **101** (1986) 453–61.
83. Fish, R. H., Browne, L. E. & Bergot, B. J., Pheromone biosynthetic pathways: conversion of ipsdienone to (–)-ipsdienol, a mechanism for enantioselective reduction in the male bark beetle, *Ips paraconfusus*. *J. Chem. Ecol.*, **10** (1984) 1057–64.
84. Silverstein, R. H., Rodin, J. O. & Wood, D. L., Sex attractants in frass produced by male *Ips confusus* in ponderosa pine. *Science*, **154** (1966) 509–10.
85. Harring, C. M., Aggregation pheromones of the European fire engraver beetles *Pityokteines curvidens*, *P. spinidens* and *P. voronzovi* and the role of juvenile hormone in pheromone biosynthesis. *Z. Ang. Ent.*, **85** (1978) 281–317.
86. Klimetzek, D., Köhler, J., Krohn, S. & Francke, W., Das Pheromon-System des Waldreben-Borkenkäfers, *Xylocleptes bispinus* Duft. (Col.: Scolytidae). *J. Appl. Entomol.*, **107** (1989) 304–9.
87. Naves, Y. R., Etudes sur les matières végétales volatiles LVI: sur de nouvelles cétones, les tagéténones, isolées de l'huile essentielle de *Lippia asperifolia*. *Helv. Chim. Acta*, **31** (1948) 29–32.
88. Whitten, W. M., Hills, H. G. & Williams, N. H., Occurrence of ipsdienol in floral fragrances. *Phytochemistry*, **27** (1988) 2759–60.
89. Silverstein, R. M., Rodin, J. O., Wood, D. L. & Brown, L. E., Identification of two new terpene alcohols from frass produced by *Ips confusus* in ponderosa pine. *Tetrahedron*, **22** (1966) 1929–36.
90. v. Schantz, M., Widin, K. G. & Hiltunen, R., Structures of some aliphatic monoterpenoids isolated from the essential oil of *Ledum palustre* L. *Acta Chem. Scand.*, **27** (1973) 551–5.
91. Francke, W., Bartels, J., Schmutzenhofer, H., Kohnle, U. & Vité, J. P., The odor bouquet of *Ips schmutzenhoferi* (Holzschuh) (Col.: Scol.). *Z. Naturforsch.*, **43c** (1988) 958–60.
92. Granger, R., Parset, J. & Girard, J. P., 2-Methyl-6-methylene-2,7-octadienol isolated from *Thymus vulgaris* essential oil. *Phytochemistry*, **11** (1972) 2301–5.
93. Klimetzek, D. & Francke, W., Relationship between the enantiomeric composition of α-pinene in host trees and the production of verbenol in *Ips* species. *Experientia*, **36** (1980) 1343–4.
94. Renwick, J. A. A. & Vité, J. P., Bark beetle attractants: mechanism of colonization by *Dendroctonus frontalis*. *Nature*, **224** (1969) 1222–3.

95. Naumann, C. M., Ockenfels, P., Schmitz, J., Schmidt, F. & Francke, W., Reactions of *Zygaena* moths (Lepidoptera: Zygaenidae) to volatile compounds of *Knautia arvensis*. *Entomologica Generalis*, **15** (1991) 255–64.
96. Tumlinson, J. H., Gueldner, R. C., Hardee, D. D., Thompson, A. C., Hedin, P. A. & Minyard, J. P., Identification of the four compounds comprising the boll weevil sex attractant. *J. Org. Chem.*, **36** (1971) 2616–21.
97. Phillips, T. W., West, J. R., Foltz, J. L., Silverstein, R. M. & Lanier, G. N., Aggregation pheromone of the deodar weevil, *Pissodes nemorensis* (Coleoptera: Curculionidae): isolation and activity of grandisol and grandisal. *J. Chem. Ecol.*, **10** (1984) 1417–24.
98. Bohlmann, F., Zdero, C. & Faass, U., Über die Inhaltsstoffe von *Artemisia fragrans* Willd. *Chem. Ber.*, **106** (1973) 2904–9.
99. Borden, J. H., Handley, J. R., Johnston, B. D., MacConnell, J. G., Silverstein, R. M., Slessor, K. N., Swigar, A. A. & Wong, D. T. W., Synthesis and field testing of 4,6,6-lineatin, the aggregation pheromone of *Trypodendron lineatum* (Coleoptera: Scolytidae). *J. Chem. Ecol.*, **5** (1979) 681–9.
100. Schurig, V., Weber, R., Klimetzek, D., Kohnle, U. & Mori, K., Enantiomeric composition of lineatin in three sympatric ambrosia beetles. *Naturwiss.*, **69** (1982) 602–3.
101. Klimetzek, D., Vité, J. P. & Mori, K., Zur Wirkung und Formulierung des Populationslockstoffes des Nutzholzborkenkäfers *Trypodendron* (=*Xyloterus*) *lineatum*. *Z. ang. Ent.*, **89** (1980) 57–63.
102. Bierl-Leonhard, B. H., Moreno, D. S., Schwarz, M., Forster, H. S. & Plimmer, J. A., Isolation, identification and synthesis of the sex pheromone of the citrus mealybug, *Planococcus citri* (Risso). *Tetrahedron Lett.*, **22** (1981) 389–92.
103. Meinwald, J. & Eisner, T., Pheromones VII. African monarch: major components of the hair pencil secretion. *Tetrahedron Lett.*, (1971) 3485–8.
104. Tanaka, K., Ohsawa, K., Honda, H. & Yamamoto, I., Copulation release pheromone, erectin, from the azuki bean weevil (*Callosobruchus chinensis* L.). *J. Pestic. Sci.*, **6** (1981) 75–82.
105. Jefson, M., Meinwald, J., Nowicki, S., Hicks, K. & Eisner, T., Chemical defense of a rove beetle (*Crephilus maxillosus*). *J. Chem. Ecol.*, **9** (1983) 159–80.
106. Wheeler, J. W., Avery, J., Birmingham, F. & Duffield, R. M., 8-Acetoxy-2,6-dimethyl-2-octenal, a novel mandibular gland component from *Panurginus* bees (Hymenoptera: Andrenidae). *Insect. Biochem.*, **14** (1984) 391–4.
107. Leal, W. S., Kuwahara, Y., Suzuki, T., Nakano, Y. & Nakao, H., Identification and synthesis of 2,3-epoxyneral, a novel monoterpene from the Acarid mite *Tyrophagus perniciosus* (Acarina: Acaridae). *Agric. Biol. Chem.*, **53** (1989) 295–8.
108. Leal, W. S., Kuwahara, Y. & Suzuki, T., 2(*E*)-(4-Methyl-3-pentenylidene)-butanedial, β-acaridial: a new type of monoterpene from the mold mite *Tyrophagus putrescentiae* (Acarina: Acaridae). *Agric. Biol. Chem.*, **53** (1989) 875–8.
109. Leal, W. S., Kuwahara, Y., Nakano, Y., Nakao, H. & Suzuki, T., 2(*E*)-(4-Methyl-3-pentenyl)-butenedial, α-acaridial, a novel monoterpene from the acarid mite *Tyrophagus perniciosus* (Acarina: Acaridae). *Agric. Biol. Chem.*, **53** (1989) 1193–6.
110. Ayorinde, F. O., Wheeler, J. W. & Duffield, R. M., Synthesis of dihydrocineole, a new monoterpene from the acarid mite, *Caloglyphus rodriquezi* (Arachnida: Acari). *Tetrahedron Lett.*, **25** (1984) 3525–8.
111. Olagbemiro, T. O. & Staddon, B. W., Isoprenoids from metathoracic scent gland of cotton seed bug *Oxycarenus hyalinipennis* (Costa) (Heteroptera: Lygaemidae). *J. Chem. Ecol.*, **9** (1983) 1397–412.
112. Baeckström, P., Bergström, G., Bjorkling, F., Hui-Zhu, H., Högberg, H.-E., Jacobsson, U., Guo-Qiang, L., Löfquist, J., Norin, T. & Wassgren, A.-B.,

Structures, absolute configurations, and syntheses of volatile signals from three sympatric ant-lion species, *Euroleon nostras, Grocus bore* and *Myrmeleon formicarius* (Neuroptera: Myrmeleontidae). *J. Chem. Ecol.*, **15** (1989) 61–80.
113. Vidari, G., de Bernardi, M., Pavan, M. & Raggozzino, L., Rose oxide and iridodial from *Aromia moschata* L: (Coleoptera: Cerambycidae). *Tetrahedron Lett.*, (1973) 4065–8.
114. Schulz, S., Francke, W., Edgar, J. & Schneider, D., Volatile compounds from androconial organs of Danaine and Ithomiine butterflies. *Z. Naturforsch.*, **43c** (1988) 99–104.
115. Bernardi, R., Cardani, C., Ghiringhelli, D., Selva, A., Baggini, A. & Pavan, A., On the components of secretion of mandibular glands of the ant *Lasius (Dendrolasius) fuliginosus. Tetrahedron Lett.*, (1967) 3893–6.
116. Ahlgren, G., Bergström, G., Löfquist, J., Jansson, A. & Norin, T., *trans*-Perillenal, a furanoid monoterpene from pine saw-fly, *Neodiprion sertifer* (Hymenoptera: Diprionidae). *J. Chem. Ecol.*, **5** (1979) 309–19.
117. Kunesch, G., Zagatti, P., Lallemand, J. Y., Debal, A. & Vigneron, J. P., Structure and synthesis of the wing pheromone of the male African sugar cane borer, *Eldana saccharina* (Wlk) (Lepidoptera Pyralidae). *Tetrahedron Lett.*, **22** (1981) 5271–4.
118. Büchi, G., Kovats, E. st., Enggist, P. & Uhde, G., Syntheses of rosefuran and dehydroelsholtzione. *J. Org. Chem.*, **33** (1968) 1227–9.
119. Sierra, J. R., Woggon, W. D. & Schmid, H., Transfer of Cantharidin (1) during copulation from the adult male to the female *Lytta vesicatoria* (Spanish flies). *Experientia*, **32** (1976) 142–4.
120. Eisner, T., Dugrap, M., Jacobs, R. & Meinwald, J., Necrodols: anti-insectan terpenes from defensive secretion of carrion beetle (*Necrodes surinamensis*). *J. Chem. Ecol.*, **12** (1986) 1407–15.
121. Cavill, G. W. K., Insect terpenoids and nepetalactone. In *Cyclopentanoid Terpene Derivatives*, ed. W. I. Taylor & A. R. Battersby. Marcel Dekker, New York, 1969, pp. 203–38.
122. Smith, R. M., Brophy, J. R., Cavill, G. W. K. & Davies, N. W., Iridodials and nepetalactone in the defensive secretion of the coconut stick insect *Graeffea crouani. J. Chem. Ecol.*, **5** (1979) 727–35.
123. Rowell-Rahier, M. & Pasteels, J. M., Economics of chemical defense in *Chrysomelinae. J. Chem. Ecol.*, **12** (1986) 1189–203.
124. Pagnoni, U. M., Pinetti, A., Trave, R. & Garanti, L., Monoterpenes of *Teucrium marum. Aust. J. Chem.*, **29** (1976) 1375–81.
125. McElvain, S. M., Walters, P. M. & Bright, R. D., The constituents of the volatile oil of catnip. *J. Am. Chem. Soc.*, **64** (1942) 1828–36.
126. Dawson, G. W., Griffiths, D. C., Jones, N. F., Mudd, A., Pickett, J. A., Wadhams, L. B. & Woodcock, C. M., Identification of an aphid sex pheromone in *Megoura viciae* female legs. *Nature*, **225** (1987) 614–16.
127. Bellas, T. E., Brown, W. V. & Moore, B. P., The alkaloid actinidine and plausible precursors in defensive secretions of rove beetles. *J. Insect Physiol.*, **20** (1974) 277–80.
128. Smolanoff, J., Kluge, A. F., Meinwald, J., McPhail, A., Miller, R. W., Hicks, K. & Eisner, T., Polyzonimine: a novel terpenoid insect repellent produced by a milliped. *Science*, **188** (1975) 734–6.
129. Baker, R., Herbert, R. H. & Grant, G. G., Isolation and identification of the sex pheromone of the Mediterranean fruit fly, *Ceratitis capitata* (Wied.). *J. Chem. Soc., Chem. Commun.*, (1985) 824–5.
130. Sutherland, O. R. W., Wearing, C. H. & Hutchins, R. F. N., Production of α-farnesene, an attractant and oviposition stimulant for codling moth by developing fruit of ten varieties of apple. *J. Chem. Ecol.*, **3** (1977) 625–31.

131. Vander Meer, R. K., Alvarez, F. & Lofgren, C. S., Isolation of the trail recruitment pheromone of *Solenopsis invicta*. *J. Chem. Ecol.*, **14** (1988) 825–38.
132. Bowers, W. S., Nault, L. R., Webb, R. E., & Dutky, S. R., Aphid alarm pheromone: isolation, identification, synthesis. *Science*, **177** (1972) 1121–2.
133. Burger, B. V., le Roux, M., Spies, H. S. C., Truter, V. & Bigalke, R. C., Mammalian pheromone studies IV: terpenoid compounds and hydroxy esters from the dorsal gland of the springbok, *Antidorcas marsupialis*. *Z. Naturforsch.*, **36c** (1980) 340–3.
134. Bergström, G. & Tengö, J., Studies on natural odoriferous compounds IX: farnesyl and geranyl-esters as main volatile constituents of the secretion from Dufour's gland in 6 species of *Andrena* (Hymenoptera: Apidae). *Chem. Scripta*, **5** (1974) 28–38.
135. Tengö, J. & Bergström, G., Identical isoprenoid esters in the Dufour's gland secretions of North American and European *Andrena* bees (Hymenoptera: Andrenidae). *J. Kansas Ent. Soc.*, **51** (1978) 521–6.
136. Calam, D. H., Species and sex-specific compounds from the heads of male bumble bees (*Bombus* spp.). *Nature*, **221** (1969) 856–7.
137. Svensson, B. G. & Bergström, G., Volatile marking secretions from the labial gland of North European *Pyrobombus* D. T. males (Hymenoptera, Apidae). *Insectes Sociaux*, **24** (1977) 213–14.
138. Tengö, J. & Bergström, G., *All-trans*-farnesyl hexanoate and geranyl octanoate in the Dufour gland secretion of *Andrena* (Hymenoptera: Apidae). *J. Chem. Ecol.*, **1** (1975) 253–68.
139. Fernandes, A., Duffield, R. M., Wheeler, J. W. & La Berge, W., Chemistry of the Dufour's gland secretions of North American *Andrenid* bees (Hymenoptera: Andrenidae). *J. Chem. Ecol.*, **7** (1981) 453–63.
140. Burger, B. V., Munro, Z., Röth, M., Spies, H. S. C., Truter, V., Tribe, G. D. & Crewe, R. M., Composition of the heterogenous sex attracting secretion of the dung beetle, *Kheper lamarcki*. *Z. Naturforsch.*, **38c** (1983) 848–55.
141. Regev, S. & Cone, W., Evidence of farnesol as male sex attractant of the two spotted spider mite, *Tetranychus urticae* Koch. *Environ. Entomol.*, **4** (1975) 307–11.
142. Doskotch, R. W., Cheng, H.-Y., Odell, T. M. & Girard, L., Nerolidol: an antifeedant sesquiterpene alcohol for the gypsy moth larvae from *Melaleuca leucadendron*. *J. Chem. Ecol.*, **6** (1980) 845–51.
143. Kullenberg, B., Bergström, G. & Ställberg-Stenhagen, S., Volatile components of the marking secretion of male bumble bees. *Acta Chem. Scand.*, **24** (1970) 1481–3.
144. Bergström, G. & Svensson, B., 2,3-Dihydro-6-*trans*-farnesol main component of the cephalic marker secretion of *Bombus jonellus* K. (Hym. Apidae) males. *Zoon Suppl.*, **1** (1973) 61–5.
145. Bergström, G. & Löfquist, J., Chemical basis for odour communication in four species of *Lasius* ants. *J. Insect Physiol.*, **16** (1970) 2353–75.
146. Wheeler, J. W., Rasmussen, L. E., Ayorinde, F., Buss, I. O. & Smuts, G. L., Constituents of temporal gland secretion of the African elephant *Loxodonta africana*. *J. Chem. Ecol.*, **8** (1982) 821–35.
147. Castillo, B., Del, J., Brooks, S. J. W. & Campbell, M. M., Caparrapidiol and caparrapitriol: two new acyclic sesquiterpene alcohols. *Tetrahedron Lett.*, (1966) 3731–6.
148. Baker, R., Edwards, M., Evans, D. A. & Walmsley, S., Soldier-specific chemicals of the termite *Curvitermes strictinasus* Matheas (Isoptera, Nasutitermitinae). *J. Chem. Ecol.*, **7** (1981) 127–31.
149. Zagatti, P., Kunsch, G., Ramiandrasoa, F., Malosse, C., Hall, D. R., Lester, R.

& Nesbitt, B. F., Sex pheromones of rice moth, *Corcyra cephalonica* Stainton I: identification of male pheromone. *J. Chem. Ecol.*, **13** (1987) 1561–73.
150. Nishino, C., Bowers, W. S., Montgomery, M. E., Nault, L. R. & Nielson, M. W., Alarm pheromone of the spotted alfalfa aphid, *Therioaphis maculata* Buckton (Homoptera: Aphididae). *J. Chem. Ecol.*, **3** (1977) 349–57.
151. Weinheimer, A. J., Youngblood, W. W., Washecheck, P. H., Karns, T. K. B. & Cierseczko, L. S., Isolation of the elusive (−)-germacrene-A from the gorgonian *Eunicea mammosa*; chemistry of coelenterales XVIII. *Tetrahedron Lett.*, (1970) 497–500.
152. Baker, R., Coles, H. R., Edwards, M., Evans, D. E., Howse, P. E. & Walmsley, F., Chemical composition of the frontal gland secretion of *Syntermes* soldiers (Isoptera, Termitidae). *J. Chem. Ecol.*, **7** (1981) 135–45.
153. Honda, K., Larval osmeterial secretions of the swallowtails (*Papilio*). *J. Chem. Ecol.*, **7** (1981) 1089–113.
154. Persoons, C. J., Verwiel, P. E. J., Talman, D., Nooijen, P. F. J. & Nooijen, W. J., Sex pheromones of the American cockroach, *Periplaneta americana*: a tentative structure of periplanone-B. *Tetrahedron Lett.*, (1976) 2055–8.
155. Adams, M. A., Nakanishi, K., Still, W. C., Arnold, E. V., Clardy, J. & Persoons, C. J., Sex pheromone of the American cockroach: absolute configuration of periplanone-B. *J. Am. Chem. Soc.*, **101** (1979) 2495–8.
156. Hauptmann, H., Mühlbauer, G. & Sass, H., Identifizierung und Synthese von Periplanon A. *Tetrahedron Lett.*, **27** (1986) 6189–92.
157. Biendl, M., Hauptmann, H. & Sass, H., Periplanon D_1 und Periplanon D_2—zwei neue biologisch aktive germacranoide Sesquiterpene aus *Periplaneta americana*. *Tetrahedron Lett.*, **30** (1989) 2367–9.
158. Howard, J. J., Green, T. P. & Wiemer, D. F., Comparative deterrency of two terpenoids to two genera of *Attine* ants. *J. Chem. Ecol.*, **15** (1989) 2279–88.
159. Pearce, G. T., Gore, W. E., Silverstein, R. M., Peacock, J. W., Cuthbert, R. A., Lanier, G. N. & Simeone, J. W., Chemical attractants for the smaller European elm bark beetle *Scolytus multistriatus* (Coleoptera: Scolytidae). *J. Chem. Ecol.*, **1** (1975) 115–24.
160. Prestwich, G. D. & Collins, M. S., Chemotaxonomy of *Subalitermes* and *Nasutitermes* termite soldier defense secretions. Evidence against the hypothesis of diphylatic evolution of the Nasutitermitinae. *Biochem. Syst. Ecol.*, **7** (1979) 211–18.
161. Lundgren, L. & Bergström, G., Wing scents and scent-released phases in the courtship behavior of *Lycaeides argyrognomon* (Lepidoptera: hycaenidae). *J. Chem. Ecol.*, **1** (1975) 399–412.
162. Baker, R., Borges, R., Cooke, N. G. & Herbert, R. H., Identification and synthesis of (*Z*)-(1′*S*, 3′*R*, 4′*S*)(−)-2-(3′,4′-epoxy-4′-methylcyclohexyl)-6-methylhepta-2,5-diene, the sex pheromone of the southern green stinkbug. *Nezara viridula* (L.). *J. Chem. Soc., Chem. Commun.*, (1987) 414–16.
163. Aldrich, J. A., Lusby, W. R., Marvon, B. E., Nicolaou, K. C., Hoffmann, M. P. & Wilson, L. T., Pheromone blends of green stink bugs and possible parasitoid selection. *Naturwiss.*, **76** (1989) 173–5.
164. Quilico, A., Piozzi, F. & Pavan, M., The structure of dendrolasin. *Tetrahedron*, **1** (1957) 177–85.
165. Baker, R., Briner, P. H. & Evans, D. A., Total synthesis of ancistrofuran, a defensive compound from the termite *Ancistrotermes cavithorax*. *J. Chem. Soc., Chem. Commun.*, (1978) 981–3.
166. Maurer, B. & Ohloff, G., Zur Kenntnis der stickstoffhaltigen Inhaltsstoffe von Castoreum. *Helv. Chim. Acta*, **59** (1976) 1169–85.
167. Evans, D. A., Baker, R. and Howse, P. E., The chemical ecology of termite

defense behavior. In *Chemical Ecology: Odor Communication in Animals,* ed. F. Ritter. Elsevier/North Holland, Amsterdam, 1979, pp. 213–24.
168. Heath, R. R., McLaughlin, J. R., Tumlinson, J. H., Ashley, T. R. & Doolittle, R. E., Identification of the white peach scale sex pheromone; an illustration of micro techniques. *J. Chem. Ecol.,* **5** (1979) 941–53.
169. Roelofs, W. L., Gieselmann, M., Mori, K. & Moreno, D. S., Sex pheromone chirality comparison between sibling species—California red scale and yellow scale. *Naturwiss.,* **69** (1982) 348.
170. Mori, K. & Ueda, H., Synthesis of the optically active forms of 2,6-dimethyl-1,5-heptadien-3-ol acetate, the pheromone of the Comstock mealybug. *Tetrahedron,* **37** (1981) 2581–3.
171. Roelofs, W., Gieselmann, M., Cardé, A., Tashiro, H., Moreno, D. S., Henrick, C. A. & Anderson, R. J., Identification of the California red scale sex pheromone. *J. Chem. Ecol.,* **4** (1978) 211–24.
172. Anderson, R. J., Adams, K. G., Chinn, H. R. & Henrick, C. A., Synthesis of optical isomers of 3-methyl-6-isopropenyl-9-decen-1-yl acetate, a component of the California red scale pheromone. *J. Org. Chem.,* **45** (1980) 2229–36.
173. Demole, E. & Enggist, P., A chemical study of *Burley* tobacco flavour (*Nicotiana tabacum* L.) VI. Identification and synthesis of four irregular terpenoids related to solanone, including a 'prenylsolanone'. *Helv. Chim. Acta,* **58** (1975) 1602–7.
174. Baker, R., Parton, A. H. & Howse, P. E., Identification of an acyclic diterpene alcohol in the defense secretion of soldiers of *Reticulitermes lucifugus. Experientia,* **38** (1982) 297–8.
175. Bergström, G. & Löfquist, J., Chemical congruence of complex odoriferous secretions from Dufour's in three species of ant of the genus *Formica. J. Insect Physiol.,* **19** (1973) 877–907.
176. McKibben, G. H., Thompson, M. J., Parrot, W. L., Thompson, A. C. & Lusby, W. R., Identification of feeding stimulants for boll weevils from cotton buds and anthers. *J. Chem. Ecol.,* **11** (1985) 1229–39.
177. Phelan, P. L., Silk, P. J., Northcott, C. J., Tan, S. H. & Baker, T. C., Chemical identification and behavioral characterization of male wing pheromone of *Ephestia elutella* (Pyralidae). *J. Chem. Ecol.,* **12** (1986) 135–46.
178. McDowell, P. G. & Oloo, G. W., Isolation, identification and biological activity of trail following pheromone of termite *Trinervitermes bettonianus* (Sjöstedt) (Termitidae: Nasutitermitinae). *J. Chem. Ecol.,* **10** (1984) 835–51.
179. Prestwich, G. D., Wiemer, D. F., Meinwald, J. & Clardy, J., Cubitene: an irregular twelve-membered ring diterpene from a termite soldier. *J. Am. Chem. Soc.,* **100** (1978) 2560–1.
180. Klimetzek, D., Bartels, J. & Francke, W., Das Pheromonsystem des Bunten Ulmenbastkäfers *Pteleobius vittatus* F. (Col.: Scolytidae). *J. Appl. Entomol.,* **107** (1989) 518–23.
181. Birch, M. C. & Hefetz, A., Extrusible organs in male moths and their role in courtship behavior. *Bull. ESA,* **33** (1987) 222–9.
182. Goodrich, B. S., Hestermann, E. R., Murray, K. E., Mykytowycz, R., Stanley, G. & Sugowdz, G., Identification of behaviorally significant volatile compounds in the anal gland of the rabbit *Oryctolagus cuniculus. J. Chem. Ecol.,* **4** (1978) 581–94.
183. Borden, J. H., Chang, L., McLean, J. A., Slessor, K. N. & Mori, K., *Gnathotrichus sulcatus*: synergistic response to enantiomers of the aggregation pheromone sulcatol. *Science,* **192** (1970) 894–6.
184. Borden, J. H. & McLean, J. A., Secondary attraction in *Gnathotrichus retusus* and cross attraction of *G. sulcatus* (Coleoptera: Scolytidae). *J. Chem. Ecol.,* **5** (1979) 79–88.

185. Francke, W., Pan, M. L., König, W. A., Mori, K., Puapoomchareon, P., Heuer, H. & Vité, J. P., Identification of 'pityol' and 'grandisol' as pheromone components of the bark beetle *Pityopthorus pityographus*. *Naturwiss.*, **74** (1987) 343–5.
186. Francke, W., Structural concepts in pheromone chemistry. Paper presented at the 32nd IUPAC Congress, Stockholm, 2–7 August 1989.
187. Kinzer, G.-W., Fentiman, A. F. Jr, Page, T. F., Foltz, R. L., Vité, J. P. & Pitman, G. B., Bark beetle attractants of a new compound isolated from *Dendroctonus*. *Nature*, **211** (1969) 475–6.
188. Hefetz, A., Eickwort, G. C., Blum, M. S., Cane, J. & Bohart, G. E., A comparative study of the exocrine products of cleptoparasitic bees (*Holcopasites*) and their hosts (*Calliopsis*) (Hymenoptera: Anthophoridae, Andrenidae). *J. Chem. Ecol.*, **8** (1982) 1389–97.
189. Burger, B. V., Mackenroth, W. M., Smith, D., Spies, H. S. C. & Atkinson, P. R., Chemical composition of the wing gland and abdominal hair pencil secretion of the male African sugar cane borer, *Eldana saccharina* (Lepidoptera: Pyralidae). *Z. Naturforsch.*, **40c** (1985) 847–50.
190. Hall, D. R., Corte, A., Lester, R., Nesbitt, B. & Zagatti, P., Sex pheromones of rice moth *Corcyra cephalonica* Stainton; II. Identification and role of female pheromone. *J. Chem. Ecol.*, **13** (1987) 1575–89.
191. Schildknecht, H., Neumeier, H. & Tauscher, B., Gyrinal, die Pygidialdrüsensubstanz der Taumelkäfer (Coleoptera: Carabidae) *Liebigs Ann. Chem.*, **756** (1972) 155–61.
192. Meinwald, J., Opheim, K. & Eisner, T., Gyrinidal: a sesquiterpenoid aldehyde from the defensive glands of gyrinid beetles. *Proc. Natl. Acad. Sci. USA*, **69** (1972) 1208–10.
193. Wheeler, J., Oh, S., Benfield, E. & Neff, S., Cyclopentanoid norsesquiterpenes from gyrinid beetles. *J. Am. Chem. Soc.*, **94** (1972) 7589–90.
194. Miller, J. R., Hendry, L. B. & Mumma, R. O., Norsesquiterpenes as defensive toxins of whirligig beetles (Coleoptera: Gyrinidae). *J. Chem. Ecol.*, **1** (1975) 59–82.
195. Eisner, T., Hendry, L. B., Peakall, D. B. & Meinwald, J., 2,5-Dichlorophenol (from ingested herbicide?) in defensive secretion of grasshopper. *Science*, **172** (1971) 277–8.
196. Birgersson, G., Schlyter, F., Löfquist, J. & Bergström, G., Quantitative variation of pheromone components in the spruce bark beetle *Ips typographus* from different attack phases. *J. Chem. Ecol.*, **10** (1984) 1029–55.
197. Albone, E. S., Dihydroactinidiolide in the supracaudal scent gland of the secretion of the red fox. *Nature*, **256** (1975) 575.
198. Rocca, J. R., Tumlinson, J. H., Glancey, B. M. & Lofgren, C. S., The queen recognition pheromone of *Solenopsis invicta*, preparation of (*E*)-6-(1-pentyl)-2H-pyran-4-one. *Tetrahedron Lett.*, **24** (1983) 1889–92.
199. Meinwald, J., Erickson, K., Hartshorn, M., Meinwald, Y. C. & Eisner, T., Defensive mechanism of arthropods XXIII. An allenic sesquiterpenoid from the grasshopper *Romalea microptera*. *Tetrahedron Lett.*, (1968) 2959–62.
200. Francke, W., Schulz, S., Sinnwell, V., König, W. A. & Roisin, Y., Epoxytetrahydroedulan, a new terpenoid from the hairpencils of *Euploea* (Lep.: Danainae) butterflies. *Liebigs Ann. Chem.* (1989) 1195–201.
201. Winter, M., Schulte-Elte, K.-H., Velluz, A., Limacher, J., Pickenhagen, W. & Ohloff, G., Aromastoffe der roten Passionsfrucht. Zwei neue Edulanderivate. *Helv. Chim. Acta*, **62** (1979) 131–4.
202. Prelog, V. & Ruzicka, L., Über zwei moschusartig riechende Steroide aus Schweinetestes Extrakten. *Helv. Chim. Acta*, **27** (1944) 61–6.
203. Brooksbank, B. W. L., Brown, R. & Gustafsson, J.-A., The detection of

5α-androst-16-en-3α-ol in human male axillary sweat. *Experientia,* **30** (1974) 864–5.
204. Brooksbank, B. W. L. & Haslewood, G. A. D., The estimation of androst-16-en-3-ol in human urine. Partial synthesis of androstenol and of its β-glucosiduronic acid. *Biochem. J.*, **80** (1961) 488–96.
205. Gower, D. B., 16-Unsaturated C_{19} steroids. A review of their chemistry, biochemistry and possible physiological role. *J. Steroid Biochem.*, **3** (1972) 45–103.
206. Cowley, J. J., Johnson, A. L. & Brooksbank, B. W. L., The effect of two odorous compounds on performance in an assessment-of-people test. *Psychoneuroendocrinology,* **2** (1977) 159–72.
207. Kirk-Smith, M., Booth, D. A., Carroll, D. & Davies, P., Human social attitudes affected by androstenol. *Res. Commun. Psychology, Psychiatry and Behaviour,* **3** (1978) 379–84.
208. Benton, D., The influence of androstenol—a putative human pheromone—on mood throughout the menstrual cycle. *Biolog. Psychol.*, **15** (1982) 249–56.
209. Black, S. L. & Biron, C., Androstenol as a human pheromone: no effect on perceived physical attractiveness. *Behavioral Neural Biol.*, **34** (1982) 326–30.
210. Labows, J. N., McGinley, K. J. & Klingman, A. M., Perspectives on axillary odor. *J. Soc. Cosmet. Chem.*, **34** (1982) 193–202.
211. Ayorinde, F., Wheeler, J. W., Wemmer, C. & Murtaugh, J., Volatile components of the occipital gland secretion of the Bactrian camel (*Camelus bactrianus*). *J. Chem. Ecol.*, **8** (1982) 177–83.
212. Claus, R. & Hoppen, H. O., The boar-pheromone steroid identified in vegetables. *Experientia,* **35** (1979) 1674–5.
213. Novotny, M., Harvey, S., Jemiolo, B. & Alberts, J., Synthetic pheromones that promote intermale aggression in mice. *Proc. Natl. Acad. Sci. USA,* **82** (1985) 2059–61.
214. Suzuki, T. & Mori, K., (4*R*,8*R*)-(−)-4,8-Dimethyldecanal: the natural aggregation pheromone of the red flour beetle, *Tribolium castaneum* (Coleoptera: Tenebrionidae). *Appl. Entomol. Zool.*, **18** (1983) 134–6.
215. Levinson, H. Z. & Mori, K., Chirality determines pheromone activity for flour beetles. *Naturwiss.*, **70** (1983) 190–2.
216. Meinwald, J., Thompson, W. R., Eisner, T. & Owen, D. F., Pheromones. VII. African monarch: major components of hairpencil secretion. *Tetrahedron Lett.*, (1971) 3485–8.
217. Battiste, M. A., Strekowski, L., Vanderbilt, D. P., Visnick, M., King, R. W. & Nation, J., Anastrephin and epianastrephin, novel lactone components isolated from sex pheromone blend of male Caribbean and Mexican fruit flies. *Tetrahedron Lett.*, **24** (1983) 2611–14.
218. Chuman, T., Sivinski, J., Heath, R. R., Calkins, C. O., Tumlinson, J. H., Battiste, M. A., Wydra, R. L., Strekowski, L. & Nation, J. L., Suspensolide, a new macrolide component of male Caribbean fruit fly (*Anastrepha suspensa* [Loew]) volatiles. *Tetrahedron Lett.*, **29** (1988) 6561–4.
219. Wong, J. W., Verigin, V., Oehlschlager, A. C., Borden, J. H., Pierce, H. D. Jr, Pierce, A. M. & Chang, L., Isolation and identification of two macrolide pheromones from the frass of *Cryptolestes ferrugineus* (Coleoptera: Cucujidae). *J. Chem. Ecol.*, **9** (1983) 451–74.
220. Sakan, T., Isoe, S. & Hyeon, S. B., The structure of actinidiolide, dihydroactinidiolide and actinidol. *Tetrahedron Lett.*, (1967) 1623–7.
221. Kikukawa, T., Matsumura, F., Olaifa, J., Kraemer, M., Coppel, H. C. & Tai, A., Field evaluation of chiral isomers of the sex pheromone of the European pine sawfly, *Neodiprion sertifer. J. Chem. Ecol.*, **9** (1983) 673–93.
222. Röller, H., Dahm, K. H., Sweeley, C. C. & Trost, B. M., Die Struktur des

Juvenilhormons. *Angew. Chem.*, **79** (1967) 190–1 (Int. Ed. Engl., **6** (1967) 179).
223. Attygalle, A. B., Cammaerts, M. C. & Morgan, E. D., Dufour gland secretion of *Myrmica rugulosa* and *Myrmica schencki* workers. *J. Insect Physiol.*, **29** (1983) 27–32.
224. Morgan, E. D., Chemical words and phrases in the language of pheromones for foraging and recruitment. In *Insect Communication*, ed. T. Lewis. Academic Press, London, 1984, pp. 169–94.
225. Kobayashi, M., Koyama, T., Ogura, K., Seto, S., Ritter, F. J. & Brüggemann-Rotgans, I. E. M., Bioorganic synthesis and absolute configuration of faranal. *J. Am. Chem. Soc.*, **102** (1980) 6602–4.
226. Alvarez, F. M., Vander Meer, R. K. & Lofgren, C. S., Synthesis of homofarnesenes; trail pheromone components of the fire ant, *Solenopsis invicta*. *Tetrahedron*, **43** (1987) 2897–900.

Chapter 4

Origin of Natural Odorants

RODNEY CROTEAU & FRANK KARP

Institute of Biological Chemistry, Washington State University, Pullman, WA 99164, USA

1 INTRODUCTION

Only substances that have a molecular weight below about 400 and an appreciable vapor pressure at room temperature can be perceived as having odor. The spectrum of odorants is thus limited to relatively small, neutral organic compounds, including undissociated acids and nitrogenous bases.[1] Relatively few organic acids are sufficiently volatile to contribute to natural aromas. Acetic (vinegary), propionic (goaty), butyric (spoiled butter), and lactic (sauerkraut) acids are odorous at relatively high concentration.

Historically, aroma research was largely directed toward the isolation, identification, and cataloging of specific odor-producing constituents from the complex mixtures produced by nature. This work continues, and many important aromas from foodstuffs, beverages, smoking materials and flowers have only recently been separated and the constituents positively identified. The cost of natural isolates, coupled to the often difficult separation of the desirable component(s), has stimulated the development of chemical syntheses of many substances from more abundant raw materials, including petrochemical feedstocks. Synthetically derived fragrances in many cases have superseded those produced by living organisms.

The more recent emphasis on 'natural' as opposed to 'artificial' ingredients has, however, refocused attention on the origin of natural odorants and on the possibility of improving biological methods for the production of these materials. Recent advances in analytical techniques and instrumentation, the use of radiolabeled tracers, the development of instructive biomimetic syntheses and chemical model reactions, and the exploitation of cell-free enzyme systems, have greatly facilitated our understanding of the origin of natural odorants via the enzymatic, and occasionally non-enzymatic, processes that occur in microbial, animal and plant cells. Of the natural fragrances, those of plant origin are certainly the most structurally diverse. In spite of this diversity, and the potential biogenetic possibilities, most odor-bearing compounds appear to be formed by way of relatively few, often overlapping, metabolic pathways that form the subject of this chapter.

2 PRODUCTS OF LIPID AND AMINO-ACID METABOLISM

Aliphatic aldehydes, ketones, lactones and related compounds are among the most widely distributed of natural odorants, and are often major contributors to fruit aromas. Many such compounds, associated for example with the essence of banana, apple and pear, are neither produced during growth nor present at harvest. Rather, they arise during a short ripening period marked by a climacteric rise in respiration and the onset of cellular catabolic processes. The biogenesis of a large number of these volatile aliphatic types can be rationalized by invoking a reasonably small number of metabolic pathways in which non-volatile fatty acids and amino-acids serve as the most important precursors.

Aliphatic Compounds

Aroma-bearing aliphatic compounds are considered to be produced via three major pathways: (1) lipoxygenase-catalyzed conversion of polyunsaturated fatty acids into C_6 and C_9 aldehydes and related substances, and into C_{12} and C_9 oxo-acids; (2) conversion of fatty acids, via α- and β-oxidation, into short-chain acids, aldehydes and ketones, and related metabolites; (3) conversion of amino-acids to their corresponding C_{n-1} acyl derivatives via transamination and oxidative decarboxylation.[2]

The major fatty acids of plant glyceryl lipids are the even-numbered, saturated derivatives from C_{12} to C_{18} and the unsaturated C_{18} derivatives (oleic, linoleic and linolenic acids), all of which are products of primary metabolism derived via the well-known acetyl coenzyme A–malonyl coenzyme A pathway. Free fatty acids exist at very low levels in intact plant cells, and must first be released from the corresponding glycerides and phosphoglycerides by enzymatic hydrolysis (i.e. by the action of lipases and phospholipases). Unsaturated fatty acids (primarily linoleic and linolenic) may then become substrates for lipoxygenase, this enzymatic oxygen insertion reaction ultimately leading to the formation of C9- and C13-hydroperoxides (Fig. 1). The lipoxygenase reaction mimics radical-initiated autoxidation; however, the enzymatic process is both regio- and stereo-specific, the selectivity depending on the enzyme source. Hydroperoxide lyases then cleave these reactive compounds, the cleavage of the C9-hydroperoxide leading to unsaturated C_9 aldehydes and 9-oxononanoic acid, and the cleavage of the C13-hydroperoxide leading to C_6 aldehydes and the corresponding 12-oxododecenoic acid. Lyase activity has been demonstrated in a variety of plants, including cucumber, tomato, pear and apple. The lyase from pear fruit is specific for the C9-hydroperoxide, whereas that from tomato fruit is specific for the C13-hydroperoxide. The lyase from cucumber fruit accepts both isomers.[3]

Hexanal and *trans*-2-nonenal originate from linoleic acid ($C_{18:2}$), as illustrated in Fig. 1, while *trans*-2-hexenal and nonadienal arise from linolenic acid

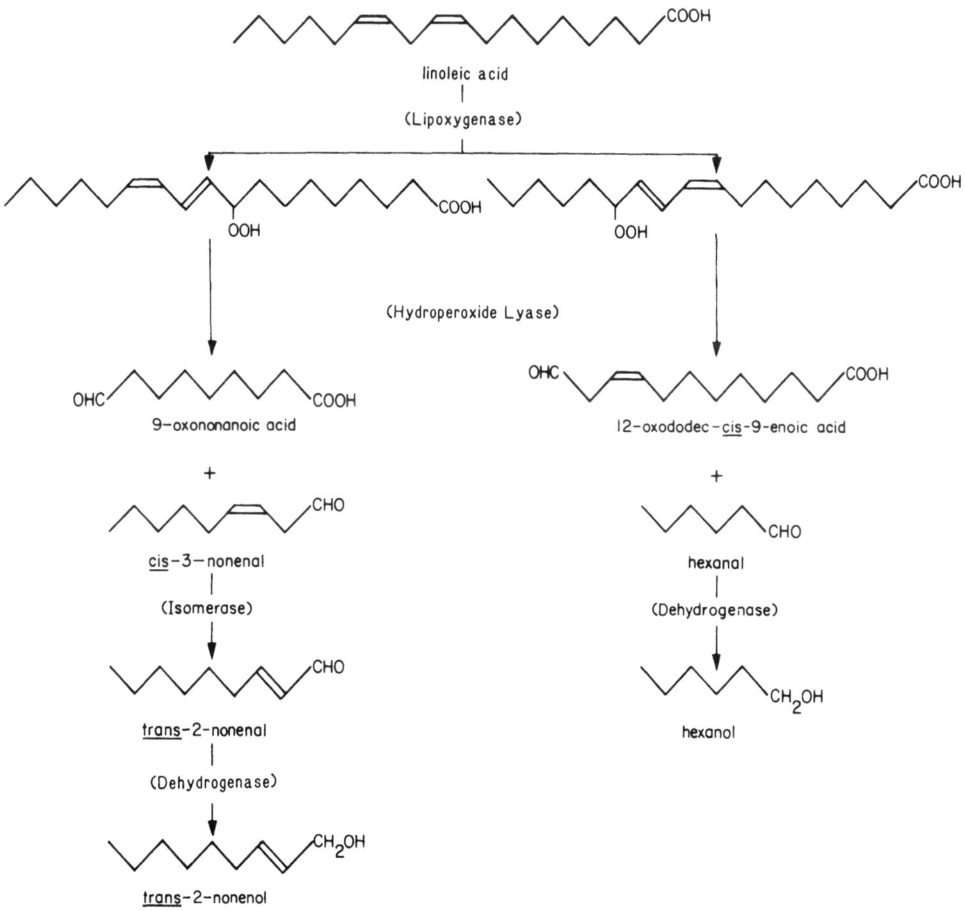

Fig. 1. Enzymatic sequence for the formation of C_6 and C_9 aldehydes and alcohols from linoleic acid.

($C_{18:3}$).[4] Alcohols are derived from these aldehydes by the action of alcohol dehydrogenase(s). The C_6 aldehydes and alcohols are largely responsible for the characteristic odor of green leaves, and they contribute to the aroma of a variety of fruits.[3] The 3-enals are converted by an isomerase to the conjugated 2-enals. Such isomerization of (3*Z*)-hexenal to (2*E*)-hexenal yields 'leaf aldehyde'. The former provides 'leaf alcohol' upon reduction. Nonenal and nonadienal, and their derived alcohols, have been described as melon, cucumber or violet-like.[5] Species-dependent variations in the type and quantity of available fatty acids, and the specificity of lipoxygenases, lyases, isomerases and dehydrogenases are responsible for the diverse mixtures of C_6 and C_9 aliphatics encountered in plant volatiles.[2]

An alternative means of generating volatile, chain-shortened products from the fatty acids of fruits and vegetables is the β-oxidation pathway. This general scheme is initiated by conversion of the fatty acid to the corresponding acyl coenzyme A derivative by the action of acyl–thiol ligases (acyl CoA synthetase, acyl thiokinase). The acyl chain then undergoes sequential desaturation at C2–C3, hydration of the double bond to the C3(β)-hydroxy derivative, and oxidation to the corresponding C3(β)-ketoacyl chain. In the final step of this sequence, the enzyme thiolase, with the assistance of coenzyme A, cleaves the β-keto derivative to yield acetyl-coenzyme A and the acyl-coenzyme A with the chain now shorter by two carbons.[6] Repetition of the β-oxidation cycle would ultimately cleave the chain to acetate units. Both saturated and unsaturated chains may be oxidized by this route, although in the latter case the participation of ancillary reduction and/or isomerization step(s) is required to convert unsaturated chains to normal β-oxidation intermediates.[7] The resulting chain-shortened fatty acyl-coenzyme A esters may be reduced to the corresponding aldehydes by the action of acyl-coenzyme A reductases, and the aldehydes in turn converted to alcohols by dehydrogenases. The alcohols thus formed may be esterified with acyl coenzyme A derivatives by the action of transacylase. A generalized scheme for these transformations is provided in Fig. 2, in which the production of such compounds as ethyl octanoate and octyl acetate is illustrative of the diverse set of volatile metabolites which could be so generated.[2]

Odd-chain products are formed from the more common even-chain fatty acids by the process of α-oxidation in which flavin-dependent enzymes decarboxylate the fatty acid, via the α-hydroperoxy intermediate, to yield the aldehyde shortened by one carbon. Reduction of the aldehyde may occur and the resulting alcohol may enter the esterification cycle as illustrated (Fig. 2). Alternatively, oxidation of the aldehyde to the acid could lead to the production of odd-chain esters, including propionates and valerates.[6,8] It is obvious that the combination of α- and β-oxidation, when coupled to reduction and esterification steps, is capable of generating an extremely wide and complex range of volatile, odor-bearing metabolites.

β-Oxidation and alternate modes of lipid oxygenation are also thought to be involved in the production of additional odorants in plants, including the

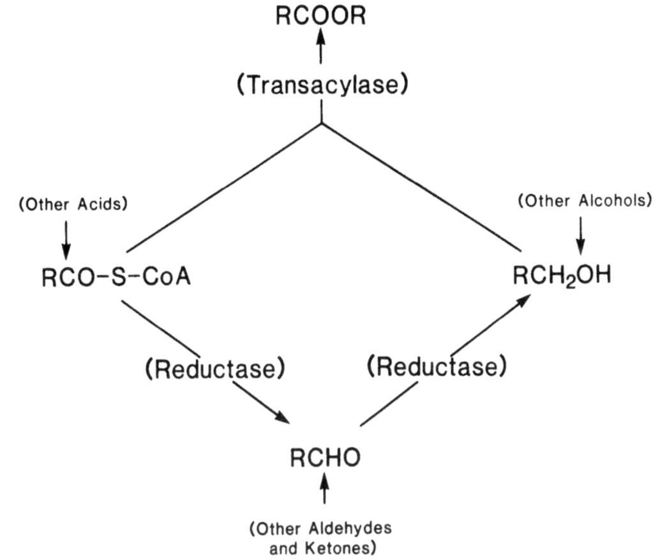

Fig. 2. Pathway for the biosynthesis of short-chain aldehydes, alcohols, and esters from acyl CoA.

methyl ketones and lactones. Aliphatic methyl ketones are characteristic of the essential oil of rue (*Ruta graveolens*). Most prominent are 2-nonanone and 2-undecanone (and the corresponding secondary alcohols) which are considered to arise from C_{10} and C_{12} acyl chains in the course of β-oxidation. Thus, following conversion to the β-ketoacyl chain, decarboxylation yields the odd-chain methyl ketone (Fig. 3).[9] The γ- and δ-lactones, characteristic of pineapple, peach, apricot, strawberry, mango and coconut,[8] undoubtedly arise

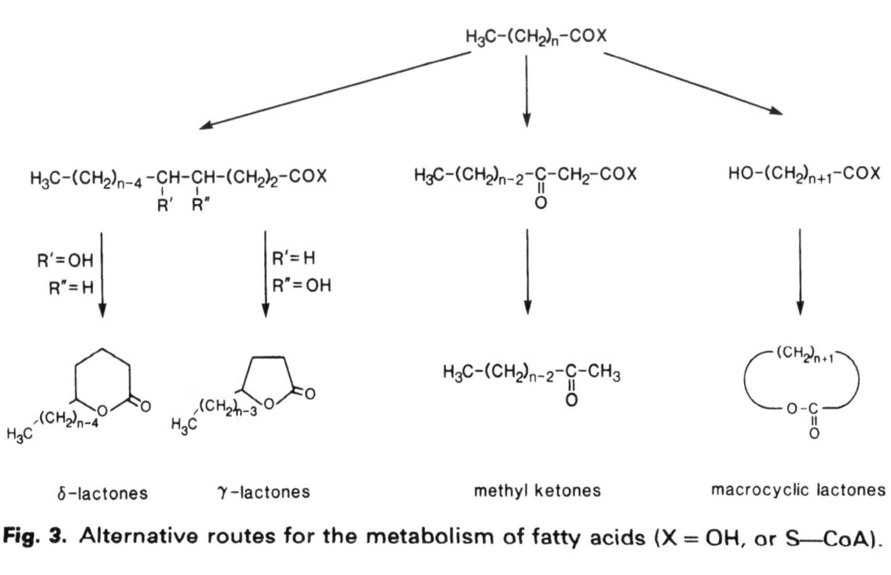

Fig. 3. Alternative routes for the metabolism of fatty acids (X = OH, or S—CoA).

from intermediates of fatty acid β-oxidation, or the biosynthetic elongation process. It seems likely that C4-unsaturated chains undergo hydration of the double bond, or epoxidation and reduction, to give the C4- and C5-hydroxyacyl chains which are subsequently lactonized to the corresponding γ- and δ-lactones (Fig. 3). Macrocyclic lactones, such as those found in galbanum, angelica root and ambrette seed,[10] are probably formed by ω- (or $\omega - 1$)-hydroxylation of the corresponding odd or even acyl chain, followed by internal esterification via a transacylase-type enzyme (Fig. 3). The aroma-producing phthalides (bicyclic γ-lactones bearing a fused cyclohexane ring), typical of the essential oils of certain Umbelliferae,[11] are probably formed by the polyketide pathway rather than by oxidative modification of normal fatty acids.

Short branched-chain aldehydes, alcohols and esters are also important aroma compounds of ripening fruits and vegetables produced during the climacteric rise in respiration. Such branched-chain derivatives arise from branched amino-acids by transamination and oxidative decarboxylation to the shortened acyl-coenzyme A derivative (Fig. 4).[2] Cellular levels of free leucine and valine may increase several-fold during the climacteric period, and may serve as precursors of 3-methylbutyl and 2-methylpropyl derivatives, respectively.[12] The aroma of field-ripened tomatoes is attributed to esters of 3-methyl-1-butanol, derived, ultimately, from leucine. Thus, leucine aminotransferase converts the amino acid to the α-keto acid (α-ketoisocaproic acid) which is decarboxylated to 3-methylbutanoyl CoA. The coenzyme A ester may either be a substrate for transacylation or be reduced to the corresponding aldehyde and alcohol (isoamyl alcohol). Valine and isoleucine may be transformed to branched esters, aldehydes and alcohols by the same general pathway, and alanine and aspartic acid are similarly metabolized by mitochondrial transaminases.[13]

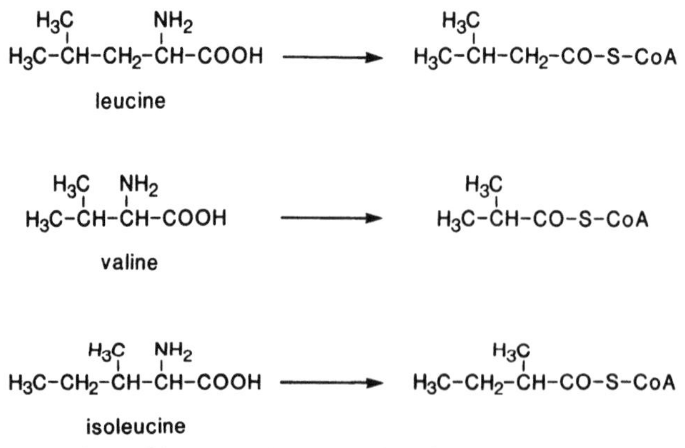

Fig. 4. Formation of branched acyl chains from common amino-acids.

The Jasmone Group

Jasmone was discovered as a principal aroma constituent of jasmine in 1899. This cyclopentenone derivative has a fruity, celery-like odor which, upon dilution, becomes sweet-floral and blends well with all floral scents.[14] Of the numerous compounds in jasmine flower oil, jasmone, methyl jasmonate and jasmine lactone (Fig. 5) are thought to carry the principal jasmine fragrance, while comprising but a very small fraction of the fresh weight of the flowers.[15]

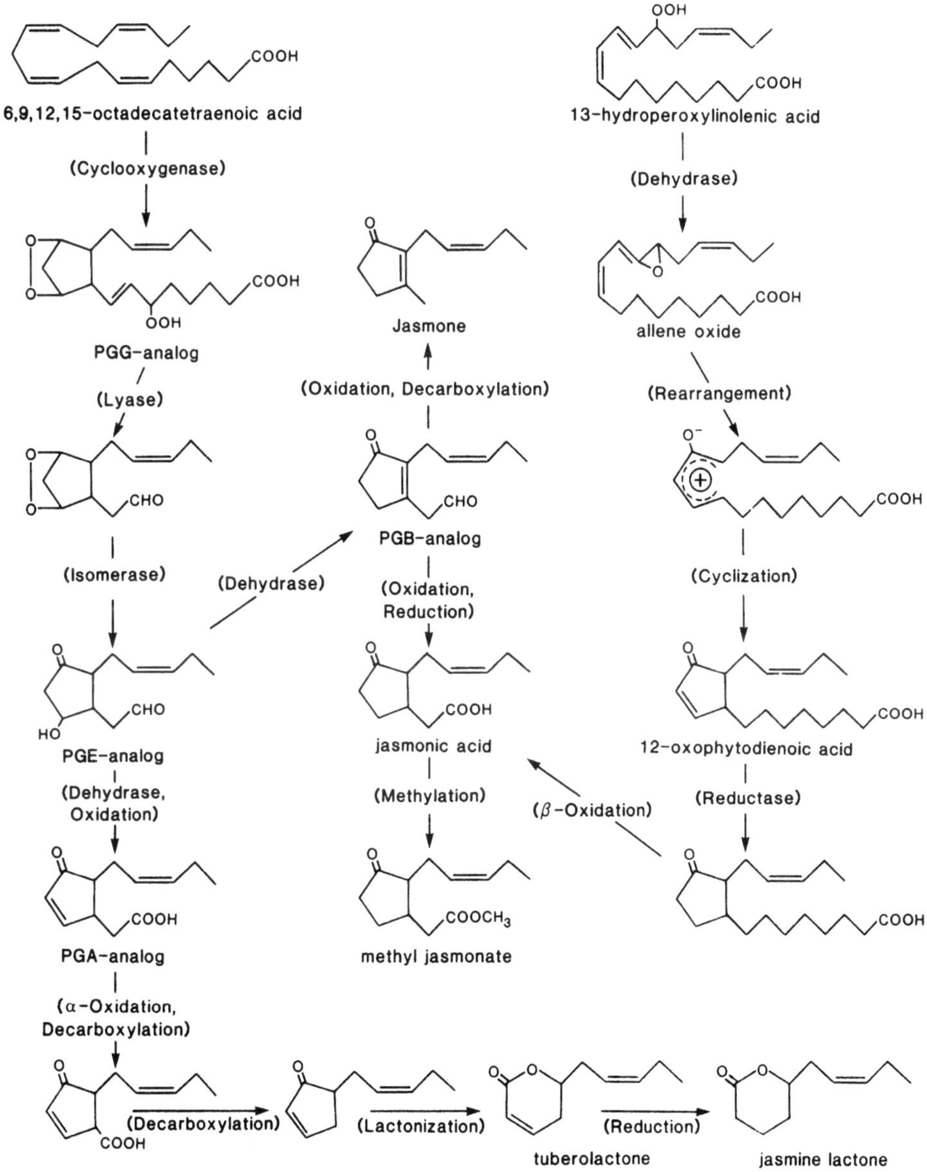

Fig. 5. Alternative routes to jasmonoid compounds.

All proposed biosynthetic schemes for the formation of jasmonoid compounds begin with polyunsaturated fatty acids. The extensive studies by Vick & Zimmerman[16] have led to a proposal for the formation of jasmonic acid which serves as a plant growth regulator (promotes senescence). This pathway begins with the C13-hydroperoxide derivative of linolenic acid (via lipoxygenase; see above) which is dehydrated (via hydroperoxide dehydrase of the chloroplast[17]) to a short-lived allene oxide (Fig. 5). The latter may undergo hydrolysis to ketols or spontaneous rearrangement and cyclization to 12-oxo-10,15-phytodienoic acid, a key intermediate of the pathway which, by reduction of the endocyclic double bond and β-oxidative 'carboxyl-end-shortening', yields jasmonic acid (Fig. 5). Methylation of this product, presumably by *S*-adenosylmethionine, could provide methyl jasmonate.

An alternative scheme, bearing analogy to prostaglandin formation in animals, has been put forward by Demole.[15] This scheme involves chain-shortening from the methyl end of the cyclopentane derivative and requires a most unusual[18] polyunsaturated acid, 3,6,9,12-octadecatetraenoic acid, as precursor. As a final variant, again analogous to prostaglandin biosynthesis, 6,9,12,15-octadecatetraenoic acid (formed by C6-desaturation of linolenic acid) could, via the action of a cyclo-oxygenase, yield an endoperoxide analog of prostaglandin G_2 (Fig. 5; PGG-analog). Hydroperoxide lyase could then afford the carboxyl-end-shortened product in a single step. Isomerization of the endoperoxide (in a manner analogous to the transformation of prostaglandin H to prostaglandin E) could provide the hydroxycyclopentanone (Fig. 5; PGE-analog) which by dehydration (in a manner analogous to the conversion of prostaglandin E to prostaglandin B) yields the α,β-unsaturated ketone (Fig. 5; PGB-analog). A straightforward oxidation–reduction sequence could thus provide jasmonic acid with methylation as before. In a similar vein, decarboxylation of the α,β-unsaturated ketone (an enzymatic reaction precedented in plant paraffin formation) could give rise to (*Z*)-jasmone directly. The origin of jasmine lactone is rather more difficult to rationalize; however, if the hydroxycyclopentanone (PGE-analog) were dehydrated (in a manner analogous to the transformation of prostaglandin E to prostaglandin A) to the alternate conjugated ketone (PGA-analog), an α-oxidation–decarboxylation sequence could accomplish the removal of the carboxymethyl function. Lactonization (a reaction well documented in terpene metabolism)[19] would thus generate tuberolactone and trivial reduction of the endocyclic double bond would yield jasmine lactone (Fig. 5). Although this hypothetical sequence does not involve oxophytodienoic acid as an intermediate, it does account for the formation of all relevant jasmonoid compounds and it employs reaction types that have direct enzymatic precedent in other biological systems.

Phenylpropanoids

The characteristic C_6–C_3 skeleton of the fragrant phenylpropanoids originates from phenylalanine or tyrosine, the aromatic amino-acid end-products of the shikimic acid pathway.[6] Phenylalanine ammonia-lyase (PAL) removes ammonia from phenylalanine to produce *trans*-cinnamic acid, and this enzyme is generally considered to mark a transition from primary to secondary metabolism in plants (similar lyase action on tyrosine yields *p*-coumaric acid) (Fig. 6). Cinnamic acid is not an important odorant, but rather serves as an important precursor of numerous other aromatic metabolites including aldehydes, alcohols, lactones and esters.[20] The methyl, ethyl, benzyl and cinnamyl esters have pleasant and long-lasting aromas, and all occur naturally in various plants.

Hydroxycinnamic acids are obtained by sequential aromatic hydroxylations, giving the common 4-, 3,4-, and 3,4,5-substitution patterns. Cinnamate-4-hydroxylase is a specific mixed-function oxidase of the cytochrome P-450 type that introduces the *p*-hydroxy function (Fig. 6). The resulting *p*-coumaric acid

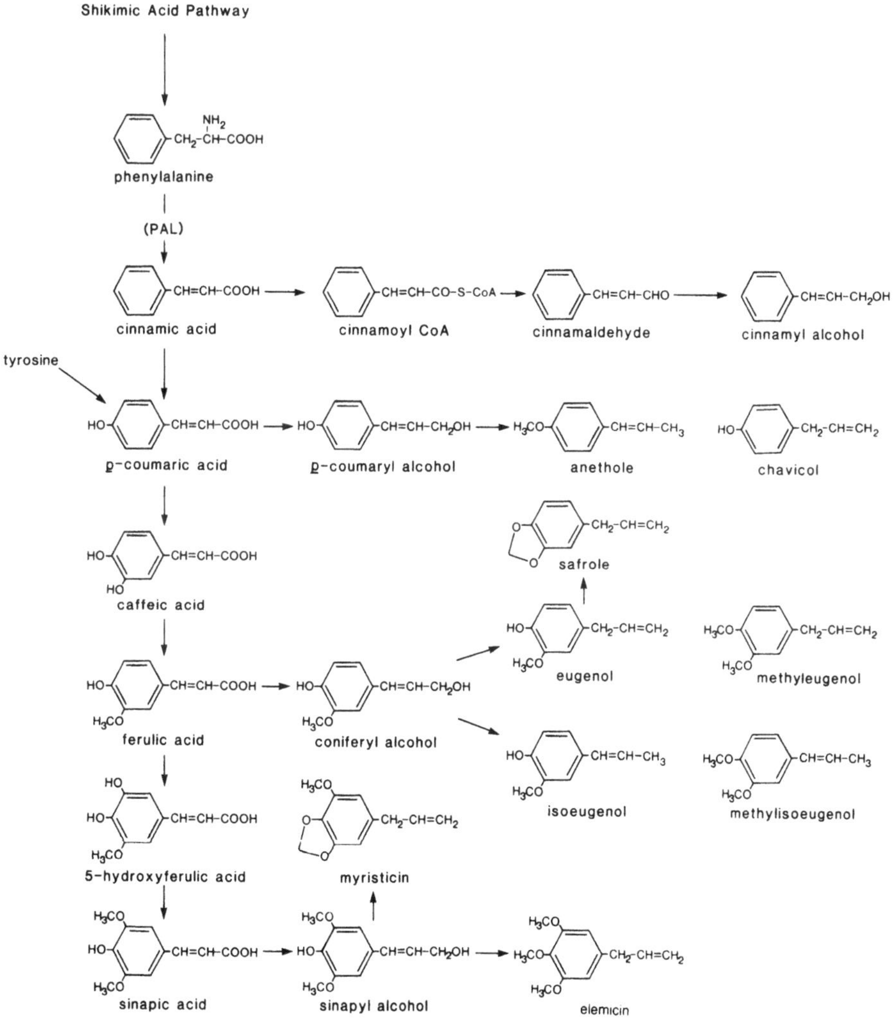

Fig. 6. Conversion of phenylalanine to phenylpropanoids.

may be further hydroxylated to ferulate by another mixed-function oxidase, or possibly via superoxide or enzymes of the phenolase or peroxidase type.[20] The C5-hydroxylation of ferulic acid is accomplished by a distinct cytochrome P-450 oxygenase.[21] The phenolic hydroxyl functions may be converted to the corresponding methoxy groups by transmethylation (*O*-methyltransferase) involving *S*-adenosylmethionine. Although the arrangement of structural types illustrated in Fig. 6 may imply a 'metabolic grid' in the transformation of phenylpropanoids, the actual sequence and positional specificity of the reactions is considered to be tightly regulated in most species. The resulting hydroxy and methoxy acids may undergo subsequent metabolism, much like the corresponding acyl derivatives (see Fig. 2), to provide aldehydes, alcohols and esters. As with the corresponding acyl derivatives, thermodynamic considerations appear to require activation of the carboxyl group, and this is accomplished via formation of the coenzyme A or glucose ester.[22] Cinnamaldehyde is an important contributor to the aroma of cinnamon bark and cassia oils. Cinnamyl alcohol has been described as warm-balsamic, floral and sweet, resembling cinnamon oil and many flower types. It occurs in both hyacinth and styrax.[23] Diverse aliphatic and aromatic acids may form esters with various aromatic alcohols, which are constituents of essential oils or resins, such as coniferyl benzoate in *Styrax* benzoin, cinnamyl acetate in cinnamon bark and sinapyl acetate in *Citrus bergamia.*[20] As mentioned above, aliphatic and aromatic esters of cinnamate are widespread in the plant kingdom. The coenzyme A esters of cinnamate and related aromatic acids may also be elongated via malonyl coenzyme A, providing a wide range of additional products. Piperine, the pungent principle of pepper, is derived by this pathway.[6]

Coumarins can be regarded as a special type of cinnamate ester. Following *o*-hydroxylation of *trans*-cinnamic acid to *o*-coumaric acid[20] and glycosylation, the product is stored in the vacuoles of fresh plant tissue where it may undergo light-dependent isomerization. On disruption of this compartmentation, for example by wounding or desiccation, the glycoside is hydrolyzed and the *cis*-aglycone spontaneously cyclizes (Fig. 7). Coumarin is described as possess-

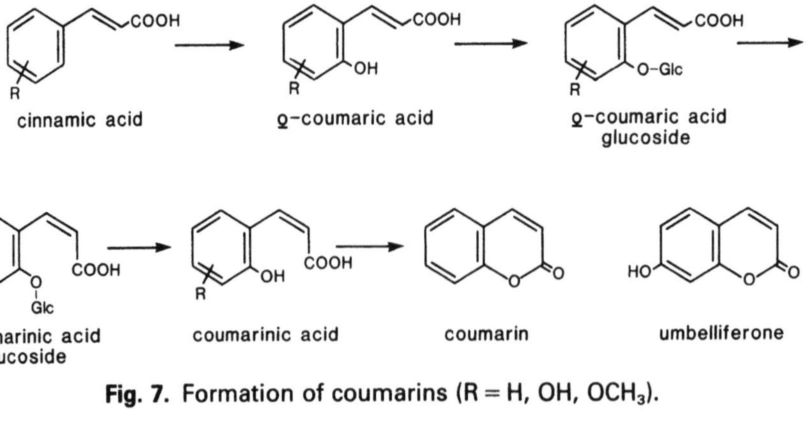

Fig. 7. Formation of coumarins (R = H, OH, OCH_3).

ing a sweet, grass- or hay-like odor with a spicy note, and is characteristic of the aroma of drying woodruff (*Asperula odorata*), sweet clover (*Melilotus officinalis*) and *Anthoxanthum odoratum*.[6] Additional substituents on the parent *o*-coumaric acid will give rise to the corresponding substituted coumarins, such as the hydroxycoumarins (e.g. umbelliferone) and the methoxycoumarins of orchids (*Dendrobium*). Synthetic coumarin, used formerly in artificial vanilla flavors, is now used mostly in soaps, detergents and industrial odorants.[14]

The substituted cinnamyl alcohols serve as precursors of the various allyl and propenyl phenols (Fig. 6). Although these transformations can be viewed as a combination of dehydration and reduction, the mechanism of the reaction sequence is uncertain and alternative routes have been proposed.[24,25] Nevertheless, the origin of eugenol, methyleugenol, isoeugenol and methylisoeugenol from coniferyl alcohol (via ferulate), anethole and chavicol from *p*-coumaryl alcohol, and elemicin from sinapyl alcohol, seems clear from the substitution patterns alone; and biosynthetic evidence is available in some cases.[26] Of the allylphenols, the sharp, spicy odor of eugenol is easily recognized as reminiscent of cloves. It is the main component of clove, cinnamon leaf, pimento and bay-leaf oils. Methyleugenol, described as 'herbaceous–spicy, dry–musty with slight carnation undertones',[23] frequently accompanies eugenol and also occurs in laurel leaf and citronella oils. An interesting variation on this class of phenylpropanoids is the occurrence of the methylenedioxy derivatives (e.g. myristicin, safrole, apiole) formed by ring closure of the corresponding hydroxymethoxy compounds (Fig. 6). Allyl and propenyl derivatives of this type occasionally co-occur, such as myristicin and isomyristicin in nutmeg, and anethole and methylchavicol in anise and fennel.[27] Genetic analysis of *Perilla frutescens* hybrids has provided suggestive evidence for the organization of pathways leading to methyleugenol, myristicin, dillapiole and elemicin;[28] however, the lack of fundamental knowledge of the origin of these products prevents any detailed assessment of the regulation of phenylpropanoid metabolism.

Phenylpropanoid precursors (C_6–C_3) give rise to a very wide range of compounds that are, strictly speaking, no longer phenylpropanoid, but rather can be regarded as derivatives of benzoic acid (C_6–C_1) or phenylacetic acid (C_6–C_2). Side-chain degradation of the variously substituted cinnamates by removal of one carbon (perhaps by α-oxidation) or two carbons (by β-oxidation of the coenzyme A ester or alternative means) would provide the respective phenylacetic acid or benzoic acid with maintenance of the substitution pattern. Although such side-chain cleavage reactions are often considered 'catabolic' transformations, the chain-shortened products clearly must serve as biosynthetic precursors of related aldehydes, alcohols and esters. Thus, the origin of benzyl acetate (jasmine and ylang ylang oil),[29] methyl salicylate (oil of wintergreen), benzyl salicylate (40% of *Ocotea teleiandra* oil)[14] and piperonal (odor of heliotrope) from the corresponding parent acid can be readily envisioned (Fig. 8), although knowledge of the mechanisms involved is very

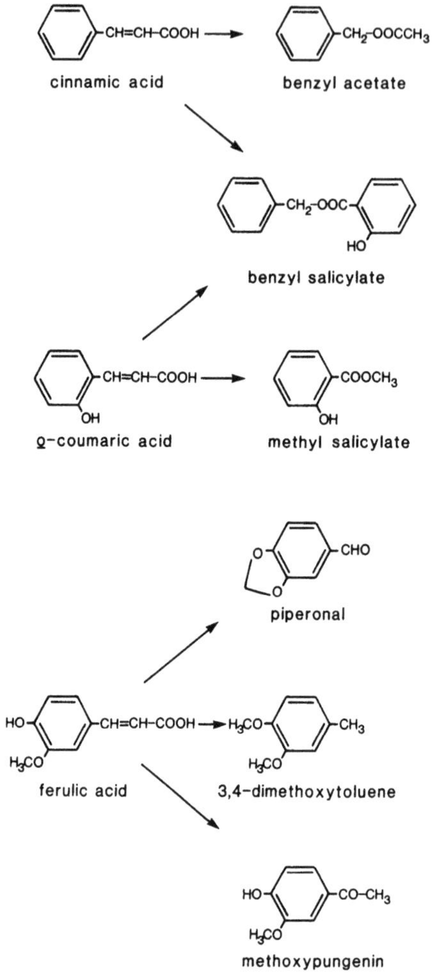

Fig. 8. Additional metabolites of cinnamic acid.

limited.[30] Other interesting groups of chain-shortened derivatives are the methoxytoluenes (such as 3,4-dimethoxytoluene of banana), probably formed from ferulic acid by combination of α-oxidation and decarbonylation or decarboxylation reactions, and the acetophenones (such as methoxypungenin), likely derived by decarboxylation of the β-ketoacid corresponding to ferulate (Fig. 8).

Vanillin and Related Metabolites

Of the very large family of phenylpropanoid odorants, vanillin, which is largely responsible for the familiar fragrance of *Vanilla* and *Orchis*,[6] has received the most attention with regard to biosynthetic origin. Labeling studies have confirmed the transformation of phenylalanine, cinnamic acid and ferulic acid to vanillin in the *Vanilla* bean [in which the product is sequestered as glucovanillin (β-D-glucopyranosyloxy-3-methoxybenzaldehyde)], thus providing a broad outline of the biosynthetic pathway that involves loss of two carbons from the acid side-chain (Fig. 9).[31] Since vanillic acid is only poorly incorporated into vanillin, and benzoic acid is not incorporated at all into the corresponding aldehyde, it would appear that the free acid is not the immediate product of the chain-shortening reaction. A possible route, as noted in an earlier section, could involve the conversion of ferulic acid to the coenzyme A ester (via ATP and CoA), followed by one round of β-oxidation to provide vanilloyl coenzyme A directly. Reduction of this activated derivative would then yield vanillin, which undergoes glycosylation to the storage form in the fresh bean and is later released during post-harvest treatment. Salicylaldehyde is probably formed by similar reaction sequence from *o*-coumaroyl coenzyme A.[32] An alternative, 'non-oxidative' route has also been described in which the parent acid is hydrated to the β-hydroxy derivative, followed by cleavage to the aldehyde with release of acetic acid (Fig. 9).[33] The question of the absolute requirement for coenzyme A activation in the reactions of side-chain cleavage remains open; both routes may operate, depending on the species and the particular C_6–C_3 precursor.[20] The origin of substituted benzoate compounds is further complicated by the likelihood that some derivatives originate by quite different biosynthetic pathways (i.e. gallic

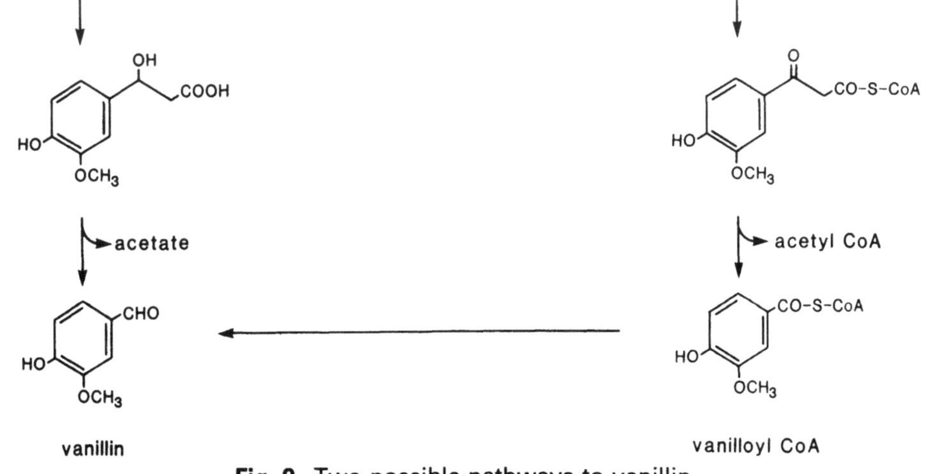

Fig. 9. Two possible pathways to vanillin.

acid from dehydroshikimic acid) or by peroxidative degradation of more complex phenolics (i.e. di- and tri-hydroxybenzoates from flavonoids), and by the observation that, at least in some species, benzoic acid is efficiently converted to benzaldehyde and benzyl alcohol, presumably following activation of the free acid.[32,34] Benzaldehyde may occur as the mandelonitrile aglycone of cyanogenic glycosides, and is released following hydrolysis and the action of a lyase.[35]

3 MONO AND SESQUITERPENOIDS

The monoterpenoids constitute a very large group of natural products that are known for their aroma properties and find wide use in the flavor and fragrance industries. Monoterpenoids are minor components of many fruit and vegetable aromas, but are major components of citrus and conifer essences and, of course, the essential oils of herbs and spices. Many plants that produce monoterpenes accumulate these volatile metabolites in specialized oil glands or resin ducts. The majority of monoterpenoids are constructed by the 'head-to-tail' fusion of isoprene units derived from mevalonic acid (ultimately from acetyl coenzyme A). This biosynthetic pattern is typified in the action of geranyl pyrophosphate synthase, which mediates the condensation of dimethylallyl pyrophosphate with isopentenyl pyrophosphate to form geranyl pyrophosphate, the progenitor of the monoterpenes (Fig. 10). This synthase appears to be strictly associated with monoterpene biosynthesis since the enzyme is confined to the oil glands in monoterpene-producing species and is essentially absent in leaf tissue of non-producers such as potato and wheat.

The acyclic oxygenated monoterpenes, geraniol, nerol, linalool, citral and citronellol, are among the most abundant in nature and arise more or less directly from geranyl pyrophosphate by isomerization and/or redox processes (Fig. 10). (−)-Linalool constitutes about 50% of American basil oil, while (+)-linalool forms 70% of coriander fruit oil. Citral, a mixture of geranial and neral, is an important component of lemon essence. Acyclic olefins such as myrcene and the isomeric ocimenes are also derived from geranyl pyrophosphate. These compounds are associated, particularly, with the essential oils of the Compositae and Lamiaceae families. Myrcene constitutes over 60% of the oil of hops.[11]

Most monoterpenoids are cyclic, and are based upon a relatively small number of skeletal themes that occur as many simple derivatives, positional isomers and stereochemical variants. The monoterpene cyclases catalyze the multistep isomerization and cyclization of geranyl pyrophosphate to the parent compounds of the various structural types (Fig. 10). Following initial ionization of the geranyl precursor and isomerization to an enzyme-bound linalyl pyrophosphate intermediate, *transoid* to *cisoid*-rotation about C2–C3 overcomes the original impediment to the cyclization of the geranyl form and promotes a second ionization (of the linalyl intermediate) with C1–C6

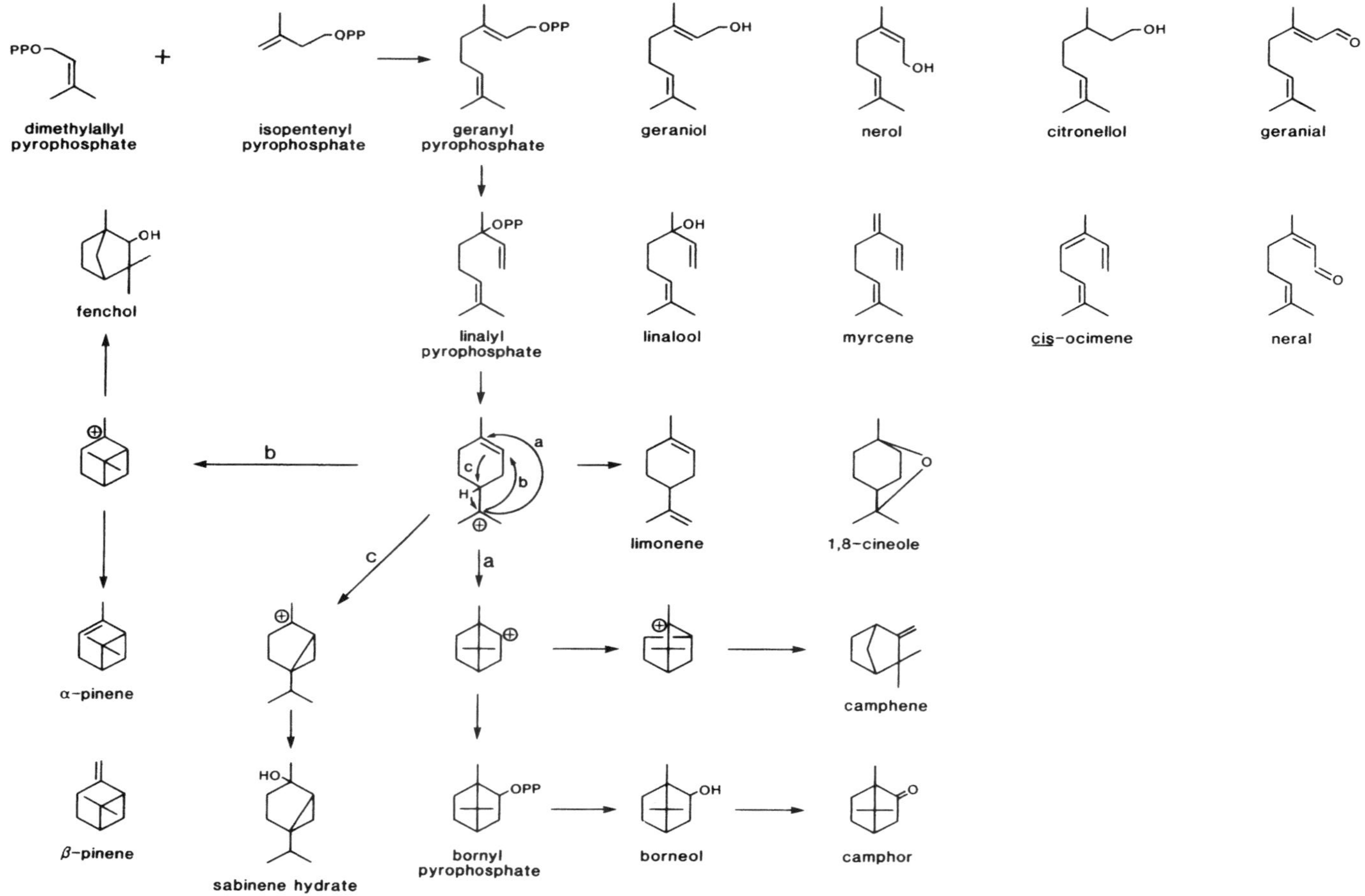

Fig. 10. Biosynthesis and isomerization–cyclization of geranyl pyrophosphate (OPP indicates the pyrophosphate moiety).

cyclization to the α-terpinyl cation. This universal intermediate may undergo deprotonation to an olefin (e.g. limonene) or be captured by a nucleophile, or it may undergo additional cyclizations, hydride shifts or rearrangements before the cationic reaction is terminated.[36] Over 50 different cyclases probably exist. Some of these enzymes generate a single product, while others can produce several skeletal types or positional isomers and derivatives of the same skeletal class. Enantiomers of a given product are generated by distinct cyclases.

In the biosynthesis of camphor, geranyl pyrophosphate is first cyclized to bornyl pyrophosphate, which is hydrolyzed (via a specific phosphatase) to borneol and oxidized to the ketone (Fig. 10, pathway a). *endo*-Fenchol cyclase from fennel produces the alcohol product directly via rearrangement of a pinyl cation and water capture (Fig. 10, pathway b). Similarly, 1,8-cineole cyclase forms the cyclic ether characteristic of *Eucalyptus*, while sabinene hydrate cyclase from marjoram produces both *cis*- and *trans*-isomers of this tertiary alcohol (Fig. 10, pathway c). Separate olefin cyclases from sage produce the (+)- and (−)-enantiomers, respectively, of the pinenes, camphene and limonene. (+)-Limonene may account for over 90% of the oil of *Citrus*, whereas the (−)-enantiomer is more abundant in *Mentha* and *Pinus* species.

The olefinic parent compounds produced by the cyclases are often further modified by oxidative transformations. Limonene, for instance, may undergo cytochrome P-450-catalyzed allylic hydroxylation at C3, C6, or C7 and subsequent oxidation to the corresponding carbonyl compound (Fig. 11).[37]

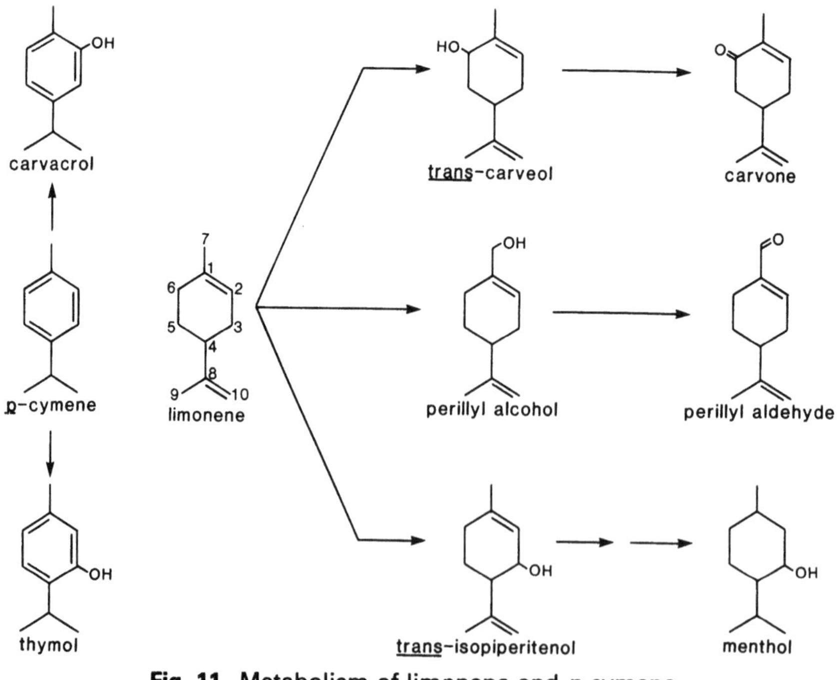

Fig. 11. Metabolism of limonene and *p*-cymene.

Perillyl aldehyde (possessing a spicy-cumin odor) is produced in this way, as is carvone. (−)-Limonene gives rise to (−)-carvone, characteristic of spearmint, whereas (+)-limonene yields the enantiomeric (+)-carvone that produces a quite different aroma, typical of caraway and dill. Isopiperitenol, generated by C3-oxygenation, undergoes subsequent transformations to afford menthol, the primary constituent of peppermint oil. This 'allylic oxidation–conjugate reduction' pathway is widespread in essential oil plants, and is also important in the metabolism of bicyclic monoterpenes of the thujane and pinane type. The aromatic olefin *p*-cymene is also selectively hydroxylated to give the phenols thymol and carvacrol (Fig. 11), aroma compounds of particular significance in thyme, oregano and *Monarda* species.

The sesquiterpenoids (C_{15}) contain one more isoprene unit than the monoterpenoids, which allows for much greater diversity in construction pattern. Nearly 200 skeletal types, containing up to four carbocyclic rings, are known; yet, while many sesquiterpenoids are of importance as antibiotics, phytoalexins and toxins, the aroma-bearing compounds of the essential oils are associated with a limited number of carbon skeletons. Among those having noteworthy aroma properties are: the acyclic compound farnesol, occurring naturally in cabreuva oil and musk grains, and having a very delicate floral fresh-green note reminiscent of lily-of-the-valley; nerolidol and its esters, which are described as having a mild but long-lasting floral odor;[14] cedrol, which is isolated from cedarwood oil and bears a woody odor; caryophyllene, an olefin found in numerous essential oils, with a spicy, acidic, woody odor reminiscent of vetiver;[23] nootkatone, found in grapefruit and possessing a very intense citrus aroma; 'vetiverol' (a complex mixture containing, *inter alia*, the ketones α-vetivone, β-vetivone and khusimone) with a balsamic, earthy fragrance, isolated from the grass *Vetiveria zizanoides*; and patchoulol, with a distinct woodiness characteristic of patchouli (*Pogostemon cablin*)[23] (Fig. 12).

trans,trans-Farnesyl pyrophosphate is considered to be the progenitor of the sesquiterpenoids and is derived by the addition of an isoprene unit to geranyl pyrophosphate via prenyl transferase (Fig. 12). Farnesol is obtained by enzymatic hydrolysis of this precursor, and nerolidol by preliminary isomerization to nerolidyl pyrophosphate followed by hydrolysis (analogous to the formation of linalool). Cyclization in the sesquiterpene series is initiated by ionization of the allylic pyrophosphate and involves electrophilic attack of the cationic center on the central or distal double bond of the precursor. The formation of 10- or 11-membered rings by addition to the distal double bond may occur from farnesyl pyrophosphate directly without preliminary isomerization to nerolidyl pyrophosphate; however, formation of six- or seven-membered rings requires the isomerization step to overcome the stereochemical impediment due to the *trans*-C2–C3 double bond of the farnesyl precursor (Fig. 12). Further internal additions, hydride shifts, methyl migrations or other rearrangements may occur before termination of the cationic reaction sequence by deprotonation or water capture.[38] The sesquiterpene cyclization mechanisms can be seen to resemble those of monoterpene biosynthesis.

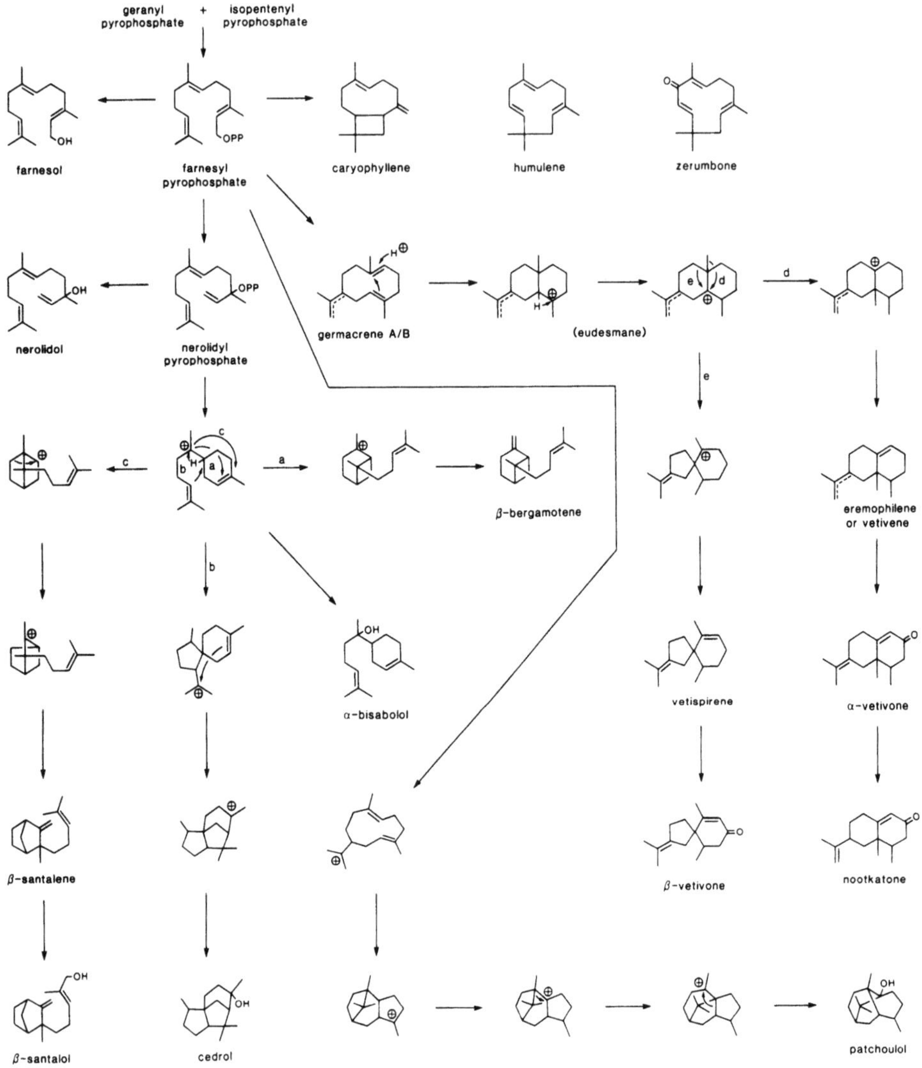

Fig. 12. Conversion of farnesyl pyrophosphate to sesquiterpenoids.

Indeed, the sesquiterpene analogs of most common monoterpenes have been reported (e.g. bisabolol, bergamotene); however, the C_{10} and C_{15} analogs seldom co-occur in the same species. Secondary oxidative transformations of the sesquiterpenes may also occur, often resulting in extensive modification of the carbon skeleton.[39]

Bisabolyl derivatives, while not noted for their fragrance properties, represent one of the simplest sesquiterpene types derived by cyclization to the central double bond (Fig. 12).[40] α-Bisabolol occurs in the flowers of camomile (*Matricaria chamomilla*) and, while possessing a weak but agreeable odor, is used mainly as a fixative.[41] The cyclization to *trans*-β-bergamotene (Fig. 12, pathway a) can be seen as the analog of the pinene cyclization in the monoterpene series.[42] Cedrol and the santalols can be derived from a bisabolyl intermediate as well; however, the biosynthetic details have yet to be demonstrated. Thus, a hydride shift in the bisabolyl cation (Fig. 12, pathway b) could lead to a spirane intermediate with subsequent closure to the second five-membered ring to provide the cedrane skeleton. Simple water capture of the cation will yield cedrol. The cyclization to santalene (Fig. 12, pathway c) is analogous to the formation of camphene in the monoterpene series (cf. Fig. 10), and allylic hydroxylation will afford β-santalol in a straightforward manner (Fig. 12).

Cyclization to the distal double bond generates a 10-membered ring (germacrane) or 11-membered ring (humulane) derivative (Fig. 12). Germacrane-type sesquiterpenoids are of minor significance as odor compounds, but they serve as intermediates in the biosynthesis of several important odorants. Proton-initiated cyclization of germacrene A or B leads to the eudesmane skeleton, which by combination of hydride and methyl migration (Fig. 12, pathway d) yields an eremophilene or vetivene.[40] Allylic oxygenation yields α-vetivone (from germacrene B, isopropylidene substituent) or the isomeric nootkatone (from germacrene A, isopropenyl substituent). Rearrangement of the eudesmane system with ring contraction (Fig. 12, pathway e) generates the vetispirane skeleton, and allylic oxygenation of the corresponding olefin affords β-vetivone. A very complex series of rearrangements leads from the initially formed germacryl cation to the tricyclic patchoulyl system, and this reaction sequence is terminated by water capture to provide patchoulol (Fig. 12).[43] It is interesting to note that, whereas many of the pathways discussed in earlier sections involve relatively few enzymes acting on numerous potential substrates, the mono and sesquiterpenes are derived from only two substrates (geranyl and farnesyl pyrophosphates) by the action of numerous enzymes that employ variations on a single reaction mechanism.

Ionones and Damascenones

The unusually potent ionones, damascenones and related aroma compounds are produced by the degradation of high-molecular-weight terpenoids (carotenoids; C_{40}) which in themselves have no odor or taste. These carotenoid breakdown products are found in a variety of vegetables and fruits (especially berries) and, in surprising quantities, in tea and tobacco. The biosynthetic head-to-head coupling of geranylgeranyl pyrophosphate yields the yellow to red carotenoid pigments (Fig. 13) which are characteristic of corn, carrot and tomato, and which are also found in leaves where they function as photoprotectants and supplementary light receptors in photosynthesis.[44] The extensive network of conjugated double bonds of the carotenoids makes these compounds especially susceptible to oxidative decomposition that is accompanied by bleaching of the chromophore. Among the many degradation

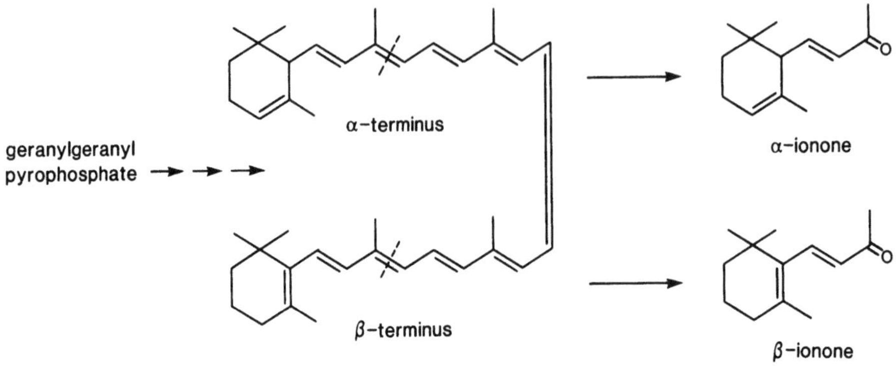

Fig. 13. Conversion of carotenes to ionones.

products of the carotenoids, the ionones and damascenones are perhaps the most important. Some 70 years ago, ionone was observed to arise from carotene by autoxidation, and the role of enzyme action and molecular oxygen in an alternative means of generating ionone from this precursor was subsequently demonstrated.[11]

During senescence or fruit ripening, lipoxygenases, peroxidases and other oxidases of the organelles (plastids, mitochondria) are released and employ molecular oxygen, either directly or indirectly via co-oxidative reactions, to cleave the polyene chain of the carotenoids. The resulting volatile trimethylcyclohexene derivatives may contain 9, 10, 11, or 13 carbon atoms, and all bear an allylic oxygen function in the side-chain (Fig. 13).[45] Dioxygenases, similar to those involved in the cleavage of β-carotene to retinal (a component of the visual pigment rhodopsin in animals),[6,46] may also be involved in carotenoid degradation in plants. Non-enzymatic oxidation also plays a role, particularly in post-harvest processing such as the curing of tobacco and tea. Although all in-chain double bonds are vulnerable to oxidative attack, cleavage at C9–C10 is favored to produce ionones (C_{13}). Both α- and β-ionone (Fig. 13) are characterized by an intense floral aroma which, upon dilution, resembles the odor of violets.

The acceleration of carotenoid breakdown by the addition of linoleic acid led to the formulation of a role for lipoxygenase in the cleavage reaction. The process is attributed to a coupled oxidation–reduction (co-oxidation) involving hydroperoxides and other reactive intermediates of the lipoxygenase reaction in mediating the cleavage.[13,47,48] Such intermediates of the co-oxidation process can also be generated by lipid autoxidation. Tea flavanols are oxidized to *o*-quinones by catechol oxidase and serve either directly or indirectly (via reaction with unsaturated fatty acids) in the co-oxidative cleavage of carotenoids.[49] α-Carotene (α/β-termini) gives rise to α-ionone, whereas cleavage of β-carotene (β/β-termini) yields β-ionone.[48,50] Although non-volatile carotenoid oxidation products can be stored as glycosides,[51] it is not known whether ionone derivatives may be sequestered in this form.

Although *β*-ionone was reported as a component of *Boronia* in 1927,[14] it was only in 1970 that damascenone was isolated from rose oil, and later also found in geranium. The damascenones and related damascones are thought to arise by the oxidative breakdown of allenic carotenoids, such as neoxanthin, by mechanisms similar to those in the production of ionones (Fig. 14).[52] Biogenetic routes from ionones to damascones, via allenic intermediates, have also been proposed,[52] but these seem more cumbersome. Although structurally related to the ionones, the damascenones are not violet-like, but rather confer the green, sweet 'ripe-like' odor of pears, blackberries and raspberries.[45] Grape skins contain a non-volatile glycosidic precursor of damascenone that results in the production of this odorant during post-harvest processing.[53] *α*-Damascone occurs naturally in tea aroma, and bears a potent floral, fruity odor that becomes intensely rose-like on dilution. *β*-Damascone, by comparison, possesses tobacco and tea-like notes.[23] The related megastigmatrienones, which contain a keto function on the ring, have been described as 'the heart of the tobacco aroma' in which damascenones are also found. The trimethylcyclohexenones are also characteristic of tobacco, and occur in saffron as well.[46] The production of damascenones and related metabolites in tobacco, via carotenoid breakdown, is most abundant during cell-wall degradation accompanying loss of cellular function; however, living cells may contain glycosylated forms of these products[46] as a result of prior metabolism.

Natural 'violet-like' odorants apparently related to the ionones were isolated from orris root oil in 1893. These 'irones' (6-methylionones) are C_{14} ketones which are produced slowly in the stored rhizome. It was the search for this

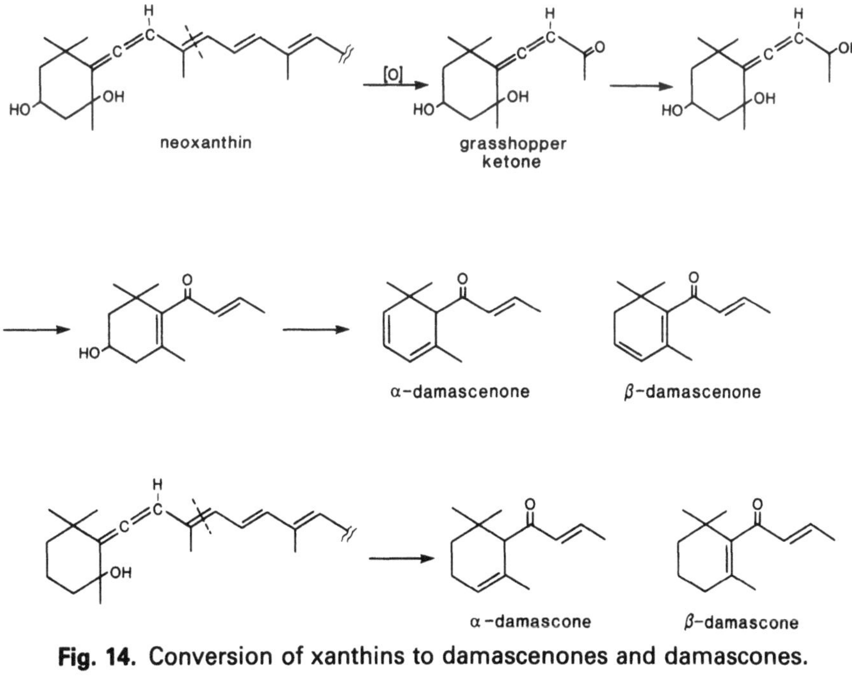

Fig. 14. Conversion of xanthins to damascenones and damascones.

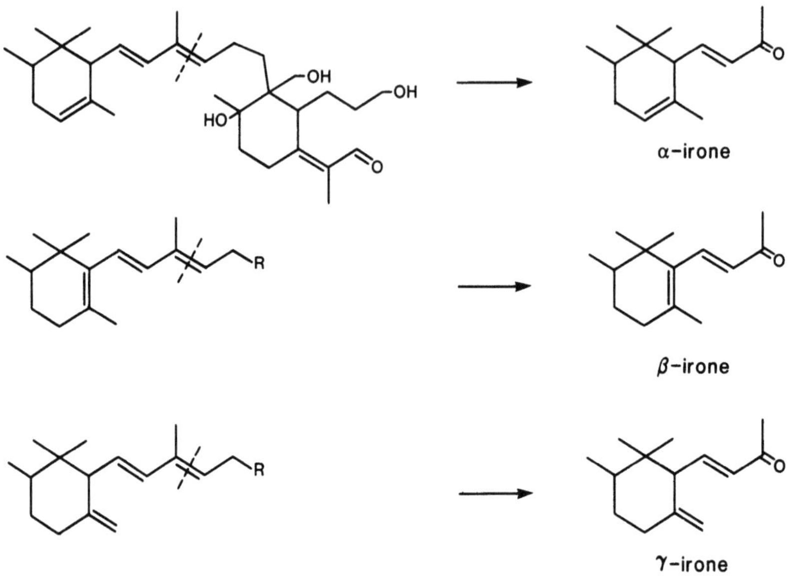

Fig. 15. Conversion of modified triterpenoids to irones.

violet odor that led to the discovery of the ionones. Although the irones were originally thought to be derived via carotenoid breakdown, extracts of the rhizomes were shown to contain a series of methyl-substituted triterpenoids which bore a striking resemblance to the irones (Fig. 15). These triterpenoids, called iridals or cycloiridals, are formed from the C_{30} olefin squalene which, in turn, is derived by head-to-head condensation of two farnesyl pyrophosphate chains. The structures are highly modified by oxygenation and desaturation, and methylation of the terminal double bond initiates the cyclization to produce the unique ring structure.[54] Oxidative decomposition of these C_{31} precursors could generate the irones,[55] much in the same way that ionones and related compounds are produced from carotenoids.

Aroma compounds derived from acyclic isoprenoids are also known. The primary tomato carotenoid, lycopene, contributes aroma-bearing methyl ketones by oxidative processes, while the grassy, cinnamon-like odor of 6-methyl-3,5-heptadien-2-one probably originates from the phytol side-chain of chlorophyll.

4 CONCLUSIONS

There is tremendous variety in both the source and structure of natural odorants. Some of these aroma compounds are catabolic breakdown products derived from primary metabolites such as acyl lipids, amino-acids, and higher isoprenoids. Many other aroma substances are 'secondary products', such as phenylpropanoids and monoterpenes, which appear to be not essential to the metabolism of the producing organism, but which may have ecological functions as attractants, phytoalexins, etc. These compounds are formed by extensions and offshoots of primary metabolism, and often employ minor pathways of limited distribution in the plant kingdom.

Aroma production may occur only during a specific phase in the life cycle of a particular tissue (postclimacteric fruit ripening, senescence, and post-harvest processing of leaves) and is often a result of rapid metabolic changes leading to the 'mixing' of formerly compartmentalized enzymes and substrates. In other cases (e.g. monoterpenes), aroma formation is constitutive, with production and accumulation taking place throughout the life cycle.

The same or similar substrate may undergo a variety of metabolic transformations to yield very different end-products (for example, polyunsaturated fatty acids may be cleaved in a number of ways to produce short-chain aliphatic and alicyclic compounds). Alternatively, the same reaction on a variety of substrates can lead to distinct, though similar, products, as in the oxidative degradation of higher isoprenoids to ionones and irones. In the case of phenylpropanoids, relatively few enzyme types lead from cinnamate to a host of metabolites. At the extreme, a single substrate, geranyl pyrophosphate, gives rise to a large number of products via an equally large number of enzymes (cyclases) that employ the same basic reaction mechanism. It is thus difficult to extract common features that would, in any significant way, unify the origin of natural odorants. Unity lies only in the ability of the olfactory system to perceive products of such diversity.

Although the metabolic pathways described here have been studied in disparate species, often chosen for practical reasons, it should be understood that they often co-occur to produce exotic fragrance blends. Flowers, for example, typically produce essences by a mixture of the diverse pathways outlined. The main components of *Gardenia* flower oil are α-farnesene, ocimene, linalool, 3-hexenyl tiglate, jasmine lactone and hexenyl and methyl benzoates. This mixture clearly originates via several distinct metabolic pathways.[56] Orchids, as well, are noted for the complexity of volatiles produced.[57] *Boronia megastigma* concrete is a natural blend of β-ionone, dodecyl acetate, methyl jasmonate and heptadecene, as well as several monoterpene olefins. Some clones are dominated by a cineole-like sesquiterpene, whereas others characteristically contain a fragrant series of esters. The long-chain acetates may act as sex pheromones (attractants) for the night-flying moths that serve as pollinators of this species.[58]

Although it is the sensual response to fragrance chemicals that has appealed to us through the ages and that fuels much of our continued interest in the origin of odorants, the generation of natural aromas ought also to be viewed from an overall metabolic, physiological and even ecological perspective. Constraints to chemical syntheses will foster this biological approach, to which this chapter forms a brief introduction.

NOTE

For recent work on the biosynthesis of vanillic acid, in which ferulate is not implicated as an intermediate, see Funk, C. & Brodelius, P. E., Phenylpropanoid metabolism in suspension cultures of *Vanilla planifolia* Andr. II. Effects of precursor feeding and metabolic inhibitors. *Phytochemistry*, in press.

REFERENCES

1. Amoore, J. E., Odor theory and odor classification. In *Fragrance Chemistry*, ed. E. T. Theimer. Academic Press, New York, 1982, pp. 27–76.
2. Croteau, R., Biogenesis of flavor components: volatile carbonyl compounds and monoterpenoids. In *Postharvest Biology and Biotechnology*, ed. H. O. Hultin & M. Milner. Food and Nutrition Press, Westport, 1978, pp. 400–32.
3. Hatanaka, A., Kajiwara, T. & Sekiya, J., Fatty acid hydroperoxide lyase in plant tissues: volatile aldehyde formation from linoleic and linolenic acid. In *Biogeneration of Aromas*, ACS Symposium Series No. 117, ed. T. H. Parliment & R. Croteau. American Chemical Society, Washington, DC, 1986, pp. 167–75.
4. Anderson, R. A., Hamilton-Kemp, T. R., Fleming, P. D. & Hildebrand, D. F., Volatile compounds from vegetative tobacco and wheat obtained by steam distillation and headspace trapping. In *Biogeneration of Aromas*, ACS Symposium Series No. 117, ed. T. H. Parliment & R. Croteau. American Chemical Society, Washington, DC, 1986, pp. 99–111.
5. Gardner, H. W., Oxidation of lipids in biological tissue and its significance. In *Chemical Changes in Food During Processing*, ed. T. Richardson & J. W. Finley. AVI Publishing, Westport, 1985, pp. 177–203.
6. Luckner, M., *Secondary Metabolism in Microorganisms, Plants and Animals*. Springer-Verlag, Berlin, 1984.
7. Schulz, H. & Kunau, W.-H., Beta-oxidation of unsaturated fatty acids: a revised pathway. *Trends Biol. Sci.*, **12** (1987) 403–6.
8. Tressl, R. & Albrecht, W., Biogenesis of aroma compounds through acyl pathways. In *Biogeneration of Aromas*, ACS Symposium Series No. 117, ed. T. H. Parliment & R. Croteau. American Chemical Society, Washington, DC, 1986, pp. 114–33.
9. Forney, F. W. & Markovetz, A. J., The biology of methyl ketones. *J. Lipid Res.*, **12** (1971) 383–95.
10. Mookherjee, B. D. & Wilson, R. A., The chemistry and fragrance of natural musk compounds. In *Fragrance Chemistry*, ed. E. T. Theimer. Academic Press, New York, 1982, pp. 433–94.
11. Ohloff, G., Recent developments in the field of naturally-occurring aroma components. *Fortschr. Chem. Org. Naturst.* **35** (1978) 431–527.
12. Tressl, R. & Drawert, F., Biogenesis of banana volatiles. *J. Agr. Food Chem.*, **21** (1973) 560–5.
13. Eskin, N. A. M., *Plant Pigments, Flavors and Textures*. Academic Press, New York, 1979.
14. Bedoukian, P. Z., *Perfumery and Flavoring Synthetics*. Elsevier, Amsterdam, 1967.
15. Demole, E. P., The fragrance of jasmine. In *Fragrance Chemistry*, ed. E. T. Theimer. Academic Press, New York, 1982, pp. 349–96.
16. Vick, B. A. & Zimmerman, D. C., Biosynthesis of jasmonic acid by several plant species. *Plant Physiol.*, **75** (1984) 458–61.

17. Vick, B. A. & Zimmerman, D. C., Pathways of fatty acid hydroperoxide metabolism in spinach leaf chloroplasts. *Plant Physiol.*, **85** (1987) 1073–8.
18. Simpson, T. J., The biosynthesis of polyketides. *Natural Product Reports*, **4** (1987) 339–76.
19. Croteau, R., Biosynthesis and catabolism of monoterpenoids. *Chem. Rev.*, **87** (1987) 929–54.
20. Gross, G. G., Biosynthesis and metabolism of phenolic acids and monolignols. In *Biosynthesis and Biodegradation of Wood Components*, ed. T. Higuchi. Academic Press, Orlando, 1985, pp. 229–71.
21. Grand, C., Ferulic acid 5-hydroxylase: a new cytochrome P-450-dependent enzyme from higher plant microsomes involved in lignin synthesis. *FEBS Lett.*, **169** (1984) 7–11.
22. Dewick, P. M., The biosynthesis of shikimate metabolites. *Natural Product Reports*, **5** (1988) 73–97.
23. Hall, R., Klemme, D. & Nienhaus, J., *Guide to Fragrance Ingredients*. Johnson Publications, London, 1985.
24. Manitto, P., Monti, D. & Gramatica, P., Biosynthesis of phenylpropanoid compounds. Part 1. Biosynthesis of eugenol in *Ocimum basilicum* L. *J. Chem. Soc., Perkin Trans. I* (1974) 1727–31.
25. Senanayake, U. M., Wills, R. B. H. & Lee, T. H., Biosynthesis of eugenol and cinnamic aldehyde in *Cinnamomum zeylanicum*. *Phytochemistry*, **16** (1977) 2032–3.
26. Klischies, M., Stöckigt, J. & Zenk, M. H., Biosynthesis of the allylphenols eugenol and methyleugenol in *Ocimum basilicum* L. *J. Chem. Soc., Chem. Comm.* (1975) 879–80.
27. Vickery, M. L. & Vickery, B., *Secondary Plant Metabolism*. University Park Press, Baltimore, 1981.
28. Koezuka, Y., Honda, G. & Tabata, M., Genetic control of phenylpropanoids in *Perilla frutescens*. *Phytochemistry*, **25** (1986) 2085–7.
29. Theimer, E. T., Benzene-derived cyclic carbinols. In *Fragrance Chemistry*, ed. E. T. Theimer. Academic Press, New York, 1982, pp. 269–83.
30. Gross, G. G., The phenolic acids. In *The Biochemistry of Plants*, Vol. 7, ed. E. E. Conn. Academic Press, New York, 1981, pp. 301–16.
31. Zenk, M. H., Biosynthesis of C_6–C_1 compounds. In *Biosynthesis of Aromatic Compounds*, ed. G. Billek. Pergamon Press, Oxford, 1966, pp. 45–60.
32. Zenk, M. H., Pathways of salicyl alcohol and salicin formation in *Salix purpurea* L. *Phytochemistry*, **6** (1967) 245–52.
33. French, C. J., Vance, C. P. & Towers, G. H. N., Conversion of *p*-coumaric acid to *p*-hydroxybenzoic acid by cell free extracts of potato tubers and *Polyporus hispidus*. *Phytochemistry*, **15** (1976) 564–6.
34. Croteau, R., Biosynthesis of benzaldehyde, benzyl alcohol and benzyl benzoate from benzoic acid in cranberry (*Vaccinium macrocarpon*). *J. Food Biochem.*, **1** (1977) 317–26.
35. Conn, E. E., Cyanogenic compounds. *Ann. Rev. Plant Physiol.*, **31** (1980) 433–51.
36. Croteau, R., Biosynthesis of cyclic monoterpenes. In *Biogeneration of Aromas*, ACS Symposium Series No. 117, ed. T. H. Parliment & R. Croteau. American Chemical Society, Washington, DC, 1986, pp. 134–56.
37. Karp, F. & Croteau, R., Role of hydroxylases in monoterpene biosynthesis. In *Bioflavour '87*, ed. P. Schreier. W. de Gruyter, Berlin, New York, 1988, pp. 173–98.
38. Croteau, R., The biosynthesis of terpene compounds. In *Fragrance and Flavor Substances*, ed. R. Croteau. D&PS Verlag, Pattensen, 1980, pp. 13–36.
39. Croteau, R. & Cane, D. E., Monoterpene and sesquiterpene cyclases. In *Methods in Enzymology*, Vol. 110, ed. J. H. Law & H. C. Rilling, Academic Press, New York, 1985, pp. 383–405.

40. Cane, D. E., Biosynthesis of sesquiterpenes. In *Biosynthesis of Isoprenoid Compounds*, Vol. I, ed. J. W. Porter & S. L. Spurgeon, John Wiley, New York, 1981, pp. 283–374.
41. Herout, V., Sesquiterpene alcohols. In *Fragrance Chemistry*, ed. E. T. Theimer. Academic Press, New York, 1982, pp. 222–65.
42. Cane, D. E., McIlwaine, D. B. & Harrison, P. H. M., Bergamotene biosynthesis and the enzymatic cyclization of farnesyl pyrophosphate. *J. Am. Chem. Soc.*, **111** (1989) 1153–4.
43. Croteau, R., Munck, S. H., Akoh, C. C., Fisk, H. J. & Satterwhite, D. M., Biosynthesis of the sesquiterpene patchoulol from farnesyl pyrophosphate in leaf extracts of *Pogostemon cablin* (Patchouli): mechanistic considerations. *Arch. Biochem. Biophys.*, **256** (1987) 56–68.
44. Torssell, K. B. G., *Natural Product Chemistry*. John Wiley, New York, 1983.
45. Weeks, W. W., Carotenoids: A source of flavor and aroma. In *Biogeneration of Aromas*, ACS Symposium Series No. 117, ed. T. H. Parliment & R. Croteau. American Chemical Society, Washington, DC, 1986, pp. 157–66.
46. Enzell, C., Biodegradation of carotenoids—an important route to aroma compounds. *Pure Appl. Chem.*, **57** (1985) 693–700.
47. Firn, R. D. & Friend, J., Enzymatic production of the plant growth inhibitor, xanthoxin. *Planta*, **103** (1972) 263–6.
48. Grosch, W., Laskawy, G. & Fischer, K.-H., Aroma compounds formed by enzymatic co-oxidation. In *Aroma Research, Proceedings of the International Symposium on Aroma Research*, *Zeist*, ed. H. Maarse & P. J. Groenen. Pudoc, Wageningen, 1975, pp. 75–83.
49. Sanderson, G. W. & Graham, H. N., On the formation of black tea aroma. *J. Agr. Food Chem.*, **21** (1973) 576–85.
50. Sanderson, G. W., Co, H. & Gonzalez, J. C., Biochemistry of tea fermentation: the role of carotenes in black tea aroma formation. *J. Food Sci.*, **36** (1971) 231–6.
51. Kodama, H., Fujimori, T. & Kato, K., Isolation of a new terpene glucoside, 3-hydroxy-5,6-epoxy-β-ionyl-β-D-glycopyranoside from flue-cured tobacco. *Agric. Biol. Chem.*, **45** (1981) 941–4.
52. Isoe, S., Katsumura, S. & Sakan, T., The synthesis of damascenone and β-damascone and the possible mechanism of their formation from carotenoids. *Helv. Chim. Acta*, **56** (1973) 1514–16.
53. Braell, P. A., Acree, T. E., Butts, R. M. & Zhou, P. G., Isolation of non-volatile precursors of β-damascenone from grapes using charm analysis. In *Biogeneration of Aromas*, ACS Symposium Series No. 117, ed. T. H. Parliment & R. Croteau. American Chemical Society, Washington, DC, 1986, pp. 75–84.
54. Marner, F.-J., Gladtke, D. & Jaenicke, L., Precursors of irones and dihydroirones. In *Progress in Terpene Chemistry*, ed. D. Joulain. Editions Frontières, Gif-sur-Yvette, 1986, pp. 193–203.
55. Krick, W., Marner, F.-J. & Jaenicke, L., On the stereochemistry of natural irones, dihydroirones, and their precursors. *Helv. Chim. Acta*, **67** (1984) 318–24.
56. Anonis, D. P., *Gardenia* in perfumery. *Perf. Flavorist.*, **8** (1983) no. 5 31–7.
57. Patt, J. M., Rhoades, D. F. & Corkill, J. A., Analysis of the floral fragrance of *Platanthera stricta*. *Phytochemistry*, **27** (1988) 91–5.
58. Davies, N. W., Volatile constituents of *Boronia megastigma* flowers. *Perf. Flavorist.*, **8** no. 6 (1984) 3–8.

Chapter 5

A Consideration of Some Psychological and Physiological Mechanisms of Odour Perception

AVERY N. GILBERT

Roure Inc., Teaneck, NJ 07666, USA

&

✝MORLEY R. KARE

Monell Chemical Senses Center, Philadelphia, PA 19104, USA

1 INTRODUCTION

Those to whom it falls to summarize the progress of a scientific discipline are often tempted to interpret an accelerating publication rate as a positive indicator of intellectual advance. Were we to subscribe to this approach we should be able to give a very rosy picture indeed of scientific achievement in odour perception and its attendant psychological and physiological mechanisms. Examination of the MEDLINE database of medical and life sciences research literature reveals a substantial increase in the annual scientific output having to do with the nose, odour, chemoreception and related topics (Fig. 1).[1] From 1966 to 1987 the yearly figure rose from 750 to over 2500 papers. The somewhat broader BIOSIS database shows olfactory research pouring forth at the numbing rate of 4000 papers a year in 1987. Sceptics might view these figures as reflecting inflation in scientific publishing in general. Yet the same analysis[1] showed that as a proportion of the total, olfactory papers grew from 0·4% in 1965 to 0·8% by 1987.

These impressive statistics may signal real advances in certain areas.

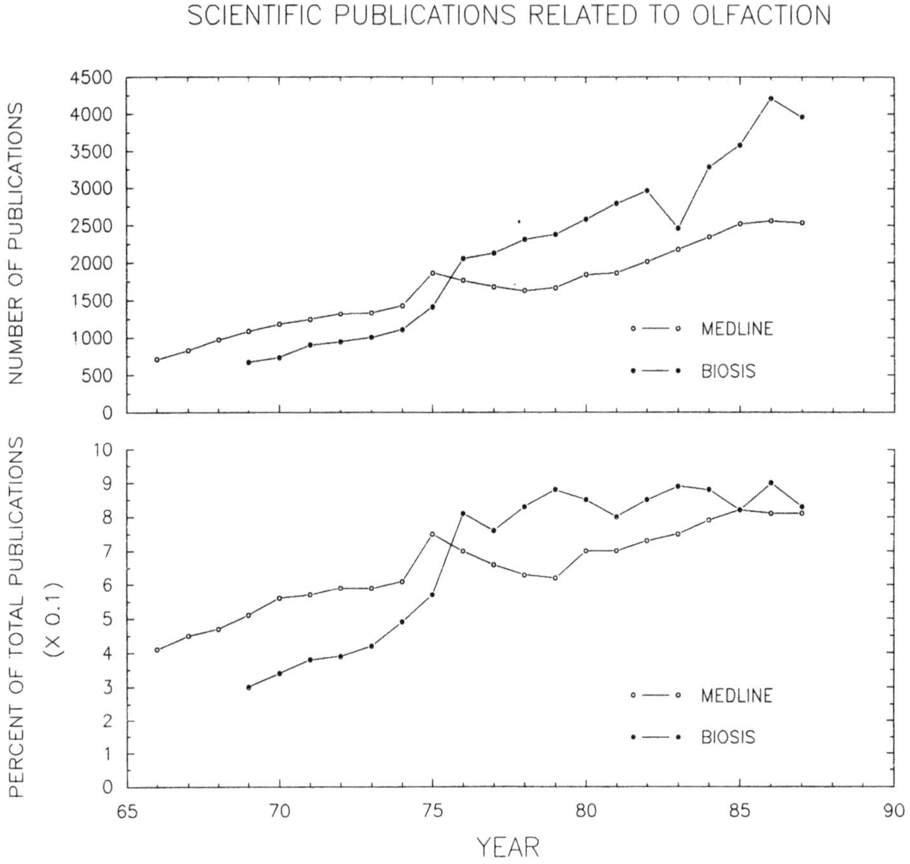

Fig. 1. Number of olfactory-related scientific publications contained in BIOSIS and MEDLINE (above) and that number as a percentage of all papers listed (below), from Ref. 1.

Outstanding among them are the discovery and genetic sequencing by Margolis of a marker protein specific for olfactory neurons,[2,3] and a neuron-specific, growth-related, calmodulin-binding, phosphoprotein (B50/GAP43) in olfactory receptor neurons.[4] Progress has also been made in identifying, at the molecular level, the events of olfactory transduction. Depolarization of the receptor neuron following binding of an odorant to the putative membrane-bound receptor protein is now thought to occur either by a receptor-complex-mediated opening of ion conductance channels in the membrane, or by a cyclic nucleotide cascade leading to depolarization through phosphorylation of channel proteins. These topics, which fall outside the range of the present discussion, have been reviewed in exemplary fashion elsewhere,[5–8] and are treated in Chapter 18 in the present volume. For a dissenting review that questions whether any of the evidence adduced to date can be said to implicate a specifically olfactory receptor, see Ref. 9.

Such advances will doubtless contribute one day to a more nearly complete understanding of all the facets of olfactory perception. Yet even the science of visual perception, which most authorities take to be far better understood than that of olfaction, provides an equivocal model. Photopigment biophysics contributes only tangentially to the phenomenology of colour vision, and even less, in turn, to other fundamental aspects of visual perception such as depth perception. It may have no bearing at all on neurological aspects such as visual tracking. So, too, little may have been revealed of the features most characteristically attributed to olfactory perception by the on-rushing torrent of recent transduction publications.

2 SNIFFING BEHAVIOUR

We begin our review and appraisal with the behaviour that is the prerequisite for olfactory perception in man and other mammals—the sniff. Exactly how long and how large in volume is the typical odour-sampling sniff, if indeed there is a 'typical' sniff?

Youngentob *et al.*[10] set out to characterize sniffing patterns in rats quantitatively using a still-air odour generator coupled to a pneumotachograph, which their subjects were induced to use by means of an operant-conditioning paradigm. Prior to their experiment, the investigators assembled a list of 52 parameters of sniffing, ranging from the elementary (sniffing bout duration) to the arcane (time ratio position of the peak flow rate in an inspiratory sniff). Of these, 12 parameters yielded coefficients that could be combined in a multiple linear regression equation (or 'Sniffing Index') to describe more powerfully than any single descriptor how the rats' sniffing varied across test conditions. The index did, in fact, discriminate between trials with different odorants, and between trials with different concentrations of one odorant.

Though some of these findings may remain specific to the species and to the odour detection tasks employed, certain general findings were evident. Sniffing bouts consisted of alternating inspirations and expirations, with the duration, volume and average flow rate of single sniffs increasing through the course of a bout, and the total inspiratory volume of a bout averaging 2·5 times the total expiratory volume. The number of inspiratory and expiratory sniffs increased with decreasing odorant concentration and when the inhaled air changed from odourized to non-odourized.[10]

Youngentob *et al.* pointed out that sniffing patterns may alter odour perception 'by influencing the spatio-temporal access of odorant molecules to the olfactory mucosa'. Specifically, they suggested that sniffing could cause more inspired air to reach the olfactory mucosa high in the nasal passage, while also minimizing the sorption taking place en route over non-olfactory mucosa. As an initial, fine-grained description of sniffing behaviour this study is a landmark effort. It also serves fair warning to those who would oversimplify, or worse still, ignore, the role of respiratory variables in olfactory perception.

A somewhat different tack was taken by Laing in an important series of experiments with humans,[11–13] in which a narrower range of sniff parameters was examined; the values of these parameters were not free but were constrained by the experimenter. Subjects sampled odours from 16-litre Perspex sampling chambers fitted with an oxygen mask whose inlet housed a hot-wire anemometer that recorded the physical characteristics of the subject's sniffing. Laing[12] found that a single natural sniff produces better results in odour threshold and intensity tests than either an unlimited number of sniffs, or a subject's deepest single sniff.

Optimal perception of odour intensity was achieved with sniffs between 0·39 and 0·64 s duration,[13] values much larger than the 0·18 s of odour presentation previously reported as the mean required for detection.[14] Another experiment by Laing[15] revealed that near-perfect identification of seven odorants presented at moderate intensity could be obtained using a single natural sniff. By restricting the duration of a single sniff to one of four values in yet another experiment, Laing established that a 0·42 s duration provided sufficient odour quality information to permit accurate identification. A final experiment indicated that sniff duration interacts with odorant concentration to produce various levels of odour identification performance. At relatively high concentrations, performance was near-perfect regardless of sniff length (a 'ceiling effect'), whereas at intermediate concentrations longer sniffs improved the rate of correct identification. Because this latter effect was seen only across a small range of concentrations, Laing stated that its contribution to error rate would be minute under most circumstances.

Taken as a whole, Laing's work relating the physical parameters of sniffing to olfactory perception sets the stage, nearly a century after Zwaardemaker's pioneering efforts, for the design of an olfactometer and a testing protocol tailored to the chief psychophysical realities of human respiratory and odour sampling behaviour. Approaches such as this, based on human-factors engineering, are more likely to produce consistent data and data with applicability beyond the circumscribed objectives of traditional laboratory olfactometry. Indeed, as Laing[15] comments, some of the notoriously high variances associated with olfactometric determinations of odour thresholds by various methods and laboratories may be due to the essentially arbitrary flow rates and sampling parameters that have been employed.

Yet even as this fundamental work is to be applauded, a note of caution might be raised. Both Youngentob *et al.* and Laing interpret their demonstrations of sniff-dependent perception in purely physiochemical terms, namely odorant concentration and flow rates to and across the olfactory mucosa. Although such parameters are doubtless involved, some consideration ought to be given to the psychological and information-processing demands of the various tasks before ascribing performance differences in the 170–640-millisecond range entirely to water–air partition coefficients and the fluid mechanics of scented air in the turbinal passages. An example of the possible role of attentional factors can be found in Laing's own work[15] when sniff durations were confined to 0·2, 0·5, 1·0 or 2·0 s by synchronizing inhalations to the sound of a buzzer. Odour identification scores at the shortest duration were significantly lower in the first set of trials, but equivalent in subsequent trials. As Laing suggested, perhaps it was initially difficult for some subjects to accommodate the competing demands of odour identification and timing of inhalation.

The importance of cognitive factors in odour judgements was underscored by results from Youngentob *et al.*,[16] who used a still-air odour generator linked to a face mask by a Teflon cannula into which nylon-mesh monofilament screens could be inserted. By varying the mesh size of the screens Youngentob *et al.* produced four calibrated resistive loads to inspiratory flow. Magnitude estimates of the perceived effort of sniffing against these loads varied directly with the log–log transformed respiratory resistance. Magnitude estimation of the strengths of four suprathreshold concentrations of ethyl butyrate proved to be sensitive to inspiratory resistance: whereas perceived intensity increased with concentration, it decreased with increasing resistance, an effect seen at all concentrations but the very highest.

As a pneumotachograph had been fitted to the end of the still-air odour generator, Youngentob *et al.* could characterize physical parameters of their subject's sniffing, including sniff duration, volume, average and peak flow rate of an inspiratory sniff, and total sampling time. These parameters showed little variation across the 16 combinations of odour concentration and resistance to flow, implying a fair degree of uniformity in the amount of physical stimulus reaching the olfactory epithelium in each combination of experimental variables. Nevertheless, perceived intensity decreased with increasing resistance. As their subjects could accurately estimate the effort of sniffing against each of the resistive loads, Youngentob *et al.* were led to endorse the perceptual constancy model of olfaction proposed by Teghtsoonian *et al.*[17,18] which holds that the perception of odour intensity is related to the vigour of the sniff producing it.

Lest others too hastily apply these findings to clinical cases of hyposmia induced by nasal obstruction, Youngentob *et al.* cautioned that their experimentally induced resistive loads were applied extranasally, and that clinical intranasal obstruction might alter not only resistance to airflow but the pattern and distribution of airflow within the nose.

3 ANATOMY AND ODOUR PERCEPTION

Indeed, the apparently straightforward empirical question of how much inhaled air actually passes over the olfactory mucosa and at what rate, is a topic on which curiously little progress has been made, despite a wealth of available instrumentation and a clearly enunciated need for the data. The gross anatomy of the human nasal passages is irregular and variable from person to person, yet it is still far less convoluted and intricate than that of other mammals, including those most frequently studied in olfactory research. So fundamental an issue as whether intranasal airflow is laminar or turbulent has yet to be resolved. Apart from structural constraints imposed by bone and cartilage, airflow may be affected by vascular engorgement of the respiratory epithelium, which though thin itself can produce large airflow changes when vasodilation takes place at the narrow (2 to 3 mm wide) aperture of the nasal valve. An excellent summary of the innervation and pharmacology of nasal respiratory function (including vascularization and secretion) is provided by DeLong & Getchell.[19]

Consideration of how patency of the nasal airways might modulate odour perception frequently devolves to an uncritical mention of the phenomenon that has come, on somewhat slender objective grounds, to be known as the 'nasal cycle'. This term is often, though incorrectly, attributed to Kayser,[20] who was among the first in modern times to observe systematically that spontaneous alternations in left and right nostril airflow are due to the vascular physiology of the epithelium. An idealized version of the nasal cycle is presented in Fig. 2. Thus conceived, it has proved extremely popular in the clinical literature of rhinology, though Gilbert & Rosenwasser[21] found that the data offered in support of the periodicity of these alternations are seldom of

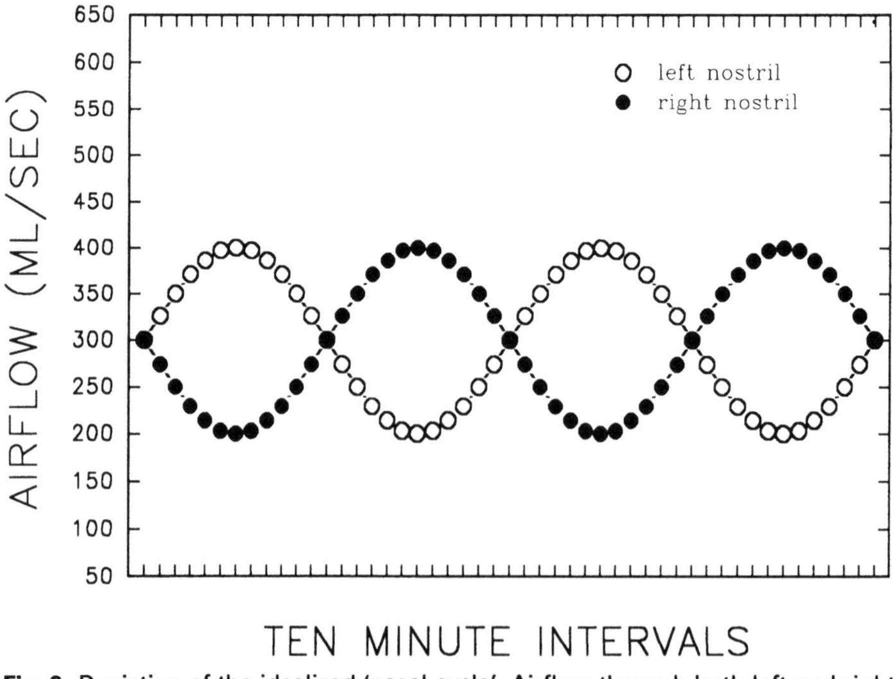

Fig. 2. Depiction of the idealized 'nasal cycle'. Airflow through both left and right nostrils is rhythmic and of the same period length, but the two nostrils are 180° out of phase.

sufficient sampling density or duration to allow even tentative application of the statistical techniques required for confirmation.

Gilbert and Rosenwasser[21] attempted to remedy this situation by observing airflow with active anterior rhinometry at a frequency (once every 10 minutes) and at an observation length (8 hours) sufficient to apply autocorrelation to the resulting time series. Only the most tenuous evidence for periodicity was found, and then only when the statistical criteria were substantially relaxed in favour of finding rhythmicity. Even then, the incidence of a putative nasal cycle among the test subjects was far below that expected from the clinical literature, a discrepancy easily resolved, given the tendency of human observers to see pattern in essentially stochastic data.

On the other hand, Gilbert & Rosenwasser[21] did find that left and right nostril airflow values were often reciprocal (negatively correlated at a significant level), if not necessarily periodic. This fairly robust tendency towards reciprocity, coupled with a conflation of the concepts of periodicity and reciprocity, may account for the persistent belief in the 'nasal cycle'. Gilbert[22] (Fig. 3) has confirmed these earlier conclusions using twice the sampling rate (i.e. observation once every 5 minutes).

If such rapid (<5 min) switchovers in internostril airflows are indeed a ubiquitous feature of nasal physiology, there arise several implications for olfactory research methods. The foremost of these is whether determination of detection thresholds, taken under otherwise highly constrained experimental conditions, might yield dubious values if taken without concomitant nasal airflow measurements. Here lies another possible source of the notoriously large variances in detection thresholds mentioned above (low within-subject reliability in this measurement has long frustrated researchers, and airflow changes could explain it). A second implication of this work[22] is that attempts to link nasal patency to periodic psychophysiological functions such as EEG[23] or cognitive tasks[24] may founder if premised on the rhythmicity of nasal physiology.

An intriguing and often overlooked feature of the olfactory system is its unilateral connectivity, i.e. the fact that neurons in the olfactory epithelium of the left and right nasal passages send their axons exclusively to the ipsilateral member of the paired olfactory bulbs, and that while these structures are connected by minor collateral branching, the central projections of the olfactory tracts display nowhere near the massive crossing-over of sensory afferents that characterize the visual and auditory systems. These pathways project via commissures to the opposite hemisphere only after synapsing in the olfactory cortex.[25]

In rats, which share this anatomical arrangement, the commissures are not fully developed before a post-natal age of 18 days. Consequently, in rat pups below this age, behavioural preference for an odour acquired during operant training with a single naris temporarily plugged is evident only when the trained side is open during testing. When pups trained via a single naris before 18 days are tested after this age (after the commissures are mature), the

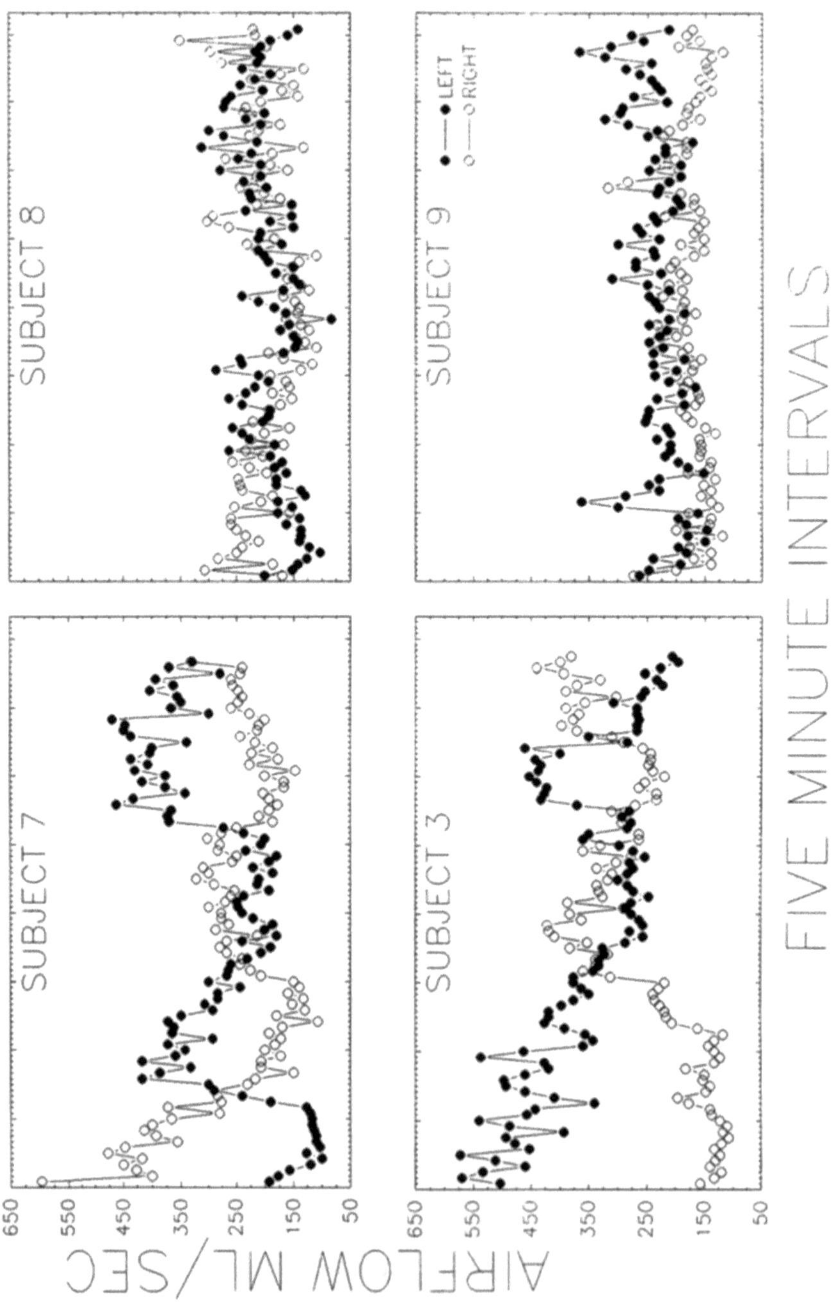

Fig. 3. Nasal airflow data for four actual subjects selected to illustrate reciprocity (subject 3), periodicity (subject 8), the coincidence of both (subject 7), and the absence of both (subject 9); from Ref. 22.

untrained naris has access to the memory on the opposite (trained) side, and the animal performs just as well with only its untrained naris open.[26–28] Surgical section of the commissural pathways makes this interhemispheric transfer of information once again impossible.

These developmentally immature rats are, in effect, the functional equivalent of human commissurotomy ('split brain') patients, whose abilities to detect and identify odorous stimuli display much the same peculiarities demonstrable in visual and stereognostic modalities.[29,30] In these patients, unilaterally delivered odours can be named only when presented to the left nostril, whereas odours presented only to the right nostril can be identified non-verbally by the patient pointing selectively to an associated object.

The ipsilateral innervation of the olfactory system has been put to ingenious use in a pair of studies that also demonstrate the utility of a neuropsychological approach to odour perception. Bellas *et al.*[31,32] examined patients with brain damage who exhibit a disorder termed 'unilateral neglect', in which the patient fails to respond to stimuli in space and on the side of the body contralateral to the damaged cerebral hemisphere. In almost all instances the right hemisphere (with damage especially in the right parietal lobe) is affected, with resultant neglect of the left side of body and space. By presenting a different, easily identifiable odour to each nostril simultaneously and asking the patient to name the odour and the side on which it was presented—a variant of the diagnostic procedure employed in other modalities—Bellas *et al.* could identify the neglected nostril. Right hemisphere lesion patients with left unilateral neglect failed to respond to olfactory stimulation of their left nostril. Analogous testing with stimuli that elicit a pungent sensation via the trigeminal nerve produced a similar result, despite the fact that trigeminal inputs are contralaterally innervated. Together these findings corroborate a recent model of unilateral neglect, namely, that it is not a defect in sensory input from the side of the body contralateral to the site of damage; rather it is primarily the result of a disordered internal representation of body and space, independent of sensory input. The findings of Bellas *et al.*[31,32] support the latter view with respect to olfaction, and emphasize how cognition mediates olfactory sensory input, even under these extraordinary circumstances.

An even more striking demonstration of dissociations among the neuropsychological components of odour perception is provided by a study[33] of the celebrated bilateral medial temporal lobe resection patient identified in the literature only as case 'H.M.' H.M. is frequently examined because his surgery, done to mitigate pharmacologically uncontrollable seizures, resulted in a persistent and profound anterograde amnesia (inability to form long-term memories). Eichenbaum and colleagues were interested in H.M. because at surgery most of his pyriform cortex in the uncus was excised, leaving only some components of the first cranial nerve projection intact, in particular the olfactory tubercle and adjacent pyriform cortex on the basal surface of the frontal lobe. The location of the excised tissue, and of the trauma incurred by the surgery, suggested that H.M. would suffer massive olfactory deficits.

Surprisingly, his performance on a series of tests demonstrated a sparing of function with a remarkable dissociation of certain olfactory perceptual capacities. Specifically, he could detect weak odours, distinguish odours by concentration, and adapt to strong odours, but was severely impaired at odour quality discrimination. This deficit was not traceable to his well-known memory impairment. The results indicate that the medial temporal lobe plays a critical role in the perception of odour quality, but that both this and the temporal olfactory pathway are necessary for odour discrimination and identification.[33]

4 HANDEDNESS AND ODOUR PERCEPTION

Despite the striking ipsilateral connectivity of olfactory afferent pathways, the potential interaction of handedness and side of nose in odour perception has been largely unexplored. Given that certain emotional and cognitive processes have asymmetric hemispheric representation, an odorant might be perceived differently depending on the nostril (and hence the hemisphere) of stimulation. Unfortunately, studies on olfactory perception have failed to control for sex and handedness in a single, comprehensive experimental design. Furthermore, most studies of olfactory lateralization have been based on clinical populations,[29,33–37] and have reached conflicting conclusions regarding its extent and direction. The few attempts to relate handedness to olfactory function in normal, intact subjects were also inconclusive. Toulouse & Vaschide[38,39] tested 64 men, women and boys, and found that 87% showed a left nostril superiority for detection and recognition of camphor. Interestingly, all five instances of right nostril superiority for detection and recognition occurred in sinistral or ambidextrous subjects. In contrast, Toulouse & Vaschide reported higher right nostril sensitivity for ammonia. Koelega[40] found no internostril difference in detection thresholds for *n*-amyl acetate in dextral men and women, but he did not test sinistrals [addendum to reprint]. Youngentob *et al.*[41] found superiority for detection thresholds of *n*-butanol in male sinistrals and dextrals in the nostril ipsilateral to the preferred hand, but they did not test women.

Gilbert *et al.*[42] had dextral and sinistral subjects of both sexes rate monorhinically presented pairs of odours for dissimilarity. Multidimensional scaling of the data revealed a sex difference in consistency of odour comparisons, as well as an internostril asymmetry related to handedness, namely there were significantly greater internostril asymmetries in mapping among dextrals than sinistrals. This latter result implies a second-order effect of handedness on normal olfactory perception.

One possible account of these results is that olfactory function is lateralized at the cerebral level. The fact that male dextrals had the (non-significantly) highest degree of internostril asymmetry corroborates previous interpretations of sex differences in cerebral functional asymmetry.[43] Suggestive results were also obtained by Ehrlichman, who had dextral male subjects rate four odorants on a hedonic scale.[44] Three experimental variants involving active and passive sniffing yielded no left versus right nostril differences for pleasant odours, but unpleasant odours were perceived as more unpleasant when presented to the right nasal passage. This finding suggests a lateralization for negative but not positive olfactory experience.

An alternative to the central lateralization account is that some degree of asymmetry in the peripheral structures of the nose is related to handedness. Unlike the former alternative, such peripheral factors are not well documented in humans[21,22,41] but have been demonstrated in rats.[26,45] This work emphasizes the need to pinpoint where along the olfactory pathway, from external nares to cortex, this structural or functional asymmetry emerges.

5 OLFACTION IN DEMENTIA AND NORMAL AGING

The rudimentary state of neuropsychological approaches to olfactory perception is glaringly apparent in recent attempts to define a possible olfactory deficit associated with Alzheimer's disease (AD). Attention to this issue has been spurred by clinical reports of olfactory deficits in AD, by findings of characteristic AD neuropathology localized to regions subserving olfaction,[46–49] and by a profusion of theories regarding a nasal route of entry for a putative toxin or pathogen. Gilbert[50] summarized the difficulties in defining olfactory deficits in AD, chief among them how to assess and characterize smell function in the demented individual adequately, and the degree to which smell deficits are specific to AD rather than to dementias in general.

Most early reports dealt with the parameters of odour recognition or odour memory; olfactory thresholds in AD patients were determined only in later studies. Doty *et al.*[51] found elevated thresholds for phenylethyl alcohol in 15 AD patients. Although these investigators claimed priority for this result, others[52] had previously documented detection deficits in 18 AD patients using eucalyptol, citral and prunolide.

These results contrast with those from two other studies. One used Amoore's pyridine test kit to quantify thresholds in 10 AD patients.[53] Because seven patients fell in the normal range, it was concluded that 'primary sensory function in olfaction is not impaired in early AD'. A second study[54] found no threshold difference between 18 AD patients and elderly controls for amyl alcohol, but the patients showed elevated thresholds for cinnamon oil. Although choice of test odorants, methodological details and patient selection might account for some of these discrepancies, the weight of the evidence indicates that olfactory thresholds in AD patients are often, though not invariably, elevated. These deficits are superimposed on the decline in olfactory performance that occurs in normal aging.[55]

Neuropsychological studies of AD confirm higher-order deficits in olfactory perception. Methodological advances include larger sample sizes, better age-matching of controls and the use of neurological controls. More sophisticated approaches to olfactory perception (signal detection methods, multiple tests of olfactory function, and concurrent non-olfactory tasks) constitute further methodological refinements. For example, Moberg *et al.*[56] compared both early and late stage AD ($N = 42$) and Huntington's Disease (HD) patients ($N = 38$) to equal numbers of healthy controls on olfactory, verbal and visual recognition memory tests. Signal detection analysis revealed that early AD patients ($n = 7$) were impaired on all three tasks, but especially so on olfactory recognition, while early HD patients ($n = 17$) were selectively impaired only on olfactory recognition. Both advanced disease groups showed deficits on all tests, but again the olfactory task was especially affected. These differences remained significant even after the greater difficulty of the olfactory task was accounted for statistically. Knupfer & Spiegel[52] compared AD patients with vascular dementia patients and controls on tests of olfaction that included detection threshold, recognition memory (with variable time delays), and identification (matching odours to names and confrontation naming). The groups were significantly different on all measures. AD patients were more impaired than vascular dementia patients, who in turn were more impaired than controls. In addition to determining olfactory thresholds, Rezek[54] compared mildly demented AD patients and elderly controls on two odour identification tasks (confrontation naming and matching odours to written names). AD patients performed more poorly than controls on both tasks. Confrontation naming was the more difficult task for both patients and controls. Finally, two research teams used a commercial, microencapsulated odour identification task to compare AD patients with age-matched controls.[51,53] Both teams found deficits on this higher-order test in the AD patients.

Does a primary perceptual deficit account for the poor performance of AD patients on higher-order olfactory tests? This interpretation, offered by Doty *et al.*,[51] is intuitively pleasing but subject to both theoretical and empirical objections.[57] Theoretically, odour naming and recognition might remain intact despite elevated thresholds.[50] After all, a relation between odour intensity and odour identification performance in healthy subjects has yet to be demonstrated (except in the limiting case of subthreshold concentration). The situation can be likened to turning down the brightness on a television screen—the picture remains identifiable over a large range of intensity. Empirical results are also at odds with Doty's interpretation. Knupfer & Spiegel[52] selected test odour concentrations based on each patient's detection threshold. Deficits in odour recognition and identification remained despite compensatory increases in stimulus concentration. Furthermore, Moberg *et al.*[56] documented higher-order deficits even after successfully screening patients for a primary perceptual deficit. Higher-order deficits were found in the absence of detection threshold deficits in two other studies.[53,54] Elsewhere, patients with focal cerebral excisions had normal detection thresholds to *n*-butyl alcohol despite suffering from large odour identification deficits.[58] In sum, higher-order olfactory impairment does not result from elevated detection thresholds.

Is severity of dementia related to the extent of olfactory impairment? In both AD and HD patients, severity of dementia on the Mini Mental State (MMS) exam was correlated with poorer olfactory performance.[56] In another comparison of AD patients, vascular dementia patients and age-matched controls, MMS scores correlated to some degree with within-group olfactory performance.[52] However, olfactory performance differed sharply between groups, while MMS scores were very similar. Therefore the severity of dementia (irrespective of underlying disease) was not strongly associated with olfactory performance. When still other AD patients were classified as Stage 1, 2 or 3 based on the criteria of Cummings, stage of dementia was not related to either detection threshold or odour identification.[51] Similarly, no odour discrimination differences were found between mildly and moderately demented AD patients, despite differences on an analogous visual task.[59] Some caution may be advised in drawing conclusions from these four studies. Only one study[52] controlled for differences in task difficulty. Without independent assessment of task difficulty, conclusions about specificity of deficits remain tentative.

In sum, recent studies have confirmed earlier work by documenting a variety of olfactory impairments in AD patients. Yet because of a lack of standardized clinical tests of smell,[60] and given the variability in measuring and defining cognitive deterioration, the functional specificity and time course of such impairments remain unclear.

Of course, any olfactory deficits in AD necessarily take place in the context of the changes in odour perception due to normal aging. There are few, if any, longitudinal studies of aging and olfaction. The longitudinal approach, although it has certain methodological advantages, tends to minimize age differences because of selective subject attrition.[61] Cross-sectional analyses of olfaction are more common, yet these may overstate age differences because of cohort effects.[61] Among cross-sectional designs, two-sample comparison between youthful and elderly subjects (e.g. that by Stevens & Cain[62,63]), is the least informative. Only a decade-by-decade analysis can provide estimates of onset and rate of decline.[64]

Threshold determinations, estimates of strength or hedonic quality, similarity judgements and stimulus identifications are some of the measures available to characterize olfactory function. Yet most studies of aging and olfaction make use of only one such measure (see the summary by Chauhan *et al.*[65]). Reliance on a single testing method compromises broader conclusions, as it has done regarding olfactory deficits in AD.[50] The use of a variety of olfactory measures and odorants in a single examination is more satisfying. This multifaceted approach was the strategy of the Smell Survey undertaken in collaboration with the National Geographic Society and its membership publication *National Geographic.* Detection, intensity estimates, hedonic judgements and identification ability were recorded in response to six microencapsulated odorants.[66]

The results[55] showed that odour perception, on average, declines with age. However, this decline is by no means uniform, across either subjects, odorants

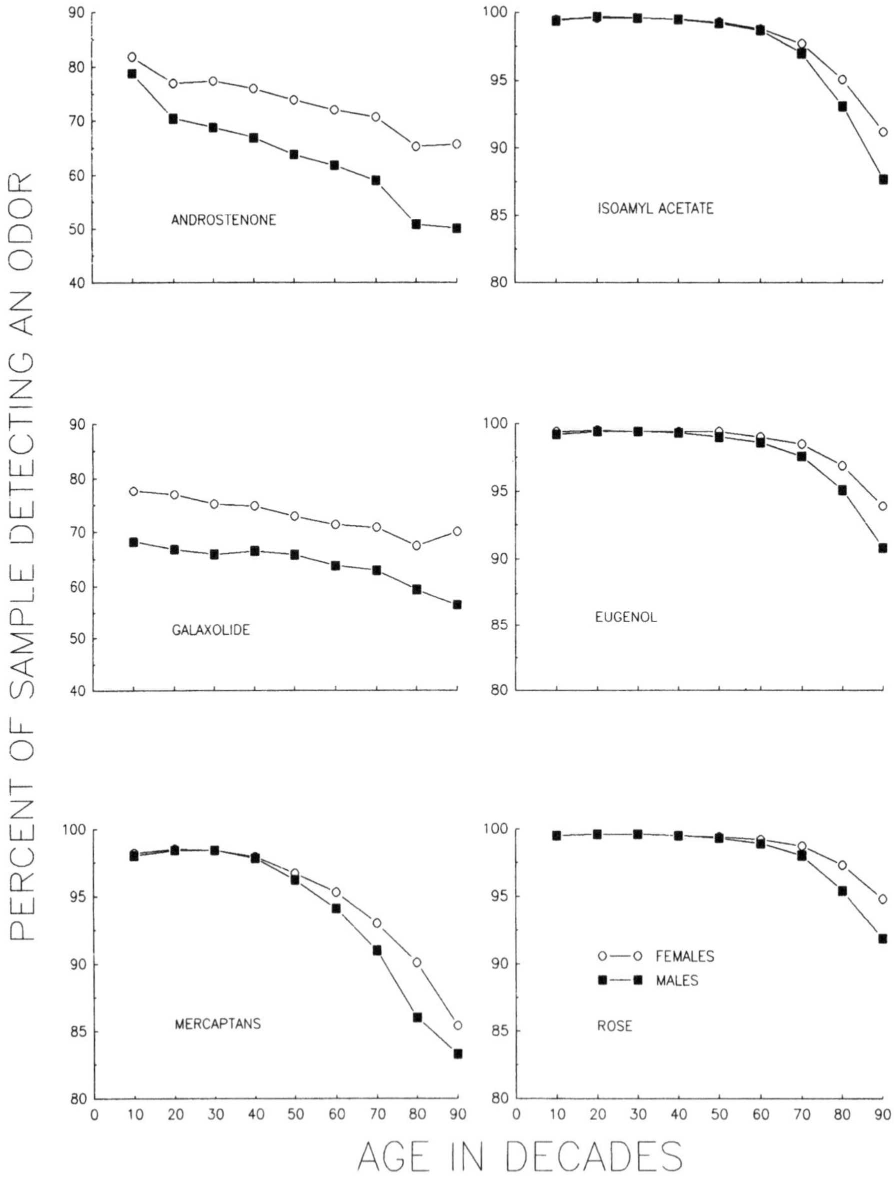

Fig. 4. The percentage of *National Geographic Smell Survey* respondents able to detect an odour, plotted as a function of age (collapsed by decade). Note the odorant-specific shape of the age–response curves. From Ref. 55.

or measures of response (Fig. 4). Age-related changes in olfactory function are distinctly heterogeneous.

Rowe & Khan[67] noted that research on aging emphasizes differences between age groups, while ignoring the substantial differences within age groups. This approach creates a 'gerontology of the usual', and neglects important psychosocial and physiological factors extrinsic to aging. In order to understand the aging process better, as well as the factors that promote resistance to it, it will be necessary to explore the central fact of heterogeneous olfactory response.

Heterogeneity is evident in the structure and function of the aging brain. Neuronal cell loss is general in some brain regions, and highly selective in others. Neurofibrillary tangles show a predilection for certain cell groups, and are rare or absent in others.[68] Neuropsychological assessment of intellectual function points to a decline beginning in the seventh or eighth decade, with a more pronounced decline evident in the ninth decade. Yet while some intellectual abilities become impaired, others are spared or, in fact, improve throughout the life-span.[61]

In the olfactory system there appear to be processes that militate for and against the preservation of function. Although the number of nerve fibres in the olfactory tract and cells in the olfactory bulb declines with age,[69–71] it is also known that the olfactory system replaces its neuroepithelial receptor cells.[72,73] There is evidence of morphological change in the olfactory system with repeated exposure to odorants.[74] Thus there is an anatomical basis for suspecting that factors extrinsic to aging can modulate olfaction. The elderly participate with gusto in many gourmet societies, and it is reported that perfumers and wine-makers improve with age. Conversely, many prescription medications are known to affect taste and smell;[75] usage may contribute to variability among the elderly. A diminished metabolic rate lowers the caloric needs of the elderly and consequently their food intake, an effect often amplified by limited physical activity. Such reduction in appetite can be misinterpreted as a loss of response to the olfactory qualities of food (see references in Ref. 76). To concentrate solely on average deficits exhibited by a population leads to a picture of ineluctable decline; to explore what sustains the perfumer and debilitates the medicated patient may lead to insight and improved quality of life.[55]

Given the presumption in favour of the heterogeneous effects of aging on the nervous system, Cain's claim that age-related deficits in olfaction are uniform across odours and concentrations throughout the life-span[62,63] is puzzling. Methodological shortcomings in the aging studies by Cain and his colleagues[62,77] have been described by others.[65] It is sufficient to note here that were 'uniformity' to be documented in age-related olfactory decline, the result would stand in contrast to every other known sensory system.

The elderly are of course only one end of a developmental continuum, and the study of olfaction in infants and children presents a host of interesting questions regarding the origin, acquisition, and valence of early odour preferences and aversions. Despite the relevance of these topics to developmentalists, until recently few undertook active experimentation with odours, leaving by default the presumption that children under the age of five years do not show adult-like differential responses to smells of varying hedonic quality. Recent work suggests that this view survived because experimentalists did not account adequately for the cognitive capacities of their young subjects. Schmidt & Beauchamp[78] employed a pair of stuffed toys representing characters on Sesame Street, a popular children's television programme. These toy creatures were, respectively, benevolent and malevolent, and 18 children averaging 35 months of age were quite willing to assign odorants to one or the other character based on the odorants' hedonic value. The children's evaluations of nine odorants were closely similar to those made by adults asked to designate their preferences in the same manner, thereby putting to rest not only the notion that three-year-olds are not possessed of demonstrable odour preferences, but also the notion that any such preferences are at odds with those of adults.

6 MEMORY, ODOUR PERCEPTION AND HEDONIC RANKING

There exist other widely shared beliefs about odour perception that, although less explicit than those found in the literature on children's sense of smell, are none the less powerful constraints shaping the course of research. One such tacit belief is the prevailing view, usefully designated the 'Proustian' perspective, that odours are powerfully intertwined with memory, to the extent that this connection is the dominant feature of interest in the psychological landscape behind the nose. Nearly every after-dinner speaker on perfumery has felt the compulsion to allude to the Overture to *Swann's Way*,[79] and it has become something of a shibboleth in the research literature as well, referenced in testimony to the powerful effects of early experience on later odour memory. It is a telling statement that a work of fiction should serve this function, rather than a series of citations to rigorous experiments. Despite the common assertion (e.g. by Engen[80,81]) that experience with an odour critically affects hedonic response, the evidence is meagre. Although odour exposure in the course of a laboratory experiment altered pleasantness ratings in a subsequent laboratory assessment,[82] there is no empirical evidence linking measurable odour preferences to prior, naturally occurring exposure.

Indeed, recent odour-memory work has, in general, made this relationship appear much more quotidian than the ornate references to Proust might imply. Memory for odours, it turns out, displays features common to all memory: it fades with time and can be interfered with by intervening events.[83] Rubin *et al.*[84] found that when odours, photographs and names of objects were used to spark autobiographical memory, the similarities between these types of stimuli exceeded the differences, again vitiating any privileged link between odour and memory. Odour-evoked autobiographical memories, though not older, were less likely to have been evoked recently.[84] This finding may explain folk belief in the evocative power of odour: odour-evoked memories may be relatively less accessible, making their recollection a novel and hence a noteworthy event.

Ehrlichman & Halpern[85] had subjects report memories cued by single words in the presence of unpleasant, neutral or pleasant odours, and later rated their memories for happiness. Pleasant odour produced a significantly greater percentage of happy memories than did unpleasant odour, demonstrating that olfactory cues can bias memory toward hedonically congruent content. Whether this effect was produced by odour-induced changes in memory rating rather than in actual memory retrieval, or by experimental demand (it's ungrateful to recall a bad memory if the experimenter gives you a good odour), requires further investigation.

Schleidt *et al.*[86] surveyed German and Japanese subjects about their memories for freely recalled odours, the associated hedonic valence (pleasant or unpleasant), and contextual information. The German subjects recalled more odours than the Japanese (9·1 versus 5·9 per subject), but the proportion of unpleasant and pleasant odours was similar (53% and 47%, respectively), as was the proportion classifiable into such categories as 'civilization', 'food and drink', etc. The investigators were impressed by the relative conformity in hedonic evaluation in certain categories, namely an aversion to the smell of rot and excrement, and a preference for the smell of plants and fresh water. The odours of familiar persons (e.g., friends and family) were judged pleasant, whereas those of strangers were not.

Pangborn *et al.*[87] obtained responses from subjects in 16 countries to 22 actual, microencapsulated odours. They also noted general agreement in hedonic rankings of the stimuli across cultures, with occasional sharp departures—for example, a marked distaste for methyl salicylate in England. Principal component analysis revealed regional patterns of preference and aversion. Because most of the test odorants were food-related, Pangborn *et al.* were inclined to attribute these regional patterns to differences in traditional food and flavouring habits. Notable national differences in response to actual microencapsulated odorants have been observed elsewhere.[66] Various historical and cultural accounts of such differences have been offered in piecemeal fashion, but few investigations of culture and history have been as comprehensive as Corbin's[88] treatise on social attitudes and the role of odour in France.

7 RESPONSE TO ODOURS

Underlying much writing on the generality of human odour response is the notion that cross-cultural commonalities, or 'universals' as it is often phrased, are evidence for an innate, adaptive, inflexible and reflex-like response presumably shaped by natural selection.[86] One investigator (Steiner) has gone so far as to refer to a 'naso-facial reflex' in characterizing the customary facial displays made in response to pleasant and unpleasant odours.[89–91] Although Steiner's experimental evidence on this point has yet to be replicated by independent researchers, the social determinants of facial displays to odour were explored in a study using the videotaped responses of female subjects.[92] Raters viewing the videotapes identified the hedonic valence of the test odorants with 76% accuracy when the subjects posed a facial response to odours real or imagined, but the accuracy rate dropped to 37% when the subject's response was spontaneous (covertly videotaped). Spontaneous facial displays to odour were characterized by little facial movement and near-impassivity. Still, subjects could easily produce, and raters easily classify, the 'typical' faces made in response to pleasant, unpleasant and neutral odours. These results pose a direct challenge to the nativist view of odour response as an innate and stereotypical reflex. The existence of a set of emblematic facial expressions, produced and interpreted uniformly across cultures, speaks as much to social convention as it does to biological dictate.

Lost in the whirl of cultural 'universals' and physiological constraints has been scholarly treatment of personality traits as determinants of response to fragrance. That Patrick Suskind's grotesque, cartoon-like protagonist Grenouille in the novel *Perfume: The Story of a Murderer*[93] can be taken seriously as a psychological portrait of the olfactive life, rather than as a stand-in for a historicopolitical subtext, speaks to the lack of credible work. Over 20 years ago, Carsch[94] proposed an approach for relating perfumes to personality. Although written without benefit of later statistical analytical techniques, Carsch's approach was multivariate and multidimensional. The psychological effects of fragrance could be located along tranquilizing/stimulating and neutral/erogenic dimensions, and the perfume wearer was characterized according to personality, temperament and dress, with each variable containing component descriptive axes. Despite the intrinsic appeal of a relationship between perfume and personality, there is little empirical work on the topic. Filsinger *et al.*[95] had 195 subjects complete Eysenck's well-validated, 24-item Introversion–Extraversion Scale, and then rate four odorants for strength, familiarity and pleasantness. Factor analysis and principal component analysis of the data produced a first factor on which loaded most of the items of the Introversion–Extraversion Scale. Factors 2, 3 and 4 corresponded to olfactory strength, familiarity and pleasantness, respectively. The clear separation of olfactory and personality measures was interpreted as evidence that these responses operate in 'different arenas of psychological functioning'.

The great majority of psychological studies on odour response deal with the conscious processing of olfactory information, despite the fact that an argument can be made for odour perception being a phenomenon largely at the periphery of conscious thought, only marginally accessible to self-awareness. Thus it is of particular interest to consider how smells affect our sleeping life. Trotter *et al.*[96] presented pleasant or unpleasant odour for a minute or more to sleeping subjects shortly after the onset of REM sleep. Upon being awakened from this state of dream sleep, subjects reported the content of their dreams. An average of 19% of these dreams contained olfactory references, a rate of incorporation similar to that seen in research on other forms of external stimulation. Incorporation of the odour stimulus into dream content occurred in 27% of pleasant odour trials, and 11% of malodour trials. Trotter *et al.* adopted a conservative approach to their results, noting that the small number of trials makes interpretation of this discrepancy rather tenuous. We, however, feel no such compunction, and cite the folklore surrounding the scented pillow as justification enough to encourage self-experimentation among our readers as they pursue sweet dreams and deep insight into the psychology of perfume.

REFERENCES

1. Wysocki, C. J., Scientific smells. *Focus on Fragrance* (Jan. 1989) 3–4.
2. Margolis, F. L., A brain protein unique to the olfactory bulb. *Proc. Nat. Acad. Sci. USA*, **69** (1972) 1221–4.
3. Rogers, K. E., Dasgupta, P., Gubler, V., Grillo, M., Kweh-Goodall, Y. S. & Margolis, F. L., Molecular cloning and sequencing of cDNA of olfactory marker protein. *Proc. Nat. Acad. Sci. USA*, **84** (1987) 1704–8.
4. Verhaagen, T., Oestreicher, A. B., Gispen, W. H. & Margolis, F. L., The expression of the growth associated protein B50/GAP43 in the olfactory system of neonatal and adult rats. *J. Neurosci.*, **9** (1989) 683–91.
5. Snyder, S. H., Sklar, P. B. & Pevsner, J., Molecular mechanisms of olfaction. *J. Biol. Chem.*, **263** (1988) 13971–4.
6. Lancet, D. & Pace, U., The molecular basis of odor recognition. *Trends Biochem. Sci.*, **12** (1987) 63–6.
7. Anholt, R. R. H., Primary events in olfactory reception. *Trends Biochem. Sci.*, **12** (1987) 58–62.
8. Bruch, R. C., Kalinoski, D. L. & Kare, M. R., Biochemistry of vertebrate olfaction and taste. *Ann. Rev. Nutr.*, **8** (1988) 21–42.
9. Dionne, V. E., How do you smell? Principle in question. *Trends Neurosci.*, **11** (1988) 188–9.
10. Youngentob, S. L., Mozell, M. M., Sheehe, P. R. & Hornung, D. E., A quantitative analysis of sniffing strategies in rats performing odor detection tasks. *Physiol. Behav.*, **41** (1987) 59–69.
11. Laing, D. G., Characterisation of human behaviour during odour reception. *Perception*, **11** (1982) 221–30.
12. Laing, D. G., Natural sniffing gives optimum odour perception for humans. *Perception*, **12** (1983) 99–117.
13. Laing, D. G., Optimum perception of odor intensity by humans. *Physiol. Behav.*, **34** (1985) 569–74.
14. Stuiver, M., Biophysics of the sense of smell. PhD thesis, Groningen University, The Netherlands, 1958.
15. Laing, D. G., Identification of single dissimilar odors is achieved by humans with a single sniff. *Physiol. Behav.*, **37** (1986) 163–70.
16. Youngentob, S. L., Stern, N. M., Mozell, M. M., Leopold, D. A. & Hornung, D. E., Effect of airway resistance on perceived odor intensity. *Am. J. Otolaryngol.*, **7** (1986) 187–93.

17. Teghtsoonian, R., Teghtsoonian, M., Berglund, B. & Berglund, U., Invariance of odor strength with sniff vigor: An olfactory analogue to size constancy. *J. Exp. Psy.: Hum. Perc. Perf.*, **4** (1978) 144–52.
18. Teghtsoonian, R. & Teghtsoonian, M., Perceived effort in sniffing: the effects of sniff pressure and resistance. *Perception Psychophys.*, **31** (1982) 324–9.
19. DeLong, R. E. & Getchell, T. V., Nasal respiratory function—vasomotor and secretory regulation. *Chem. Senses*, **12** (1987) 3–36.
20. Kayser, R., Die exacte Messung der Luftdurchgängigkeit der Nase. *Archiv. Laryngol. Rhinolog.*, **3** (1895) 101–20.
21. Gilbert, A. N. & Rosenwasser, A. M., Biological rhythmicity of nasal airway patency: a re-examination of the 'nasal cycle'. *Acta Otolaryngol.*, **104** (1987) 180–6.
22. Gilbert, A. N., Reciprocity versus rhythmicity in spontaneous alternations of nasal airflow. *Chronobiol. Int.*, **6** (1989) 251–7.
23. Werntz, D. A., Bickford, R. G., Bloom, F. E. & Shannahoff-Khalsa, D. S., Alternating cerebral hemispheric activity and the lateralization of autonomic nervous function. *Human Neurobiol.*, **2** (1983) 39–43.
24. Klein, R., Pilon, D., Prosser, S. & Shannahoff-Khalsa, D., Nasal airflow asymmetries and human performance. *Biol. Psychol.*, **23** (1986) 127–37.
25. Brodal, A., *Neurological Anatomy in Relation to Clinical Medicine*, 3rd edn. Oxford University Press, New York, 1981.
26. Kucharski, D., Johanson, I. B. & Hall, W. G., Unilateral olfactory conditioning in 6-day-old rat pups. *Behav. Neur. Biol.*, **46** (1986) 472–90.
27. Kucharski, D. & Hall, W. G., Developmental change in the access to olfactory memories. *Behav. Neurosci.*, **102** (1988) 340–8.
28. Schwob, J. E. & Price, J. L., The development of lamination of afferent fibers to the olfactory cortex in rats, with additional observations in the adult. *J. Comp. Neurol.*, **223** (1984) 203–22.
29. Gordon, H. W., Olfaction and cerebral separation. In *Hemispheric Disconnection and Cerebral Function*, ed. M. Kinsbourne. C. C. Thomas, Springfield, IL, 1974, pp. 137–54.
30. Gordon, H. W. & Sperry, R. W., Lateralization of olfactory perception in the surgically separated hemispheres of man. *Neuropsychologia*, **7** (1969) 111–20.
31. Bellas, D. N., Novelly, R. A., Eskenazi, B. & Wasserstein, J., The nature of unilateral neglect in the olfactory sensory system. *Neuropsychologia*, **26** (1988) 45–52.
32. Bellas, D. N., Novelly, R. A., Eskenazi, B. & Wasserstein, J., Unilateral displacement in the olfactory sense: a manifestation of the unilateral neglect syndrome. *Cortex*, **24** (1988) 267–75.
33. Eichenbaum, H., Morton, T. H., Potter, H. & Corkin, S., Selective olfactory deficits in case H.M. *Brain*, **106** (1983) 459–72.
34. Eskenazi, B., Cain, W. S., Novelly, R. A. & Friend, K. B., Olfactory functioning in temporal lobectomy patients. *Neuropsychologia*, **21** (1983) 365–74.
35. Eskenazi, B., Cain, W. S., Novelly, R. A. & Mattson, R., Odor perception in temporal lobe epilepsy patients with and without temporal lobectomy. *Neuropsychologia*, **24** (1986) 553–62.
36. Henkin, R. I., Comiter, H., Fedio, P. & O'Doherty, D., Defects in taste and smell recognition following temporal lobectomy. *Trans. Am. Neurol. Assoc.*, **102** (1977) 146–50.
37. Rausch, R., Serafetinides, E. A. & Crandall, P. H., Olfactory memory in patients with anterior temporal lobectomy. *Cortex*, **13** (1977) 445–52.
38. Toulouse, E. & Vaschide, N., L'Asymétrie sensorielle olfactive. *Compt. Rend. Hebd. Soc. Biol.* (14 Oct. 1899) 785–7.
39. Toulouse, E. & Vaschide, N., L'Asymétrie sensorielle olfactive. *Rev. Philos.*, **49** (1900) 176–86.

40. Koelega, H. S., Olfaction and sensory asymmetry. *Chem. Senses Flavour*, **4** (1979) 89–95.
41. Youngentob, S. L., Kurtz, D. B., Leopold, D. A., Mozell, M. M. & Hornung, D. E., Olfactory sensitivity: is there laterality? *Chem. Senses*, **7** (1982) 11–21.
42. Gilbert, A. N., Greenberg, M. S. & Beauchamp, G. K., Sex, handedness, and side of nose modulate odor judgments. *Neuropsychologia*, **27** (1989) 505–11.
43. McGlone, J., Sex differences in human brain asymmetry: a critical survey. *Behav. Brain Sci.*, **3** (1980) 215–63.
44. Ehrlichman, H., Hemispheric asymmetry and positive–negative affect. In *Duality and Unity of the Brain: Unified Functioning and Specialization of the Hemispheres*, ed. D. Ottoson. Plenum Press, New York, 1987, pp. 194–206.
45. Heine, O. & Galaburda, A. M., Olfactory asymmetry in the rat brain. *Exp. Neurol.*, **91** (1986) 392–8.
46. Ohm, T. G. & Braak, H., Olfactory bulb changes in Alzheimer's disease. *Acta Neuropathol.*, **73** (1987) 365–9.
47. Reyes, P. F., Golden, G. T., Fagel, P. L., Fariello, R. G., Katz, L. & Carner, E., The prepiriform cortex in dementia of the Alzheimer type. *Arch. Neurol.*, **44** (1987) 644–5.
48. Mann, D. M. A. & Esiri, M. M., The site of the earliest lesions of Alzheimer's disease. *New Engl. J. Med.*, **318** (1988) 789–90.
49. Talamo, B. R., Rudel, R., Kosik, K. S., Lee, V. M.-Y., Neff, S., Adelman, L. & Kauer, J. S., Pathological changes in olfactory neurons in patients with Alzheimer's disease. *Nature*, **337** (1989) 736–9.
50. Gilbert, A. N., The neuropsychology of olfaction in Alzheimer's disease. *Neurobiol. Aging*, **7** (1986) 578–9.
51. Doty, R. L., Reyes, P. F. & Gregor, T., Presence of both odor identification and detection deficits in Alzheimer's disease. *Brain Res. Bull.*, **18** (1987) 597–600.
52. Knupfer, L. & Spiegel, R., Differences in olfactory test performance between normal aged, Alzheimer and vascular type dementia individuals. *Internat. J. Geriat. Psychiatry*, **1** (1986) 3–14.
53. Koss, E., Weiffenbach, J. M., Haxby, J. V. & Friedland, R. P., Olfactory detection and recognition in Alzheimer's disease. *Lancet*, i, **14** (March 1987) 622.
54. Rezek, D. L., Olfactory deficits as a neurologic sign in dementia of the Alzheimer's type. *Arch. Neurol.*, **44** (1987) 1030–2.
55. Wysocki, C. J. & Gilbert, A. N., The National Geographic Smell Survey: the effects of age are heterogeneous. In *Nutrition and the Chemical Senses in Aging: Recent Advances and Current Research Needs*, ed. C. Murphy, W. S. Cain, D. M. Hegsted. *Ann. NY Acad. of Sci.*, **561** (1989) 12–28.
56. Moberg, P. J., Pearlson, G. D., Speedie, L. J., Lipsey, J. R., Strauss, M. E. & Folstein, S. E., Olfactory recognition: differential impairments in early and late Huntington's and Alzheimer's diseases. *J. Clin. Exp. Neurol.*, **9** (1987) 650–64.
57. Greenberg, M. S. & Gilbert, A. N., Le système olfactif dans la maladie d'Alzheimer. *Alzheimer Actual.*, **28** (December 1988) 6–9.
58. Jones-Gotman, M. & Zatorre, R. J., Olfactory identification deficits in patients with focal cerebral excision. *Neuropsychologia*, **26** (1988) 387–400.
59. Koss, E., Olfactory dysfunction in Alzheimer's disease. *Devel. Neuropsychol.*, **2** (1986) 89–99.
60. Bruyer, R., Neuropsychologie du traitement de l'information olfactive: revue bibliographique. *Psycholog. Belg.*, **28** (1988) 1–21.
61. Hochanadel, G. & Kaplan, E. Neuropsychology of normal aging. In *Clinical Neurology of Aging*, ed. M. L. Albert. Oxford University Press, New York, 1984, pp. 231–44.
62. Stevens, J. C. & Cain, W. S., Age-related deficiency in the perceived strength of six odorants. *Chem. Senses*, **10** (1985) 517–29.

63. Stevens, J. C. & Cain, W. S., Old-age deficits in the sense of smell as gauged by thresholds, magnitude matching, and odor identification. *Psychol. Aging*, **2** (1987) 36–42.
64. Cohen, M. M. & Lessell, S., The neuro-ophthalmology of aging. In *Clinical Neurology of Aging*, ed. M. L. Albert. Oxford University Press, New York, 1984, pp. 313–44.
65. Chauhan, J., Hawrysh, Z. J., Gee, M., Donald, E. A. & Basu, T. K., Age-related olfactory and taste changes and interrelationships between taste and nutrition. *J. Am. Dietet. Assoc.*, **87** (1987) 1543–50.
66. Gilbert, A. N. & Wysocki, C. J., The Smell Survey results. *National Geographic*, **172** (1987) 514–25.
67. Rowe, J. W. & Kahn, R. L., Human aging: usual and successful. *Science*, **237** (1987) 143–9.
68. Kemper, T., Neuroanatomical and neuropathological changes in normal aging and in dementia. In *Clinical Neurology of Aging*, ed. M. L. Albert. Oxford University Press, New York, 1984, pp. 9–52.
69. Liss, L. & Gomez, F., The nature of senile changes of the human olfactory bulb and tract. *Arch. Otolaryngol.*, **67** (1958) 167–71.
70. Smith, C. G., Age incidence of atrophy of olfactory nerves in man. *J. Comp. Neurol.*, **77** (1942) 589–95.
71. Bhatnagar, K. P., Kennedy, R. C., Baron, G. & Greenberg, R. A., Number of mitral cells and the bulb volume in the aging human olfactory bulb: a quantitative morphological study. *Anat. Rec.*, **218** (1987) 73–87.
72. Graziadei, P. P. & Graziadei, G. A., Neurogenesis and neuron regeneration in the olfactory system of mammals. I. Morphological aspects of differentiation and structural organization of the olfactory sensory neurons. *J. Neurocytol.*, **8** (1979) 1–18.
73. Graziadei, G. A. & Graziadei, P. P., Neurogenesis and neuron regeneration in the olfactory system of mammals. II. Degeneration and reconstitution of the olfactory sensory neurons after axotomy. *J. Neurocytol.*, **8** (1979) 197–213.
74. Panhuber, H. & Laing, D. G., The size of mitral cells is altered when rats are exposed to an odor from their day of birth. *Develop. Brain Res.* **34** (1987) 133–40.
75. Schiffman, S. S., Taste and smell in disease. (First of two parts.) *New Engl. J. Med.*, **308** (1983) 1275–9.
76. Kare, M. R. & Brand, J. G. (eds), *Interaction of the Chemical Senses with Nutrition*. Academic Press, Orlando, FL, 1986.
77. Stevens, J. C., Platinga, A. & Cain, W. S., Reduction of odor and nasal pungency associated with aging. *Neurobiol. Aging*, **3** (1982) 125–32.
78. Schmidt, H. J. & Beauchamp, G. K., Adult-like odor preferences and aversions in three-year-old children. *Child Devel.*, **59** (1988) 1136–43.
79. Proust, M., *Swann's Way*, trans. C. K. S. Moncrieff. Vintage/Random House, New York, 1970.
80. Engen, T., *The Perception of Odors*. Academic Press, New York, 1982.
81. Engen, T., Remembering odors and their names. *Am. Scientist*, **75** (1987) 497–503.
82. Cain, W. S. & Johnson, F., Lability of odor pleasantness: influence of mere exposure. *Perception*, **7** (1978) 459–65.
83. Walk, H. A. & Johns, E. E., Interference and facilitation in short-term memory for odors. *Perception Psychophysics*, **36** (1984) 508–14.
84. Rubin, D. C., Groth, E. & Goldsmith, D. J., Olfactory cueing of autobiographical memory. *Am. J. Psychol.*, **97** (1984) 493–507.
85. Ehrlichman, H. & Halpern, J. N., Affect and memory: effects of pleasant and unpleasant odor on retrieval of happy and unhappy memories. *J. Pers. Soc. Psych.*, **55** (1988) 769–79.

86. Schleidt, M., Neumann, P. & Morishita, H., Pleasure and disgust: memories and associations of pleasant and unpleasant odours in Germany and Japan. *Chem. Senses*, **13** (1988) 279–93.
87. Pangborn, R. M., Guinard, J.-X. & Davis, R. G., Regional aroma preferences. *Food Qual. Pref.*, **1** (1988) 11–19.
88. Corbin, A., *The Foul and the Fragrant: Odor and the French Social Imagination*. Harvard University Press, Cambridge, MA, 1986.
89. Steiner, J. E., The gustofacial response: observation of normal and anencephalic newborn infants. In *Oral Sensation and Perception: Development in the Fetus and Infant*, ed. J. F. Bosma. NIH–USDHEW, Bethesda, MD, 1973, pp. 254–278.
90. Steiner, J. E., Innate, discriminative human facial expressions to taste and smell stimulation. *Ann. NY Acad. Sci.*, **237** (1974) 229–33.
91. Steiner, J. E., Further observations on sensory motor coordinations induced by gustatory and olfactory stimuli. *Israel J. Med. Sci.*, **12** (1977) 1231.
92. Gilbert, A. N., Fridlund, A. J. & Sabini, J., Social and hedonic determinants of facial displays to odor. *Chem. Senses*, **12** (1987) 355–63.
93. Suskind, P., *Perfume: The Story of a Murderer*, transl. J. E. Woods. Knopf, New York, 1986.
94. Carsch, G., An experimental design for relating personality to perfumes. *J. Soc. Cosmet. Chem.*, **18** (1967) 521–6.
95. Filsinger, E. E., Fabes, R. A. & Hughston, G., Introversion–extraversion and dimensions of olfactory perception. *Percept. Motor Skills*, **64** (1987) 695–9.
96. Trotter, K., Dallas, K. & Verdone, P., Olfactory stimuli and their effects on REM dreams. *Psychiatr. J. Univ. Ottawa*, **13** (1988) 94–6.

III TRAPPING AND MEASURING OF ODOURS

Odours are composed of volatile materials in the gas phase. These are perceived when they enter the nose, and the sensation occurs almost immediately. Odours can be trapped by collecting these materials from a given volume over a given period of time. The analysis of a short odour impression, however, is still at the limit of the actual possibilities because of the very minute amounts of trace components of mixtures, which can have a significant impact on the sensation produced.

Of course, quite some progress has been made since the first analysis of odorous liquids, but the analysis of odours remains a methodological problem which is challenging to the scientists involved. The reader will, therefore, not be surprised to find in the following chapters a collection of microanalytical techniques and of sophisticated sensory measurements.

One of the important applications of these analytical efforts is represented by the study of natural scents. It is not only contributing to an improved understanding of the fine-tuning of nature's mechanisms, but it also leads to the creation of reconstitutions which faithfully reproduce specific odour experiences.

Chapter 6

The Measuring of Odors

NORBERT NEUNER-JEHLE & FRANZ ETZWEILER
Givaudan Forschungsgesellschaft AG, CH-8600 Dübendorf, Switzerland

1 APPROACHES TO ODOR MEASUREMENT—BASIC CONSIDERATIONS

Among human senses, smell is probably the most elusive, and our knowledge about odor-related phenomena is accordingly incomplete. Any progress at the scientific level towards a better understanding of what makes something odorous or, at the technical level, to know why for instance a perfume performs better in one rather than in the other application is directly connected with the availability of precise odor-measuring methods. Whether such methods exist, how they work and what kind of information they supply, will be the subject of this chapter.

To facilitate the understanding of the principal ways in which the problem of odor measurement can be approached, let us start with a short excursion into some fundamentals of odor perception.

What is Odor and How is it Perceived?

Odor is basically the result of an interaction between a chemical stimulus (in the following text called the odorant) and an olfactory receptor system causing biological and psychophysical effects in a living organism. Before we smell something, a series of consecutive events that will be only briefly discussed here must take place. More details can be found in various monographs[1–3] and in Chapter 18 of this book.

The molecules of nearly all chemical substances can act as odor stimuli if they are sufficiently volatile to be present in air. Since most odorous substances are liquids or solids, the release of odorant molecules from the liquid or solid state into the surrounding air is a prerequisite for any material to become odorous.

During inhalation, a fraction of air reaches the olfactory epithelium in the upper part of the nose, where the sensory cells (receptor cells, olfactory neurons) are located. While passing over the mucous layer that covers the olfactory epithelium, odorant molecules are partly dissolved in the aqueous phase and migrate through the mucus where they meet with the receptor sites. They are located at the hair-like endings (cilia) of the sensory cell (dendrites) which protrude into the mucous layer.

During the contact between odorant molecules and receptor cells the olfactory stimulation and signal transduction process, which is not yet completely understood, takes place:[4–6] odorant molecules are supposed to react specifically with binding proteins at the receptor cell membrane. During this interaction a series of complex biochemical processes takes place, in the course of which electrical receptor potentials are built up and rhythmic bursts of action potentials are elicited.

The resulting electrical impulses are conducted from the peripheral end of the receptor cells via the nerve fibres (axones) to the olfactory bulb, where the central-nervous processing of the signals starts. Via synaptic connections these signals are further transmitted to other parts of the brain, where the primary chemosensory information is translated into physiological and behavioral effects and finally, by comparison with memory contents, into a conscious experience, which we call olfactory perception.

Principal Approaches

In odor perception, three different dimensions are involved: the physico-chemical dimension of the stimulus, the biological dimension of the receptor system, and the psychical dimension of the perceiving individual. Each of them is defined by a set of measurable parameters. This opens various ways to odor measurement, which can be allocated accordingly to one of the three basic methodical approaches, to the physico-chemical, physiological or psychophysical (sensory) approach.

Odor-Relevant Parameters in Different Dimensions

In the **sensory dimension**, the primary parameters are odor quality and intensity. Both are not directly accessible to objective measurement, but psychophysical methods can be used to characterize an odor and to quantify its strength on the basis of subjective judgments.

The lack of objective methods allowing the measuring of odors in a direct way has prompted many researchers, especially in the second half of this century, to look for alternative methods that could be used to measure an odor indirectly via objectively measurable physiological or physico-chemical parameters. Whether the latter can be used for measuring odors depends not only on the precision of measurement, but also on the questions of how closely these parameters are related to the odor attributes and whether these relations are known to us.

In the **physiological dimension**, a number of primary and secondary processes are initiated upon stimulation, running inside and outside the olfactory system. The most prominent measurable physiological parameters are the changes of electrical potentials in the olfactory system, their amplitudes and frequencies. Whereas both parameters are known to correlate with odor intensity, their relation to odor quality is extremely complex and far from being understood.

In the **physico-chemical dimension**, the fundamental parameters are the chemical structure of the odorant and its concentration. For single odorants, strong relations between the physico-chemical and the sensory parameters can be grouped into the following three pairs:

(1) odor quality/structure relations,
(2) odor intensity/concentration relations,
(3) odor intensity/structure relations,

whereas

(4) odor intensity/concentration relations in mixtures

are to be considered separately.

Basic Relations Between Sensory and Physico-chemical Parameters

(1) The relations between an odorant's structure and its odor have been examined by many scientists[7–10] and are also the subject of Chapter 9 in this book. Therefore, we can restrict our considerations to a few basic facts.

Although it is generally accepted that size, shape and polar properties of an odorant molecule are the most relevant factors determining its odor properties and, although for single structural classes strong relations have been established empirically,[11] there is so far no universal theoretical concept of general validity that would allow prediction of an odorant's odor from its structure. As a consequence, there is no general way to measure odor quality by measuring the structural parameters of the odorant or by measuring single or multiple structure-dependent molecular properties, although many attempts in this direction have been undertaken in the past.[12–14]

However, the chemical structure and odor quality of nearly all materials used in perfumes or present in natural scents are empirically known to us. Therefore, chemical analysis and structure identification offer a way, not really to measure, but at least to specify an odor in terms of analytical data on the qualitative and quantitative composition of the odorous materials.

(2) In contrast to the odor quality/structure relation, the relation between odor intensity and odorant concentration is less complex and well-defined. It follows the general psychophysical power law

$$I = kc^n$$

$$\log I = \log k + n \log c$$

which was formulated in 1957 by Stevens and relates subjectively perceived intensity I to stimulus concentration c. Both n and k are constants for a given odorant. From its logarithmic form it can be deduced that for a given odorant a linear relationship exists between the perceived intensity and the odorant's concentration in log/log coordinates as shown in Fig. 1. The increase of odor intensity with concentration is represented by a straight line whereby n is the slope, indicating how fast odor intensity rises with concentration, and k is the

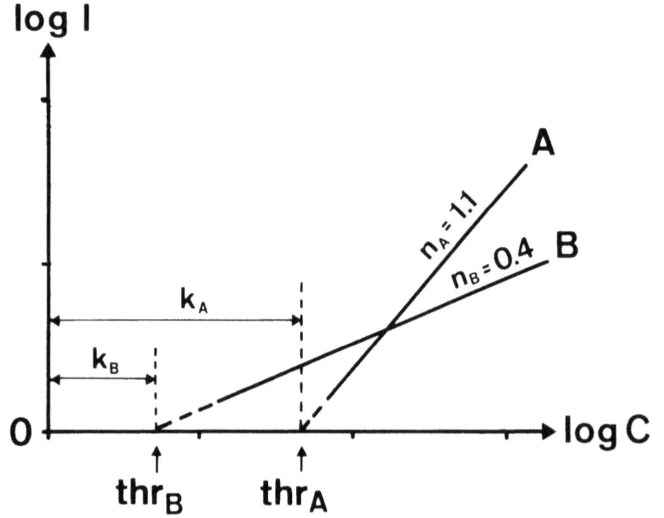

Fig. 1. Stevens' psychophysical power function applied to the odor intensity/concentration relation of two different odorants, A and B. When plotted in logarithmic coordinates, the function is represented by straight lines of different slopes (e.g. $n_A = 1{\cdot}1$ and $n_B = 0{\cdot}4$). Theoretically, the odorants' thresholds (thr_A and thr_B) correspond to the concentration where the lines cut the abscissa when extrapolated to odor intensity zero.

intercept of this line on the abscissa. The intercept defines the odorant's threshold concentration where odor intensity changes from not perceived to just being perceived. An exponent equal to 1, greater or less than 1 specifies that an odorant's perceived intensity increases at the same, at a higher or at a lower rate than its concentration, respectively.

(3) A strong but unknown relationship also exists between odor intensity and the odorant's structure, above-mentioned constants n and k being the parameters which characterize quantitatively this relation. For different odorants, the values for n are reported to vary in a wide range from 0·1 to over 1 with typical values being between 0·2 and 0·7.[15–17] For k, which specifies an odorant's threshold, differences from odorant to odorant of more than six powers of 10 are found in the literature.[18,18a]

Although many attempts have been made by several scientists in the past 25 years to establish a quantitative correlation between odor intensity and an odorant's structure or any structure-dependent objectively measurable molecular properties,[19] our knowledge is still very incomplete.

However, both parameters n and k can be determined experimentally by olfactometric methods and, if they are known for a given odorant, its odor intensity can be quantified on the basis of Stevens' power law by measuring the odorant's concentration.

(4) Odor intensity/concentration relations in mixtures are of special interest, since most odors to which we are exposed in our daily life, including perfumes or any natural scent, are complex mixtures. The fact that the odor of a mixture of odorants is not simply the sum of odors of the single components has been well known for a long time. This has been the subject of many examinations over the last decades and even today persists to be a main topic in odor research.[20,21]

If we want to use physico-chemically measurable parameters, such as the odorant's structures and concentrations, to specify the sensory property of a mixture, we should know how the single components interact among each other olfactorily.

From the theoretical point of view, little is known about how and where mixtures are processed by the olfactory system. It is not yet clear to what extent the physico-chemical properties of the stimuli (e.g. air/mucus partition and adsorption coefficients) have an influence on the question,[22,23] whether and how substantial processing occurs at the peripheral olfactory system (e.g. through competitive or non-competitive binding at the receptor membrane) and what mechanisms of interaction exist in the central part of the olfactory system through special neural circuitries and different response properties of cells at different neural levels in the brain.

Empirically, there is a great variety of experimental data referring to odor interaction phenomena. Some of them were reviewed by Laing.[24] Little evidence exists demonstrating that the odor quality of a mixture differs completely from that of the individual components. However, with regard to odor intensity, all sorts of interactive effects (e.g. additivity, synergism and suppression) and their dependence on the type of odorants, on the concentration levels and on mixing ratios, have been reported.

It must be stated critically that many experiments were carried out with liquid mixtures or solutions in which pure chemical or physico-chemical interactions occur. This leads to incorrect assumptions on the concentrations of the odorants in the vapor phase and, therefore, to incorrect conclusions. Only by using highly developed olfactometric techniques, precisely controlled experimental conditions and adequate analytical measuring methods will progress in this area be possible.

Many attempts have been made to develop mathematical models that could predict the odor intensity of a mixture. Berglund in 1973, for instance, used the relative length of vectors in a parallelogram of forces (vector summation model) in which the angle between the vectors reflects the degree of odor similarity, to describe the perceived intensities of components A and B individually and of AB in a binary system.[24a] For multicomponent systems, various other models were developed which were critically reviewed by Frijters[25] and recently by Laffort.[26]

In the next three sections, the more fundamental aspects of different methods of odor measurement, grouped into the physico-chemical, physiological and sensory approaches, will be reviewed in general. A more detailed discussion of some selected measuring techniques will follow.

Physico-chemical Approach

On the stimulus side, the primary parameters relevant for odor perception are the chemical structure and quantity of the odorant. Both are accessible to objective measurements by physico-chemical methods such as quantitative chemical analysis.

Ideal would be a method which allowed identification of the odorant's structure and determination of the number of molecules directly at the receptor site where odor stimulation occurs. However, this is not possible in practice.

A less ideal, but more realistic, approach is quantitative analysis of odorants in the air that reach the olfactory epithelium in the nose. However, since the odorants have to be dissolved in the aqueous medium of the mucous layer before reaching the cell membrane, the final composition of the odorant mixture at the receptor can still be considerably different from that in the inhaled air, depending on the odorant's air/water and aqueous/lipid phase distribution behavior.

An even more pronounced difference must be expected if, not the odorous air, but the odor-emanating material itself is analyzed, because this material often contains considerable amounts of components that are not sufficiently volatile to vaporize into the air. Although these components never contribute to the odor and although they make the analytical work more difficult, classical chemical analysis of the odorous material itself represents in many cases a complementary approach to odor measurement and sometimes turns out to be indispensable.

Classical Chemical Analysis of the Odorous Material

The odorous material may be a perfume, an essential oil or any substrate that has been obtained by steam distillation, vacuum transfer or solvent extraction from plant material or from a perfumed consumer product. Today, we have an impressive arsenal of analytical instruments such as capillary gas chromatographs coupled to specific detectors like mass spectrometers or infrared spectrometers at our disposal, allowing the analysis of almost any complex mixture within a short time. The methods and instrumental details of modern chemical analysis will not be discussed here.

But why should an odorous liquid sample be analyzed, if its composition is completely different from what we smell in the air? There are some sound reasons to do so:

—Especially low-volatility components with very low odor thresholds are emanated into the air in minute amounts only and may reach concentration levels at which they can be perceived olfactorily, but cannot be detected analytically in the air. Their detection in the substrate, however, does not pose any problem.

—In certain cases, odorant concentrations in the liquid sample or in a solution can be used to calculate the corresponding concentrations in the air, if the volatility of the odorants is known.

—Wherever analytical data are needed as a starting basis for the reconstitution of an odor—an important aspect in perfume creation which is the subject of Chapter 7—analysis of the odorous material itself, complementary to headspace analysis, has proved to be of great value.

One of the central parameters in olfaction is the odorant's volatility. Therefore, the various analytical approaches to odor measurement cannot be discussed without entering into the basic physico-chemical relations by which volatility is defined and without considering some of the main factors that are influencing it. A more detailed description can be found in special textbooks and reviews.[27–29]

Factors Influencing Volatility

If a chemical in its **pure liquid or solid state** in a closed system at a fixed temperature is exposed to air, a certain amount of the odorant vaporizes from the liquid or solid phase into the surrounding air (headspace). Its concentration in the air increases continuously and remains at a constant level as soon as the odorant has reached an equilibrium state in its gas/liquid (gas/solid) phase distribution. This concentration in the air is substance-dependent and can be expressed either in terms of pressure units (vapor pressure in pascals or torricellis) or in terms of concentration units (headspace concentration in micrograms per liter air) and represents a quantitative measure for an odorant's volatility in the pure state.

If the odorant is not in its pure state, but part of an **odorant mixture**, for instance an essential oil or a perfume, or if the odorant is dissolved in a liquid or solid product matrix, its evaporation from the condensed into the vapor phase depends not only on the vapor pressure of the pure odorant but also on its concentration in the mixture. For a closed system in the state of phase distribution equilibrium the relations are quantitatively described by Raoult's law:

$$p_i = p_i^0 x_i$$

p_i being the partial pressure of substance i in the air above the mixture, p_i^0 its vapor pressure in the pure state and x_i its concentration in the liquid mixture in terms of its mole fraction (molar dilution factor).

For **diluted systems**, e.g. a solution of an odorant in a solvent, the phase distribution is better described by Henry's law:

$$p_i = Kx_i$$

where K is the air/liquid partition coefficient, which is a constant at fixed temperature for a given system (solute/solvent).

However, both laws are valid only for a closed system with 'ideal behavior'. In practice, most systems represent non-ideal mixtures or solutions, owing to the fact that physico-chemical interactions take place between the molecules of the mixture, in the liquid as well as in the vapor phase. Consequently, the actual partial pressure p_i of a given component may be considerably higher or lower than calculated on the basis of the distribution laws cited above, depending on whether repellent or attracting forces acting on the molecules of a given component in a given system predominate.

To make the distribution laws valid for non-ideal systems also, the introduction of correction factors, so-called activity and fugacity coefficients, is necessary. The first one takes the molecular interactions in the liquid into account, the second those in the vapor phase. Their strict application leads to rather complex mathematical functions[30] which could be used, at least theoretically, to calculate the precise composition of the vapor phase above a perfume mixture, if the vapor pressures of the pure odorants and the respective correction factors for non-ideal behavior were known. Normally these factors are not available, although they can be measured experimentally. Since these coefficients are temperature- and concentration-dependent they have strong validity only for the given system for which they had been determined.

Therefore, in practice, this theoretical approach of calculating vapor phase concentrations (partial pressures) for a liquid mixture on the basis of the above distribution laws is limited to rough estimations for which activity and fugacity coefficients are neglected.

However, we must be aware that very often odor-emanating samples are far away from representing homogeneous mixtures or true solutions. The presence of solid particles or droplets due to phase separation, the formation of micelles or simply the fact that an odorant or a perfume is part of a solid matrix (e.g. a soap base or detergent powder) can have a tremendous influence on the odorant's concentration in the air. Its calculation on the basis of the phase distribution laws would not even give approximate results, as will be shown with some practical examples in Section 3. Other factors, such as adsorption or special interactions of solutes in the boundary layers of **non-homogeneous systems**, may play an additional and possibly dominating role.

The various aspects of how, for instance, the volatility of flavor substances depend on the mostly multiphasic nature of food matrices, have been discussed by Land.[31] Some of the relations can also be applied to comparable systems in the fragrance field.

Quantitative Headspace Analysis

Chemical analysis of the odor-emanating substrate often gives a first reliable basis, at least qualitatively, of what may be present in the air and, in some cases, the data may also serve for a semi-quantitative estimation of the vapor-phase composition. However, the calculation of precise vapor-phase data from the analytical data of the liquid sample using the phase distribution laws is not applicable, for reasons in part already mentioned above:

—This calculation requires vapor pressure data and, since most systems represent 'non-ideal' mixtures or solutions, activity coefficients. Even if vapor pressure data were available, activity coefficients are normally not.

—Many systems are far from being true mixtures; rather, they represent multiphasic (liquid or solid or liquid/solid) mixtures.

—Wherever an odor in a natural environment has to be measured, it is usually in the open rather than in the closed system so the equilibrium state between condensed and vapor phase leading to vapor phase saturation cannot be reached. Thus, the odorant concentrations in the headspace are always considerably below those predicted by the distribution laws.

Therefore, the most direct access to odor measurement in the physicochemical approach is quantitative analysis of the vapor phase (headspace) above an odorous material. With quantitative headspace analysis, there is no need to consider the many factors that influence an odorant's volatility and, if it is done in the right way—methods and instrumental aspects will be discussed in Section 2—the method offers an objective and reliable basis for measuring odors.

Artificial Noses

Dravnieks[32] defined an artificial nose as an instrument 'that would inspect samples of odorous air and report the intensity and quality of an odor without the intervention of the human nose'. The ideal odor-measuring instrument was supposed to be one that functioned in the same way as a natural olfactory system. Therefore, the first attempts to construct such an instrument were undertaken in the 1950s and 1960s, after scientists had developed various theories of how odor stimulation could work.

In 1950, Tanyolac started studying the effect of odorants in an air flow passing over different surfaces. He observed changes in surface tension following adsorption of odorants and used this principle in his 'electro-odocell' for odor measurement.[33] Some years later, Moncrieff developed an instrument[34] where the heat of adsorption that was generated when an odorant from the air was adsorbed onto a film-coated thermistor was used as measuring parameter. Dravnieks[32] applied sensor plates coated with a thin film of various biochemicals and measured the changes of contact potentials occurring when odorants were adsorbed on the film and formed charge-transfer complexes with the biomolecules.

These are only a few examples from a variety of different systems that had been developed at that time.[32,35] One of the main problems in developing artificial noses of this type is the identification of the odorants' properties or attributes which are really relevant to odor quality and intensity, and to find the physico-chemical techniques to measure them.

Other, more modern concepts are based on gas sensor arrays. They consist of a series of different chemical sensors with special layers of inorganic or organic materials (semiconductors) onto which gases are adsorbed, resulting in a change of conductivity or electrical charge. Since response characteristics of the sensors in the array differ from each other, their output signals, after being digitalized and processed by pattern-recognition techniques in a micro-computer, can be used to identify the type and concentration of gases. After calibrating the system, it will finally allow classification of odors.

The use of integrated sensor arrays in olfaction was already postulated by Persaud & Dodd in 1982.[36] Recently, Abe *et al.*[37] described an automated odor-sensing system where eight semiconductor gas-sensor elements and a pattern-recognition technique were applied. Its capacity for identifying odors was reported to be still improvable by using sènsor units of higher quality. A similar sensor array system in combination with cluster separation procedures was successfully applied by Shurmer *et al.*[38] to identify smokes from different brands of tobacco. As an alternative to pattern recognition techniques, Gardner & Hines[39] described an 'artificial neural network technique' as a novel approach of analysing signals from an electronic nose, a procedure which is supposed to be closer to the real signal processing as it occurs in an olfactory system.

In spite of recent progress in this area, none of the various systems which have been described in the literature have so far reached a quality standard that would really deserve the name 'artificial nose'. Although further improvements in this field can be expected, the human olfactory system can be imitated to a very limited extent only (needless to say), because odor sensation is not simply the result of a complex physical interaction between the stimulus and the olfactory system, but has also a mental dimension which is not accessible by whatsoever sophisticated electronic devices may be invented.

Physiological Approach

The application of electrophysiological measuring techniques for the study of odor perception in animals has a long tradition and goes back to the early 1950s when Adrian, Ottoson and Tucker started their pioneering research in this area. A review of this and some of the later work can be found in several proceedings and books.[40–44] Slow changes of electrical potentials representing the summated receptor potentials that are elicited by odor stimulation can be recorded, for instance from the olfactory epithelium of vertebrates ('electro-olfactogram') or from an insect's antenna ('electro-antennogram') using DC-measuring techniques, as shown in Fig. 2. With AC-recording, the rhythmic changes of action potentials from single receptor cells or from neurons in the brain are measurable.

It has been shown by many experiments that the amplitudes of the slow potential curve and the frequencies of action potential changes increase/decrease non-linearly with stimulus concentration in striking similarity to sensory responses following the psychophysical power function. Therefore, electrophysiological responses can be used to measure odor intensity perception in animals.

In contrast, odor quality is supposed to be somehow encoded in the olfactory bulb, where a marked convergence of a high number of sensory neurons to a lower number of secondary neurons takes place. From measurements of electrophysiological responses in the human olfactory bulb, Hughes *et al.* found some evidence that each odor produces many different frequency components but with a distinct component of highest amplitude depending on the type of odorant.[45] However, the mechanisms of encoding and decoding in the brain are not yet known and, therefore, no correlation between measurable electrophysiological parameters and odor quality is established, so far.

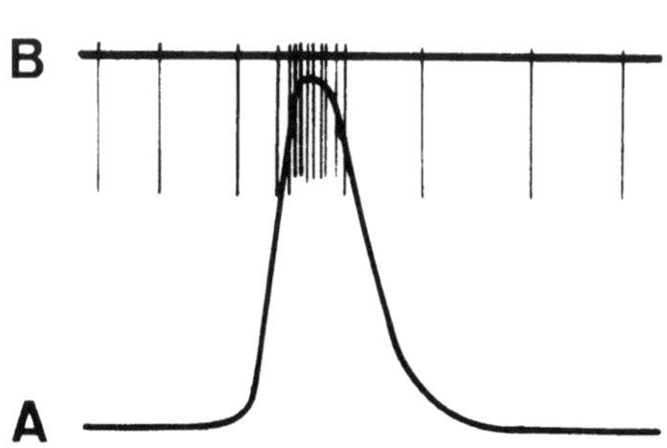

Fig. 2. Electrical records received from the frog when exposed to *n*-butyl alcohol (schematically, after K. B. Döving, cited in Ref. 1). A, DC recording of the slow electrical potential from the olfactory epithelium (electro-olfactogram); B, simultaneous AC recording of electrical pulses from a single neuron in the olfactory bulb.

Now, what does this mean with regard to human olfaction? Apart from a few experiments that have been carried out with human subjects,[45] electrophysiological techniques using implanted electrodes are generally not applicable to humans.

In recent years, several attempts have been made to investigate the relations between odor sensation and cerebral electrical activity. A promising approach was described by Van Toller.[46,47] He used EEG techniques to record the cortical electrical activities in humans during odor stimulation and transformed them into 'brain electrical activity topographical maps'. Separation of odor-evoked electrical signals from the background noises that arise from non-olfactory stimulations and other brain activities seems to be a major problem and it is still an open question whether computer programs can be developed that will solve this problem.

It has been known for a long time that stimulation of most human senses, as a consequence of psychosomatic effects, can also provoke secondary physiological processes or muscle activities, e.g. changes of skin temperature, skin conductivity, respiratory or pupillary reflexes. Steiner *et al.*[48] have shown that different odorants produce different degrees of pupil enlargements, and in addition that the response profiles are subject-dependent. Correlation between these measurable reflexes and the concentration of the odorant was not found and correlation with odor quality in the hedonic dimension was observed only for a few subjects. In another study,[49] the suitability of skin conductivity as a measurable parameter for quantifying odor perception was investigated and compared with that of the pupillary reflex method. Whereas with the latter method no quantitative correlation was found between stimulus concentration and the measurable response, the skin conductivity method opened some interesting aspects with regard to such a correlation, not at the routine but at the research level (Fig. 3). Nevertheless, none of the above-mentioned physiological or electrophysiological methods, although of great interest in odor perception research, has yet proved its applicability as a routine technique for the measurement of odors.

Sensory Approach

Any sensory approach to odor measurement is based on recording the response of a perceiving individual, either an animal or human, upon stimulation. An animal can easily be trained to react to odor perception with a typical behavior which can then be observed and registered.[44,50]

In the context of perfumes, however, it is not the animal's but the human's perceptual response that is the center of our interest. Therefore, further discussion will be restricted on sensory measuring methods with humans. In man, odor perception finally occurs in the mental sphere leading to conscious experience and as such it is not accessible to direct observations or objective measurement. However, it is exactly this conscious experience which allows a human being to 'analyze' an odor and to give a verbal description of what he

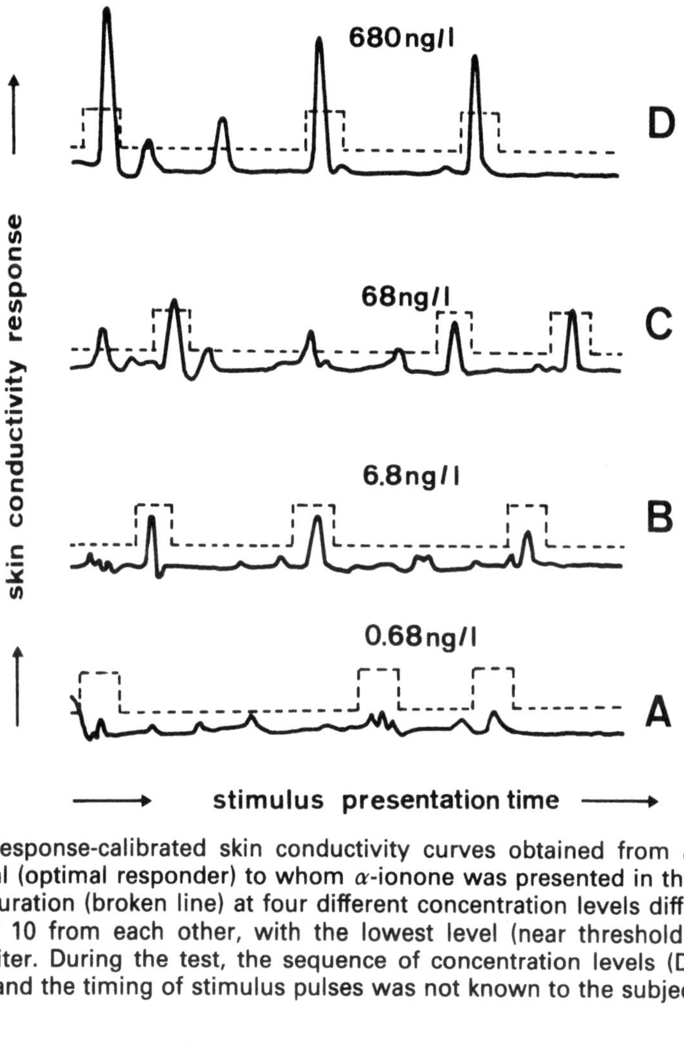

Fig. 3. Response-calibrated skin conductivity curves obtained from a selected individual (optimal responder) to whom α-ionone was presented in three pulses of 10 s duration (broken line) at four different concentration levels differing by a factor of 10 from each other, with the lowest level (near threshold) being at 0·68 ng/liter. During the test, the sequence of concentration levels (D, C, A, B) and the timing of stimulus pulses was not known to the subject.

or she perceives. This, of course, is yielding much more information than any observation of behavioral reactions could ever give.

One of the major problems in odor measurement is that sensory perception depends not only on the species of the organism but also on the perceiving individual within one species and even on the actual state of this individual at a given time. In other words, different individuals perceive one and the same odor stimulus in a different way, not only with regard to odor quality but also to odor intensity. Thus, a human being's perceptual judgment holds only for that subject with its very specific biological and psychical conditions at the moment at which the perception takes place.

The substantial individual variations associated with sensory measuring methods call for statistical methods of data gathering and data treatment using a panel of several individuals which should be representative of that part of the population to which the data are intended to be applied.

A second requirement to improve consistency and reliability of data from sensory analysis is the use of optimized and standardized measuring methods. These include sample preparation, handling and presentation to the subjects for sensory analysis under strictly controlled experimental conditions. This area will be discussed thoroughly in Section 2, under 'Olfactometric techniques'.

Before entering into experimental details of sensory analysis, the more general aspects of what can be measured with sensory methods will be shortly reviewed.

Classification of Odor Quality

The human olfactory system is said to be capable of discriminating many different odors. The actual number is not known but varies considerably among individuals and depends on the subject's experience and the method used for sensory analysis.

Odor quality can be assessed or classified by verbal description: modern statistical methods of descriptive sensory analysis require the use of a standardized vocabulary instead of spontaneous intuitive everyday language or the perfumer's special terminology. Such a standard vocabulary should be concise but large enough to cover all relevant odor notes. Each term should be precisely defined either by a descriptive word with widely accepted meaning or, more ideally, by a reference material (e.g. a pure chemical) to avoid the problems of semantics.[51]

The development of standardized descriptive techniques goes back to 1752 when Linnaeus proposed a simple odor classification system consisting of seven classes. Other systems were established later on by Zwaardemaker in 1895 and by Amoore, Schutz and Harper in the 1950s and 1960s, being similar in principle but somewhat different in vocabulary and the number of the classes.[2]

Already in the 1970s, scientists active in the flavor field started to connect odor descriptors with intensity scaling on a numerical basis or on an unstructured graphic scale ('multidimensional scaling'), attempting to improve the accuracy of the method and to establish 'quantitative descriptive analysis'.[52]

The basic idea of multidimensional scaling is to match a given odor to a given word-list of odor attributes or to a set of appropriate standard reference odors. Thereby, one estimates on an intensity scale the degree to which each attribute may be relevant for describing a given odor or the degree of similarity of the odorous sample in comparison to a set of well-defined reference odors.[53,54]

After testing, the panel's means of category scale values can be calculated and converted into relative distances in a multidimensional odor similarity space.

In the past decade, multidimensional scaling was used more and more, especially in flavor analysis. This was possible because of the enhanced availability of computers and programs for statistcal data treatment and graphical data presentation. More about this can be found in the literature.[52,55]

Psychophysical scaling methods were also used to investigate the relationship between the odor of single odorants and their binary[56] or multicomponent mixtures,[57] respectively.

Today, descriptive methods of odor classification are applied in different areas—in research, in product development, in quality control as well as in market research. There is a great variety of different methods in use depending on the specific requirements of respective applications and almost every fragrance and flavor house or food company has its own system.

Odor classification is of the utmost importance in research for the study of structure/odor relations, for instance. Any progress in this area depends to a great extent on the consistency of data obtained from a trained panel that is capable of managing the odor classification of a great number of chemically well-defined and 'olfactorily pure' fragrance chemicals by using a standardized method of odor description in combination with multidimensional scaling.

As an alternative to descriptive methods, several attempts have been made to measure odor quality, or to be more precise, the similarity of different odors, by indirectly measuring the response sensitivity of a perceiving individual with regard to these odors. Cheesman & Mayne[58] and Köster,[59] for instance, used the effects of cross-adaptation, Amoore & Venstrom[60] the occurrence of anosmia, to gain information on odor similarity from intensity measurements. Another interesting approach in this direction was recently proposed by Müller,[61] who examined the panel correlation diagrams obtained from odor threshold measurements with odorants of similar and dissimilar odor notes. None of these approaches, however, has led to a method generally applicable for measuring or classifying odor quality as an alternative to descriptive methods, so far.

Odor Intensity Measurement

The three main questions around which odor intensity measurement usually centers can be formulated in the following way:[54]

—What is the minimum quantity of an odorant to become perceptible (absolute threshold)?

—What is the minimum difference in odorant concentration that can be perceived as distinct (differential threshold)?

—How does perceived odor intensity increase with the concentration of an odorant (suprathreshold odor intensity slope)?

One distinguishes two kinds of **absolute odor thresholds**, detection and recognition thresholds, depending on whether an odor at the threshold level is just detected as being different from odorless or recognized for its typical odor. In general, detection thresholds are reported to be somewhat lower than recognition thresholds and normally the threshold values mentioned in the literature are related to the former, if not specified otherwise.

From the principles of psychophysics, it follows that an odor threshold is not a precisely fixed substance-dependent constant but it is the mean of many single values obtained from a series of repeated measurements carried out with one or several individuals. Its determination is based on a statistical concept because the probability of a subject perceiving an odor increases with the concentration of the odorant, following a steep curve as shown schematically in Fig. 4. Accordingly, an odorant's threshold value is defined, by convention, as the concentration at which a subject (or a group of subjects) gives a positive response in 50% of the trials.

One of the simplest procedures of threshold determination, called the method of limits, is based on this concept and consists of a multiple presentation of an odorant to the subject at different concentration levels, differing usually by a factor of two. After a number of ascending and/or descending series of trials, the average concentration at which the odorant shifts from being perceived to not perceived, or conversely, is calculated. This average concentration is considered to be the subject's individual threshold. Methods and experimental aspects will be discussed in more detail in Section 2 under 'Olfactometric techniques'.

From experience with psychophysical measurements, it is well known that correct detection of a stimulus, especially a weak one, does not depend only on the subject's sensitivity and capacity to distinguish an odor signal from the background noise, which may arise from various internal or external sources acting on the sensory system. There are also a number of non-sensory factors,

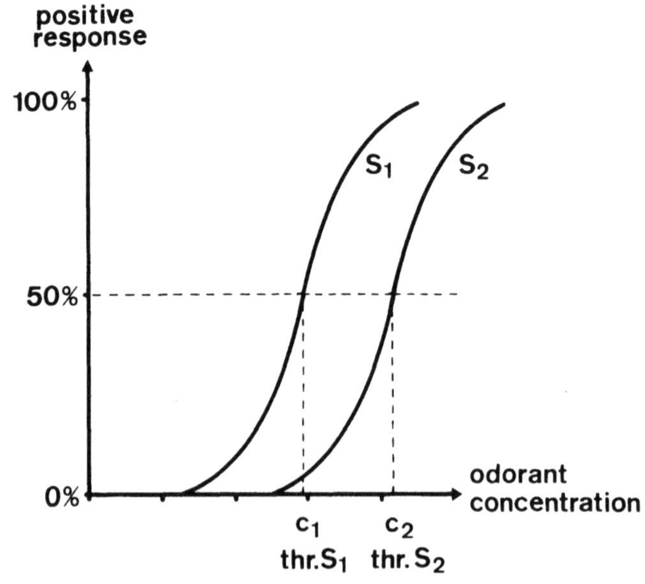

Fig. 4. Theoretical curves reflecting the odor detection probability (percentage of positive responses) in the threshold range for dependence of stimulus concentration for two subjects S_1 and S_2 (Schematic diagram after E. P. Köster [Ref. 54]). According to the subjects' different sensitivities, their individual thresholds (positive responses in 50% of the trials) are at different concentrations c_1 and c_2, respectively.

called response biases, which influence the subject's decision to answer the question of whether a stimulus is present or not, with yes or no. Such response biases can be connected, for example, with the subject's interpretation of the task instructions, his level of attention, his motivation for the task, the degree to which he expects the occurrence of a positive perception within a sequence of consecutive stimulus presentations, and with various other factors.

A statistical approach to tackling the detection problem was introduced by Green & Sweets in 1966.[62] This concept, called signal detection theory, allows isolation and evaluation of the effects on the subject's performance resulting from sensory capacity and from response biases. More about the application of this concept to sensory measurement can be found elsewhere.[54,63]

The **differential threshold**, also called the discrimination threshold or the just-noticeable difference,[54] is defined as the difference in physical intensity (concentration) of an odorant that can just be detected by the subject in 50% of the trials.

The human nose is said to be not as good at perceiving small differences in odor strength as it is at perceiving small differences in odor quality. For ethyl mercaptan, for example, a factor of three in concentration was reported to give a just-noticeable difference.[64] However, typical differential thresholds for most odorants are found to be in the range 15–30%,[3,65] and even lower values, down to 5%, were supposed to be detectable when more precisely controlled conditions of stimulus presentation and sensory measurement were used.[66]

Usually, discrimination thresholds are measured by a paired comparison method. A standard stimulus is presented with a series of reference stimuli of different concentrations from which some are one to several steps below and some above the concentration level of the standard sample. The subject is asked to indicate whether he perceives the reference samples to be stronger or weaker than the standard sample. The various factors influencing the results and the various methods of deriving differential thresholds from experimental data were described by Köster.[54]

The capacity of the human nose to perceive **suprathreshold odor intensities** ranges over more than four powers of 10 of an odorant's concentration from the threshold concentration up to the saturation point, where a further increase of the odorant's concentration does not lead to a further increase of perceived intensity. Within this range, the odor intensity/concentration relation follows Stevens' power law and can be represented in a double-logarithmic plot by a straight line with the slope n, as was shown in Fig. 1 (Section 1). Position, slope and flattening of this line are odorant-dependent characteristics which can be determined experimentally by measuring perceived odor intensities over the whole concentration range.

Basically, there are three different methods of quantifying odor intensity. In category scaling, the odorous sample is rated by being assigned to descriptive categories (e.g. very weak, weak, medium, strong, very strong). The method of magnitude estimation (ratio scaling) uses a scale of numbers of equal ratio from one scale unit to the next. The test samples, which may consist of a series of different dilutions of the odorant under investigation, are assigned to these scale numbers according to the ratio of their perceived odor intensities. In reference scaling, the test sample is compared with a series of different concentrations of a reference odorant (e.g. *n*-butan-1-ol) and the subject's task is to identify that reference sample which best matches the odor intensity of the test sample.

Each method has its advantages and disadvantages and there are numerous variations for special applications.[54,65,67]

As expected, individual differences in odor intensity rating in the suprathreshold concentration range are often rather large, too. Tuorila[68] assigned them to two main factors, the genuine perceptual differences between individuals and the different mode of handling numerical scores in magnitude estimation. Interestingly, inter-individual variations of sensitivity in the suprathreshold range seem not to correlate with those found at the threshold level. The main factors that influence odor intensity judgments and the measurement of odor intensity characteristics in the suprathreshold range have been exhaustively reviewed by Hyman.[69]

Odor Values

Instead of measuring effectively perceived odor intensities, an odorant's odor intensity at any given concentration above threshold can be related to its threshold concentration and expressed in multiples of it. This leads to the 'odor value' (OV) of a substance, which is defined as the quotient of its actual concentration and its threshold concentration, both usually in terms of nanograms per liter in air (or in another medium):

$$\mathrm{OV} = \frac{\text{actual concentration of odorant}}{\text{threshold concentration of odorant}}$$

This numerical measure, which indicates how many times the actual concentration of an odorant exceeds its threshold concentration, was already used by Rothe & Thomas in 1963 and somewhat later by Guadagni: they named it 'aroma value' and 'odor unit number', respectively. Other names that can be found in the literature are 'odor intensity unit', 'threshold odor number' or 'odor value'.[70] Originally, odor values were intended to be used for assessing the relative importance of single components that contribute to the total odor or flavor of a mixture, but other authors[71] later applied them explicitly as a quantitative measure to specify an odorant's intensity and to calculate the odor intensity of mixtures.

The odor value concept is based on the assumption that it provides a quantitative measure for an odorant's odor intensity response, not only at threshold level but also at suprathreshold concentrations. Of course, this is not correct for several reasons:

—From the psychophysical power function, it follows that odor intensity does not increase linearly with substance concentration. For most odorants, odor intensity was found to increase at a lower rate than concentration ($n < 1$). Therefore, concentration ratios from which odor values are derived cannot be a numerical equivalent for odor intensities. The doubling of an odorant's concentration with a typical value of $n = 0{\cdot}5$, for instance, increases the odor intensity only by a factor of 1·4.

—Since the odor intensity of different odorants increases with concentration at different rates, two odorants can be perceived with different odor intensities, although they may be present at concentrations at which their odor values are identical.

—Odor values, used to quantify the contributions of single odorants to the total odor of a mixture, do not account for possible interactions that may result in odor quality and intensity of a component being altered or masked by other components in the mixture.

In spite of these limitations, the odor value can be regarded not as an absolute, but at least as some sort of relative measure for odor intensity. The odor value concept still represents an approximative, fast and easy-to-handle method for quantifying an odorant's odor intensity on the basis of its actual concentration and its threshold value. With this approach, time-consuming and less precise sensory measurement in the suprathreshold range can be reduced to odor threshold measurement which has to be done only once for a given odorant and delivers fairly reproducible data. The rest of the work consists of measuring the odorant's actual concentration, which can be done with high precision by quantitative headspace analysis.

However, the point should be clearly made that only by using correctly determined headspace concentrations (vapor pressures) and odor threshold data can a successful application of this concept be expected. After discussing the various aspects and limitations of the odor value concept, Frijters[72] concluded that 'the better these limitations are appreciated the more helpful this concept can be'.

Comparison of Different Approaches

From what has been outlined in Section 1 under 'Physiological approach', it can be derived that physiological methods, although they may remain of great interest to odor perception research, are not applicable so far as routine methods to odor measurement in the context of human olfaction. Therefore, they will not be discussed further.

In the following, we concentrate on the two other approaches, physicochemical and sensory, each of them showing a clear superiority over the other depending on the kind of information that is sought.

Sensory methods are to be preferred, for instance, whenever an integral judgment on odor quality or intensity is needed and it does not make any difference whether the odor comes from one single odorant or from a complex mixture. The main strength of the sensory approach lies in the fact that we are directly measuring what we are looking for. Often, it is only after performing sensory measurement that the results obtained from non-sensory measurement can be confirmed. For instance, whether or not an odor-relevant part of the fragrance present in a detergent powder or fabric softener remains on the laundry after the washing process may be derived from physico-chemical (analytical) data, but always needs a final confirmation by sensory evaluation. The weakness of sensory methods is based on the fact that it is only possible to obtain subjective data, which are known to exhibit considerable variations. Unlike instrumental detectors, human beings used as detectors in odor measurement cannot be calibrated with a standard sample to give one and the same 'correct' response.

Physico-chemical methods, in contrast to the sensory ones, deliver objectively measurable data which are fairly precise and reproducible, provided the methods are used in the correct way. Minute changes, for instance in the composition of a complex odorant mixture leading to an alteration of the odor note of a fragranced product, are detected, and often the sources for such odor effects can be identified unambiguously by headspace analysis in the shortest time. Or, returning to the above-mentioned example of fragrance substantivity on laundry, only the physico-chemical approach allows precise measurement of which of the fragrance constituents succeed in remaining on the laundry after washing, and to what degree.

On the other hand, physico-chemical data must somehow be translated into olfactory data. Quantitative relations between the parameters in both dimensions are often not, or only approximately, known. Therefore, it is in this transformation step that the precision of objectively measured data with regard to their sensory relevance is lost.

The above considerations underline the fact which is clear to everybody who is active in this area, that in most cases a **combination** of physico-chemical and sensory methods is mandatory if meaningful results from odor measurement are expected.

Today, headspace and olfactometric techniques are probably the most advanced measuring methods representing a solid basis for such an integrated approach. Both are used not only by chemists in various industrial and other research centers, but increasingly also by perfumers and flavorists in the creation and application laboratories of the leading companies worldwide in the fragrance and flavor creation or processing industry.

This is why further discussion of the methodological and instrumental aspects of odor measuring methods will focus exclusively on these two key issues in the following section.

2 INSTRUMENTATION AND METHODOLOGY

Headspace Sampling and Measuring Techniques

Within the scope of physico-chemical methods, headspace sampling obviously represents the most direct access to odor trapping and measurement and has many advantages over other sampling methods such as solvent extraction, steam distillation, etc., the latter delivering odor samples for analysis which are considerably different in quantitative composition from what we are really smelling when breathing in a vapor sample from the air space (headspace) above any odorous material or substrate.

Although the method has been known for more than 30 years (one of the first reports on headspace analysis was published in 1959[73]) its wide application was hampered in the beginning by its lack of precision and detection sensitivity. With the enormous improvements in gas-chromatographic techniques and with the development of sample enrichment methods, headspace analysis became a standard technique for the investigation and quantification of volatile organic materials, in general.

Today, the method is routinely applied in many areas, among the most prominent being medicine, food and beverages, food packaging, cosmetics and toiletries, polymers and environmental samples, as can be seen from the references cited in various reviews.[30,74,75] Whereas in the flavor field headspace techniques have been used for many years,[76] surprisingly they have received much less attention in the fragrance field.

It is not within the scope of this contribution to discuss the many aspects and experimental details of headspace analysis. A detailed description can be found in some excellent monographs[28,76–79] and reviews.[29,30,80] However, for a better understanding some general principles of headspace sampling and trapping should be briefly reviewed before we concentrate on more specific questions, namely:

—what the requirements with regard to odor measurements are, and
—how they can be implemented into an experimental procedure.

Headspace Sampling

Headspace sampling can be achieved by different methods, using either the static or the dynamic approach.

In the **static approach**, the sample to be investigated is kept in a closed system at constant temperature. Any volatile substance present in the liquid or solid phase will slowly pass from the liquid into the vapor phase until each substance has reached a constant vapor phase concentration. At this point, phase distribution equilibrium exists and the substance concentrations in both phases remain constant according to the phase distribution laws, mentioned in Section 1.

In the **dynamic approach**, air is passed through or over the liquid or solid sample at constant temperature in an open system. Again volatile substances vaporize from the sample into the gas phase. But they never reach their constant vapor saturation, as they do in a closed system under phase equilibrium, because the vapor phase above the sample is continuously being replaced by fresh air. As a consequence, the high volatiles are slowly stripped out from the sample and their concentration in both phases decreases continuously with time to approach zero asymptotically. The dynamics of this process depends on several parameters such as flow rate, sample surface, diffusion rate, etc., from which the stripping rate can be calculated.[29]

Headspace Trapping

Headspace trapping can be performed either with or without sample enrichment.

In the simplest version **without sample enrichment**, some milliliters of headspace are sucked from a closed sample container, kept at a constant temperature, into a gas-tight syringe and injected directly into the gas chromatograph. Most headspace apparatus on the market is based on this principle. Today, the instruments are fully automated and all the different steps, from sample preparation under static conditions at a preselected temperature, headspace sampling and subsequent sample injection into the gas chromatograph, up to the final reporting of the analytical results, are integrated in one system.[81,82]

Although such systems today have reached a considerably high level of accuracy and reproducibility, their application remains restricted because of unsatisfactory sensitivity to heavy volatiles. This is due to the fact that only a limited volume of headspace (1–3 ml) can be directly injected into capillary columns without separation loss, unless additional cryogenic focusing techniques in combination with multiple headspace injections, as described by Wylie,[74] are used. To improve the method's sensitivity, headspace sampling is usually done at elevated temperature and vapor pressure to increase the headspace concentrations. However, for many applications, the sensitivity, especially for the heavy volatiles, remains unsatisfactory. Furthermore, additional drawbacks often connected with headspace sampling at elevated temperature (e.g. condensation at cold spots, decomposition, chemical reactions) cannot be tolerated.

Therefore, headspace trapping **with sample enrichment** must be used prior to analysis, whenever for sensitivity reasons large volumes of odorous air have to be sampled. This is achieved by sucking or pushing a large amount of headspace from the sample container, operating either in the static or in the dynamic mode, through a trap, where all substances present in the vapor phase are quantitatively frozen out or adsorbed on a sorbent. In a subsequent step, the volatiles can be desorbed from the trap either by heating, while purging the trap with a carrier gas, or by extracting the sorbent with solvent.

Cooled traps have serious disadvantages, such as unacceptable partial losses of headspace constituents due to fog formation, or interruption of the vapor flow from water-rich material which leads to formation of ice within the trap. Today, therefore, the use of sorbent is generally preferred, although the problem of breakthrough which depends on the nature and quantity of the sorbent, the geometry and temperature of the trap, the flow rate and sampling time, must be carefully controlled.[27,79] A review of the variety of materials that can be used as sorbent was given by Núñez *et al.*[30] Among the most commonly used ones are charcoal, Porapak, Chromosorb, Amberlite XAD-resins and

Tenax. There is no 'ideal' sorbent that could be used universally for all applications. Each sorbent has its own characteristics and the selection should be adapted to the respective application.

Heat desorption of the enriched headspace sample from the sorption trap can only be used if sorbent and volatiles from the headspace are thermostable. The method, in combination with cryogenic focusing at the gas-chromatographic column inlet, has a clear advantage with regard to sensitivity, because complete sample transfer into the capillary column without separation loss is possible. Disadvantages of the method are risk of thermal decompositions, incomplete desorption of heavy volatile or polar headspace components, ghost peaks from the trap, blocking of the column inlet by frozen water from humid samples, and the fact that from one headspace sampling experiment only one chromatogram can be obtained.

Today, for many applications, solvent desorption of the traps, especially in combination with improved GC injection techniques, is generally preferred. Apart from avoiding all the above-mentioned disadvantages, solvent desorption has further advantages: it is a simple device, it is easy to handle and aliquots of the sample extract can be used in different ways to carry out sniffing tests or to run supplementary analyses with GC/MS or other spectroscopic methods. In some cases, however, problems can occur with highly volatile headspace components, if an interfering solvent peak in the gas chromatogram makes their correct quantitation impossible.

Dynamic headspace sampling in combination with trapping techniques is frequently used for trace analysis of volatile organic materials present in a liquid or solid non-volatile matrix or in water. A stripping gas (air or nitrogen) continuously purges the sample under investigation and the gaseous effluent passes a sorption trap where enrichment of the volatile solutes occurs. An efficient variation of this 'purge and trap' method was introduced by Grob,[83] generally known as the 'closed loop stripping' technique. After having passed the trap, the stripping gas is not wasted but it is recycled in a closed loop, thus eliminating the problem of sample contamination by large quantities of carrier gas whose trace impurities would also be enriched in the trap.

Analytical results obtained with 'purge and trap' techniques may be used to support qualitative headspace analysis (detection of trace constituents), but cannot be directly correlated with a perceived odor since they do not reflect the true quantitative composition of the vapor phase above an odorous sample.

Specific Requirements with Regard to Odor Measurement

If analytical data from the vapor phase above an odorous sample are to be correlated with the perceived odor, specific requirements must be respected which clearly restrict the range of adequate headspace techniques that can be used for odor measurement.

The **volume of one sniff** by which an odor is perceived is different from individual to individual: a volume of about 0·5 liter can be regarded as average. This, at least, should be the minimum quantity the headspace method must be able to handle. If heavy volatile odorants with extremely low odor threshold values are involved, the quantity of them present in 1 liter of headspace may still be large enough to contribute to the odor, but too small to enable their analytical detection. Therefore, sampling of even larger volumes than 1 liter should be possible and an adequate sample enrichment device is indispensable.

Another important issue is the **temperature**. Since smelling is performed at room temperature and normal pressure, headspace sampling and trapping should be done under comparable conditions. Elevated temperatures and pressures, as already mentioned, must be avoided since they increase the risk of chemical reactions and decompositions of the odorants during headspace sampling that lead to changes in the headspace composition. This also means that thermal desorption of the headspace sample from the trap should be avoided.

The question of **static versus dynamic headspace sampling** must be addressed, too. In a natural environment, smelling an odorous sample, be it a flower, a perfume on a strip or a piece of soap, corresponds to something between static and dynamic headspace sampling. In a completely open system, the vapor phase around the sample is inhomogeneous. The odorant's concentration nearest to the surface is almost as high as in a static system at vapor saturation, but decreases rapidly with increasing distance from the surface. Furthermore, the actual headspace concentration profile may be influenced by air movements. It is not surprising that odor intensity measurements done in an open system either by sniffing or by headspace analysis do not yield reproducible quantitative results since there are too many uncontrolled experimental parameters (distance, air turbulence, etc.) on which they depend.

Therefore, wherever possible, quantitative headspace analysis should be based on static or quasi-static headspace sampling—'quasi-static' because, in practice, as soon as headspace is taken off from a closed system, phase distribution will be slightly shifted away from its equilibrium state by the fresh air which enters the sample container continuously and replaces the saturated vapor phase that was sucked away during headspace sampling.

Instrumental Design and Experimental Procedures for Quantitative Headspace Analysis

With the special requirements formulated above for application of headspace techniques to odor measurement in mind, the authors began in 1975 to develop their own method. Headspace trapping was based on the technique published by Grob.[83] Special attention was paid to adapting the method to a wide range of different areas in fragrance application,[84] and later on some modifications were introduced to optimize further its accuracy and reproducibility for a special application.[85]

The method uses headspace sampling (usually 0·5–1 liter) taken from a closed system under quasi-static conditions at room temperature, thus avoiding any temperature or pressure stress. Sample trapping and enrichment is performed with a microfilter using Porapak Q as sorbent. After filter extraction, an aliquot of the extract is analyzed quantitatively by gas chromatography using capillary columns.

The experimental set-up is shown schematically in Fig. 5. It consists of a thermostated sample container, a dual trap and a pumping (sucking) device.

The **headspace container** of 1·2 liter volume can accommodate samples of different size and shape (perfume, soap, detergent, tissue, etc.). To prevent any contamination from outside, air enters through a charcoal filter mounted on the headspace container, while headspace is continuously sucked from the container during sampling and trapping. Being made from silica glass, the charcoal filter can be easily reactivated by heating it up in the flame of a burner.

A water-filled buret, constructed according to the principle of the Mariotte's flask, serves as a **sucking device**. Headspace sampling starts as soon as the water in the buret is allowed to flow out into a water reservoir. Thus, vacuum is generated in the buret above the liquid, which causes the headspace to be sucked in from the sample container through the connection line via the microfilter. Because of the special construction of the buret, the flow rate of the water and hence the headspace sampling flow are absolutely constant and can be adjusted to any value by varying the difference in height, ΔH. Sucking, i.e. headspace sampling, is stopped automatically as soon as the water level in the buret equals the height of the inlet tube at the water reservoir. The precise headspace volume can be determined for calibration purposes by weighing the amount of out-flowing water.

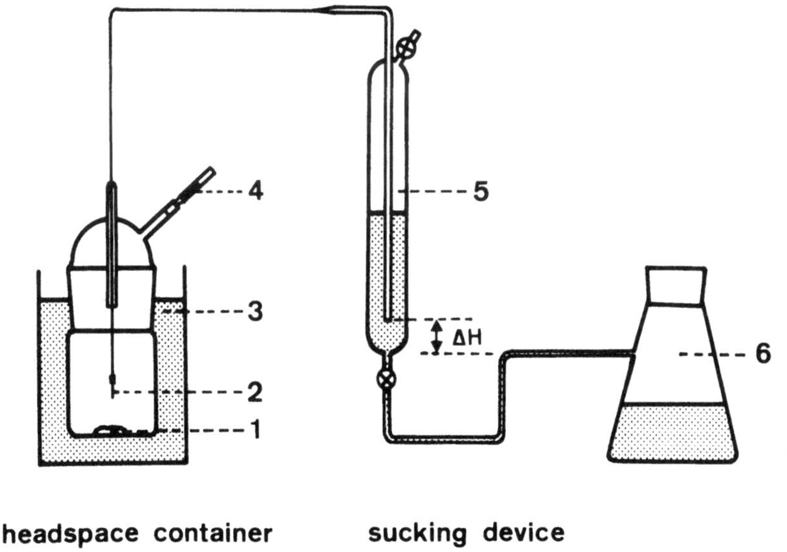

Fig. 5. Schematic diagram of the headspace sampling apparatus consisting of a headspace container and sucking device. 1, Sample; 2, headspace microfilter; 3, thermostated bath; 4, air filter; 5, buret; 6, water reservoir.

Trapping is performed by mounting two microfilters in series at the end of the sucking tube, positioned near the sample in the headspace container. The first filter is expected to trap all headspace constituents quantitatively. The second one serves as a control, making it possible to check whether any breakthrough of substances has taken place. If so, the problem can easily be solved by taking more sorption material or by sampling less headspace volume.

The construction of a headspace filter (microfilter) is shown schematically in Fig. 6A. For reasons of its large specific surface area and its favorable desorption and cleaning properties, Porapak Q (50–80 mesh) is chosen as sorbent.[84] The filters can be cleaned easily by rinsing with solvent and are reusable over years without losing their activity.

Desorption of headspace components from the filter needs a special device (see Fig. 6B) which allows a certain amount of solvent (between 20 and 100 μl

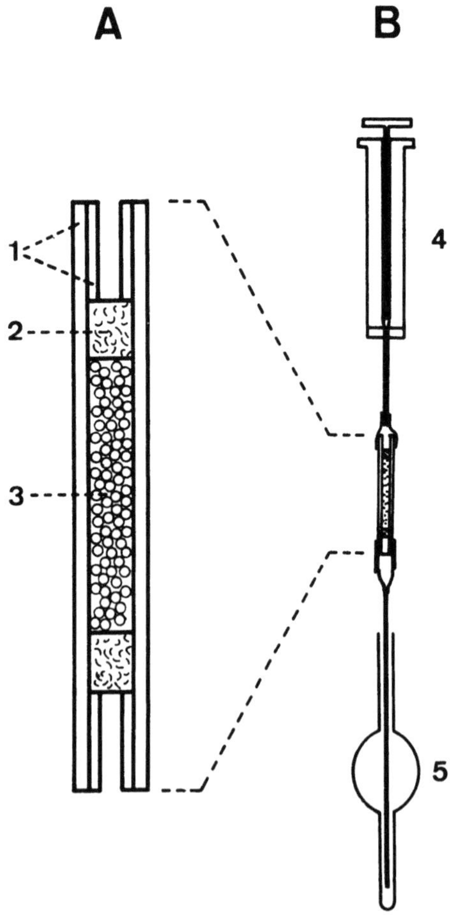

Fig. 6. Headspace microfilter. A, Construction; B, solvent extraction device. 1, Glass capillaries; 2, glass wool; 3, Porapak Q; 4, syringe for solvent injection; 5, ampoule for collecting and mixing the filter extract.

depending on the filter dimensions and on the detection sensitivity requirements) to be pressed very slowly through the filter. If only an aliquot is used for analysis, the volume of the resulting extract, which it is important to know, can be determined gravimetrically. A second extraction of the filter with the same or another solvent makes it possible to check whether quantitative desorption was achieved. Attention must also be paid to the homogeneity of the extract, if only a part of it is used for analysis. Reliable quantitative results will be obtained only after thorough mixing.

For **quantitative determination** of the headspace composition, high-resolution gas chromatography is the method of choice. Depending on the detection sensitivity required in the respective application, a smaller or larger aliquot of the total extract can be injected onto the capillary column using on-column injection techniques or, in the case of larger extract aliquots, special injection techniques.[86,87] Quantitation of the single components present in the headspace filter extract is performed by the external standard calibration method. Their concentration in the original headspace can be calculated (in terms of nanograms/liter) from the headspace volume sucked through the microfilter and the aliquot of the filter extract injected into the gas chromatograph.

Sensitivity and Reproducibility of the Method

Sensitivity of the method depends on various experimental conditions such as the headspace sample volume, the volume of the filter extract, the aliquot of extract used for analysis, the type of capillary column and the type of detector. As a typical result for routine analysis (1 liter headspace trapped on the microfilter, desorbed with 20 μl solvent, from which 2 μl are injected into the gas chromatograph equipped with a flame ionization detector), the presence of single components down to a level of 1 ng per liter headspace can normally be detected without problems. For higher detection sensitivity, special conditions are required.

Another criterion for the efficiency and reliability of the method is how reproducible the measured data are. This again depends on the experimental conditions used in the respective application. In favorable cases (e.g. perfume mixture, 1 liter headspace, 20 μl extract volume, 2 μl injected into the gas chromatograph corresponding to an absolute amount of not less than 50 ng for single constituents), relative standard deviations were found to be about 5%. However, in less favorable cases (e.g. 10 liter headspace volume from dry laundry, absolute amounts of heavy-volatile fragrance constituents around 1 ng/liter) higher relative standard deviations up to 30% must be expected. This, however, is still far better than the standard deviation (reproducibility) of intensity data obtained with sensory methods.

With regard to odor measurement, it is evident that even the highest detection sensitivity which can be performed with headspace analysis today may not be sufficient if an odorant with extremely low odor threshold (e.g. below 0·1 ng/liter) is present in the headspace at concentrations below this level. For such a case, there is still a deficiency in detection sensitivity of the headspace-analytical approach by a factor of at least 10–100 compared with the sensory approach.

Olfactometric Techniques

Dravnieks, one of the pioneers in olfactometry, defined an olfactometer as 'an instrument for preparation and delivery of an odor stimulus to a chemoreceptor system'.[88]

The first olfactometer was already built in 1895 by Zwaardemaker. Since that time numerous olfactometers have been developed. Almost every scientist active in odor perception research built his own instrument. A description of the most typical instrumental designs can be found in reviews published by Moncrieff[35] and by Dravnieks.[88]

Whereas in the earlier period, up to the 1970s, the development and use of olfactometric techniques was predominantly the domain of scientists in chemoreception and psychophysics, in the 1980s especially people working in the air pollution control area became involved. They contributed considerably to the standardization of olfactometric equipment and procedures.[89–93] The method was also widely used in flavor research,[24,45,72,94,95] whereas its application in fragrance research was rather modest.

The many instrumental and methodological aspects of olfactometric techniques will not be discussed exhaustively since this has already been done at full length by other authors.[88] The focus of the following section will rather be directed on selected areas with special interest for odor threshold measurement. The discussion begins with a brief review of the basic principles of sample generation and handling, followed by a critical examination of some aspects directly connected with the quality of olfactometric results. Finally, the instrumental design of a state-of-the-art olfactometer that the authors developed and used successfully for odor threshold determination over many years, will be presented.

Basic Principles of Instruments and Methods in Olfactometry

According to the definition cited above, an olfactometer has to generate an odorous air sample, to dilute it with odorless air, and to present the diluted air samples under controlled conditions to a panelist whose response with regard to odor intensity perception is recorded.

If the test sample is not yet an odorous gas, but a liquid or solid, the first step is the **generation** of a concentrated odorous air sample. This can be done, in analogy to headspace sampling, either by a static or by a dynamic procedure.

The static procedure uses a closed container into which either a known amount of an odorant is completely evaporated by heat or where a liquid or solid sample of a pure odorant or an odorant mixture is kept in the phase equilibrium state to reach vapor phase saturation. If the vapor pressure of the odorant is known, its concentration in the vapor phase can be calculated as outlined in Section 1.

In the dynamic approach the odorous air sample is obtained by allowing carrier gas (air or nitrogen) to pass through or over the sample, thus generating an odorant-saturated air flow. This is the method most frequently used. To achieve a completely saturated vapor phase, special conditions concerning the flow rates of the carrier gas and the sample distribution on a large surface in the sample generator are required.[88]

Dilution of the odorant-saturated air sample can be achieved in two ways, either by transferring a certain volume of saturated vapor into a fixed volume of odorless air (batch dilution) or by mixing a well-defined flow of odor-saturated air with a well-defined flow of odorless air (dynamic dilution). The best way of generating samples of different dilutions by the dynamic method is to keep the flow of the diluting odorless gas constant and to add different flows of the odorous gas (with defined odorant concentration) by means of an adequate splitter device. A variant of this principle that avoids the use of valves within the instrumental segments in contact with odorous air—thus reducing the risk of adsorption effects—was first introduced by Dravnieks in 1975.[88] Today, dynamic dilution olfactometers of this type have become the standard instrumentation for precise odor intensity measurement.

For **presentation** of an odorous sample to the panelist, a large variety of different delivery systems have been developed and only a few can be mentioned here.

The simplest device, often used in combination with static sample preparation and dilution, is a closed flask which has to be opened by the panelist shortly before sniffing.

The presentation device can also be a hood around the subject's head or even a specially ventilated room to which a constant flow of odorous air is delivered.

In some studies a direct leak-tight connection is used between the olfactometer and the subject's nose by inserting tubes into the nostrils or placing the nose in a glass mask.

In the past two decades, especially for odor threshold measurement, the use of sniffing ports or sniffing masks delivering sufficient air flow to allow normal sniffing without mechanical stress to the nostrils was preferred. This type of olfactometer/nose interfacing has proved to be a good compromise on the one hand, to guarantee a fairly good control of the odorant's mass flow into the nostrils and, on the other hand not to be too far away from the panelist's natural mode of breathing.

The last step in the olfactometric procedure is the **odor perception testing** with the human subject as detector. Here, the panelist's task can be completely different, depending on whether odor thresholds or odor intensities in the suprathreshold range are to be determined:

—In the suprathreshold range, the panelist has to rate the odor intensity of a test sample within a given numeric scale (magnitude estimation) or he has to allocate its intensity to the most similar intensity of a reference sample within a set of reference samples of gradually differing odor intensities (reference scaling).

—The panelist's task in odor threshold determinations, however, is reduced to a simple 'yes' or 'no' decision. A series of test samples, usually differing by a factor of two in dilution from one to the next sample, are presented to the panelist either in an ascending or descending order of concentration. He has to identify that odor sample which he first or last perceives, depending on whether an ascending or descending order, respectively, is used.

—Without applying special precautions, the reliability of the panelist's answer depends on various psychological factors, such as expectation, imagination, motivation, guessing, and so on.

To reduce such interferences, a special sample presentation technique called 'forced choice triangle test' has been introduced. In this test, one diluted odorous sample and two odorless samples (blanks) are presented simultaneously to the panelist at each dilution level. The panelist has to find out which of the three samples is the odorous one and differs from the two others. If the forced choice triangle test is repeated several times with the same test sample but with different positioning of the odorous sample versus the blanks, the level of confidence for correct answers can be considerably improved.

—Since odor threshold is defined as the odorant concentration at which a panelist just perceives a faint odor in 50% of the trials, one usually determines the panelist's individual threshold concentration by calculating the geometrical mean from the correctly identified higher and of the adjacent lower concentration level at which he failed to perceive the odor.[96]

—From a set of individual thresholds, finally, a mean threshold value representative for the panel as a whole (group threshold) can be derived; this will be discussed later in the context of data treatment.

Today's Principal Instrumental Approaches

Most olfactometers developed during the last decades are based on one of the above-mentioned principles or on a combination of them. Their difference in size and complexity is impressive, ranging from the simplest devices up to rather complicated and huge machines.

A selection of three types of olfactometric equipment which are predominantly in use today will be mentioned here.

In the simplest approach, known as the **'flask or beaker method'**, the sample, usually a solution of odorous material, is transferred into a closed flask or into a beaker covered with a watch glass and after sample equilibration it is presented to a panelist for sniffing. A set of vapor phase dilutions can be easily obtained by adding different amounts of an odorless solvent in known proportions to the original sample.

The method has some serious drawbacks. Precise vapor phase concentrations cannot be calculated from the concentrations in the solution without knowing the odorant's vapor pressure, diffusion rate and activity coefficient in a given solvent, as outlined in Section 1 under 'Factors influencing volatility'. Furthermore, the delivery of the vapor sample to the panelist's nose while sniffing on the stagnant vapor phase in the flask is not well defined.

Because of its simplicity, however, the method is still very popular, especially in flavor analysis of food and beverages where smelling *and* tasting are often encountered.[68,97] Recently, it has also found application in the clinical area as a standardized method to quantify the olfactory defects in patients as a diagnostic tool[98,99] and to control the success of therapeutic measures.

Although the beaker method is generally considered to be less precise, it has also been used in the fragrance field, for instance to measure the psychophysical power function parameters of musk chemicals.[100]

Whenever more precise odor intensity data are required, **dynamic dilution olfactometry** is the method of choice. The test sample can be ambient air, sampled continuously or batchwise from an odor-emitting source, or it can be an odorant-saturated gas delivered from a sample generator. Diluted samples are obtained by mixing variable flows of odorous air into a constant flow of odorless air. The dilution ratios of the diluted vapor samples delivered to the olfactometer/nose interface are usually determined by flow rate measurements. The dynamic dilution olfactometer equipped with an appropriate forced-choice triangle presentation device has become the equipment most widely used in odor threshold measurement today.

The lack of standardization of instrumental parameters and experimental procedures, however, is still a problem and certainly one of the reasons why various attempts to make such an instrument commercially available[101–103] have so far had little success. A state-of-the-art instrument of this type will be described later in this section (see p. 189).

A gas chromatograph equipped with an outlet splitter and a sniffing device represents another, completely different approach in olfactometry. With the enormous evolution of capillary gas chromatography in the past decade, the **GC sniffing technique** has found wide application and must be regarded today as an indispensable tool for every laboratory dealing with odor measurement. Odor threshold measurement can be performed simply by injecting different dilutions of a sample, e.g. in an ascending order of concentrations, into the gas chromatograph until the panelist can perceive the substance at the sniffing port at the time when the detector indicates its elution from the GC column or at the correct retention time if the concentration of the eluting substance is below the detection limit.

Dravnieks & O'Donnell[104] used this technique as early as 1971, as a 'pulse-type' stimulus delivery device for measuring odor thresholds, which they found to be comparable with those determined by classical methods, if the odorant's time/concentration profile in the gas-chromatographic effluent was considered in the calculation. Similar good agreement was also reported by Christoph & Drawert,[95] who applied this technique to the study of odor threshold/structure relations. Only recently the method has also found application in demonstrating the individual variations of odor sensitivity in humans.[105]

One main limitation of the GC sniffing technique is that it cannot be used for thermally unstable substances. Other disadvantages are that application of the forced-choice presentation technique is not possible and that the number of trials must be limited in view of the fact that each stimulus presentation at the same or at different concentration levels needs a time-consuming GC run of at least 10–30 minutes, depending on the odorant's volatility.

On the other hand, the method's advantages must not be underestimated. Well-defined sample dosaging and continuous monitoring of the sample composition guarantee an optimal control on the quantity and purity of the odorant shortly before it enters the panelist's nose. Moreover, minute samples that could never be handled in a classical olfactometer can be investigated by this way.

The complementary character of the GC sniffing versus the classical olfactometric technique also becomes evident when investigating odorant mixtures. Only with the latter, olfactometric data of the mixture as a whole can be obtained, whereas with the GC sniffing technique the odor intensities of individual constituents are recorded separately which, however, cannot simply be summed up to the mixture's integral odor intensity.

Critical Methodological and Experimental Aspects

Considering the large variation of threshold values published by different authors for one and the same odorant using different methods and instruments—some of them differ from each other by several powers of ten—the many problems and pitfalls encountered in olfactometric measuring techniques become evident.

Such differences cannot be attributed solely to the differences in the individual sensitivity of the panelists. Inadequate concepts, instruments and non-standardized experimental conditions and procedures in use are the major causes for the observed variations. It is impossible to discuss all the different sources of error and the factors which impair the consistency of olfactometric data. Most of them have already been described in detail by various authors.[96,106] Further discussions will therefore be restricted to some special aspects to which—in the authors' opinion—more attention should be paid.

The **odorant's purity** plays an important role in the study of odor/structure relations, for instance. Wherever odor quality and intensity of an odorant are to be related with its defined chemical structure, a check for its purity by qualitative and quantitative analysis is mandatory.

However, a liquid or solid sample often seems to be fairly pure, whereas its vapor phase contains considerable amounts of impurities, especially when they are more volatile than the pure odorant. The method of choice for purity testing is therefore headspace sampling and analysis by capillary gas chromatography. The gas chromatograph should be equipped with an outlet splitter and sniffing device, since only by sniffing each peak eluting from the gas chromatograph is one able to check whether or not one of the present impurities contributes to the typical odor of the whole sample. If not, there is no need to worry about a higher analytical purity.

This 'olfactory purity testing' should always be done prior to the start of olfactometric measurements. Some examples of how the method can yield rather unexpected and surprising results in practice will be shown in Section 3 (see p. 195).

Incorrect determination of the **concentration** of the odorant at the olfactometer outlet is another source of error for olfactometric data. Usually, the concentration is calculated from the odorant's vapor pressure and from the dilution ratios that were applied in the olfactometer, assuming that vapor saturation is achieved in the sample generator.

Incorrect vapor pressure data from the literature, incomplete saturation of the odorant in the vapor phase, or incorrect dilution ratios, which are usually based on flow rate measurements, directly impair the precision of calculated concentration values.

One of the most severe problems in olfactometry in this context is the odorant's **adsorption** to the surfaces of the inner walls of the olfactometer. Even if the main sources of strong adsorption, such as valves, greased joints, sealing gums, plastic tubes, oily films, organic residues and so on, are eliminated and only inert and clean materials are used, the phenomenon of adsorption can never be completely excluded. There is always a dynamic process of adsorption/desorption at the surface which is in contact with odorous air. In the state of equilibrium, a certain fraction of odorant is always adsorbed on the surface. As soon as the odorant concentration in the dilution device is changed from higher to lower levels, for instance, a certain amount of odorant is desorbed from the surface into the air until a new state of equilibrium is reached. Therefore, prior to the equilibrium state the true concentration in the olfactometer effluent is somewhat higher than calculated from the dilution ratio. The same holds true in the opposite direction, if the odorant concentration is changed from a low to a higher level. Adsorption may then lead to temporarily lower levels than the calculated ones.

The only way to become truly independent of the many factors which can cause incorrect assumptions about the actual odorant concentrations at the olfactometer outlet is **direct quantitative analysis** of the **olfactometer's effluent** at the sniffing port. This proves not always to be a simple task when we consider the very low concentrations of some odorants at the threshold level. However, trapping and sample enrichment at the olfactometer outlet with subsequent quantitative headspace analysis, as described earlier in this section, is certainly the best method of solving the problem. Furthermore, it indicates whether the olfactometer operates correctly or whether anything is going wrong. This can be seen directly from the degree of conformity of measured data with those calculated on the basis of dilution ratios. The present authors implemented this method of analytical control at the olfactometer outlet in their standard procedure for odor threshold determinations more than 10 years ago and are still convinced that it is indispensable.

Among the factors influencing the results of human **intensity perception**, some are related to the experimental conditions of stimulus presentation (e.g. flow rates, sequence of concentration changes) and others to the size and quality of the panel or to the data handling and statistical treatment. Improved recommendations have recently been formulated by Hangartner.[93] As to the **flow rates**, it is widely accepted that they should be as close as possible to the conditions of normal breathing. Inhalation rates for humans are reported to be at about 30 liter/min,[107,108] and to reach maximum values up to 44 liter/min.[109] It has been clearly demonstrated that flow rates influence odor intensity perception at suprathreshold as well as at threshold levels.[96,106] This shows the need for standardization. For dynamic dilution olfactometers where the odorous air is presented to the panelist via a set of masks or funnels using the forced-choice triangle technique, a flow rate of about 10–15 liter/min is recommended.[106] It should be large enough to bathe the nose with an ample flow without disturbing the panelist by too high a flow velocity. Flow rates that are too low lead necessarily to an additional dilution of the presented sample flow with ambient air to achieve a rate up to that corresponding to the panelists' normal sniff volume. In this case, the odorant concentration in the air flow entering the panelist's nose is lower (more highly diluted) than the originally adjusted value at the olfactometer outlet. Therefore, threshold determinations carried out with too low sample flow rates necessarily lead to higher threshold values than with high flow rates.

In our experience, flow rates around 10 liter/min, or even slightly below that, are more convenient and more readily accepted by most panelists than higher flow rates. They usually feel less exposed to a stressing situation and nearer to normal conditions of smelling. However, in calculating threshold values, the additional dilution by ambient air required to allow the panelist to sniff at his normal inhalation rate (about 30 liter/min) must be considered. Therefore, flow rates should always be specified; otherwise threshold data which were determined under different conditions cannot be compared with each other.

The **mode of stimulus presentation** in threshold measurement is a subject of controversial discussion. The different dilution steps can be presented in ascending, descending or random order. The adaptation problem, due to desensitization of the panelist with a strong sample prior to a weak sample, is the argument most often cited in the literature against the descending or random order. Random presentation may have the additional disadvantage of requiring more experimental runs compared with ascending or descending presentation order. Therefore, the method of stimulus presentation in ascending order is generally recommended.[54]

However, in the authors' experience, more consistent threshold data can be obtained when a descending rather than an ascending presentation order is used. It turned out, for instance, that many panelists perceived indistinct odors, such as woody notes, rather badly when they started with subthreshold concentration levels. The advantage of starting with higher concentrations is that the panelist can memorize the odor which he must subsequently detect in the rather difficult range near the threshold concentration.

The risk of adaptation effects related to the descending method can be minimized when sniffing tests are started with concentration levels not higher than about six to seven binary steps above the group threshold level, and when the concentration levels presented to the panelist are reduced faster than by the factor of only two in the upper range where odor perception poses no problem. The risk of adaptation is further reduced when the forced-choice triangle technique is applied where the panelist's nose is more often exposed to odorless than to odorous air samples.

Furthermore, the forced-choice triangle sample presentation technique can be arranged as a 'double-blind test', where neither the panelist nor the panel-leader knows the correct position of the odorous sample within the triangle. In this manner, even an unconscious influence from the panel-leader to the panelist can be entirely eliminated.[110]

The **composition of the panel** may have an even greater influence on the olfactometric results than presentation techniques. To summarize some of the authors' experience in this area: to obtain consistent data, the panelist's motivation, for instance, was found to be more important than his level of training. He should be interested in his work and he should be allowed to repeat a trial when he feels he could do better. The panelist should possess normal acuity to odorants and an average physical and mental health which in our experience is more relevant to odor perception performance than the frequently cited factor, age.[105]

Wherever the results of odor measurement are finally intended to be used for developing new fragrances or perfumes, the panel should reflect the average of the whole population. Therefore a well-balanced team with respect to sex and age should be aimed at. Anosmic panelists, however, must be excluded.

As to the panel size, from the statistical point of view a large panel of more than 20 people is desirable. However, a large panel requires more trials and more experimental work, which should also be taken into consideration. A panel size of between 10 and 20 individuals was found to be an acceptable compromise for odor threshold measurements.

The last step in odor threshold determination is **gathering raw data** from the individuals of a panel, and **data treatment**.

In the literature, numerous procedures are described which indicate how to gather olfactometric raw data and how to calculate individual or group threshold data. Some of them have been compared in detail by Dravnieks *et al.*[111]

One of the general problems is that different procedures yield different results. Thus, very often published threshold data cannot be compared with each other unless precise specifications on the measuring procedure and data treatment are given, allowing some sort of conversion.

A further problem is how to select from among the known procedures the one yielding correct and reliable results at reasonable expense (limited number of experiments; simple data handling and treatment).

As mentioned above, the threshold of an individual is defined as the concentration being perceived in 50% of the trials within a series of stimulus presentations. Some of the methods for calculating the group threshold are based on the assumption that the percentage of the population showing a positive response with increasing stimulus concentration follows a normal distribution when concentrations are indicated in logarithmic units. However, as can be seen from the histograms presented in Fig. 8 (p. 193), this assumption does not always hold true.

Because of the non-ideality of the response distribution and for some other statistical reasons whose explanation would exceed the frame of this chapter, we favor the standard data treatment procedure recommended in the guideline VDI 3881,[92,112] with some minor modifications. The main difference between the two methods lies in the determination of the subject's threshold.

Both methods use the forced-choice triangle technique with stimulus presentation in binary steps of concentrations in ascending and descending order, respectively.

In the procedure recommended in the VDI guideline, each concentration level within a test cycle of sample presentations in an ascending concentration order is tested only once, and the test cycle is repeated three times. A test cycle is terminated as soon as the panelist has correctly distinguished the stimulus from the blanks at two consecutive concentration steps. The mean value between the first correctly identified concentration and the concentration below is supposed to be the subject's threshold for that respective test cycle. The probability that the panelist has produced 'correct' answers by guessing and that his 'true' threshold lies several concentration steps higher, is 1:9. This, of course, can have a considerable influence on the subject's calculated mean threshold value. Therefore, the panelist's response data should at least be corrected for guessing.[93]

In our procedure, the subject's threshold is defined as the lowest concentration within a cycle of forced-choice triangle presentations in a descending order of concentrations where the panelist has given three correct answers within three consecutively repeated trials at the same concentration level. The probability that the panelist's true threshold lies one or two concentration steps higher is 1:27 or 1:729, respectively, and its influence on the final results can be neglected.

In both procedures, group threshold values are derived from the individual thresholds by calculating their geometric mean, the standard deviation and the confidence interval, as described in the VDI guideline mentioned above. The data obtained from the former procedure (according to the VDI guideline) represent 50-percentile values, those from the later one 95-percentile data. Since binary concentration steps are used in both methods, they differ theoretically by a factor of $\sqrt{2}$ and can easily be converted into each other.

Data treatment should also include calculation of other means such as the median and the mode since, in addition to the geometric mean, they give a more realistic picture of the statistical distribution of individual response characteristics of the panel for a given odorant.

A State-of-the-Art Olfactometer

It is primarily the olfactometer's **instrumental design** which determines the reliability of experimental results in odor measurement. A modern state-of-the-art instrument must meet a number of requirements. The most important ones are:

—Adsorption/desorption effects within the odorant-containing flow lines must be minimized and all parts should be easy to clean.
—The dynamic range of vapor phase dilution must be large enough to cover the whole range of fragrance raw materials showing large differences in volatility and odor strength.
—The instrument must allow the analytical control of the odorant's real concentration at the sniffing port.
—The olfactometer sniffing device must be equipped in such a way that forced-triangle presentation technique combined with fully automated random channel settings is possible.

An instrument that meets the above-mentioned requirements is shown schematically in Fig. 7. After being saturated in the sample container (1), a

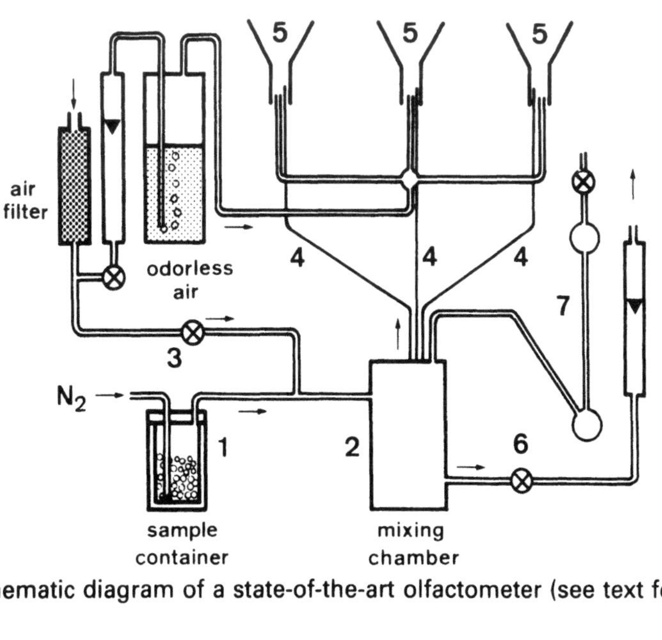

Fig. 7. Schematic diagram of a state-of-the-art olfactometer (see text for details).

flow-regulated stream of nitrogen at between 1 and 1000 ml/min is fed to the mixing chamber (2) where it can be diluted with up to 10 liter/min of filtered air via a valve (3). From the mixing chamber one part of the diluted odorous air is allowed to flow via a set of three identical steel capillaries (4) to the three sniffing funnels (5) and another part is passed via the valve (6) to the ventilation. The flow rate through the capillaries, which determines the dosage of odorous air from the mixing chamber into the sniffing funnels, depends on the one hand on the capillaries' diameter and length, which are kept constant, and on the other hand on the degree of opening of the valve (6). The latter is variable and allows regulation of the pressure in the mixing chamber and thus of the flow rates through the capillaries. Flow rates can be directly read from the pressure measuring device (7) that is connected to the mixing chamber and calibrated in flow rate units.

Special attention was paid to the following instrumental features:

—The final dilution of the odorous air sample occurs in the glass funnels by flushing them permanently with odorless air at a flow rate of 8 liter/min. The flow of odorous air into each of the funnels can be switched on/off without using valves and without interruption of flow by making use of a special technique.[113] By pressing a button, the capillaries delivering odorous air to the funnels can be moved pneumatically from position 'on', where their endings are positioned inside the funnels, to position 'off', where the endings are positioned outside the funnels and where their effluents are sucked away.

—To keep adsorption/desorption effects at a minimum, plastic tubes and valves are avoided. Only stainless steel or glass is used for all parts of the olfactometer, including fittings which are in contact with the odorous air sample. Unlike most conventional olfactometers, the stimulus concentration within the olfactometer during a threshold measurement session remains constant since final dilution for sample presentation to the panelist's nose is performed by dosing different amounts of odorous air in binary steps from 1 to 256 ml/min directly into the sniffing funnels where adsorption/desorption effects are negligible.

The olfactometer can be easily cleaned of a previously measured odorant by dismantling those parts of the olfactometer which were in contact with the odorant and by heating them with a burner or in an oven.

—With a minimum dilution ratio of 1:32 and a maximum at 1:8 000 000, the dynamic range of the olfactometer is 1:250 000. An extension to a virtually unlimited dynamic range is possible, simply by cooling the sample container with a thermostat, thus lowering the vapor pressure of the odorant.

—The concentration of the odorant in the sniffing device is measured analytically by sampling a known volume from the capillary effluent into a headspace filter according to the method described earlier in this section. Thus, the odorant's actual concentration at the sniffing port is determined.

—Forced-choice triangle presentation is achieved by a special automated channel setting device where only one of the three glass funnels is connected with the odorant-delivering capillary, whereas the other two are flushed with odorless air only and serve as 'blank' channels. By pressing a button the channel setting is changed automatically at random in such a way that neither the panelist nor the panel-leader knows the new position of the stimulus-containing funnel relative to the blank funnels. Thus, forced-choice triangle stimulus presentation under true double-blind testing conditions is possible. Whether the panelist identifies the odorant-delivering channel correctly or not is indicated automatically by lamp signals upon pressing one of the buttons assigned to each channel.

The **standard procedure** for threshold determination can be subdivided in four steps: olfactory purity testing of the sample; sensory testing at the olfactometer with the panel; determination of the threshold concentration by headspace analysis; and data treatment.

—As outlined above, olfactory purity testing is indispensable prior to starting any work with the olfactometer. It begins with collecting headspace from the pure odorant and determining the odorant's vapor pressure by injecting a known amount of headspace into the gas chromatograph. By sniffing the eluting peaks at the GC outlet, the presence of odor-relevant minor or trace constituents can be detected. If this testing is done stepwise with increasing dilutions of the headspace sample, down to the concentration level where none of the headspace constituents can be perceived any more, a rough estimation of the threshold values of the odorant and of the impurities is obtained.

—After his approximate sensitivity has been tested by the panel leader, the panelist starts sniffing at the olfactometer at a concentration level at which he perceives the odorant at medium intensity. After three correct answers in three consecutive trials at the same level, stimulus concentration is decreased by a factor of two to the next lower level, and so on, until the panelist has reached his threshold level.

—As soon as the panel session is terminated, the sample flow in the olfactometer is set to a concentration level nearest to the panel's geometric mean, and the real concentration in the olfactometer effluent is measured by headspace sampling from each of the three capillaries. If the measured values differ by more than 50% from the theoretical value that has been calculated from the odorant's vapor pressure and the dilution ratio in the olfactometer, the measurements are repeated to check for experimental errors.

—The group threshold is calculated according to the procedure described above including standard deviation, confidence limit, median and mode.

It is rather futile to discuss the method's **accuracy or reliability** by comparing, for instance, the threshold values obtained with the method described above with those that have been published. It is a well-known fact that odor threshold values reported in the literature for one and the same odorant sometimes differ from one another by a factor of several powers of ten.[18a] As long as different instrumental parameters, experimental procedures and methods of data treatment are used in olfactometry, any conformity among data must be considered to occur more or less by chance. Therefore, the only way to judge the reliability of a method is to check the reproducibility of data received from repetitive measurements.

In Table 1, individual thresholds and group threshold data of valeric acid are presented. These were obtained from six repeated panel sessions carried out over a period of two months with two individuals A_1 and A_2 (Table 1(a)) and with whole panel A (Table 1(b)). It reveals that a subject's individual threshold can vary from one session to another (intra-individual variations) by a factor of up to 2 (corresponding to one dilution step) from the calculated geometric mean $\bar{x}_g$.

The variation of group thresholds within panel A, which is indicative of the reproducibility of the method, is relatively small, as can be seen from Table 1(b). This group threshold differed only slightly when the measurement was repeated one year later at another location, even with another panel (panel B) and another panel leader, using another olfactometer (but of identical design) and different analytical equipment for the determination of the concentration of the odorant.

Table 1
Individual and group threshold data of valeric acid

(a) Individual variations

	Individual threshold values from six sessions (ng/l)						*Mean $\bar{x}_g$ (ng/l)*	*Log s*	*95% confidence limit (ng/l)*
Subject A_1	0·27	0·40	0·69	0·11	0·39	0·36	0·32	0·27	0·17–0·62
Subject A_2	0·55	0·20	0·69	0·44	0·39	0·18	0·36	0·24	0·20–0·64

(b) Group threshold variations

		Group mean $\bar{x}_g$(ng/l)	*Log s*	*95% confidence limit (ng/l)*	*Median (ng/l)*	*Mode (ng/l)*	*Number of panelists*
Panel A	session 1	0·21	0·32	0·11–0·40	0·19	Trimodal	12
	session 2	0·23	0·25	0·11–0·47	0·20	Trimodal	5
	session 3	0·20	0·32	0·13–0·33	0·17	0·17	14
	session 4	0·20	0·31	0·13–0·30	0·22	0·22	14
	session 5	0·15	0·45	0·08–0·30	0·20	Bimodal	12
	session 6	0·26	0·22	0·18–0·40	0·18	0·18	9
Panel B	(1 session)	0·28	0·54	0·18–0·42	0·24	0·12	37

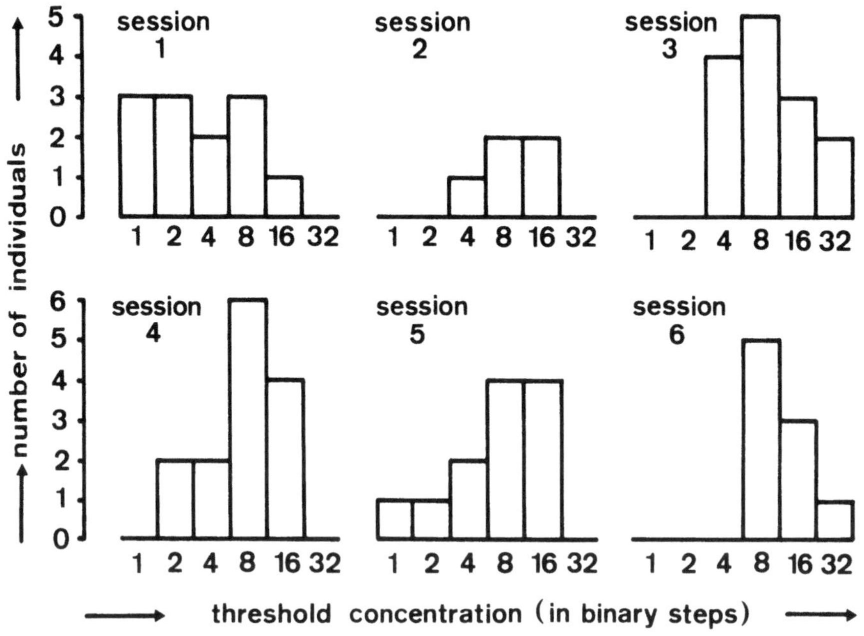

Fig. 8. Panel distribution histograms representing the number of individuals whose threshold values were found to be at one of the six concentration levels when presenting valeric acid in binary steps from 1 to 32 to the panel A during six consecutive panel sessions (refer to data in Table 1).

The importance of panel techniques and statistical treatment of olfactometric data becomes clear when one is considering the large differences of individual thresholds (inter-individual variations) within a panel. The distribution of individual threshold responses over the range of single concentration levels in the course of six subsequent sessions with panel A is shown in a histogram (Fig. 8). Furthermore, it shows that the characterization of an odorant by the shape of its panel distribution must be handled with care.

The significantly larger standard deviations obtained for the measurements from panel B compared with those from panel A may be explained by the fact that panel B was a newly created team with no experience in threshold measurement either on the panel's or on the panel leader's side, whereas the members of panel A were experienced and well motivated to give their best.

The data presented demonstrate, however, that fairly consistent and reproducible threshold data can be obtained with adequate olfactometric techniques, even at different places with different panels and instruments, if at least the instrumental design, the experimental procedure and the method of data treatment are strictly standardized.

3 PRACTICAL APPLICATIONS

Headspace and olfactometric techniques have been applied for many years to measure all kinds of odors, especially in the area of environmental protection and in flavor and food analysis, as can be seen from various reviews.[75,94,114–116]

Comparatively little has been reported on the use of these techniques in the fragrance field and the few publications are almost exclusively restricted to headspace measurements for investigating, for example, the behavior of fragrance constituents in soap,[117,118] or in detergent powder,[119] and their emanation from the fabric after washing,[84] from cosmetic products[120,121] or from air fresheners.[122] Since the early 1980s, headspace techniques were also used extensively for exploring and discovering the secrets of natural scents, a fascinating topic that will be treated thoroughly in Chapter 7. Reports on practical applications of olfactometric techniques and the use of odor intensity characteristics in the fragrance field are almost non-existent, although as early as 1970 Appell published a book entitled *Physical Foundations in Perfumery*,[123] in which he gave a comprehensive survey of how physico-chemical and sensory methods can be used in the perfumer's daily work. Since that time, only a few papers have been published dealing with practical applications, for instance of intensity power function data[124] and odor values[125,126] of fragrance chemicals, or dealing with quantitative measurement of sensory and physical parameters to be used for developing fragrances with functional properties.[127]

However, this does not at all reflect what is actually going on in some specialized laboratories of the fragrance industry, where the importance of modern techniques for quantifying odors was recognized long ago.

In the late 1970s the authors had already started to complement pure headspace measurements by olfactometric techniques and they began to explore—in close cooperation with professionals on the perfume creation and application side—how an integral approach based on physico-chemical and olfactometric methods could be applied in different areas. The main features of this concept and some practical examples illustrating its usefulness will be the subject of this last section.

Data Collection of Odor-Relevant Measurable Parameters

Characterization of fragrance raw materials by measuring and collecting their physico-chemical and olfactometric data was recognized to be—after method development—the second, but equally important, step in creating a reliable basis which makes it possible to learn more about fragrance behavior in different applications.

Therefore, vapor pressures, water solubilities, odor thresholds and odor values of more than 800 fragrance chemicals, most of them representing raw materials from the perfumer's shelf (the rest being research chemicals), were determined over a period of many years.

Numerical values of the lower and upper limits for each of these parameters, taken from the above-mentioned, fairly representative, data collection, are shown in Table 2. They illustrate the large dynamic range in each category of data and thus their potential for quantifying the odor-relevant properties of fragrance raw materials.

Table 2
Odor-relevant measurable parameters

Measurable parameter	*Measuring method*	*Approximate numerical range (lower–upper limit)*[a]	*Practical application*
Vapor pressure	Quantitative headspace analysis	0·05–50 000 µg/liter 1 : 10^6	Measure for fragrance volatility, diffusivity
Odor threshold	Olfactometry	0·002–2 000 ng/liter 1 : 10^6	Indication of concn below which odor contribution is negligible
Odor value	Calculated from headspace concn and threshold concn	100–10 000 000 1 : 10^5	Approximate numerical measure of an odorant's odor intensity and substantivity
Water solubility	Direct measurement of aqueous solution with special GC technique	Low ppm–totally miscible with water	Indication of the odorant's behavior in aqueous or water-related media

[a] Representing average values of lower and upper limits found for frequently used chemicals—figures taken from Givaudan's data collection of more than 800 fragrance chemicals.

These parameters, for instance, can be used as a **screening method** in fragrance research, allowing objective comparison, on the basis of quantitative data, of new research chemicals with old ones or with existing fragrance chemicals on the perfumer's shelf. Thus, within a given class of odor notes, the chemicals that are more potent with regard to their odor impact can be recognized faster and more reliably, and further product development can be concentrated on fragrance chemicals with a better chance of becoming a 'hit'.

A further application field is fragrance creation, in which these data are used for **classifying fragrance raw materials** with respect to their odor-relevant properties, such as volatility, diffusivity, odor strength or substantivity in aqueous, non-aqueous or other specific applications. Perfumers have quickly learned to deal with such data and there is no doubt that this approach facilitates the perfumer's difficult task of finding not only an optimal solution for the artistic aspects of perfume creation, but also of coping with the difficult technical problems of fragrance performance sometimes encountered in special applications. Practical examples will be given later.

Olfactory Purity Testing

Olfactory purity testing provides another means of characterizing fragrance materials, delivering valuable information on their odor properties not only with regard to intensity but also to quality.

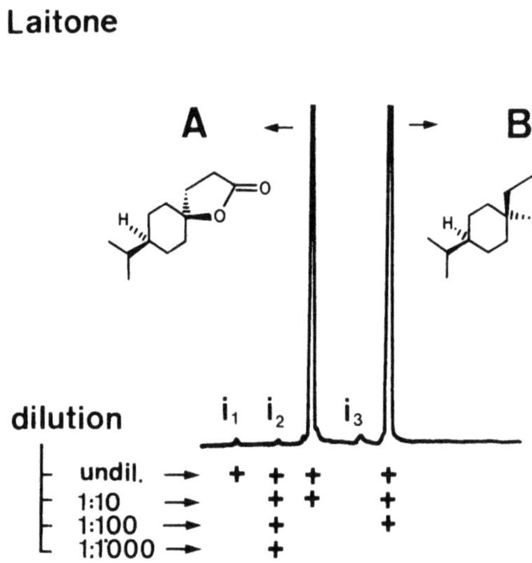

Fig. 9. Analytical and olfactory purity of Laitone® characterized by its gas chromatogram and by the results obtained with the sniffing technique. The number of + signs below each peak in the gas chromatogram indicates how many times the original headspace sample could be diluted by a factor of 10 without losing the distinct odor perception for that respective component. At the dilution level 1 : 1000, for instance, the trace component i_2 was no longer detected by the GC detector but it was still clearly detectable with the nose at the correct retention time. Olfactory purity testing was carried out independently with three panelists.

From the analytical point of view, a fragrance chemical may appear to be fairly pure but must possibly be considered impure if impurities which contribute significantly to the product's odor are present in the headspace.

An example is given in Fig. 9, which represents the gas chromatogram of Laitone®, a new fragrance chemical with an interesting lactonic, creamy and floral odor note.[128] Although the liquid consists of 99·7% of two stereoisomeric forms A and B, olfactory purity testing reveals the presence of two impurities, i_1 and i_2, at the 0·1% concentration level which contribute significantly with a similar odor note to the product's total odor. The relative importance with regard to their odor impact becomes evident when olfactive purity testing is repeated with stepwise diluted headspace samples of Laitone®. Whereas at a dilution of 1:10 impurity i_1, at 1:100 stereoisomer A, and at the next dilution step also stereoisomer B, cannot be perceived any more, impurity i_2 is still perceptible and becomes odorless only after further dilution (1:10 000) of the original sample.

By micropreparative isolation and structure analysis the structure of this highly active impurity was found to be the ethyl analog of Laitone®. The chemical was synthesized and after product development promoted to a new fragrance chemical. It turned out to be an interesting variant of Laitone®, thanks to its similar but not identical odor note and its considerably higher odor impact in comparison with the original product.

Another impressive example may be mentioned here which demonstrates that not only the odorant's stereoisomeric purity but also its chirality has to be taken into account when talking about olfactory purity. This is of special importance in exploring and establishing precise odor/structure relations.

In the course of the routine screening of research products we recently tested a product with a pleasant and powerful woody amber-like note for its analytical and olfactory purity. The sample, a crude reaction product obtained from a multistep chemical synthesis, turned out to be a complex mixture of various constituents with closely related structures, as can be seen in Fig. 10. Olfactory purity testing revealed that one of the constituents (peak P_4) determines predominantly the odor of the sample. After micropreparative isolation and structure analysis, it became clear that the constituent in question is a pure stereoisomer with a defined structure representing a racemic mixture of two enantiomers. In view of the component's extremely low odor threshold (around 3 pg/liter air) we wondered whether the threshold values of the two enantiomers would be different. Complete separation of a purified sample into

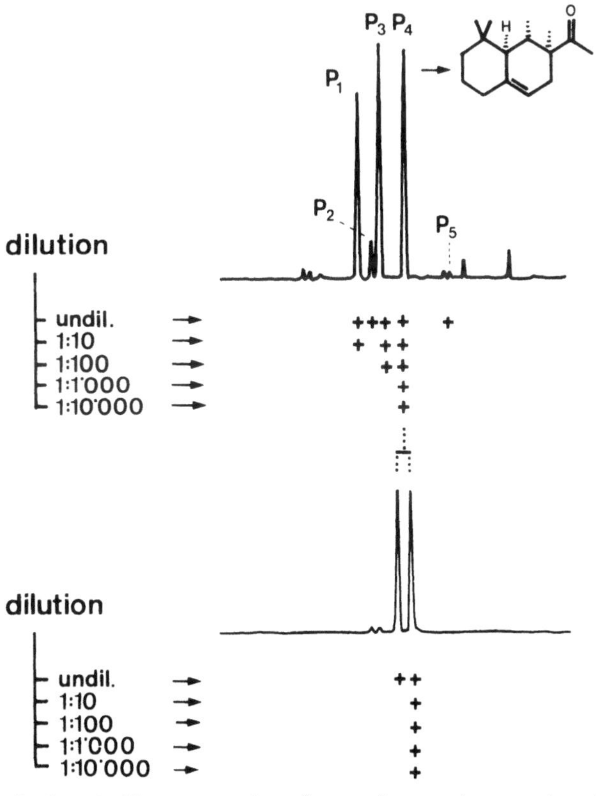

Fig. 10. Analytical and olfactory purity of a crude reaction product from chemical synthesis separated by gas chromatography on DB-WAX as liquid phase (upper chromatogram) and of a purified sample of stereoisomer P_4 separated on a chiral phase (permethylated β-cyclodextrin dissolved in OV-1701) into the enantiomers (lower chromatogram).

its enantiomers on a chiral phase (see Fig. 10, lower part) in combination with olfactory purity testing revealed that only one of the enantiomers is virtually responsible for the odorants' extremely high odor impact, its odor threshold concentration being at least 10 000 times below that of the other enantiomer. Similar differences in the odor thresholds were also found for the enantiomers of the structurally related components P_1 and P_3 of the mixture shown in the upper chromatogram of Fig. 10, indicating that such a high enantioselectivity of odorants—at least in certain structural classes—may not be the exception but rather the rule.

In the literature, differences in odor quality and odor intensity have been described for the enantiomers of various odorants.[129–132] The above-mentioned example, however, represents one of the few cases, to our knowledge, in which such a striking difference in the threshold value of highly pure enantiomers has been quantitatively detected.

The list of examples could be continued *ad libitum* with other well-known fragrance chemicals. The examples mentioned, however, may already illustrate sufficiently the importance of olfactory purity testing in different areas:

—The method turns out to be a rich source for discovering new chemicals with high odor impact that make it possible, for instance, to reduce the dosage of fragrance chemicals in a perfume formulation and in a consumer product without necessarily losing odor intensity.

—Wherever odor quality or intensity data from existing fragrance chemicals or new research chemicals are intended to be used for deriving odor/structure relations, olfactory purity testing is mandatory.

—Quality control is another application where the method, in combination with quantitative headspace analysis, can contribute to objective measurement of the sensory difference between two qualities of the same product or to overcoming an off-odor problem occasionally appearing in the chemical production or in the stability testing of fragrance materials in a consumer product.

Odorant/Matrix Interactions

As outlined in Section 1, the degree to which single odorants of a perfume emanate from the fragranced product into the surrounding air depends not only on their vapor pressures and concentrations in the product, but to a great extent also on the nature of molecular interactions between the odorant molecules in the perfume mixture and even more on the interactions between the odorants and the components of the product matrix.

Since in most cases it is not obvious whether the perfume is really dissolved in the matrix, whether it is part of an emulsion or whether it is simply deposited as droplets or as a liquid film on the surface of solid matrix particles—very often the real status of the perfume/matrix mixture is somewhere in between—any theoretical approach to calculating the odorant's concentration in the headspace is extremely difficult, if not impossible.

However, quantitative information on how the perfume in a fragranced product is influenced by the product matrix can be obtained experimentally from headspace measurements. In a systematic study, we recently tested a great number of fragrance chemicals with different structures and physicochemical properties in a series of various products, perfume formulations, different sorts of creams, detergent powders, softeners and so on. Some of the results are shown in Fig. 11, in which measured headspace concentrations of six selected fragrance chemicals, present in equal amounts in the model mixture and in the different products, are represented by a histogram arranged from the left to the right according to the decreasing vapor pressure of the odorant. The figure clearly demonstrates that fragrance emanation from the perfume mixture decreases almost linearly with the vapor pressure of the odorant. In the case of fragranced products, however, this relationship seems to be superimposed more (detergent powders and softeners) or less (creams) by specific odorant/matrix interaction effects, which in part can be correlated, for instance, with the water solubilities of the odorants.

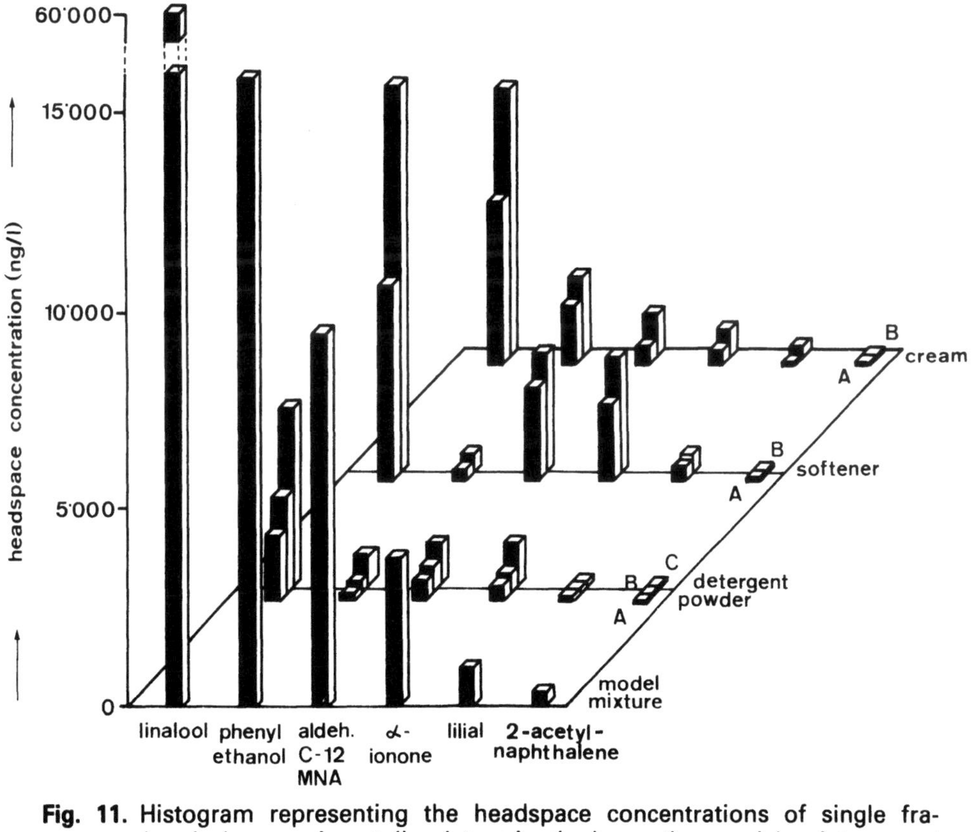

Fig. 11. Histogram representing the headspace concentrations of single fragrance chemicals experimentally determined above the model mixture and above different kinds (A, B and C) of detergent powders, softeners and creams after being fragranced with the model mixture.

To make matrix effects quantifiable, a so-called matrix factor can be determined experimentally. This is defined as the quotient of an odorant's measured concentration in the headspace above the fragranced product (matrix) and its theoretical headspace concentration, calculated on the basis of vapor pressure and dosage in the matrix assuming that the odorant is completely dissolved in the matrix and that no intermolecular interactions take place. Thus, the matrix factor is a numerical measure indicating how much an odorant's actual headspace concentration deviates from its theoretical value owing to the fact that the odorant/matrix system is not an ideal mixture.

Some matrix factors, experimentally determined for the above-mentioned perfume mixture, are compiled in Table 3. The factors found for single fragrance chemicals in the perfume mixture range from 0·7 to 1·9, indicating that they deviate from ideal behavior (factor 1·0) only to a small extent in both directions. In cream, however, these factors are much higher, which means that the odorant's headspace concentrations in reality are between 8 and 68 times higher than theoretically predicted. For detergent powders and softeners, even larger deviations of a factor of up to 190 and 269, respectively, have been determined. By comparing the matrix factors in Table 3 in the horizontal and vertical line, their dependence on two main parameters, the structure of the odorant and the nature of the matrix, becomes apparent. This opens new possibilities for rational interpretations and possible correlations with other measurable parameters of the matrix (chemical composition, water content, particle size, shape, etc.) as well as of the fragrance chemical (polarity, functional groups, solubilities, etc.), which cannot be discussed here in detail.

Further exploration of matrix effects will certainly make a substantial

Table 3
Matrix factors experimentally determined for single fragrance chemicals in differen product matrices

Product matrix	*Sub-group*	*Matrix factors of selected fragrance chemicals*					
		Linalool	*Phenyl-ethanol*	*Aldehyde C-12 MNA*	*α-Ionone*	*Lilial*	*2-acetyl-naphthalene*
Perfume mixture		1·1	1·4	1·2	0·7	1·3	1·9
Cream	A w/o	8	15	8	8	17	41
	B o/w	14	19	16	14	36	68
Detergent powder	A	20	13	36	42	47	55
	B	28	16	52	52	68	81
	C	52	44	89	128	109	190
Softener	A	43	12	139	156	193	156
	B	82	19	184	234	269	211

Abbreviations: w/o, water-in-oil; o/w, oil-in-water.

contribution to our knowledge of the behavior of fragrance materials in specific applications and therefore deserves our special attention.

Adaptation of a Perfume Formula

It can be a challenging job in perfume application to adapt the performance of a perfume which was originally developed for a single product alone, to another product or to a series of different products within a whole product line. Instead of tackling the task conventionally by tedious trial-and-error tests, headspace measurement in combination with the odor value concept can be used to solve the problem in a relatively simple and elegant way.[126]

The principle of this approach is best explained by the following model experiment. Two different cosmetic products, cream and shampoo, were fragranced with a model mixture at dosage levels which were empirically chosen by the perfumer to obtain comparable odor intensities from the two products. Sensory testing revealed that both odor intensity and quality of the two products were distinctly different, as can be expected because of the matrix effects discussed in the preceding section. To quantify this difference objectively, the fragrance emanation from cream and shampoo was measured by headspace analysis. Headspace data were transformed into odor values and plotted on a logarithmic scale, as shown in Fig. 12(A). Remarkable differences in the odor values of the two products could be observed. For instance, the odor value of α-ionone increases by a factor of 3·3 and that of linalyl acetate by a factor of 7·6 from cream to shampoo. As a consequence of these changes, the ratio of odor values between the two mixture components with the highest odor impact, α-ionone and γ-undecalactone, is shifted too, their ratio being 1:3·7 in cream and only 1:1·1 in shampoo. These findings were well in line with the results from the sensory test, which indicated a higher odor intensity and a different odor note for the shampoo in comparison with the cream.

To adapt the odor characteristics of the model perfume in the shampoo to those in the cream, the concentrations of single fragrance constituents in the shampoo were corrected by linear adjustment according to the differences of their concentrations found in the headspace. This simple one-step procedure led to a fairly good match of the odor value histograms for cream and shampoo, as shown in Fig. 12(B). The success of this approach was also confirmed in the sensory test.

Fragrance Performance in Functional Applications

The perfume in a consumer product has a well-defined role and must meet very different requirements. One important requirement is that the fragrance displays its specific action during the whole period of the product's use. That means, for instance, that the fragrance in an air freshener has to be still active after a couple of weeks and that a piece of soap should not only pervade a bathroom with its pleasant scent but it should also convey a substantial part of

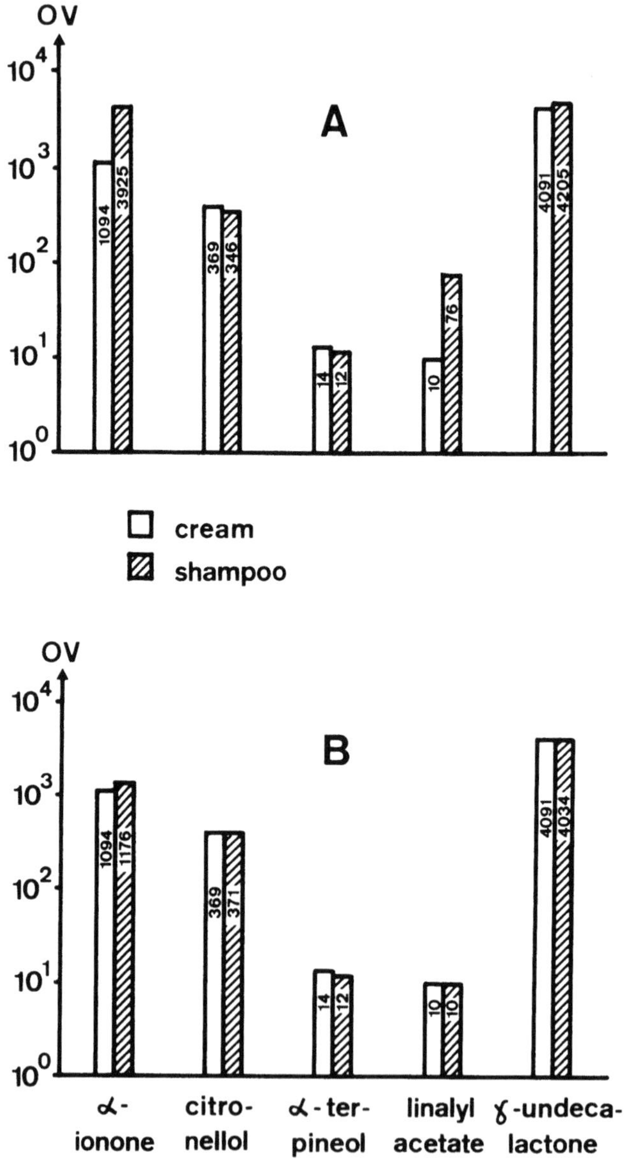

Fig. 12. Histogram representing the differences in odor values of single fragrance chemicals in the headspace of cream and shampoo. A, After the products had been fragranced with the model mixture at dosages empirically chosen by the perfumer (0·8% in cream and 0·5% in shampoo); B, after the dosages of single fragrance chemicals in the shampoo had been corrected by a one-step linear adaption.

it to the user's skin. If the perfume performs well, it is said to be diffusive and substantive. Classical examples where substantivity plays an important role are detergent powders and fabric softeners. When these products are used, the perfume is expected to remain active through the various stages of laundering: during the washing process, while removing the laundry from the machine, while drying, ironing and storing the clothes in the cupboard and finally also while wearing them. Every perfumer well knows that it is not a simple task to achieve fragrance substantivity and diffusivity on dry laundry, for example.

So far, the classical approach to solving this kind of problem has been systematic testing of single fragrance raw materials in appropriate applications, in combination with sensory analysis. This, however, is a tedious and time-consuming method if one considers the large number of fragrance raw materials and the fact that they have to be tested in many products since the latter differ in composition from one supplier to the other and each supplier changes the composition of his products from time to time.

Today, a more scientific approach is preferred to tackle these problems. It is based on the rather trivial assumption that such parameters as vapor pressure, water solubility and odor threshold must be the dominating factors determining the behavior of a fragrance chemical, for instance its diffusivity and substantivity, in different stages of the product's use. Headspace and olfactometric measuring techniques deliver both (a) the data collection on the basis of which diffusive and substantive raw materials can be selected and (b) the tool for quantifying the success of using them. The latter aspect may be exemplified with regard to fragrance substantivity on laundry by the following model experiment.

Terry cloth was washed in a washing machine according to a standard procedure which included the addition of a fabric softener in the final step of the washing program. This softener was fragranced with the same model mixture as used in the experiment for determining matrix factors in various consumer products (see the section on 'Odorant/matrix interactions', above). After washing, the fragrance emanation from wet and dry laundry was measured by the headspace technique. Some of the data are represented in Fig. 13 in odor value units to allow correlation with sensory results. In the plot, fragrance chemicals are arranged from left to right in order of decreasing vapor pressures. Thus, it can be seen immediately that the more volatile fragrance chemicals (e.g. benzyl acetate, linalool and α-ionone) do not contribute to the odor of dry laundry. Obviously, they are lost during the 24 h drying period, due to their relatively high evaporation rate. In the headspace of wet laundry, however, they reach high concentrations and accordingly high odor values, except for benzyl acetate which—in spite of its high volatility—reaches only a low odor value due to its high threshold.

From the results of this model experiment it becomes clear that only chemicals with moderate to low volatility (e.g. Lilial®, Fixolide®, benzyl salicylate and Musk Ketone®) have a real chance of not being lost during drying of the laundry. Although they are present in the headspace of dry

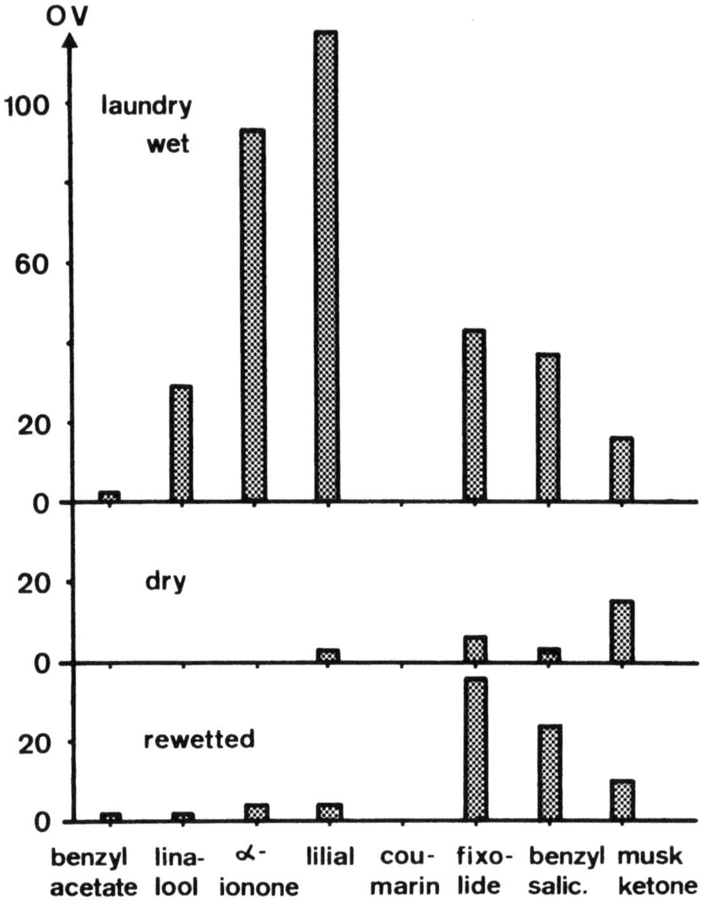

Fig. 13. Odor values of selected fragrance chemicals emanating from wet, dry and rewetted laundry (terry cloth) after washing and treatment with fabric softener in a washing machine. The softener was fragranced to 0·2% with a model perfume consisting of 24 fragrance chemicals in equal amounts. Dry laundry was rewetted with the same quantity of water which was lost during the drying of wet laundry.

laundry in low concentrations only, they can be considered to be fairly substantive, thanks to their favorably low odor thresholds.

Coumarin, although having a moderate volatility and a very low odor threshold, turns out not to be substantive either on wet or on dry laundry. This exceptional behavior can be easily explained by its relatively high water solubility leading to its removal when the aqueous softener solution is sucked off from the laundry in the washing machine. An interesting effect can be shown when dry laundry is rewetted with water. Its odor emanation is represented in Fig. 13. It can be seen that some of the fragrance chemicals which are believed to be lost during drying reappear. This is a clear indication that they are in fact still present on the dry laundry and that only strong adsorption on the surface of the fabrics prevents them from emanating into the headspace. After rewetting, they enter into competition with water and are therefore partly desorbed from the surfaces.

The evaluation of the complete data obtained from these experiments revealed a number of yet other important factors having a marked influence on the substantivity behavior of fragrance chemicals on laundry, for instance the presence of special functional groups in the fragrance molecules, the chemical nature of the fabric and other factors. They cannot be discussed here in detail.

When comparing the odor values obtained from wet, dry and rewetted laundry by headspace measurements with the results obtained from the same pieces of laundry by sensory testing, a fairly good correlation was observed. A strong and highly diffusive odor was perceived from the wet laundry and only a faint odor from the dry laundry, whereas after rewetting dry laundry its odor intensity was found to increase distinctly. This is in good agreement with the measured data and demonstrates again the usefulness of the odor value concept.

This example demonstrates clearly that, to achieve fragrance substantivity on laundry, the perfume in a detergent powder or fabric softener must contain, at least in part, fragrance chemicals which are not or are only slightly soluble in water and which exhibit high odor values. High odor values in combination with low vapor pressures (only possible with extremely low odor thresholds) are an additional prerequisite to guarantee high substantivity on dry laundry, for instance. In contrast, high odor values in combination with high vapor pressures are the typical characteristics of highly diffusive fragrance materials.

It is obvious that every perfumer who has empirically developed his own successful strategy of creating highly performing perfumes for specific applications uses, more or less consciously, the above-mentioned relationships. With the availability of headspace and olfactometric techniques, however, this strategy now becomes objectively controllable, quantifiable and—on the basis of an existing data collection of fragrance raw materials—predictable.

4 CONCLUSIONS

It was the authors' intention to restrict the detailed discussion of odor-measuring methods to the two selected techniques, quantitative headspace analysis and threshold olfactometry, and furthermore to stress the importance of using them combined in an integral approach.

Although the few practical examples described in the preceding section cover only a small fraction of the possible applications in the area of fragrance research and perfume creation, they may illustrate sufficiently the potential and also the limitations of these techniques.

In the analytical approach one limitation becomes impressively clear to the research chemist. It happens again and again that he experiences real fireworks from the most exciting odors while sniffing at the outlet of the gas chromatograph, although the flame ionisation detector running at its highest sensitivity level only draws a noisy but straight baseline. What a disappointment—or better, what a challenge—for the analyst, who never ceases to strive for further improvements of his analytical tools.

The merging of headspace analytical data with odor threshold data, on which the odor value concept is based, includes another limitation. More precise information on odor intensity and thus a better predictability of an odorant's performance could be expected, if the concept were not restricted to threshold data. Further refinement of the concept by measuring and implementing odor intensity data from the suprathreshold range, e.g. the slope and flattening of an odorant's dose/response curve, can be regarded as another challenge.

There is no doubt that future progress in both areas, instrumental analysis and olfactometry, will further improve our capability of measuring and exploring this most fascinating phenomenon—odor.

REFERENCES

1. Moncrieff, R. W., *Odours.* William Heinemann Medical Books, 1970, pp. 10–36.
2. Harper, R., *Human Senses in Action.* Churchill Livingstone, Edinburgh, 1972, pp. 237–47.
3. Burdach, K. J., *Geschmack und Geruch.* Verlag Hans Huber, Bern, Stuttgart, Toronto, 1988, pp. 14–22.
4. Cagan, R. H. & Kare, M. R., *Biochemistry of Taste and Olfaction.* Academic Press, New York, 1981.
5. Dodd, G. H., The molecular dimension in perfumery. In *Perfumery, the Psychology and Biology of Fragrance* Chapman & Hall, London, New York, 1988, 38–46.
6. Lancet, D., Chen, Z., Ciobotariu, A., Eckstein, F., Khen, M., Heldman, J., Ophir, D., Shafir, J. & Pace, U., Toward a comprehensive molecular analysis of olfactory transduction. In Proceedings of the Congress Olfaction and Taste IX, ed. S. D. Roper & J. Atima. *Ann. NY Acad. Sci.*, **510** (1987) 27–32.
7. Amoore, J. E., *Perfum. Essent. Oil Rec.*, **43** (1952) 321.
8. Wright, R. H., *J. Appl. Chem.*, **4** (1954) 611.
9. Davies, J. T., *J. Theor. Biol.*, **8** (1965) 1.
10. Beets, M. G. J., *Structure–Activity Relationships in Human Chemoreception.* Elsevier Applied Science Publishers, London, 1978, pp. 207–42.
11. Jurs, P. C., Ham, C. L. & Brügger, W. E., Computer-assisted studies of chemical structure and olfactory quality using pattern recognition techniques. In *Odor Quality and Chemical Structure,* ACS Symposium Series No. 148, ed. H. R. Moskowitz & C. B. Warren. American Chemical Society, Washington, DC, 1981, pp. 143–59.
12. Dravnieks, A., Comparison of theories on relations between odor parameters and other properties of odorants. In *Theories of Odor and Odor Measurement,* ed. N. N. Tanyolac. Unwin Bros, Old Woking, UK, 1968, pp. 11–46.
13. Laffort, P., A model of the olfactory mechanism based on chromatographic data. In *Structure–Activity Relationships in Chemoreception,* ed. G. Benz *et al.* Information Retrieval, London, 1976, pp. 186–95.

14. McGill, J. R. & Kowalski, B. R., Intrinsic dimensionality of smell. *Anal. Chem.*, **49** (1977) 596–602.
15. Cain, W. S. & Moskowitz, H. R., Psychophysical scaling of odors. In *Human Responses to Environmental Odors,* ed. A. Turk *et al.* Academic Press, New York, 1974, pp. 1–32.
16. Patte, F., Etcheto, M. & Laffort, P., Selected and standardized values of suprathreshold odor intensities for 110 substances. *Chem. Sens. Flav.*, **1** (1975) 283–305.
17. Cain, W. S., Physicochemical characteristics and supraliminal odor intensity. *Percep. Psychophys.*, **9** (1971) 478.
18. Van Gemert, L. J., Belz, R. & Van Straten, S., Compilation of odor threshold values in air. Supplement V, Report No. A 84.220/090070, Central Institute for Nutrition and Food Research TNO, Zeist, 1984.
18. (a) Devos, M., Patte, F., Rouault, J., Laffort, P. & Van Gemert, L. J., *Standardized Human Olfactory Thresholds.* IRL at Oxford University Press, Oxford–New York 1990.
19. Greenberg, M. J., Dependence of odor intensity on the hydrophobic properties of molecules, a quantitative structure odor intensity relationship. *J. Agric. Food Chem.*, **27** (1979) 347–52.
20. Roper, S. D. & Atema, J. (eds), Proceedings of the Congress Olfaction and Taste IX. *Ann. NY Acad. Sci.*, **510** (1987).
21. Laing, D. G., Cain, W. S., McBride, R. L. & Ache, B. W. (eds), *Perception of Complex Smells and Tastes.* Academic Press, 1989, New York, London.
22. Laing, D. G., Coding of chemosensory stimulus mixtures. In Proceedings of the Congress Olfaction and Taste IX, ed. S. D. Roper & J. Atema. *Ann. NY Acad. Sci.*, **510** (1987) 61–6.
23. Laing, D. G., The role of physicochemical and neural factors in the perception of mixtures. See ref. 21, pp. 189–205.
24. Laing, D. G. & Panhuber, H. In *Progress in Flavor Research,* ed. D. G. Land & H. F. Nursten. Applied Science Publishers, London, 1982, pp. 27–46.
24. (a) Berglund, B., Berglund, U., Lindvall, T. & Svensson, L. T., A quantitative principle of perceived intensity summation in odor mixtures. *J. Exp. Psych.*, **100** (1973) 29–38.
25. Frijters, J. E. R., Psychophysical models for mixtures of tastants and odorants. In Proceedings of the Congress Olfaction and Taste IX, ed. S. D. Roper & J. Atema. *Ann. NY Acad. Sci.*, **510** (1987) 67–78.
26. Laffort, P., Models for describing intensity interactions in odor mixtures: a reappraisal. See ref. 21, pp. 205–23.
27. Cole, R. H. & Coles, J. S., *Physical Principles of Chemistry.* W. H. Freeman, San Francisco, 1964.
28. Hackenberg, H. & Schmidt, A. P., *Gas Chromatographic Headspace Analysis.* Heyden & Son, London, 1977.
29. Drozd, J. & Novák, J., Headspace gas analysis by gas chromatography. *J. Chromatogr.*, **165** (1979) 141–65.
30. Núñez, A. J., Gonzáles, L. F. & Janák, J., Preconcentration of headspace volatiles for trace organic analysis by gas chromatography. *J. Chromatogr.*, **300** (1984) 127–62.
31. Land, D. G., Some factors influencing the perception of flavor-contributing substances in food. In *Progress in Flavor Research,* ed. D. G. Land & H. F. Nursten. Applied Science Publishers, London, 1982, pp. 53–66.
32. Dravnieks, A., Approaches to objective olfactometry. In *Theory of Odor and Odor Measurement,* ed. N. N. Tanyolac. Unwin Bros, Old Woking, UK, 1968, pp. 373–89.
33. Tanyolac, N. N., The electro-odocell for odor measurement and surface effect. In

Surface Effects in Detection, ed. J. I. Bregman & A. Dravnieks. Spartan Books, Washington, DC, 1965.
34. Moncrieff, R. W., An instrument for measuring and classifying odors. *J. Appl. Physiol.,* **16** (1961) 742.
35. Moncrieff, R. W., *The Chemical Senses.* Leonard Hill, London, 1967, pp. 208–44.
36. Persaud, K. & Dodd, G. H., Analysis of discrimination mechanisms in the mammalian olfactory system using a model nose. *Nature,* **299** (1982) 352–5.
37. Abe, H., Yoshimura, T., Kanaya, S., Takahashi, Y., Miyashita, Y. & Sasaki, S., Automated odor-sensing system based on plural semiconductor gas sensors and computerized pattern recognition techniques. *Anal. Chim. Acta,* **194** (1987) 1–9.
38. Shurmer, H. V., Gardner, J. W. & Chan, T. S., The application of discrimination techniques to alcohols and tobaccos using tin oxide sensors. *Sensors and Actuators,* **18** (1989) 359–69.
39. Gardner, J. W. & Hines, E. L., Integrated sensor array processing in an electronic nose. *IEE Coll.,* **130** (1988) 7/1–4.
40. Ottoson, D., Generation and transmission of signals in the olfactory system. In *Olfaction and Taste, Proceedings of the First International Symposium, Stockholm, Sept. 1962,* ed. Y. Zolterman. Pergamon Press, Oxford, 1963, pp. 35–44.
41. Tucker, D., Olfactory, vomeronasal and trigeminal receptor responses to odorants. In *Olfaction and Taste, Proceedings of the First International Symposium, Stockholm, Sept. 1962,* ed. Y. Zolterman. Pergamon Press, Oxford, 1963, pp. 45–70.
42. Schneider, D., Basic problems of olfactory research. In *Theories of Odor and Odor Measurement,* ed. N. N. Tanyolac. Unwin Bros, Old Woking, UK, 1968, pp. 201–12.
43. Gesteland, R. C., Technique for investigating single unit activity in the vertebrate olfactory epithelium. In *Methods in Olfactory Research,* ed. D. G. Moulton, A. Turk & J. W. Johnston. Academic Press, London, 1975, pp. 269–322.
44. Moulton, D. G., The simultaneous measurement of neural response and performance in rats and rabbits involved in an odor detection task. In *Methods in Olfactory Research,* ed. D. G. Moulton, A. Turk & J. W. Johnston. Academic Press, London, 1975, pp. 407–29.
45. Hughes, J. R., Hendrix, D. E. & Wetzel, N., Evidence from the human olfactory bulb for the frequency component hypothesis. In *Theories of Odor and Odor Measurement,* ed. N. N. Tanyolac. Unwin Bros, Old Woking, UK, 1968, pp. 87–114.
46. Van Toller, S. & Reed, M. K., Brain electrical activity topographical maps produced in response to olfactory and chemosensory stimulation. *Psychiatry Res.,* **29** (1989) 429–30.
47. Van Toller, S., Emotion and the brain. In *Perfumery, the Psychology and Biology of Fragrance,* ed. S. Van Toller & G. H. Dodd. Chapman & Hall, London, 1988, pp. 121–46.
48. Steiner, W., Hanisch, E. & Schwarz, D., Geruchserlebnis und Pupillenerweiterung—eine experimentelle Untersuchung. *Parfümerie und Kosmetik,* **58** (1977) 189–96.
49. Rioult, M., Etzweiler, F., Müller, P. M., Neuner-Jehle, N. & Schudel, P., Givaudan Forschungsgesellschaft AG, Dübendorf, 1986, unpublished.
50. Myers, L. J. & Pugh, R., Threshold of the dog for detection of inhaled eugenol and benzaldehyde determined by electroencephalographic and behavioral olfactometry. *Am. J. Vet. Res.* **46** (1985) 2409–12.
51. Piggott, J. R. & Canaway, P. R., Finding the word for it—methods and uses of descriptive sensory analysis. In *Flavor 81, Proceedings of the 3rd Weurman Symposium, Munich, April 1981,* ed. P. Schreier. Walter de Gruter, Berlin, 1981, pp. 33–46.

52. Pangborn, R. M., A critical review of threshold intensity and descriptive analysis in flavor research. In *Flavor 81, Proceedings of the 3rd Weurman Symposium, Munich, April 1981,* ed. P. Schreier. Walter de Gruter, Berlin, 1981, pp. 3–32.
53. Gregson, R. A. M. & Mitchell, M. J., Odor quality similarity scaling and odor–word profile matching. *Chem. Senses and Flavor,* **1** (1974) 95–101.
54. Köster, E. P., Human psychophysics in olfaction. In *Methods in Olfactory Research,* ed. D. G. Moulton, A. Turk & J. W. Johnston. Academic Press, London, 1975, pp. 345–74.
55. Schiffman, S. S., Characterization of odor quality utilizing multidimensional scaling techniques. In *Odor Quality and Chemical Structure,* ACS Symposium Series No. 148, ed. H. R. Moskowitz & C. B. Warren. American Chemical Society, Washington, DC, 1981, pp. 1–22.
56. Moskowitz, H. R., Psychophysical scaling and optimization of odor mixtures. In *Odor Quality and Chemical Structure,* ACS Symposium Series No. 148, ed. H. R. Moskowitz & C. B. Warren. American Chemical Society, Washington, DC, 1981, pp. 23–56.
57. Dravnieks, A., Bock, F. C. & Jarke, F. H., Sensory structure of odor mixtures. In *Odor Quality and Chemical Structure,* ACS Symposium Series No. 148, ed. H. R. Moskowitz & C. B. Warren. American Chemical Society, Washington, DC, 1981, pp. 79–92.
58. Cheesman, G. H. & Mayne, S., The influence of adaption on absolute threshold measurements for olfactory stimuli. *Q. J. Exp. Psychol.,* **5** (1953) 22–29 and **8** (1956) 8–14.
59. Köster, E. P., Adaption and cross-adaption in olfaction. Thesis, University of Utrecht, The Netherlands, 1971.
60. Amoore, J. E. & Venstrom, D., Correlation between stereochemical assessments and organoleptic analysis of organic compounds. In *Olfaction and Taste II,* ed. O. O. Hayashi. Pergamon Press, Oxford, 1966.
61. Müller, P. M., Measurable properties of odors and their application for structure odor relationships (SORs). In *Proceedings of the 11th Int. Congress of Ess. Oils,* New Delhi, November 1989, ed. S. C. Bhattacharyya, N. Sen & K. L. Sethi, Oxford & IBM Publishing Co. Ltd., New Delhi, 1989, pp. 25–39.
62. Green, D. M. & Sweets, J. A., *Signal Detection Theory and Psychophysics.* Wiley, New York, 1966.
63. Schiffman, H. R., *Sensation and Perception—an Integrated Approach.* Rutgers, The State University; John Wiley & Sons, New York, 1976.
64. Wright, R. H., *The Science of Smell,* British Columbia Res. Council, University of British Columbia, Vancouver, Canada; George Allen & Unwin, London, 1964, p. 50.
65. Turk, A., Switala, D. & Thomas, S. H., Suprathreshold odor measurements by dynamic olfactometry. In *Proceedings of the 72nd Annual Meeting of the Air Pollution Control Association, Cincinnati, Ohio, June 1979,* **72** (1979) pp. 3–22.
66. Cain, W. S., Differential sensitivity for smell: noise at the nose. *Science,* **195** (1977) 796–8.
67. Moskowitz, H. R., Dravnieks, A., Cain, W. S. & Turk, A., Standardized procedure for expressing odor intensity. *Chemical Senses and Flavor,* **1** (1974) 235–7.
68. Tuorila, H., Suprathreshold odor intensity assessment: individual variation in scaling. In *Flavor 81, Proceedings of the 3rd Weurman Symposium, Munich, April 1981,* ed. P. Schreier, Walter de Gruter, Berlin, 1981, pp. 53–62.
69. Hyman, A. M., Factors influencing the psychophysical function for odor intensity. *Sens. Proc.,* **1** (1977) 273.
70. Frijters, J. E. R., Some psychophysical notes on the use of the odor unit number.

In *Progress in Flavor Research,* 1982, ed. D. G. Land & H. F. Nursten. Applied Science Publishers, London, 1982, pp. 47–51.
71. Guadagni, D. G., *Correlation of Subjective–Objective Methods in the Study of Odors and Taste.* ASTM 440, Philadelphia, USA, 1968, p. 36.
72. Frijters, J. E. R., A critical analysis of the odor unit number and its use. *Chemical Senses and Flavor,* **3** (1978) 227–33.
73. Mackay, D. H. M., Lang, D. H. & Berdick, M., Analysis of volatiles in breath. *The Toilets Goods Association,* **32** (1959) 45–51.
74. Wylie, P. L., Headspace analysis with cryogenic focusing: a procedure for increasing the sensitivity of automated capillary headspace analysis. *Chromatographia,* **21** (1986) 251–8.
75. McNally, M. E. & Grob, R. L., Static and dynamic headspace analysis: a review, part one. *Am. Lab.,* **17** (1985) 20–33; part two, ibid. 106–20, **17** (1985) 106–20.
76. Charalambous, G. (ed.), *Analysis of Food and Beverages.* Academic Press, New York, 1978.
77. Kolb, B. (ed.), *Applied Headspace Gas Chromatography.* Heyden & Son, London, 1980.
78. Joffe, B. V. & Vittenberg, A. G., *Headspace Analysis and Related Methods in Gas Chromatography.* Wiley & Sons, New York, 1982.
79. Poole, F. P. & Schuette, S. A., *Contemporary Practice of Chromatography.* Elsevier, New York, 1984.
80. Joffe, B. V., New applications of gas-chromatographic headspace analysis. *J. Chromatogr.,* **290** (1984) 363–75.
81. Kolb, B., Application of gas-chromatographic headspace analysis for the characterization of non-ideal solutions by scanning the total concentration range. *J. Chromatogr.,* **112** (1975) 287–95.
82. Wylie, P. L., Comparing headspace with purge and trap for analysis of volatile priority pollutants. *J. Am. Water Works Assoc.,* **80** (1988) 65–72.
83. Grob, K., Organic substances in potable water and in its precursor, part one: methods for their determination by gas–liquid chromatography. *J. Chromatogr.,* **84** (1973) 255–73.
84. Etzweiler, F., Neuner-Jehle, N. & Senn, E., Einsatz der quantitativen Headspace Analyse zur Untersuchung des Verhaltens von Riechstoffen in Seifen und Detergentien. *Seifen–Öle–Fette–Wachse,* **106** (1980) 419–27.
85. Etzweiler, F., Senn, E. & Neuner-Jehle, N., Eine einfache Methode zur Bestimmung von Dampfdrücken. *Ber. Bunsenges. Phys. Chem.,* **88** (1984) 578–83.
86. Etzweiler, F., Modification of a packed gc injector to a versatile capillary injector. *HRC & CC,* **8** (1985) 436–43.
87. Grob, K., Jr & Kuhn, S., Speed of temperature increase when using large retention gaps in capillary gas chromatography. *J. Chromatogr.,* **301** (1984) 1–9.
88. Dravnieks, A., Instrumental aspects of olfactometry. In *Methods in Olfactory Research,* ed. D. G. Moulton, A. Turk & J. W. Johnston. Academic Press, London, 1975, pp. 1–61.
89. Mannebeck, H., Zur Problematik der Olfaktometrie und ihre Anwendung zur Bestimmung von Geruchsemissionen. *Gesundheits-Ingenieur,* **98** (1977) 313–17.
90. Johnson, R. A., A system for field measurement of environmental odor intensities and odorant concentrations based on the ASTM odor reference method. In *Proceedings of the 71st Annual Meeting of the Air Pollution Control Association, Houston, Texas, June 1978,* pp. 3–16.
91. Homans, W. J., Kohler, H., Egbert, P. & Bardtke, D., Bestimmung von Geruchsschwellenwerten mit dem Olfaktometer TO-4 in Kombination mit dem Flammenionisationsdetektor. *Staub-Reinhalt. Luft,* **38** (1978) 217–21.
92. Olfaktometrische Technik der Geruchsschwellen-Bestimmung—Grundlagen. VDI-Richtlinien—Entwurf VDI 3881, Blatt 1, November 1983.

93. Hangartner, M., Hartung, J., Paduch, M., Pain, B. F. & Voorburg, J. H., Improved recommendations on olfactometric measurements. *Environmental Technol. Lett.*, **10** (1989) 231–6.
94. Teranishi, R., Buttery, R. G. & Guadagni, D. G., Some properties of odoriferous molecules. In *Flavor 81, Proceedings of the 3rd Weurman Symposium, Munich, April 1981,* ed. P. Schreier. Walter de Gruter, Berlin, 1981, pp. 133–43.
95. Christoph, N. & Drawert, F., Olfactory thresholds of odor stimuli determined by gas-chromatographic sniffing technique. In *Topics in Flavor Research, Proceedings of the Int. Conference, Freising, April 1985,* ed. R. G. Berger, S. Nitz & P. Schreier. H. Eichhorn, Marzling-Haugenham, FRG, 1985, pp. 59–77.
96. Dravnieks, A. & Jarke, F. H., Odor threshold measurement by dynamic olfactometry—significant operational variables. In *Proceedings of the 72nd Annual Meeting of the Air Pollution Control Association, Cincinnati, Ohio, June 1979,* **72** (1979), pp. 3–21.
97. Meilgaard, M. C. & Reid, D. S., Determination of personal and group thresholds and the use of magnitude estimation in beer flavor chemistry. In *Progress in Flavor Research,* ed. D. G. Land & H. F. Nursten. Applied Science Publishers, London, 1982, pp. 67–73.
98. Knupfer, L. & Spiegel, R., Differences in olfactory test performance between normal aged Alzheimer and vascular-type dementia individuals. *Int. J. Geriatric Psychiatry,* **1** (1986) 3–14.
99. Takagi, S. F., Standardized olfactometries in Japan—a review over ten years. *Chemical Senses,* **14** (1989) 25–46.
100. D'Andrea, C. L., Psychophysical characterization of musk chemicals. *Chemical Senses and Flavor,* **1** (1975) 359–69.
101. Randebrock, R. E., Über den Einsatz eines Olfaktometers. *J. Soc. Cosmet. Chem.,* **21** (1970) 289 and *Parfumerie und Kosmetik,* **67** (1986) 10–24.
102. Lachenmayer, U. & Kohler, H., Untersuchungen zur Neuentwicklung eines Olfaktometers. *Staub.-Reinhalt. Luft,* **44** (1984) 359–62.
103. Kronen, H., Olfaktometrie—ein Begriff, der zum Lachen reizt? *Labo* (Nov. 1986) 86–9.
104. Dravnieks, A. & O'Donnell, A., Principles and some techniques of high resolution headspace analysis. *J. Agr. Food. Chem.,* **19** (1971) 1049–56.
105. Marin, A. B., Acree, T. E. & Barnard, J., Variations in odor detection thresholds determined by charm analysis. *Chemical Senses,* **13** (1988) 435–44.
106. Jann, P. R., A survey of dynamic olfactometer design. In *Proceedings of the 77th Annual Meeting of the Air Pollution Control Association, San Francisco, June 1984,* **77**, 5—paper 84–93.4, 1984, pp. 2–13.
107. Engen, T., *The Perception of Odors.* Academic Press, New York, 1982.
108. Laing, D. G., Optimum perception of odor intensity by humans. *Physiology & Behavior,* **34** (1985) 569–74.
109. Laing, D. G., Characterization of human behavior during odor perception. *Perception,* **11** (1982) 221–30.
110. Dravnieks, A. & Prokop, W. H., Source emission odor measurements by a dynamic forced choice triangle olfactometer. *JAPCA,* **25** (1975) 28–36.
111. Dravnieks, A., Schmidtsdorff, W. & Meilgaard, M., Odor thresholds by forced-choice dynamic triangle olfactometry: reproducibility and methods of calculation. *JAPCA,* **36** (1986) 900–5.
112. Hangartner, M., Sensorische Geruchsmessung. *Umweltschutz/Gesundheitstechnik,* **1** (1987) 20–3.
113. Patent Application filed by Givaudan.
114. Termonia, M. & Guns, M., Characterization of malodorous volatiles in air using olfactometry and computer assisted capillary gas-chromatography–mass spectrometry. *Int. J. Environ. Anal. Chem.,* **20** (1985) 69–84.

115. Werkhoff, P., Bretschneider, W., Herrmann, H. J. & Schreiber, K., Fortschritte in der Aromastoff-Analytik. *Labor Praxis,* (1990) 51–8.
116. Grosch, W., Fischer, K. H., Ulrich, F. & Gasser, K., Fleischbrühe und Fleischextrakt—Zusammensetzung und qualitätsbestimmende Inhaltsstoffe. In *Lebensmittelqualität: Wissenschaft und Technik,* ed. R. Stute, VCH, Weinheim, 1989, pp. 47–63.
117. Sfiras, J. & Demeilliers, A., Etude par chromatographie gazeuse de la vapeur odorante émise par une savonette parfumée. *Recherches,* **19** (1974) 259–69.
118. Burrell, J. W. K., *J. Soc. Cosmet. Chem.,* **25** (1974) 325.
119. Palazzo, A., Conton, C. & Toninelli, G., Détermination du parfum dans une poudre détergente. *Parfums, Cosmétiques, Arômes,* **24** (1978) 43–8.
120. Klimes, I. & Lamparsky, D., Einsatz der Headspace—Methode zur Bestimmung von Riechstoffen in Kosmetika. *Parfümerie und Kosmetik,* **59** (1978) 407–12.
121. Cozzoli, O., Gasparoli, A., Melis, S. & Mignini, E., Applicazione della gaschromatografia spazio di testa all'analisi del profumo nei formulati cosmetici. *La Rivista Italiana delle Sostanze Grasse,* **60** (1983) 59–64.
122. Esteve, N., Hogg, J. & Julia, E., Tecnica para la concentracion y analisis de los espacios de capeza de los productos de limpieza domestica perfumados. In *Comun. Jorn. Com. Esp. Deterg., 13 Assoc. Invest. Deterg. Tensioactivos Afines, Barcelona, 1982,* pp. 109–30.
123. Appell, L., *Physical Foundations in Perfumery.* Norda/Schimmel Int., New York; published by American Perfumer and Cosmetics, 1031 South Boulevard, Oak Park, IL 60302, Dec. 1970.
124. Moskowitz, H. R., Chandler, J., Moldaweir, R. & Laterra, R., Psychophysical measurement as a tool for perfumery and the cosmetic industry. *J. Soc. Cosmet. Chem.,* **30** (1979) 91–104.
125. Ohloff, G., Importance of minor components in flavors and fragrances. *Perfumer & Flavorist,* **3** (1978), no. 1, 11–22.
126. Etzweiler, F., Neuner-Jehle, N. & Müller, P. M., Application of odor values in perfuming cosmetic products. Poster presentation at the 15th IFSCC Congress, London, Sept. 1988.
127. Warren, C. B., Development of fragrances with functional properties by quantitative measurement of sensory and physical parameters. In *Odor Quality and Chemical Structure,* ACS Symposium Series No. 148, ed. H. R. Moskowitz & C. B. Warren. American Chemical Society, Washington, DC, 1981, pp. 57–77.
128. Fragrance Chemical, manufactured by Givaudan (see US patent 4 519 944 (priority August 25, 1982).
129. Beets, M. G. J., *Structure Activity Relationships in Human Perception,* Elsevier Applied Science Publishers Ltd., London 1978, pp. 127–42.
130. Ohloff, G., Chemistry of odour stimuli, *Experientia,* **42** (1986) 271–9.
131. Christoph, N. and Drawert, F., Olfactory thresholds of odour stimuli determined by gaschromatographic sniffing technique; Structure activity relationships. In *Topics in Flavour Research,* Proceeding of the International Conference Freising-Weihenstephan, April 1985, ed. R. G. Berger, S. Nitz & P. Schreier, H. Eichhorn, Marzling—Haugenham, FRG, 1985, p. 71.
132. Werkhoff, P., Bretschneider, W., Güntert, M., Hopp, R. & Surburg, H., Chiral Analysis in Flavour and Essential Oil Chemistry, part B—Direct enantiomer resolution of (E)-α-ionone and (E)-α-damascone by inclusion gaschromatography. In Flavour Science and Technology, ed. Y. Bessière, A. F. Thomas, John Wiley & Sons, Chichester, 1990, pp. 33–6.

Chapter 7

Trapping, Investigation and Reconstitution of Flower Scents

ROMAN KAISER
Givaudan Forschungsgesellschaft AG, CH-8600 Dübendorf, Switzerland

1 INTRODUCTION

Up to the middle of the 19th century, natural extracts of scented flowers/plants, and to a certain extent of animal secretions, had been the only raw material used for the creation of fragrances. No wonder that chemists working in the fragrance industry started to investigate these 'essential oils' (used as a generic term and including steam-distilled oils, concretes, absolutes, resinoids and related products) at the dawn of modern organic chemistry in the later 19th century. The progress in 'essential oil' research achieved during the following decades was a great source of motivation and stimulation for both the chemist and the perfumer, and many of the still commonly used fragrance chemicals can be traced back to this period. Today, the perfumer's palette comprises, besides 300–400 'essential oils',[1,2] around 3000 regularly used synthetic products.[3,4]

As recently as 30–40 years ago the isolation and structural elucidation of natural substances had to be performed almost exclusively by chemical means requiring large quantities of starting materials and involving time-consuming methods. As a result, only main constituents were discovered at that time and, therefore, progress was relatively slow. The introduction of gas chromatography and the development of spectroscopic methods revolutionized the investigation of 'essential oils'. Suddenly it became possible to detect olfactorily relevant trace constituents and a new era began just in time to cope with a special challenge of the late 1960s/early 1970s—the scarcity of certain 'essential oils', their uncontrollable price fluctuations, and dermatoxic aspects occasionally connected with such natural products. As a result of joint efforts between analytical, synthetic and olfactory research, many fragrance companies have now had, for years already, a comprehensive series of synthetic reconstitutions of such 'essential oils' on the perfumer's shelf. The still-increasing consumption data of the natural products clearly demonstrate that these reconstitutions are not thought to compete with or even to replace the originals. They are rather designed to improve the quality of fragrances in which the natural products cannot be used for price reasons.

However, many highly attractive flower/plant scents, e.g. those of lily-of-the-valley, honeysuckle species, linden blossom and many tropical flowers including a great variety of orchid species, cannot be captured in adequate quality and on a commercial scale as an 'essential oil', because olfactorily important constituents are destroyed or disturbing artefacts are formed during their manufacture. Furthermore, many of these flowers/plants would not be available in sufficient quantities, or at least the manufacturing costs of such 'essential oils' would be astronomically high. Fortunately, knowledge of natural product chemistry and the sensitivity of the physico-chemical methods in the mid-1970s had reached such a level that the analytical investigation of the scents emitted by these flowers/plants—their so-called headspace—could be taken into consideration. In fact, a long-term program had been initiated at that time within Givaudan Research to study and subsequently to reconstruct those olfactorily attractive natural scents which are not available as an 'essential oil' of adequate quality. It may be assumed that many other fragrance houses simultaneously became involved in this field.

At that time, several techniques had already been described in the literature for obtaining sufficient material for the analytical investigation of such diluted vapor phases. In most cases, however, these methods were applied to the investigation of volatile flavor components of natural food products or of biologically active compounds in orchid fragrances. Since then, the application of all types of headspace technique has been an important topic in flavor research (cf. references cited in Ref. 5), while only a comparatively small number of papers dealing with the composition of the headspace of living flowers/plants has appeared until now. They will be reviewed in the following subsection according to the type of headspace technique applied.

2 SURVEY OF EXPERIMENTAL APPROACHES

Trapping of Flower Scents

The First Motivation

The scents of flowers are of more than aesthetic interest and importance; they are directly involved in the preservation of the respective plant species. By attracting specific pollinators, in most cases insects, they assure the reproduction of species which have developed highly unique and complex pollination mechanisms and, over a long span of time, they have governed the mutual evolution of the interdependent plant and its pollinator. One of the most striking examples of such plant–insect interactions is that involving the male euglossine bees of the American tropics and the orchids they pollinate (cf. Ref. 6 and references cited therein). The male bees visit these orchid flowers not to be rewarded with food, but to collect the extremely intense aromatic floral fragrances and to use them in their own reproductive biology, probably as precursors of their own sex pheromones.

It was just this highly interesting and complex 'male euglossine syndrome' of scented neotropical orchids that gave the stimulus for the first application of headspace techniques to the investigation of flower scents. Although many of these orchid scents are also highly 'attractive' to the human nose, these first headspace experiments have been performed by biologists and botanists, not by chemists working in the fragrance field.

Direct Injection of Headspace Vapor

This technique started in 1966 with a paper by Dodson & Hills[7] on the gas chromatography of orchid fragrances. They placed the orchid inflorescence in a plexiglass chamber at the time of optimal odor production. After an equilibration/saturation time of 30 min, 10 ml of the headspace gas was withdrawn and directly injected into the gas chromatograph. At that time none of the constituents was identified, but it could be demonstrated that these orchid fragrances were composed of many compounds.

In 1968, Hills *et al.*[8] published the continuation of this work with a comprehensive list of orchid species investigated by this method. They were able to identify tentatively some of the major constituents by co-injections, comparing relative retention times and olfactory evaluations of the respective peaks eluted from the column via an effluent splitter. It may be astonishing, but this direct headspace injection technique permitted reasonable gas chromatograms to be obtained, even on packed columns. The reason for this is the extremely high fragrance production of the orchid species investigated—it may be up to 50 times higher than in the usual cases of fragrant flowers.

Although these first results of the investigation of flower scents were a starting point rather than a breakthrough, they should have earned more attention among chemists involved in fragrance and flavor chemistry.

Low-Temperature Trapping

The direct analysis of the headspace vapor is a rapid method; however, the amount of minor and trace constituents is far from sufficient for their subsequent investigation, e.g. by the combination of gas chromatography and mass spectrometry. As a consequence, an enrichment step has to be inserted. One way to accumulate volatile compounds from any natural substrates is to use a cold trap which may have the form of a 'cool finger', a glass spiral or a glass capillary. The cold trap can be kept at 0°C, −70°C or −170°C using ice, solid carbon dioxide/ethanol or liquid nitrogen, respectively. Water is inevitably condensed by this method and, therefore, the condensed volatiles have to be recovered afterwards by extraction with a suitable solvent.

In 1973, Mack & Köpsel[9] used low-temperature trapping to capture and to subsequently investigate (GC/MS) the scents of lily-of-the-valley and lilac. In fact, this was the first paper published by chemists relating to the fragrance industry, which describes the application of headspace trapping to the investigation of perfumistically important flower scents. In one experiment, freshly picked flowers were placed in a modified desiccator, whence the volatiles were purged with humified nitrogen and condensed in traps at −70°C. After 8 h, the condensate was extracted with Frigen 11® and, after concentration, investigated by GC/MS. In a second experiment, the flowers were exposed to vacuum-degassing for 8 h at 10^{-1} torr and the volatiles trapped as above. This investigation was certainly a first step in realizing an old olfactory dream, to reproduce synthetically the incomparable scent of lily-of-the-valley. Although a series of olfactorily interesting compounds were identified for the first time, no real odor impact compound was found for muguet. This situation has not changed since then. In fact the scent of lily-of-the-valley seems to be based on a highly delicate equilibrium of fragrant compounds also occurring in many other scented flowers.

Seven years later, Komaki[10] used the same principle of purging and low-temperature trapping to capture the flower scent of *Osmanthus fragrans* and compared the results of GC/MS investigation with those obtained from a steam distillate prepared from fresh flowers of the same bush. All constituents identified have also been found in an absolute of Chinese origin.[11]

Finally, Joulain[12] described the application of vacuum degassing and low-temperature trapping in connection with the identification of structurally and olfactorily interesting constituents in the trapped headspace fragrance of the flowers of *Spartium junceum* L., *Narcissus poeticus* L., *Lonicera caprifolium* L., *Syringa vulgaris* L., *Pittosporum tobira* Ait., *Robinia pseudoacacia* L., *Wisteria sinensis* D.C., *Lilium candidum* L., *Olea europaea* L., *Philadelphus coronarius* L. and *Coronilla emerus* L.

Isolation/Enrichment by Sorption

Sorption is another alternative for the preconcentration of volatile components over a period of time, some of its variants having found wide application. One such variant (*a*) is to place the sorption agent directly in the headspace of a scented substrate or to bring it in direct contact with the substrate, a form of 'enfleurage', which may be considered as a static method. Another, dynamic, variant (*b*) is closer to low-temperature trapping and consists of sweeping or drawing the headspace vapor from the natural substrate through a sorption agent kept in a glass tube or capillary.

(a) Modified 'enfleurage technique'. Aiming at somewhat larger quantities of fragrant material than are available by direct injection of the headspace vapor, in 1973 Holman & Heimermann[13] described a technique using oil-impregnated glass fiber paper to collect the scents of orchids. The principle is the same as that of the very traditional enfleurage technique, which is still used to a certain extent to produce very luxurious flower extracts (e.g. Tubereuse absolue pommade). To collect the fragrance, they placed such impregnated strips (5 cm × 15 cm, 200 mg Convalex-10 per piece) together with the flower—without injuring the plant—in a size-adapted glass flask for one to seven days. To analyze the adsorbed scent, the strip was rolled tightly and placed in the adapted injection port of the gas chromatograph connected with a mass spectrometer, and the material was transferred to the cooled column by heating for 10 min at 200°C. In the experiment described in more detail, they exposed six flowers of *Neofinetia falcata* during four days to an impregnated strip; it was estimated to have adsorbed 2 mg of scent during this time. As a result, linalool, methyl benzoate and caryophyllene could be identified as main constituents. Just to illustrate briefly the progress achieved since then in the sensitivity of the analytical methods, the scent of the same orchid species would have to be trapped for only 1 h, for example, on 5 mg charcoal, to obtain about 30–50 μg of fragrant material, which is enough for many GC/MS measurements and complementary investigations.

Applying exactly the same technique, two years later, the same authors together with Thien[14] investigated the flower odors of eight species of *Magnolia* native to the United States. These analytical results have been used to discuss taxonomic aspects of the genus *Magnolia*. One of these species especially—*Magnolia grandiflora*—is often cultivated as an ornamental tree and is highly appreciated for its beautiful creamy-white flowers, which emit an extremely delicate and distinct scent. Interestingly, this Magnolia fragrance contains a series of bicyclic monoterpenes seldom found in flower scents.[15] Two of them—verbenone and isopinocamphone—had already been identified at that time.

Another type of modified 'enfleurage technique' was described by Bergström and colleagues in 1973[16] and in 1980.[17] They used 10% silicon high vacuum grease on Chromosorb W to extract the volatiles, for example from the labella of *Ophrys* flowers. The extracted material was recovered mainly in two ways.

1. A precolumn was filled with a small portion (0·5–1 ml) of the sorbent containing the volatiles and placed directly in the splitless injection system of the GC, where the volatiles were desorbed under heating onto the GC column.

2. The scents were evaporated from the sorbents and condensed under vacuum in a glass spiral at −70°C. The condensate was recovered by rinsing it with diethyl ether.

A third type of 'enfleurage technique' especially designed for field experiments was described by Joulain[15] in 1986 and applied to the trapping of the flower scents of *Sambucus nigra* L., *Ligustrum ovalifolium, Ligustrum vulgare* L., *Ailantus glandulosa* Desf., *Magnolia grandiflora* L. and *Trachelospermum jasminoides*. In such an experiment, around 100 g of freshly picked flowers are spread on the grid of a stainless steel device. In the first stage, a stream of ambient air, supplied by an unheated portable hair-drier, blows through the flowers on the grid. The entrained volatiles are then trapped on a layer of purified fat placed above the grid. After two hours the fat is collected and, in a second laboratory stage, the volatiles are recovered by the low-temperature trapping[12] already discussed.

(b) Packed precolumn sorption/sorption traps. Referring to results described by Silverstein & Rodin,[18] Bergström[16] and his co-workers in 1973 were apparently the first to use precolumn tubes to study floral fragrances. They redesigned the inlet system of their gas chromatograph to accept the precolumn collection tube, where the volatiles were heat-desorbed onto the cold GC column. Some years later, Nilsson[19] used this principle to study the flower scent of *Platanthera chlorantha.* The disadvantage inherent to all trapping techniques involving heat desorption is that the entire sample is used in one injection, leaving no material for further experiments.

Two years later, Sirikulvadhana *et al.*[20] applied Poropak Q, which was described by Jennings *et al.*[21] as a useful sorbent for aroma volatiles, to capture the scent of a Floribunda rose (Spanish Sun) and a Hybrid Tea rose (Forever Yours). One to two dozen long-stemmed rose buds were placed in a glass chamber and humidified air was swept through it and the Poropak trap attached with a flow rate of 150 ml/min for five days. The volatiles adsorbed were backflushed with nitrogen at 100–110°C and collected in a glass capillary at −70°C. Samples for subsequent GC and GC/MS analyses were taken from the capillary by syringe. The chromatograms obtained on an 88-m capillary column coated with Carbowax 20M showed a remarkable separation and, although not all of the constituents could be identified by GC/MS at that time, a serious basis for the investigation of flower scents was laid.

Poropak Q was also used by Tsuneya *et al.*[22] in 1977 to trap the headspace volatiles of Gardenia flowers (*Gardenia jasminoides* Ellis. f. *ovalifolia*). In this case, the volatiles accumulated by sweeping 260 g of flower petals during 75 min (3 liters of headspace gas) were directly transferred by heat desorption onto the cooled capillary column and subsequently investigated by GC/MS. For completeness and as a comparison, Tsuneya *et al.* also investigated the absolute sample obtained from the petals of the same species. Among 60 constituents identified, ocimene, linalool, methyl benzoate and the tiglates of methanol, ethanol, *n*-hexanol and *cis*-3-hexenol proved to be major constituents of the headspace sample while α-farnesene and *cis*-jasmine lactone have been described as the dominating constituents of the absolute.

Although activated charcoal is an extremely powerful adsorbent which possesses optimal thermal and chemical stability and which is not deactivated by water, it was only used in a few cases up to the mid-1970s to trap the volatiles of flavor substrates. The main fears have been of incomplete recovery by solvent extraction and of thermal decomposition caused by applying heat desorption; however, these have been greatly overcome by the development by Grob[23] and by Grob & Zürcher[24] of small traps holding only 1–2 mg of charcoal. The adsorbed volatiles can easily be recovered by extraction with 10–50 μl of carbon disulfide, thus avoiding a further concentration step prior to the investigation by GC/MS. The successful preparation, handling and extraction of such traps is thoroughly described by the same authors.[23,24]

In 1977, Kaiser & Lamparsky[25] used this type of charcoal trap to complete their investigation of a hyacinth absolute and identified in the headspace of the living flower some olfactorily important constituents not found in the absolute. Especially worth mentioning are the 2,6-dimethyl-3(*E*),5(*E*),7-octatrien-2-ol and its 5(*Z*)isomer, which have been identified since then by the same authors in a great variety of flower scents. These trienols will be discussed below in more detail. In 1980, the same authors[26] discussed some structurally interesting nitrogen-containing compounds identified in the flower scents of two *Lonicera* species, *Citrus aurantium* and *Hedychium flavum*, and trapping on charcoal was later applied by them to the headspace fragrance of *Freesia hybr.* and *Syringa vulgaris*[27] and *Jasminum sambac.*[28] Activated charcoal has been used by Kaiser for the trapping of many other olfactorily important flower scents. In 1987, he discussed night-scented flowers especially[29] and in 1988 he gave an overview on olfactory and chemical characteristics of floral scents.[30] Since no proceedings of these two congresses have been published, some of these topics will be treated later in this review.

In 1983, Williams & Whitten[6] reported on the progress they had achieved since 1968[8] in analyzing floral scents of orchids characterized by the 'male euglossine syndrome'. Instead of withdrawing 10 ml of headspace vapor from the collection chamber adapted in size and shape to the inflorescence, in 1982 they started to use[31] a trapping tube laden in the first part with Tenax and in the second part with charcoal.[6,31] The adsorbed volatiles were recovered with the aid of a special heat-desorbing device and subsequently condensed in a glass capillary by cooling with liquefied nitrogen. One year later, Zhu Liang-feng *et al.*[32] discussed their GC/MS investigation of the flower fragrance of *Hedychium coronarium* Koen., which was trapped on XAD-4 resin, an adsorbent widely used in water analysis. They identified 30 constituents in the beautiful 'white-floral' fragrance of this member of the Zingiberaceae.

Beside Poropaks, Chromosorb and activated charcoal, Tenax GC has become quite popular, not only in flavor research—in a few cases it has also found application for the trapping of flower scents. De Pooter *et al.*[33] used this adsorbent for the preconcentration of the scent of *Lonicera periclymenum*. The volatiles were recovered by thermal desorption onto a cooled storage tube. In 1986[34] and 1989,[35] Mookherjee *et al.* compared the scents of a series of living flowers trapped on Tenax with those of the corresponding so-called 'dead' flowers. Some aspects of these investigations will be discussed later in this review. Also very recently, Patt *et al.*[36] applied Tenax as well as charcoal to the trapping of the floral fragrance of *Platanthera stricta,* a North American relative of the Eurasian species *Platanthera chlorantha* investigated by Nilsson[19] using Poropak Q.

Although not directly applied to the investigation of flower scents, the methodology described by Bicchi *et al.*[37] in 1987 and 1988 should also be mentioned. They used fused silica open tubular traps coated with a 15 μm film of dimethylsilicone to capture volatiles emitted by living plants. The trap is then installed in a multidimensional capillary GC system and the volatiles are thermally desorbed, cold-trapped and analyzed on the analytical column.

This review on methods applied to the headspace investigation of flower scents should not be concluded without drawing attention to the 'comparison of techniques for the isolation/enrichment of plant volatiles in the analysis of *Ophrys* orchids' published by Bergström *et al.*[17] in 1980. Although *Ophrys* scents are hardly of importance to the fragrance industry, these principle considerations and comparisons are certainly still valid and may be applied in the investigation of any plant volatiles.

Complementary Methods

It is well understood that the total amount of volatiles present in the plant tissue at the time of headspace trapping cannot be considered as the 'liquid phase'. The relative quantity of a compound in the scent depends not only on its vapor pressure but also on its rate of interaction with other constituents of plant tissue such as water, lipids, carbohydrates, proteins and other non-volatiles, and last but not least on the permeability of the cell membranes. Nevertheless, the complementary/comparative investigation of a micro-extract of the fragrant plant tissue—if available at all for this purpose—often facilitates the estimation of the quantitative data in the 'liquid phase' to reproduce finally the respective headspace composition (cf. Section 5).

Such micro-extracts may be obtained by extracting the flower(s) with highly purified solvents such as hexane, fluorocarbons or liquid carbon dioxide. Jennings[38] presented a convenient laboratory system using liquid carbon dioxide to prepare 'solvent-free' extracts of such natural substrates. Since injuring the plant/flower may initiate enzymic reactions, possibly leading to the rapid formation of volatiles which occur only in traces or not at all in the original scent, the substrate should be exposed immediately to the solvent, which may contain an enzyme-inactivating agent such as methanol. However, such micro-extracts often contain considerable amounts of high-boiling materials or non-volatiles, which should be removed prior to their investigation by GC/MS by applying mild clean-up procedures.

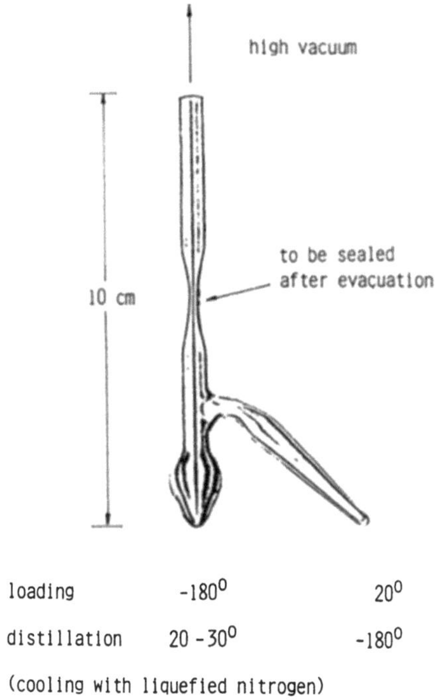

Fig. 1. Apparatus for microdistillation of natural extracts. Temperatures are in °C.

High-vacuum transfer, a very mild type of distillation, is often the method of choice, which may be performed in a size-adapted version (cf. Fig. 1) of the apparatus described by Weurman *et al.*[39] In such a clean-up experiment, the microdistillation apparatus is loaded with the minute amount of 'concrete' obtained by solvent extraction (0·5–5 mg) and cooled with liquid nitrogen. The whole system is evacuated to about 0·01 torr and sealed under high vacuum by melting off the inlet section. Cooling with liquid nitrogen is now changed to the trapping section and, therefore, allows an extremely mild distillation even at room temperature. After 1–6 h—depending on the temperature applied—the material condensed in the trapping section is rinsed with an adequate portion of hexane (10–500 μl) and subsequently investigated by GC/MS.

Simultaneous steam-distillation–extraction, originally described by Likens & Nickerson[40] and later modified in many ways,[41] is one of the most widely used isolation techniques for separating volatiles from non-volatile material. However, the suitability of this method for particular substrates should be carefully evaluated, since the formation of artefacts has to be taken into consideration. To minimize artefact formation, it is highly recommended that the plant material should not be used directly, but a solvent extract thereof. As described by Schmaus & Kubeczka,[42] special attention has to be given to substrates rich in linalyl acetate, which partially rearranges under these conditions. This applies in general to tertiary allylic terpene alcohols, which may undergo dehydration or allylic rearrangement.[43] Degradation products of carotenoids may also be formed, as illustrated by comparing headspace samples of various color types of *Freesia hybr.* with corresponding extract samples obtained by distillation–extraction.[27] These aspects to be considered are more comprehensively discussed in a paper published by Schreier[44] in 1987. To prevent the formation of such artefacts Schmaus & Kubeczka[42] developed a dynamic vapor phase stripping technique, which allows separation of the volatiles from the non-volatiles of a micro-extract.

The apparatus for simultaneous distillation–extraction should be size-adapted in such a way that additional concentration of the solvent containing the volatiles is not necessary any more. Godefroot *et al.*[45] described a system in which the volatiles are collected in only 1 ml samples. Another microscale apparatus for steam distillation has been developed by Bicchi *et al.*[46]

3 BIOLOGICAL ASPECTS TO BE CONSIDERED

Sample preparation is certainly one of the critical steps in the investigation of natural scents and neither a sophisticated instrumental technique nor a high level of data-processing can ever compensate major errors in these early analytical steps. If one has the intention of capturing, investigating and subsequently reconstructing a flower scent at its prime, however, not only physico-chemical but also biological aspects connected with scent emanation should be carefully taken into consideration.

The high diversity in flower scents and the phenomenon of their diurnal variation typical of so many species reflect the high diversity in pollination biology.[47] Mrs Louise B. Wilder,[48] a great lover of scented flowers, paraphrased the biodynamic aspects of scent emanation—as they appear to the human nose—by the following words:

> Some day-scented flowers give out little or no perfume after sunset while others seem then to increase their emanations. Not infrequently too it is surprising to note that the fragrance given out by day is of quite a different quality from that given out by night. The aromatics of the Honeysuckle by day and by night are not at all the same, and the white Petunia, whose day-scent falls short of being agreeable, gives forth a refined and delicious perfume after sunset . . .

Being aware that the quality and quantity of a scent emitted by a botanically well-defined species of a flower/plant may strongly depend on phenomena such as maturity, biological rhythm and environmental factors, the experimental parameters of headspace trapping have to be very flexibly adapted to the individual cases, as summarized in the following, by the method of adsorption on activated charcoal and subsequent solvent elution.

Scented flowers which still display the original fragrance profile for 0·5–2 hours after cutting—this sampling time is normally sufficient—may be placed in a suitable glass chamber of appropriate volume (Fig. 2).

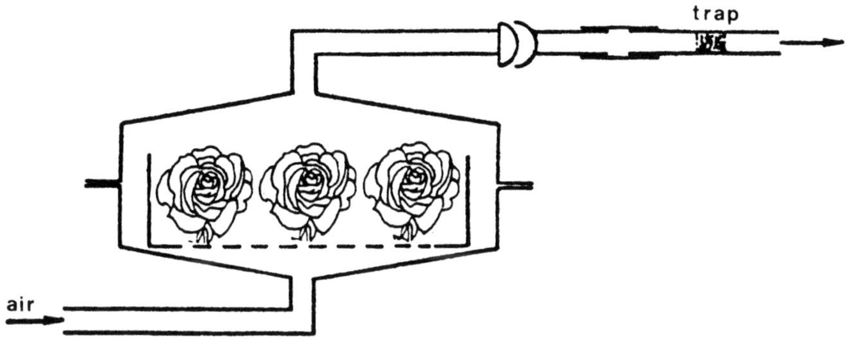

Fig. 2. Trapping the headspace of cut flowers emitting the original scent for an adequate time.

The emitted scent is collected on the charcoal trap (1–10 mg, with a second trap in series to control a possible breakthrough) by passing air through the system (30–80 ml/min, 0·5–2 h, depending on species). Afterwards, the headspace preconcentrated in this way (1–150 μg) can be recovered by extraction with carbon disulfide (10–50 μl, depending on trap size) and the sample thus obtained is subsequently investigated by GC/MS. Compared with the heat desorption often applied today, solvent extraction has the advantage of having material available for further investigation, e.g. for sniffing at the outlet of a capillary column to judge the olfactive importance of constituents hardly or not characterizable in a regular GC/MS run. Furthermore, thanks to the enormous improvement in the sensitivity of NMR spectroscopy and the development of procedures for preparative capillary gas chromatography (e.g. Refs 49, 50) it has become possible to isolate new compounds occurring in adequate concentrations directly from such headspace trappings and to elucidate their structure by NMR.

However, if the quality and quantity of the floral fragrance changes rapidly after cutting, if the rhythmicity in fragrance emission is to be studied or if the investigation is performed with rare species, the headspace has to be collected without disconnecting the flower from the plant, by sucking or sweeping the air around it through the trap (Fig. 3). This can be done either with an entire inflorescence (Fig. 3, left-hand side) or even with a single flower as illustrated by the fascinating example of *Angraecum sesquipedale* (Fig. 3, right-hand side), an orchid native to Madagascar, which displays exclusively during the night its beautiful 'white-floral' scent of the lily type. In the latter case, the trap is placed as near as possible to the part of the flower showing the highest odor emanation. *Angraecum sesquipedale* is also an impressive example demonstrating how strongly a floral scent may depend on the maturity of the individual flower.[30] The first GC trace shown in Fig. 4 represents the scent accumulated during the first night after inflorescence, the second GC trace that accumulated during the subsequent night.

When trying to capture a natural scent at its prime, diurnal rhythms in the

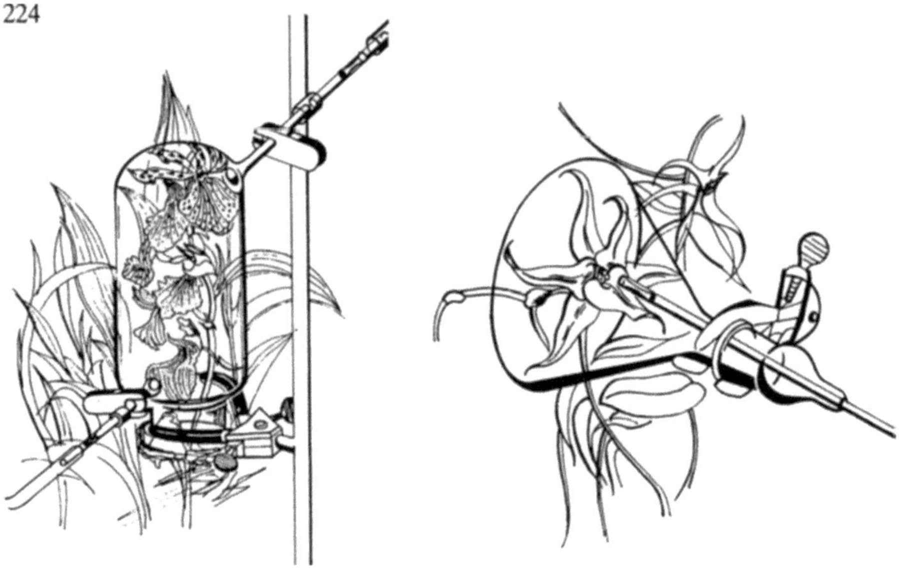

Fig. 3. Trapping the headspace of an entire inflorescence (left) or of a single flower (right) still attached to the plant.

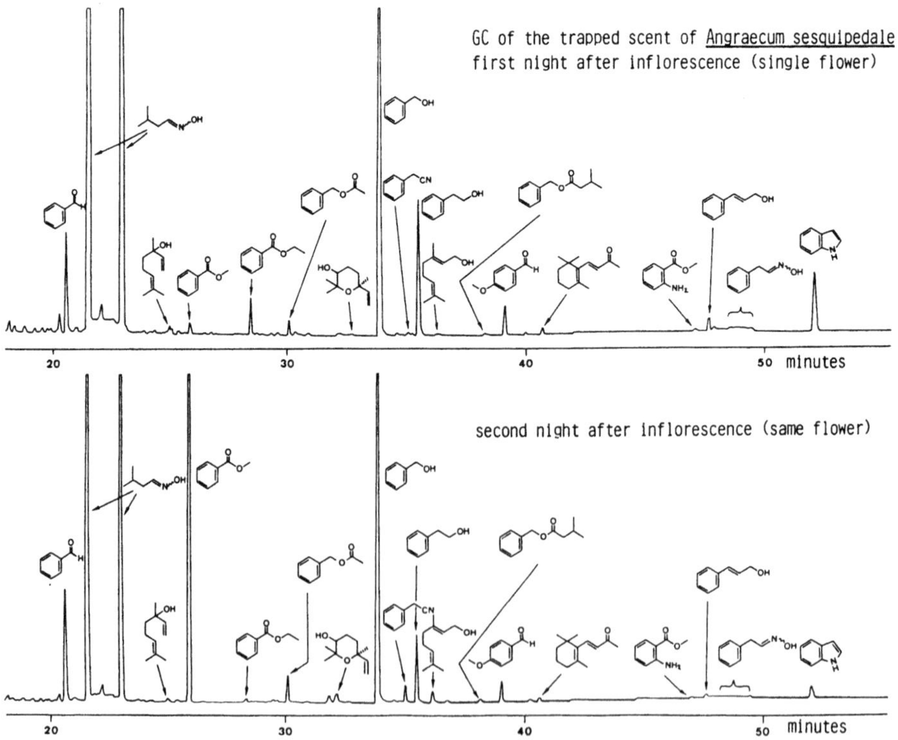

Fig. 4. GC of the trapped scent of *Angraecum sesquipedale*: UCON 50 m × 0·3 mm i.d., 60–190°C at 2°C/min.

emanation of certain constituents have also to be taken into consideration. For example, the well-known hyacinth (*Hyacynthus orientalis* Hort.) shows a pronounced rhythmicity in the emanation of the 2,6-dimethyl-3(*E*),5(*E*),7-octatrien-2-ol and its 5(*Z*) isomer (in hyacinth *ca.* 10:1), having its maximum between 6 p.m. and 10 p.m. when observed in Switzerland during the second part of February (Fig. 5).[30] The time when the relative amount of these

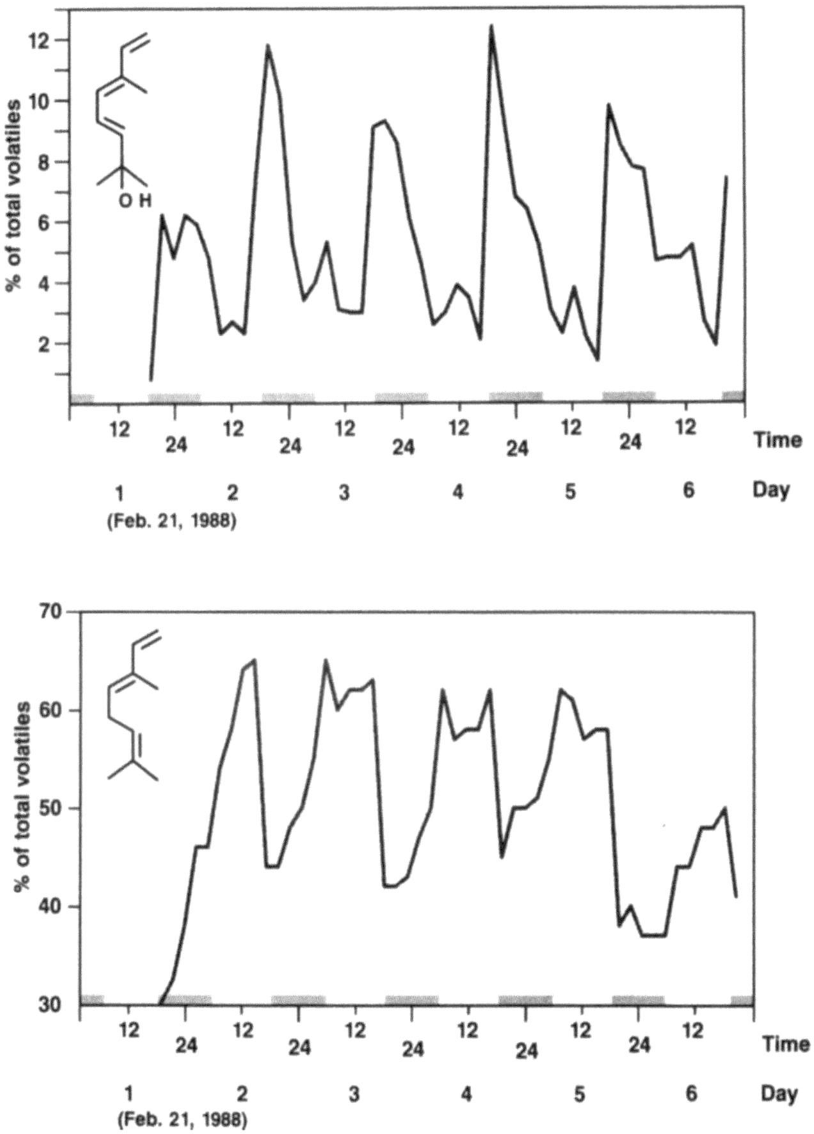

Fig. 5. Diurnal changes in the content of 2,6-dimethyl-3,5,7-octatrien-2-ol (3(*E*),5(*E*):3(*E*),5(*Z*) ~10:1) and (*E*)-ocimene in the fragrance emitted from *Hyacinthus orientalis* Hort. (white).

trienols starts to increase dramatically coincides with the sunset, as was learned from experiments performed one and two months later. Interestingly, their precursors—the (*E*)-ocimene and its (*Z*)-isomer (approx. 10:1 in hyacinth)—show more or less anticyclic behavior.

Although diurnal variations of scents emitted by flowers are so obvious for many species, the scientific investigation of this phenomenon is the topic of only a few publications. In 1960, Overland[51] described how the opening and closing of flowers of *Cestrum nocturnum* (*Solanaceae*) occurs in a cyclic manner both in constant light and in constant darkness. The strong odor, which occurs in nature at night and was previously believed to be a direct result of darkness, has been shown to exist as an endogenous rhythm, occurring under constant conditions of temperature and light, as well as in darkness. Although in nature the odor cycle is synchronized by the environment so as to have a period of roughly 24 h, in constant conditions temperature affects the cycle by lengthening it at low temperature and reducing it at high temperature. Comparable results have been obtained for both intact and cut flowers.

In 1978, studying pollination ecology and adaptation in *Platanthera chlorantha,* Nilsson[19] showed that some of the fragrance compounds such as caryophyllene emanate continuously from the flowers while others, e.g. methyl benzoate, are characterized by distinct periodicity.

In a recent publication, Matile & Altenburger[52] discussed the rhythms of fragrance emission in flowers in a broader sense. Applying trapping on activated charcoal over periods of three hours and subsequent solvent extraction as summarized above, they investigated the diurnal changes in quality and quantity of the floral fragrances of *Odontoglossum constrictum* Lindl., *Citrus medica* L., *Hoya carnosa* R. Br. and *Stephanotis floribunda* Brongs. Whereas even in detached flowers of *O. constrictum* and *C. medica* rhythmicity could be observed for up to four cycles, flowers of *H. carnosa* showed this phenomenon only when attached to the plant. *Hoya carnosa* is described as an example of nocturnal emission of volatiles as illustrated in Fig. 6 by the major headspace constituents isopentanol, β-pinene, cineole and methyl salicylate. Despite the marked overall decline of fragrance (Fig. 6), the diurnal changes of scent production with sharp maxima at approx 3 a.m. could be followed for up to four days. In agreement with the olfactory evaluation by nose, between noon and 6 p.m. practically no volatiles were present in the headspace of the enclosed flower. The synchronism of emission regarding the five major components of fragrance is remarkable. Yet the quantitative composition of fragrance changes in the course of a night, and the olfactory perception of the *Hoya* fragrance at midnight is therefore different from that at 6 p.m.

In the case of *Stephanotis floribunda,* they found as many as seven diurnal cycles of rhythmic emission. This very attractively scented species represents an example of marked rhythmicity not only with regard to the quantity but also to the quality of fragrance produced. As shown by Matile & Altenburger in

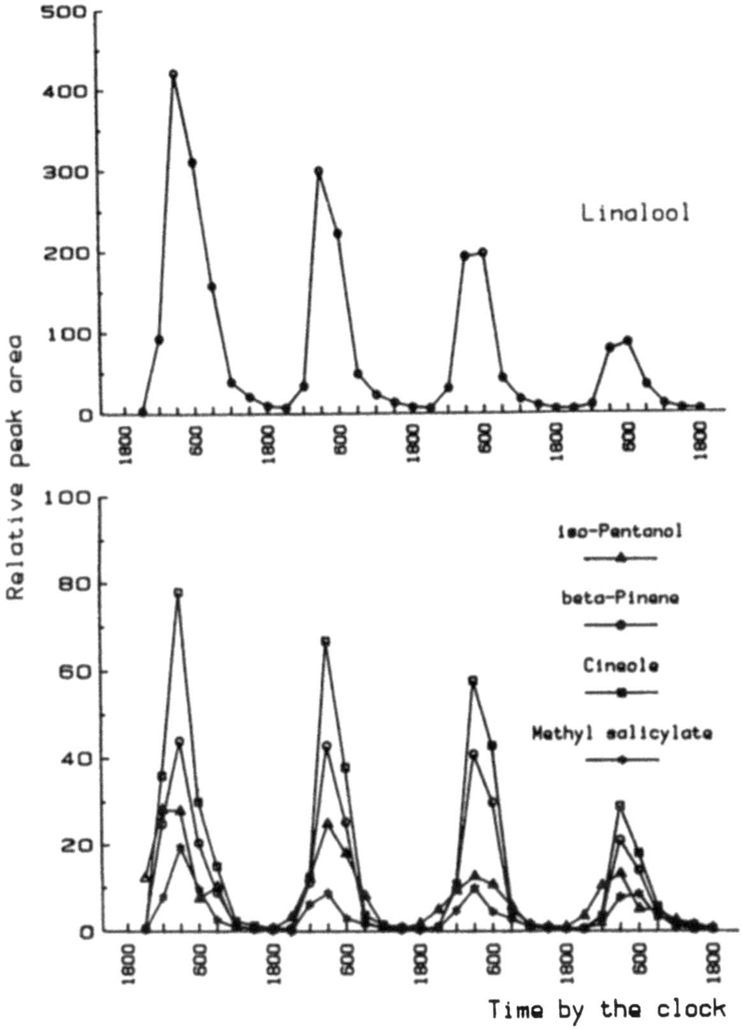

Fig. 6. Rhythmic emission of volatiles in flowers of *Hoya carnosa* (Matile & Altenburger[52]).

Fig. 7(*a*) the maxima of the three components methyl benzoate, linalool and 1-nitro-2-phenylethane do not coincide. In Fig. 7(*b*) the relative abundances of these three major volatiles are given for each sampling period. The maxima for methyl benzoate and linalool occur at midnight whereas the percentage of 1-nitro-2-phenylethane is highest at noon.

In the same year Altenburger & Matile[53] described how the nocturnal emission of fragrance from flowers of *Hoya carnosa* occurs according to an endogenous circadian rhythmicity. This was demonstrated by the continuation of rhythmical emission for two to three cycles under conditions of permanent

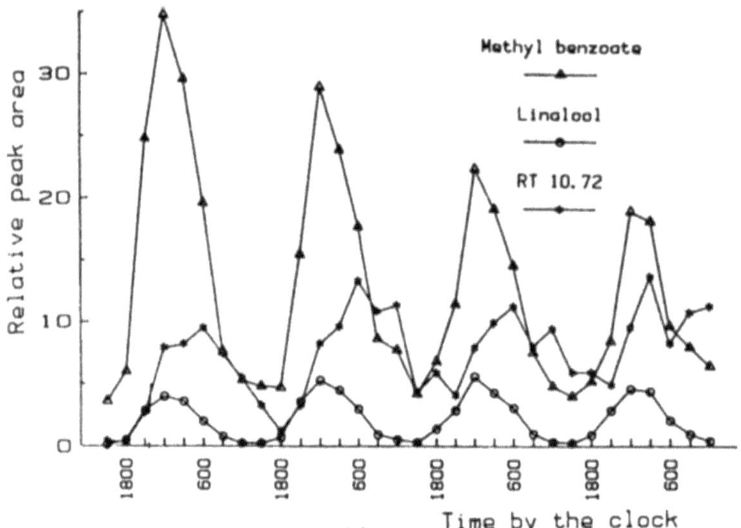

(a)

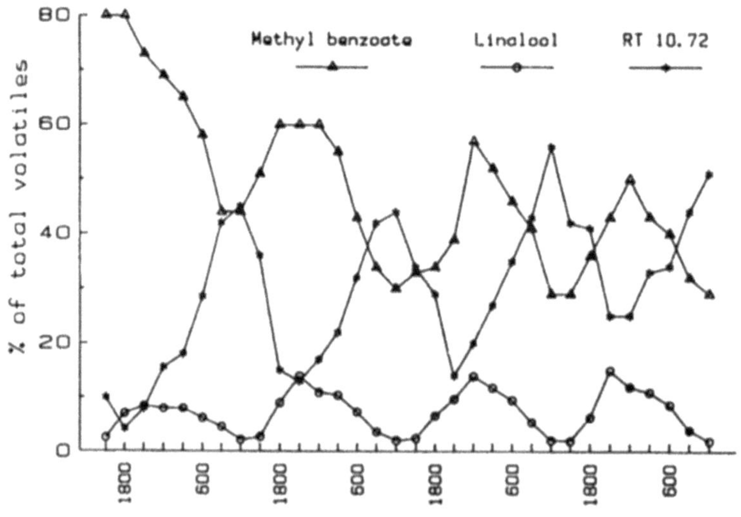

(b)

Fig. 7. (*a*) Diurnal changes of volatiles collected from the headspace of a flower of *Stephanotis floribunda*. Only four out of seven cycles altogether analyzed are shown. The compound with retention time (RT) 10·72 min has tentatively been identified as 1-nitro-2-phenylethane.[52] (*b*) Rhythmic changes of fragrance composition in *Stephanotis floribunda* in the course of four diurnal cycles.[52]

illumination. The free-running period was approximately 29 h under these conditions. The inversion of light/dark cycles caused the entrainment of the next emission peak by approximately 12 h, whether the inversion was started with a light period or with a dark period prolonged for 12 h. These studies on rhythmicity in fragrance emission illustrate the possibilities that future applications of advanced headspace techniques might not only help us to understand the real mystery of a flower scent but also to enormously enlarge our knowledge of biological phenomena of the utmost importance to the preservation of the respective species.

Concluding this section on biological aspects to be considered in connection with the trapping of flower scents, two recent papers published by Mookherjee *et al.*[34,35] should be mentioned. They compared the composition of the fragrance emitted from living flowers still attached to the plant with those of picked flowers, which they call 'dead flowers'. In side-by-side experiments, they placed the living and the 'dead' flowers in suitable glass flasks equipped with a Tenax trap. The flasks were purged with air for 6–12 h, then the trapped volatiles were desorbed onto the GC column and subsequently investigated by GC/MS. The authors discussed the quantitative differences for some compounds found in the fragrance of living and 'dead' flowers of *Jasminum grandiflorum, Freesia, Syringa vulgaris,* 'Easter lily',[34] Tea rose, *Narcissus tazetta* and *Osmanthus fragrans.*[35] Unfortunately, they did not mention if the cut flowers had any water support during the 6–12 h experiment or if the living as well as the dead flowers have been investigated at the same time after anthesis. As an example of the different flower scents investigated by this comparative approach, Table 1 shows the major differences in the composition of the scent of living and picked 'Yellow Tea Rose' (JFK) flowers as reported by Mookherjee *et al.*[35] They stressed, for example, the extreme difference in the content of *cis*-3-hexenyl acetate, amounting to about 21% of the total volatiles of the living Tea rose and only present at 5% in the scent of the picked rose.

However, these findings do not agree with some results discussed by Kaiser[30] in 1988 concerning the comparison of the diurnal changes in the content of *cis*-3-hexenyl acetate and related volatiles in the fragrance emitted from a

Table 1
Major differences[a] between living and picked Yellow Tea Rose (JFK) flowers[35]

Compound	*Living flowers*	*Picked flowers*[b]
cis-3-Hexenyl acetate	20·67	5·39
Hexyl acetate	8·40	4·26
Phenylethyl alcohol	5·73	3·30
Phenylethyl acetate	5·50	1·53
3,5-Dimethoxytoluene	9·96	18·58
α-Elemene	—	4·07
Geranylacetone	2·17	—
Dihydro-β-ionol	—	2·62
Isocaryophyllene	0·30	2·12
α-Farnesene	5·83	2·96

[a] Data are given as area-normalized percentages.
[b] Air-purged.

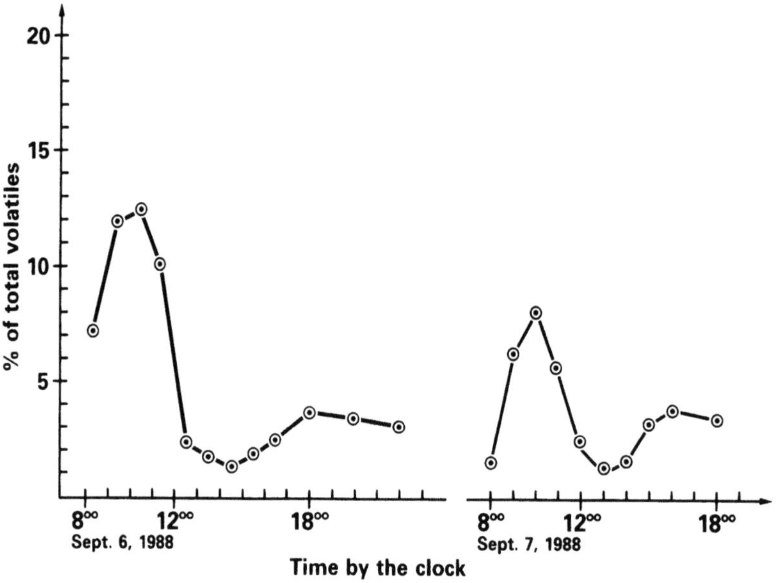

(a)

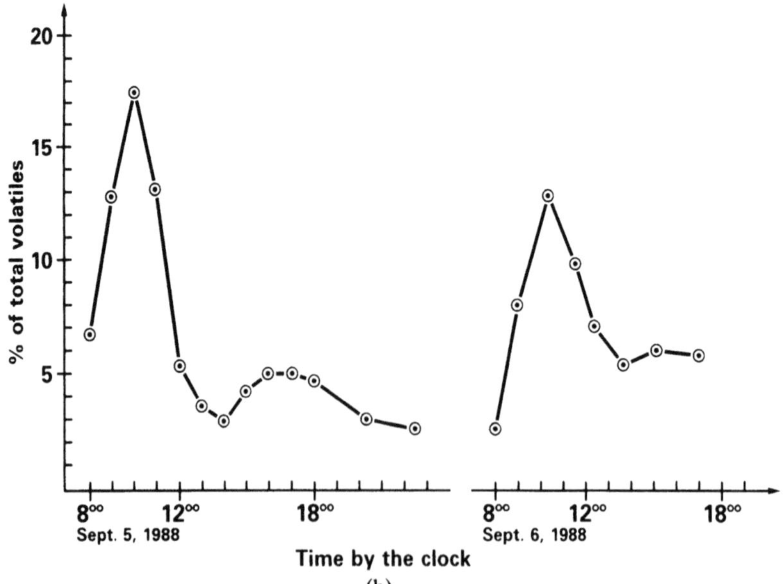

(b)

Fig. 8. (*a*) Diurnal changes in the content of *cis*-3-hexenyl acetate in the fragrance emitted from a flower of the Hybrid Tea rose 'Fragrant Cloud' still attached to the plant. (*b*) Diurnal changes in the content of *cis*-3-hexenyl acetate in the fragrance emitted from a cut flower of the same cultivar.

flower of the Hybrid Tea rose 'Fragrant Cloud' still attached to the plant with that of a cut flower of the same plant. These comparative experiments—in one case the flower still attached to the plant and placed in a glass vessel adapted in size and shape, in the other case the cut flower having water support and placed in the same type of vessel—were performed under field conditions (5–7 September 1988; clear days) with flowers of comparable maturity (one day after anthesis). The fragrant air was sucked with a flow rate of approximately 60 ml/min through the charcoal traps (5 mg) attached to the vessels; these traps were exchanged every hour. Total adsorption of the volatiles in the first trap was demonstrated by attaching a second consecutive trap. As shown in Fig. 8(*a*), the relative abundance of *cis*-3-hexenyl acetate increases dramatically during the morning to its maximum around 10 a.m. and decreases afterwards within a short time to its minimum. The same behavior was found for related volatiles, e.g. *n*-hexyl acetate (maximum concentration approximately 2.2%) and *trans*-2-hexenyl acetate (maximum concentration approximately 4.6%). Interestingly, a rhythm showing the same characteristics exists also for the cut flower (Fig. 8(*b*))—even during the second day of observation—and the term 'dead' seems not to be appropriate to describe this phenomenon. As experimentally verified, the differences in the relative amounts of *cis*-3-hexenyl acetate for the attached and the cut flower at a defined time definitely lie within the range of differences observed for individual flowers of comparable maturity still attached to the same plant. The

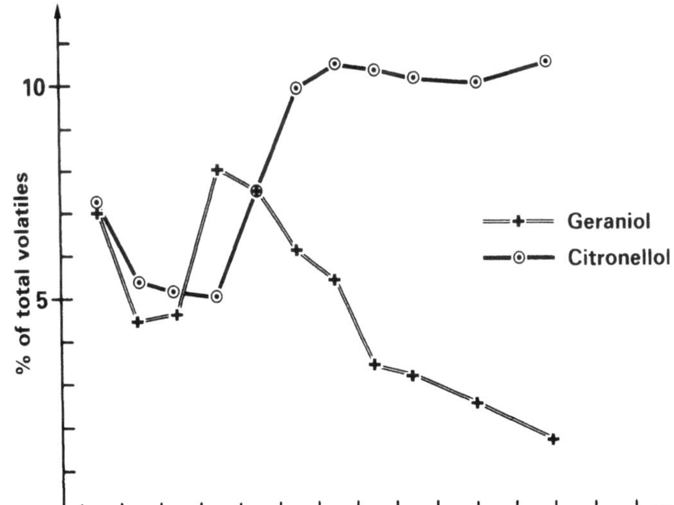

Fig. 9. Diurnal changes in the content of geraniol and citronellol in the fragrance emitted from a flower of the Hybrid Tea rose 'Fragrant Cloud' still attached to the plant (the same flower as was used for measurement shown in Fig. 8(*a*), 6 September 1988).

total amount of fragrance captured on the charcoal traps during the sampling periods of 1 h have been comparable for both the attached and the cut flower. It gradually increased from approximately 30 μg between 8 a.m. and 9 a.m. to its maximum of approximately 100 μg between 11 a.m. and 12 noon and decreased afterwards to approximately 15 μg between 5 p.m. and 6 p.m.

It has already been shown by Matile & Altenburger,[52] (cf. Fig. 7), and it is recognizable in many cases just by smelling at the flower, that the maxima of relative abundance of individual components do not have to coincide. While geraniol, in the case of 'Fragrant Cloud', shows an emission dynamic similar to *cis*-3-hexenyl acetate (compare Fig. 9 with Fig. 8(a)), citronellol and (for example) phenylethyl alcohol (not shown in Fig. 9) reach the highest content in the headspace in the later afternoon. Other constituents, e.g. germacrene D, appear to be emitted more or less continuously.

4 NEW OR UNUSUAL COMPOUNDS IDENTIFIED IN FLOWER SCENTS

2,6-Dimethyl-3,5,7-octatrien-2-ol

The first new natural products identified in the headspace of fragrant flowers might have been the 2,6-dimethyl-3(*E*),5(*Z*),7-octatrien-2-ol (**1a**) and its 5(*E*) isomer **1b** described by Kaiser & Lamparsky[25] as occurring in the scent of hyacinth. These two trienols normally occurring in nature in a ratio of 1:10 are practically always accompanied by smaller amounts of the corresponding dehydration products **2a** and **2b** in a similar ratio.

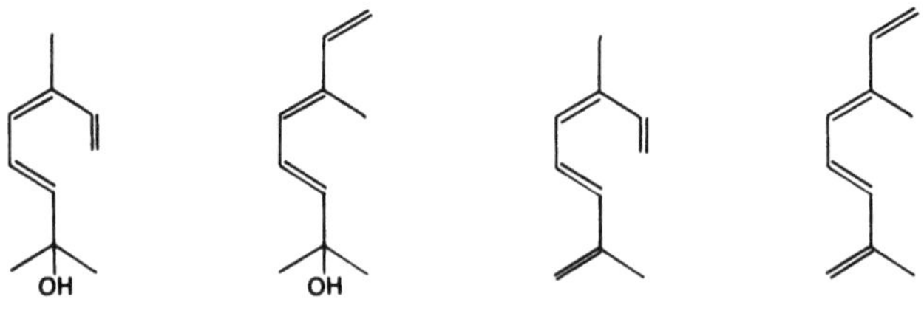

1a 1b 2a 2b

As already discussed (cf. Fig. 5), **1a** and **1b** are emitted from flowers of *Hyacinthus orientalis* in a pronounced diurnal rhythm having the maximum in the late evening. A comparable rhythm in the emanation of **1a/1b** could be observed when investigating the time-dependence of the floral scents of *Hoya carnosa* and *Stephanotis floribunda.*[30] In all three cases the emanation of **1a/1b** starts to increase dramatically at dusk, reaching the maximum of relative abundance for *Hyacinthus orientalis* around 8 p.m. (20 February: 8–12% depending on cycle, 1–4% during day; cf. Fig. 5), for *Stephanotis floribunda* around 9 p.m. (2 May: 2–4%, 0·1–0·3% during day) and for *Hoya carnosa* around 3 a.m. (27 April: 0·1–0·3%, only traces during day). The maximum of relative abundance of **1a/1b** coincides in the cases of *S. floribunda* and *H. carnosa* with the maximum of total fragrance production.[52]

These trienols **1a/1b**—characterized by a highly attractive sweet floral odor—show a surprisingly widespread occurrence among night-active as well as day-active scented flowers, as illustrated by some examples summarized in Table 2 (identified in flower scents between 1977 and 1989; R. Kaiser, unpublished). Due to their high sensitivity, however, they are normally only found in the original scent—the headspace.

Amino-acid derivatives and other nitrogen-containing compounds

A series of new compounds of flower scents, which may be considered as being derived from the corresponding amino-acids, has been reported by Kaiser & Lamparsky[26] in 1980. They identified in the trapped headspace of two *Lonicera* species and in that of *Hedychium flavum* the (E) and (Z) isomers of isobutyraldoxime (**3**), 2-methylbutyraldoxime (**4**) and 3-methylbutyraldoxime (**5**), all being accompanied in the concentration range of trace or minor constituents by the corresponding nitriles and nitro compounds. Although these oximes **3–5** as well as their derivatives are not responsible for the basic fragrance character, they contribute to the peculiar olfactory image of the living flower. As discussed later by Kaiser in 1987[29] and 1988,[30] the oximes **3–5** as well as phenylacetaldoxime (**6**) are frequently found in the trapped

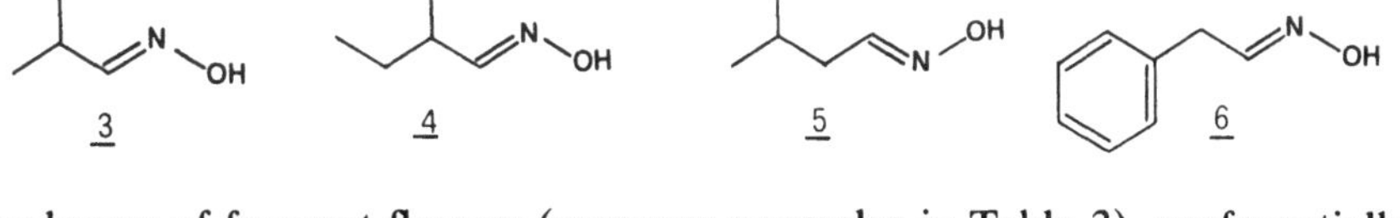

headspace of fragrant flowers (compare examples in Table 3), preferentially in those of flowers showing the syndromes associated with moth pollination. The oxime **6** was reported for the first time as a natural product by Sakurai *et al.*[54] in 1979; they identified it in the flower absolute of *Citrus unshiu*. Later, in 1986, Joulain[12] described it as occurring together with other known nitrogen-containing constituents in the headspaces of *Spartium junceum, Lonicera caprifolium* and *Robinia pseudoacacia.*

Based on the investigations of Underhill *et al.*[55,56] into the biosynthesis of mustard oil glucosides, it has to be assumed that **6** represents a metabolite of phenylalanine. In analogy, isobutyraldoxime (**3**) may be regarded as a

Table 2
Examples of the occurrence of **1a/1b**

Source of flower scent[a]	*Family*	*Abundance*[b]
Polianthes tuberosa	Amarylidaceae	++
Trachelospermum jasminoides	Apocynaceae	+++
Hoya carnosa	Asclepiadaceae	++
Stephanotis floribunda	Asclepiadaceae	+++
Dolichothele longimamma	Cactaceae	++
Selenicereus hamatus	Cactaceae	+
Sambucus nigra	Caprifoliaceae	++
Rhododendron flavum	Ericaceae	+++
Freesia hybrida hort.	Iridaceae	++
Iris germanica	Iridaceae	++
Robinia pseudoacacia	Leguminosae	++
Hyacinthus orientalis hort.	Liliaceae	++++
Lilium longiflorum hort.	Liliaceae	++
Magnolia grandiflora	Magnoliaceae	++
Michelia alba	Magnoliaceae	+
Michelia champaca	Magnoliaceae	++
Osmanthus fragrans	Oleaceae	+++
Angraecum compactum	Orchidaceae	++
Angraecum sesquipedale	Orchidaceae	+
Brassavola digbyana	Orchidaceae	+++
Brassavola glauca	Orchidaceae	+++
Cattleya labiata	Orchidaceae	++
Dendrobium chrysotoxum	Orchidaceae	+
Himantoglossum hircinum	Orchidaceae	++
Odontoglossum constrictum	Orchidaceae	+++
Oncidium tigrinum	Orchidaceae	++++
Cyclamen purpurascens	Primulaceae	++
Reseda odorata	Resedaceae	+
Citrus aurantium	Rutaceae	++
Boronia megastigma	Rutaceae	++
Dictamnus albus	Rutaceae	++
Datura arborea	Solanaceae	++++
Daphne japonica	Thymelaceae	++
Daphne mezereum	Thymelaceae	+
Tilia cordata	Tiliaceae	+++
Tilia platyphyllos	Tiliaceae	++++
Hedychium gardnerianum	Zingiberaceae	+++

[a] Refers to headspace samples accumulated on charcoal; concentration may show pronounced time-dependence.
[b] ++++, Main constituent; +++, major constituent; ++, minor constituent; +, trace constituent.

Table 3
Examples of the occurrence of the oximes **3–6** in flower scents[a]

Species investigated	**3**	**4**	**5**	**6**
Lonicera japonica	—	b	a	—
Lonicera periclymenum	—	b	d	—
Lonicera caprifolium	a	b	c	—
Hedychium flavum	a	a	b	—
Hedychium gardnerianum	a	a	b	a
Coffea arabica	a	b	a	—
Nicotiana alata	a	a	b	—
Nicotiana suaveolens	a	b	c	—
Michelia champaca	—	b	b	b
Citrus aurantium	—	a	—	b
Convallaria majalis	—	—	—	b
Gardenia taitensis	—	—	—	b
Angraecum sesquipedale	—	a	d	b

[a] a, 0·01–0·1%; b, 0·1–2%; c, 2–5%; d, >5% in the trapped headspace. Concentration may show pronounced time-dependence.

metabolite of valine and, correspondingly, the oximes **4** and **5** as metabolites of isoleucine and leucine, respectively.

Extracts of scented flowers rich in the oximes **3–6** often contain whole series of further oximes and derivatives thereof as minor or trace constituents.[26,57] They may be generated by transoximination of the first-formed oximes **3–6** with carbonyl compounds of the same substrate, and subsequent reactions.

Phenylacetaldoxime (**6**), which occurs practically always together with the corresponding nitrile and nitro compound, is frequently accompanied by its *O*-methyl ether (**7**), e.g. as in the flower scents of *Gardenia taitensis* (0·3–0·7%), *Citrus aurantium* (0·05–0·15%) and *Spartium junceum* (0·05–0·15%)[29] as well as in that of *Leptactinia senegambica* (Karo-Karounde).[58] The odor of **7** is reminiscent of a combination of phenylacetaldehyde, phenylpropionaldehyde and phenylethyl methyl ether.

Another widely spread nitrogen-containing compound is 2-aminobenzaldehyde (**8**) identified by Joulain[12] as a major constituent in the flower scents of *Robinia pseudoacacia* (>20%), *Spartium junceum* (10–20%), *Pittosporum tobira* (1–5%) and *Philadelphus coronarius* (~10%). In the last of

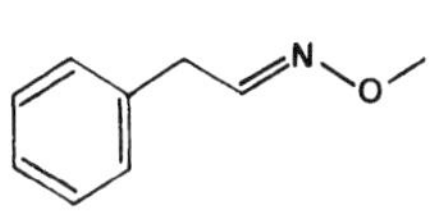

7 (E/Z)

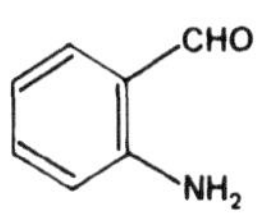

8

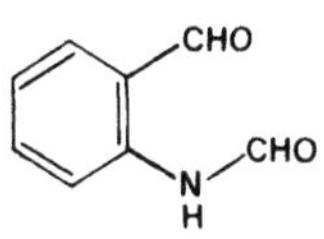

9

these scents, the corresponding *N*-formyl derivative **9** was also found as a minor constituent. 2-aminobenzaldehyde (**8**) shows a typical 'spring flower' odor and seems to be of interest as a fragrance product.[59]

Besides the oximes **3–6**, another series of compounds derived from amino-acids attracts interest. Joulain[15] identified in the scent of *Sambucus nigra* L., which appears both attractively floral and somewhat fetid, the methyl ester of valine (**10**, ~0·3%) and isoleucine (**11**, ~7%) as well as the imines **14** (~0·15%), **15** (~0·1%) and **16** (~0·15%) derived from **11**. To complete this type of naturally occurring volatiles, the methyl and ethyl esters of *N,N*-dimethylleucine (**12** and **13**) identified in fig leaf absolute (*Ficus carica* L.)[60] may be mentioned. Although only present in trace amounts, the components **10**, **11** and **14–16** are described as contributing to the peculiar odor character of the fresh flower.

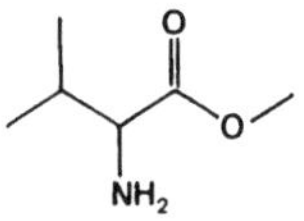
10

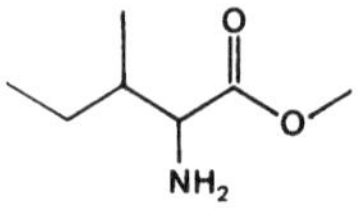
11

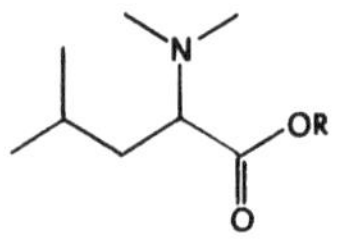
12 R = CH_3

13 R = C_2H_5

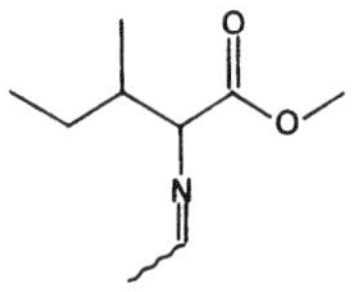
14

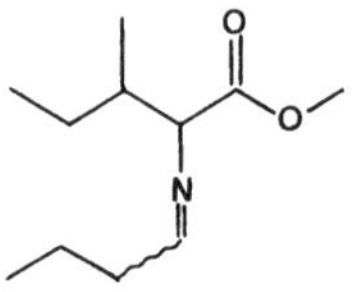
15

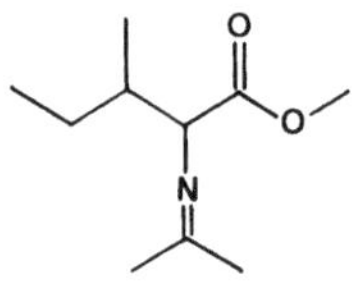
16

Unusual Oxygenated Monoterpenes

Oxygenated bicyclic monoterpenes, mostly characterized by herbaceous, woody, camphoraceous and balsamic odor aspects, are rather unusual in flower scents. Two such compounds—isopinocamphone (**17**, 9%) and verbenone (**18**, 18%)—together with a high amount of *cis*-jasmone (27%) have already been identified (in 1975) by Thien *et al.*[14] in the delightful fragrance of *Magnolia grandiflora*. Twelve years later, Joulain[15] confirmed these results and additionally identified in the same flower headspace myrtenol (**19**, 1%), myrtenal (**20**, trace), *trans*-verbenol (**21**, trace), chrysanthenone (**22**, trace) and *cis*-verbanone (**23**, trace).

Indeed, it was very stimulating to learn from nature that such high amounts of jasmone and bicyclic monoterpenes perform so harmoniously in a flower scent. The only scent comparably rich in this type of compounds is that of *Philadelphus coronarius,* in which Joulain[15] identified as much as 46% of

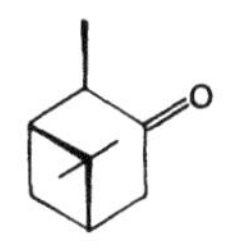

17

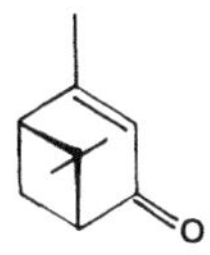

18

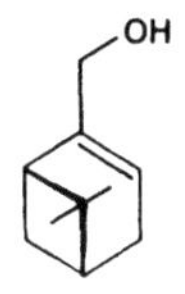

19

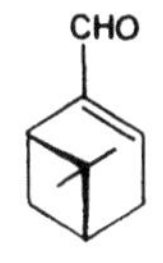

20

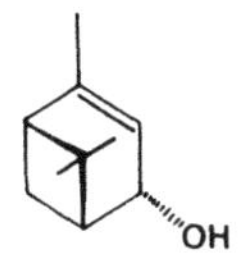

21

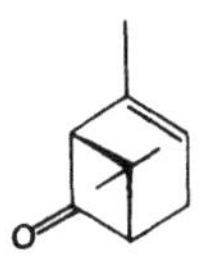

22

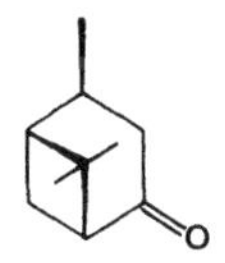

23

24

25

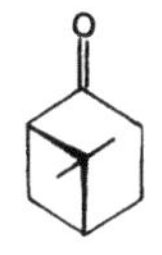

26

trans-myrtanal (**24**) together with its *cis* isomer **25** (~2%) and nopinone (**26**, ~5%) as well as **19** and **20** in trace amounts.

Constituents known as Insect Pheromones and Structurally Related Compounds

Besides the bicyclic monoterpenes discussed above and other constituents, Kaiser[29] identified, in the scent of *Philadelphus coronarius,* the two isomers of chalcogran (**27a/b**), which were discovered by Francke *et al.*[61] as principal aggregation pheromones of the bark beetle *Pityogenes chalcographus* (L.). This result appears remarkable, since the compounds **18–21** including the *cis* isomer of **21** belong to the most widespread oxygenated terpenes found in bark beetles.[62] Interestingly, the chalcogran **27a/b** could also be found at about 0·1% in the trapped scent of *Dracaena fragrans,* which contains at about 0·3% another pair of diastereomers known as pheromones—(*E*)-7-methyl-1,6-dioxaspiro(4,5)decane (**28a**) accompanied by its (*Z*) isomer (**28b**) in a ratio of about 10:1 (R. Kaiser, 1985, unpublished). These spiroketals (**28a/b**) have been identified by Francke *et al.*[63] in the pentane extract of workers of *Paravespula vulgaris* L.

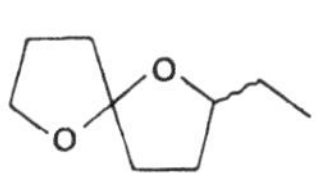

27a/b

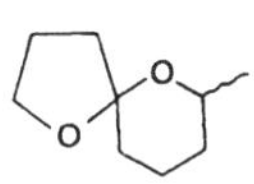

28a/b

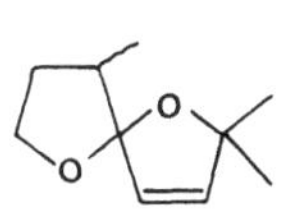

29a/b

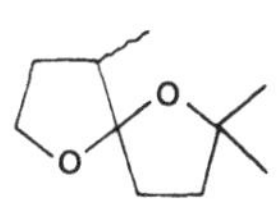

30a/b

In this context, the identification of the diastereomers of the two structurally related spiroketals **29a/b** and **30a/b** in geranium oil[64] attracts interest.

Furthermore, an interesting cyclic ortho ester (**32**), which may be considered as an oxa analogue of brevicomin (**34**), could be identified by Kaiser[28] in the trapped headspace of a Philippine type of *Jasminum sambac.* Brevicomin (**34**) is an aggregation pheromone of several scolytid bark beetles of the genus *Dendroctonus.*[65]

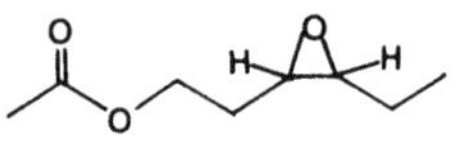

31

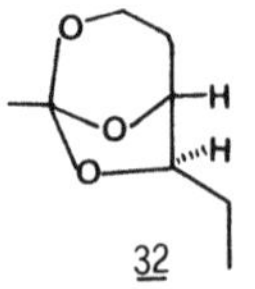

32

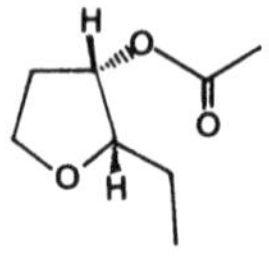

33

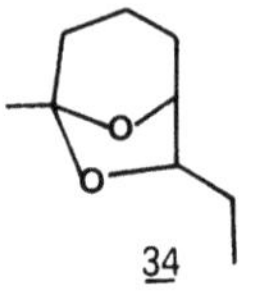

34

The ortho ester **32** represents the intermediate in the rearrangement of (*Z*)-3,4-epoxyhexan-1-yl acetate (**31**) to the *trans*-2-ethyl-3-acetoxy-tetrahydrofuran (**33**), both products being present in the same flower scent. Both **31** as well as **33** are characterized by interesting, very diffusive, green herbaceous odors. The formation of the epoxide **31** and its rearrangement products **32** and **33** has to be considered as a characteristic aspect in the biogenesis of volatiles of *Jasminum sambac,* which distinguishes the latter from *J. grandiflorum.*

Two Interesting Homoterpenes

In 1986, Maurer *et al.*[66] described the 4,8-dimethyl-1,3(*E*),7-nonatriene (**35b**) and the 4,8,12-trimethyl-1,3(*E*),7(*E*),11-tridecatetraene (**38d**) as minor constituents of cardamon oil. As discussed by Kaiser in 1987,[29] these two structurally interesting homoterpenes, characterized by the same molecular weight and a very similar mass fragmentation as perillene and dendrolasine, respectively, might be nearly as widespread in flower scents (Table 4) as the tertiary ocimenols **1a** and **1b** already discussed (cf. Table 2).

A typical example of the natural occurrence of **35b** and **38d** is in the scent of *Selenicereus hamatus,* a representative of the Cactaceae family native to south-east Mexico and the Lesser Antilles, which blooms after sunset with very large and fragrant flowers that last for one night only. The GC curve shown in Fig. 10 gives a partial overview of the composition of the trapped headspace (5-mg charcoal traps) of *Selenicereus hamatus,*[29,30] from which the main constituent **35b** was isolated some years ago by preparative capillary GC, finally allowing its structural elucidation as well as that of **38d**. These two hydrocarbons (**35b** and **38d**) are accompanied in this flower scent by the corresponding 3(*Z*) isomers **35a** and **38c** in a ratio of approx. 50:1, both showing MS practically identical with **35b** and **38d**.

Knowing the fragmentation pattern of **35b** made it possible to elucidate two derivatives thereof as minor constituents—2(3)-epoxy-2,6-dimethyl-6(*E*),8-nonadiene (**36b**) and 2,6-dimethyl-3(*E*),6(*E*),8-nonatrien-2-ol (**37b**)—which have been prepared as outlined in Fig. 11.

Epoxidation of the readily accessible mixture of **35a**/**35b**[67] with peracetic acid led to the two epoxides **36a** and **36b** in a ratio of 2:3, the main product (**36b**) being identical with the compound found in *S. hamatus*. Cleavage of the epoxy group in **36b** finally gave the allylic alcohol **37b**, characterized by an attractive 'Lily-of-the-valley'-type odor. The tertiary ocimenols **1a** and **1b**, also present in this natural scent as well as in many other flower scents containing

Table 4
Examples of the occurrence of **35b** and **38d** in flower scents

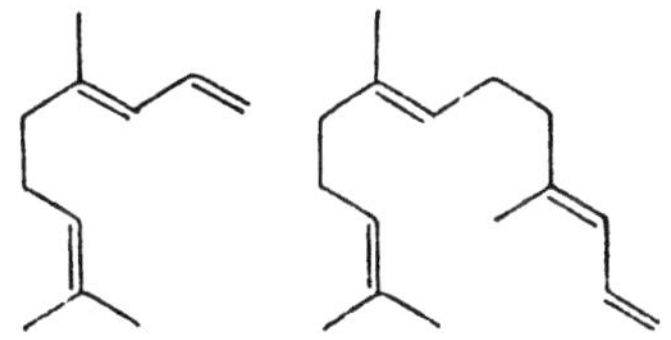

Source of flower scent	Abundance[a] **35b**	Abundance[a] **38d**
Selenicereus hamatus (Cactaceae)	++++	++
Magnolia liliiflora nigra (Magnoliaceae)	++++	+
Osmanthus fragrans (Oleaceae)	+++	+
Robinia pseudoacacia (Leguminosae)	+++	+
Philadelphus coronarius (Saxifragaceae)	+++	+++
Lilium longiflorum (Liliaceae)	++	+++
Trachelospermum jasminoides (Apocynaceae)	++	+
Dracaena fragrans (Agavaceae)	+	++
Daphne cneorum (Thymelaeaceae)	+	++
Plumeria alba (Apocynaceae)	+	+++
Hoya carnosa (Asclepiadaceae)	+	++++
Aerangis friesiorum (Orchidaceae)	+	++++

[a] ++++, Main constituent; +++, major constituent; ++, minor constituent; +, trace constituent; refers to headspace samples accumulated on charcoal.

the hydrocarbons **35a**/**35b**, are accessible by application of the same reaction sequence to (*Z*)- and (*E*)-ocimene.

In analogy to the C_{11}-trienes **35a** and **35b**, the four isomers of 4,8,12-trimethyl-1,3,7,11-tridecatetraene (**38a**–**38d**; Fig. 12) are accessible by application of the same type of Wittig reaction to a corresponding mixture of farnesal isomers, and the two naturally occurring isomers **38c** and **38d** (in nature approximately 1:50) may be obtained by column chromatography of the synthetic mixture and subsequent preparative GC.

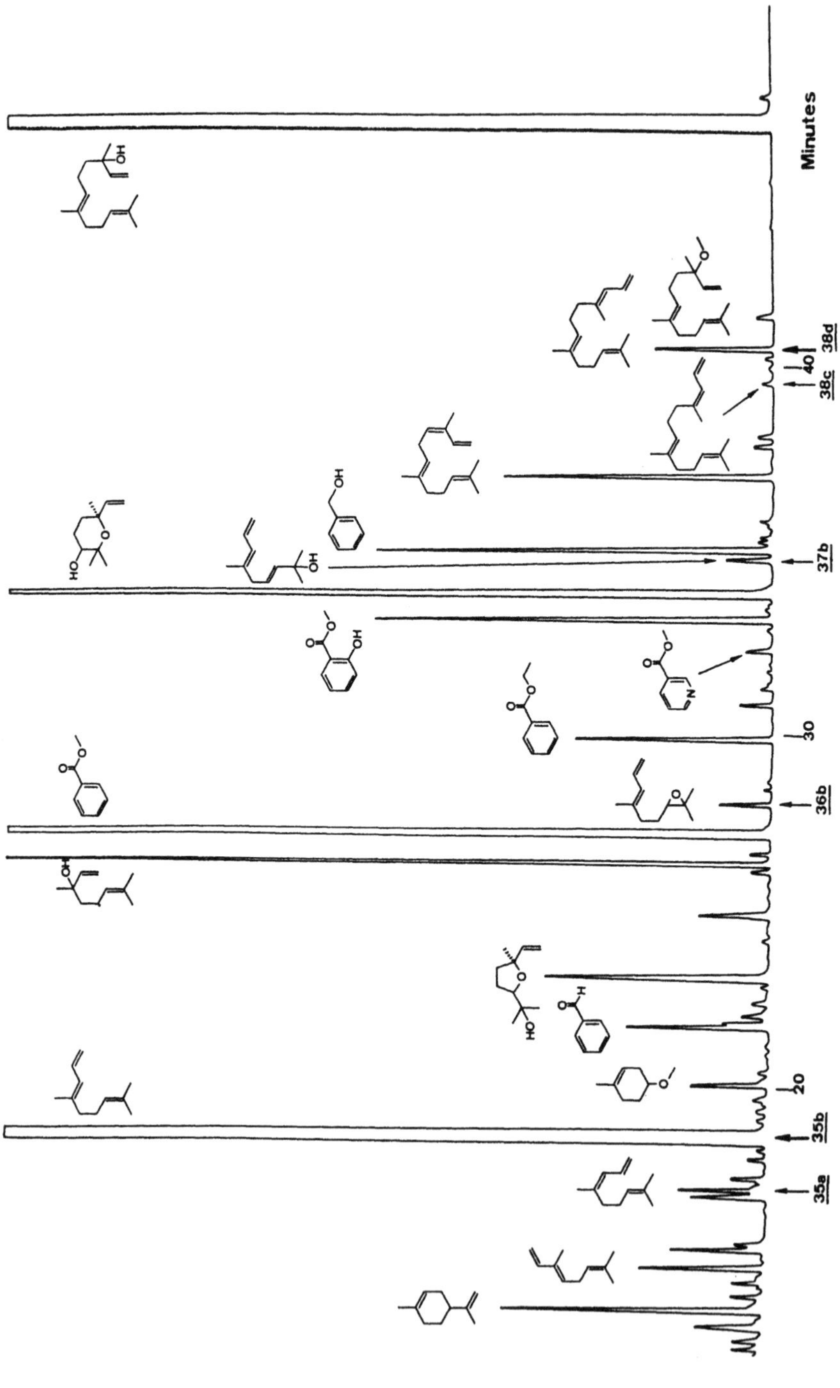

Fig. 10. GC of the trapped scent of *Selenicereus hamatus*: UCON 50 m × 0·3 mm i.d., 60–190°C at 2°C/min.

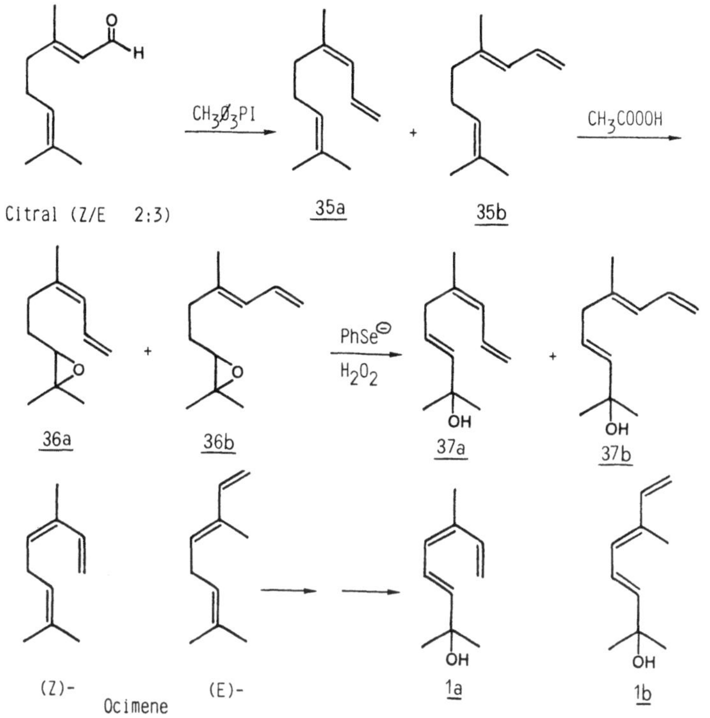

Fig. 11. Synthesis of the homoterpenes **35a/35b–37a/37b**.

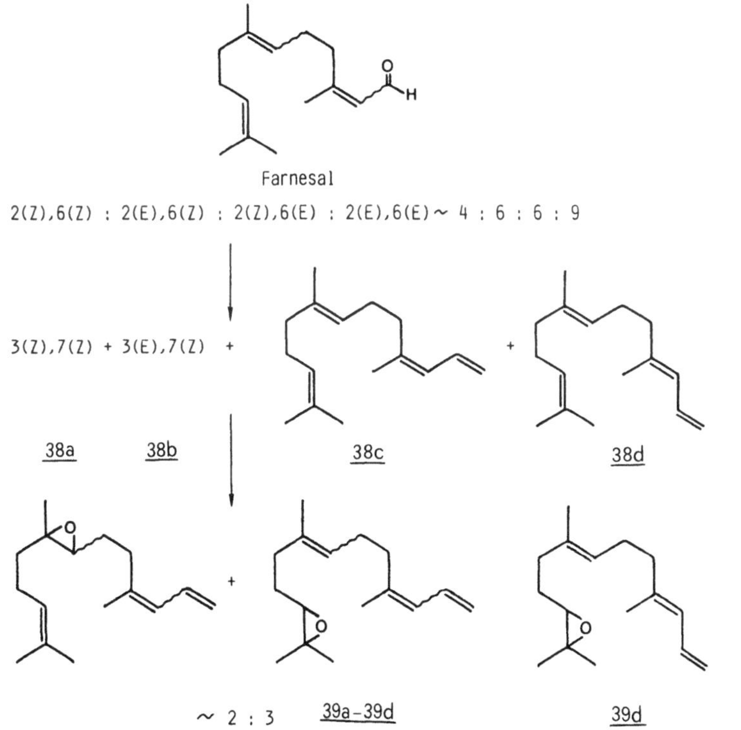

Fig. 12. Synthesis of the homoterpenes **38c/38d** and the epoxide **39d**.

As can be seen from Table 4, which illustrates the widespread occurrence of **35b** and **38d**, the tetraene **38d** represents, for example, the main constituent in the trapped headspace of *Aerangis friesiorum*, a highly fragrant night-scented orchid native to Kenya and characterized by the typical syndromes associated with moth pollination. A minor constituent of this orchid fragrance showed the same retention data and the same mass spectrum as the epoxide **39d**, which was accessible from the epoxidation product of **38a–38d** by column chromatography and preparative GC.[29]

Many of the flower scents listed in Table 4 contain as major constituents the acyclic terpene alcohols nerolidol and geranyllinalol, which may be considered as biological precursors for the hydrocarbons **35b** and **38d**. In analogy to the biosynthesis of secologanin from loganin (Ref. 68 and references cited therein) and 1-alkenes from fatty acids,[69] the C_{11}-homoterpene **35b** was postulated to be formed by enzymic abstraction of an allylic hydride ion in nerolidol (Fig. 13) and subsequent fragmentation.[29] Correspondingly, geranyllinalol would be transformed to the C_{16}-homoterpene **38d**.

Very recently, Boland & Gäbler[70] could give the experimental evidence for this hypothesis by conversion of [^{2}H]nerolidol in *Hoya purpureofusca, Magnolia liliiflora nigra, Robinia pseudoacacia* and *Philadelphus coronarius.*

Dehydrogeosmin in the Scent of Some Species of Cactaceae

Among the natural products with extremely low threshold values figures 1,10-dimethyl-*trans*-9-decalol (**40**)—the so-called geosmin—which emits a very

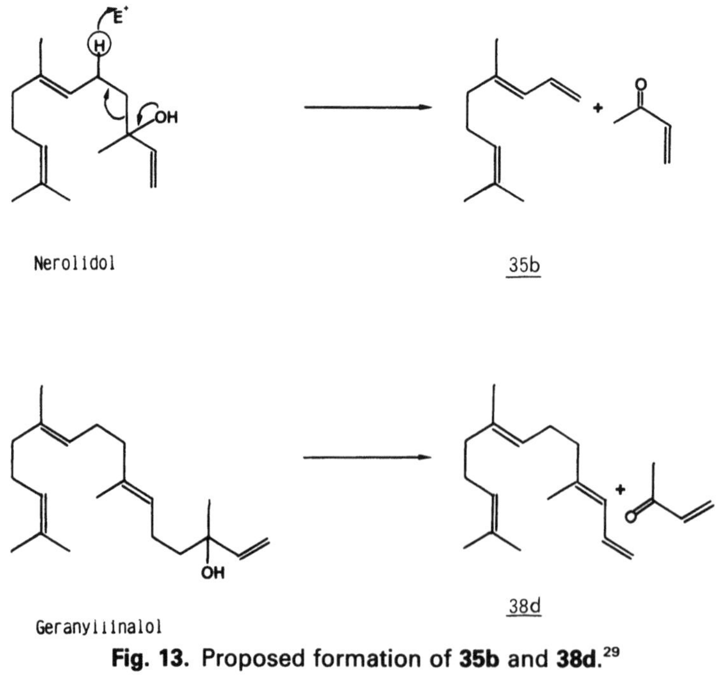

Fig. 13. Proposed formation of **35b** and **38d**.[29]

strong earthy-musty odor typical of freshly plowed soil. Geosmin (**40**), probably derived from an eudesmane sesquiterpene, is a metabolite of many actinomycetes and of several blue-green algae and is known as an off-flavor contaminant of water supplies as well as a trace constituent of several foodstuffs (compare the review given in Ref. 71). In the course of a broad olfactory evaluation of fragrant flowers, Kaiser recognized the characteristic musty–earthy odor typical of geosmin, most surprisingly in the flower scents of a series of species belonging to the genera, for example, of *Rebutia, Sulcorebutia, Dolichothele* and *Mammillaria.*[72] Searching for the compound responsible for this most unusual olfactory aspect in a flower scent, the headspace of a single flower of *Rebutia marsoneri* Werd. native to north Argentina was trapped on charcoal (5 mg) during the time of maximal opening (noon to 2 p.m.). Indeed, the sample thus obtained (approximately 6 μg) contained 35% of a compound characterized by a mass fragmentation indicative of a dehydrogeosmin. Based on the key fragment *m/z* 126 (100%) considered to originate from a retro-Diels–Alder fragmentation process, the structure of the dehydrogeosmin **41** (Fig. 14) could be proposed, which was subsequently confirmed by synthesis as described by Kaiser & Nussbaumer.[72]

In the course of this investigation, the dehydrogeosmin **41** was also identified in the flower scent of *Dolichothele longimamma* (D.C.) Br. et R. (approx. 5·5%) native to central Mexico and in that of *Sulcorebutia kruegeri* (Card.) Ritt. (approx. 7·6%), native to Bolivia. According to the olfactory evaluation of many additional species—**41** is easily detectable by nose down to the level of trace constituents—this compound seems to have a broad distribution within the genera mentioned and related ones; interestingly, the flowers of species characterized by the distinct odor of **41** are all of yellow coloration. It is certainly most striking that the flower scents of such representatives of the Cactaceae family growing under extreme dry and hot conditions are olfactorily dominated by a compound of extreme musty–earthy character which—for the human nose—is always associated with moist/damp places. It would not be too

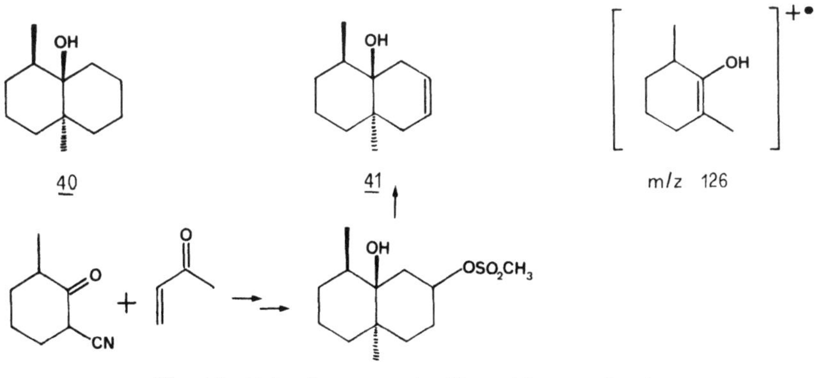

Fig. 14. Dehydrogeosmin **41** and its synthesis.

surprising if this new natural product **41** were of significant importance to the pollination biology of such Cactaceae.

5 RECONSTITUTION OF A FLOWER SCENT

An excellent basis—probably even a requirement—for the successful reconstitution of a flower scent is certainly the experience accumulated during the investigation and reconstitution of 'essential oils', which are normally available in defined quality and unlimited amounts and for which the determination of the quantitative data offers no special problems.

This is in contrast to the scent of a living flower/plant to be investigated and subsequently reconstituted, which is often characterized by a time-dependence of its olfactory appearance and which is normally only available in a quantity of 1–100 μg as a preconcentrated headspace sample. These limitations can only be compensated by choosing an interdisciplinary approach from the very beginning. Diurnal changes in the quality and quantity of fragrance emission have to be followed and described verbally in a reproducible manner so as to determine the optimal range of time for headspace trapping and to have at least a comprehensive verbal description of the scent, if the flower should not be available at the stage of reconstitution.

Much attention has to be given not only to optimal experimental conditions during the investigation of the trapped scent by GC/MS and complementary methods, but also to trace constituents hardly or not characterizable by these analytical methods which, however, might be of significant olfactory importance to the total fragrance. That means that every investigation of a trapped scent has to start with a careful olfactive evaluation of all peaks eluted from at least two capillary columns of appropriate polarities, finally resulting in a comprehensive 'olfactogram', which has to be used as a completion of the analytical results during the reconstitution of the respective scent. Considering the interdisciplinary character of such projects, it appears to be a necessary requirement that the analyst involved has also passed through a serious training in perfumery. Nevertheless it is certainly of utmost importance that the perfumer assigned to the final reconstitution work participates from the very beginning.

Furthermore, a major problem connected with headspace trapping in general lies in relating the amount of a substance measured in the trapped vapor phase to the amount actually present in the original matrix. This applies especially to flower scents, which are normally characterized by a high complexity of constituents covering a wide range of volatilities and polarities and which, in practice, may never be trapped under conditions of 'equilibrium with the liquid phase'; for such scents, the quantitative data can hardly be calculated by applying approaches such as that described by Kolb.[73] Especially critical is the estimation of the quantities in the liquid phase for the less volatile components with low vapor pressure, often only present as trace constituents in the trapped headspace; therefore, comparison with the quantitative data of a micro-extract of the respective flower—if available for this purpose—is strongly recommended for a better approximation (cf. 'Complementary methods' in Section 2). On the other hand, headspace techniques are often the only methods of investigating rare flowers in a nondestructive manner if no chance of preparing such micro-extracts exists.

A very pragmatic approach to overcome this dilemma is a well-developed comparison file of the scents of regularly available fragrant flowers, comprising the GC profiles and quantitative data of the trapped scents as well as of the corresponding micro-extracts. Based thereon indicative 'correlation factors' may be established for critical constituents, facilitating the estimation of the amounts needed in the 'liquid phase' finally to mimic the natural scent to a first approximation.

The decision at which stage of the project the reconstitution of the envisaged scent should be initiated may be problematical. Based on cost/benefit considerations it appears reasonable to have an adequate overlap of the research and reconstitution phases.

There is certainly not just one way to approach the reconstitution of a natural scent; the principal considerations, however, might always be the same. The process starts with the question of which of the (say) 150 compounds identified are necessary to reproduce the flower scent satisfactorily. There is no doubt that thinking in terms of odor values is of utmost importance for the selection of compounds, as illustrated by Ohloff[74] for the reconstitution of Bulgarian rose oil. However, main and major constituents of rather low odor values also have to be considered very carefully, because they might influence the evaporation profile of the final reconstitution in the way envisaged and they might act as harmonizers. It appears reasonable that the first compounding trial should be a simple reflection of the qualitative and estimated quantitative data. This means that it should cover as far as possible all olfactorily relevant major and minor constituents in the quantities as estimated, for example, by applying the pragmatic approach described above. Selecting these products, attention has to be given to their olfactory purity—they should not contain significant by-notes of inappropriate quality—as well as to the possibility that the racemic product may differ in odor quality and/or intensity from the pure enantiomer normally occurring in the natural scent. A subsequent series of trials, which might be laborious but inevitable, aims at the quantitative composition which mimics the basic odor profile of the fresh flower at its best.

Having reached the basic skeleton, some principle considerations also important for the reconstitution of 'essential oils'[75] have to be taken into account. What is the acceptable price for the future reconstitution? Are all products used in the basic skeleton justifiable, considering their cost/benefit ratio? Have all constituents an adequate stability for the intended application? What about legal and/or safety restrictions? To fulfill all these requirements some of the constituents may have to be replaced by olfactorily related, unproblematical products. To do this without sacrificing too much on the olfactory quality is a demanding undertaking, which needs the creativity and imagination of the experienced perfumer.

As a reward for these efforts, the effects of olfactorily relevant trace components may now be examined, which often make it possible to impart distinctive and striking facets. Simultaneously, preliminary application tests in all types of end-uses have to be initiated and, depending on their results, some additional adjustments may be necessary.

As a last exciting challenge, science, technology and artistry have to meet once again for a final judgment and questioning. Is there still the possibility that some of the unidentified minor or trace constituents could contribute importantly to the natural fragrance? Is the supply of new products developed for this purpose guaranteed? Are all legal and safety aspects clarified? Does the reconstitution really reflect the flower scent at its prime? Does it evoke similar emotional feelings? Does it also include olfactory shadings probably arising from the immediate natural surroundings? Does its evaporation profile agree with expectation? If the answer to all these questions is 'yes', the time has come to prove that all the efforts have not been '*l'art pour l'art*' but '*l'art pour dollar*'.

REFERENCES†

1. Arctander, S., *Perfume and Flavor Materials of Natural Origin.* S. Arctander, Elizabeth, NJ, USA, 1960.
2. Rassmann, H. W., Rohstoffe in der Riech- und Geschmackstoffindustrie. *Dragoco Report,* **36** (1989) 95–115.
3. Arctander, S., *Perfume and Flavor Chemicals.* S. Arctander, Elizabeth, NJ, USA, 1969.
4. Bauer, K. & Garbe, D., *Common Fragrance and Flavor Materials.* VCH Verlagsgesellschaft, Weinheim, FRG, 1985.
5. Schreier, P., *Chromatographic Studies of Biogenesis of Plant Volatiles.* Hüthig, Heidelberg, Basel, New York, 1970, pp. 1–9.
6. Williams, N. H. & Whitten, W. M., Orchid floral fragrances and male euglossine bees: methods and advances in the last sesquidecade. *Biol. Bull.,* **164** (1983) 355–95.
7. Dodson, C. H. & Hills, H. G., Gas chromatography of orchid fragrances. *Am. Orchid Soc. Bull.,* **35** (1966) 720–5.
8. Hills, H. G., Williams, N. H. & Dodson, C. H., Identification of some orchid fragrance components. *Am. Orchid Soc. Bull.,* **37** (1968) 967–71.
9. Mack, H. & Köpsel, M., Flower oils of lily of the valley and lilac. *Parf. Kosm.,* **54** (1973) 233–7.
10. Komaki, R., The constituents of the essential oil from Osmanthus flower. In *Proc. 8th Internat. Congress of Essential Oils, Cannes, 1980*, ed. FEDAROM, Grasse, 1982, pp. 394–400.
11. Kaiser, R. & Lamparsky, D., Volatile constituents of Osmanthus absolute. In *Essential Oils,* ed. B. D. Mookherjee & C. J. Mussinan. Allured, Wheaton, IL, USA, 1981, pp. 159–92.

† This reference list covers literature published up to Spring 1989 only.

12. Joulain, D., Study of the fragrance given off by certain springtime flowers. In *Progress in Essential Oil Research,* ed. E.-J. Brunke. W. de Gruyter, Berlin, 1986, pp. 57–67.
13. Holman, R. T. & Heimermann, W. H., Identification of components of orchid fragrances by gas chromatography–mass spectrometry. *Am. Orchid Soc. Bull.*, **42,** (1973) 678–82.
14. Thien, L. B., Heimermann, W. H. & Holman, R. T., Floral odors and quantitative taxonomy of *Magnolia* and *Liriodendron. Taxon,* **24** (1975) 557–68.
15. Joulain, D., The composition of the headspace from fragrant flowers. *Flavour and Fragr. J.* **2** (1987) 149–55.
16. Bergström, G., Studies on natural odoriferous compounds. *Chem. Scripta,* **4** (1973) 135–8.
17. Bergström, G., Appelgren, M., Borg-Karlson, A.-K., Groth, I., Strömberg, S. & Strömberg, St., Studies on natural odoriferous compounds. *Chem. Scripta,* **16** (1980) 173–80.
18. Silverstein, R. M. & Rodin, J. O., Insect pheromone collection with absorption column. *J. Econ. Entomol.*, **59** (1966) 1152–4.
19. Nilsson, L. A., Pollination ecology and adaption of *Platanthera chlorantha. Bot. Not.*, **131** (1978) 35–51.
20. Sirikulvadhana, S., Jennings, W. G. & Vogel, G., Collection of flower aroma concentrates for gas chromatographic analysis. *Int. Flavours Food Addit.*, **6** (1975) 126–8.
21. Jennings, W. G., Wohleb, R. & Lewis, M. J., Gas chromatographic analysis of headspace volatiles of alcoholic beverages. *J. Food Sci.*, **37** (1972) 69–71.
22. Tsuneya, T., Ikeda, N., Shiga, M. & Ichikawa, N., GC/MS analysis of Gardenia flower volatiles. In *Proc. 7th Internat. Congress of Essential Oils, Kyoto, 1977.* 1979, pp. 454–7.
23. Grob, K., Organic substances in potable water and its precursor. *J. Chromatogr.*, **84** (1973) 255–327.
24. Grob, K. & Zürcher, F., Stripping of trace organic substances from water; equipment and procedure. *J. Chromatogr.*, **117** (1976) 285–94.
25. Kaiser, R. & Lamparsky, D., Nouveaux constituants de l'absolue de jacinthe et leur comportement olfactif. *Parf. Cosm. Arômes,* no. 17 (1977) 71–9.
26. Kaiser, R. & Lamparsky, D., Constituants azotés en trace de quelques absolues de fleurs et leurs headspaces correspondants. In *Proc. 8th Internat. Congress of Essential Oils, Cannes, 1980,* ed. FEDAROM, Grasse, 1982, pp. 287–94.
27. Lamparsky, D., Headspace technique as a versatile complementary tool to increase knowledge on constituents of domestic or exotic flowers and fruits. In *Essential Oils and Aromatic Plants,* ed. A. Baerheim Svendsen & J. J. C. Scheffer. Martinus Nijhoff/Dr W. Junk, Dordrecht, 1985, pp. 79–92.
28. Kaiser, R., New volatile constituents of *Jasminum sambac* (L.) Aiton. In *Flavors and Fragrances: a World Perspective,* Proc. 10th Internat. Congress of Essential Oils, Fragrances and Flavors, Washington, DC, 1986, ed. B. M. Lawrence, B. D. Mookherjee & B. J. Willis. Elsevier Science Publishers, Amsterdam, 1988, pp. 669–84.
29. Kaiser, R., Night-scented flowers, not only attractive to moths . . . Paper presented at EUCHEM 1987, 'Semiochemicals in the Plant and Animal Kingdoms', Angers, France, October 1987.
30. Kaiser, R., Olfactory and chemical characteristics of floral scents. Paper presented at the 196th National ACS Meeting, Symposium on Progress in Essential Oil Research, Los Angeles, CA, September 1988.
31. Williams, N. H. & Whitten, W. M., Identification of floral fragrance components of *Stanhopea embreei* and attraction of its pollinators to synthetic fragrance compounds. *Am. Orchid Soc. Bull.*, **51** (1982) 1262–6.

32. Zhu Liang-feng, Lu Bi-yan, Chung-shih Tang, Jian Ke & Kang Zhi-quan, Application of XAD-4 hydrophobic resin and GC/MS/DS to study the headspace of *Hedychium coronarium* Koen. In *Proc. 9th Internat. Congress of Essential Oils, Singapore, 1983.* Technical Paper, Book 3, pp. 127–8.
33. De Pooter, H. L., Coolsaet, B. A., Dirinck, P. J. & Schamp, N. M., GLC of the headspace after concentration on Tenax GC and of the essential oils of apples, fresh celery, fresh lovage, honeysuckle and ginger powder. In *Essential Oils and Aromatic Plants,* ed. A. Baerheim Svendsen & J. J. C. Scheffer. Martinus Nijhoff/Dr W. Junk, Dordrecht, 1985, pp. 67–77.
34. Mookherjee, B. D., Trenkle, R. W., Wilson, R. A., Zampino, M., Sands, K. P., & Mussinan, C. J., Fruits and flowers: live vs dead—which do we want? In *Flavors and Fragrances: A World Perspective.* Proc. 10th Internat. Congress of Essential Oils, Fragrances and Flavors, Washington, DC, 1986, ed. B. M. Lawrence, B. D. Mookherjee & B. J. Willis. Elsevier Science Publishers, Amsterdam, 1988, pp. 415–24.
35. Mookherjee, B. D., Trenkle, R. W. & Wilson, R. A., Live vs dead, Part II. A comparative analysis of the headspace volatiles of some important fragrance and flavor raw materials. *J. Ess. Oil Res.,* **2** (1989) 85–90.
36. Patt, J. M., Rhoades, D. F. & Corkill, J. A., Analysis of the floral fragrance of *Platanthera stricta. Phytochemistry,* **27** (1988) 91–5.
37. Bicchi, C., D'Amato, A., David, F. & Sandra, P., Direct capture of volatiles emitted by living plants. *Flavour Fragr. J.,* (a) **2** (1987) 49–54; (b) **3** (1988) 143–53.
38. Jennings, W., Vapor-phase sampling, *HRC & CC,* **2** (1979) 221–4.
39. Weurman, C., Groenen, P. J. & van Gemert, L. J., Experiments on 'high-vacuum transfer' in food odor research. *Nahrung,* **14** (1970) 607–16.
40. Likens, S. T. & Nickerson, G. B., Detection of certain hop oil constituents in brewing products. *Proc. Am. Soc. Brew. Chem.,* (1964) 5.
41. Bemelmans, J. M. H., Isolation and concentration from the product phase. In *Isolation, Separation and Identification of Volatile Compounds in Aroma Research,* ed. H. Maarse & R. Belz. Akademie-Verlag, Berlin, 1981, pp. 4–36.
42. Schmaus, G. & Kubeczka, K. H., The influence of isolation conditions on the composition of essential oils containing linalool and linalyl acetate. In *Essential Oils and Aromatic Plants,* ed. A. Baerheim Svendsen & J. J. C. Scheffer. Martinus Nijhoff/Dr W. Junk, Dordrecht, 1985, pp. 127–33.
43. Weston, R. J., Composition of essential oil from leaves of *Eucalyptus delegatensis. Phytochemistry,* **23** (1984) 1943–5.
44. Schreier, P., Isolierung von Aromastoffen.—Entscheidender Schritt bei der Analyse von Lebensmittelaromen. *Lebensmit. Gericht. Chem.,* **41** (1987) 25–34.
45. Godefroot, M., Sandra, P. & Verzele, M., New method for quantitative essential oil analysis. *J. Chromatogr.,* **203** (1981) 325–35.
46. Bicchi, C., D'Amato, A., Nano, G. M. & Frattini, C., Improved method for the analysis of small amounts of essential oils by micro-distillation followed by capillary gas chromatography. *J. Chromatogr.,* **279** (1983) 409–16.
47. Hess, D., *Die Blüte.* Ulmer, Stuttgart, 1983.
48. Wilder, L. B., In *The Fragrant Garden.* Dover, New York, 1974, p. 149. (Republication of *The Fragrant Path.* Macmillan, 1932.)
49. Wassgren, A.-B. & Bergström, G., Revolving fraction collector for preparative capillary gas chromatography in the 100-μg to 1-ng range. *J. Chem. Ecol.,* **10** (1984) 1543–50.
50. Etzweiler, F., Computer controlled micropreparative isolation and enrichment from GC capillary columns and subsequent sample handling. *HRC & CC,* **11** (1988) 449–56.
51. Overland, L., Endogenous rhythm in opening and odor of flowers of *Cestrum nocturnum. Am. J. Bot.,* **47** (1960) 378–82.

52. Matile, P. & Altenburger, R., Rhythms of fragrance emission in flowers. *Planta,* **174** (1988) 242–7.
53. Altenburger, R. & Matile, P., Circadian rhythmicity of fragrance emission in flowers of *Hoya carnosa* R. Br. *Planta,* **174** (1988) 248–52.
54. Sakurai, K., Toyoda, T., Muraki, S. & Yoshida, T., Odorous constituents of the absolute from flowers of *Citrus unshiu* Marcovitch. *Agric. Biol. Chem.,* **43** (1979) 195–7.
55. Kindl, H. & Underhill, E. W., Biosynthesis of mustard oil glucosides: *N*-hydroxyphenylalanine, a precursor of glucotropaeolin and a substrate for the enzymatic and non-enzymatic formation of phenylacetaldehyde oxime. *Phytochemistry,* **7** (1968) 745–56.
56. Matsuo, M., Kirkland, D. F. & Underhill, E. W., 1-Nitro-2-phenylethane, a possible intermediate in the biosynthesis of benzylglucosinolate. *Phytochemistry,* **11** (1972) 697–701.
57. Kaiser, R., New volatile constituents of the flower concrete of *Michelia champaca* L. In *Proc. 11th Internat. Congress of Essential Oils, Fragrances and Flavours, New Delhi, November 1989*, Vol. 4, ed. S. C. Bhattacharyya, N. Sen & K. L. Sethi. Oxford & IBH Publishing Co., New Delhi, Bombay, Calcutta, pp. 1–13.
58. Joulain, D. & Laurent, R., The absolute from Karo-Karounde flowers. In *Flavors and Fragrances: A World Perspective.* Proc. 10th Internat. Congress of Essential Oils, Fragrances and Flavors, Washington, DC, 1986, ed. B. M. Lawrence, B. D. Mookherjee & B. J. Willis. Elsevier Science Publishers, Amsterdam, 1988, pp. 607–25.
59. Johnson, P., Anwendung der Headspace Erkenntnisse in der Kreation. *Seifen–Oele–Fette–Wachse,* **113** (1987) 173–5.
60. Kaiser, R., New natural products of structural and olfactory interest identified in fig leaf absolute (*Ficus carica* L.). In *Progress in Essential Oil Research,* ed. E.-J. Brunke. W. de Gruyter, Berlin, 1986, pp. 227–39.
61. Francke, W., Heemann, V., Gerken, B., Renwick, J. A. A. & Vité, J. P., 2-Ethyl-1,6-dioxaspiro[4.4]nonane, principal aggregation pheromone of *Pityogenes chalcographus* (L.). *Naturwissenschaften,* **64** (1977) 590–1.
62. Francke, W. & Vité, J. P., Oxygenated terpenes in pheromone systems of bark beetles. *Z. Ang. Ent.,* **96** (1983) 146–56.
63. Francke, W., Hindorf, G. & Reith, W., Alkyl-1,6-dioxaspiro[4.5]decanes—a new class of pheromones. *Naturwissenschaften,* **66** (1979) 618–19.
64. Kaiser, R., (5*R**,9*S**)- and (5*R**,9*R**)-2,2,9-Trimethyl-1,6-dioxaspiro[4.4]non-3-ene and their dihydro derivatives as new constituents of geranium oil. *Helv. Chim. Acta,* **67** (1984) 1198–203.
65. Silverstein, R. M., Brownlee, R. G., Bellas, T. E., Wood, D. L. & Browne, L. E., Brevicomin. Principal sex attractant in the frass of the female western pine beetle. *Science,* **159** (1968), 889–91.
66. Maurer, B., Hauser, A. & Froidevaux, J.-C., (*E*)-4,8-Dimethyl-1,3,7-nonatriene and (*E,E*)-4,8,12-trimethyl-1,3,7,11-tridecatetraene, two unusual hydrocarbons from cardamom oil. *Tetrahedron Lett.,* **27** (1986) 2111–12.
67. Pattenden, G. & Weedon, B. C. L., Carotenoids and related compounds. Synthesis of *cis*- and di-*cis*-polyenes by reactions of the Wittig type. *J. Chem. Soc.* (*C*), 1968 1984–2006.
68. Tietze, L.-F., Secologanin, eine biogenetische Schlüsselverbindung—Synthese und Biogenese der Iridoid- und Secoiridoid-glykoside. *Angew. Chem.,* **95** (1983), 840–53; *Int. Ed. Engl.,* **22** (1983) 828.
69. Ney, P. & Boland, W., Biosynthesis of 1-alkenes in higher plants. *Eur. J. Biochem.,* **162** (1987) 203–11.
70. Boland, W. & Gäbler, A., Biosynthesis of homoterpenes in higher plants. *Helv. Chim. Acta,* **72** (1989) 247–53.

71. Ohloff, G., Recent developments in the field of naturally-occurring aroma components. In *Progress in the Chemistry of Organic Natural Products,* Vol. 35, ed. W. Herz, H. Gisebach & G. W. Kirby. Springer, Wien/New York, 1978, pp. 483–4.
72. Kaiser, R. & Nussbaumer, C., Δ^6-*trans*-1,10-Dimethyl-*trans*-9-octalol (dehydrogeosmin), a novel compound occurring in the flower scent of various species of *Cactaceae. Helv. Chim. Acta,* **73** (1990), 133–9.
73. Kolb, B., Quantitative aspects of flavour analysis by equilibrium and dynamic headspace gas chromatography with capillary columns. In *Essential Oils and Aromatic Plants,* ed. A. Baerheim Svendsen & J. J. C. Scheffer. Martinus Nijhoff/Dr W. Junk, Dordrecht, 1985, pp. 3–21.
74. Ohloff, G., Importance of minor components in flavors and fragrances. *Perfumer & Flavorist,* **3** (1978) no. 1, 11–22.
75. Kastner, D., Naturprodukt und Rekonstitution in der Parfümerie. *Parf. Kosm.,* **59** (1978) 223–9.

IV CLASSIFICATION OF ODOURS

The classification of odours and odorant materials is an important prerequisite for fragrance creation as well as for studies of odour perception and action.

Classifications are rarely a final goal, either in art, or in science and technology. Usually they become meaningful only in view of creations to be completed, questions to be answered, or problems to be solved. It is, therefore, not surprising that an abundance of corresponding classifications exist, and it cannot be the aim of the following chapters to enumerate all the approaches. Rather, an attempt will be made to demonstrate the usefulness of classifications in connection with the creation of perfumes and the relationships between chemical structure and odour perceived.

Chapter 8

Empirical Classification of Odours

MAURICE THIBOUD

L. Givaudan & Cie S.A., CH-1214 Vernier-Genève, Switzerland

1 INTRODUCTION

Numerous studies have been dedicated to the classification of odours, and more specifically to the phenomena of the perception, identification and classification of odorous molecules (see Chapters 3, 4, 5, 6, 7 and 9). The purpose of this chapter is not to summarize and analyse these studies. It will try to explain, empirically, how perfumers describe and classify the odours of the natural and synthetic raw materials and of the perfumes they create, and how and why they use these raw materials when they compose a perfume, based on their knowledge and experience.

2 WHAT IS A FRAGRANCE?

In the first part of this book, Edmond Roudnitska gives us a detailed description of the art of perfumery and the creation of a perfume, particularly the prestigious perfumes destined for extraits and eau de toilette. However, in summary, we can say that fragrances are living works of art that change with the times. By its very essence, a fragrance evokes nature, a specific impression or ambience. Like symphonies, fragrances are composed of notes and chords. Using these chords, these notes, the perfumer interprets a client's needs as closely as possible. Following his inspirations—a whim, an impulse—the perfumer creates what is for him the ultimate expression of dream and function. His inspiration springs from two basic concepts.

There is the classic concept: natural and synthetic raw materials are artistically used for their intrinsic olfactory value. These scents evoke flowers, fruits, spices, precious woods, or animals. The beauty and excellence of these fragrances are the perfumer's first source of inspiration. Ingeniously blended with amber, musk, mosses, through a thousand different nuances, they slowly transform into a masterpiece of harmony, rich in mystery, subtlety and warmth, which underlines and emphasizes a woman's charm, her elegance or her natural freshness or stresses a man's virility, his sportmanship or his feelings. This is how fine fragrances have played their classic role for at least a century.

Using the same fragrances, the same notes, the perfumer can compose compounds which evoke or create a specific ambience, underline a functional product's specific attributes and confer to it a unique personality. We call this the *ambience concept*: conceived to fire the imagination; to create an intimate, cosy atmosphere; to transform bathrooms into environments of fresh, relaxing and natural scents; to add touches of softness and cleanliness to fabrics and household cleaners; or to achieve stimulating effects on urges and passions.

Given the different ways fragrances are perceived by individuals, a perfume very often embodies both the classic and the ambience concept.

Examples

The importance of fragrances in creating certain ambience concepts can be illustrated as follows.

Rose	Besides its olfactory note, well known by everyone, the rose odour reflects softness, femininity, sensitiveness.
Spicy notes	Examples such as clove, coriander and cinnamon express exoticism, light, warmth, even sometimes arrogance.
Violet notes	Bring elegance and distinction.
Lily-of-the-valley	With its flowery, fresh, green odour brings spring, morning dew, youth, lightness, delicacy.
Woody notes	Confer soft (Sandalwood), warm (Vetiver), voluptuous (patchouli) effects.
Balsamic and animal notes	Can be soothing, mysterious, sweet and sensual.
Fruity notes	Bring an acid, mellow but also voluptuous, glamorous shade.

Lemon odours	Express freshness, sun, beach, sub-tropical comfort. Associated with green notes, lemon may indicate sea, wind, waves and, with some spices, we can even reach the Caribbean Islands.

From this summary and the examples cited we can derive two possible descriptions or classifications:

—An *objective* description or classification based on the olfactory note of the raw material.
—A *subjective* description or classification based on the ambience created or brought by these very same products.

3 DEFINITION OF 'OBJECTIVE' AND 'SUBJECTIVE' ADJECTIVES

Since every classification system requires a description of the materials to be classified, we shall begin by defining 'objective' and 'subjective'. This classification is instinctively made by the perfumer on two different levels: the first is objective and emanates from knowledge which every individual has; the second is subjective and individual. The synthesis of the objective and subjective descriptions provides us with the empirical classifications of raw materials.

The definition of an objective adjective is simple because it reflects references well known to all. However, subjective adjectives can be divided into three categories:

(a) *Subjective quality*: such as fresh, woody, flowery, spicy, citrus-like, clear, clean, etc. These allow us to differentiate between two different qualities or origins of the same product or between two products with the same tendency.
(b) *Subjective ambience*: such as oriental, summery, sensual, young, sporty, feminine, masculine.
(c) *Subjective functional*: such as refreshing, cleansing, softening, skin care, calming, or stimulating, etc.

We shall now explain how the perfumer has established these different categories.

The descriptions of odours are based exclusively on olfactory memory, on knowledge of raw materials and on experience. We shall see how experience can influence and modify an olfactory description which seemed correct during the first evaluation of a substance in its pure state.

4 KNOWLEDGE OF ODOURS IN GENERAL

Of course, even before his professional training, the perfumer knows the odour of oranges, tangerines, grapefruit, anise, mint, pepper, clove, cinnamon, pine, thuja, honey, chocolate, caramel, vanilla, peach, coconut, thyme, smoke, leather, etc.—in short, all the odours we have known since our childhood and which we have kept in our memory. Obviously, the number of odours or products vary according to the lifestyle and environment of the perfumer before his professional schooling.

We also recognize certain flowers by their odour: carnation, lily-of-the-valley, lilac, rose, honeysuckle, jasmine, freesia, lavender, etc. Here again, the environment of each individual is important, just as are his olfactory interests and sensitivity.

The initiation to odours starts at a very early age, probably at the same time that speaking starts, and stays engraved in the mind. It may even be at this moment that the interest in odours awakens in the individual and, most definitely, in future perfumers, who unconsciously memorize them as well as the memories and emotions they engendered.

On the basis of this fundamental knowledge, perfumers discover and build their awareness of other raw materials during the course of their studies. All future olfactory descriptions will rely on a profound knowledge of natural raw materials.

5 KNOWLEDGE OF RAW MATERIALS

Natural Raw Materials

The progressive discovery of odours can be compared with the study of a new language: we improve our language, and the way we express ourselves in a foreign language, as fast as our vocabulary develops. For example, knowing the odour of lavender and camphor, the perfumer will describe the odour of rosemary—when he discovers it for the first time—as reflecting lavender and camphor. Then, later, when discovering the odour of aspic, it will be likened to lavender or rosemary before becoming an adjective of reference itself.

In other words, the process of description begins with an association of well-known odours; then the odours themselves form an adjective. During his training, the perfumer will notice that the odour of rose essence and rose absolute are different and that they are different from the odour of the flower. The adjective will thus be rounded off with the mode of extraction or its origin (Turkish rose, essence . . .). The perfumer will also differentiate odours according to the purity of their qualities. To this end, he supplements the objective adjective with other adjectives we shall call subjective *quality* adjectives.

Isolated Products and Nature-Identical Synthetics

When the perfumer proceeds to pure substances from natural sources or to nature-identical synthetic products he will first use the objective adjectives which arise from his knowledge of complex natural products, in particular those natural raw materials which constitute his precise reference base.

At first sight, this description may seem simple: take the odour of the natural raw material as a base from which the principal constituents are derived. Thus, linalool will be described as smelling like the essence of rosewood, eugenol as clove, irone as orris, geraniol, citronellol and phenylethyl alcohol as rose, benzyl acetate as jasmine, citral as lemon, cinnamic aldehyde as cinnamon, etc.

Although it would be easy to add to this list, it quickly becomes more difficult to say that linalool and linalyl acetate smell like lavender, bergamot, coriander or petitgrain, even though they are principal constituents. This is true for the simple reason that other constituents of natural essences—even at a smaller dosage—have a preponderant role in the odour of these essences.

At this stage, the perfumer takes the odour of the purified natural or synthetic materials and uses them as new adjectives of reference. Linalool, its acetate and other linalyl odours will constitute a new reference for him.

Artificials

The perfumer will at first use the closest adjectives of natural and synthetic products for new synthetic substances which do not exist in nature.

Then, the note of these products will also become an adjective (see the examples in Tables 1 and 2, e.g. Lilial). On the basis of the odour of the products which have become references, sample families will be created such as aldehydes, salicylics, etc.

6 DEFINITION AND INTERPRETATION OF ODOURS BY THE PERFUMER

The more raw material odours the perfumer knows, the greater his memorized vocabulary will be. He will use this vocabulary to describe the odour of raw materials which are new to him or which he has forgotten. This is where he will perceive the importance of a first and second subjective description which will enable him to develop his description and—if necessary—to give the elements of difference between two similar olfactory products (Tables 1 and 2).

These subjective adjectives will also arise out of the perfumer's experience and environment. The odours of vanilla, caramel or chocolate can evoke pastries, sugar will give rise to the adjective 'soft'. The adjective 'green' evokes everything associated with the green found in nature, for example leaves, grass, green fruits, etc. The note of freshness is undoubtedly linked to a

Table 1
Some examples of definition and interpretation of odours by perfumers

Name	*Odour descriptions*
Ethyl heptanoate	Ethereal, aldehydic, fruity, dry
p-tert.-Butylcyclohexyl acetate	Floral, woody, fresh, strong
Decahydro-β-naphthyl acetate	Floral, fruity, woody, green (jasmine)
Dimethyl phenylethyl carbinol acetate	Floral, sweet, rosy, fresh
Dihydromyrcenol	Agrestic, lime, fresh, strong
Lilial	Floral, fresh, strong, sweet (direction lilac, lily-of-the-valley)
Lyral	Floral, fresh, sweet, tenacious
Cedramber	Woody, amber, dry, diffusive
Rosalva	Fresh, rosy, fatty, aldehydic
Sandalore	Woody, warm, sweet, strong (direction sandalwood)
Canthoxal	Floral, anisic, green–sweetish
Cashmeran	Musky, woody, flowery, camphoraceous
Hedione	Jasmine, floral
Floropal	Woody, citrus, fruity (grapefruit)
Dynamone	Amber, leather, balsamic, fixative properties

Table 2
Some examples of definitions and interpretation of odours as above, but for natural products

Name	*Odour description*
Cananga essence	Floral, sweet, medicinal, balsamic
Ylang-Ylang essence	Floral, strong, warm, medicinal
Ylang Ylang absolute	In addition to the above terms: sweet
Olibanum essence	Woody, dry, terpenic, spicy
Olibanum odoresin	More balsamic, sweet
Iris absolute	Orris, warm, woody, powdery, diffusive
Iris concrete	Orris, sweet, woody, flowery, slightly fruity, tenacious
Jasmine absolute (Grasse)	Jasmine, floral, fresh
Jasmine absolute (Egypt)	Jasmine, floral, sweet, warm
Rose absolute (France)	Rosy, very rich, tenacious, slightly green
Rose absolute (Turkey)	Rosy, rich, warm, honey, slightly spicy
Clary sage absolute	Agrestic, amber, rich, tenacious
Clary sage essence (France)	Agrestic, warm, amber, musky
Thyme absolute	Agrestic, tobacco, hay
Thyme essence	Agrestic, fresh, green

sensation of volatility. In fact, one notices that products which have a note of freshness are generally more volatile than others. However, not all volatile products can be described as fresh. Many esters are not fresh but fruity, for example amyl acetate.

Nevertheless, all these adjectives deal with the description of the various raw materials, but not with their definition in terms of usefulness for creating and composing. Other definitions give the raw materials a position in the palette of the perfumer for the time when he will have to choose the elements which will allow him to respond or translate his creative idea. These definitions (for which adjectives of the third category (c) are used) will depend solely on his intuition and conception. For example, certain perfumers base their creations on the study of forms (see Chapter 1) such as round, square, pointed, flat, etc., and others on colours. Others base their creations on music, for example dull, strident, booming notes, or major or minor tones. Still others use paintings or locations (castles, ballrooms, theatres . . .), or even landscapes.

Depending on the future use of the perfume, these definitions will be complemented by a functional adjective which will reinforce the functional element of the end-product. Amongst all these definitions, the perfumer uses the raw materials which give his compositions the characteristics needed to reinforce the image of the end-product (cream, milk, bubble bath, detergent, soap, etc.).

Using these adjectives, the perfumer can establish tables which include both the objective and subjective adjectives he may need to position the raw materials one against the other in his palette. This is demonstrated by the examples in Tables 3–12.

Numerous tables of this kind already exist in several other publications and are used by the professional perfumer to position a new raw material in comparison with others. However, he will do so according to his personal preferences and establish his own distinctions and families. Above all, when using a new raw material, the perfumer can establish the description of its odour. The behaviour of a raw material in a mixture is dependent on the mixture and its constituents in terms of proportion and of end-use (soap, detergent, etc.). We could draw a comparison with the behaviour of an individual who is different depending on whether he is alone or in company, depending on which company, on his location and on his situation. We can also come back to language, to the differences of the meaning of a word according to the construction of the sentence or the context of the story told or described.

The behaviour analysis of raw materials, depending on the mixtures, gives the perfumer the opportunity to find another definition of the odours, e.g. he will realize—according to the proportion and the future use of the perfume—that raw materials can be 'modifying', 'binding', the 'main body', 'marking', 'fixing', and so on.

The professional perfumer rarely consults these tables except to inform sales departments and perfume trainees. The fact that they do not consult these

Raw material	Objective adjectives																					Subjective adjectives: Quality							Subjective adjectives: Ambience							Subjective adjectives: Functional	
	Agrestic	Aldehydic	Amber	Aromatic	Balsamic	Camphoraceous	Citrus	Coumarinic	Floral	Food-related	Fruity	Green	Hay-like	Herbaceous	Honey	Liquorice	Musky	Rosy	Spicy	Tobacco-like	Woody	Bitter	Diffusive	Metallic	Rich	Strong	Tenacious	Terpenic	Active	Fresh	Masculine	Natural	Sporty	Sweet	Warm	Medicinal	Stimulating
Natural																																					
Carrot seed essence	■		☒							■			☒													■									■		
Cistus essence (Spanish)	■		■		■										■												☒								☒		
Clary sage essence	■		■					☒					☒				■			☒					☒				☒		☒	☒	☒		■		☒
Coriander essence	■			☒															■		■	☒	☒						☒	■			☒				☒
Cumin essence	■			■						☒		☒		☒			☒		☒			■				■		☒			☒	☒			☒		
Mugwort essence	■			☒		■						☒							☒			☒				■				☒						■	☒
Myrtle essence	■			☒	■	■			■										☒				☒					☒	☒		☒		☒				☒
Roman chamomile essence	■										■							☒					■						☒	☒		☒				■	
Thyme essence	■			☒								■	☒						☒	☒			☒							■			☒			☒	
Wormwood essence	■								☒	☒	☒	☒	☒	■					☒			■						☒	☒			☒				☒	☒
Synthetic																																					
Amyl benzoate	■								■			■																						■	☒		☒
Cetone V	■										■									☒	☒			■	☒	■						☒					
Methyl nonyl ketone	■	■		☒			■				■	☒	☒									☒	☒					☒	☒	☒							☒

Key:
■ Main (or more important) odour characteristic
☒ Secondary note (or according to dosage)

Table 3
Adjectives applied to 'agrestic' raw materials

	Objective adjectives														Subjective adjectives												
															Quality				Ambience					Functional			
Raw material	Agrestic	Animal	Balsamic	Cinnamic	Citrus	Floral	Fruity	Honey	Mushroom	Rosy	Spicy	Tobacco	Vanilla	Woody	Dry	Rounded	Terpenic	Weak	Oriental	Powdery	Sensual	Sweet	Warm	Medicinal	Skin-care	Smoothing	Softening
Natural																											
Benjoin			■										■						X	X		■	■		X	X	
Copaiva			■			■			■					X	X			X				X		X			
Opoponax			■		X						■			■	X		■		X	X							
Peru			■										■		X				X	X		X	■				
Styrax		■	■			■																■				X	X
Tolu			■													X			X	X	X	■	■	X	X	X	X
Synthetic																											
Benzyl cinnamate			■	X		■		X			■											■			X	X	X
Cinnamyl propionate			■	■			■															■					
Ethyl cinnamate			■	X			■	■							X							■		X			
Eugenyl phenylacetate			■					X			■	■	■		X				X	X			X	X			
Isoeugenyl benzoate			■								X		■					■	X	X	X	■	X		X		
Methyl cinnamate			■	X			■												X	X		■	X	X	X	X	X

Table 4
Adjectives applied to 'balsamic' raw materials

	Raw material	Natural				Synthetic							
		Cedarwood essence Virginia	Patchouli essence	Sandalwood essence	Vetiver essence Bourbon	Ambroxane	β - Methylional	Cedramber	Madrox	Methyl cedryl ketone	Methyl ionone (s)	Oxyoctalin formate	Sandalore
Objective adjectives	Agrestic	X	X					X		X			
	Amber			X		■		■	■				X
	Animal					■							
	Aromatic		X				X						
	Balsamic			■									X
	Camphoraceous		■									■	
	Chypre		X										
	Earthy		■		■								
	Floral										■		
	Fruity												
	Green				X					X			
	Honey						■						
	Leathery										X		X
	Minty												
	Musky			X					■				
	Orris			X			■			■	■		X
	Phenolic												
	Pine	X						X					
	Spicy		X		X		X						
	Terpenic							X					
	Tobacco	X		X			X			X	X		X
	Woody	■	■	■	■	■	■	■	■	■	■	■	■
Subjective adjectives: Quality	Bitter						X						
	Diffusive		X		X	X	X	■		■	X		
	Dry	X		■	■	X		■		X		■	
	Fixative					■							
	Hard												
	Powerful		X		X	X	■	X			X	■	■
	Rich												
	Tenacious	■		■		X			■	■	X		
Subjective adjectives: Ambience	Fresh									X			
	Masculine	X			X	X	X	X		X			
	Oriental		X	X		X	X	X			X		X
	Powdery			X		X	X				■	X	X
	Sporty	X			X			X		X			
	Sweet	■			■						X		■
	Warm		X		X	X	X	X			X		■
Subjective adjectives: Functional	Clean	X						X		X	X		X
	Medicinal		■										

Table 5
Adjectives applied to 'woody' raw materials

		Synthetic: Eugenol	Synthetic: Cinnamyl acetate	Synthetic: Cinnamic aldehyde	Synthetic: Acetyl isoeugenol	Synthetic: Acetyl eugenol	Natural: Pepper (black) essence	Natural: Nutmeg essence	Natural: Ginger	Natural: Clove bud essence	Natural: Cinnamon leaf essence	Natural: Cardamom essence	Natural: Allspice/Pimento berry
Objective adjectives	Agrestic	X	■					X			X	■	
	Aromatic			■			■	X			X	X	
	Balsamic				X								■
	Camphoraceous											■	
	Floral				■	■				X			
	Food-related								■		X		
	Fruity									■			
	Herbaceous												
	Phenolic	■				X				X			
	Spicy	■	■	■	■	■	■	■	■	■	■	■	■
	Terpenic						■	■	■				
	Vanilla				X								
	Woody		■			X	■		■				■
Subjective adjectives – Quality	Diffusive			■					X	X	X		X
	Hard	■											
	Powerful	■		X						■	■		
	Rich							X				■	
	Weak				■	■							
Subjective adjectives – Ambience	Active	X		X			X		X	X	X	X	X
	Exotic	X							X			X	
	Masculine	X					X		X	X	X	X	X
	Natural												
	Oriental	X			X		X	X		X	X	X	
	Powdery				■					X		X	
	Rigorous/Harsh			X		X			X				X
	Sensual						X	X		X		X	
	Sporty	X					X		X	X	X		X
	Sweet		■		■	■		X					■
	Warm	X		■	X		X	■	X	■	■	X	■
Subjective adjectives – Functional	Clean	X			X	X				X	X		X
	Medicinal	X				X				X		X	X

Table 6
Adjectives applied to 'spicy' raw materials

	Raw material	Natural				Synthetic							
		Jonquil absolute	Mimosa absolute	Narcissus absolute	Tuberose absolute	Benzyl acetate	Gardenol	Hedione	Hydroxycitronellal	Lilial	Lyral	Methyl N-methylanthranilate	Terpineol
Objective adjectives	Agrestic	☒	■										☒
	Balsamic			☒									
	Chypre			☒									
	Citrus											■	
	Floral	■	■	■	■	■	■	■	■	■	■	■	■
	Fruity	☒			■	■	■	☒				■	
	Green	■					■			☒			☒
	Hay-like			■									☒
	Herbaceous		■							☒			☒
	Honey				■								
	Iris / Orris	☒									☒		
	Jasmine	☒	☒		☒	■		■					☒
	Lilac		☒			☒		☒		☒	☒		■
	Milky				☒								
	Mossy												
	Lily-of-the-valley		☒		☒	☒		☒	■	☒	■		☒
	Orange											☒	
	Pine												■
	Rosy								☒				
	Spicy	☒											
	Tobacco												
Subjective adjectives: Quality	Bitter	■		■									
	Diffusive					☒	☒	☒				☒	
	Dry												☒
	Hard						☒						
	Harsh						☒						
	Powerful						■			■			
	Rich	☒	■		☒								
	Tenacious								☒	☒	■		
Subjective adjectives: Ambience	Feminine	☒			☒				☒				
	Fresh	■				■	☒	■	■	■	■	■	■
	Lively	☒			☒	☒	☒					☒	☒
	Natural	☒			☒				☒			☒	☒
	Powdery	☒	☒								☒		
	Sensual			☒	☒						☒		
	Sweet		☒		☒			■	■	■	☒	☒	
	Tender	☒	☒		☒				☒				
	Warm		☒	☒	■	☒					☒		
	Young	☒			☒	☒	☒			☒		☒	☒
Subjective adjectives: Functional	Medicinal						☒						
	Skin-care	☒	☒	☒	☒					☒	☒		
	Smoothing	☒	☒						☒	☒	☒		
	Softening									☒			☒

Table 7
Adjectives applied to 'floral' raw materials

Raw material	Objective adjectives														Subjective adjectives: Quality					Subjective adjectives: Ambience							Subjective adjectives: Functional		
	Agrestic	Aldehyde	Balsamic	Citrus	Floral	Fruity	Green	Honey	Lemon	Minty	Rosy	Spicy	Terpenic	Woody	Bitter	Diffusive	Earthy	Fatty	Metallic	Active	Fresh	Lively	Natural	Sporty	Sweet	Young	Clean	Humidifying	Refreshing
Natural																													
Elemi essence	☒		■	☒			☒					■	■			☒					■		☒	☒				☒	☒
Petitgrain Bergamot essence	☒			■	☒	☒				☒					■				■	☒	■	☒		☒		☒			☒
Synthetic																													
Anther							■	■						■		☒					■								
l-Carvone	☒									■		■				■					■		☒		☒				☒
Citronellyl acetate						■					■										■		☒		☒			☒	
Decatone	☒			☒		■	☒		■					■		☒				☒	■			☒		☒	☒		☒
Dihydromyrcenol	■			☒	☒				■							■				☒	■	☒	☒	☒		☒	☒		
Geranyl acetone				☒	■		☒				■										■				■	☒	☒	☒	
Menthanyl acetate	☒				■					■				■	☒	☒				☒	■		☒	☒			☒		
Octyl acetate		☒			☒	■	■											■			■							☒	☒
Rosalva		■									■					☒		■		☒	■							☒	
Terpenyl formate				■	■		■							☒	☒		☒			☒	■			☒			☒		

Table 8
Adjectives applied to 'fresh' raw materials

Adjectives		Raw material (Synthetic)	Allyl amylglycolate	Allyl cyclohexylpropionate	Amyl hexanoate	Butyl butyrate	Citronellyl isobutyrate	Ethyl malonate	Ethyl maltol	Ethyl methylphenylglycidate (strawberry aldehyde)	Nootkatone crist.	Rhodinyl isobutyrate	Undecalactone (peach)
Objective adjectives		Agrestic		X									
		Aldehydic	■										
		Animal			■								
		Balsamic						X	X				
		Citrus		X	X			X			■		
		Floral					X			■		X	
		Food-related		X	X	X		X	X				X
		Fruity	■	■	■	■	■	■	■	■	■	■	■
		Green	■	X	X						■	X	
		Herbaceous	X		■								
		Honey						X	X				
		Jasmine											X
		Lactonic						■		X			
		Lavender					X						
		Lemon		X									
		Milky											■
		Rosy					■					■	
		Spicy							X				
		Tobacco							X	X			
		Woody									■		
Subjective adjectives	Quality	Diffusive	X	X	X	■	X				X	X	
		Ethereal			X	X	X			X			
		Fatty			■		X						■
		Fixative							■				X
		Harsh	X		X								
		Metallic	X										
		Powerful	■	■					■	■	X		■
		Rich										X	
		Tenacious											X
	Ambience	Fresh	X	X	X	■	X	■		■	X		
		Natural								X			
		Oriental	X	X	X				X	X	X		X
		Sweet		■		■	■		■	X		■	X
		Warm					X		X		X		X
	Functional	Refreshing	X	X	X			X			\	X	
		Skin-care					X			X			X
		Smoothing					X			X		X	X
		Softening								X		X	X
		Stimulating	X	X	X			X			X	X	

Table 9
Adjectives applied to 'fruity' raw materials

		Raw material										
		Natural						Synthetic				
		Bergamot essence	Lemon essence	Lime essence	Mandarin essence	Orange Florida essence	Grapefruit essence	Bergamyl acetate	Citral	Dimethyloctenone	Linalyl acetate	Tetrahydrolinalyl acetate
Objective adjectives	*Agrestic*	X						■				
	Aldehydic		X			X			■			
	Citrus	■	■	■	■	■	■	■	■	■	■	■
	Fruity		■	X	■	■	X			■	■	■
	Herbaceous	■						X				
	Lavender							X			■	X
	Lemon	X	X			X	X		X			
	Lime		X	X		X	X		X			
Subjective adjectives — *Quality*	*Bitter*	X		X	X		■		X		X	
	Diffusive		X	X					X	■	X	
	Powerful						X		X			X
	Terpenic			■			X					
Subjective adjectives — *Ambience*	*Active*	X	X	X			X		X			X
	Fresh	■	■	■	■	■	■	■	■	■	■	■
	Lively		X		X		X		X			
	Natural					X			X		X	
	Sporty	X	X	X			X		X	X	X	X
	Young	X	X	X			X		X		X	
Subjective adjectives — *Functional*	*Clean*	X	X	X			X	X	X		X	X
	Stimulating	X	X	X			X		X			X

Table 10
Adjectives applied to 'citrus' raw materials

		Natural: Geranium Bourbon essence	Natural: Rose essence (Bulgarian)	Natural: Rose absolute (French)	Synthetic: Citronellol	Synthetic: Damascenone	Synthetic: Geraniol	Synthetic: Geranyl phenylacetate	Synthetic: Nerol	Synthetic: Phenylethyl alcohol	Synthetic: Rhodinol	Synthetic: Rhodinyl formate	Synthetic: Rose oxide
Objective adjectives	Aromatic				X							X	X
	Earthy												■
	Floral	X	X	X	X	X	X			X	X	X	
	Fruity					■	■						
	Green	■	X							X		■	■
	Herbaceous	X			X			■	X		X		X
	Honey		■					■					
	Lemon								X				
	Marine								■		X		
	Minty	■									■	X	
	Rosy	■	■	■	■	■	■	■	■	■	■	■	■
	Spicy		X			■							
	Tobacco		X										
	Woody				X								
Subjective adjectives – Quality	Dry											X	X
	Fatty				■								
	Fixative												
	Fresh	X	X		■				■		X	■	■
	Powerful	■				■						■	X
	Rich		■	■							X		
	Rounded						X	X		X			
	Tenacious			■									
	Weak				X		X						
Subjective adjectives – Ambience	Feminine		X	X				X		X	X		
	Natural	X	X				X		X	■	X		
	Powdery		X	X				X			X		
	Sweet			X	X		■	■		■	■		
	Tender		X	X			X	X		X	X		
	Warm	X	■	X				X			X	X	
Subjective adjectives – Functional	Skin-care		X	X			X	X		X	X		
	Smoothing		X	X			X	X		X	X		
	Softening				X		X	X		X	X		

Table 11
Adjectives applied to 'rosy' raw materials

		Raw material: Natural — Galbanum essence	Natural — Violet leaf absolute	Synthetic — Acetal R	cis-3-Hexenyl acetate	cis-3-Hexenyl formate	cis-3-Hexenyl cis-3-hexenoate	cis-3-Hexenyl isobutyrate	cis-3-Hexenyl salicylate	Isocyclocitral	Nonadienal	Stemone	Viridine
Objective adjectives	Agrestic	X		X		X		X	X	X		X	
	Aldehydic				X		X				X		
	Balsamic								■				
	Earthy	X							X			X	
	Floral		■	■					X		X		■
	Food-related	■											
	Fruity		■	■	■		X	■			■		
	Green	■	■	■	■	■	■	■	■	■	■	■	■
	Herbaceous	X			■	X	X	X	■	■		X	X
	Iris/Orris		X								X		
	Leafy		X		X	■	X						
	Lemon									X			
	Lilac			X	X								X
	Metallic			■			■						
	Minty									X			
	Mossy											X	■
	Orange												
	Rosy			X									X
	Spicy	■										X	
	Terpenic	■											
	Woody								X				
Subjective adjectives — Quality	Bitter	X	X	X				X		X		X	
	Dry					■			X			X	
	Fatty				X		X				X		
	Powerful	X	X	X	■	■	■	■	X	■	■	■	X
	Rich		■										
	Tenacious												
Subjective adjectives — Ambience	Active	X	X	X			X		X		X		
	Fresh	X	X	X	X	X	X	X	X	■		■	■
	Natural	X	X		X	X	X				X	■	
	Sweet							■	■				
Subjective adjectives — Functional	Clean	X		X		X	X	X	X	X			
	Cleansing	X	X	X			X		X	X		X	X
	Humidifying		X		X	X	X				X		X
	Refreshing						X			X			

Table 12
Adjectives applied to 'green' raw materials

tables systematically ensures the originality of each perfumer, leading to different interpretations of a given theme which reflect the creator's style. Of course, the professional perfumer will consult them when confronted with a new raw material, to obtain an idea of the olfactory note. However, as soon as he has found the unique character of this new raw material, it will instinctively find its place in his palette according to the criteria described.

7 CLASSIFICATION OF PERFUMES

Here again, we find the two main themes, an objective and a subjective classification. The objective classification is based on the use of adjectives which correspond to the raw materials which give the olfactory note to the mixture. This classification comprises well-known families: flowery, aldehydic perfumes, chypre and ferns (fougère), etc. The subjective description—which is the most important since the role of perfume is to convey a message or to stimulate the imagination—uses the subjective adjectives of categories (b) and (c).

As before, the perfumes will successively become references and serve as adjectives for new perfumes. In any case, the dream of every perfumer is to find the perfume which will create an adjective, because this would be an incontestable sign of success.

According to the perfume houses and organizations, the perfumes sold on the market are classified in families (adjectives like floral, fresh, etc.; see Figs 1 and 2), or compared one by one (see, for example, Figs 3 and 4), in terms of what they evoke or in terms of colours or forms.

8 CONSUMER DEFINITION AND INTERPRETATION OF ODOURS

All the descriptions and classifications we have just seen come from the knowledge of the perfumer and from his contacts and discussions with colleagues and associates. However, perfumes are intended for people who have nothing to do with this profession. Therefore, the perfumer must keep the consumer in mind, i.e. the consumer's olfactory perception. Various studies are geared to this end. These studies have the purpose of establishing how consumers perceive and translate odours. They have different names, such as 'fragrance mapping', and require a sophisticated data system to balance and analyse results. In simple terms, these tests consist of giving the participants blind raw materials and perfumes and asking them to perceive and translate these odours by choosing from a selected list of adjectives or by offering these adjectives themselves. The best tests combine both methods. The results of these studies, which the perfumer must interpret, are added to the definitions he had before. They may sometimes alter his own definitions, allowing him to tailor his approach better.

It must be noted that the results of one study on a given product, raw material or perfume will vary according to the user, environment (soap, extrait, etc.), ethnic group, religion, climate, tradition, etc. For example, to tell an inhabitant of a far-east country that a perfume is oriental will most likely evoke nothing in him.

These studies based on raw materials lead directly to subjective adjectives; that is, the same odours of raw materials which evoke objective adjectives from the perfumer will evoke subjective ones from the public. It is up to the perfumer to interpret the results and find the corresponding note in his palette (Figs 5–8).

9 CONCLUSIONS

All perfumers and every person in the industry use the same vocabulary and references in these cases:

—the name of the raw material;
—the name of the successful perfume;
—the trend (in such and such a segment);
—the technical and physico-chemical properties of raw materials.

However, each perfumer will instinctively define:

—the olfactory note;
—the adjective of quality;
—the evoked ambience;
—the suggested or reinforced functional idea.

All these definitions deal with raw materials and perfumes.

It is these definitions—particular to each perfumer—upon which the latter interprets and adapts:

—market studies;
—his style of creation (form, colour, music, etc.).

How the perfumer interprets and classifies odours depends on his style, as do all the other established rules and methods: the description of odours, everything which touches the unconscious, the imagination, sensitivity, taste and traditions—the manner is never precise.

Like all the other arts, the creation of a perfume is a means of expression and communication. This communication depends on how the perfumer expresses himself when mixing odours. It will vary according to the perfumer, his degree of experience, the perfume he creates, and the market.

Finally, to make one last comparison with language, be it in a foreign language or in one's mother-tongue, the more the individual understands the exact meaning of words and their impact on the receiver, the better he will be able to express himself according to the criteria which have been established.

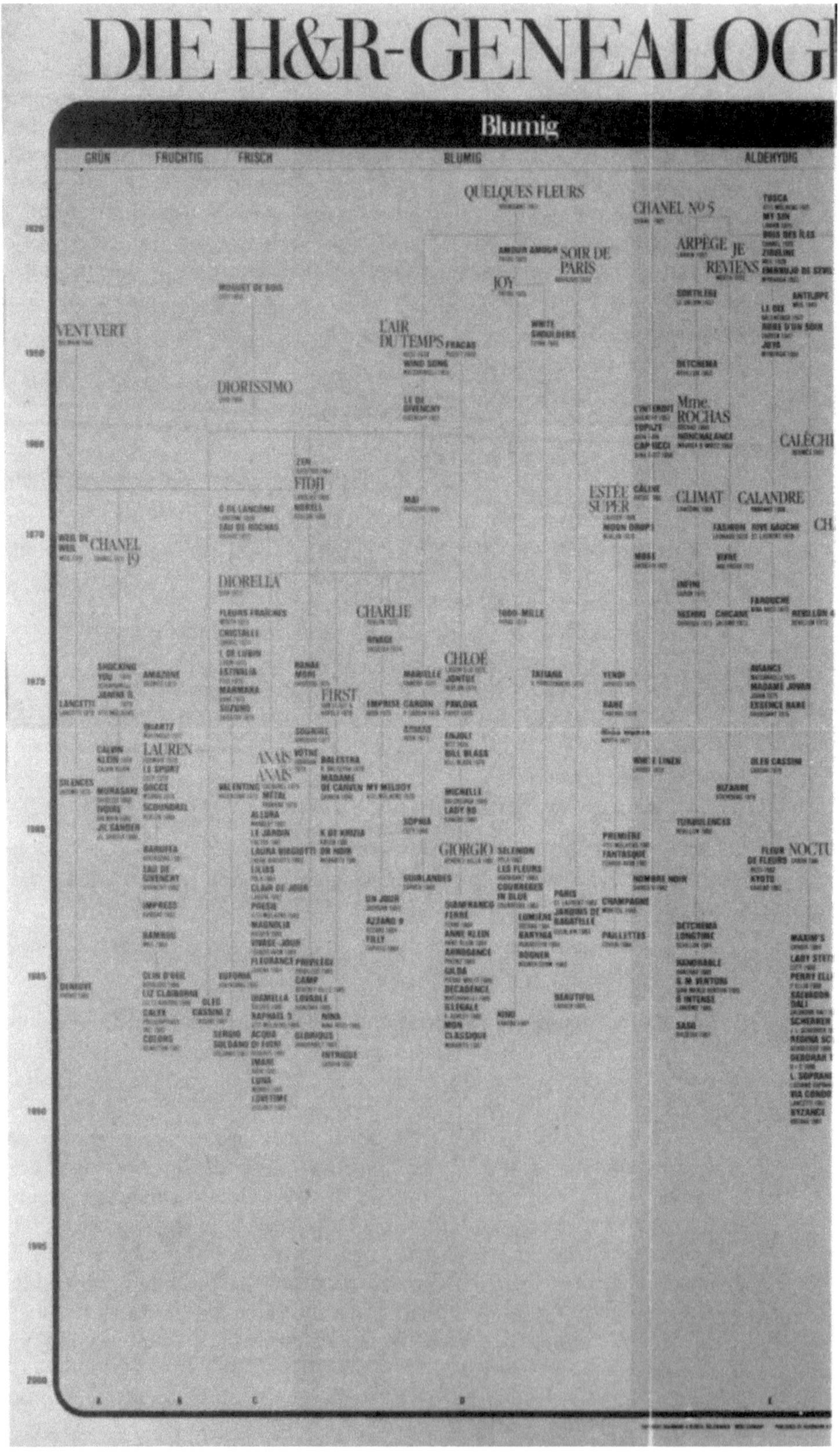
DIE H&R-GENEALOGI
Blumig
GRÜN
FRUCHTIG
FRISCH
BLUMIG
ALDEHYDIG
QUELQUES FLEURS
CHANEL Nº 5
ARPÈGE
JE REVIENS
SOIR DE PARIS
JOY
VENT VERT
L'AIR DU TEMPS
DIORISSIMO
Mme. ROCHAS
CALÈCHE
FIDJI
ESTÉE SUPER
CLIMAT
CALANDRE
CHANEL 19
DIORELLA
CHARLIE
CHLOÉ
FIRST
LAUREN
ANAÏS ANAÏS
GIORGIO
1920
1950
1960
1970
1975
1980
1985
1990
1995
2000

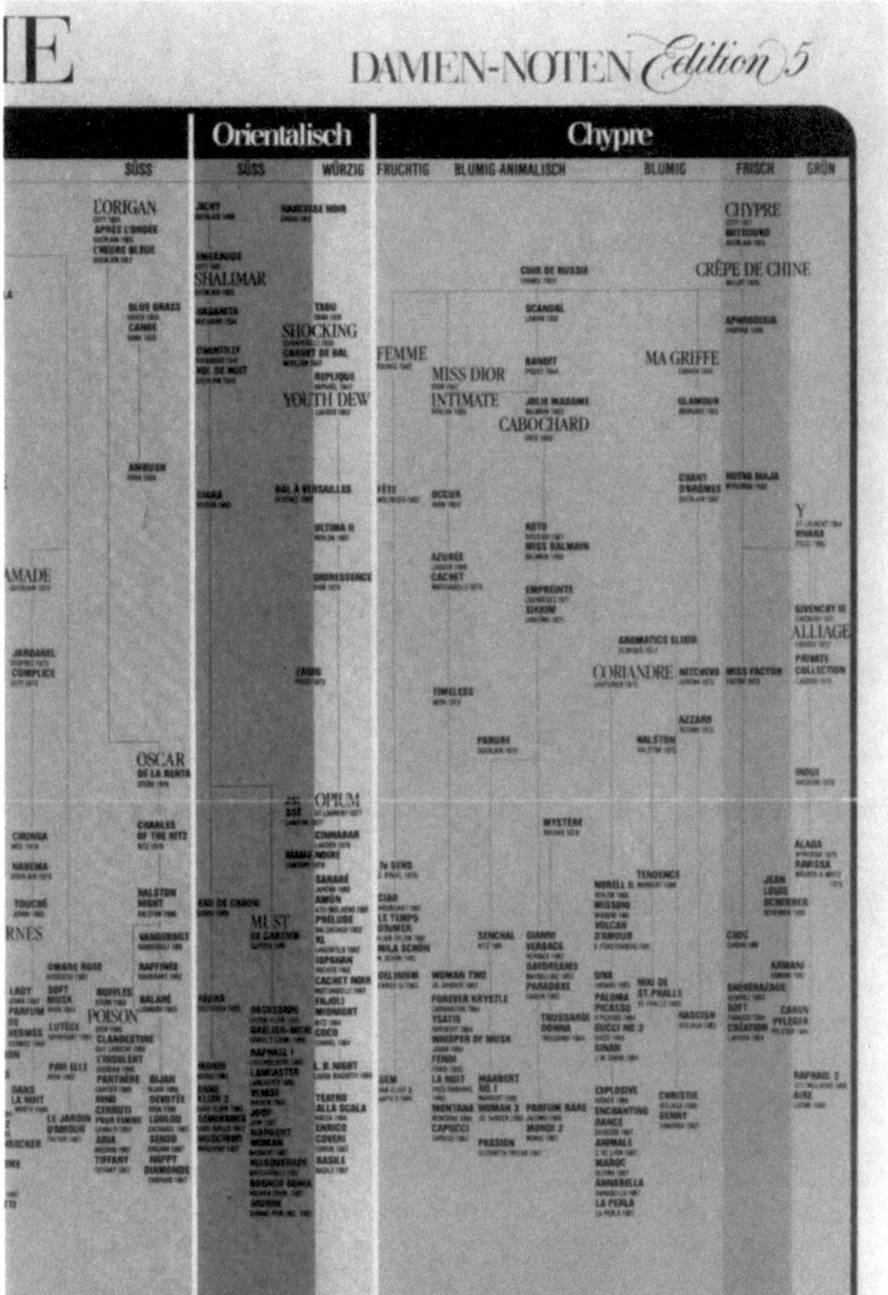
DAMEN-NOTEN Edition 5
Orientalisch
Chypre
SÜSS
SÜSS
WÜRZIG
FRUCHTIG
BLUMIG-ANIMALISCH
BLUMIG
FRISCH
GRÜN
L'ORIGAN
SHALIMAR
SHOCKING
YOUTH DEW
CHYPRE
CRÊPE DE CHINE
FEMME
MISS DIOR
INTIMATE
CABOCHARD
MA GRIFFE
ALLIAGE
CORIANDRE
OSCAR
OPIUM
MUST
POISON

Fig. 1

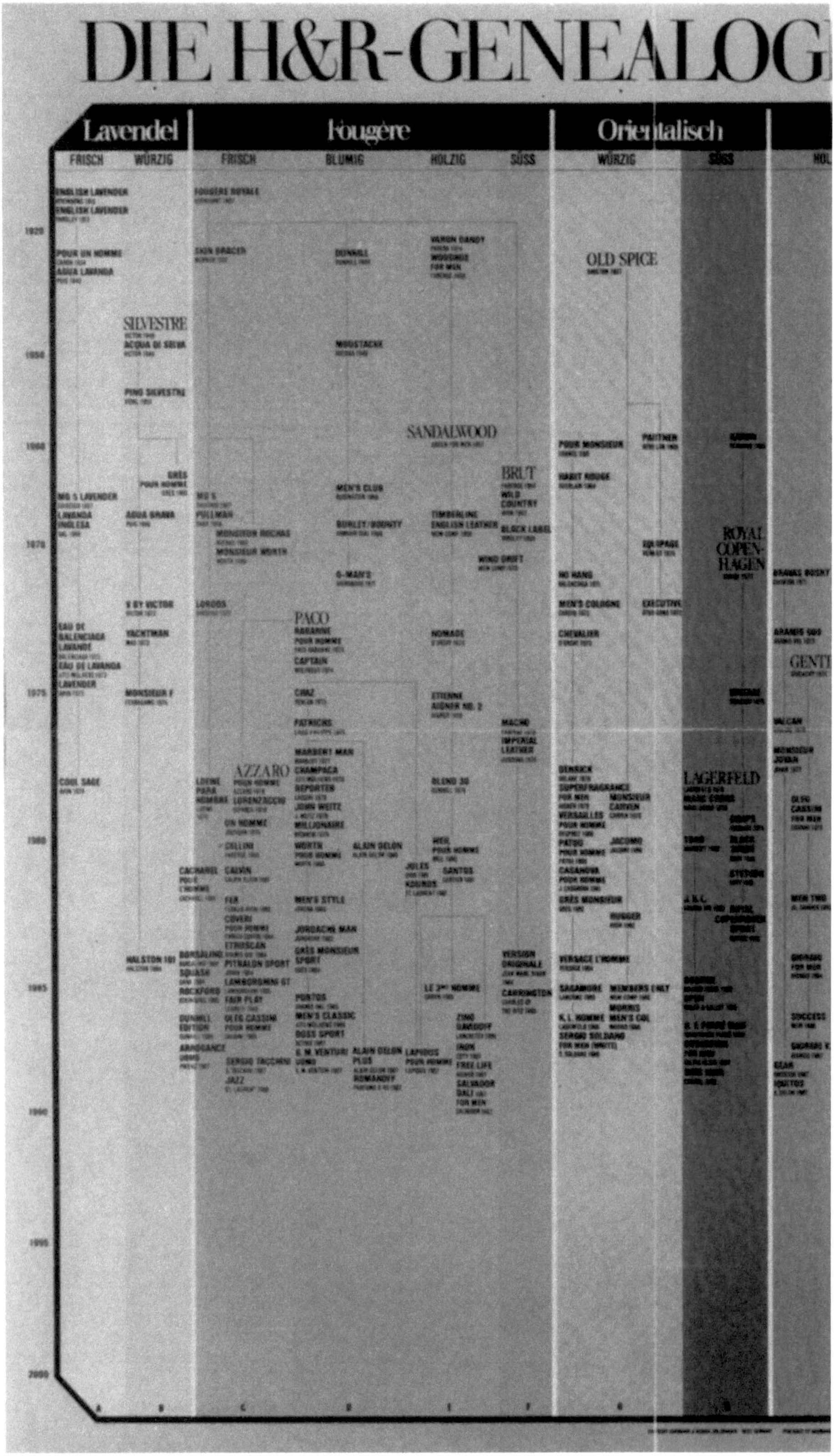
DIE H&R-GENEALOG
Lavendel
Fougère
Orientalisch
FRISCH
WÜRZIG
FRISCH
BLUMIG
HOLZIG
SÜSS
WÜRZIG
SÜSS
OLD SPICE
SILVESTRE
SANDALWOOD
BRUT
ROYAL COPEN-HAGEN
PACO
AZZARO
LAGERFELD

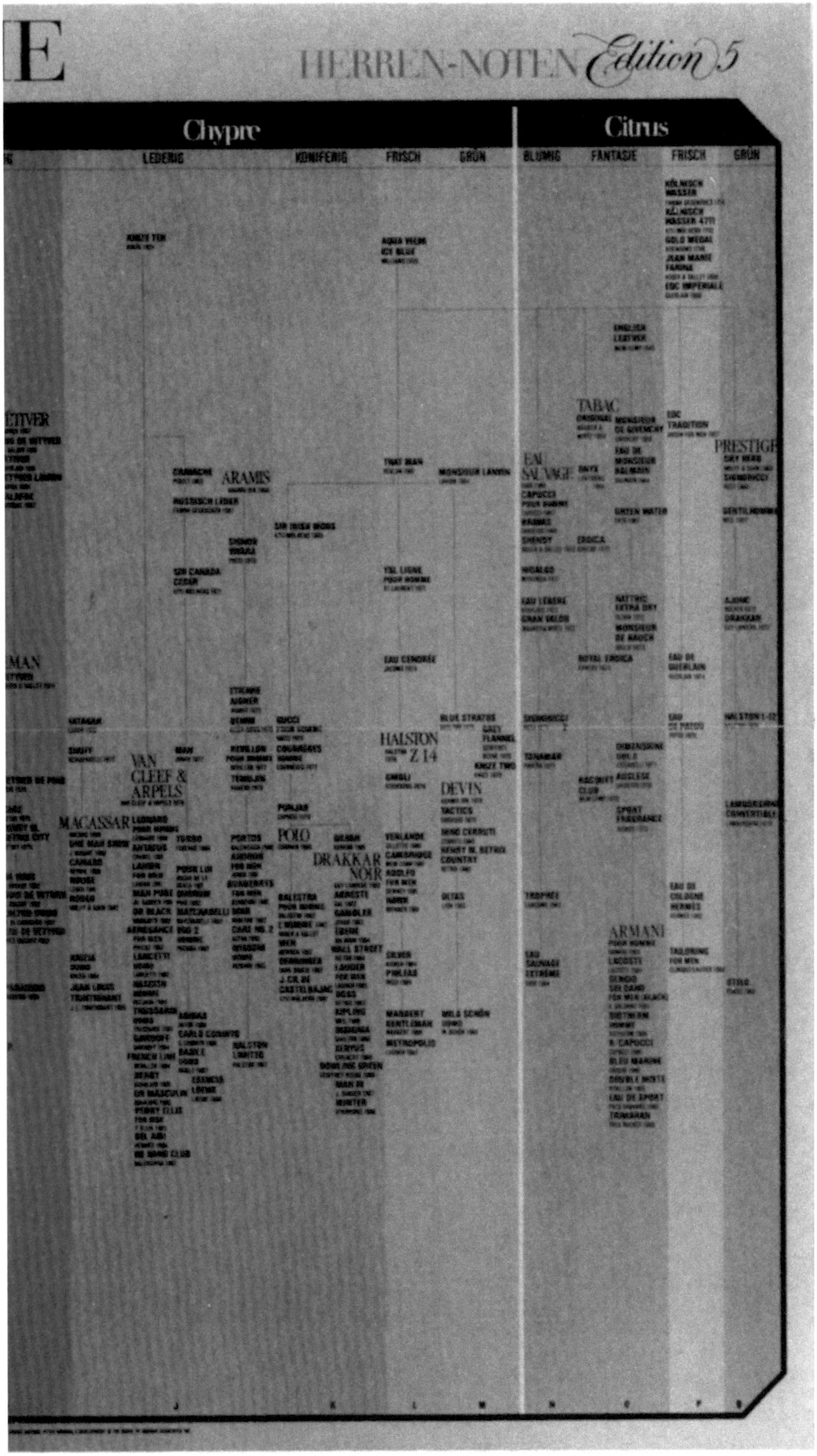
HERREN-NOTEN Edition 5
Chypre
Citrus
LEDERIG
KONIFERIG
FRISCH
GRÜN
BLUMIG
FANTASIE
FRISCH
GRÜN
ARAMIS
TABAC
EAU SAUVAGE
PRESTIGE
HALSTON Z 14
VAN CLEEF & ARPELS
MACASSAR
POLO
DRAKKAR NOIR
DEVIN
ARMANI

Fig. 2

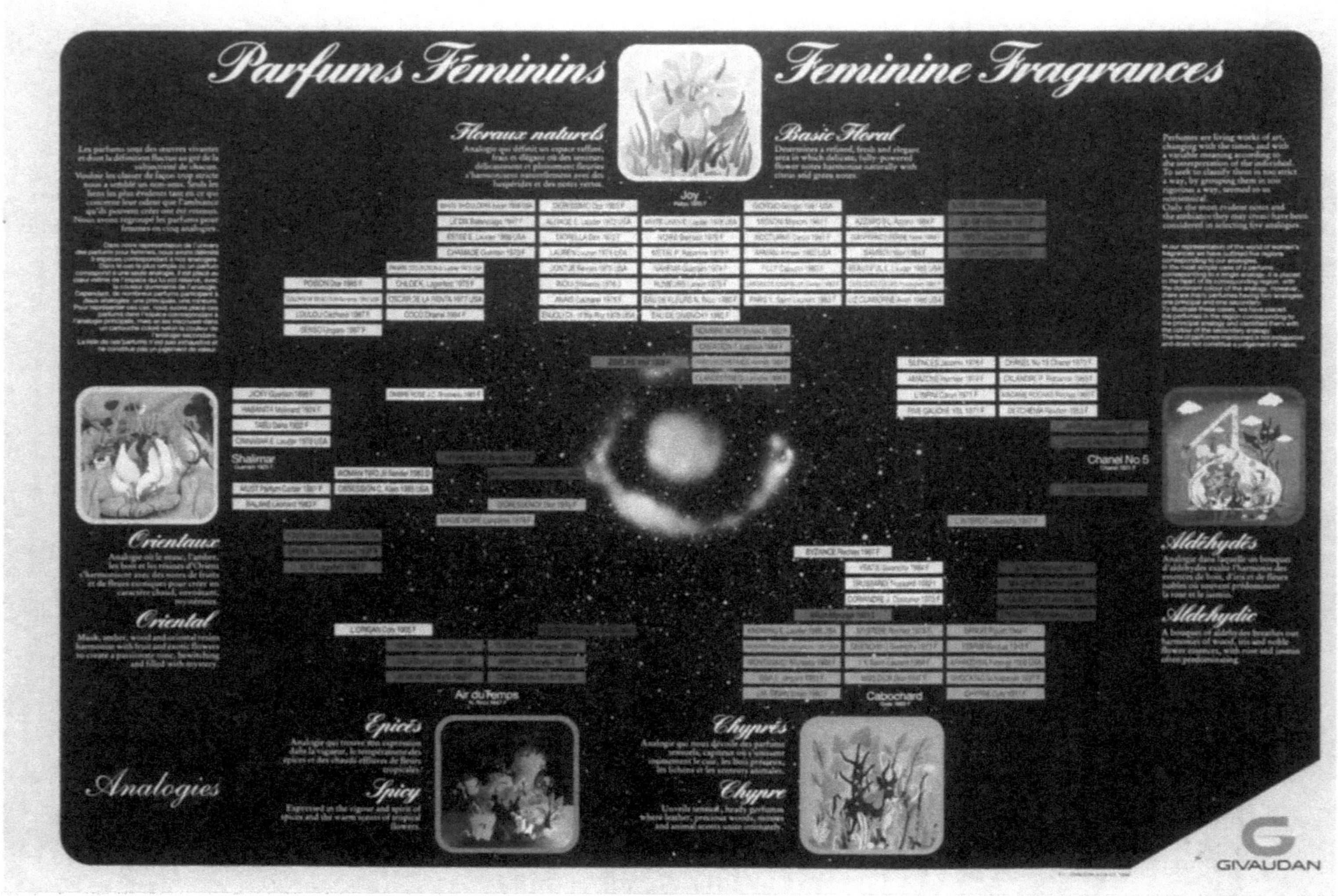
Parfums Féminins
Feminine Fragrances
Floraux naturels
Basic Floral
Joy
Shalimar
Chanel No 5
Orientaux
Oriental
Aldéhydés
Aldehydic
Air du Temps
Cabochard
Epicés
Spicy
Chyprés
Chypre
Analogies
GIVAUDAN

Fig. 3

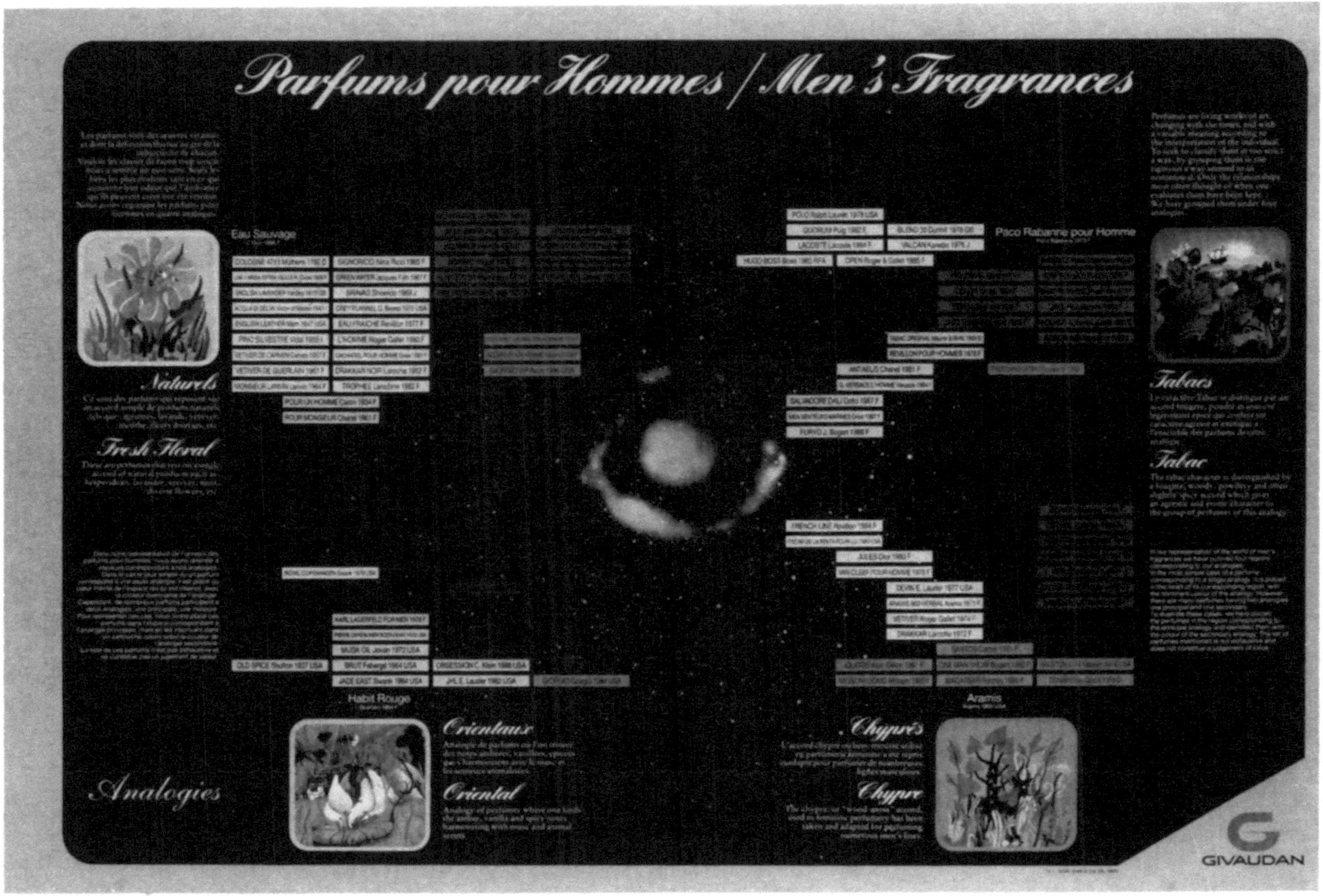
Parfums pour Hommes / Men's Fragrances
Eau Sauvage
Paco Rabanne pour Homme
Naturels
Fresh Floral
Tabacs
Tabac
Habit Rouge
Aramis
Orientaux
Oriental
Chypres
Chypre
Analogies
GIVAUDAN

Fig. 4

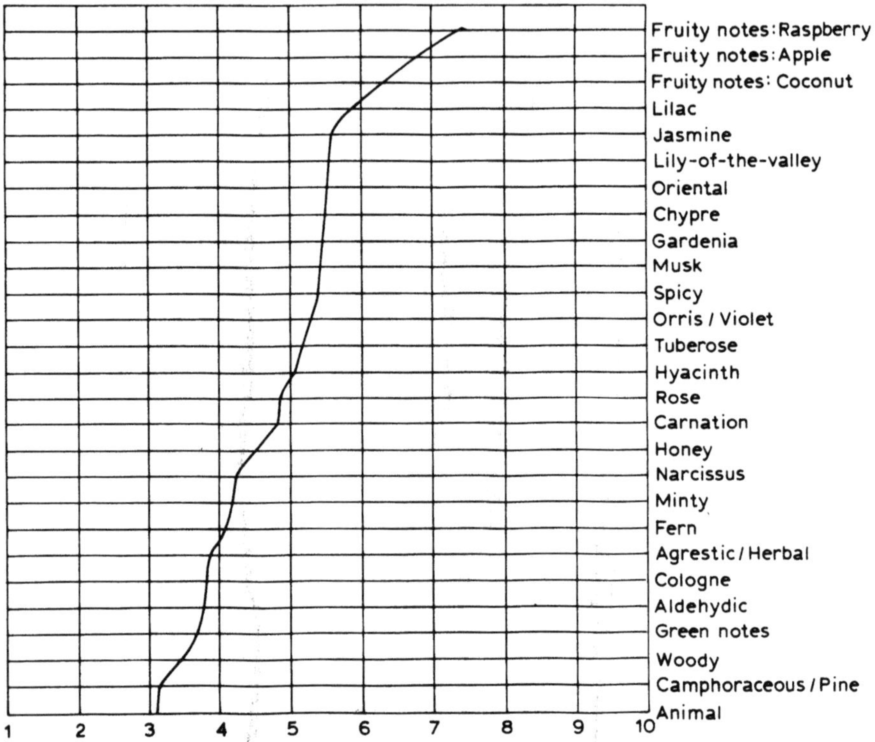

Fig. 5. Sweet attributes defined by oils: —, sweetness, on a scale of 1 to 10 where 1 = not at all sweet and 10 = very sweet.

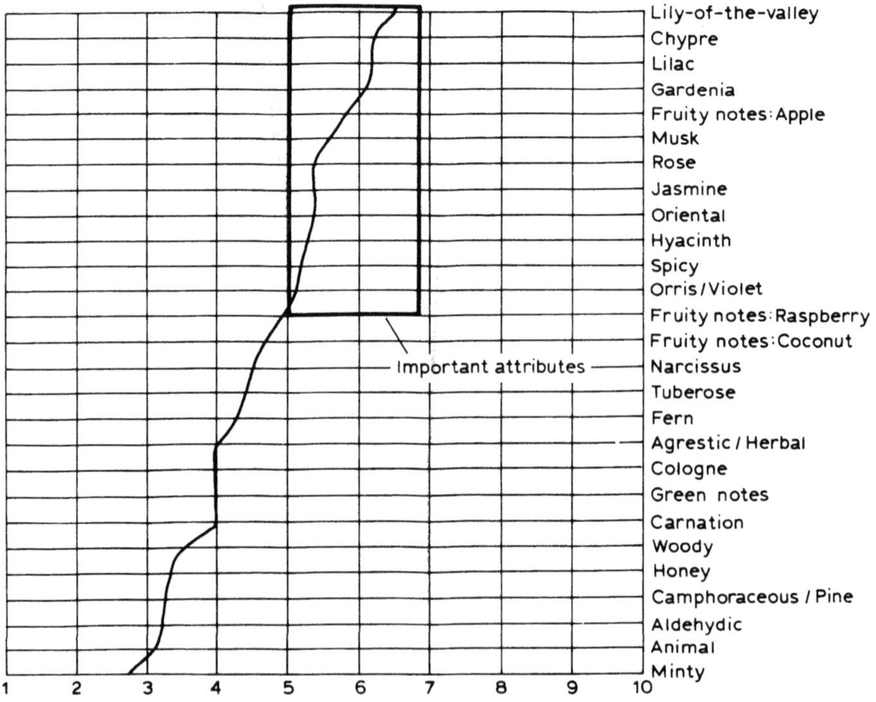

Fig. 6. Feminine attributes defined by oils: —, femininity, on a scale of 1 to 10. Attributes scoring above 5 are significant (outlined by the box).

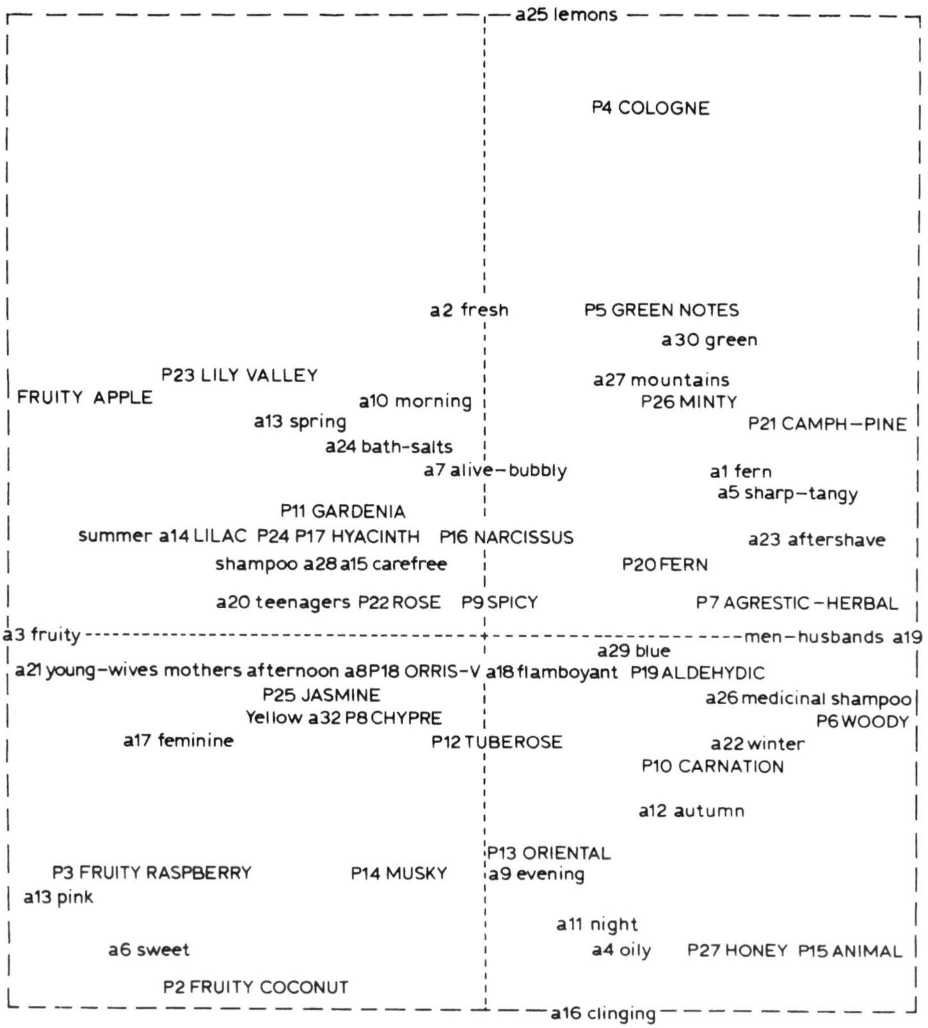

Fig. 7. Fragrance Mapping Exercise. Compiled results from USA, Spain, Japan, UK, Brazil.

10 GLOSSARY OF ADJECTIVES USED IN PERFUMERY†

Aldehydic. For the perfumer, 'aldehydic' notes and modern aldehydic fragrances are those which contain fatty aldehydes in sufficient concentration to contribute to the character of a fragrance. Some of the most famous perfumes in modern times have used aldehydes in daring and original ways. But aldehydes alone do not make up a complete blend. An aldehydic fragrance is

† We thank Mr J. Hadorn and his team for the development of this glossary, which has shown its utility in the Perfumery School (Fragrance Training Centre) at Givaudan.

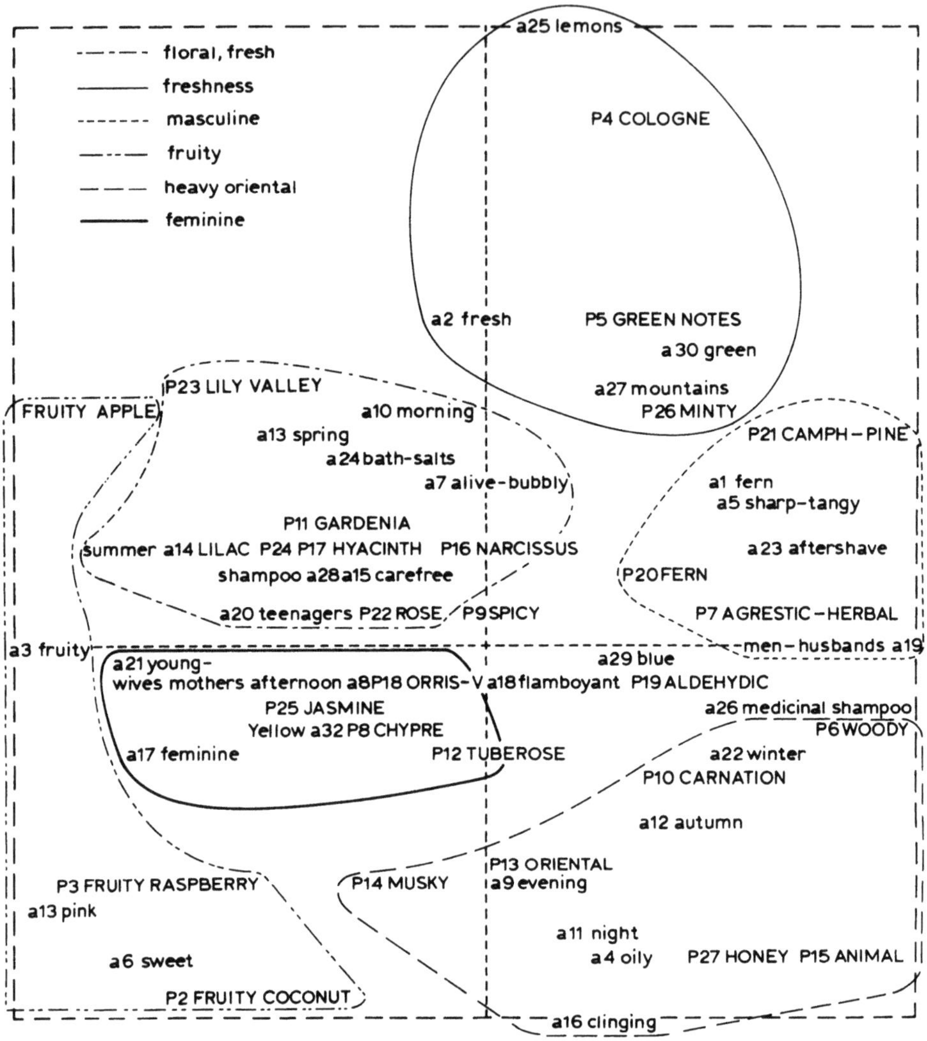

Fig. 8. Fragrance Mapping Exercise. Compiled results, from USA, Spain, Japan, UK, Brazil, showing olfactive categories as listed below.

'a floral bouquet harmonized with a complex of fatty aldehydes which contribute to the fragrance blend'.

Animal. The only animals of practical interest to the perfumer are the musk deer, beaver, civet and whale. Therefore, an 'animal' odour is 'one that is reminiscent of either musk Tonkin, castoreum, civet or ambergris and contains a warm, vibrant nuance'.

Balsamic. Balsams are natural exudations from trees and plants. Although balsams are resinous masses, they differ from resins: they contain high percentages of benzoates and cinnamates. A 'balsamic' odour has 'a mild, sweet, vanillic note with a slightly woody background'. Examples are the balsams of Tolu and Peru.

Bitter. 'Bitter' is one of the four basic taste sensations identified by human tastebuds. Its definition and usage often carry unpleasant connotations: people refer to 'bitter weather', a 'bitter debate' or to a 'bitter taste left in the mouth'. Yet some bitters, such as bitter chocolate or Angostura bitters, are pleasant.

Although flavourists largely agree on what is or what is not bitter, perfumers do not. One author in the perfumery industry uses the word as generally descriptive of all vegetable materials with the exception of flowers and balsams. Although there is a relationship between taste and odour, not all materials with a bitter taste have a bitter odour, and not all materials with a bitter odour have a bitter taste. Nevertheless, products with a bitter odour generally affect the olfactive nerve similarly to the way in which products with a bitter taste affect tastebuds.

Body. 'Body' refers to a 'collective whole or totality' as well as to 'fullness or richness', such as in a musical tone. Perfumery has adopted the latter definition, describing body as 'the main fragrance theme', of which the heart is a part.

Citrus. The meaning of 'citrus' is clear. The term applies to a group of fruits including orange, lemon, lime, mandarin, tangerine, bergamot and others. It is one of the few descriptive terms that is universally defined within the perfume industry. A citrus odour is one that is 'reminiscent of citrus fruits'.

Classical. 'Classical' relates to the ancient Greek and Roman worlds, particularly their literature, art, architecture and ideals. The achievements of these civilizations are respected for having retained meaning and relevance over many centuries. They have stood the test of time.

Classical fragrances are also considered time-tested. These fragrances are well balanced, artistic creations that parallel the classical world's emphasis on symmetry and proportion. After long public acceptance, such fragrances become classics.

New fragrance developments that are similar in type and effect are considered classical because of their relationship to a time-tested standard.

Cloying. The perfumer can use the word 'cloying' to describe one of two odour effects: (1) extreme or excessive sweetness; or (2) the ability of a scent to linger long after contact.

Cool. Because 'cool' has acquired many meanings in addition to its reference to temperature, no precise definition is possible for the perfumer. However, a cool scent often suggests a 'clean, fresh, outdoor quality with possibly green, citrus–mint undertones'.

Depth. In the past, 'depth' has been defined by perfumers as a fragrance quality incorporating 'low-toned' chemicals of a heavy or unspectacular odour, but this definition acknowledges insufficiently the profound impression that some odours of depth can leave with a user. Therefore, an odour with depth is one that is 'rich, full-bodied and profound'.

Diffusive. There is wide concurrence on the use of 'diffusive'. It refers to the ability of a fragrance to spread quickly when released from the bottle.

Dry. Like the beverage industry, perfumery uses 'dry' to describe the absence of sweetness. A number of odours suggest dryness, including woody, grassy and ferny. There are also some scents that have a powdery character which also gives an impression of dryness.

Earthy. To a perfumer, an 'earthy' smell is necessarily one that resembles the earth. But what does the earth smell like? One perfumer has suggested that it is 'the peculiar aroma of freshly turned soil represented in perfumery by a tincture of ambergris'. Ambergris itself is said to have an earthy, musty, somewhat stale smell. Thus, earthy is defined as 'the musty, stale smell of freshly turned soil'. The best example is vetiver oil.

Ferny. Many plants are classified as ferns. But ferns do not have flowers and are therefore limited in their potential to supply odours. However, they do have a uniquely green note. When perfumers describe an odour as 'ferny', they are referring to this 'special green quality punctuated with a woodiness from the stem'.

Flat. Like a beverage that has lost its effervescent sparkle, a 'flat' fragrance is one that is devoid of stimulating and interesting qualities: it 'lacks lift, diffusiveness, character and distinction'.

Floral. It is possible that 'floral' will never be used more specifically than according to the current practice. The word always applies to the odour of flowers in general, but at times can refer to a particular flower and at other times refers to the combined fragrances of several flowers.

Fresh. Some maintain that 'fresh' pertains to 'an effect introduced into a fragrance by the use of a citrus oil or certain aldehydes'. But the failure to describe the effect points to the difficulties associated with defining the term. There are professional perfumers who use 'fresh' when referring to the green note of recently cut leaves, the clean and invigorating scent of early morning air, or perhaps an ozonic smell. Thus, a fresh fragrance is usually considered 'an invigorating odour, reminiscent of the outdoors and typified by green, citrus notes'.

Fruity. It is sometimes mistakenly presumed that 'fruity' refers only to citrus notes and that the two terms are synonymous. In fact, fruity applies to many varieties of fruits, each with its own odour type. Red fruits, such as berries, have odours that are fruity but not citrus. The same could be said of allyl caproate's pineapple note, or amyl formate's blackcurrant–plum characteristics. In truth, fruity can be applied to any odour 'suggestive of any of the edible fruits'.

Fungal. In everyday usage, 'fungal' often has an unpleasant connotation because of its allusion to fungus plants. Perfumers use the word only in reference to the aroma of a specific fungus, usually the mushroom. Therefore, a fungal odour is one that is suggestive of moulds, mushrooms and fungi, best illustrated by one perfumer's description of the odour of hydratropic aldehydic dimethyl acetal: 'reminiscent of mushrooms'.

Grassy. Although 'grassy'—like ferny and herbaceous—refers to green, perfumers use the term for a specific type of green note: freshly cut grass. Consequently, a grassy odour is best defined as being 'green and leafy with a slight touch of sweetness reminiscent of freshly cut grass'.

Green. Though 'green' is a colour, its colour does not describe a fragrance. Green has been defined as 'having abundant foliage', or meaning 'full of life and vigour', or as representing something 'not fully processed or treated', or simply 'unripened'.

But what does the perfumer mean when using the word 'green'? One authority has said the word refers to freshly cut leaves or vines, such as violet leaf absolute. Other perfumers have interpreted it as fresh, youthful, bright or like Spring. Yet all agree that green generally refers to a 'fresh, leafy scent'.

Harsh. 'Harsh' describes something 'disagreeable to taste or smell' or 'sharply unpleasant or rigorous'. When applied to perfumery, harsh describes a 'crude, unbalanced, rough, pungent odour'.

Hay. 'Hay' is normally defined as grass that has been mown and cured for fodder. But hay has its own special odour. To the perfumer, it is the smell of sun-dried grass or the smell of alfalfa stored in a barn. Some have said that it has suggestions of clover and fern. More precisely, it is a 'sweet clover odour reminiscent of coumarin'.

Heady. 'Heady' is not often used by perfumers since the word cannot be restricted to specific fragrances. But when invoked by the perfumer, it applies to an odour that is 'powerful, stimulating and intoxicating'.

Heart. Chemists use 'heart' to denote a very important fraction of a chemical derived through distillation and other processes. However, the chemist's use of the word should not determine the perfumer's. The heart of a perfume is its central and decisive part, 'the main constituent, core or base of a composition which gives it character'.

Heavy. The word has many meanings in addition to weight. Some definitions cite harshness and oppressiveness, others excessive richness. One source described a 'heavy' odour as 'one which requires time and effort to be recognized'. An example is musk xylol: although its odour is pleasant, physical effort, however momentary, is required for appreciation. These types of materials usually have high boiling points. Although many perfumers concur with the general descriptions, most limit their definitions of a heavy fragrance to one which is 'generally sweet and balsamic'.

Herbaceous. Though once applied to all green crops, the word 'herb' currently refers to a plant which is wholly or partly used as a food or medicine. Because there are many herbs, each having a different odour quality, it is necessary for perfumers to identify a quality of odour that is common to most if not all herbs.

'Herbaceous' has been described 'as a medicinal or phenolic note, combined with a coarse grassiness' which is often pungent. But the essential oils of thyme, hyssop, calamus and chamomile all illustrate the herby note in different ways. For example, calamus has a meaty, fatty note that is neither medicinal nor phenolic. It is therefore difficult to define 'herbaceous' more precisely than to say that it is 'grassy-green, spicy and somewhat medicinal'.

Honey. 'Honey' can come from a variety of sources such as alfalfa, pine and orange blossoms. Yet all types of honey share certain fragrance characteristics. They tend to be 'sweet, heavy, and syrupy with a waxy background'.

Leafy. 'Leafy' is one of a variety of green notes potentially present in a fragrance.

Leather. The 'leather' odour is a very popular scent imparted to the hides of animals during the tanning process. The leather odour is pungent and smoky, but with a slight sweetness. The perfumer can achieve the effect with either oil of birch tar, cod oil, isoamylquinoline or *p*-tert-butylphenol.

Lift. 'Lift' has been defined as the conversion of a fragrance from a heavy to a light effect through the use of citrus oils and synthetics. However, when used by the perfumer, the word has a more specific meaning concerning the top note of a given composition. To add lift to a blend is to give it life and brilliance. The term also refers to the diffusiveness of a blend. Lift therefore refers to a 'brilliant top note with wide diffusiveness'.

Light. In contrast to heavy odours, 'light' odours are said to be easily and quickly recognized in materials with low boiling points, or in 'products of ester structure'. Perfumers generally refer to light fragrances as neither sweet nor cloying with a fresh note that is predominant.

Liqueur. 'Liqueur' is difficult to specify because of the many fragrances and tastes (e.g. anisette, kummel, cherry) it represents. Its scent is usually 'sweet and fruity with a flavour top note and alcoholic overtones'.

Mellow. 'Mellow' refers to something fully matured, mild and pleasing. In perfumery, a well rounded or smooth perfume is often called mellow. Mellow fragrances tend to be aged, balanced, smooth and rich.

Modern. The 'modern' era of perfumery began at the start of the 20th century. At that time, synthetic aromatic chemicals, such as aldehydes, came into use. New fragrance blends were created that differed from the individual floras and floral blends that previously had been popular.

One source contends that the word is used 'loosely to describe fragrance compositions containing aliphatic aldehydes to the extent that they are observable in the top note. This nomenclature came about when the aliphatic aldehydes were first introduced in perfumery'.

'Modern' has also been applied to those blends recently created for which the perfumer used newer materials developed by research chemists.

Mossy. Although there are many mosses and lichens, the perfumer is interested primarily in oak moss and tree moss. In addition, several species of seaweed are usable in perfumery. Despite the fact that the odours of these products are generally classified as 'mossy', there are definite differences among their odours. 'Mossy' is a generic term, but its meaning is not as broad as either floral or woody. Its odour is earthy, green, herbaceous and ferny, but is primarily 'suggestive of the aromatic lichens and mosses, particularly oak moss and tree moss'.

Mushroom. 'Mushroom' and 'fungal' are nearly interchangeable. 'Mushroom' is more specific, while 'fungal' implies additional odour characteristics such as mouldiness or mossiness. A mushroom odour tends to be 'pungent, musty with some earthy-green notes'.

Musty. The perfumer uses the word 'musty' in ways that closely resemble its common usage: an odour that is 'mouldy, damp, and possibly even fungal, reminiscent of the atmosphere in an unventilated room'.

Nutty. When describing a fragrance as 'nutty', the perfumer has two basic characteristics in mind: the hint of oiliness generally associated with crushed nuts, and the bitter almond note exemplified by benzaldehyde. There are, however, many different types of nuts, each having its own taste and smell.

Resinous. When referring to the odours of gums exuded from trees, particularly pines and other evergreens, a perfumer describes them as 'resinous'. Resins have highly distinct notes peculiar to themselves. However, a majority have a somewhat harsh piny note from the turpentine oil usually present.

Rich. A 'rich' fragrance is one that possesses depth and fullness of body.

Rounded. A perfumer rounds out a perfume by adding the finishing touches to the creation. As in music, a 'rounded' fragrance is balanced to perfection. The first whiff immediately suggests balance, smoothness and harmony.

Sharp. A 'sharp' fragrance is 'strong, penetrating and often pungent'. A penetrating quality can be found in any of the aliphatic aldehydes used in perfumery. In addition, 'a combination of loud, coarse, aromatic chemicals in a cheap fragrance occasionally produces a mildly penetrating effect that can be classified as sharp'.

Smoky. 'Smoke' has long been associated with perfumery. The word 'perfume' is itself derived from the Latin word *fumus*, meaning smoke. In all probability, the first perfumes were smoke offerings. In addition, ancient religious rites often included the burning of odoriferous woods resulting in sweet-smelling smoke.

Smooth. A 'smooth' perfume—a perfectly blended one—is harmonious, balanced and rounded. There is a complete absence of harshness in a smooth fragrance. The tones of a smooth scent are quiet and even.

Spicy. 'Spicy' is one of the seven basic classifications of perfumes. These scents derive their spicy characteristics from the ingredients present in the formulation, or from the spicy notes of some of the flower oils used. Cinnamon, cloves and ginger are a few of the spices used in perfumery. Among the basic chemicals, eugenol and isoeugenol have spicy notes; among the flowers, carnation and gillyflower have spicy nuances. All spicy fragrances share a common pungency and sharpness that stimulate the senses of smell and taste.

Sweet. Like 'bitter', 'sweet' is a term common to tastes and flavours. However, there are two applications of 'sweet' in perfumery: it can describe a rich fragrance with the ambrosial characteristics of a sweet taste; it can also refer to the perfumer's changes in a fragrance which make the odour smoother or softer when it is too dry, leafy or bitter.

Thin. A 'thin' fragrance is one that lacks body and richness.

Tobacco. There are two types of 'tobacco' odour. One is the odour of freshly cut tobacco leaves brought in from the field. In this condition, tobacco has a green, leafy scent. But the consumer is more familiar with the second form that tobacco odour takes after it has been cured, aged and cased. For perfumers, a tobacco odour is primarily pungent—from the drying, curing and casing processes—with a light touch of background green.

Top note. The 'top note' is the 'initial fragrance impression, the first odour perceived when smelling a fragrance product'. It usually consists of the most volatile portion of the composition if it is to make a quick impression on the user.

Velvety. To the professional perfumer, a 'velvety' fragrance is one that is soft and smooth, lacking harsh chemical notes.

Warm. Although 'warm' is applied to a variety of circumstances and conditions, most perfumers think of a warm fragrance as being 'stimulating, full-bodied, rich and possessing depth'.

Woody. There are several woods used in perfumery, including sandalwood, rosewood and cedarwood, together with the roots and barks of trees, perhaps even cinnamon and vetiver. Yet the perfumer does not refer to all of these when using the word 'woody'. It is a generic term referring to a family grouping in which all members have an odour type considered common to the entire group. Not every odour is the same, though certain similarities and relationships are present. In some cases members of the woody group do not originate from wood, but instead derive their woody odours from other sources. However, in all cases, woody fragrance notes are suggestive of aromatic woods and roots.

Chapter 9

Chemical Classification and Structure–Odour Relationships

GÜNTHER OHLOFF, BEAT WINTER & CHARLES FEHR

Firmenich S. A., Research Laboratories, CH-1211 Geneva, Switzerland

1 INTRODUCTION

The physical and chemical properties required of a suitable sensory active molecule are determined by the location, molecular architecture and physiological medium of the chemoreceptor. Clearly sensory activity in human olfaction is exclusively associated with volatile molecules. The higher limit of molecular weight found for odorants is around 300 (for some examples, see Refs 1–3). Chemical reactivity of a ligand has little if any direct connection with olfactory activity since odorant molecules are uncharged and hardly require metabolic activation. Nevertheless, several molecular requirements must be met. Thus most odorous molecules contain both a strong hydrophobic and a relatively weak polar region. The latter, usually termed the 'osmophore',[4] is associated with a functional group such as carbonyl, hydroxyl, occasionally an ether or a limited variety of heteroatomic homologues. However, the presence of a functional group is not a *conditio sine qua non* for receptor interaction. Even alkanes can have distinctive odours.

According to present-day knowledge, the triggering of a sensory impulse occurs by direct contact between the odorant and the constituents of the receptor membrane in the olfactory cilia.[5] A signal is produced via the depolarization of the membrane and electronically conducted to the olfactory bulb, analysed in the brain and then confronted with stored-up recognition patterns. Thousands of different odorant molecules can be discriminated.

Olfaction is considered to be a bimolecular process involving the interaction of an airborne molecule with a complementary site of a receptor system. A most probable molecular model for stimulant interaction and signal transduction has been developed (Fig. 1).[6,7] The initial and most important event in the biochemical stimulation process is the binding of the odorant molecule to the membrane-bound receptor protein R, which seems to be an allosteric enzyme[8] consisting of a variable (v) and a constant region (c).[6] Non-covalent bonding leads to a reversible complex, activating the GTP binding-protein (G) and triggering a cascade of enzyme reactions. cAMP, through adenylate cyclase (C) activation, causes phosphorylation of protein kinase (P) inducing a change in ion-channel protein (IC) and ion gating. Consequently the neuronal membrane is depolarized and the output signal transmitted to the central nervous system via the olfactory bulb.

These biochemical aspects of olfaction are supported by the fact that odour sensation can reach a high degree of specificity and sensitivity. A low olfactory threshold and high diversity in odour discrimination are the result. Structure–

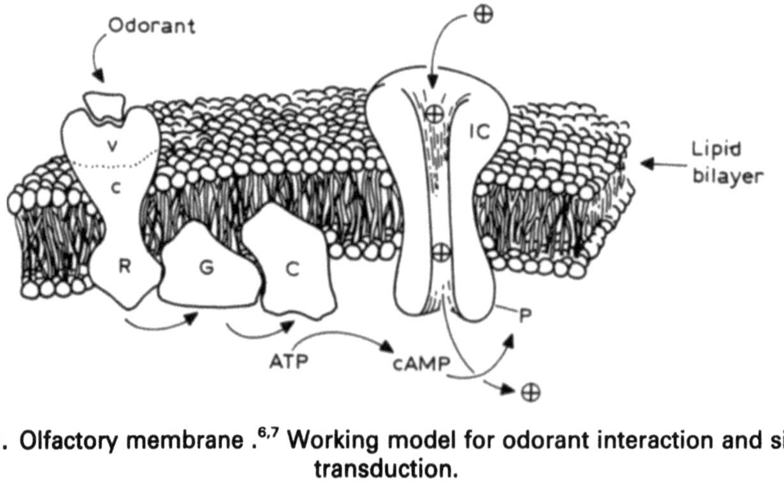

Fig. 1. Olfactory membrane.[6,7] Working model for odorant interaction and signal transduction.

activity relationships (SAR) can serve as important instruments in the investigations of processes on membrane-bound receptor proteins and can at the same time support the chemical characterization of active sites.

Progress on the olfactory recognition process in this field will be shown for four different odour qualities and concepts of mechanisms at the molecular level are developed. The option for the four basic odours, i.e. ambergris, 'steroid scent', sandalwood and musk, is not made arbitrarily. On the contrary, because there is strong evidence for an overlap in tonalities of some typical compounds of the four odour classes, the target of the investigation is directed towards the common molecular features responsible for the odour similarities.

2 AMBERGRIS ODORANTS

The fragrance of ambergris is due to an oxidative degradation of the main constituent in this animal secretion, the triterpene alcohol (−)-ambrein (**1**).[9] The fully developed scent of aged ambergris tincture consists of several different odour qualities which can be easily discerned by aroma experts and described as:[10]

1. wet mossy forest soil;
2. strong tobacco;
3. balsamic, sandalwood-like;
4. warm animal tonality of musk;
5. seaweed, ocean;
6. fecal.

Among the compounds obtained in the early chemical elucidation work were the tricyclic ether **2** and its hydrogenation product **3**, both possessing typical ambergris odour.[11,12] The diastereoisomer **4** has been reported as inactive.[13]

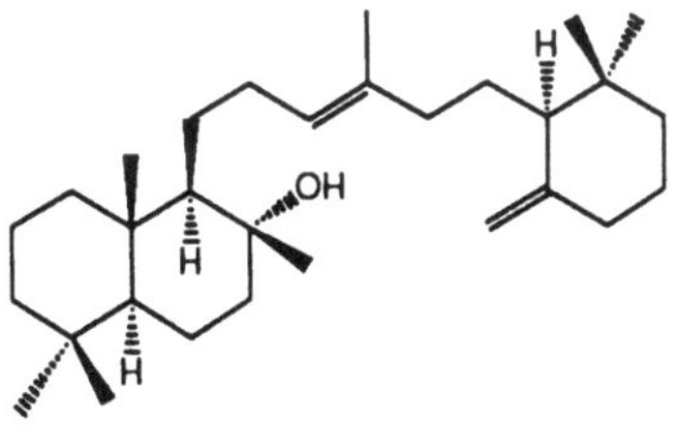

1

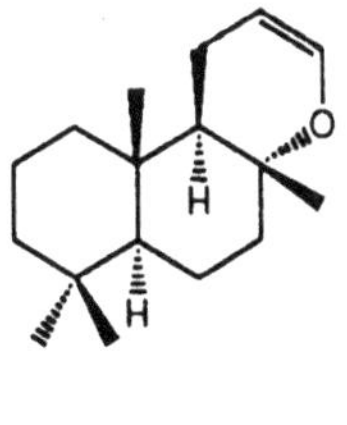

2

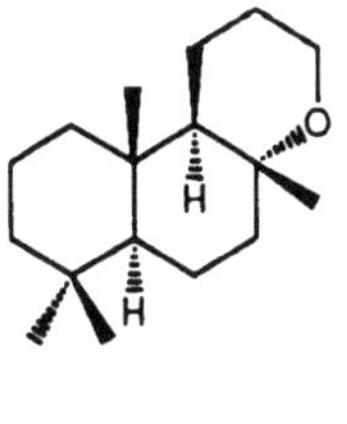

3

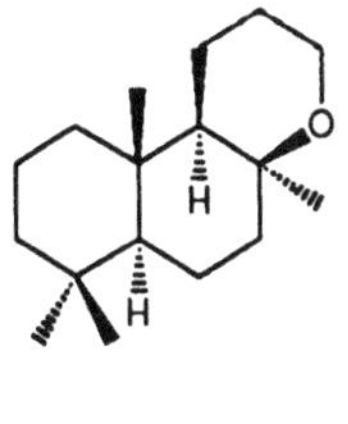

4

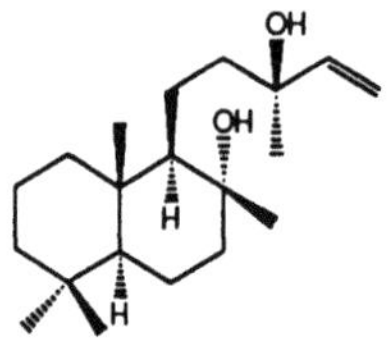

5

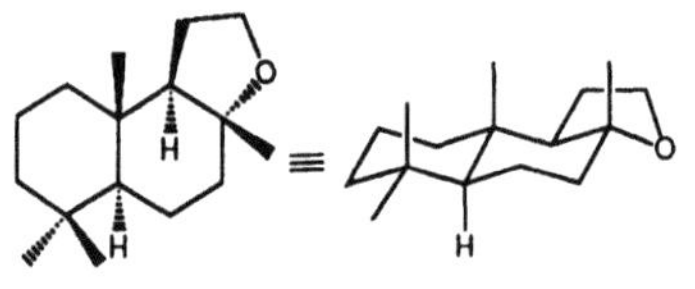

6

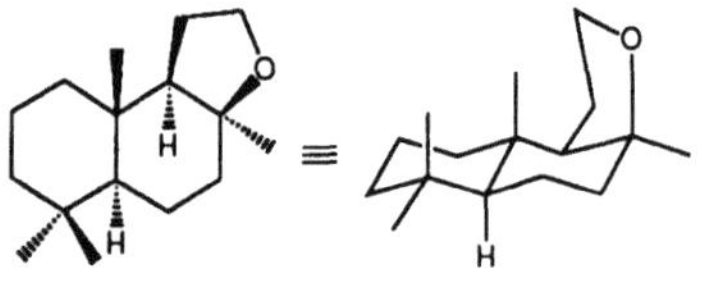

7

Another source of ambergris odorants was found in the labdanoid diterpene (−)-sclareol (**5**), which can be transformed into the tricyclic ethers **6** and **7**.[14,15a,15b] Diastereoisomer **6**, known as Ambrox®, was later identified in the odoriferous constituents of tincture of ambergris.[16] Both diastereoisomeric ethers **6** and **7**, which develop a rich aroma matching the first four tonalities mentioned above, are key ambergris-type fragrances used in perfumery. A further compound with an intense odour of ambergris, the tetracyclic ketal **9**, was discovered among the oxidation products of the related labdane derivative (+)-manool **8**.[17,18] In high concentration **9** also develops a urine and perspiration odour. Remarkably enough, ethers **4** and **10**, diastereomers of **3** and **9** respectively, are odourless.[13] Compounds **3**, **6** and **9** thus might be considered as prototypes of ambergris odorants, both from the point of view of chemical structure and of olfactory activity. They have in common a rigid *trans*-decalin skeleton, which limits conformational freedom and variations in function of the molecular environment and allows a relatively reliable model of their three-dimensional structure to be made. Moreover, their chirality has the same absolute configuration, which might be a decisive factor in biologically active compounds.

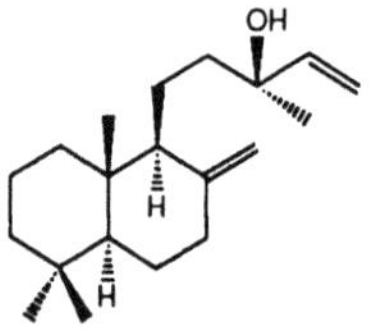

8

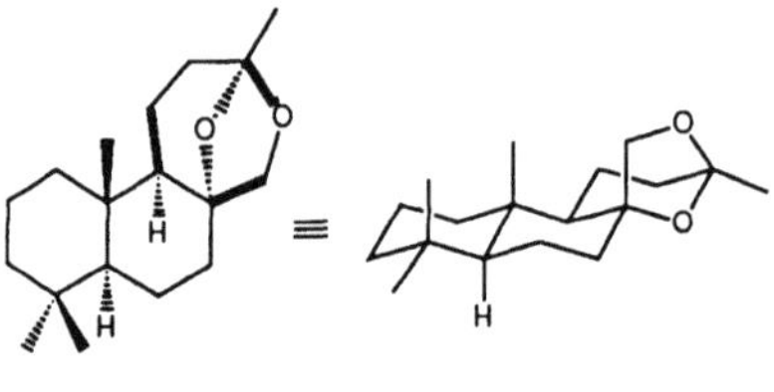

9

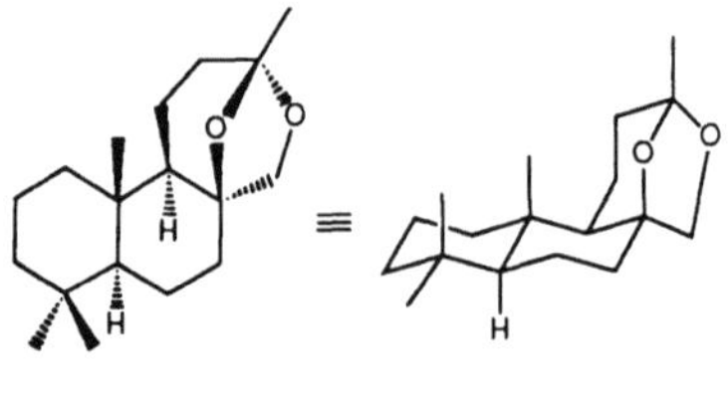

10

These facts, coupled to the growing demand for ambergris-type odorants and dwindling natural sources, have stimulated extensive investigations of analogues of these compounds, and we shall now point out some aspects of the structure–activity relationships that have emerged from the wealth of knowledge on ambergris-type compounds.[9,19]

Enantioselectivity

Enantioselectivity of odour perception in the group of ambergris-type odorants has been examined in several cases.[13] As a result, only the C_{18} acetal **9** and the tricyclic ether **3** possess an ambergris-like odour, whereas their enantiomers *ent*-**9** and *ent*-**3** were found far less active. Conversely, while compounds **4** and **10**, diastereoisomers of **3** and **9** respectively, are odourless, their enantiomers *ent*-**4** and *ent*-**10** exhibited odours, although distinct from ambergris. These pairs of inactive/active enantiomers were the first recorded examples of complete enantiospecificity in odour perception.[13] In the case of the enantiomers of **6**, only rather small differences in their odours were found, both quantitatively and qualitatively. The derivative *ent*-**6** has a higher detection threshold value (2·4 versus 0·2 ppb for **6**, measured in water) and, with its accentuated note of sandalwood, lacks the strong and warm musk-like animal note of **6**.[20]

The differences in odour properties found for the enantiomeric pairs mentioned above, as well as for other pairs,[21,22] and which are based on only three-dimensional structural features, reflect the chiral nature of the human receptor system and the diastereoselective nature of its interactions with odorants or with other bioactive compounds.

Diastereoselectivity

The odour response as a function of the diastereoisomerism of substrates has been well studied,[23a,23b] particularly in the ambergris series. The structural difference between diastereoisomers lies in their three-dimensional configuration, but influences all their physico-chemical parameters; the resulting effect for each one is a combination of all these factors, which act in concert during the interaction with the receptors, and which are not easily singled out. The large number of compounds which have been obtained by structural modification of compounds **3**, **6** and **9** have provided information about the relationship between sensory activity and possible interaction with a complementary receptor system.[9] The diastereospecificity of the ambergris-type fragrance has been elaborated and summarized in the 'triaxial rule of odour sensation',[24a,24b] the molecular model of which is shown in Fig. 2. Thus, the odour of this class appears to be connected with the *trans*-decalin skeleton. A multi-point interaction between the stimulant molecule and the complementary receptor site seems to be associated with an axial orientation of the substituents R′, R″ and R_a, whereby oxygen functions can be attached at one of the critical

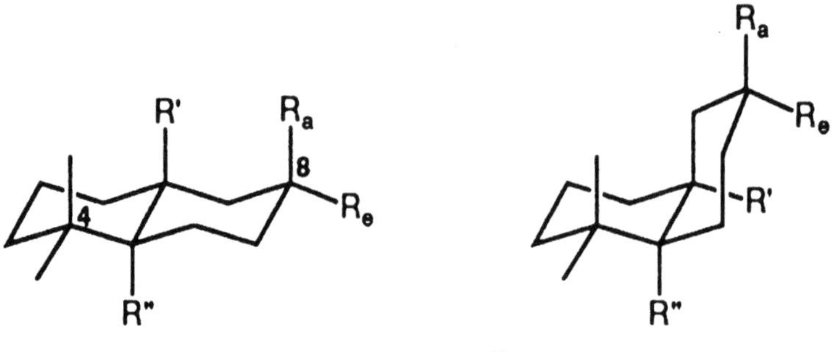

A B

Fig. 2. Triaxial rule of odour sensation. Schematic representation of the relationship between the decalin ring system and ambergris odour.[9]

positions (R′, R″, R_a or R_e), as shown in model **A**. Ambergris odour disappears in its diastereoisomers related to the *cis*-decalin ring system,[23] as represented by model **B** (Fig. 2). In addition to the already-mentioned pairs of diastereoisomers **3/4** and **9/10**, diastereoselectivity of odour perception is illustrated by selected examples listed in Table 1.[9] These are related to the

Table 1
Ambergris-like odorants and their diastereoisomers

Active	*Inactive*
11 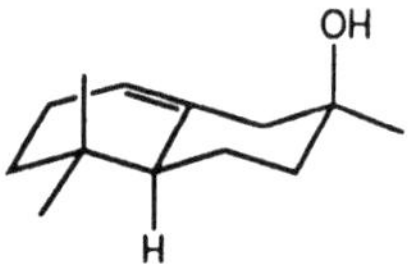	**12** 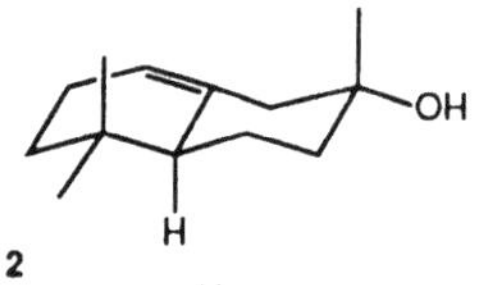
13 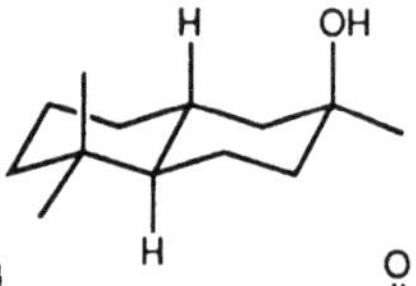	**14** 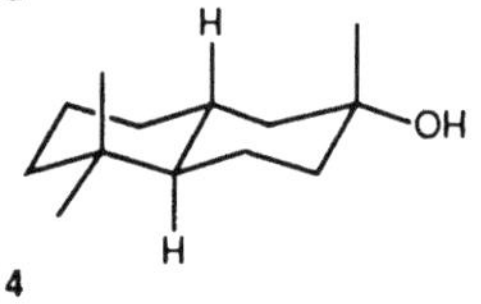
15 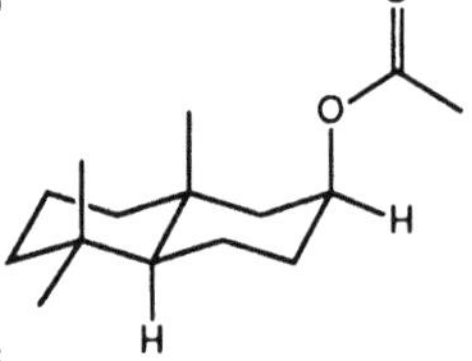	**16** 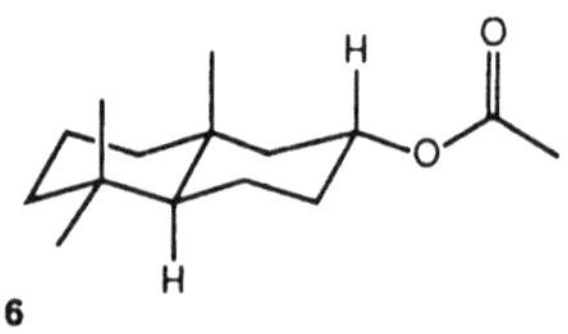
17 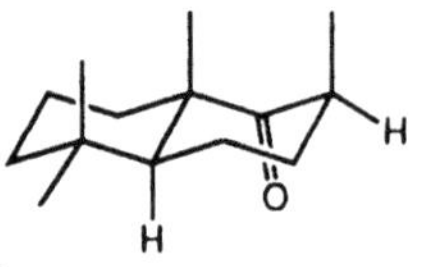	**18** 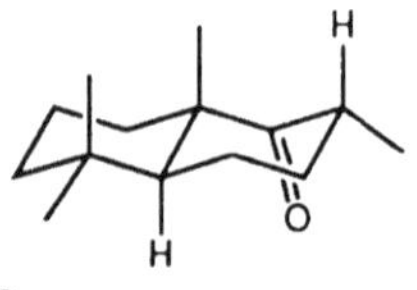

isomeric α-ambrinols **11** and **12**, their dihydro derivatives **13** and **14**, to Polywood® **15** and its odourless diastereoisomer **16**, as well as to the ketones **17** and **18**.

In the following discussion we shall try to focus on physico-chemical aspects which have emerged from studies of analogues in the group of ambergris-type compounds, in particular the structure-dependent parameters related to electronic, hydrophobic (lipophilic) and hydrophilic properties.

Electronic Properties

The relationship between electronic properties of ambergris-type compounds and their olfactory activity has been examined by Vlad and co-workers,[25a,25b] who were not satisfied with their results based on only structural and stereochemical features. Their analysis of 10 compounds revealed that in all structures there is a LUMO (or a nearby unoccupied molecular orbital) with a large contribution from the atomic orbital of axial, tertiary, allylic and other hydrogen atoms which are near-neighbours to the oxygen atom; moreover, the active compounds have two hydrogen atoms providing a major contribution to the LUMO (acceptor molecular orbital) and which form with an oxygen atom a so-called 'ambergris triangle' of certain dimensions. Illustrations of the 'ambergris triangle' are given in Fig. 3 for Ambrox® **6**, its di-demethyl analogue **19** and the two spiro ethers **20** and **21**.

However, an indication that the results of the quantum chemical calculations of the electronic structural features can be replaced by a simple analysis of

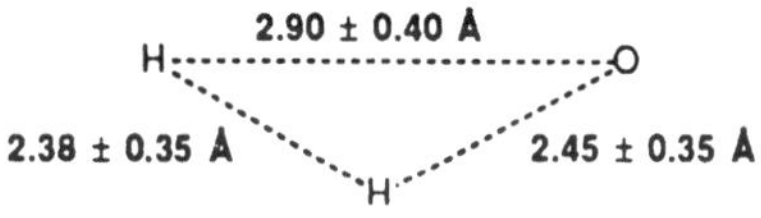

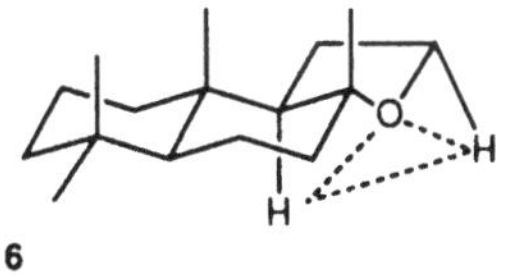

6

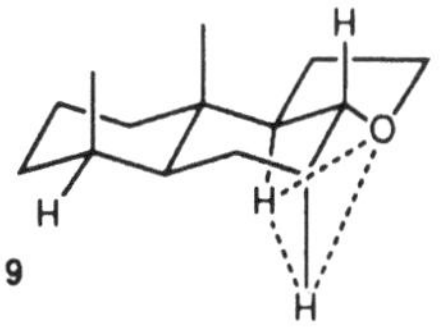

19

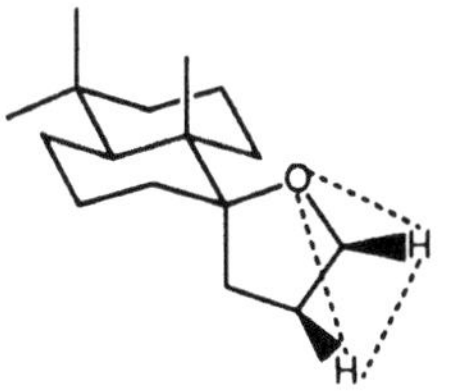

20

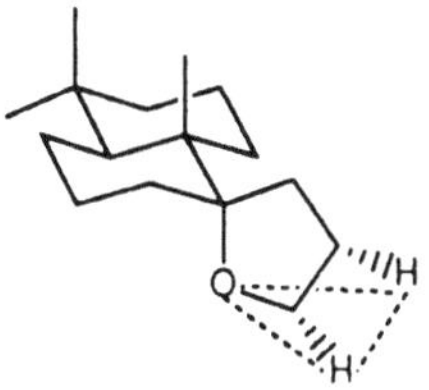

21

Fig. 3. Triangle of ambergris odorants.

geometrical features (presence or absence, in the compound under consideration, of a triangle corresponding to the 'ambergris triangle') reduces the importance of electronic features to the profit of geometrical/structural features and may lead to ambiguous conclusions. An example of the predominance of steric factors over electronic ones is shown by compounds **20** and **21**; both have an 'ambergris triangle', but only **20** is active.

Lipophilicity

Lipophilicity is a complex molecular property encoding volume (or cavity) and polarity as the two major contributions. The volume term expresses non-directional hydrophobic interactions and reflects both the steric demands (hydrophobic binding) and limits (steric hindrance) of the interacting species. Several studies of ambergris-type compounds have been examined under this aspect.

For example, two racemic demethyl analogues **19** and **22** of Ambrox® (**6**) have been prepared,[26–28] and a decrease in the odour intensity together with the appearance of side-notes was observed in relation to the replacement of methyl groups by hydrogen atoms. A more recent study (see Table 2) has also

Table 2
Nor-derivatives of Ambrox® (**6**) and its diastereoisomers

Increased ←—— Activity ——→ Decreased

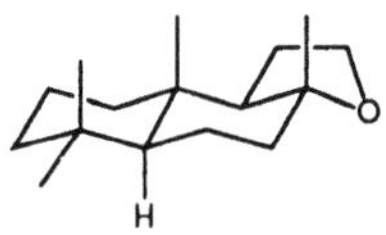
6

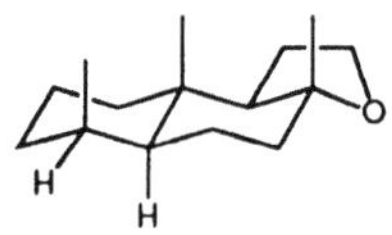
22

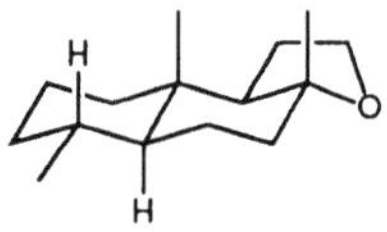
23

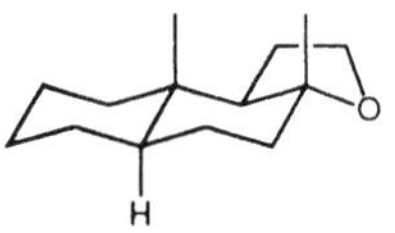
24

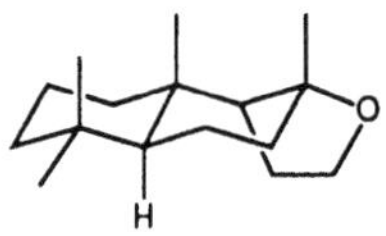
25

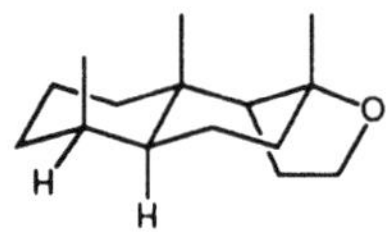
26

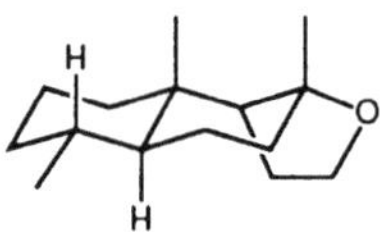
27

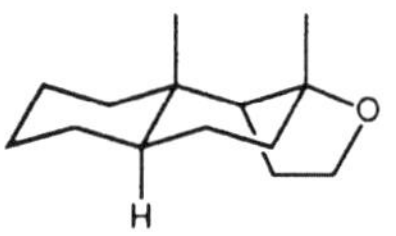
28

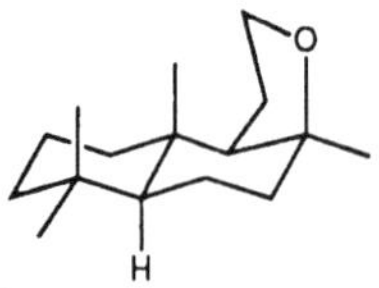
7

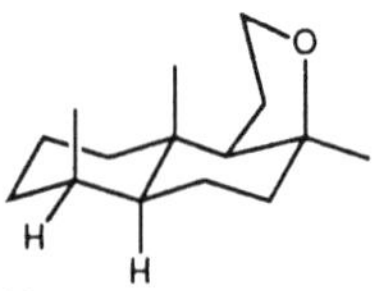
29

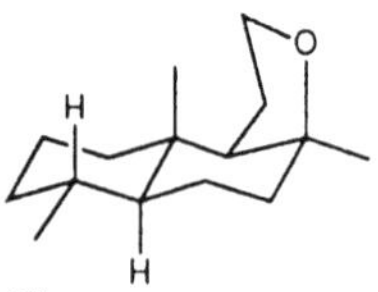
30

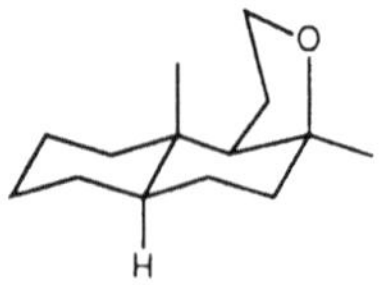
31

examined in detail the change in intensity and quality of the olfactory activity in function of the removal of methyl groups in the apolar part of Ambrox® (**6**) and its two diastereomers **25** and **7**.[20] In the case of optically active **22**, the result obtained with the corresponding racemate (*vide supra*) was confirmed; moreover, this investigation showed that the strongest and most typical sensory activity in these series was associated with an accumulation of the three axial methyl groups in the 4-, 8- and 10-positions (as shown in **6**, **22**, **25** and **26**).

Table 3
Accessible surface area of the ether oxygen in Ambrox® (**6**) ambra oxide (**3**) and their analogues

Active		*Inactive*	
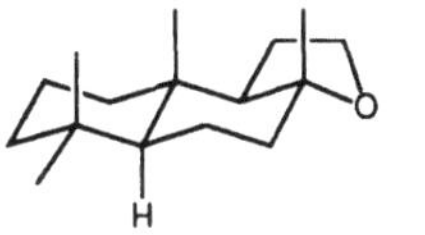 6	12.8 Å^2		
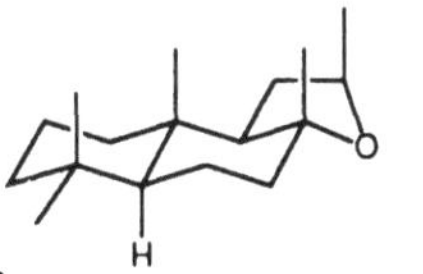 32	8.8Å^2	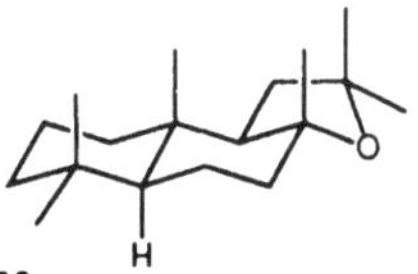 33	5.5Å^2
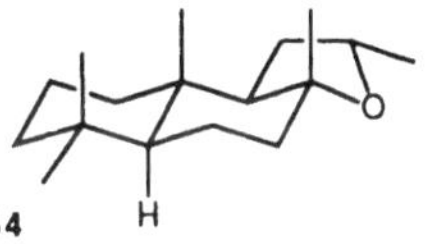 34	7.8Å^2		
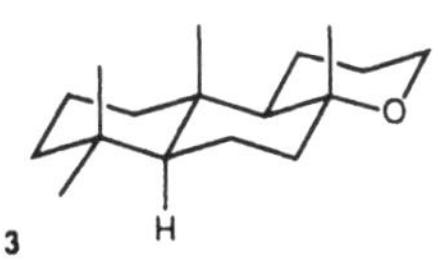 3	8.3Å^2	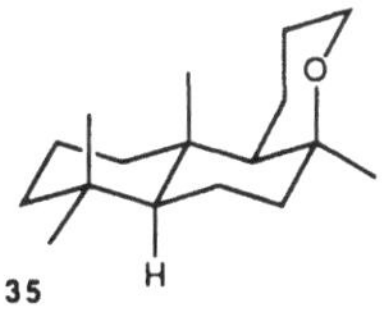 35	5.3Å^2
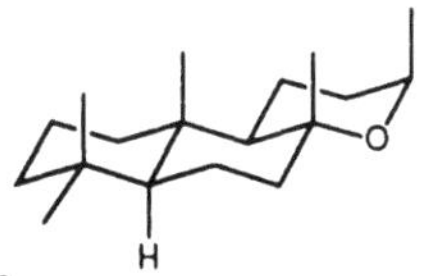 36	7.2Å^2	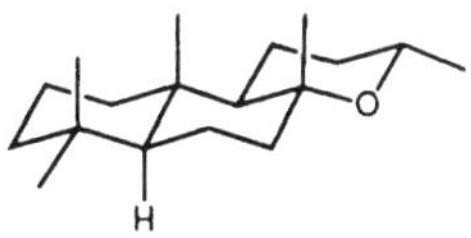 37	2.9Å^2
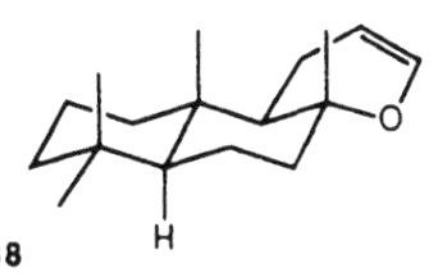 38	9.3Å^2	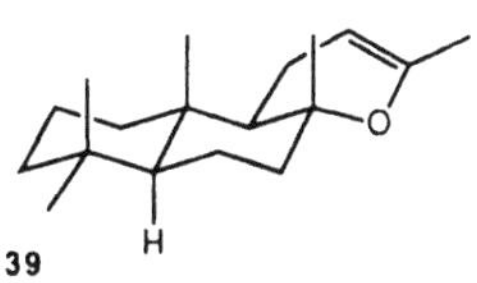 39	3.4Å^2

Hydrophilicity

The affinity of a compound for water is related to the polar interaction and in particular to hydrogen bonding. Both electronic and steric factors play a role in such interactions of the polar part of a compound with water, or with any polar entity in a receptor system. Thus, the steric accessibility of the functional group has been qualitatively shown to be a decisive factor in several examples of musk odorants.[29] In the case of ambergris-type compounds, a study of several monofunctional compounds has related activity/inactivity to steric accessibility of the oxygen atom, as quantitatively measured by its accessible surface area.[30a,30b] As examples of this correlation, some analogues of Ambrox® (**6**) and ambra oxide (**3**) are shown in Table 3.

The balance between lipophilicity and hydrophilicity as an index of molecular properties can be determined by the partition coefficient, log *P*(*n*-octanol/water). The measured or calculated values have been applied with some success in the area of QSAR of biologically active compounds, in particular pharmaceuticals, but the *n*-octanol/water reference system is probably more related to transport phenomena than to the 'real' situation that prevails at the sites of olfactory reception. Although some correlations have been found between log *P* and threshold data in taste and in olfaction,[31] it should be kept in mind that 'no single molecular property is sufficient to determine the odour of a molecule'.[23a]

3 STEROID ODORANTS

Odoriferous steroids (Fig. 4) can also be considered in the light of the 'triaxial rule',[21] especially in view of the fact that the stereochemistry of rings A and B is suspected to be chiefly responsible for specific odour release.[32] 5α-Androst-16-en-3α-ol (**40**) has a strong and lasting animal musk-type and sandalwood-like odour, whereas its 3β-epimer (**41**) is odourless. The former was first isolated from pig testes,[33] and both, together with ketone **44**, an odoriferous urine constituent, were then found in both urine and arm-pit perspiration of human beings. These compounds have been considered as possible sexual chemical messengers, and there has even been talk of the 'likelihood of human pheromones'.[34] Furthermore, the 3α-alcohol (**40**) and the corresponding ketone (**44**) have been identified as pheromones of sexually mature pigs; on the other hand the 3β-epimer (**41**) has been found to be devoid of any aphrodisiac activity.[35] Alcohols **42** and **43**, with a *cis* junction between rings A and B, appear to lack a characteristic odour altogether.[32] From the examination of about 50 such steroid derivatives a fairly comprehensive picture of the receptor activity of the 'steroid scent' has been obtained.[21]

(+)-Androsta-4,16-dien-3-one (**45**) is known to have a pronounced urine and perspiration odour with a very low threshold concentration of 1 ppb.[21] When its 'unnatural' enantiomer, *ent*-**45**, was submitted for testing to 40 people, none

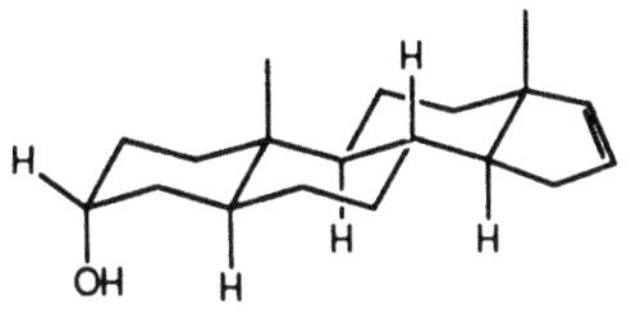

40

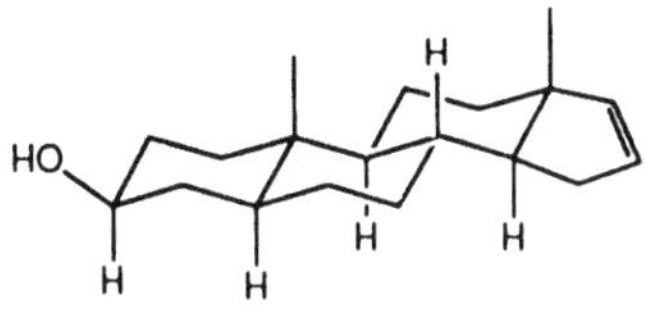

41

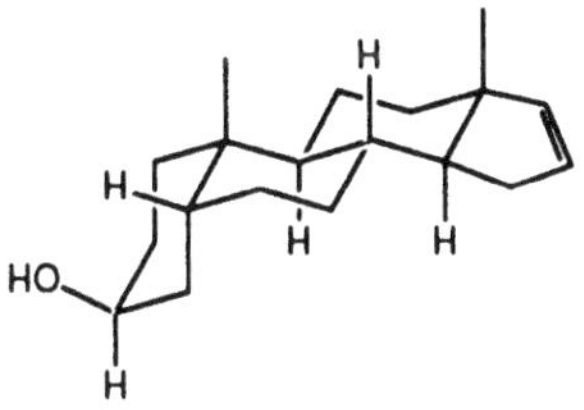

42

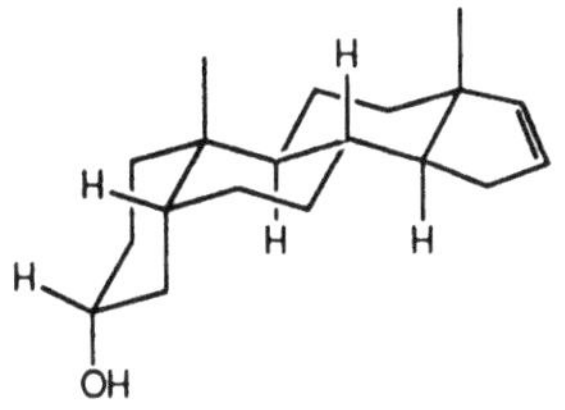

43

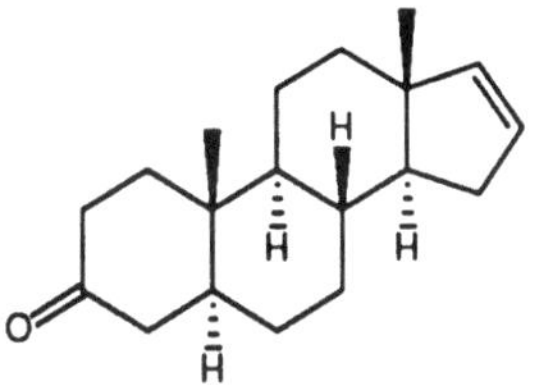

44

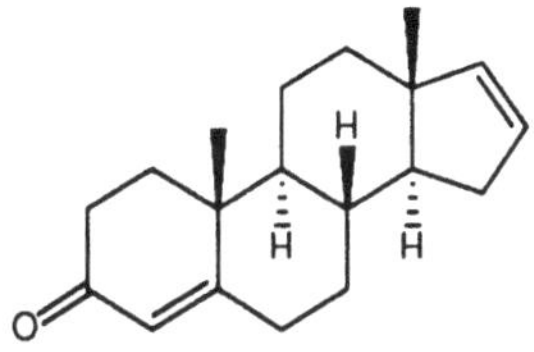

45

Fig. 4. Molecular representatives of steroid odour.

of them was able to detect an odour of any kind. Of the panel, 75% were likewise unable to detect any odour when confronted with the unnatural enantiomeric ketone *ent*-**44** and alcohol *ent*-**40**. The remaining 25% ascribed to ketone *ent*-**44** a very weak urine musk type odour, whose lower threshold value was estimated to be 10^6 times higher than that of the 'natural' epimer **44** (0·62 ppb), and to alcohol *ent*-**40** an exceedingly weak musk odour.

Structurally unrelated types of compounds (Fig. 5) may show similar odour characteristics, but, so far, on the basis of molecular structure, no satisfactory explanation for this phenomenon has yet been found.[29] A particularly remarkable example is that of the diastereoisomeric ketones **46** and **47**.[36] The *cis* compound **47** possesses the same penetrating urine odour as the steroid **44**, while its *trans* epimer **46** is odourless.[37] If one attempts to draw a parallel between the structures of the two ketones **46** and **47** and the corresponding steroid **44** on the basis of the size and shape of their molecular peripheries, one reaches the conclusion that it is compound **46** and not its epimer **47** which ought to exhibit a 'steroid-type' odour. The situation becomes yet more complicated as we consider the lower homologues **48** and **49**. Here it is the *trans* epimer **48** which has an odour identical to the *cis* compound **47** and it is the *cis* epimer **49** which is odorless.[38]

One explanation is based on conformational considerations occasioned by the gem-dimethyl group. From force field calculations it can be deduced that it is only the nor-compound **48** which is able to assume an elongated seco-

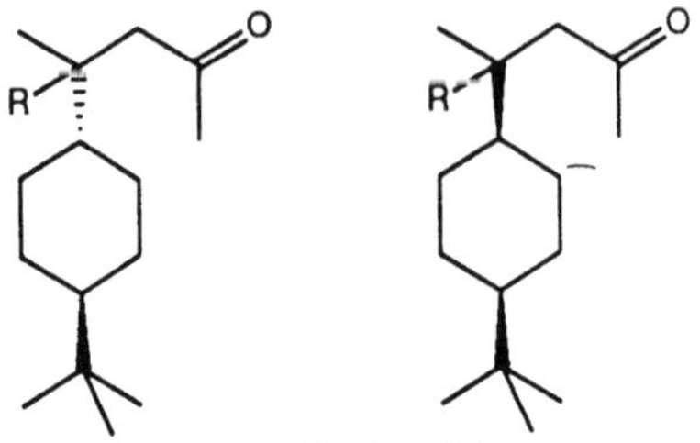

46 : R = CH_3 **47** : R = CH_3
48 : R = H **49** : R = H

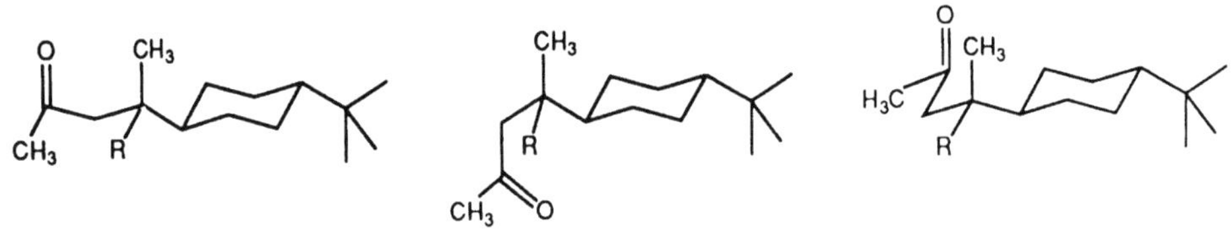

46a: R = CH_3 **46b**: R = CH_3 **46c**: R = CH_3
48a: R = H **48b**: R = H **48c**: R = H

47b : R = CH_3 **47a** : R = CH_3
49b : R = H **49a** : R = H

Fig. 5. Conformers of stereoisomeric 4-(4′-*tert*-butylcyclohexyl)4-methyl-2-pentanone (R = CH_3) and 4-(4′-*tert*-butylcyclohexyl)-2-pentanone (R = H).

steroidal conformation such as **48a**, whereas its dimethyl homologue **46** prefers a bent-chain conformation such as **46b** or **46c**. For the *cis* compound **47** a chair conformation such as **47a** is unlikely for reasons of steric hindrance, and most probably this compound exists in a twist form such as **47b** which would be more amenable to a steroid-like receptor. On the other hand, in the monomethyl analogue **49**, a chair conformation appears to be favoured in which the side chain bearing a carbonyl group would be most likely to assume an axial conformation and thus result in a non-steroidal shape such as **49a**.[40] Ketones **46–49** are the first examples in the history of olfaction where conformational analysis has been used to explain specific odour release without any information about the molecular characteristics of the active site within the sensory nerve cell membrane.

4 SANDALWOOD ODORANTS

Substances With Sandalwood Odour

The characteristic and prized fragrance of sandalwood is mainly due to the sesquiterpene alcohols (+)-α-santalol (**50**) and (−)-β-santalol (**51**), which are

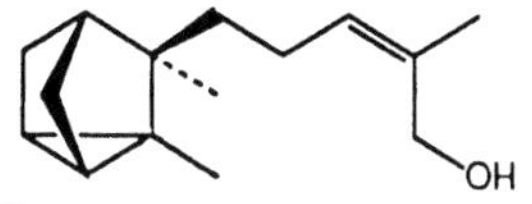

50

(+)-α-santalol

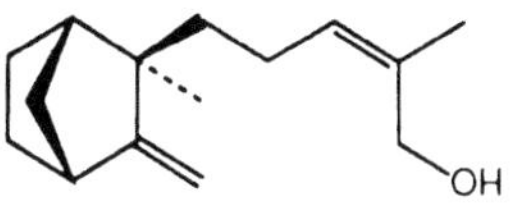

51

(−)-β-santalol

the major components (45–47% **50**, 20–30% **51**) of East Indian sandalwood oil.[39] The structures of **50** and **51** were elucidated by Semmler[40] and Ruzicka & Thomman,[41] respectively, whilst the geometry of the side-chain double bond in both sesquiterpene alcohols **50**[42] and **51**[43] was ascertained more recently. The absolute configuration of **50** was established by stereoselective partial synthesis starting from optically active (+)-camphor,[42,44,45] whilst that of **51** was defined by mechanistic considerations and ORD data.[46,47] In comparing highly purified samples of **50** and **51**, a relatively weak, slightly woody odour reminiscent of α-cedrene was associated with **50**, whereas the typical sandalwood odour, with powerful woody, milky and urinous tonalities, was attributed to **51**.[39,48]

The structurally related compounds **52–57** (Fig. 6) have been identified recently among the minor constituents of the sesquiterpene fraction of East Indian sandalwood oil,[48] however, their odoriferous activity can hardly be

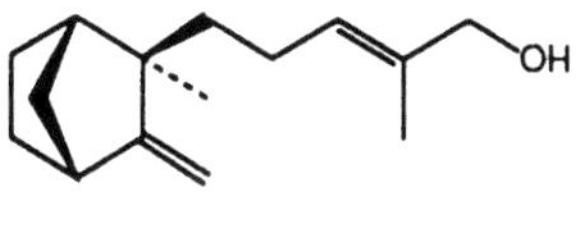

52

(−)-*(E)*-β-santalol

medium strength, woody to medicinal

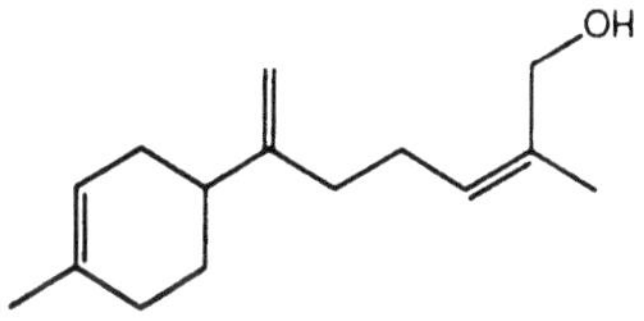

55

(Z)-lanceol

weak, woody

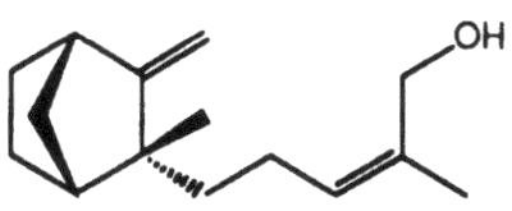

53

(+)-*(Z)*-*epi*-β-santalol

weak, woody

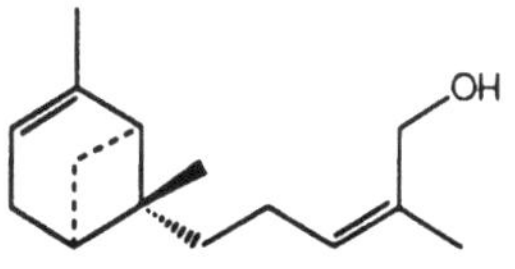

56

(−)-*(Z)*-α-*trans*-bergamotol

bright, somewhat woody, waxy, reminiscent of agrumen

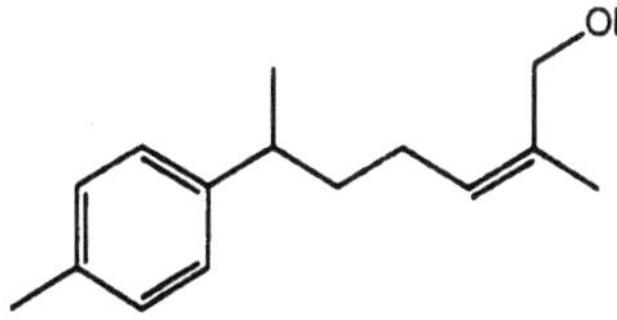

54

(Z)-nuciferol

weak, woody

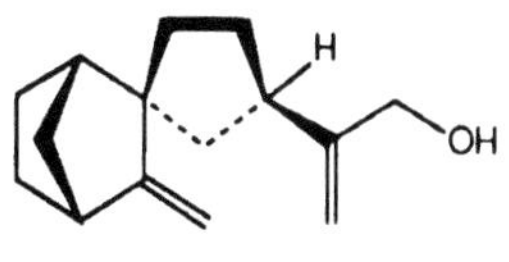

57

spiro-santalol

weak, woody

Fig. 6. Trace sesquiterpene alcohols from East Indian sandalwood oil.

Table 4
Synthetic sesquiterpene alcohols derived from santalols

Sandalwood odour	*No sandalwood odour*
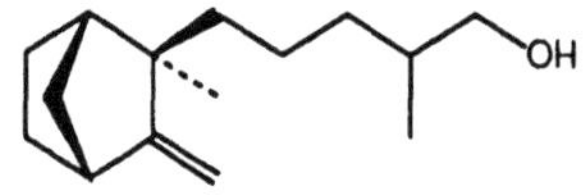 58	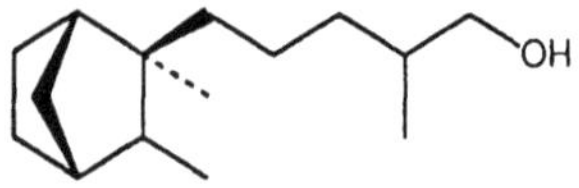 61
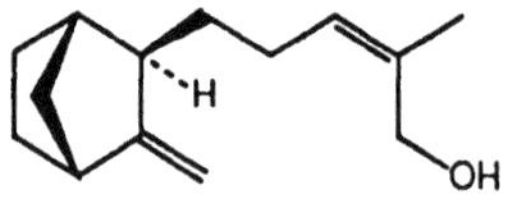 59	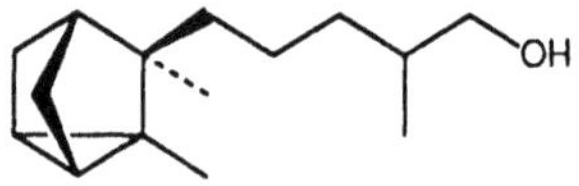 62
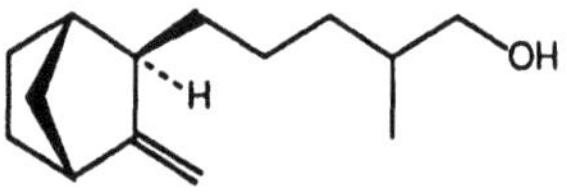 60	

63

α-photosantalol A

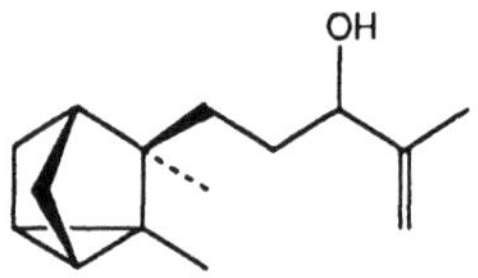

66

α-photosantalol B

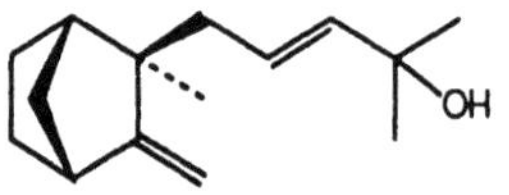

64

β-photosantalol A

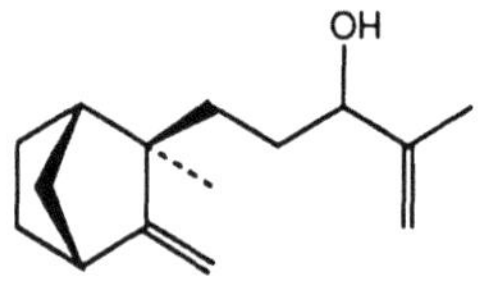

67

β-photosantalol B

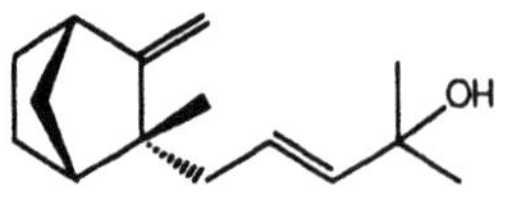

65

epi-β-photosantalol A

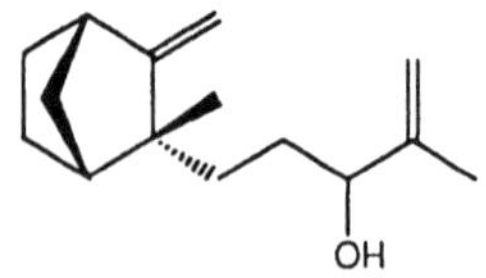

68

epi-β-photosantalol B

Fig. 7. Photosantalols.

compared with that of **51**, indicating that changes in relative stereochemistry and geometry of the side-chain, as well as in the ring system, have a profound effect on the basic odour. Meanwhile, synthetic work towards β-santalol (**51**) had provided the racemic compounds **58–60** (Table 4), with typical sandalwood odour, and demonstrated that side-chain-saturated β-santalol **58**[49,50] and the demethyl analogues **59** and **60**[43] were also active. In contrast, the totally saturated β-santalol **61** and the side-chain-saturated α-santalol **62** were inactive.[51]

Among the photosantalols **63–68** (Fig. 7), obtained by sensitized photo-oxidation of the corresponding santalenes, Mookherjee[52] described **63–65** as having a strong sandalwood aroma, **65** being the strongest, whereas **66–68** had no prominent woody odour. On the other hand, Kaiser & Lamparsky[53] reported that all four isomers **63**, **64**, **66** and **67**, had odours reminiscent of the corresponding santalols. Of the related compounds **69** and **70**, obtained by photo-oxidation of (−)-*trans*-α-bergamotene, and having similar side-chains, only **69** was found to possess a sandalwood-like odour.[54]

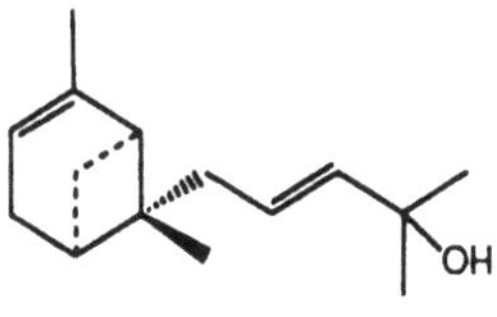

69

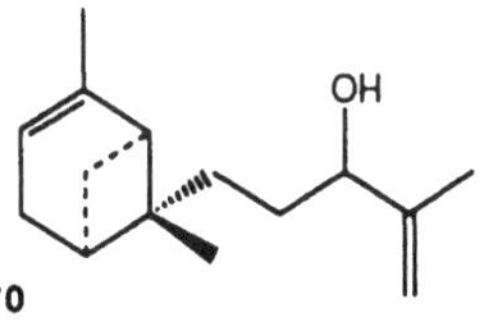

70

An important group of synthetic odorants with sandalwood-like scent is constituted by the terpenyl cyclohexanols, the odoriferous properties of which were discovered by Weissenborn[55] in the 1940s and disclosed by Byers.[56] The detailed structural elucidation of the two isomers **71** and **72**, exclusively responsible for the typical sandalwood odour in the complex mixture **71–84**, was a major achievement.[57,58] It clearly demonstrated the regio- and diastereoselectivity of the human receptive system for this class of odorants, as illustrated in Table 5. All the other products **73–84** have a very weak or no woody tonality. As derivatives of the bicyclo[2.2.1]heptane (norbornane) skeleton, compounds **71** and **72** belong to the same structural class as the santalols, but they are considerably less flexible in their hydrophilic part.

In attempts to find more readily available analogues of the santalols or of the active terpenyl cyclohexanols, a large number of compounds based on the norbornane skeleton, and some based on the bicyclo[3.1.1]heptane skeleton, have been prepared. Examples with well-defined structures, juxtaposing the active compounds **85**,[59] **86**,[60] **87**,[61] **88**[62] and **89**[60] with the weak-smelling or even inactive analogues **90**,[59] **91**,[39] **92**,[61] **93**[62] and **94**[60] are shown in Table 6, illustrating how small changes in the molecular structures can affect the odour impression of sandalwood.

Another large group of synthetic sandalwood-type odorants, which has gained considerable commercial importance, is derived from readily available

Table 5
Regio- and diastereo-isomeric terpenyl cyclohexanols

Strong sandalwood odour

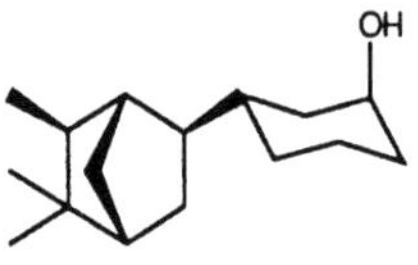

71 erythro

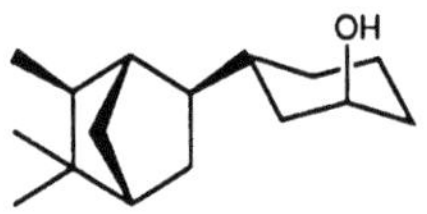

72 threo

Weak or no sandalwood odour

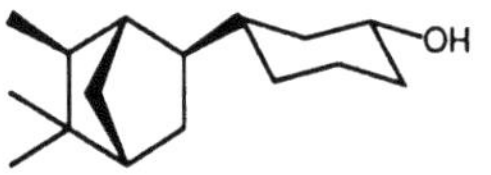

73

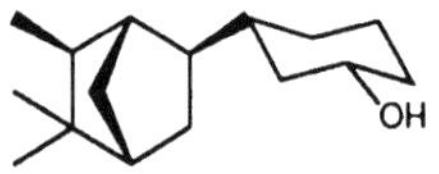

74

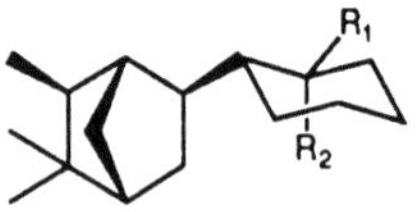

75 : R_1 = OH; R_2 = H
76 : R_1 = H; R_2 = OH

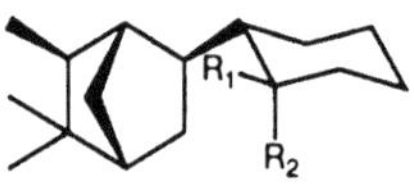

77 : R_1 = OH; R_2 = H
78 : R_1 = H; R_2 = OH

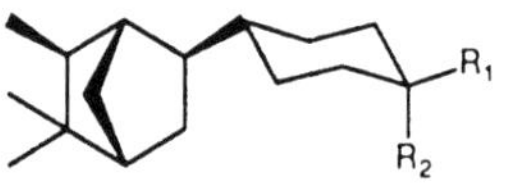

79 : R_1 = OH; R_2 = H
80 : R_1 = H; R_2 = OH

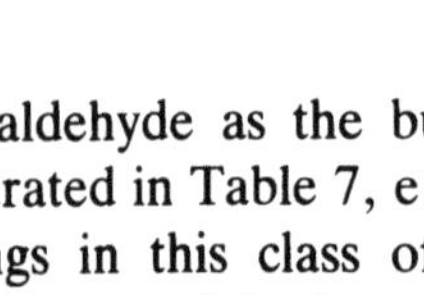

81 : R_1 = OH; R_2 = H
82 : R_1 = H; R_2 = OH

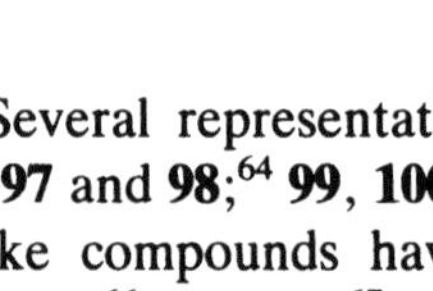

83 : R_1 = OH; R_2 = H
84 : R_1 = H; R_2 = OH

campholenic aldehyde as the building block. Several representatives of this class are illustrated in Table 7, e.g. **95** and **96**;[63] **97** and **98**;[64] **99**, **100** and **101**.[65] Recent findings in this class of sandalwood-like compounds have revealed some extremely powerful odorants, e.g. **102** and **103**,[66] and **104**.[67] The odour of compound **104** was reported to be about 400 times more intense than the odour of α-santalol (**50**) and more than 500 times as intense as its diastereoisomer **105**, as measured by their odour threshold values.[67] The ring-saturated analogues **106**[39], **107**[39] and **108**[39] and **109** (Schulte-Elte, K. H., Müller, B. & Pamingle, H. (1985) unpublished) have more or less completely lost their sandalwood activity.

Related to the class of campholenyl derivatives is the synthetic sandalwood odorant Ionosantalol® (**110**; Fig. 8). Its side-chain-saturated analogue **111**,

Table 6
Analogues of santalol

Sandalwood odour	*Weak or no sandalwood odour*
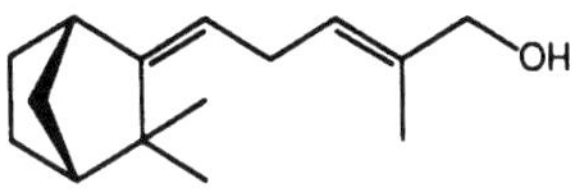 **85**	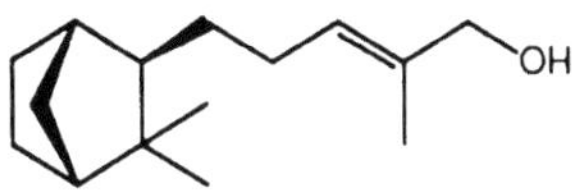 **90**
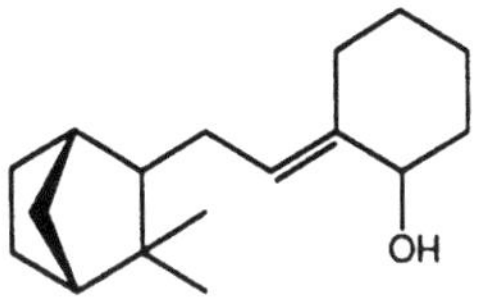 **86**	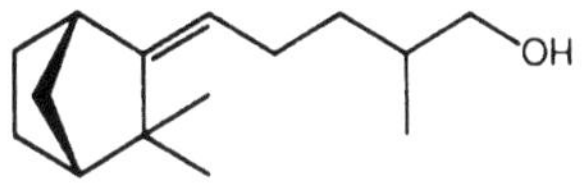 **91**
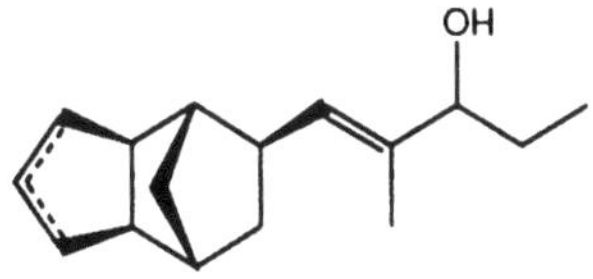 **87**	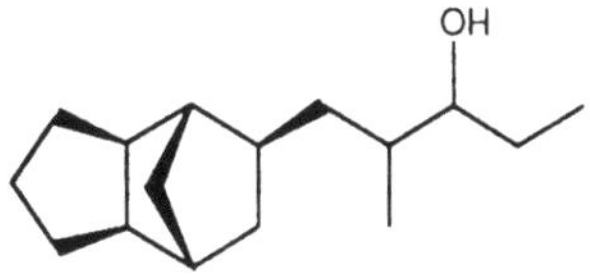 **92**
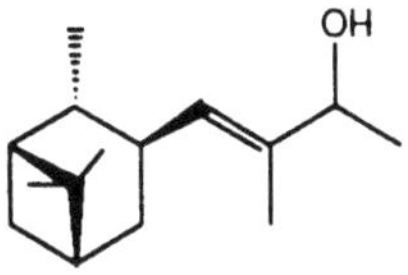 **88**	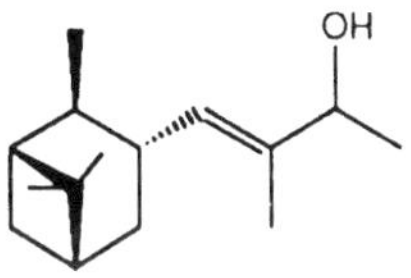 **93**
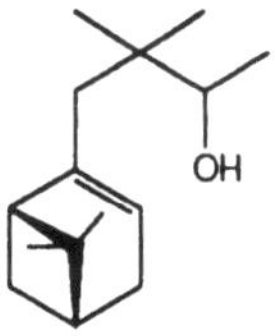 **89**	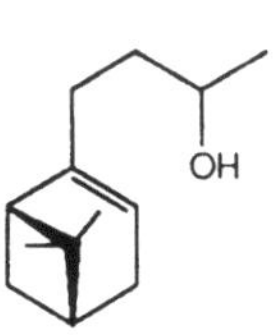 **94**

however, is nearly odourless.[68] Structurally related to the latter two compounds is a diastereoisomeric mixture of the completely saturated ionol derivative 1-(2,2,6-trimethylcyclohexyl)-3-hexanol, known under the trade name Timberol®.[69] The *trans* isomer Norlimbanol® (**112**)[70] has been found to be the carrier of the odoriferous activity, which is essentially woody in nature, with warm animal, musk and ambergris-like subnotes and a rather strong urinous tonality; none of these odour qualities is perceived from its *cis* isomer (**113**).

A particular group of synthetic sandalwood aroma chemicals with a

Table 7
Sandalwood-like odorants based on the campholenic skeleton

Sandalwood odour		*Weak or no sandalwood odour*
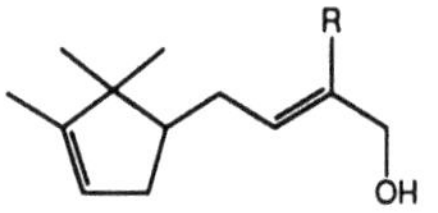 95 : R = CH_3 Sandacore® 96 : R = C_2H_5 Bacdanol®	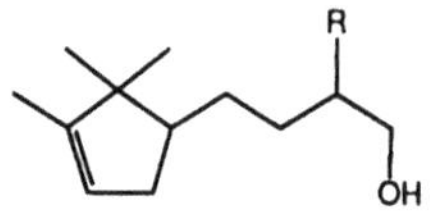 97 : R = CH_3 Brahmanol® 98 : R = C_2H_5	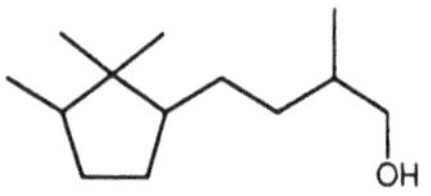 106
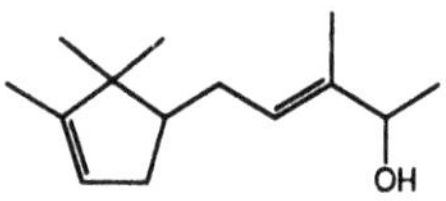 99	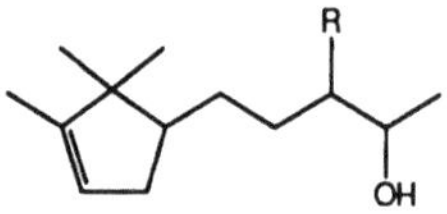 100 : R = CH_3 Sandalore® 101 : R = C_2H_5	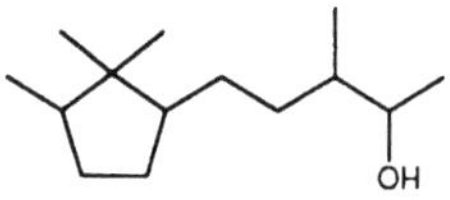 107
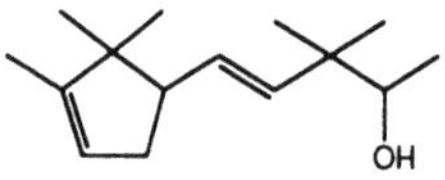 102 Polysantol®	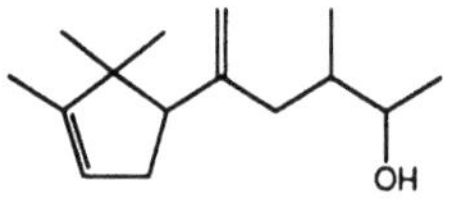 103	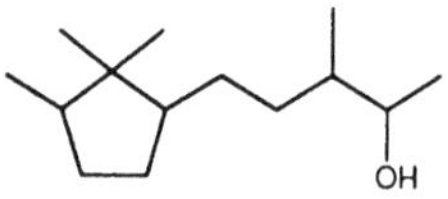 108
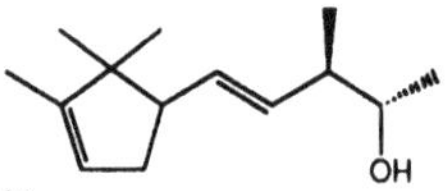 104	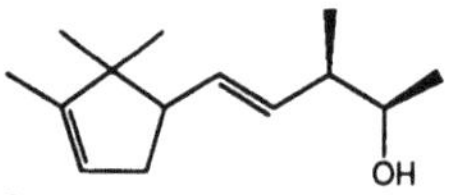 105	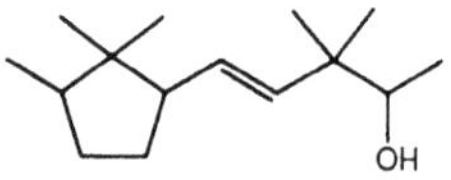 109

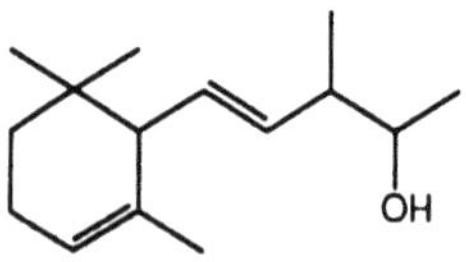

110 Ionosantalol®

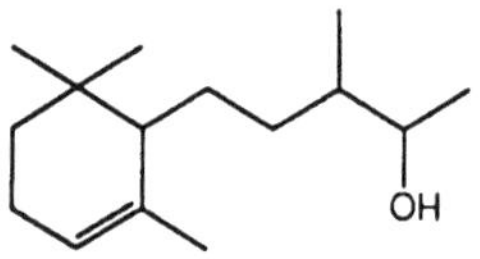

111

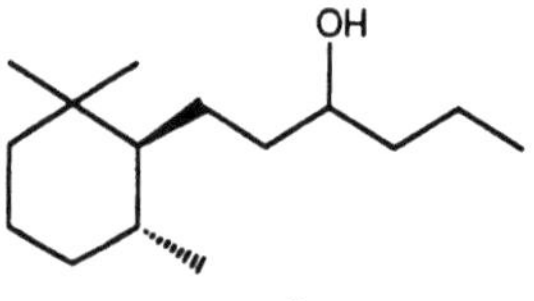

112 Norlimbanol®

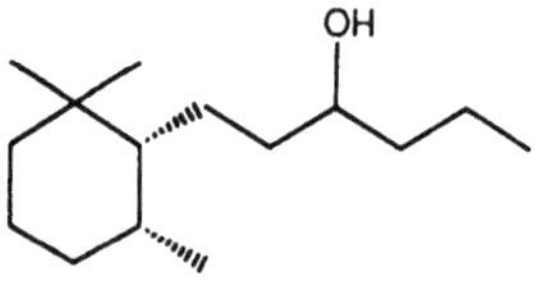

113

Fig. 8. Sandalwood-like odorants based on the ionol skeleton.

Table 8
Sandalwood-like odorants based on *p-tert*-butylcyclohexane derivatives

Sandalwood odour	*Weak or no sandalwood odour*
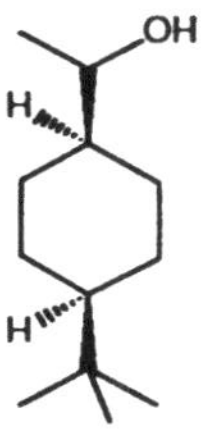 **114**	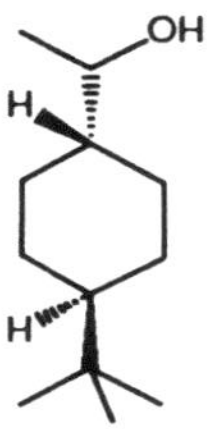 **121**
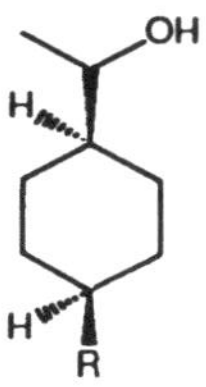 R = 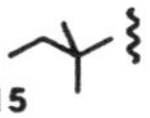 **115** 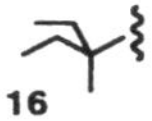 **116** 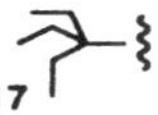 **117**  **118**  **119**  **120**	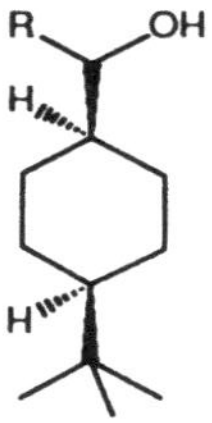 **122** R = H; **123** R = C_2H_5; **124** R = $CH(CH_3)_2$ 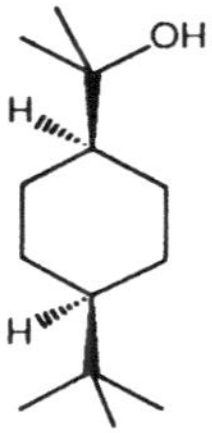 **125**

relatively rigid skeleton, which has also contributed to a more precise definition of the structural requirements for sandalwood odour, is constituted by the cyclohexane derivatives **114–125**.[71,72] Only the *cis* isomers **114–120** are active. Their corresponding *trans* isomers (e.g. compound **121**) are all devoid of sandalwood activity. As illustrated in Table 8, several *cis* compounds (**122–125**) are also inactive, indicating that odoriferous activity also depends on the molecular environment of the hydrophilic OH group.

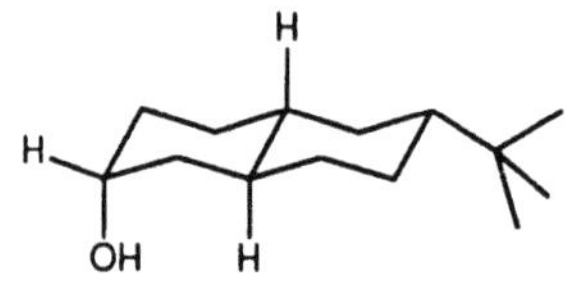

126

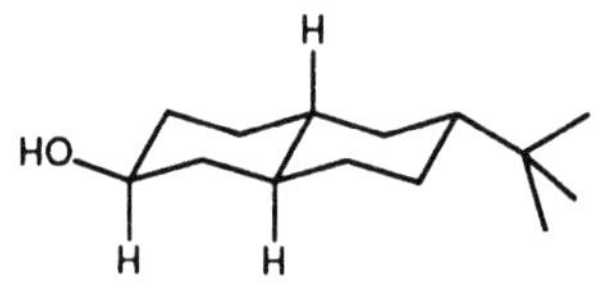

127

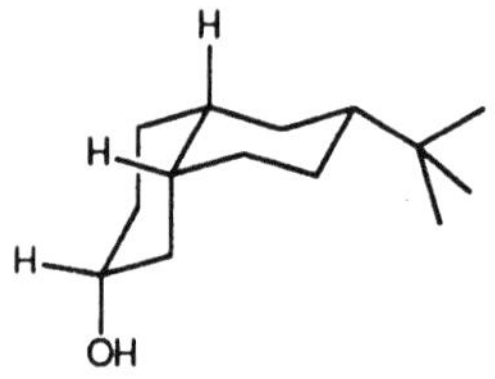

128

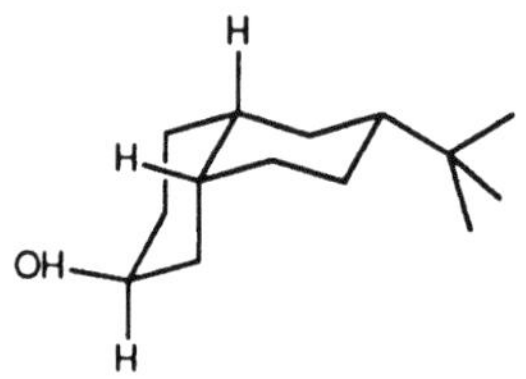

129

Finally, one of the most rigid odorants of this class known today, compound **126**,[73] has a structure based on the *trans*-decalin skeleton and was derived in analogy to the odoriferous 5α-androst-16-en-3α-ol **40**, which (to some subjects) has a smell reminiscent of sandalwood. The diastereoisomers **127–129**, however, have no sandalwood odour, a result comparable with the corresponding odourless diastereoisomers **41–43**. Only the axial alcohol **126** fulfils the requirements of the 'triaxial rule' of odour sensation.[9]

Structure–Activity Relationships (SAR)

Noting the great similarity between the structures of the synthetic 3-terpenylcyclohexanols **71/72** and the natural α-santalol **50** (Fig. 9), in 1970 Aulchenko & Kheifits[74] proposed the following structure–odour correlation: 'The sandal odour, in varying degrees, is characteristic of all compounds containing a hydroxyl group which have a "rigid" and bulky alkyl radical in direct proximity to atom C-(5) and which are capable of assuming a configuration similar to the convolute conformation of natural *cis*-santalols.' Based on this pertinent assumption, they predicted a sandalwood odour for the *cis*-β-santalol **51** without knowing the stereochemistry of the double bond, for the dihydro-β-santalol **58** and for compounds **66** and **67**, as well as the absence of a sandalwood odour for the (*E*)-β-santalol **52** and its α-isomer. Nevertheless, **52** shows some woody tonality of medium odour strength.

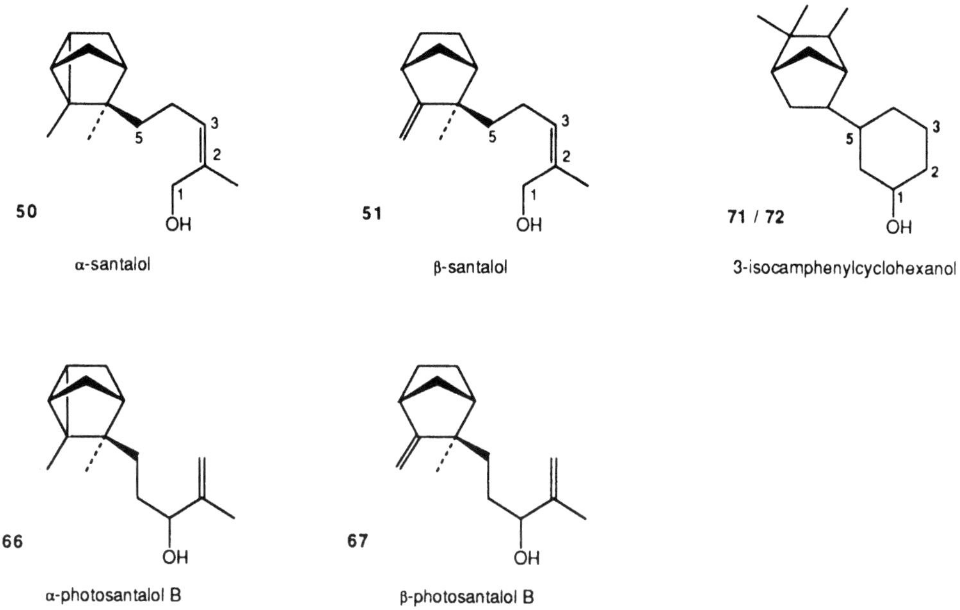

Fig. 9. Molecular shape analysis of sandalwood odour.

In 1981, Brunke & Klein[61] extended this concept by pointing to the high degree of molecular similarity between the bicyclic derivatives **130** and **131** and some useful sandalwood aroma chemicals like *β*-santalol (**51**), the campholenyl derivatives **97** and **100**, and even the acyclic compound **132**[75] when drawn with their chain in a common 'chair-like' conformation (Fig. 10).

They summarized the relationships between chemical structure and sandalwood odour as follows: 'A sandalwood odour seems to be displayed by

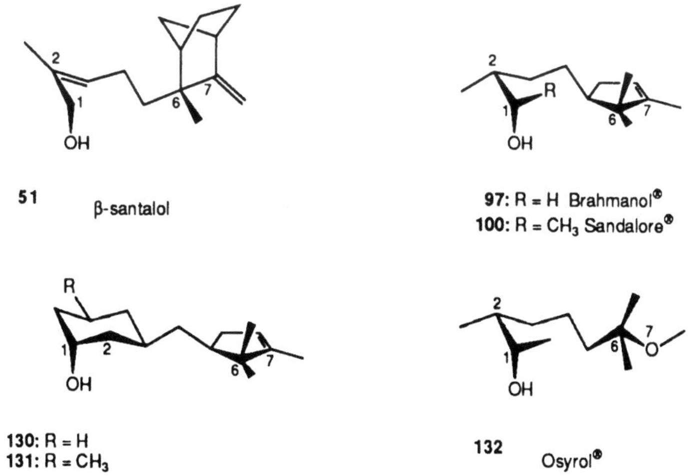

Fig. 10. Molecular shape analysis of sandalwood odour.

molecules with 13–17 carbon atoms, which possess a hydroxy group at a certain distance to a voluminous substituent. This substituent may be a polycyclic, monocyclic or aliphatic system with ramifications at C-atom 2 and 6, and a double bond or a cyclopropane ring at C-7.' At the same time, Naipawer *et al.*[76] gave a detailed account of their structure–odour studies and formulated some necessary structural requirements ('sandalwood rule') for compounds having a sandalwood odour, distinguishing the following effects.

—*Molecular weight*: The number of carbon atoms should be between 12 and 16 with an ether oxygen atom being considered equivalent to a carbon atom.

—*Functional groups*: The molecule should contain a free hydroxyl group.

—*Molecular geometry*: The hydroxyl group and a highly substituted or quaternary carbon atom should be situated in the molecular framework in such a manner that at least one 'reasonable' conformation exists in which the hydroxyl oxygen atom and the highly substituted carbon atom are approximately 4 Å apart.

—*Electronic effects*: If an electron-donating substituent (e.g. an ether, double bond or cyclopropyl group) is bonded to the quaternary carbon atom, the sandalwood odour may be favourably modified. Also, if the carbon chain between the hydroxyl group and the highly substituted carbon atom contains an sp^2-hybridized carbon atom, then the sandalwood odour may be affected in some manner.

—*Substituent effects*: When the hydroxyl group is part of an acyclic side-chain, the carbinol atom and the carbon atom alpha to it may bear alkyl substituents.

Regarding the molecular geometry effect cited above, the proposed distance requirement of *ca.*4 Å between the hydroxyl oxygen atom and a highly substituted carbon atom was mainly based on the comparison of the relatively rigid sandalwood odorants **71** and **114**, with their inactive diastereomers **73** and **121** having larger corresponding distances (Table 9).

In the light of more recent findings, in particular the extremely rigid sandalwood odorant **126**, the corresponding distance requirement should be extended to a higher value of *ca.*6·5 Å (Table 9); however, a comparison of **126** with **129** or **121** indicates that this geometrical parameter is not a sufficient condition for sandalwood odour; Witteveen & van der Weerdt[73] considered the axial hydroxyl group and the flat-shaped ring system of the active compound **126** as an important prerequisite for the odour release. On the contrary the inactive diastereoisomer **129** has a bent topology.

Recently, Beyer *et al.*[77,78] have compared compounds **71** and **126** in one study and *β*-santalol (**51**) with compounds **130** and **126** in another, using computer-aided molecular modelling techniques. The most probable conformations for the flexible molecules were determined by empirical force field (molecular mechanics) calculations. A pairwise comparison and fitting of geometries showed, as expected, a common structural element in the relative

Table 9
Molecular geometry of rigid sandalwood-type compounds: distances measured on Dreiding models

Sandalwood odour	*No sandalwood odour*
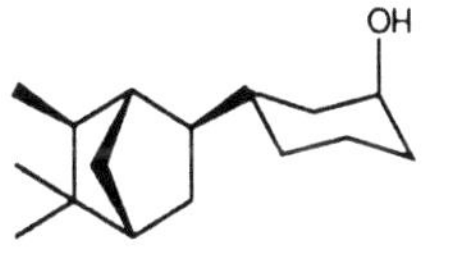 71 erythro	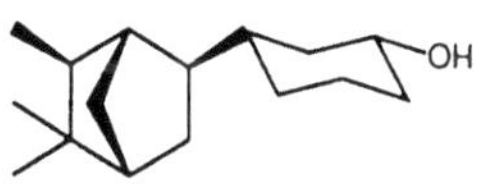 73
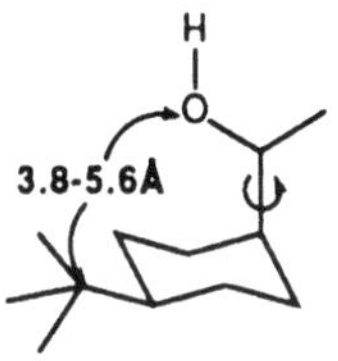 114	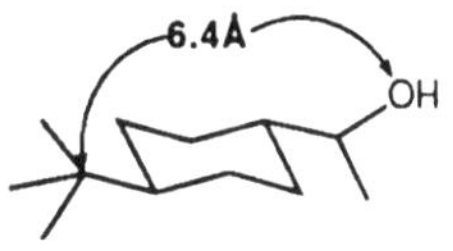 121
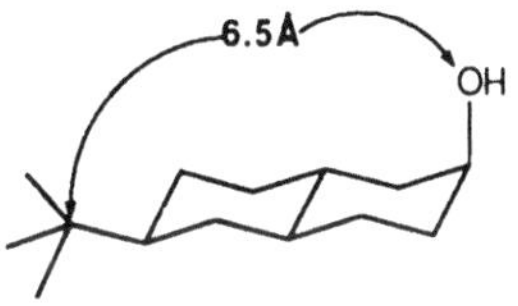 126	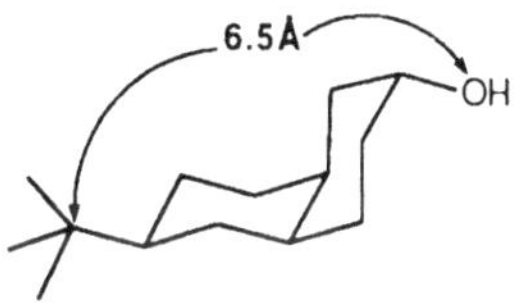 129

arrangement of the carbinol function and a quaternary carbon atom or a fictive centre.

Concerning the dependence of sandalwood odour on the absolute configuration of chiral substrates, surprisingly little information has been published. Nevertheless, during work carried out at Firmenich, significant differences in odoriferous activity between enantiomers in the series of campholenic aldehyde derivatives have been noticed (Schulte-Elte, K. H., Müller, B. & Pamingle, H. (1985) unpublished).

In conclusion, it appears that the rigid compound **126**, although ideal for structure–odour relationship studies, is not the most active substance; as mentioned above, three of the most powerful sandalwood odorants known today are compounds **102**, **103** and **104**, having a considerable degree of conformational freedom. This seems to indicate that the human chemoreceptors responsible for the sandalwood odour impression interact more efficiently with molecules having some flexibility. Further refinements in the delineation of the structural requirements for sandalwood odour are to be expected, if the vigorous activity in synthetic sandalwood chemistry of the last decade, reviewed in 1988 by Naipawer,[79] continues in the future.

5 MUSK ODORANTS

The term 'structure–odour relationship' implies that certain compounds belong both to the same structural *and* olfactory class. However, very often, within a given structural class,[80] and even between two enantiomeric compounds,[21,22] the most diverse odours can be detected; inversely, a well-defined odour can be found within a large variety of very different structural types. This is precisely the case with musk odorants, which contain such diverse structural classes as the macrocyclic musks (e.g. muscone **133**),[29,81] the widely applied aromatic musks (e.g. musk ketone **134**, Tonalid® or Fixolide®, **135**),[29,82] and the small group of steroid-musks (e.g. androstenol, **40**)[81] (Fig. 11).

It would be futile to speculate on molecular parameters common to these three structural classes. Firstly, such parameters would necessarily be too vague for an accurate structure description and would not allow discrimination between potential musks and non-musks in a SAR study (see below). Secondly, it should be made clear that, although the musk odour is very characteristic—it is referred to as warm, sensual, animal, natural—there are substantial odour differences between the different musk categories and sometimes even between representatives of the same structural class.

Nitro-free aromatic musks represent a structurally and olfactorily well-defined class of compounds and today several hundred structurally related compounds of different odour strength are known, including also a large number of non-musks. This unique situation for SAR studies, coupled with the perfumistic value of nitro-free aromatic musks, has incited us to review past and present trends in the design of new, strong musks belonging to this family.

In the context of SAR studies Beets[29] and Theimer & Davies[82] have established the most important structural requirements ('musk rules') for musk odour in the class of nitro-free aromatic musks.

1. 14–20 C atoms (optimum at C_{16}–C_{18}).
2. Two quaternary centres (or at least one quaternary and one tertiary centre) attached in *ortho* or *meta* manner to the aromatic system.

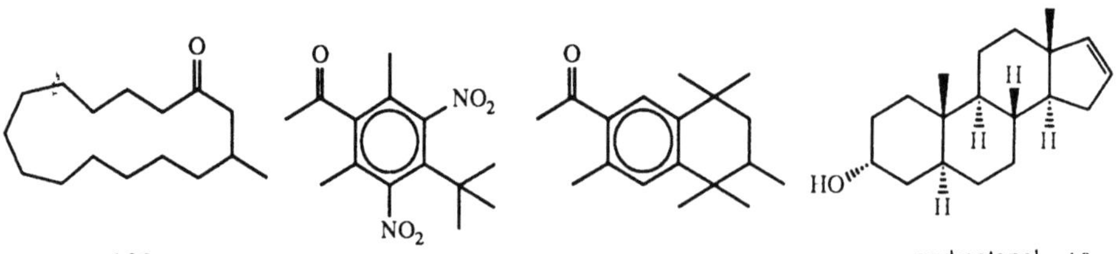

133 134 135 androstenol 40

Fig. 11. Musk odorants. Commercial products: **133**, muscone (Firmenich); **134**, musk ketone (Givaudan); **135**, Tonalid® (Polak's Frutal Works) or Fixolide® (Givaudan).

3. An acyl group in the aromatic nucleus. In certain cases the acyl group can be replaced by an ether[83] or a nitrile function.[84]
4. Sterically unhindered position of the functional group.
5. Orientation of the molecular dipole axis.
6. Closely packed structure.

These rules, which allow a qualitative prediction on whether or not a compound of a specific structure possesses a musk odour (Fig. 12) are of great value because they also indicate discriminative, negative molecular descriptors. For instance, a molecule with two quaternary centres *para*-positioned in the aromatic nucleus (contrary to rules 2 and 4), as shown in type **C**, or with an acyl group placed next to a quaternary centre (types **D**, **E**; contrary to rules 4 and 5) is expected to be either a non-musk or a weak musk odorant.

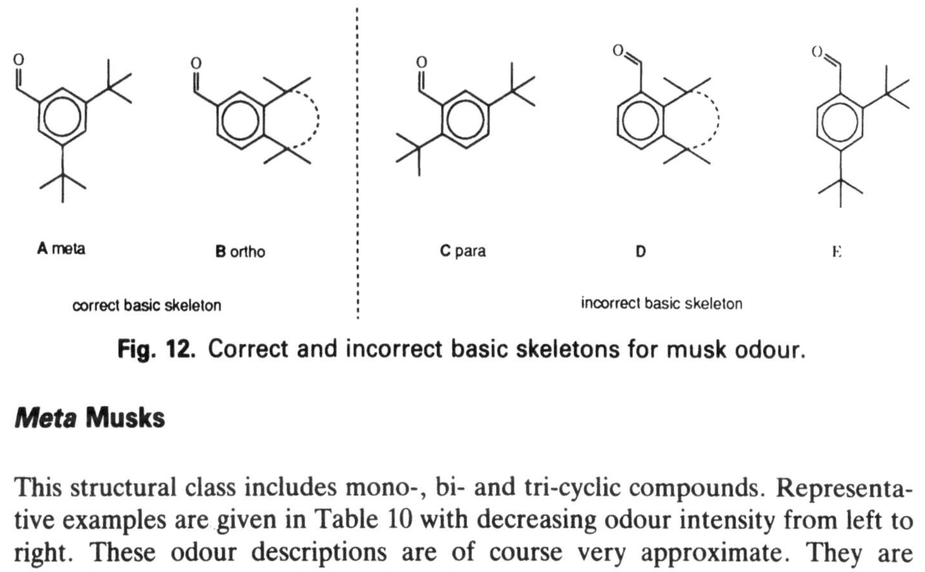

Fig. 12. Correct and incorrect basic skeletons for musk odour.

Meta Musks

This structural class includes mono-, bi- and tri-cyclic compounds. Representative examples are given in Table 10 with decreasing odour intensity from left to right. These odour descriptions are of course very approximate. They are either based on comparisons of performance in a perfume or detergent or taken from the literature (mainly Refs. 29, 82, 85). The molecules in Table 10 are oriented always the same way with respect to the tertiary or quaternary substituents (first priority) and—if possible—to the polar group (second priority). Such a graphical representation is very helpful for the design of new strong musks, as it gives a detailed picture of structure/odour intensity changes (not only of musks and non-musks) in a most transparent manner. From this table the following observations can be made: Most active musk compounds have the basic skeleton **A**. Aldehydes **142** and **143** represent exceptions—the functional group is placed next to the quaternary substituent (type **E**) but, as expected, all ketones of types **C** or **E** (**140**, **150**, **164**, **170**, **171**) are weak or odourless. This may be due to greater conformational mobility and a larger accessible polar surface (mainly Refs. 29, 30) for aldehydes as compared with ketones. Structurally, phenol **160** with musk-like character does not fit at all

Table 10
Representative examples of *meta* musks[a]

	Strong (s)/ Very strong (vs)	*Medium (m)/ Fairly strong (fs)/Musk-like (ml)*	*Very weak (vw)/Weak (w)/ Non-musk (nm)/Odourless (o)*
Monocyclic		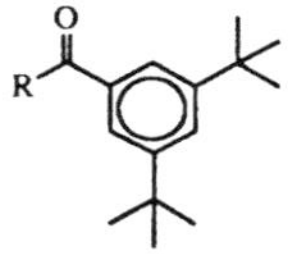 R = H 136 (m) R = Me 137 (m) R = Et 138 (m) 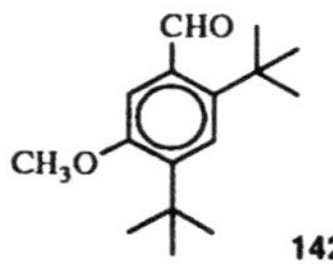 **142** (m) 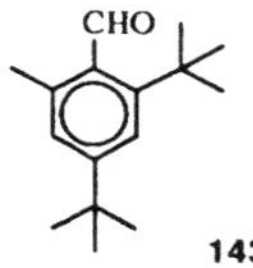 **143** (m)	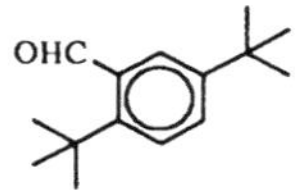 **139** (w, nm) 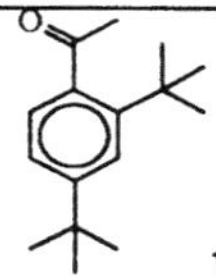 **140** (o) 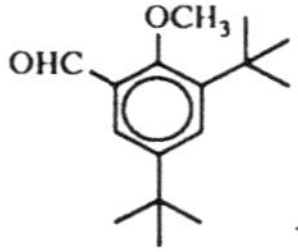 **141** (o) 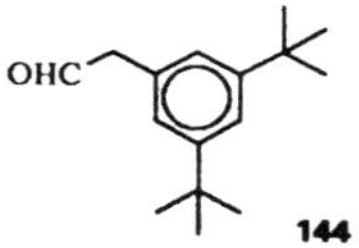 **144** (w, nm) 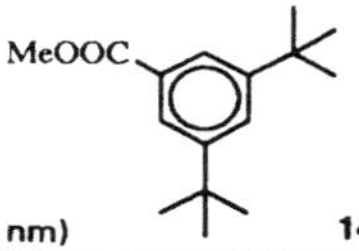 **145** (o) 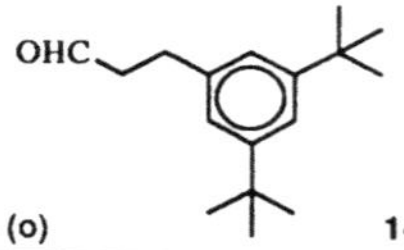 **146** (o)
Bicyclic	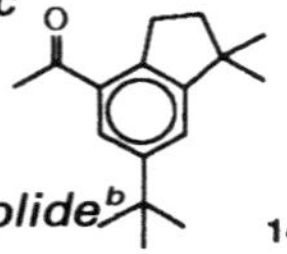 *Celestolide*[b] **147** (s) 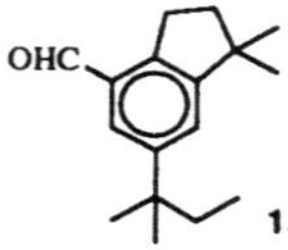 **152** (s)	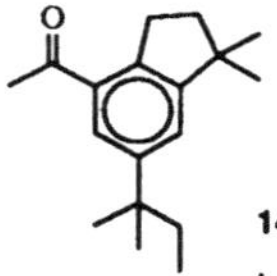 **148** (fs) 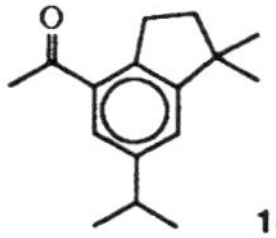 **149** (fs) 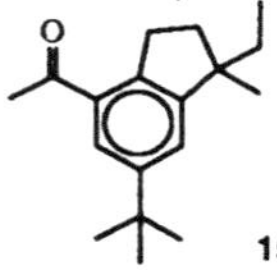 **153** (fs) 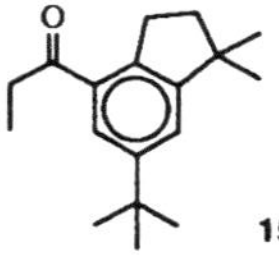 **154** (fs)	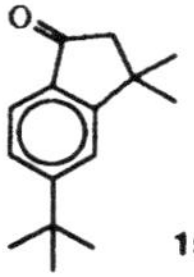 **150** (nm) 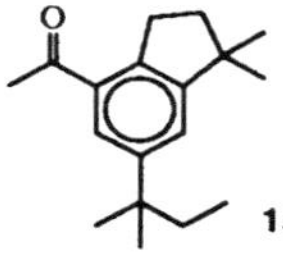 **151** (w) 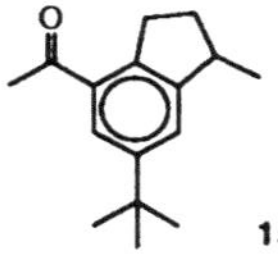 **155** (vw) 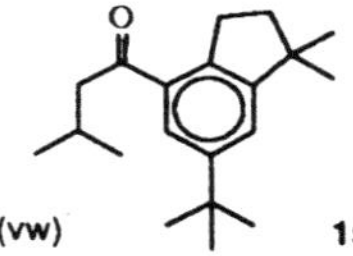 **156** (vw) 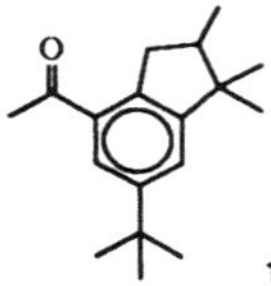 **157** (vw)

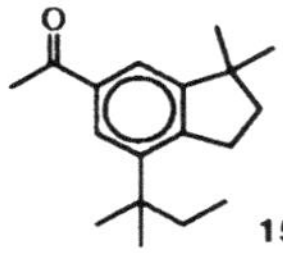
158 (s)

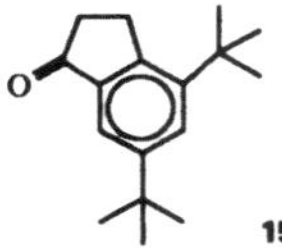
159[c] (m)

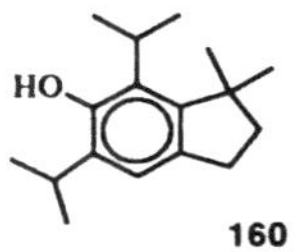

160 (ml)

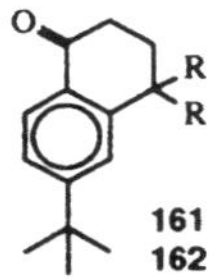

161 R = H (o)
162 R = CH_3 (w)

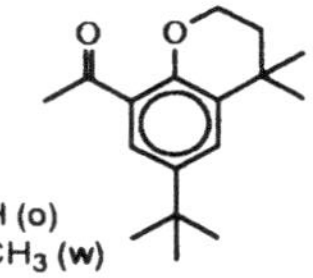
163 (vw)

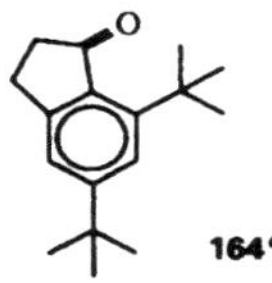
164[c] (o)

Tricyclic

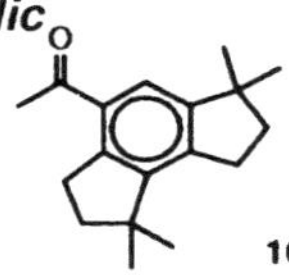
165 (s)

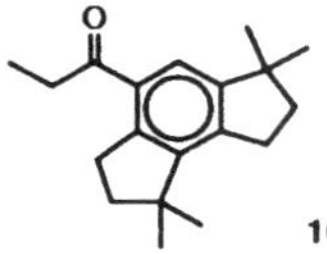
166 (m)

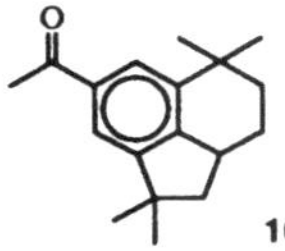
169 (s,ml)

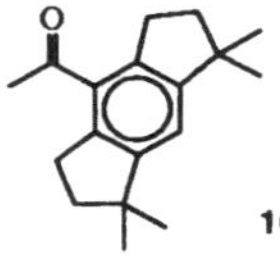
167 (w)

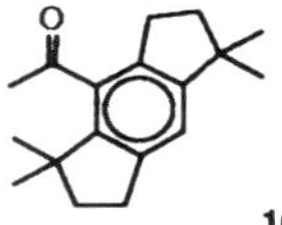
168 (nm)

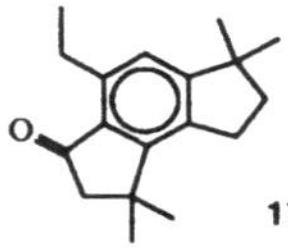
170 (w)

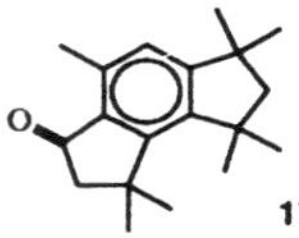
171 (o)

[a] Taken from earlier reviews (Refs 29, 82, 85, 86), if not otherwise stated.
[b] Celestolide® commercialized by International Flavors and Fragrances.
[c] Fehr, C. & Galindo, J. (unpublished).

Table 11
Representative examples of *ortho* musks[a]

Indan musks

Strong (s)/ Very Strong (vs)	*Medium (m)/ Fairly strong (fs)/ Musk-like (ml)*			*Very weak (vw)/ Weak (w)/ Non-musk (nm)/ Odourless (o)*		
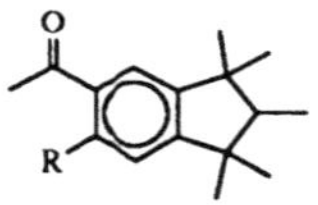	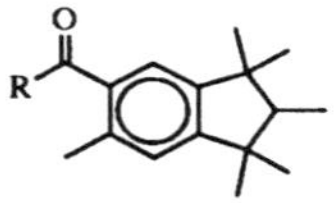	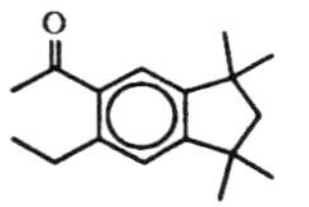	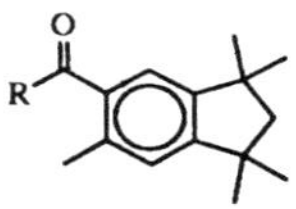	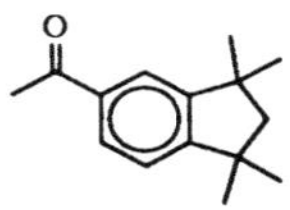	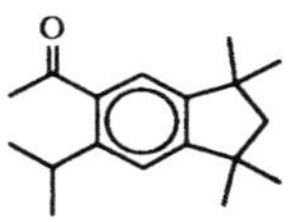	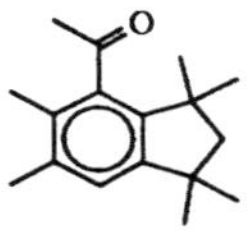
172 R = CH_3 (s) **Phantolide®**[b] **173** R = Et (s)	**174** R = H (fs) **175** R = Me (m) **176** R = Et (m)	**177** (m)	**178** R = H (m) **179** R = Et (ml)	**180** (vw)	**181** (o)	**182** (vw)
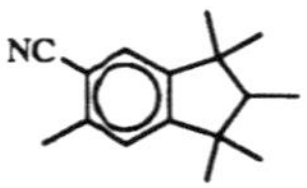	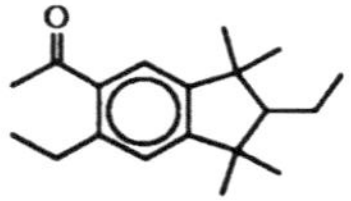	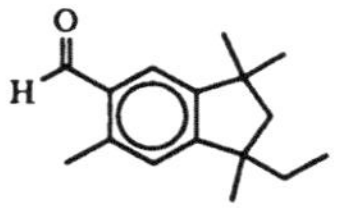		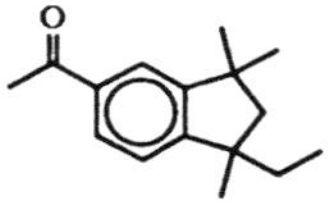	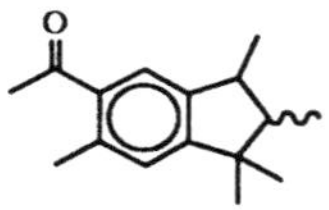	
183	**184** (m)	**185** (m)		**186** (w)	**187** (w)	
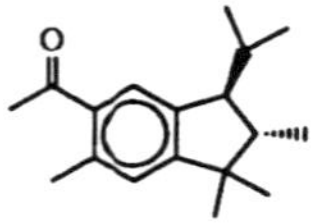	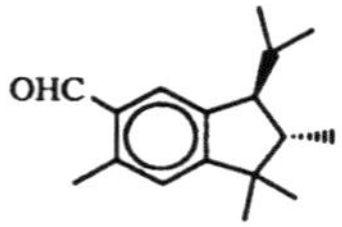	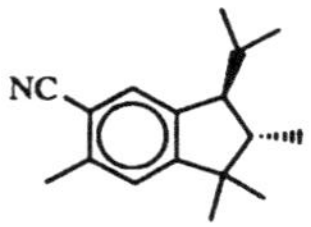		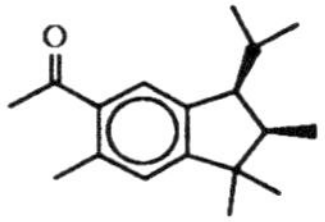		
188 (s) **Traseolide®**[b]	**189** (ml)	**190** (ml)		**191** (ml)		

Tetralin musks

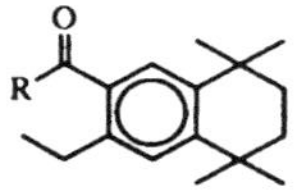

192 R = H[b] (s)
193 R = CH_3 (s)

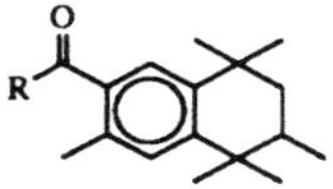

202 R = H[d] (vs)
135 R = CH_3 (vs)
Tonalid®[b]
Fixolide®

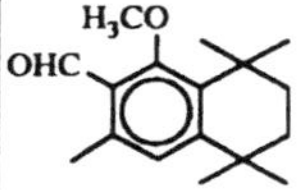

211 (s)

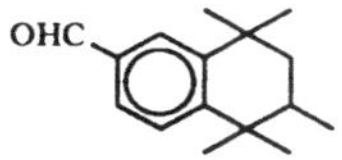

194[c] (s, ml)

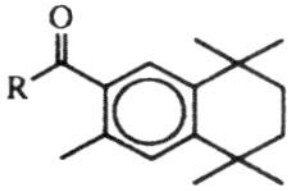

195 R = H (m)
196 R = CH_3 (m)

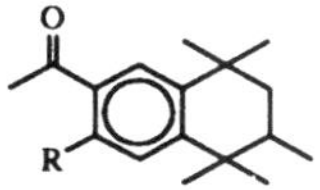

203 R = H (m)
204 R = Et (m)

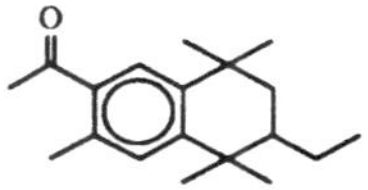

205[e] (m)

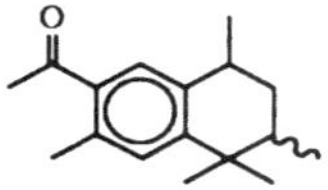

206 (m)

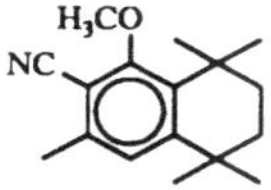

212 (ml)

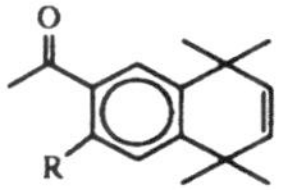

213 R = H (ml)
214 R = CH_3 (ml)

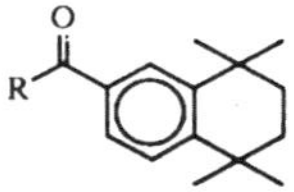

197 R = H (w, nm)
198 R = CH_3 (o)

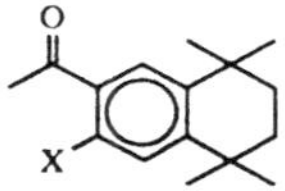

199 X = OMe (o)
200 X = Cl (nm)

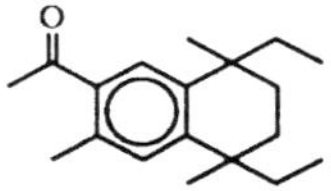

201 (o)

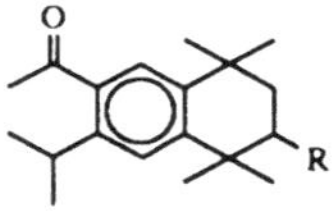

207 R = H (vw)
208 R = CH_3 (nm)

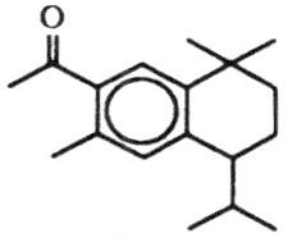

209 (o)

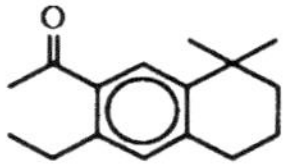

210 (nm)

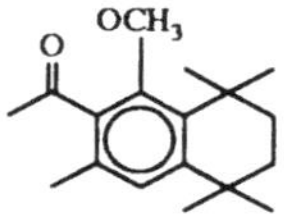

215 (o)

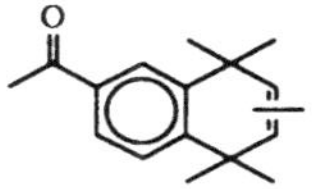

216 (o)

(*continued*)

Table 11—*contd.*

Tricyclic indan or tetralin musks

Strong (s)/ Very Strong (vs)	*Medium (m)/ Fairly strong (fs)/ Musk-like (ml)*			*Very weak (vw)/ Weak (w)/ Non-musk (nm)/ Odourless (o)*	
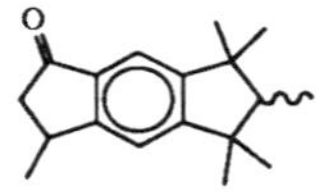 **217** (s)	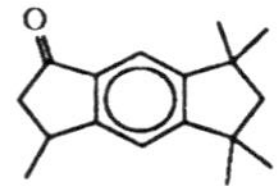 **218** (m)	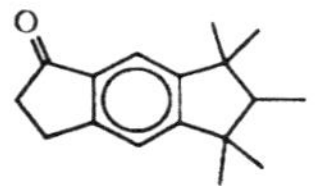 **219** (s)	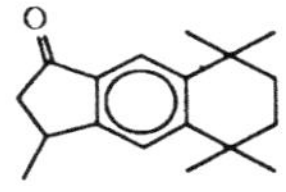 **220** (m)	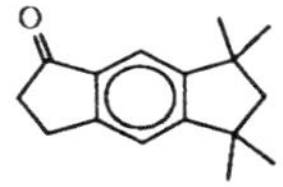 **221** (w)	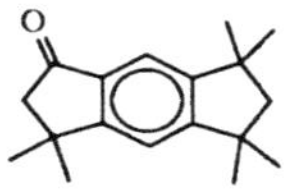 **222** (o)
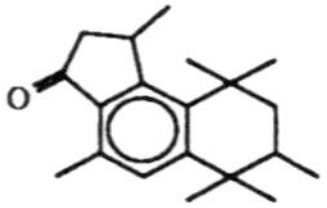 **223** (s)	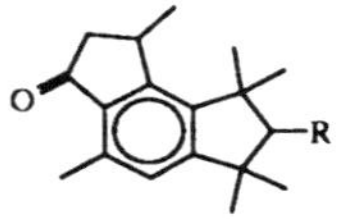 **224** R = H (m) **225** R = CH_3 (m)	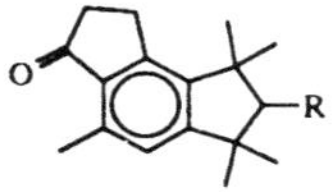 **226** R = H (m) **227** R = CH_3 (m)		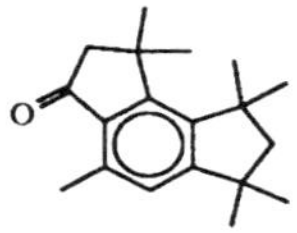 **171** (o)	
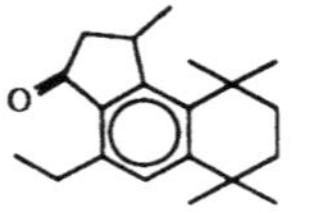 **228** (s)	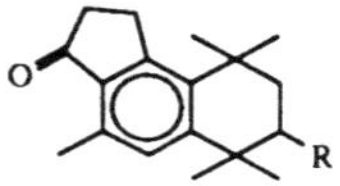 **229** R = H (m) **230** R = CH_3 (m)	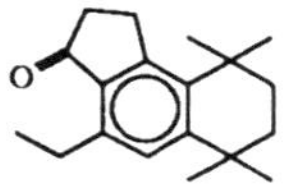 **231** (m)	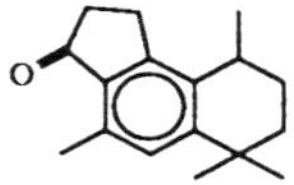 **232** (m)	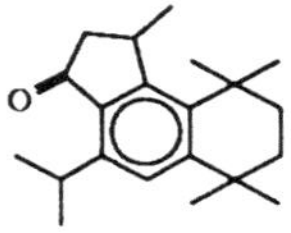 **233** (w)	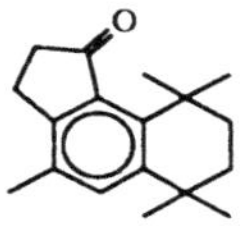 **234** (o)

[a] See Table 10 footnote *a*.
[b] Commercial products: Phantolide **172** (Polak's Frutal Works). Traseolide **188** (Quest), Tonalid **135** (Polak's Frutal Works) = Fixolide (Givaudan).
[c] Ref. 93.
[d] Ref. 90a.
[e] Ref. 94.

with substructure **A**. The strongest *meta* musks (**147**, **152**, **158**, **165**) are indan musks, of which Celestolide® (**147**) is commercialized. The monocyclic musks are all much weaker and less persistent. This demonstrates the importance of a closely packed structure (rule 6). In the same context it is instructive to compare Celestolide® (**147**) with the homologues **148**, **153** and **154**. Replacement of a methyl group by ethyl leads to a decrease of odour strength. The same tendency is noticed when a quaternary carbon atom is replaced by a tertiary C (cf. **149**, **155**). In the category of non-musks, one finds the '*para* compounds' (**139**, **168**), an ester (**145**), two aldehydes which are not benzylic (**144** and **146**), difunctional structures (**141** and **163**, but cf. **142**), and a methyl homologue of Celestolide® (**147**), ketone **157** which would have been expected to have a more pronounced musk odour. Ketone **171**, a hybrid of *ortho* and *meta* musks, is odourless (substructure of type **E**). Ketone **167** is a fairly weak musk. This is due to the alkyl substitution of both positions adjacent to the carbonyl group.

Ortho Musks

Inspection of Table 11 shows that almost all active compounds have the correct basic skeleton **B**. Compounds **188–191**,[87] **206** and **232** are structurally closely related and fulfil rule 2. Although this substructure **B** is found to be necessary, it is not a sufficient criterion for the creation of musk odour. Indeed, many examples of weak or odourless compounds are of type **B**, but have no β C=O substituent (**180**, **186**, **197**, **198**, but cf. **194**, **203**, **213**), or too voluminous alkyl groups (**181**, **201**, **207**, **208**, **215**, **216**, **222**, **233**). Incorporation of methyl groups into the basic skeleton appears to be very important for the creation of a densely packed structure and thus of a strong musk (rule 6). Whereas **180** is almost odourless, the methyl homologue **178** is a typical musk and Phantolid® (**172**) with two supplementary methyl groups is a strong musk. Exactly the same tendency is observed with the tetralins (Fig. 13) and the tricyclic musks (cf. **221**, **219**, **218**, **217** and **230**, **223**). Substitution of the β C=O position by ethyl can also lead to strong musks (e.g. **173**, **192**, **193**, **228**). However, some of these products are suspected of being neurotoxic.[88,89] Interestingly, both rigid ketones **217** and **223** (with opposite orientations of the C=O group—when superimposed as drawn) are strong musks. Compounds **183**,[90b] **190**[87] and **212**[91] represent nitrile musks. Finally, most difunctional tetralins are non-musks (**199**, **200**, **215**). Methoxyaldehyde **211**,[92] with its shielded methoxy

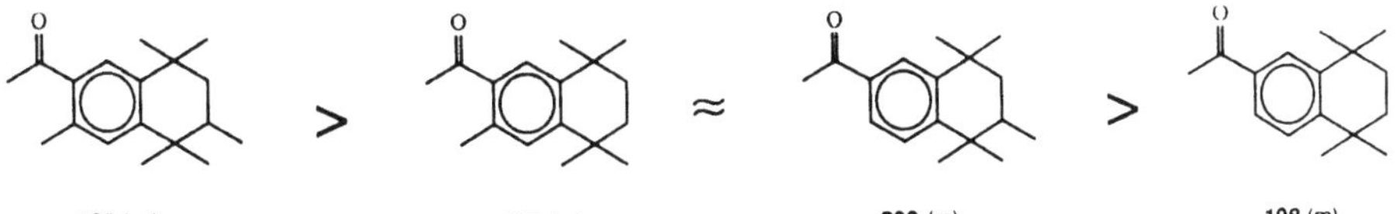

135 (vs) > 196 (m) ≈ 203 (m) > 198 (m)

Fig. 13. Tetralins.

group and accessible aldehyde function, constitutes a noteworthy exception. It is a strong musk, whereas the related ketone **215** is odourless.

With this large number of examples given, we wanted to demonstrate how even subtle structural changes within a well-defined subclass can cause dramatic changes in odour profiles; however, a comparison of the different subclasses with each other can also be very instructive. Indeed, it appears that the strongest and most tenacious musks belong to the class of *ortho* musks (in particular tetralins). These compounds are characterized by a densely packed structure which allows a very high molecular weight (C_{18} to C_{20}). On the other hand, the monocyclic *meta* musks represent the least densely packed structures and tolerate not more than 16 carbon atoms. This family contains only musks of medium intensity and low tenacity. The bi- and tri-cyclic *meta* musks (up to 18 C atoms) occupy an intermediate position, both in odour intensity/tenacity and compactness of structure.

Recent Studies

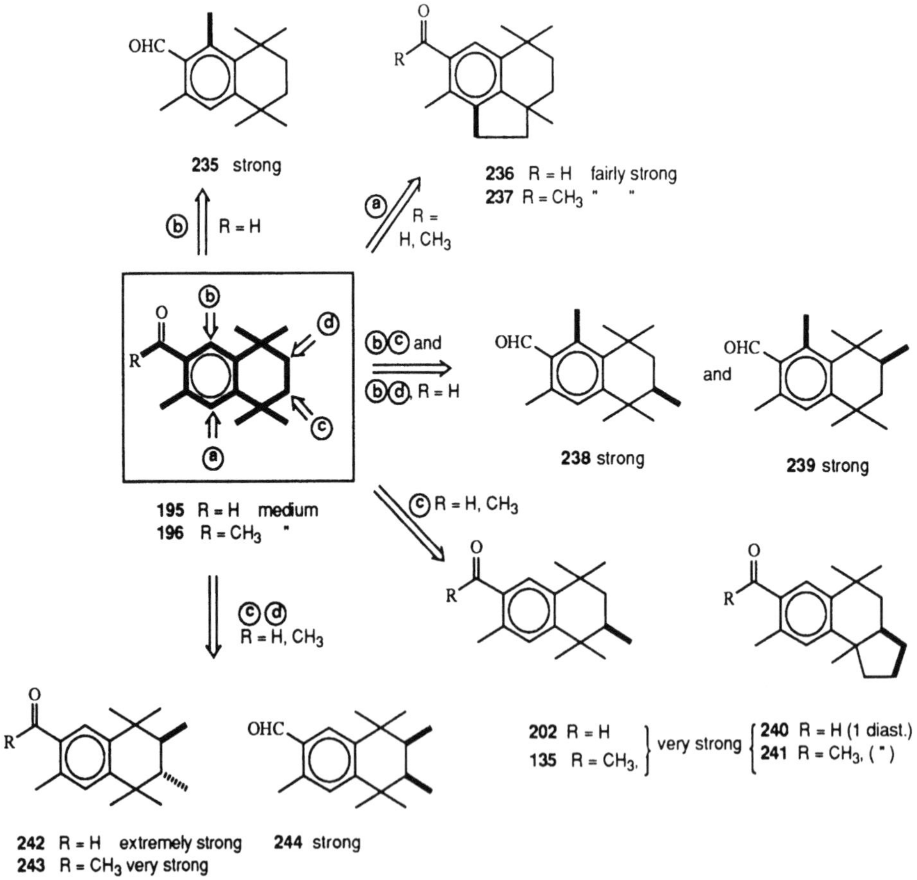

Fig. 14. Highly alkyl-substituted new strong musks.

The most promising basic skeleton for the design of a new strong musk is thus certainly structure **195** (or **196**) (Fig. 13). This fact, coupled with the observation that high alkyl substitution in a closely packed spherical structure is important for the creation of musk odour (see above), prompted us to undertake a new more detailed SAR study on tetralin musks. Indeed, we have discovered several new strong musks, **235–244** (Fig. 14)[95] Aldehyde **242** is much stronger than Tonalid® **135** and represents the most powerful aromatic musk known up to now. It is informative to compile also the structurally related weak or inactive compounds **245–259** (Fig. 15).[95]

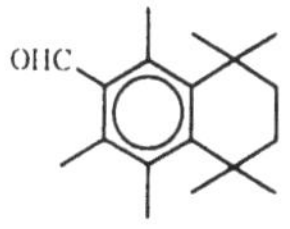

245 (w)

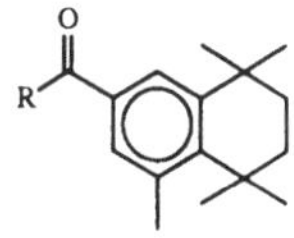

246 R = H (w)
247 R = CH_3 (w)

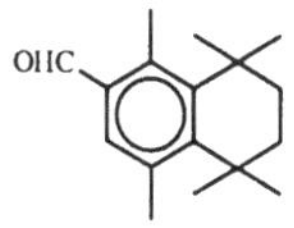

248 (w)

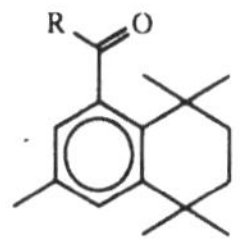

249 R = H (w)
250 R = CH_3 (w)

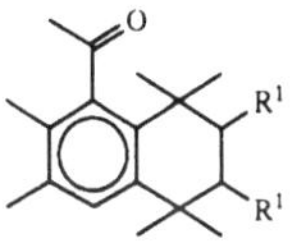

251 $R^1 = R^2 = H$* (o)
252 $R^1 = H$, $R^2 = CH_3$* (o)
253 $R^1 = CH_3$, $R^2 = H$* (o)

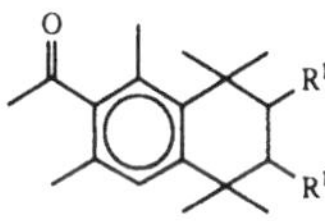

254 $R^1 = R^2 = H$* (o)
255 $R^1 = H$, $R^2 = CH3$* (o)
256 $R^1 = CH_3$, $R^2 = H$* (o)

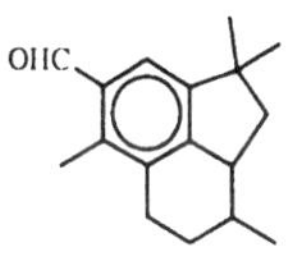

257 (o) (1 diast.)

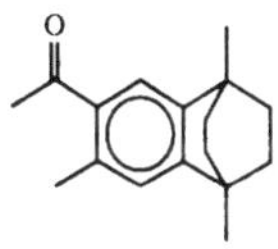

258 (o) *

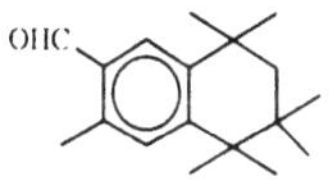

259 (o) *

Fig. 15. Weak or inactive compounds. * Fehr, C. (unpublished).

Based on our study, it is possible to draw schematically the maximal molecular dimensions for tetralin musks. The proposed general structures **F** and **G** (Fig. 16) are the result of superpositions of practically all known tetralin musks (Table 11; Fig. 14). Most compounds having 16–19 C atoms (possibly 20 C atoms) and matching the general formulae **F** or **G** can be expected to give rise to musk odour.

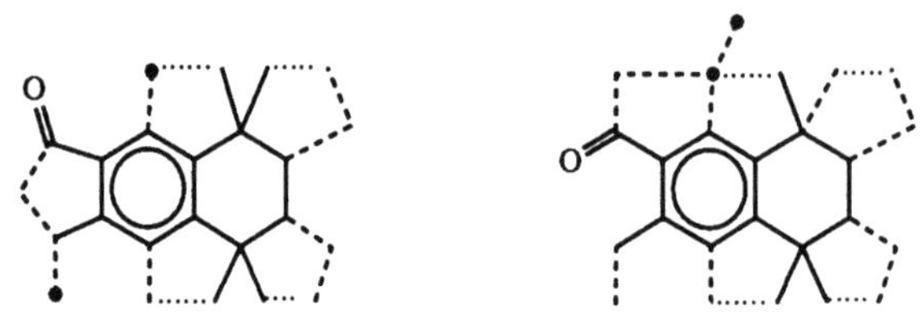

F G

Fig. 16. Schematic representation of basic tetralin musk skeleton and possible extensions to maximal molecular dimensions. ——, Basic skeleton; - - -, · · ·, maximal molecular dimensions. Restrictions: •, only for cyclopentanones and aldehydes; · · ·, only when part of a ring (**201** is odourless).

Evidently the indan musks can also accommodate themselves to this matrix, as well as other yet unexplored systems expected to lead to the discovery of new musk odorants.

Configuration of Cyclohexane Ring

For an optimal space filling (spatial arrangement) it appears that the substituents at the lipophilic end of the molecule should be in the *trans* configuration (Fig. 17). The regularly distributed methyl groups around the dipole axis in **242** lead to a powerful musk; the *cis* diastereoisomer **244** is clearly weaker, whereas the closely related analogue **259** (Fig. 15) is odourless. A similar relation is found with the indan musk Traseolide® (**188**).[87]

242 **188**

Fig. 17. Diastereoselectivity and musk odour.

The odour response is not only dependent on the diastereoisomeric, but also on the enantiomeric control of the lipophilic part of the molecule, as the two enantiomerically pure forms of **135** (Fig. 18) show quite different organoleptic properties.[96]

(S)-**135**

"strongly musky odour with a good fixation property"

(R)-**135**

"light and sweet aromatic odour"

Fig. 18. Enantioselectivity and musk odour.

In principle, these phenomena are well known;[21,22] none the less, for the class of tetralin musks we are intrigued by the fact that the odour depends less on the orientation of the C=O group than on the correct configuration in the cyclohexane ring. In an attempt to interpret this highly sensitive nature of the lipophilic part of the tetralin musks, one may envisage that the interactions between this portion of the molecule and the receptor is more selective than the polar bonding. Indeed, the strongest musks (**242**, **243**, **202**, **135**) have flexible C=O groups which can accommodate themselves to an optimal orientation. This striking relation between lipophilicity and odour intensity (for another example: compounds '**14**' and '**1**' cited in Ref. 70 may also be ascribed to the role of odorant-binding proteins. These proteins, which have been discovered in the mucosa of bovines and rats,[97] seemingly carry or concentrate odorants. Therefore, it is plausible that a more efficient transport through an aqueous medium is assured for molecules having, at least locally, pronounced lipophilic character. Compounds **195**, **202** and **242**, which are similar both topologically and stereoelectronically, possess increasingly strong musk odours. A correlation between musk odour intensity and desorption rate, and thus the ease with which a molecule passes from an aqueous layer into another phase (gas or lipid), was already established by Theimer & Davies in 1967.[82]

Carbonyl Group Environment

Our hypothesis, that increasing the lipophilicity of a known musk compound can lead to new compounds of stronger odour intensity, may also be extended—within certain limits—to the C═O group environment. Whereas **195** (Table 11) is a musk of medium odour intensity, **235** which possesses an additional CH_3 group at position 'b' (Fig. 14) is a strong musk. This contrasts with the generally accepted concept of a sterically unhindered C═O group.[29,82,85,86,98] However, the sterically more congested methyl ketone **254** is odourless, as would be expected (see also **255**, **256**). We conclude therefore that, for a musk odorant, the carbonyl system should not be prevented from adopting coplanarity with the aromatic ring; nevertheless, from the musk compounds known hitherto, nothing can be deduced about the most favourable conformation of the polar group for a strong interaction with a receptor site. Clearly, additional β-C═O substituents also exert an influence on the polarity and the global shape of the molecule. Recently, a deviation of the C═O group of *ca.* 55° from the π-plane was postulated for an efficient hydrogen bonding between a C═O lone pair and a hypothetical H-donor on the basis of the fact that **198** possesses no musk odour, whereas **196** has typical musk character (Table 11).[99,100] This hypothesis merits comment. Firstly, several musks are known where the C═O group is part of a cyclic system (Table 11). These clearly show that both coplanar orientations are compatible with musk odour (when superimposed as shown; **217**, **223**, Table 11). Secondly, replacement of the C═O group by a nitrile (*sp* lone pair) does not generally provoke extinction of the musk odour[84,87,90b,91] (**183**, **190**, **212**, Table 11) and, thirdly, **194** and **203** possess strong musk-like and medium musk odour although the *ortho* position of the C═O groups are not methylated.

Other Systems

In addition to the carbonyl group, the nitrile[84] (e.g. in **183**, **190**, **212**) and the ether function[83] (e.g. in **260**, **261**) can also act as good osmophoric groups for musk odorants. Galaxolide® (**261**), commercialized by International Flavors and Fragrances, is the strongest representative of the class of isochroman musks. The normethyl derivative **260** is a strong musk of weaker intensity, and the isomers of **261** with the methyl group at C-1 or C-3 are odourless. These cases again convincingly illustrate the importance of a correct molecular shape

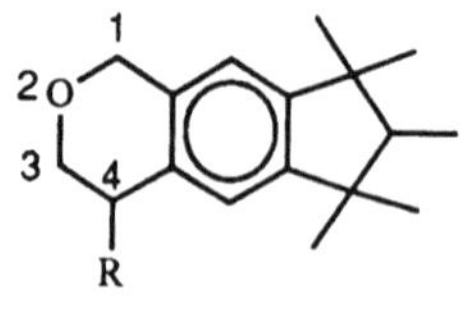

260 R = H (s)
261 R = CH_3 (s) Galaxolide®

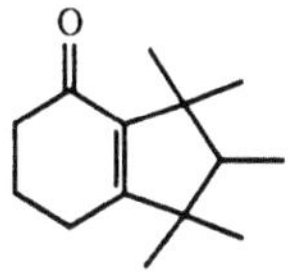

262 Cashmeran®*

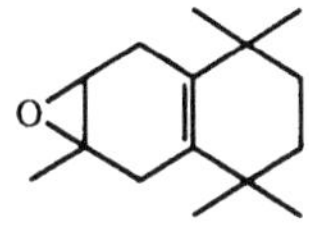

263

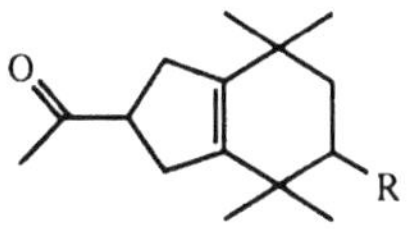

R = H **265**
R = CH_3 **266**

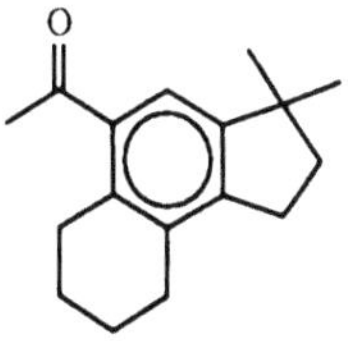

267

+

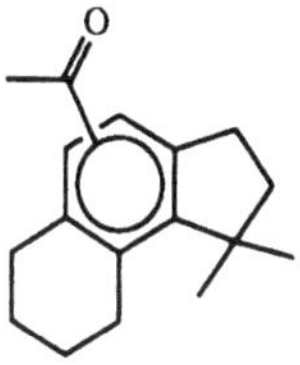

268

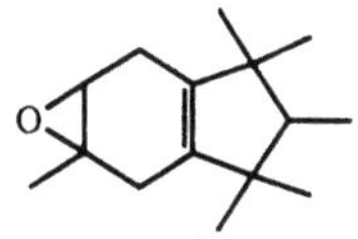

264

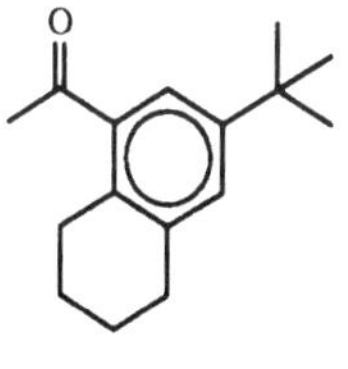

161 (o)

Fig. 19. Representative examples of structurally divergent musks. * Commercialized by International Flavors and Fragrances.

coupled with the 'fine tuning' of polarity and accessibility of the hydrophilic portion of the molecule. In addition, several other musks have been discovered which are structurally less closely related to the 'classical' aromatic musks. Some representative examples are listed in Fig. 19. Whereas **263**, **264**,[101] **265** and **266**[102] possess typical musk notes, the odour of Cashmeran® (**262**)[103] can be denoted as musk-like, but not typically musky. The mixture **267/268** has been stated to possess a noble musk odour.[104] This is in apparent contradiction with the rules (see above). The question arises whether an annulated cyclohexane ring can replace a *tert*-butyl group (cf. **158** and **267**; **147** and **268**). The long-known structurally related ketone **161** is odourless, as expected.

Computer-Assisted SAR Studies

With respect to computer-assisted SAR studies, statistical methods have been applied to select the most significant molecular descriptors for musk odour.[98,105,106] The goal was to elucidate general structural features encompassing all classes of musks; however, appreciable prediction rates could only be achieved within a well-defined subclass and the most relevant substructure

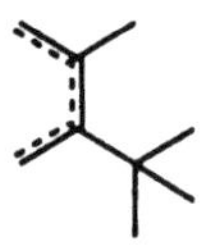
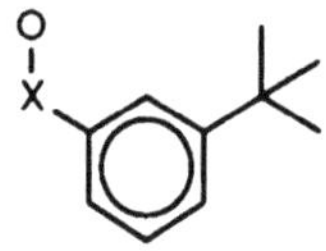

(H) (I) X = C or N

(H) for the classification of musks and non-musks was found intuitively. In another study,[99,100] a substructure (**I**), common to the nitro- and nitro-free aromatic musks, has been proposed; however this substructure does not permit accurate predictions, due to its lack of discrimination. Moreover, computer-assisted classification into only two groups (musks and non-musks) is not very useful for the design of highly active compounds and can even be misleading when the file for musks also contains all the compounds which are only weakly active. In an attempt to define the skeletal requirements for aromatic musks Jurs and co-workers[98] measured the distances between the polar hetero-atom and the two quaternary centres. Optimal dimensions of this triangle (cf. **J**)

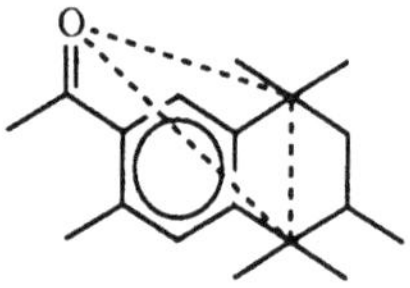

(J)

were introduced as a valuable descriptor. Although expressed differently, this parameter was already recognized long ago by Beets (cf. Ref. 29) (rule 5). There is also a doubt about the chosen C=O conformation. In the light of the complex structure–odour variations found even within a well-defined subclass (e.g. tetralin-musks), it is almost impossible to find a computer-assisted method of high predictive value. As admitted by Jurs *et al.*,[106,107] the rate of success of a statistical method largely depends on the test molecules chosen. If the applied method correctly predicts both Cashmeran® (**262**) and Tonalid® (Fixolide®) (**135**) as musk odorants[100] it would probably misclassify most of the non-musks cited in this chapter.

6 CONCLUSION

The four sensory impressions, ambergris (Ambrox® **6**), urine (5α-androst-16-en-3-one, **44**), sandalwood (β-santalol, **51**) and musk (compounds **133** and **135**), typified by the compounds listed in brackets, exhibit distinct structure–activity relationships. As we have tried to show in this chapter, small structure modifications can influence decisively the odour quality and potency.

There is no doubt that the odour of a compound is a global molecular property, whereas molecular structure can be seen in many ways as an arrangement of various substructural elements. It becomes more and more apparent that the three-dimensional disposition of selected substructural elements of a given compound is decisive for a specific odour release. The high degree of structural specificity for odour release, translated in terms of odour quality, implies the presence of highly organized active sites and supports the assumption that there are protein receptor molecules.

The recognition process in the initial step leads to a reversible neuroactive complex between the odorant and the receptor molecule; the latter, in order to bind the substrate, must possess a complementary stereoelectronic arrangement of binding sites and steric barriers. Many examples in this chapter have illustrated how critical these conditions are.

The low energy interactions implied by the ground-state complexation of odorants require a high structural organization through multiple binding. In this respect, it is interesting to note that the strongest sandalwood and aromatic musk odorants have structures allowing some flexibility and that highly rigid analogues are often weaker. This can be explained in terms of a more facile receptor–substrate conformational accommodation, resulting in a more efficient binding.

Nevertheless, it should be kept in mind that the receptor molecules also have some flexibility and can accommodate the highly rigid ambergris and steroid odorants with high selectivity and a remarkable efficiency, if the intense odour response elicited by several compounds of these classes can be taken as an indicator. Perhaps a key compound is the steroid odorant androstenol (**40**), which can evoke simultaneously and to various degrees the odours of musk, of sandalwood, of urine and of certain aspects of ambergris. Indeed, some substructural elements common to compounds of these different classes of odours can be recognized in **40**, which remains however an almost unique case.

Generally, as illustrated in this chapter, it is by carefully examining and modifying congeneric series of odorants, i.e. families of compounds structurally closely related, that possible relationships between chemical structure and odour may be discovered and tested, and become useful guidelines for the search of new odorants.

DEDICATION

This chapter was written *in memoriam* Edgar Lederer who certainly was not only highly competent but also very interested in the field of fragrant chemicals and relationships of structure and odour.

REFERENCES

1. Carpenter, M. S. & Easter, W. M., Nitro Musks. II. *tert*-Amyl homologs of the commercial musks. *J. Org. Chem.*, **16** (1951) 618.
2. Carpenter, M. S. & Easter, W. M., Nitro Musks. IV. Miscellaneous experiments. *J. Org. Chem.*, **19** (1954) 87.
3. Demole, E. & Wüest, H., Synthèses stéréoselectives de deux trioxydes $C_{18}H_{30}O_3$

stéréoisomères, d'ambréinolide et de sclaréol-lactone à partir de dérivés du (+)-manool. *Helv. Chim. Acta,* **50** (1967) 1314.

4. Rupe, H. & Majewsky, K. v., Notizen. I. Über osmophore Gruppen. II. Über die Darstellung von Diazoimiden (Triazoverbindungen). *Ber. Dtsch. Chem. Ges.,* **33** (1900) 3401.
5. Rhein, L. D. & Cagan, R. H., Role of cilia in olfactory recognition. In *Biochemistry of Taste and Olfaction,* ed. R. H. Cagan & M. R. Kare. Academic Press, New York, 1981, p. 47.
6. Lancet, D., Vertebrate olfactory reception. *Ann. Rev. Neurosci.,* **9** (1986) 329–55.
7. Anholt, R. R., Primary events in olfactory reception. *Trends Biochem. Sci.,* **12** (1987) 58.
8. Dodd, G. H., Biochemical aspects of coding and transduction in the primary olfactory neurones. *Drugs Res.,* **28** (1978) 2362.
9. Ohloff, G., The fragrance of ambergris. In *Fragrance Chemistry,* ed. E. T. Theimer. Academic Press, New York, 1982, p. 535.
10. Ohloff, G., Chemie der Geruchs- und Geschmackstoffe. In *Fortschritte der Chemischen Forschung,* Vol. 12, Part 2, ed. A. Davidson *et al.* Springer-Verlag, Heidelberg, 1969, p. 185.
11. Ruzicka, L. & Seidel, C. F., Über die flüchtigen Bestandteile des grauen Ambra. 2. Mitt. Über ein Oxyd $C_{13}H_{22}O$, einen Oxyaldehyd $C_{17}H_{30}O_2$ und ein Keton $C_{13}H_{20}O$. *Helv. Chim. Acta,* **33** (1950) 1285.
12. Collin-Asselineau, C., Lederer, E., Mercier, D. & Polonsky J., Sur quelques produits d'oxydation et de dégradation de l'ambréine; obtention de substances à odeur ambrée. *Bull. Soc. Chim. Fr.,* **17** (1950) 720.
13. Ohloff, G., Vial, C., Wolf, H. R., Job, K. Jégou, E., Polonsky, J. & Lederer, E., Stereochemistry–odor relationships in enantiomeric ambergris fragrances. *Helv. Chim. Acta,* **63** (1980) 1932.
14. Stoll, M. & Hinder, M., *Odeur et constitution III.* Les substances bicyclohomofarnésiques. *Helv. Chim. Acta,* **33** (1950) 1251.
15. (a) Hinder, M. & Stoll, M., *Odeur et constitution IV.* Sur les époxydes hydroaromatiques à odeur ambrée. *Helv. Chim. Acta,* **33** (1950) 1308.
15. (b) Hinder, M. & Stoll, M., Odeur et constitution IX. Préparation de substances bicyclohomofarnésiques stéréoisomères. *Helv. Chim. Acta,* **36** (1953) 1995.
16. Mookherjee, B. D. & Patel, R. R., Isolation and identification of volatile constituents of tincture of ambergris. *7th Int. Congr. Essent. Oils, Kyoto, 7–11 Oct. 1977*, p. 479.
17. Schenk, H. R., Gutmann, H., Jeger, O. & Ruzicka, L., Odeur et constitution III. Les substances bicyclohomofarnésiques. *Helv. Chim. Acta,* **37** (1954) 543.
18. Scheidegger, U., Schaffner, K. & Jeger, O., Über die Struktur und Umwandlungen von zwei stereoisomeren Riechstoffen $C_{18}H_{30}O_2$ aus Manool. *Helv. Chim. Acta,* **45** (1962) 400.
19. Vlad, P. F., The chemistry of perfumes based on labdane diterpenoids. *Russ. Chem. Rev.,* **51** (1982) 644.
20. Ohloff, G., Giersch, W., Pickenhagen, W., Furrer, A. & Frei, B., Significance of the geminal dimethyl group in the odor principle of Ambrox®. *Helv. Chim. Acta,* **68** (1985) 2022.
21. Ohloff, G., Maurer, B., Winter, B. & Giersch, W., Structural and configurational dependence of the sensory process in steroids. *Helv. Chim. Acta,* **66** (1983) 192.
22. Pickenhagen, W., Enantioselectivity in odor perception. In *Flavor Chemistry, Trends and Developments,* ACS Symposium Series No. 388, ed. R. Teranishi, R. G. Buttery & F. Shahidi. American Chemical Society, Washington, DC, 1989, p. 152.
23. (a) Ohloff, G., Chemistry of odor stimuli. *Experientia,* **42** (1986) 271.

23. (b) Ohloff, G. & Giersch, W., Structure–activity relationships in odor perception of drimane derivatives. *Croat. Chem. Acta,* **58** (1985) 491.
24. (a) Ohloff, G., Relationship between odor sensation and stereochemistry of decalin ring compounds. In *Gustation and Olfaction, an International Symposium. Geneva, June 1970.* Academic Press, London, New York, 1971, p. 178.
24. (b) Ohloff, G., Giersch, W., Schulte-Elte, K. H. & Vial, C., Zur Stereochemie von 1-Dekalon-Derivaten und ihren oxaanalogen Verbindungen. *Helv. Chim. Acta,* **59** (1976) 1140.
25. (a) Bersuker, I. B., Dimoglo, A. S., Gorbachov, M. Y., Koltsa, M. N. & Vlad, P. F., Structural and electronic origin of ambergris odor of cyclic compounds. *Nouv. J. Chim.,* **9** (1985) 211.
25. (b) Gorbachev, M. Y, Dimoglo, A. S., Bersuker, I. B., Vlad, P. F. & Koltsa, M. N., The role of electronic and structure factors in the origin of amber odor in cyclic compounds. *Theor. Eksp. Khim.,* **22** (1986) 355; *Chem. Abstr.,* **105** (1986) 158591q.
26. Wolf, H., Mätzel, U., Brunke, E.-J. & Klein, E., Stereoselektive Synthese von (±)-4α-Desmethyl-Ambrox und (±)-4α,8-Didesmethyl-Ambrox. *Tetrahedron Lett.* (1979) 2339.
27. Brunke, E.-J., Modifizierte Ambra-Riechstoffe I. *Dragoco Report,* **26** (1979) 276.
28. Brunke, E.-J., Ambra und Ambra-Riechstoffe II. *Dragoco Report,* **27** (1980) 9.
29. Beets, M. G. J., *Structure–Activity Relationships in Human Chemoreception.* Applied Science Publishers, London, 1978.
30. (a) Winter, B., Ring-opened analogues of Ambrox®: synthesis and structure–odour relationships. *Helv. Chim. Acta,* **72** (1989) 1278.
30. (b) Winter, B., QSAR in olfaction: ambergris-type odorants. In *QSAR: Quantitative Structure–Activity Relationships in Drug Design, Proc. 7th European Symposium on QSAR, Interlaken, Switzerland, 5–9 September 1988*, ed. J. L. Fauchère. Alan R. Liss, New York, 1989, p. 401.
31. Greenberg, M. J. In *Odor Quality and Chemical Structure,* ACS Symposium Series No. 148, ed. H. R. Moskovitz & C. B. Warren. American Chemical Society, Washington, DC, 1981, p. 177.
32. Prelog, V., Ruzicka, L., Meister, P. & Wieland, P., Steroide und Sexualhormone. (113. Mitt.). Untersuchungen über den Zusammenhang zwischen Konstitution und Geruch bei Steroiden. *Helv. Chim. Acta,* **28** (1945) 618.
33. Prelog, V. & Ruzicka, L., Untersuchungen über Organextrakte. (5. Mitt.). Über zwei moschusartig riechende Steroide aus Schweinetestes-Extrakten. *Helv. Chim. Acta,* **27** (1944) 61.
34. Comfort, A., Likelihood of human pheromones. *Nature,* **230** (1971) 432.
35. Reed, H. C. B., Melrose, D. R. & Patterson, R. L. S., Androsten steroids as an aid to the detection of estrus in pig artificial insemination. *Br. Vet. J.,* **130** (1974) 61.
36. Beets, M. G. J. & Theimer, E. In *Ciba Foundation Symposium on Taste and Smell in Vertebrates,* ed. G. E. W. Wolstenholme & J. Knight. A. Churchill, London, 1970, p. 313.
37. Theimer, E. T., Yoshida, T. & Klaiber, E. M., Olfaction and molecular shape. Chirality as a requisite for odor. *J. Agric. Food Chem.,* **25** (1977) 1168.
38. Ohloff, G., Giersch, W., Thommen, W. & Willhalm, B., Conformationally controlled odor perception in 'steroid-type' scent molecules. *Helv. Chim. Acta,* **66** (1983) 1343.
39. Brunke, E.-J. & Klein, E., Chemistry of sandalwood fragrance. In *Fragrance Chemistry,* ed. E. T. Theimer. Academic Press, New York, 1982, p. 397.
40. Semmler, F. W., Zur Kenntnis der Bestandteile ätherischer Öle. (Konstitution der α-Santalol-und α-Santalen-Reihe: Die Konstitution der Sesquiterpenalkohole und Sesquiterpene.) *Ber. Dtsch. Chem. Ges.,* **43** (1910) 1893.

41. Ruzicka, L. & Thomman, G., Polyterpene und Polyterpenoide XCIII. Über die Konstitution des β-Santalols und des β-Santalens. *Helv. Chim. Acta,* **18** (1935) 355.
42. Lewis, R. G., Gustafson, D. H. & Erman, W. F., Total synthesis of α-santalol. *Tetrahedron Lett.,* (1967) 401.
43. Kretschmar, H. C. & Erman, W. F., The total synthesis and geometric configuration of *dl*-β-santalol. *Tetrahedron Lett.* (1970) 41.
44. Guha, P. C. & Bhattacharyya, S. C., Santalol series. I. Separation of the santalols and the santalenes. *J. Indian Chem. Soc.,* **21** (1944) 261.
45. Corey, E. J., Kirst, H. A. & Katzenellenbogen, J. A., A stereospecific total synthesis of α-santalol. *J. Am. Chem. Soc.,* **92** (1970) 6314.
46. Brunke, E.-J., Hammerschmidt, F.-J. Struwe, H., (+)-Epi-β-Santalol Isolierung aus Sandelholzöl und Partialsynthese aus (+)-α-Santalol. *Tetrahedron Lett.,* **21** (1980) 2405.
47. Brunke, E.-J., Böhme, A. & Struwe, H., Absolute Konfiguration von (−)-β-Santalol, (+)-epi-β-Santalol und (*E*)-(−)-β-Santalol. *Liebigs Ann. Chem.* (1982) 1105.
48. Brunke, E.-J., Schatkowski, D., Struwe, H. & Tumbrink, L., Bergamotol and spirosantalol—New constituents of East Indian sandalwood oil. In *Flavours and Fragrances: a World Perspective,* ed. B. M. Lawrence, B. D. Mookherjee & B. J. Willis. Elsevier, Amsterdam, 1988, p. 819.
49. Fanta, W. I. & Erman, W. F., Novel use of borate protective groups in organic synthesis. A facile synthesis of dihydro-β-santalol. *Tetrahedron Lett.* (1969) 4155.
50. Fanta, W. I. & Erman, W. F., Studies related to the synthesis of (±)-dihydro-β-santalol. *J. Org. Chem.,* **37** (1972) 1624.
51. Herout, V. Jarolim, V. & Pliva, J., *Chem. Listy,* **50** (1956), 1271; *Chem. Abstr.,* **51** (1957) 296c.
52. Mookherjee, B. D., Kamath, V., Patel, R. R. & Shuster, E. J., German Patent 2 604 397 (9 Sept. 1976); *Chem. Abstr.,* **86** (1977) 95875b.
53. Kaiser, R. & Lamparsky, D., Neue mono- und sesquiterpenoide Inhaltsstoffe des Lavendel-bzw. Lavandinoeles. *Tetrahedron Lett.* (1977) 665.
54. Schulte-Elte, K. H. (Firmenich S.A.), Swiss Patent 622 946 (prior 22 Aug. 1977); *Chem. Abstr.,* **95** (1981) 169545d.
55. Weissenborn, A., GDR. Patent 5051 (24 Nov. 1942), publ. 18 Sept. 1954.
56. Byers, J. R., Jr, New woody-type odors. *Am. Perf. Essential Oils Rev.,* **49** (1947) 483.
57. Demole, E., Synthèses et relations entre constitution chimique et odeur dans la série des terpényl-3-cyclohexanols. *Helv. Chim. Acta,* **47** (1964) 1766.
58. Demole, E., Synthèses stéréospécifiques et étude organoleptique comparée de onze triméthyl-2,2,3-*exo*-norbornyl-5-*exo*-cyclohexanols racémiques. *Helv. Chim. Acta,* **52** (1969) 2065.
59. Buchbauer, G., Zur Synthese von Isosantalol und δ-Santalol. *Monatsh. Chem.,* **109** (1978) 289.
60. Góra, J., Kotarska-Markowicz, L. & Gibka, J., Synthesis of some new sandalwood type aroma chemicals from alpha pinene or camphene. In *IXth Int. Congr. Essential Oils, 13–17 March 1983, Singapore,* Book 5, Essential Oil Technical Paper, 45.
61. Brunke, E.-J. & Klein, E., Lower homologues and analogues of the odoriferous principles in ambergris, musk and sandalwood oil. In *Essential Oils,* ed. B. D. Mookherjee & C. J. Mussinan. Allured Publishing Corp., Wheaton, 1981, p. 83.
62. Trenkle, R. W. & Mookherjee, B. D., Generation of novel sandalwood odorous compounds via hydroformylation of α- and β-pinene. In *IXth Int. Congr. Essential Oils, 13–17 March 1983 Singapore,* Essential Oil Technical Paper, Book 5, 57.

63. Mühlstädt, M., Feustel, G., Hermann, M. & Dollase, W., East German Patent 68 936 (20 Sept. 1969); *Chem. Abstr.*, **72** (1970) 125008b.
64. Klein, E. & Brunke, E.-J., British Patent 2 024 208 (9 Jan. 1980); *Chem. Abstr.*, **93** (1980) 137899q.
65. Naipawer, R. E. & Easter, W. M. (Givaudan, L. et Cie S.A.), US Patent 4 052 341 (4 Oct. 1977); *Chem. Abstr.*, **88** (1978) 22229a.
66. Schulte-Elte, K. H., Müller, B. & Pamingle, H. (Firmenich S.A.). European Patent 155 591 (25 Sept. 1985); *Chem. Abstr.*, **105** (1986) 191435q.
67. Naipawer, R. E. (Givaudan, L. et Cie S.A.), European Patent 203 528 (3 Dec. 1986); *Chem. Abstr.*, **106** (1987) 175828k.
68. Naegeli, P. (Givaudan, L. et Cie S.A.), US Patent 4 046 716 (6 Sept. 1977); *Chem. Abstr.*, **87** (1977) 200931n.
69. Klein, E. & Rojahn, W., (Dragoco, Holzminden), German Patent 28 07 584 (30 Aug. 1979).
70. Schulte-Elte, K. H., Giersch, W., Winter, B., Pamingle, H. & Ohloff, G., Diastereoselektivität der Geruchswahrnehmung von Alkoholen der Iononreihe. *Helv. Chim. Acta,* **68** (1985) 1961.
71. Shaffer, G. W. & Purzycki, K. L. (Givaudan, L. et Cie S.A.), German Patent 2 605 610 (26 Aug. 1976); *Chem. Abstr.*, **85** (1976) 192236f.
72. Shaffer, G. W. & Purzycki, K. L. (Givaudan, L., et Cie S.A.), German Patent 2 804 075 (3 Aug. 1978); *Chem. Abstr.*, **89** (1978) 179608s.
73. Witteveen, J. G. & van der Weerdt, A. J. A., Structure–odour relationships of some new synthetic sandalwood aroma chemicals. Synthesis and olfactive properties in a series of bicyclo[4.4.0]decan-3-ols. *Rec. Travaux Chim. Pays-Bas,* **106** (1987) 29.
74. Aulchenko, I. S. & Kheifits, L. A., Investigating the composition of Santalydol. *Am. Perfum. Cosmet.*, **85** (1970) 37.
75. Jones, B. N., Ansari, H. R., Jagger, B. G. & Janes, J. F. (Bush Boake Allen Ltd), German Patent 2 255 119 (30 May 1973); *Chem. Abstr.*, **79** (1973) 41918e.
76. Naipawer, R. E., Purzycki, K. L., Shaffer, G. W. & Erickson, R. E., A structure–odor relationship for sandalwood aroma chemicals. In *Essential Oils,* ed. B. D. Mookherjee & C. J. Mussinan. Allured Publishing Corp., Wheaton, 1981, p. 105.
77. Beyer, A., Wolschann, P., Becker A., Pranka, E. & Buchbauer, G., Conformational calculations on odoriferous molecules of sandalwood, I. The search for the odoriferous principle of sandalwood oil. *Monatsh. Chem.*, **119** (1988) 711.
78. Beyer, A., Wolschann, P., Becker A., Buchbauer, G. & Winiwarter, S., Conformational calculations on sandalwood odour molecules, II. Force field investigations on β-santalol. *Flav. Fragr. J.*, **3** (1988) 173.
79. Naipawer, R. E., Synthetic sandalwood chemistry—a decade in review. In *Flavours and Fragrances: a World Perspective,* ed. B. M. Lawrence, B. D. Mookherjee & B. J. Willis. Elsevier, Amsterdam, 1988, p. 805.
80. Vial, C., Thommen, W. & Näf, F., Structure–activity relationship in ambergris-type woody odorants possessing a hydronaphthalene-skeleton. *Helv. Chim. Acta,* **72** (1989) 1390.
81. Mookherjee, B. D. & Wilson, R. A., The chemistry and fragrance of natural musk compounds. In *Fragrance Chemistry,* ed. E. T. Theimer. Academic Press, New York, 1982, p. 433.
82. Theimer, E. T. & Davies, J. T., Olfaction, musk odor, and molecular properties. *J. Agr. Food Chem.*, **15** (1967) 6.
83. Heeringa, L. G. & Beets, M. G. J., (Internat. Flavor & Fragrances) US Patent 3 360 530 (26 Dec. 1967).
84. Kulka, K. (Fritzsche Dodge & Olcott, Inc.), US Patent 3 910 853 (7 Oct. 1975); *Chem. Abstr.*, **84** (1976) 73972n.

85. Wood, T. F., *Chemistry of the Aromatic Musks. Givaudanian,* Givaudan Corporation, 1968, p. 1.
86. Wood, T. F., Chemistry of synthetic musks, II, Benzoid musks. In *Fragrance Chemistry,* ed. E. T. Theimer. Academic Press, New York, 1982, p. 509.
87. Boelens, H., Traas, P. C. & Takken, H. J. (Naarden Int.), NV NL-002076 (1977); Musks in perfumery. *Perfumer and Flavorist,* **5** (1980) 39.
88. Gressel, Y., Troy, W. R. & Foster, G. V., *Mech. Tox. Haz. Eval.* (1980) 53.
89. Cammer, W., *Biochem. Pharmacol.,* **29** (1980) 1531.
90. (a) Polak's Frutal Works Inc., French Patent 1 392 804 (prior. 21 March 1963); *Chem. Abstr.,* **63** (1965) 6781d.
90. (b) De Simone, R. S. (Polak's Frutal Works), US Patent 4 018 719 (prior. 20 April (1976); *Chem. Abstr.,* **87** (1977) 22866k.
91. Christenson, P. A., Drake, B. J. & Willis, B. J. (Fritzsche Dodge and Olcott, Inc.), US Patent 4 483 786 (prior. 20 Nov. 1984); *Chem. Abstr.,* **102** (1984) 100615x).
92. Willis, B. J. & Zazula, T. J. (Fritzsche Dodge & Olcott, Inc.), US Patent 4 476 040 and European Patent 98 546 (prior. US 2 July 1982); *Chem. Abstr.,* **101** (1984) 72466h.
93. Gonzenbach, H. (L. Givaudan & Cie S.A.) European Patent 301 375 (prior. CH 29 July 1987); EP-301-375-A (1989); *Chem. Abstr.,* **111** (1989) 97596v.
94. Suzukamo,G., Sakito, Y. (Sumitomo Chem. Corp. Ltd) European Patent 71 006 (prior. JA 12 Aug. 1981), See also Japanese pat. applic. 58'029 721 and 58'029 738 (same priorities).
95. Fehr, C., Galindo, J., Haubrichs, R. & Perret, R., New aromatic musk odorants: design and synthesis. *Helv. Chim. Acta,* **72** (1989) 1537; C. Fehr, Swiss Patent Appln (prior. 30 June 1989).
96. Suzukamo, G. (Sumitomo Chem. Corp. Ltd.) US Patent 4 767 882 (prior. JA 11 June 1981).
97. Pevsner, J., Sklar, P. B. & Snyder, S. H., *Proc. Natl. Acad. Sci. U.S.A.,* **83** (1986) 4942; *Chem. Abstr.,* **105** (1986) 58388k.
98. Narvaez, J. N., Lavine, B. K. & Jurs, P. C., Structure–activity studies of musk odorants using pattern recognition: bicyclo- and tricyclo-benzenoids. *Chem. Senses,* **11** (1986), 145.
99. Chastrette, M. & Zakarya, D., Sur le rôle de la liaison hydrogène dans l'interaction entre les récepteurs olfactifs et les molécules à odeur de musc. *C.R. Acad. Sci. Paris,* **307** (II) (1988) 1185.
100. Zakarya, D., Recherche de relations structure–odeur de musc. Mise au point d'une nouvelle méthode G.E.S.D.E.M. (Génération et Sélection de Descripteurs et Elaboration de Motifs). PhD thesis no. 02—88, Université Claude-Bernard, Lyon 1, 1988.
101. Helmlinger D. & Pesaro, M. (Givaudan, L., et Cie S.A.), European Patent 115274 (prior. 13 Jan. 1983); *Chem. Abstr.,* **102** (1985) 62066j.
102. Klaus M., Helmlinger, D. & Pesaro, M. (Givaudan, L., et Cie S.A.), European Patent 116277 (prior. 13 Jan. 1983); *Chem. Abstr.,* **102** (1985) 61860b.
103. Hall, J. B. & Lala, L. K. (International Flavors and Fragrances Inc.), German Patent 2 220 820 (prior. 21 June 1971); *Chem. Abstr.,* **78** (1973) 124151e.
104. Takasago Perfumery, World Pat. Applic. 88'06435 (prior. 5 Mar. 1987) = European Patent 306 529.
105. Stuper, A. I., Brügger, W. E. & Jurs, P. C., *Computer Assisted Studies of Chemical Structure and Biological Function.* Wiley–Interscience, New York, 1979.
106. Jurs, P. C., Ham, C. L. & Brügger, W. E. In *Odor Quality and Chemical Structure,* ACS Symposium Series No. 148, ed. H. R. Moskowitz & C. B. Warren, American Chemical Society, Washington, DC, 1981, p. 143.

V COMPOSITORY TECHNIQUES AND APPLICATION SEGMENTS

Fragrances are usually designed in view of specific applications or application segments. Their industrial production has resulted in compository techniques which supplement aesthetic and hedonic criteria. It has also resulted in investigational methods which help to predict the performance and acceptance of perfumed products.

The corresponding information forms an important part of the core know-how of the large fragrance houses, and the related findings or innovations are hardly protected by patents. This not only because these patents would be hard to defend, but also because of ethical behaviour preventing industry from action which would contribute to an erosion of the delicate and invaluable rights which should be reserved to protect inventions that are more than just variations of the state-of-the-art.

The reader will not be surprised, therefore, to find that some of the statements in the following chapters will not be supported by very much information on detailed formulae, etc. Nevertheless, this part of the book will allow an insight into actual solutions and needs as well as an appreciation of the technologies that are actually used.

Chapter 10

Creative Perfumery: Composition Techniques

JEAN-CLAUDE ELLENA

Givaudan France S.A., F-92532 Levallois Perret Cedex, France

'In every artist, there is a poet and a workman'
Emile Zola

1 INTRODUCTION

Have you noticed how much smells give shape to a place?

When I stroll around Paris, I am assailed or seduced all the time by the odours or the perfumes all around me: the bitter smell of rotten eggs, of coal in subway stations, seats made of polyurethane, vinyl plastic flooring, handles and banisters in brushed aluminium, cold and biting; the macadam and the wifts of sweat in corridor draughts; the warm fat of escalators; the dust, dog-piddle, car fumes in the streets, the warm smell of car engines as they rush by; the smell of the plane trees, dry and hard. The Champs-Elysées suddenly become olfactive 'action paintings', with the sweeping smells of restaurants, pizzerias, fast food, vanilla ice-cream, the exhaust fumes of buses outside the Lido, the smell of chocolate praline and the bustling crowd with its very own smell akin to the bland smell of chicken, of spicy black skins, the Japanese with their aura of ambergris; warm, living, human smells, mixed with the smell of clean parquet, of leather, the sumptuous materials of the Galeries du Rond Point—but that's *Samsara, Poison, Opium, Paco*: the connoisseurs can name all of them, the public do not know the fragrance but they do know that it's different, that it is something apart from all the others. That this is *perfume.*

Odour or Perfume? The difference between the two is that perfume, *object* of emotion, of evocation, of adornment, of seduction, cult and culture . . . is the culmination of a *project,* a performance, something that has gradually taken shape, come to life, composed by the Perfume Composer, the *subject.*

The technique of olfactive composition, described below, follows the same theme: *subject, project, object,* always keeping in mind that the technique is the background of a perfume and not its finality.

2 THE SUBJECT

Knowledge of the *subject* first of all requires an extensive study of raw materials (odorant materials) and the curiosity and passion for smell.

Raw materials of natural origin are called essential oils, absolute essences, resinoids or concretes, depending on the manufacturing companies or the extraction methods used. Synthetic materials are of natural origin, like eugenol for example (the major component of clove essential oil), or they are obtained by simple or complex chemical reactions. Synthetic chemistry can also create new odours, non-existent in nature, and thus reveal olfactive sensations that are totally new.

Using this kind of fragrant material, the perfume composer makes up a collection of fragrances selected according to certain standards:

(a) ideas disguised in odorant materials;
(b) technical quality;
(c) reliability and availability.

Ideas Disguised in Odorant Materials

Nature arouses the imagination and makes it fertile: so a privileged relationship exists between the perfumer and the fragrant substances that belong to nature. Odour arouses in each of us intense affective reactions—emotional, sensorial, descriptive; it awakens within us all a certain kind of sensuality that turns the milky smell of sandalwood into the image of a woman's body, the odour of myrrh into the memory of the smell of humid earth, or bags of fresh wheat.

Emotional

Through smell we can experience desire, pleasure or an aversion, hope or fear, surprise or attention, using subjective terms which imply affective (aggressive, suffocating, violent . . .), erotic (supple, sensual, round . . .) or hedonic (young, lively, radiant . . .) sensations.

Sensorial

We also borrow words used to describe taste (sweet, salty, bitter, acid . . .), touch (harsh, soft, hard . . .) and physical qualities like volume (rich, full . . .), mass (heavy, weighty . . .), or dimension (outline, flatness, thickness . . .).

Descriptive

Our memory groups together certain odours under an olfactive common denominator, with descriptions such as:

—*Floral*: Flowers, white flowers, exotic flowers . . .
—*Fruity*: Fruits, red fruits, citrus, exotic
—*Spicy*
—*Grassy*: Aromatic, culinary, green, vegetal, hay . . .
—*Woody*: Wood, humus, roots, resins . . .
—*Burnt*: Fire, leather . . .
—*Balsam and vanilla*
—*Animal*: Animal, human, amber, musky . . .
—*Marine*: Sea, ozone, seaweed . . .
—*Mineral*: Metal, stone, earth, dust . . .
—*Abstract*: Chemical, gas . . .

Certain odours can belong to several categories. Cumin essential oil has a grassy, culinary odour but can also be compared with an animal odour or with human perspiration.

These descriptions can be completed with personal impressions, the most important information of all, which serve as reference marks, the milestones of creation.

Although it is possible to recognize millions of fragrant substances, especially in the field of chemical synthesis, it does not appear to me to be very desirable or logical to use such a large number of fragrant substances: the main reason being that it is impossible for a human being to assimilate any real knowledge of so many fragrances.

Total familiarity with a fragrant substance does not only consist of being able to recognize its name; it also involves a comprehension of the ideas disguised in the material itself.

There is no particular aesthetically ranked sequence established in odours: the noble character of a raw material, the value it represents when the perfume composer finds analogies with more costly odours, is of no interest.

How 'to use an odour best' is the determining factor: the result of an exchange, of the *sympathy* felt between the 'object', the fragrant substance, and the 'subject', the perfume composer.

Technical Qualities

Today, combined scientific research, including the disciplines

—physics and chemistry (which give us the attributes of a fragrant substance),
—neurobiology (which shows us the functioning of our olfactive organs) and
—psychology (which explains the sense of our emotions),

brings out all the technical attributes of each fragrant substance as well as of olfactive compositions.

Remanence

This term seems to me more appropriate than 'tenacity', or 'substantivity', often used in the industry.

'Remanence' comes from how quickly a substance evaporates (volatility): this is usually measured by olfactive examination at regular intervals (seconds, minutes, hours, days): blotters are impregnated beforehand with each of the fragrant substances.

Intensity

A scale can be established by determining the levels of perception on each fragrant substance, by diluting it in stages, until it no longer produces any olfactive sensation.

Volume

This is calculated by how much space the fragrant substance takes up.

A simple experiment to measure volume consists of putting a blotter previously impregnated with a drop of a diluted odorant (from 1 to 10%, according to the intensity of the odorant), inside a jar or closed container. Olfaction takes place an hour later.

Cost

The cost of a fragrant substance can vary between FF12 for a kilogram of Florida orange essence and FF30 000 for violet leaf absolute.

Each of these aspects is developed in other chapters of this book.

Reliability or Availability

Geographical and economical positions, along with production variations, will be developed in another chapter: therefore I will discuss here only the importance of a knowledge of fragrant materials, as well as the notion of olfactive standards.

Knowledge of Fragrant Materials

This had led me to choose and to limit my collection of raw materials: 400 fragrant substances made up of natural-origin, reconstituted, and synthesized products using the 'headspace' technique, with a few bases (which are simple odorant compositions) and this is already quite a lot!

Although this collection is open to any significant new discovery, *choice* is the key word, because, faced with such a large quantity of materials (more than 2000 available on the market), everyone has by far a wide enough choice. However, this is both marvellous and very disturbing, because there is never enough time to really consider each new substance and to give it a real chance.

A short statistical analysis shows that 80% of the synthesized products being used today in most of our compositions were discovered between 1880 and 1930 (Table 1)—another reason to find the time to study carefully new fragrant substances!

Olfactive Standards

What distinguishes the industry from art is that the industry creates the tools, the methods, to create and remake the same product indefinitely: every year essential oils are selected (lavandin, jasmine, vetiver . . .), which correspond to very precise olfactive standards, standards which often differ according to the company involved: the same applies to the production of synthesized products, which is also stringently controlled.

The standardization of fragrant substances is a direct consequence of the industrialization of perfumery, but it is not a necessity from a creative point of view: for the perfume composer, the artist, creates what he does not know: his tools (fragrant substances), his methods (formulation techniques) evolve to help him perceive and express the world around him.

Table 1
Synthetic and perfumery classics

	Date discovered	*First used*	*Recently used*
Vanillin	1877	*Jicky,* Guerlain, 1889	*Must,* Cartier, 1981
Ionone and methyl ionone	1892	*Vera Violetta,* Roger & Gallet, 1892	*Balahe,* Leonard, 1983
Aldehydes	1903	*Chanel No. 5,* 1921	*First,* V.C. & A., 1976
Aurantiol (Schiff bases)	1908	*Heure Bleue,* Guerlain, 1912	*Giorgio,* 1983
Hexenol	1942	*Diorissimo,* Dior, 1956	*Alliage,* E. Lauder, 1972
Dimetol	1950	*Paco Rabanne,* 1974	*Poison,* Dior, 1986
Hedione®	1962	*Eau Sauvage,* Dior, 1966	*Cristalle,* Chanel, 1974
Galaxolide	1965		*Ysatis,* Givenchy, 1986
α-Damascone	1969	*Nahema,* Guerlain, 1979	*Rumba,* Balenciaga, 1988
Iso E Super	1974	*Halston,* Halston, 1975	*Fahrenheit,* Dior, 1988

For example, the territory, the cut, the harvest, all have different olfactive effects on lavenders: lavender from Sault, Ferrassieres, Montbrun or Diois in France vary from a very delicate floral fragrance to fruity, grassy, earthy and even cut-hay or animal notes: this is the delicate problem that exists in the conciliation of art and industry.

3 THE PROJECT

For the Genevan epistemologist, Jean Piaget, there are three ways to approach creation:

—The first consists of approaching it without knowing anything at all about the chosen subject, then learning what exists . . . which would no doubt have some surprises in store unless you do the same thing that someone else has tried already!
—The second is to find out as much as possible about everything that surrounds the subject, for example by research in the fields of botany, art (music, art and literature), psychology, technique or any other subject of interest to you!
—The third is just to be as stubborn as a mule, an excellent stimulator of creativity.

Another method, complementary and more personal, is to have several things to do at the same time, because constant research creates a driving force.

Harmonious Complexes and Simple Compositions

Personally, creation does begin with a knowledge of something: for it is never predictable or predetermined, but a 'sudden stroke of genius', a host of ideas that come from old recipes that are just waiting to be rediscovered.

Using two different pretexts, I am going to take a chance in a difficult field, because creation is like a human being, it escapes you, and does not always stand up to explanation. The following paragraphs, pretexts 1 and 2, could represent shorthand notes from the perfumer's workbook.

Pretext 1: *An approach towards a 'Tomato leaf' harmony. I do not think the same colour made me choose the red mandarin essential oil, but rather its fruity, round, fresh, sweet, lively character: but mandarin smells like mandarin, so I break apart this characteristic with a grassy fragrance, green, vegetal, which will give it a leafy aspect: this will be galbanum essential oil.*

Various combinations: 1/9—2/8—3/7 . . .

The last one is what I was looking for, but the red fruit result is acid, vegetal, does not have enough presence.

Ethyl phenylglycidate (strawberry aldehyde) is too syrupy, heavy; the substance N 112 (raspberry ketone) is too powdery, sweet; blackcurrant bud absolute, fruity green, acid, seems to be the answer.

New combinations: 8/1/1—7/2/1 . . .

New attempts with the addition of bergamot essential oil for its freshness, and basil, celery, to amplify the initial note.

New combinations until the final choice.

Each successive attempt is evaluated after dilution (5% solution in 90% ethyl alcohol).

Pretext 2: *A simple composition*
Recreate the lily-of-the-valley fragrance using affinities and contrasts, and with a different quantity of each chosen fragrant substance: choice is made as a function of (a) ideas and (b) the technical quality of each odorant.

Ideas disguised in chosen fragrant substances:

—*Ylang-ylang essential oil: floral, exotic, warm, 'present';*
—*Indole: odour of animals, of naphthalene, warm, massive, powerful;*
—*Benzyl acetate: floral odour, jasmine, rising, sweet, diffusive . . .;*
—*Hedione: floral odour, jasmine, light inconsistent, supple, diffusive;*
—*Rhodinol: floral odour, rosy, sharp, clear, cold, large;*
—*Phenylethyl alcohol: floral odour, rosy, dull, cold, linear . . .;*
—*Hydroxycitronellal: floral odour, cold, neutral, clear . . .;*
—*Cetonial: floral odour, petal-like, vegetal, cold, suave, compact;*
—*Citronellyl acetate: floral odour, rosy, fruity, rising;*
—*Phenylethyl acetate: floral odour, rosy, 'faded roses', thick, heavy;*
—*Cyclamen aldehyde: floral odour, petal, clear, large;*
—*Acetal CD: floral odour, vegetal, green, heavy . . .;*
—cis-*3-Hexenol: grassy odour, 'cut grass', green leaves, hard . . . .*

Dosing: The quantity of the fragrant substances used needs to be determined in such a way that the overall olfactive image takes over from each fragrant substance (synthetic perception), with the possibility for the perfume composer to isolate an olfactive element (analytical perception) and to correct, by varying proportions or the choice of another fragrant substance, the final shape of the composition he is creating.

Choice: Ideas, the technical quality, the disposition of the fragrant substances by affinity and contrast, all have a direct influence on the perception of a perfume.

If the formula of a composition is simple, the perfume itself will also be simple: if it is complex, the perfume will also be complex. Two examples will explain this fact.

Formula no. 1: This formula is simple, made up of six fragrant substances, the 'pillars' or the framework of this composition. (The real odour of lily-of-the-valley contains several hundred different components.)

Formula no. 1

Fragrant substance	*Dosage (%)*	*Technical quality (Intensity + Volume)**
Indole	1	10+
Benzyl acetate	10	3
Hedione®	20	4
Rhodinol	20	2
Phenylethyl alcohol	15	4
Hydroxycitronellal	44	1
	100	

* Technical quality: The scale (Intensity + Volume) is reduced from 1 to 10+: it is calculated on the basis of research carried out on the different levels of perception.

Formula no. 2: Formula no. 1 is now completed by 'ornamental' fragrant substances which give various tinges to olfactive perception. ('Ornamental' fragrant substances can also become 'pillars' in another olfactive composition.)

Formula no. 2

Fragrant substance	*Dosage (%)*	*Technical quality (Intensity + Volume)*
Indol	1	10+
Ylang ylang essential oil	1	4
Cetonial	1	3
Benzyl acetate	8	3
Hedione®	15	4
Rhodinol	8	2
Citronellyl acetate	2	1
Phenylethyl alcohol	12	4
Phenylethyl acetate	2	4
Hydroxycitronellal	36	1
Cyclamen aldehyde	1	3
Acetal CD	12	1
cis-3-Hexenol	1	10+
	100	

In both cases, we have a lily-of-the-valley form, similar to one another: the fragrance perception is a function of the technical quality, the action of each fragrant substance (the rhythm), olfactive subtlety and chosen contrasts.

The floral jasmine harmonies will be underlined by the contribution of indole and not by the addition of benzyl acetate, which would make it much sweeter.

The second formula is therefore less abrupt, has more outline and the 'ornamental' substances accentuate the lily-of-the-valley shape with:

—*ylang ylang essential oil for the floral tones;*
—*the cetonial and cyclamen aldehyde for the 'petal' and vegetal characteristics;*
—*citronellyl acetate for the cold, rosy touch;*
—*phenylethyl acetate for the faded rose;*
—*acetal CD for the heavy vegetal accent;*
—cis-*3-Hexenol for the green leaf undertone.*

The Need for Perfumistic Models

In spite of numerous extraction processes on lily-of-the-valley flowers, there is no such thing as an essential oil or lily-of-the-valley absolute available: only the perfume composer can recreate the odour of this flower and its fragrance for bases and perfumes.

This has been done for bases:

—*Muguet 16* by Givaudan, composed by Reboul in 1929;
—using hydroxycitronellal, a fragrant molecule which does not exist in nature, *Muguet Longchamps* by I.F.F. in 1960, created by Arif;
—*Mayciane* by De Laire in 1953, created by Robert;

and for perfumes:

—*Muguet des Bois* by Coty in 1943;
—*Muguet du Bonheur* by Caron in 1952;
—*Diorissimo,* the Dior symbol, composed by E. Roudnitska in 1956.

The sensitivity of these perfume composers has contributed a new innovation with each creation: they have set up *models,* sometimes perfume milestones, now an essential part of our olfactive culture.

The study of these models, which have today become internationally part of prestige perfumery or toiletries, is useful in the training of a perfumer: this is made easier by the use of scientific techniques (gas chromatography, mass spectrometry) and helps in the objective analysis of each fragrance—something perfume composers of the 1960s could not achieve.

The interest of these materials is not that they make it easy to copy existing fragrances but that they give a far wider scope of technical knowledge, composition techniques, structures, new fragrant compounds and workable harmonies.

Data Processing

Compared with the man who builds up his knowledge and his structures with an *object* and a *subject,* or as function of what he sees and what he is, the computer has predetermined structures and knowledge.

Nevertheless, this instrument of knowledge and information, apart from stocking technical and economical formulae, can facilitate creativity: imagine that its prodigious memory could record the verbal perception each perfume composer has of every fragrant substance, and instead of reducing information by announcing generalities such as 'floral', 'fruity', 'green', etc., that each perfume composer has already defined, the computer could provide the particulars, the olfactive singularities of each fragrant substance: the use of such a program would open vast horizons of creativity!

Imagine too a much wider freedom in composition, where each perfume composer would be free to find new harmonies, bases, and compositions for his palette and the computer would then take care of regulating the optimal production presentation of the formula obtained from this wide variety of fragrant material.

Paradoxically, the computer does not usually have the memory of how a fragrance is put together. The review of previous experiments would help retrace the progress, the stumbling blocks, special discoveries, failures and doubts which are all important parts of creation.

The elaboration of *First* by Van Cleef & Arpels required 102 tests, each using several dozen fragrant substances including certain bases which required several dozen experiments: each test is a negative or positive contribution, anxious to come alive or remain obscure within the perfume that is to come.

4 THE OBJECT

Perfume or the Art of Consensus

We all possess an imaginary olfactive museum—a group of olfactive images, often personalized, sometimes in common with others, that the memory has put aside and reconstitutes and compares, as soon as we smell a perfume.

As we all have access either consciously or unconsciously to one or several olfactive images stocked in our olfactive memories, a fragrance triggers off one or several emotional reactions . . . a consensus.

This imaginary olfactive museum is made up of our experiences: the odours of cities, of houses, of men, of women, of a whole book of memories, odours of perfumes, attractive odours, disagreeable odours . . . but also, indirectly, of information given to us orally or found in books.

Oral information is utilized when an apprentice perfumer composes a jasmine fragrance by following his professor's directions, even though he has never smelt the blossom itself.

Books can be exemplified by *Le Parfum* by P. Süsskind, 'La Chevelure' from the *Fleurs du Mal* by Charles Baudelaire, and Chapter 7 of the second part of the *Chant du Monde* by Jean Giono, the most wonderful imaginary olfactive description ever written.

Olfactive images that we recognize in a perfume are made up of personal experiences that we either reject or revel in. Obviously, our reactions are often linked to tradition, to fashion, to certain conventions or the rules of a society.

Among the animal fragrances which enter into the composition of a perfume, the Japanese do not really appreciate civet, which evokes for them decaying meat; they prefer ambergris, which is more subtle and which evokes the odour of dried fish, or *Nuoc Mam,* which they have been familiar with for centuries.

The Scandinavians have a special preference for light floral scents, which compensate for their very long winters.

The Orientals prefer powerful, spicy fragrances, like the honey often used in local cooking.

The Germans have an attraction for the citrus ingredients of eau de Cologne, and for pine, abundantly used in thermal baths.

The combination of geranium and cinnamon smells in soap is liked by the French, e.g. the perfume used in Palmolive soap that became so popular just after the Second World War.

Old Spice, the best-selling men's fragrance in the United States, smells of Coca Cola: cinnamon, orange and lemon are its main components.

Chrysanthemum is associated to funerals and death in France while in Japan it is the symbol of the nation, violet is considered cheap, musk hippy, and lemon signifies cleanliness throughout the world.

Patchouli, the smell of mouldy oak leaves and of dried boletus, a name that is often considered negative, is the major component of modern eaux de toilette. Originally, these dried leaves protected cotton fabrics printed in India—Indian cotton—from insects: they were imported into England during the middle of the 19th century. This odour has an olfactive influence even now: fabric softeners sold in Great Britain contain an important percentage of this fragrant substance.

The French olfactive memory is illustrated by textile softeners perfumed with lavender.

Classification and Language

Perfumes by nature resist general classification. Each one is trying to be unique, and each composer tries to give his work an irreplaceable character.

Nevertheless, the human spirit needs to classify in groups, families, categories, themes, analogies . . . and so on.

This corresponds to the need to make communication, an olfactive approach, and the memorization of fragrant substances and perfumes easier: so a professional language has been established, a language which often seems rather esoteric to the uninitiated.

Today, when we see how well the majority of American and European consumers know how to analyse and describe a fragrance in very simple terms, understandable by any other user, psychology becomes a promising means of research for the comprehension of the olfactive world.

Language Psychology

Linguistic and cognitive psychology has an interest in the human being because of his complex coding system and his analysis and treatment of information. Through the knowledge of language behaviour, it tries to decipher the logic which commands visual, tactile, auditive and *olfactive* information.

Thus, we are interested in the psycholinguistic aspects of

—the organization of olfactive perception among professionals and consumers;
—their verbal coding and decoding methods;
—fruitful (or highly organized) reconnaissance areas.

5 SUBJECT AND OBJECT

To bring this chapter to conclusion, I would like to present some of my own non-objective and impassioned thoughts between the perfume composer that I am and the perfume—between the subject and the object.

I like simple, uncomplicated perfumes that envelop you, light and present, provocative, subtle and generous.

I like them to draw their own olfactive image around you, in sensual voyages, to caress your skin and blend in with it: I like them to insinuate, infiltrate and leave a mark on my memory.

I dislike nondescript perfumes, closed, heady, provocative, 'noisy' perfumes: they prevent me from being able really to appreciate them, to smell the other things around me: they form a fortress without a footbridge, that is closed up from other bodies. I do not like perfumes that are too polished, that do not have enough character, do not take any risks, that give you nothing to hold on to, that have no life in them.

Of course, there are perfumes you cannot resist, but I think passion often dies fast and you tire easily of something that takes you so much by surprise, that requires effort and attention and incites a sudden search for new pleasure: clichés make life dull and you become bored very easily.

Creation is linked to the outside world: the composer, conservative or progressive, needs the contact with the outside world to create from, to correct him, to help him progress and find new ideas.

Perfume follows very much the same process: it needs new values, to follow new beliefs, to find a new guiding light.

I can see a return to a certain academic kind of approach in olfactive formulae and the use of certain fragrant substances which date back to the beginning of the century (Schiff bases, leather, animal materials) but also I see a new expression of our times by more intense aggressive, perfume odour and the non-specificity of fragrances.

The industry has taken over from Art, but we need to find a way to make fragrances rounder and more generous, to add new vigour, to find a new balance between the rational and the emotional, between order and the element of chance, between false disorder and simplicity, and between realism and fantasy. We need to introduce some oriental spirituality into our Cartesian way of thinking and to learn to combine various cultures: in a word, to find a new harmony, a new respect for the workings of the instinct.

The art of the perfume composer is his capacity to choose and compose using his memories, his passions and his beliefs; and to write the words that will then become perfume.

The secret lies in his talent, but also in the fact that talent is something that is totally impossible to explain.

Chapter 11

Support Materials for Odorant Mixtures

BENNO STRESCHNAK

Schönwasserstrasse 259, D-4150 Krefeld 1, Germany

1 TOILET WATERS AND COLOGNES

Not including the fats and oils enriched with the fragrance of flowers in antiquity, or the mixtures of resins and balsams used to burn incense thousands of years ago, the oldest fragrance carrier is ethyl alcohol. It was used as early as the 13th century to produce Hungarian toilet water, a predecessor of eau de Cologne. Today, it constitutes the basis of all toilet waters as well as many other cosmetic products. Among the most important fragrance products are extraits, eaux de parfum, eaux de toilette and eaux de Cologne, as well as the various aftershave lotions. These toilet waters generally show an alcohol concentration of 50–90% and a perfume oil content of about 30% down to 0·5%.

During the storage period, several odorants and essential oils interact with the alcohol. This can lead to the desired perfection of the fragrance thus reaching a higher degree of maturity. However, it can also lead to an undesired change in the fragrance. The influence of oxygen and light, for instance, as well as traces of heavy metal which are found in many essential oils, can intensify or accelerate such fragrance modifications in the wrong direction.

By using additives such as antioxidants and UV absorbers, the storage stability of alcoholic toilet waters can be improved considerably. The real quality of a toilet water also depends to a considerable extent on the quality of the ethyl alcohol used. Unfortunately, good-quality alcohol is not available in every country and the perfumer needs a great deal of ingenuity to develop an adequate toilet water from a bad-smelling alcohol.

Perfume oils dissolve poorly in lotions which have a low alcohol content, such as facial tonics and aftershave lotions. Particular attention must be paid to this factor and suitable emulsifiers often have to be used.

2 SKIN CLEANSERS

Soaps

The majority of soaps are produced through the saponification of neutral fats, such as beef tallow, palm oil, coconut oil and palm-kernel oil, with sodium hydroxide. Often, antioxidizing, sequestering and superfatting agents are added to the soaps to improve their durability and skin care properties. In fragranced soaps, the dosage of perfume oil generally lies between 0·5 and 2% and can even reach 5% in luxury soaps. The quality of the fats and additives used has a considerable influence on the perfuming process. However, the quality of the soap can also be impaired by certain perfume ingredients. This applies especially to white soaps which can become discoloured by a large number of odorants. Discoloration occurs least often in soaps which have a surplus of free fatty acids for improved superfatting. Soaps that do not contain animal fats, in view of religious considerations, consist of vegetable oils and fats, e.g. rice skin oils, with a large proportion of highly unsaturated fatty acids. These soaps have an intense, unpleasant scent which is hard to conceal. Fragrancing such raw materials gives a real challenge to the perfumer.

Of all the products which can be perfumed, soap holds a special position: the cost of perfuming soap in most countries can amount to 5–10% of the sale price. This price far exceeds that of other products. More than ever, the perfumer must select the odorants needed for his composition judiciously.

In addition to bar soaps based on sodium salts of fatty acids, soaps based on tensides have also been available for quite some time. Preferably, they contain fatty alcohol sulphates, *N*-methyltaurine amides of fatty acids and monoesters of sulphosuccinic acid as active cleaning substances together with builders and superfatting agents. The pH value can be adjusted over a large range. There are fewer discoloration problems than when perfuming alkaline soaps. On the other hand, more fragrance problems may arise because the wash-active substances in these soaps have a penetrating odour which must be concealed or incorporated harmoniously into the composition.

Transparent soaps are another category of special soaps. The transparency is achieved by adding glycerine, sugar or alcohol to the soap base to prevent the fatty acid salts from crystallizing. The transparency can be improved to a certain extent by avoiding apolar odorants in a perfume composition or at least by using polar odorants predominantly. Moreover, it is helpful to work with strongly smelling substances so that the perfume oil concentration can be kept as low as possible.

Deodorant soaps, cream soaps, baby soaps and skin care soaps are special forms of bar soaps which technically do not require a special additional effort from the perfumer for his task in creating appropriate compositions. However, the perfume oils should support the declared efficiency or the specific properties, respectively.

Like bar soaps, liquid soaps can be produced from neutral fats. The saponification process requires potassium instead of sodium hydroxide. Liquid soaps contain up to 90% water and many of them are produced with synthetic wash-active substances, primarily sulphates of etherified fatty alcohols.

Pasty yellow soaps and washing pastes are intensive cleaning products. Abrasive agents are occasionally added to enhance their cleaning strength. They are usually used in workshops and no special requirements exist in terms of their fragrances. Therefore, technical difficulties do not arise in the perfuming process with respect to this special application field.

Shower and Bubble-Bath Products

These products can contain a multitude of different tensides or tenside compounds as well as the following auxiliary agents: thickeners, foam stabilizers, substances which provide a nacre sheen, preservatives, superfatting substances and 0·5–4% of perfume oil. The perfumer will have no difficulties if the structure of such products is kept simple. Problems arise when 'active agents' are added (for advertisement-related purposes). They give bubble bath and shower gel products an intense odour which is difficult to conceal.

Occasionally, the pH value of these products is adjusted to approximately 5. In such a case, the fragrance can be altered after a relatively short storage period. The esters—primarily acetates—contained in the perfume compositions are saponified. A pungent vinegar odour, especially apparent when the package is opened, develops in the products during storage. As an example, linalyl acetate, extremely sensitive in this respect, must often be replaced with terpenyl acetate, a far more stable odorant.

Bath Tablets and Salts

Of little importance in Europe, these products are very popular in Japan. The perfuming process is somewhat difficult because complications can arise regarding the odorant's stability and degree of evaporation. Since its stability relies primarily on the composition of the tablets, it is recommended that the reaction of various suitable odorants with the support materials is tested and perfume compositions are created only with those which are compatible.

3 SKIN CARE PRODUCTS

Creams and Body Lotions

These products are normally emulsions, i.e. mixtures of water and oily or fatty substances which contain between 2 and 10% of a suitable emulsifier to aid and maintain the uniform dispersion of oil in water. Skin care agents can be added to the water as well as to the fatty phase.

There are emulsions in which the fatty material exists in the form of microdroplets surrounded by water. These emulsions belong to the oil-in-water (o/w) group. Most body lotions belong to this group. They generally have a higher water content than the water-in-oil (w/o) emulsions, which are produced with a different type of emulsifier and subjectively feel more greasy. Night creams and nutritive creams as well as many all-purpose creams always belong to the w/o type.

In general, emulsions available on the market contain 0·3–0·5% of perfume oil. Today, the perfuming process is relatively simple since there is an extensive supply of appropriate raw materials exhibiting no odour or only a very weak one *per se*.

From time to time however, certain difficulties arise if a skin care product line requires all its products to emit the same fragrance. A perfume incorporated into an o/w cream can diffuse a fragrance different from that in a w/o cream. This applies to the fragrance's quality as well as to its intensity, which can be perceived differently despite equal dosages of perfume oil.

Further, the perfuming process can be negatively influenced when high concentrations of active agents are used in the skin-care product.

Gels and Similar Products

Gels are transparent systems with a high viscosity. Three gel types are available on the market:

—Water-free gels;
—Oil-free gels;
—Oil-in-water gels.

The perfuming of oil-free gels may trigger technical problems that can be solved by using emulsifiers.

Another special case in this group are so-called oils. These products are monophase systems that contain only fat-soluble raw materials, i.e. mineral oils, such as paraffin, natural fats or oils and esters of higher fatty acids with different alcohols.

The perfuming process for oils produced with large quantities of natural fats and oils can be more difficult because such natural materials very often smell much stronger than purely synthetic products, which are virtually odourless.

Sunscreen Products

These skin-care products differ principally from those described above in that they contain UV absorbers in addition to the other components of the formulation. The perfuming of the products is rarely impaired by the sunscreens. However, both the perfume oil and the perfumed product must be thoroughly tested for possible phototoxic reactions.

Deodorants and Antiperspirants

Active agents, which prevent unpleasant odours originating from bacterial decomposition of sweat can be incorporated into soaps, aerosols, roll-ons and sticks.

The most important active agents in this field are germicides, bacteriostats or biocides, which ideally combat in a selective manner those bacteria responsible for the decomposition of perspiration.

At present, the perfuming process for deodorant sprays has become relatively simple since the fluorochlorinated hydrocarbons that have been responsible for the instability of several fragrances are no longer used as propellants. Today, perfumers use very tenacious perfume oils to provide a psychological support for the effectiveness of the products through a long-lasting fragrance.

Several fragrance companies supply perfume oils which, due to their composition, allegedly provide an effective deodorant performance without having to rely on germicides.

Enzyme inhibitors represent a relatively new class of deodorant agents. These do not possess antibacterial properties but they inactivate the enzymes contributing to the decomposition of apocrine sweat.

These agents can alter perfume oils, i.e. the individual odorants in the aqueous alcoholic solutions. For this reason, it is important to conduct comprehensive stability tests with the products prior to their large-scale distribution.

Deodorant sticks often contain sodium stearate, an agent used to give them a certain consistency. Therefore, odorants which are known to discolour in soaps cannot be used in white deodorant sticks.

The most frequently used active agents in antiperspirants are aluminium salts. Diluted in a water-based solution, they turn acid and destroy many odorants, especially esters. Here again, extensive storage tests are necessary to develop stable perfume compositions.

4 HAIR CARE PRODUCTS

Liquid and Dry Shampoos

In addition to tensides or tenside mixtures, shampoos contain several additives which can influence the perfuming process. These additives are expected to have a superfatting property to prevent dandruff or to strengthen the hair. The shampoos may contain an agent for natural shine, protein hydrolysate, vitamins, lecithin, herbal extracts, preservatives, alkanolamides, zinc-pyrithione, etc.

The perfume dosage for shampoos is generally around 0·5%. Special shampoos which have a strong odour due to their active agents can contain as much as 1% of perfume oil. In contrast to many other products, shampoos must meet the fragrance expectations of individual countries to be successful. This applies especially to the Japanese market. Specialized shampoos should be perfumed according to their functions. Basically, all odorants and essential oils used in bubble-bath and shower gel products are suitable for the creation of shampoo perfume oils. Caution is advised when using easily saponifiable esters. This unfortunately includes allyl amyl glycolate, a popular and—for some perfume oil types—vital ester.

Seen from an economic viewpoint, dry shampoos are not very important. They do not contain any active cleaning substances but depend on solids such as talc, starches, silica gel, etc. These solids absorb the hair fat and dirt particles and are removed by brushing. The perfuming of powder shampoos, which contain as much as 1% of perfume, is not problematic. However, highly volatile odorants should be avoided because they evaporate much faster because of the powder's large surface.

Hair Conditioners

Conditioners and strengtheners are hair treatment products which must be rinsed out after every application. Just like shampoos, they have a considerable market value.

The composition of hair treatment products differs depending on the requirements of various hair types. Products for greasy hair include a degreasing agent, those for dry hair contain 1–2% of oils to work against dryness. Conditioners are composed of 2–3% of a quaternary ammonium salt solution which prevents the generation of static charges, of some 4–6% of fatty alcohol, of minute quantities of an o/w emulsifier and of 0·3–0·5% of perfume oil. The water content is often above 80%. In general, these products do not have a strong odour *per se* and can therefore be perfumed easily.

Hair strengtheners are similar to conditioners but contain more fatty ingredients. They too can be perfumed without difficulty.

The group of hair-styling products includes cold permanent wave and hair-straightening preparations. Despite the difference in the composition of these two product groups, the difficulties in creating an adequate perfume lies in two chemical substances, thioglycolic acid and ammonia, which are used in both products.

The selection of suitable odorants, which qualify for covering of the product's unpleasant odour, can only be made after conducting a storage test. In this case, the chemical stability of the fragrance is less important than its capability to conceal the product's odour. In the light of the high water content of hair-styling preparations, the perfume oils have to be mixed with emulsifiers.

Although the hair-styling achieved by thioglycolate has to be fixed by an oxidant (usually either hydrogen peroxide or bromate), it would be an excessive precaution to test the stability of the fragrance materials against oxidative alterations. Such slight changes go unnoticed, in contrast to the unpleasant odour which is always left on the hair by traces of the thioglycolate.

Hair Sprays

Hair sprays are used to condition and hold the hair after it has been washed and dried. They are composed of film-making, softening and cationic compounds which are dissolved in aqueous alcohol. They often also contain perfume, colorants and UV absorbers. Copolymers composed of polyvinylpyrrolidone and polyvinyl acetate are used to form the film.

Today, foams which have a hair-setting effect are more popular. They resemble liquid hair sprays in both their ingredients and functions and are made up of the same raw materials.

Since the raw materials have virtually no odour, the perfuming process is relatively simple and good results are achieved with fairly low perfume concentrations—between 0·1 and 0·3%.

Whereas gels are applied to the hair prior to styling, hair sprays are used once the hair has been styled. The film-forming agents are often the same. The dosage of perfume oil is slightly higher in sprays than in gels.

Brilliantines, used earlier primarily by men, have lost considerable ground to gel preparations. Since the raw materials used in them have little or no unpleasant odours, the perfuming process is without difficulty. However, the high water content of gels, which can amount to over 90%, requires an emulsifier for the perfume. If no emulsifier is used, an additional 0·2% of perfume is needed.

Pomades are water-free hair care products based on vaseline. Their application in Africa is still widespread and they seem to be gaining ground again in Europe. The vaseline content usually lies between 70 and 80%. In addition, pomades include paraffin, fatty alcohols and emulsifiers. Since vaseline has a relatively strong odour on its own, approximately 0·5% of perfume oil is needed to make it fragrant.

Hair Lotions

Hair lotions are products based on an aqueous alcoholic solution (normally with 40–50% alcohol) and active agents. These include antidandruff agents, vegetable extracts, and a quaternary ammonium base or film-forming component. Owing to the low alcohol content, the perfume oils have to be mixed with emulsifiers. The dosage of perfume oils lies between 0·2 and 0·4%.

Hair Colorants and Hair Bleaches

Bleaches consist primarily of a 12% hydrogen peroxide solution and stabilizers. Because of the high concentration of hydrogen peroxide, it is virtually impossible to achieve a satisfactory fragrancing since almost all the odorants are destroyed. Most hair colorants require hydrogen peroxide for the dyeing process and cannot be perfumed adequately.

Depilatories

To remove hair mechanically, wax is often applied (either directly or on linen strips) to the regions which need to be depilated. The wax sets around the hair and pulls the hair out of the follicle when it is pulled off the skin. Colophonium, beeswax and vaseline preparations are often not perfumed, although the process would not pose any difficulties.

The active ingredient in chemical depilatories is calcium thioglycolate. The products are similar to the cold permanent wave solutions but have a higher alkali content. To improve their very unpleasant odour, high perfume concentrations are required. However, even with a perfume dosage of as much as 1%, the product's odour is barely neutralized.

5 COSMETICS

Facial Make-up and Powders

The purpose of facial make-up is twofold: to cover skin blemishes and to provide it with a healthy glow. The most important raw materials in this product group are pigments. These have absolutely no influence on the perfuming process. Pigments are often incorporated into creams. The following notes apply to the perfuming process of such products.

In face powders, the particles consist primarily of talc to which white pigments (such as magnesium silicate) and kaolin, zinc stearate and colour pigments (such as iron oxide and glimmer) can be added. Preservatives are a vital element of the product. The perfume dosage usually amounts to 1%. The development of perfume oils for powders demands long years of experience. Due to the large surface of the powder components, especially of the talc, the perfume is considerably exposed to the air's oxygen. This can lead to the destruction of various odorants.

Moreover, the large surface contributes to an increased volatility of the odorants. A discoloration of the powder has often been observed. This is attributable to the fragrance components.

The situation is somewhat simpler with compressed powders. In addition to the powder particles, they contain binding agents used to press the powder into its compact form. These binding agents, namely fatty acid esters, paraffin oil, lanolin alcohols and other agents (in total some 5% of the composition) considerably reduce the volatility of the perfume and contribute to its stability. Compressed powders usually contain approximately 0·5% of perfume.

Eye Make-Up

This category includes eye shadow, mascara, eyeliner and eyebrow pencils. These products are similar in structure to the products described above; therefore the same information applies. It should be pointed out that the skin around the eyes is much more sensitive than in other regions. This calls for a very careful selection of odorants and the perfume dosage should be kept at a minimum.

Lip Care Products

None of the agents found in lipsticks has a negative effect on the perfuming process. On the other hand, it is important to remember that many odorants not only emit an odour but also have a distinctive taste. Many odorants taste bitter and should not be used for this reason. In general, lipsticks contain 1% of perfume.

6 LAUNDRY PRODUCTS

Pretreatment Powders and Sprays

A wide selection of special detergent products for the pretreatment of very soiled or stained laundry is available on the market as powders and sprays or in the form of pastes.

Most of the powdered surface-active agents are oxidants and contain percarbonate or perborate. Since the amount of oxygen required is best released at temperatures between 40 and 60°C, the bleach activator TAED (tetra-acetyl ethylene diamine) can be added to the sodium perborate to release oxygen at lower temperatures.

The oxidizing stain-removing salts are very difficult to perfume because the majority of the odorants known at present are unstable in these formulae. Soil-removing salts often contain dithionites. These products are easier to perfume because there are almost no stability problems. During the washing process, sulphur dioxide (SO_2) is released. The normal perfume dosage of 0·1–0·2% is inadequate to mask the unpleasant odour.

Spray-dried powders contain between 15–40% of non-ionic surfactants, 20–70% of hydrocarbons, and (to a certain extent in Europe) 20–35% of methylene chloride, as well as 1–15% of a propellant, namely CO_2 or propane/butane. Moreover, the products can also contain colorants, up to 0·2% of perfume, and water.

Since these products do not include any aggressive substances, almost all odorants can be used for the perfuming process. This process is only impeded by the strong odour of the surface-active agents used to remove spots.

Products available in the form of pastes are applied to highly soiled areas, for example shirt collars. Then the laundry is washed without any additional pretreatment. The pastes, often filled in convenient, ready-to-use tubes, contain a mixture of anionic surfactants with non-ionic surfactants. They can be perfumed easily with those perfume oils used in soaps and washing powders.

Powdered Heavy-Duty Detergents

Washing habits vary tremendously from region to region; therefore, the composition of washing products varies from country to country and from continent to continent. One of the reasons lies in the hardness of water, which influences the washing quality considerably. In Europe, for example, it is much higher than in the USA or in Japan.

The variety of washing powder compositions and the differences in construction of washing machines in Europe, the USA and Japan are already reasons for the diverging composition of perfume types in the above-mentioned countries.

Table 1
Percentage composition of heavy-duty detergent powders

Ingredients	*Western Europe*	*Japan*	*USA*
Anionic surfactants	5–10	5–15	0–20
Non-ionic surfactants	3–6	0–2	0–17
Ion-exchangers	20–30	5–20	10–35
Bleaching agents (mostly sodium perborate)	20–25	0–5	0–5
Bleach activators (mostly TAED)	0–2	—	—
Fabric softeners	—	0–5	0–5
Enzymes (proteases and amylases)	0·3–0·8	0–0·5	0–2·5
Perfume oil	0·15–0·4	0·1–0·25	0·1–0·3

The most important differences in the composition of washing powders which have the greatest influence on the suitability (i.e. the stability) of individual odorants can be seen in Table 1.

Modern heavy-duty detergents are normally phosphate-free. Since products that include phosphates are increasingly being removed from the market, they have not been taken into consideration here.

The surfactants have an unpleasant odour which must be concealed. The ion-exchangers, most of them zeolites, call for a high perfume dosage because they lower the intensity of the fragrance. Bleaches modify or destroy many odorants, all the more so when bleach activators are present. The peracetic acid which arises in the decomposition of TAED additionally burdens the odour of the suds with its strong acidic smell. It can rarely be adequately perfumed. Enzymes, which themselves provoke a very strong odour, lead to further problems in perfuming.

The washing powder's pleasant fragrance adheres better to laundry the lower the washing temperature. The latter differs from country to country:

Japan	Up to 25°C
USA	Up to 55°C
Western Europe	Up to 95°C

Liquid Detergents

Liquid detergents are increasingly replacing dry powdered products. The US market share already lies well above 30%. The stability in these products of the odorants and essential oils is generally better than in powders. The strong odour of liquid detergents, which results from formulation aids such as xylene sulphonates, glycols and alcohols, requires a much higher perfume dosage and can amount to up to 1%.

Special detergents for coloured textiles, wools or curtains are widespread on the European market. The perfuming of these products is generally easier than of heavy-duty detergents because they contain less aggressive ingredients. The perfume dosage is between 0·1 and 0·3%.

Softeners and Similar Products

Fabric Softeners

The active agents used in fabric softeners are dialcoyl or higher dialkyl dimethylammonium chlorides. In Europe and in the United States, mainly distearyl dimethylammonium chloride in concentrations between 2·5 and 6% is used, while in Japan some 13% of dioleyl dimethylammonium chloride is combined with 7% of ethylene glycol. The products also contain preservatives, colorants, perfume and synergists. The following synergists are patented components and can be used in quantities varying between 2 and 4%: fatty acid mono- and di-glycerides, fatty alcohols and fatty acids (C_{16}/C_{18}).

Household softeners have a pH value between 3 and 5. Unfortunately these values greatly reduce the stability of numerous odorants, especially of esters. Another difficulty in perfuming softeners stems from the relatively strong odour inherent in some of the ingredients. This can be attributed to an 0·5% content of technical isopropyl alcohol in the consumer end-products.

Softeners for commercial laundries are even more difficult to perfume because their pH value can be as low as 1·5, which excludes many odorants in fragrancing such materials. Furthermore, discoloration problems often arise during the perfuming process, leading to yet another limitation in the selection of suitable odorants.

Softener concentrates are based on the same ingredients as the regular softeners. They contain approximately three times the amount of active agents and therefore have an isopropyl alcohol content of about 1·5%. This also makes the perfuming process considerably more difficult in comparison with the normal softener products.

Tumbler Sheets

Tumblers are very common in the United States and are becoming increasingly popular in Europe. Although laundry becomes relatively soft through the drying programme's constant circular motion, additional softeners are available on the market. Tumbler sheets are among the most important products in this category. These sheets are made of the most diversified materials onto which the active agent distearyl dimethylammonium methane sulphonate is added with other agents such as niotensides, glycols and glycerides. The active agent mass is heavily perfumed—up to 7%.

During the perfuming process, attention must be given to the discoloration of the usually white sheets. Such problems may be due to the perfume oils and the water-vapour volatility of most odorants.

Acidic Softeners

These aqueous products are highly colloidal and acidic. The pH value is about 1·5. The most commonly used acid is citric acid. Several products also include quaternary ammonium compounds. Owing to the high instability of most odorants in such an acidic medium, a pleasant fragrancing—although highly desirable—is almost impossible to attain.

Other Products

Ironing sprays and mangle finishers contain waxes, esters, niotensides and fatty acids or starch derivatives, silicone oils and sometimes even quaternary ammonium compounds. These products can be perfumed without difficulty.

7 HOUSEHOLD CLEANING PRODUCTS

Dishwasher Products

Liquid hand-dishwashing detergents contain tenside compounds (primarily of the non-ionic type in Japan), xylene sulphonates, preservatives and protein derivatives to protect the skin. The perfume contents vary between 0·1 and 0·5%. In addition to citrus notes, cosmetic fragrances are used to underline the product's skin-care value.

However, two facts must be observed: (1) monoterpene hydrocarbons (main components in citrus oils) can diffuse through bottles consisting of polyethylene; and (2) several protein derivatives used for skin care have a strong odour which can only be concealed with high perfume concentrations.

Until recently, products used in dishwashing machines still contained salts of dichloroisocyanuric acid for hygienic reasons and could not be perfumed due to the active chlorine emitted from them. Modern dishwasher detergents are free from chlorine and now contain metasilicates, phosphates (only to a certain extent), soda and sodium perborate. Odorant compositions which maintain their stability against alkali and perborates can be used.

Acidic rinses are used in the last step of the dishwashing programme in machines. They contain organic acids, primarily citric acid, non-ionic tensides, alcohols and fragrances. The perfuming process is difficult because of the product's low pH value.

Neutral rinses represent another solution in this field because spots on machine-washed glassware are not—as is wrongly believed by many—attributable to calcareous residues, but to salts which are left behind after the water has evaporated. The spots can be prevented by using a combination of niotenside compounds. Thus, the water runs off smoothly and completely. These products can be perfumed easily.

All-Purpose Cleaners

Cleaners recommended for porcelain, enamel and plastic are neutral or only slightly alkaline and contain, in addition to anionic and non-ionic active substances, special fat-dissolving agents, glycols and builders. The phosphates, which were previously used as builders, have been replaced with more ecological products such as salts of organic acids, mostly of citric acid and/or gluconic acid. Xylene sulphonate is often used as an additional solubilizer.

The perfume dosage amounts to 0·1–0·3%. There are no problems of stability. However, the odours of the raw materials in certain products are penetrating and difficult to conceal.

Floor cleaners are a sub-group of the all-purpose cleaners. They are based on similar ingredients. In addition, they contain waxes and acrylic acid esters (which have a fairly strong odour) to ensure a high finish. For this reason, even a 0·3% perfume dosage is often inadequate.

A last sub-group in this section comprises the window cleaners. In addition to small amounts of tenside, glycol ether and possibly some alcohol, they contain only water. They are difficult to perfume because the perfume oils cannot be dissolved with the aid of emulsifiers. In such a case, the product would leave oily streaks on the glass. Therefore, it is recommended to use odorants either which are—at least to a certain extent—water-soluble or which have a low threshold value and can therefore be perceived even when using very small amounts of them.

WC Cleaners

These highly acidic cleaners are very difficult to perfume. As powders, they contain sodium hydrogen sulphate or, more rarely, aminosulphonic acid. As liquids, they contain aqueous hydrochloric acid or up to 9% of formic acid.

Less acidic powdered cleaners include citric acid combined with tensides and carbonates. The resulting carbon dioxide bubbles enhance the cleaner's effectiveness. Liquid products also often include citric acid or, in some cases, acetic acid combined with tensides.

These products are somewhat easier to perfume than the strongly acidic cleaners. However, suitable odorants have to be carefully selected after technical studies in the application laboratories.

In addition to citric acid and tensides, fragranced rinsing products contain sodium citrate, a highly complexing element in the formula. They should prevent resoiling by hardness and iron contents in the flushing water. They contain up to 5% of perfume oil and thus have an additional function as air fresheners.

8 ROOM DEODORANTS/AIR FRESHENERS

Aerosol Sprays

Nowadays, aerosol air fresheners are much easier to perfume than earlier because fluorochlorinated hydrocarbons are no longer being used for ecological reasons. In several countries, however, the use of propane/butane mixtures as propellants may lead to a strong off-odour which is difficult to compensate by the fragrances. The perfume dosages in this product group normally lie between 0·5 and 2%.

Gels

Gel air fresheners contain a 2–6% gellifying compound made up of agar-agar or carragheenate, 1–5% of tensides, 0–10% of dissolvents, e.g. propylene glycol and 2–10% of perfume. The rest consists of preservatives, colourants and water. There are no restrictions with regard to the choice of odorants because the products do not include any aggressive agents.

Fleeces and Ceramics

These highly absorbing materials are directly steeped in a perfume oil. To diminish the evaporation of the fragrance, it is useful to add 10–50% of a niotenside acting as a fixative agent. Furthermore, it is recommended chiefly to use tenacious odorants when composing perfume oils for this type of air freshener.

Wicks

Devices in which the diffusion and dispersion of fragrances can be regulated by pulling out a wick are very popular, especially in Japan. Here, the active part consists of a 3–10% solution of perfume oil in aqueous alcohol. Depending on the alcohol content, the perfume oil needs a more or less important amount of an emulsifier (solubilizer) to achieve its perfect dissolution.

Urinal Stones

Until recently, urinal stones consisted exclusively of *p*-dichlorobenzene (PDCB) with approximately 1% of perfume oil. The product had no cleansing effect and could only be defined as some kind of air freshener. All the newer products, which had to be developed following the ban on PDCB, do not have the same intensity and radiation of *p*-dichlorobenzene. They are composed of 5–30% of polywax, 10–90% of tensides, plus fatty acid amides. There are no restrictions on the selection of suitable odorants.

9 FINAL REMARKS

For the three main categories—toiletries, detergents and room deodorants—the support materials industrially used for the production of consumer end-products have been reviewed. All the categories mentioned ask for a carefully selected fragrancing.

If we have, for instance, a trend-setter in the fine fragrance field, it is not so easy to take the same fragrance impression to a product-line built later on with the idea of using the successful smell for further commercial developments. The task is, however, challenging and rewarding for obvious reasons. The following chapter will describe the perfumery requirements necessary to achieve it.

BIBLIOGRAPHY

Jakobi, G. & Löhr, A., *Detergents and Textile Washing*. VCH Verlagsgesellschaft, Weinheim, 1987.
Umbach, W., *Kosmetik,* Georg Thieme Verlag, Stuttgart, 1988.

Chapter 12

Perfumery Applications: Functional Products

JOHN K. FUNESTI

FlavorScents, Inc., Montvale, New Jersey 07645, USA

1 INTRODUCTION

Cleanliness has been an on-going pursuit of mankind for centuries. The earliest mention of soap for cleaning purposes appeared in the second century AD. A great deal has happened since then, as the human race became quite serious about getting—and remaining—clean. This seriousness is probably the genesis of the age-old adage, 'Cleanliness is next to Godliness'.

Not until the early part of this century was there anything but soap available for cleaning one's body or the clothes on it—and it was a rather crude soap at that. As the years progressed, however, a variety of cleaning products started appearing for a yet unsophisticated and undemanding consumer.

The evolution began with a small selection of washing products for personal use, as well as clothes, in the 1920s and 1930s. These early products were unperfumed, and allowed the soap-base odor to linger on the skin and cloth. By the 1940s, simple perfumes started to make their way into functional products. These perfumes, for the most part, relied on citronella, thyme and lemongrass oils, along with isolates and chemicals such as bornyl acetate, camphor, diphenyl oxide and terpinolene.

Although it appeared that the perfumery development of functional products was destined for fast growth, World War II stunted the growth of this industry—as it did others—as it caused monumental shortages of raw materials. It was not until the 1950s, therefore, that the business got rolling again.

The industry was up and running. The 1950s brought us expanded palettes of raw materials through the production of greater numbers of synthetic products called aromatic chemicals, which later became known simply as aroma chemicals. The fragrance industry expanded to meet the growing need of cleaning-products manufacturers, who were now creating new products at a quickened pace. All the new products were aimed at the consumers, so they could keep their personal environments, their wearing apparel, and themselves, clean.

Now the fun began, because along with this unprecedented growth came a completely new segment of the industry—creative perfumery for household products. Perfumers were not only required to become more and more imaginative, they were also faced with the challenge of mastering and manipulating raw materials through hostile base environments.

2 FRAGRANCING OF FUNCTIONAL PRODUCTS

By the mid-1970s, the fragrancing of functional products became a very important part of the industry, contributing significantly to the growth and profits of fragrance suppliers, ultimately losing their 'stepchild' image to the fine fragrance creators.

During that same decade we witnessed the larger fragrance supply houses establish creative departments of perfumers whose sole responsibility was to work in the functional-products area. Suppliers with research capabilities to develop new aroma chemicals were feeding the creative efforts of the perfumery. Voluminous and tedious stability studies of fragrance raw materials produced results which often gave direction to the research laboratory efforts to find new aroma chemicals which answered specific stability and creative needs.

At the close of the 1970s, the work emanating from the research facilities of household-products manufacturers produced aggressive bases which were fraught with problems that the household-products perfumer had to face—problems which carried into the next decade. Examples included solubilities, active integrity, oxygen and chlorine bleaches, encapsulated chlorine, fabric conditioners, enzyme additives, and so on.

As if this were not enough, the perfumer had then to consider linear fragrance diffusion and longevity for perfuming the environmental air fresheners with emphasis on continuous action. These various mechanisms created problems related more to the physical aspects of the chemistry of fragrance materials than to the interaction of a product's base nature with the fragrance itself.

The household-products perfumers needed to address all of these various criteria for their fragrance creations—often learning something from one product base and applying it to another. At all times the tenets adhered to with each project were raw material stability, availability and cost, and the marriage of these three to produce a fragrance which would aesthetically send the appropriate signal about the product to the consumer.

This chapter will discuss the specific problems encountered by the perfumer when presented with the challenge of fragrancing functional products—a category which can be further subdivided into cosmetic and household products.

Not all great fragrance ideas for functional products are born independently in the mind of the perfumer. Rather, they are inspired by what has gone before, or what is currently in the marketplace in an upscale way. Fine fragrances have in recent years been a source of ideas which inspire the creation of a functional-product perfume. This process is known as the 'trickle down' effect. Although a tried and true method of developing new functional-products perfumes, it has several obstacles that must be overcome before a successful transfer of a fine fragrance theme to a functional product can be accomplished.

Successful fine fragrances can create a polarizing effect among consumers—causing a love/hate feeling toward the fragrance. Functional products, on the other hand, particularly the household ones, cannot afford this love/hate reaction from consumers, because they must be acceptable to a broad cross-section within the consumer market. Therefore, the unique combinations which give birth to the couturier fragrance theme have to be modified, tamed and massaged by the functional perfumer to broaden the fragrance profile for greater consumer acceptance. Of course, there are always exceptions to this successful translation of polarizing fragrances to those which 'trickle down' almost exactly intact, giving consideration to cost, stability and availability. Examples of both occurrences are:

—*Bandit*, Piquet, launched in 1944. The accord was interesting, but too polarizing for acceptance in functional products.

—Grès's, *Cabochard*, introduced in 1959. Inspired by Bandit, it became a theme which later was used widely in shampoos, beauty soaps and deodorants.

—*Miss Dior*, which came along in 1947. It gave inspiration to *Intimate,* a more popular version of the 1947 creation that found its way into many cosmetics and toiletry products.

—*Chanel No. 5*, launched in the early 1920s. This set the stage for the new aldehyde floral perfumes and has trickled down into almost every known cosmetics product and a few household ones too. The accord never had to be modified to make it more acceptable to the consumer, but it did need modification to reduce costs, assure availability of the perfume materials and create stability in the various numbers of products. Sixty years after Chanel's introduction, this fragrance is still the most widely used couturier fragrance with extremely broad consumer appeal—casting almost a positive 'halo effect' over the product's acceptability. Today's suppliers know that when all else fails a *Chanel No. 5* variation will usually save the day, even in this day of *avante garde* perfuming. *L'Air du Temps* is another fragrance which has the same broad acceptance as *Chanel No. 5.*

A more recent introduction by Dior—*Poison*—has already made its way into a room air freshener, and prognosticators see this as on the way for fabric softeners.

Line extensions of couturier fragrances into functional cosmetics is the first, and the most, obvious example of the trickle-down effect, although not representative in the truest sense of this phenomenon. Today, it is expected that the lead fragrance note in the perfume or cologne will be carried throughout all the products in the line of the same name. Trickle-down in the present sense is from couturier perfume to hair spray, deodorants, bar soaps, detergents, fabric softeners, etc. In attempting to adapt a fragrance theme for functional products, perfumers must consider the following factors.

(a) Acceptability of couturier theme in product category.
(b) Technical considerations of matching diffusion, stability of both odor and color is of paramount importance. Substitution of raw materials is almost always necessary.
(c) Costs and availability of aroma chemicals and natural materials.
(d) Need for fragrance materials to comply with the IFRA and RIFM safety guidelines.

Aroma chemicals create the palette for today's modern perfumer that make it possible to translate couturier perfumes into other products. Cologne (notes) have found stability and musk notes do not discolor; these are just two examples of how aroma chemical developments in the last 30 years have become the key to the expanding world of fragranced products.

Hair products such as shampoos and hair sprays have provided fertile areas for the trickle-down effect. The desired residual character of the couturier fragrances is easily translated to these products. Popular types in shampoos are derived from fragrances like *Chloé*, *Chanel No. 5*, *Charlie* and *Lauren*. Major hair spray products utilize the fragrance types of *Anais Anais*, *L'Air du Temps* and *Chanel No. 5*.

Several of the top beauty complexion soap bars in the world have been inspired by *Arpège*, *Chanel No. 5*, *Chamade*, and *Cabochard.*

Detergent products and cleansing products in general place too many constraints on the perfumer trying to translate a couturier fragrance for the end-use. Usually the base odor, indicative of the active ingredient, and pH considerations dictate the direction of fragrances in the products. This, however, does not preclude the perfumer from being inspired by a couturier theme and using it as a starting point, modifying the theme until all constraints are dealt with successfully—usually producing a fragrance which bears slight resemblance to the original idea.

However, this does not dissuade the perfumer from saying the creation is of a certain couturier type taking into consideration all the aforementioned reasons to make changes in the odor profile.

Defying the natural order of things, there are also two variations of the 'trickle down' theory: those that 'trickle up' and 'trickle across'. An example of the 'up' phenomenon is the *Youth Dew* bath oil by Estée Lauder, which became the inspiration for the perfume oil. Another example of bath oil-to-perfume is *Dioressence* by Dior. Though these examples are rare, they do happen, giving credence to the 'trickle up' state of creativity and inspiration.

'Trickle across' has one famous example—the Johnson & Johnson Baby Powder fragrance, a powdery rose accord—which far overshadows all others and probably has become the most overdone fragrance type of the past three decades. This fragrance implies—on the very first whiff—softness and babies. All perfumers, even those with moderate experience, know how difficult, if not impossible, it is to change the mind-set of marketers and product formulation chemists away from the J&J type when discussing fragrancing baby products, bathroom deodorants or any other product which might demand a powdery fragrance type. The marketing request is usually for something new and different, but which smells like 'J&J Baby Powder'. There is a fear of limited product acceptance by the consumer if it does not have familiar notes which can always be related back to the Baby Powder note.

The popularity of J&J Baby Powder fragrance, making it one of the all-time classic blends, has been said to be responsible for *Ombre Rose*, *Baby Soft* and *Sweet Honesty*, all perfume/cologne fragrances which are examples of the 'trickle up' theory.

Fruity notes did not always enjoy the prestige of being the lead notes of many couturier fragrances as found in today's markets. Rather, these notes were complete fragrances themselves, used particularly in the USA as strong masking agents and most effective in public lavatories. These fragrances, which bordered on flavors with their intensity, were discharged into the atmosphere through mechanical means or continuous-action para blocks.

It was only in the late 1960s and early 1970s that fruity fragrances began to gain respectability in their own right by providing a whole new range of odor concepts for functional products. Shampoos by the dozens had myriad fruity notes espousing their different claims on the hair.

This fruity era 'trickled across' to many different product areas. The natural fragrances picked up along the same lines, giving way to vegetable and other food-like fragrances.

The fruity notes have 'trickled up' to modify and lead the couturier palettes of aldehydic florals, chypre, spicy, woody, etc.

The fruity notes have also 'trickled down' to detergents and fabric softeners—leaving their imprint of residual quality along with richness and masking power.

Today, the 'trickle down' phenomenon is occurring more quickly than it did in the past. The sophisticated demands of the customer for more upscale fragrances in the functional-products area, along with the availability of a greater range of aroma chemicals yielding increased stability and creativity in more product bases, have made this possible. As mentioned, *Poison* has already been translated to a room freshener.

Today's functional-products perfumer has much to be inspired by to enhance his creativity, but he needs the knowledge and experience to make successful fragrance translations. The remainder of this chapter will discuss some of the considerations in perfuming functional products.

3 LINE EXTENSIONS

The first group of functional products to be considered are line extensions to perfume/cologne. While these do present problems, they are not as severe or seemingly insurmountable as those in the household-products area.

Cost, stability and color are the prime considerations when making line extensions. Costs of extensions are proportionately less than the cologne oil. Some guidelines are as follows: the bath fragrance might be 60–75% of the cost of the cologne, the talc, hair products, baby lotions and deodorants at 50% of the cologne, with the soap interpretations coming in the 30–40% of the original cost.

Initially, as a fact-finding experiment, the cologne oil should be incorporated into all bases at appropriate levels while the perfumer, using his experience, examines the formula for obvious necessary modifications for cost and compatibility. The first exercise involving the cologne oil will help to narrow down the potential problems such as coloration and discoloration. Frequently, one lower-cost version may work in two or three line extensions. Therefore, fragrances developed from modified cologne oils for line extensions are usually more intense and used at higher levels than those in non-line-extension consumer products. Fragrances which are lower-cost versions of a cologne oil and which work in more than one line extension may tolerate the use of solvents to adjust costs where necessary.

Solvents

This is a good opportunity to say a word or two about solvents. It is no revelation that solvents are used extensively in perfuming. For example, dipropylene glycol (DPG) and to a lesser extent diethyl phthalate (DEP) are used to adjust costs downward where necessary when the potential business does not warrant significant investment of perfumer time. However, there are as many areas where solvents have secondary and equally important roles besides cost reduction, including positive effects on diffusion and blending and solubilizing certain materials. Triethyl citrate, hercolyn D, and deodorized kerosene are solvents which substantively contribute some of these effects.

Solvents provide the taming of extremely powerful materials which are too difficult to work with in concentrated form. Solvents such as DPG, phenylethyl alcohol or benzyl alcohol blend and mellow particularly strident materials such as the aliphatic aldehydes while forming more stable versions of these chemicals in the hemi-acetal state.

Solvents also contribute to the accuracy of addition of extremely powerful materials by controlling their influence on the compound by gradual additions. Their value is also very basic—that is, they are also used to dissolve solids.

Over the years, solvents have gained a negative connotation from purchasers of fragrance compounds because it is felt that they are being deprived of their full value by the presence of solvents in a compound. Whether there is 10% or 90% of solvents in the compound the fragrance should be evaluated on the performance, compatibility and price. So, if it is necessary to include solvents to achieve the price or ease of use, so be it. This could be explained to the customer. But by no means should the supplier apologize for including solvents.

Bath Oils

Usually the line extension for the cologne starts with the dispensable bath oil. This modification must retain as much of the cologne character as possible because of its high concentration (3–5%) and the fact that the fragrance envelopes the user in the bath. The consumer will be quicker to pick out differences in this form than all others. The bath environment, coupled with hot water to aid diffusion, constitutes the need for a critical interpretation of the cologne oil.

Some major considerations in formulating for this medium concern solubility as many bath oil products are formulated with mineral oil, which is well known for its perfume insolubility. Resins and crystalline materials present the most solubility problems. The resins can be replaced by their soluble or molecular distilled versions produced by a number of suppliers. The floral absolutes also have similar versions for inclusion in these modifications.

Crystalline products usually contribute the sweetness and musky notes. These can be reduced in concentration and replaced with some of the newer liquid aroma chemicals espousing these notes.

The products which normally cause discoloration in any medium are suspect here too. Phenol derivatives, vanillin, maltols, and anthranilates all need to be substituted to reduce the risk of discoloration.

The aldehydic complexes in all probability should be rebalanced upward in concentration to compensate for the reduction of strength and body necessitated by the substitution of the more expensive materials. The increased aldehydic portion will enhance the diffusion too.

The entire perfume compound must be incorporated into the mineral oil, usually with the aid of surfactants. The usual ratio is 3:1, surfactant to perfume, incorporated at the desired level. Stability studies should be conducted not only for color and odor stability, but for solubility as well. If there are indications of oil separating, then you know it has to go back to the laboratory again.

The dry bath salt has a few different problems, but solubility and coloration are not among them. The bath salt presents an oxidation problem because of the massively increased surface area of the medium. In these variations, it is prudent to use crystalline materials wherever possible to increase the fixative effect of the perfume. An antioxidant could be included to reduce the oxidation of easily oxidizable materials.

The creation of the bath salt fragrance utilizes much of the same technology and know-how used in developing fragrances for detergent powders, which will be discussed later in this chapter.

Cosmetics and Lotions

Lotions in a fragrance line are usually the next most popular item used by the consumer. For the most part, lotions are primarily body lotions, although they are also used for the face and hands. Replicating the fragrance for this line extension need not be as critical as the bath oil, because the product is primarily a functional one used to soothe or soften. The bath oil extension product, on the other hand, has the requirements of setting a mood, relaxing the user and capturing the bloom of the cologne, transferring the user into a state of well-being.

Body lotions, from a technical perfumery standpoint, have their own set of requirements. Naturally, those materials judged to be photosensitizers, sensitizers or toxic in other ways may not be used in this product which ultimately comes in contact with the skin, as do many household products and detergents.

The RIFM/IFRA safety dicta on offending fragrance materials should be observed in all the above-mentioned phases of perfumery and need not be mentioned again in this chapter, as they apply to all discussions.

In modifying the cologne fragrance for body lotion, one can usually use the bath oil rendition as a starting point for many of the problems have been dealt with satisfactorily except cost and masking of base odors. Here is a situation where a diluent may come into play. Masking of the usual fats, vegetable extracts and other cosmetic additives is achieved by the high concentration of perfume (1–2%), a level demanded by the fact that this is a line extension. Adjusting the fixative portions of the fragrance will provide more residual on skin. Odor stability is usually not difficult to achieve because the products are emulsions which restrict and protect the interaction of materials and reduce the oxidation potential. Discoloration and emulsion cracking are the big problems of fragrancing lotions and creams. Usually the same guidelines known to all perfumers must be applied as mentioned earlier regarding offending materials, which can be used at safe levels specified, or replaced entirely.

The fragrance of lotions which are individual products and not part of a line is usually more difficult because the product costs dictate lower levels of perfume and the active agents incorporated into a moisturizing base provide greater challenges for masking base odors while not being overly perfumed and expensive.

These fragrances require greater strength and residual notes to mask and be pleasant on the skin, while satisfying the discoloration and stability guidelines of upscale line extension fragrances. Many of these products are tinted, thus resolving some color problems, but the concentrations of 0·25–0·75% pose different problems. The top notes of lotion fragrances need to be emphasized in the oil state before incorporation in order to overcome the depressing effect of the emulsion. The lotion base, as with most product media (with the exception of alcohol), tends to mellow and suppress the effect of the perfume.

Perfuming of cosmetic lotions and creams for face, hands and body is an ever-increasing and challenging art for the perfumer. Learning the basics of identifying which raw materials perform well without the side effects of discoloration, phase separation or untoward residual notes on skin will certainly simplify the project for the perfumer.

Talc

The consumer prefers talc products to be white, whether the product is a line extension or a talc like J&J Baby Powder. The talc base is chemically inert and the only way it significantly affects the perfume is through adsorption. Adsorption, for our purposes, means the substrate is only on the surface of the medium and not absorbed into the medium itself. This phenomenon creates stability problems for the perfumer, particularly with citrus oils because they oxidize rather quickly. The second problem is discoloration caused by high levels of perfumes (1–2%) in line-extension fragrances. Much modification is needed to solve this problem by using decolorized natural oils or molecular distilled products. Discoloration caused by the usual offending materials is more critical in talc than in other line extensions. The quality of talc and other additives can create masking problems, e.g. poor-quality talcs have a greater preponderance of metal oxides which readily complex with some fragrance materials causing discoloration.

Adding an antioxidant such as BHT (butylated hydroxytoluene) or vitamin E helps to reduce the oxidation potential of citrus oils; a chelating agent such as citric acid reduces the discoloration potential of perfume oil.

Talcs developed for the mass market require less perfume (0·2–1·0%), thus reducing and even eliminating many of the problems.

Perfuming of talcs has always created much discussion on various theories of fixation of perfumes. Many have approached it by advancing complicated theories involving high molecular weight, low volatility, solvent action of one material on another, or differences in dielectric constant regarding fixation of more volatile materials; or the outright rejection of the existence of fixatives.

What does work is the careful study of blended and single perfume materials in a talc base under accelerated aging conditions; eliminating those which do not perform with good odor stability and longevity from those that do, to produce the many fine talc fragrances on the market today.

Aerosol Products (Pressurized and Pump Spray)

The perfumer can divide the fragrances of aerosol products into two basic areas:

(1) the aerosol line extension of a fine fragrance;
(2) the aerosol product, which is an independent stock-keeping unit (SKU) by itself.

Fragrancing for aerosol line extensions requires much rebalancing of the perfume formula in order to produce the fragrance fidelity of the initial cologne product. Usually this means readjusting the top note of the original formula to minimize the fugitive aldehydic, citrus and other notes which appear first in the normal order of smelling a fragrance, so that when the fragrance is sprayed from an aerosol, and the perfume is instantly available to smell in its entirety, these top notes do not overpower and distort the fragrance profile. The fragrance delivered to the user by means of an aerosol should resemble the original fragrance.

For aerosol adjusting, the perfumer may need to make several variations of the original creation—rebalancing not just the top notes, but the entire fragrance—to reproduce the profile of the fragrance that the consumer can associate with the original product.

Developing fragrances for products which are individual in concept and promotion, such as room sprays, deodorants, hair sprays and furniture polish, the need to replicate another form of the product does not exist and 'perfumer's license' in creativity takes over.

Certain considerations in approaching fragrance creation for aerosol/pump products become self-evident after the first few attempts when heady top notes fail to produce the desired effect of a well-rounded fragrance which is pleasing to all. However, there is much more latitude available to the creator in designing fragrances for these independent products.

This latitude of creativity does not preclude some serious considerations of stability, solubility and color. Initial testing of fragrances with the active agent and propellant are best carried out in glass containers so the problems, if any, are visible as well as olfactive. However, if the commercial container is to be metal, then metal vessels must be used in testing to determine the compatibility with the interior of the can. Ideally, cans should be emptied and cut open after accelerated aging to determine whether any corrosion problems have occurred. Fragrances dependent upon natural oils contain traces of moisture which may contribute to the corrosion. These fragrances should be modified, with synthetics replacing the naturals to eliminate the problem.

Solubility considerations of perfumes in aerosols involve the utilization of alcohols, such as ethanol, isopropanol and dipropylene glycol, to increase the interaction and mixing of the propellant and the perfume. Certain perfume materials, such as those from natural animal origin, resins, and crystalline synthetics with limited solubility, will precipitate after a few days in solution. The resultant precipitates can inactivate the aerosol by obstructing the valve. Solubilized versions of these materials can usually be found and substituted to solve the problem. Solubilized versions can mean products which are molecular-distilled or solvent-extracted. It may be necessary to create a replacement by using several materials more compatible with the system to contribute the necessary olfactive effect. It is essential to observe the critical solubility of aerosols so as not to render the product inoperable after a few months on the shelf. The best insurance against this happening is to subject the product to a freeze/thaw cycle testing for a minimum of 30 days to see if any precipitation is produced.

Discoloration in aerosols is caused by the same group of materials which cause problems in non-aerosol products. It is intensified if the final package is to be clear glass which allows the ultraviolet rays to act on such notable materials as indole, vanillin and phenols like eugenol.

Water-based, hydrocarbon-filled room sprays, making up a significant portion of the fragrance market, do not present any difficult stability problems. However, another equally important commercial product—namely antiperspirants—does cause considerable problems, especially discoloration. This is the result of the reaction between the perfume and the active ingredients. Aside from some of the more aggressive bases now used in the laundry market, antiperspirants require the greatest skill and knowledge of raw material inter-reactivity. A knowledge of those materials which are known to create color problems, along with individual testing of raw materials incorporated in the base, can readily solve the problem.

Fragrance degradation caused by reactions of perfume materials with the active ingredient of the product base is, in general, the greatest concern for the perfumer in creating fragrances for aerosols. Antiperspirants are acidic in nature owing to the presence of the complexed salts of aluminum and zinc, which of course are the active ingredients used to inhibit perspiration. These salts make the pH of the antiperspirant solution around 4, thereby rendering many ingredients insoluble and unstable. Natural oils cause many problems because of their complexity of chemical groups; it is, therefore, difficult to predict their behavior. Because odor degradation also produces significant discoloration reactions, it is important to test many of the raw materials individually in the base under pressurized conditions to know for sure which ones are acceptable. Antiperspirant bases are particularly corrosive to can liners, creating additional problems of free iron to enhance the possibility of discoloration.

Basic axioms come into play when selecting materials for stability in an antiperspirant spray, cream or roll-on. Saturated primary and secondary alcohols provide the best stability. Phenols, for the most part, should not be considered. Ketones, ethers and esters are reasonably stable. Many saturated aliphatic aldehydes exhibit moderate stability. Unsaturated aldehydes can be substituted by their corresponding nitriles. Natural oils must be tested individually, noting the supplier's method of distillation and the preservatives used, recognizing that natural products of the same origin can differ in treatment from supplier to supplier.

Other considerations when formulating for antiperspirants include the origin of the valve gasket. If it is synthetic rubber, it could produce disagreeable odors when it interfaces with the fragrance. At the very least, extensive raw material testing is required, or possibly an alternative gasket material should be used. The presence of water in the system—possibly from alcohol—can cause hydrolysis of certain ingredients as well as corrosion of the metal container. Anhydrous alcohol will minimize this problem.

Generally, aerosols are filled under a vacuum system which eliminates air from the container and precludes oxidation of perfuming materials. Aerosols not filled in this manner will present unwanted problems. Occasionally, perfumes incorporated in antiperspirant bases will cause the formation of a gel or thickening of the base solvents, rendering the product inoperable. Normally, this is attributable to the influence of solvents, and studies must be set up to evaluate the perfume in different solvents incorporated into the base.

While addressing the technical considerations of perfuming antiperspirants, the perfumer must also consider the aesthetics of the fragrance itself. As antiperspirants, and deodorants as well, are primarily directed at masking or inhibiting human body odor, the fragrance itself must be devoid of any materials that would enhance the decomposed odors of sweat. Raw materials of animal-like and fecal notes which are desirable in fine cologne are not recommended for this end-use. Fragrance notes that harmonize with the fatty aldehydic and fatty-acid-like body notes are usually the most effective in masking. Some of these are fruity, citrus blends supported by background powdery notes. Lavender notes also do well but must be screened thoroughly for untoward reactions with the base.

The preceding paragraphs are to be used as a guide, and not a restraint, in creating aerosol antiperspirants, or aerosols in general. Sometimes innovative and daring perfumes create successful fragrances by moving boldly into the area heretofore considered technically impossible, only to have combinations of unknown reactions take place which overcome all the rules. This happens sporadically in all areas of perfumery, and is discussed in detail in the next section.

4 PERFUMERY FOR HOUSEHOLD PRODUCTS

Each specific product used in the maintenance of a clean personal environment comes with its own particular criteria which need to be addressed regarding fragrance development. These fragrances must convey an unseen dimension of beauty in a bar soap, efficient cleaning power in a detergent or softness in a fabric softener, yet remain true to their original odor profile during months on the shelf. The fragrances for these products, and the many others in the household category, need to send not only an aesthetic message about the product but a functional message as well. It is important to note that, in order to achieve the dual functionality of the fragrance, the perfumer must be technically creative, utilizing all of the information available while marrying the fragrance with the untoward nuances contributed by the base.

The detergent market, surely one of the biggest, most challenging and competitive segments today, yields great financial rewards to those suppliers who successfully compete for the business. It is also a great challenge to the perfumer.

The detergent product can come in liquids, granules, or agglomerated powders, and its uses are myriad—for cleaning laundry, dishes, walls, floors, etc. The powdered and liquid forms are produced in both phosphate and non-phosphate forms. Fragrances developed for the phosphate base do not necessarily perform in the same manner in the non-phosphate base. The latter, for example, usually absorbs and suppresses many perfume ingredients, therefore changing the entire fragrance profile. The fragrance, then, must be specially formulated, taking particular care to reinforce the top note so as to duplicate the same effect in the phosphate base.

Heavy-Duty Detergents

There also exist certain fragrance translations for lines of laundry detergent products that are marketed as both the liquid and granulated heavy-duty detergents. It is rare that the same fragrance can be used in both without modifying for top notes, base coverage or residual qualities. Perfumers working on such projects quickly become aware that there is more than one way to replicate an odor profile.

Originally, when perfuming detergents, the creator had only to consider the stability of raw materials in an alkaline pH of 10 or more. It was simple; it usually meant that only tertiary esters would withstand the rigors of this pH, ruling out the more popular acetates and salicylates. The perfumer would rely more heavily on the rose alcohols, nitro-musks and macrocyclic musks, coupled with some aldehydes and natural oils. What seemed difficult 25 years ago pales in the face of today's complicated formulations, when one is confronted with many new additives such as polymers, bleaches (both oxygen and chlorinated), enzymes, sulfonic acid actives and non-ionic actives—to name but a few. Many ingredients today are of an organic nature which increases the strength of the base odor and, in turn, the problems of masking or incorporating notes into the perfume profile. There is also more demand today to create fragrances which are residual on cloth, deodorize heavily soiled clothes, or convey the softness of a fabric softener.

The challenges of detergent perfuming are great, but the rewards of a successful marriage of perfume products are even greater. Developing a new fragrance for today's high-technology detergent begins in the usual way of ruling out those materials which are readily oxidized, easily hydrolyzed at high pH or capable of reducing the activity of active ingredients by inter-reacting chemically through unsaturated bonds. The creative procedure, however, evolves by the testing and evaluation of single materials in the base. Redundant as it may seem, there is just no substitute for raw material testing in any kind of creative perfuming involving new base technology or new aroma chemicals. The new technologies incorporated into detergent bases usually invalidate the results of the more traditional aroma chemicals or natural oils. Testing of raw materials in various bases enables perfumers to establish databases of information for future projects. It is often possible to extrapolate stability data from one medium to another to help shorten development time.

The testing regimen usually involves samples of raw materials incorporated at levels similar to those of the finished fragrance and subjected to accelerated aging in ovens (35–50°C) for periods of one, two or four weeks. These are evaluated for odor variation at each interval. Clear jars are the best to use in most cases so that color changes can also be observed.

The finished fragrance should then be tested in both glass and the packaging material which best simulates the final package. The glass sample, then, becomes a control against which others can be measured olfactorily. In the case of granulated detergents it is important to see the effects of the fragrance and package on each other. It is important to use laminate or wax moisture barriers in an 'as marketed' container to ascertain the stability under various conditions of high heat and humidity.

While these aging studies are being conducted it is also important—particularly in the case of laundry detergents—to carry out actual 'in use' tests of the fragranced product in washing machines with various load cloths assembled to represent a regular household laundry load. The 'in use' test provides information on the 'bloom' of the fragrance in hot water as well as the residual left after the spin-dry cycle. Basically, it is desirable to reproduce the homemaker's laundry task precisely to simulate the consumer's conditions.

When developing fragrances designed to have a deodorizing effect on soiled clothes, it is necessary to use soiled clothes in the evaluation process. The perfumer's olfactory evaluation after the spin-dry cycle will be able to test the effectiveness of the fragrance on deodorizing soiled fabrics.

Before leaving the subject of detergents, I would like to discuss briefly perfume levels and trends. In the 1960s and 1970s, the concentration of perfume for granulated detergents and the one or two liquid detergents on the market did not exceed 0·1%, with cost being in the very low range. Through the 1980s, the trend was to use more expensive residual fragrances at higher levels, with some on the market today at 0·35% for granular products and 0·4% for liquid detergents with softener added (softergents). These usually cost about twice as much as the aforementioned perfumed compositions. The quality and strength of these perfumes has certainly improved during this period too. The competition is strong and the consumer more demanding for superior products.

Trends are made by market success. This is true for practically all fragrance products. In laundry detergents, both solid and liquid, the trend is dependent upon what covers and is stable in the base. For example, when a major soaper develops a new detergent base with upscale technology, the stability dictates what the final fragrance profile is going to be. Rarely can a perfumer in the laundry detergent area consider making a certain fragrance type without taking into consideration the stability of key materials in the base. Once the fragrance for a new detergent technology is developed, proved stable and accepted in consumer testing, the market-place determines if it will be a trend-setter or not. Assuming the new product is immensely successful, its influence on trends usually applies to the smaller manufacturer trying to piggy-back on the success of the market leader with a 'me too' perfume, even if the product technology is totally foreign.

Major manufacturers monitor their competitors' new product market progress and may be influenced to extrapolate the effect of fragrance—such as intensity, burst of fragrance in water, residual strength—on that product's success and incorporate these ancillary characteristics into their own product without even coming close to the market leader's profile. Thus, trends such as those seen in the couturier market are not found among the major players in the laundry detergent field.

Detergent fragrance evolution is dependent upon new technologies in the laundry detergent field that require more stable aroma chemicals from suppliers who do basic research in synthesizing and testing of new aromatic molecules. The successful history of functional products perfumery was made possible by the breakthroughs in aroma chemical research, giving the customer a greater variety of stable fragrances at a reasonable price.

The proliferation of lemon-type fragrances in every type of functional product, especially those with aggressive bases, is a good example of what aroma chemical research can do today, compared with what was almost impossible to achieve 25 years ago. Perfumers no longer have to depend on the saturated alcohols, terpenes and derivatives of lemon and other citrus oils to recreate the fresh lemon scent. They now have a choice of synthetically produced chemicals with greater resistance to the destructive reactivity on the molecule by active ingredients such as chlorine (hypochlorite), oxygen bleaches (perborate), chlorine salts, enzymes, zeolites and peracids. The fruit of this technology has given the perfumer greater tools. It also challenges the perfumer to be more creative in producing fragrances more pleasing in use to the consumer with improved stability for longer shelf-life.

It is interesting to note that from a base of more than 2000 raw materials available to perfumers today, approximately 60% are suitable for perfuming detergents without aggressive active ingredients, and only 10% or less are suitable for perfuming those bases with the aforementioned aggressive active ingredients.

Creative perfuming for any class of product always involves the consideration of cost and availability.

Fragrances for the leading laundry detergents produced around the world involve enormous volumes of fragrance materials. It is always prudent for the creator to know the projected sales volume of the proposed laundry product and to integrate that with the worldwide availability of the natural oils as well as the projected production of synthetic raw materials for the product. The volume of raw materials, particularly natural products, commonly used in laundry products frequently imposes restrictions on the creative perfumer. These restrictions also give rise to the reconstitution of some natural oils through aroma chemical research and perfumery know-how to capture synthetically what is not available in sufficient quantities from mother nature—supporting the old cliché that 'every problem creates an opportunity'.

These limiting factors in detergent perfumery also tend to confirm why trends do not necessarily exist in detergent fragrances as they do in fine fragrances. The market of raw materials can be cornered easily by large manufacturers, leaving key materials in short supply for competitive companies.

Fabric Softeners

Fabric softeners, the second largest and most competitive market after laundry detergents, provide one of the greatest areas of creativity for perfumers. These fragrances must not only meet stability criteria for the product, but must also convey aesthetic attributes beyond the normal detergent fragrance.

The fabric softener, existing in both liquid and sheet (non-woven) form, is the finishing touch on the laundry. It must continue to project the cleanliness of the detergent perfume while imparting its own impact on clothes. Achieving this requires great skill on the part of the perfumer, combined with a database of materials which demonstrate residual odor qualities on various kinds of fabrics. The base odor of the softener product, usually composed of quaternary cationic actives, creates its own masking problems as well.

Fabric softeners generally use high levels of perfume—from 0·25 to 0·5% and maybe higher—costing significantly more than those found in detergent powders.

Recent successful developments of fragrances for leading fabric softener products in the USA attest to the perfumers' skill in creating fragrances which make the product and the clothes smell good, but also signal an extra dimension to the consumer that the clothes are really soft. The perfume is then a major contributor to the perceived physical effect of softness. The accomplishment of this phenomenon comes from the skillful blending of perfume materials which were previously painstakingly tested one by one for stability, base coverage and substantive qualities on various fabrics usually found in loads of laundry.

Interestingly enough, if one were to examine closely the odor characteristics of the successful fabric softener perfumes, there would be an underlying theme of fruity notes, skillfully blended, which connote softness with cosmetic overtones. Also, the use of fruity notes such as peach, strawberry or apricot (to name a few) could be regarded as a trend in the laundry product area of perfumery. These notes convey softness when properly blended with other notes, as well as providing the deodorizing effect found in certain detergent products.

Light-Duty Liquid Detergents (LDL)

Light-duty liquid detergents have long been an important segment of the household cleaning products area. Fragrances for these products throughout their existence since the mid-1950s have been a variety of light florals and citrus blends. Today, as the decade of the 1990s begins, much of that remains the same, with the forerunning fragrance being lemon. Light-duty liquid detergent bases themselves do not challenge the perfumer on stability as much as they do on base odor masking without leaving a residual fragrance on the dishes. The mandate of these fragrances has always been to be fresh, clean and light, and non-conflicting with the odors connected with the cleaning chore. That is why lemon became so prevalent in the masking of these products. Evaluation of the myriad of lemon-fragranced LDLs on the supermarket shelf today will quickly show that there are many variations of lemon fragrances. Some of this variety is dictated by the creativity of the perfumer, while most of it involves the integration of the product base odor into the fragrance profile resulting in a range of lemons all doing their job of masking while projecting 'fresh' and 'clean'.

The perfumer's task of masking the active ingredients, which are responsible for cutting grease and cleaning the dishes, is best carried out by a series of tests of single materials and blends for masking with the absence of residual qualities. The pH values of LDLs are usually between 7 and 10 and do not pose any great stability problems other than discoloration. Levels of fragrance range from 0·08 to 0·2%, with costs falling in a very moderate range compared with laundry detergents and fabric softeners.

Automatic Dishwasher Detergents (ADD)

Automatic dishwasher detergents products developed for the machine dishwasher have followed the fragrance profiles found in the LDL. Lemon is still the most popular, with pine in second place. Automated dishwasher detergent products present more problems for stability considerations than LDL because they usually contain a chlorinated active ingredient to reduce spotting of glassware. Automated dishwasher detergent comes as either granulated powder or liquid, containing different forms of the chlorinated active ingredients. The granulated powder usually contains an encapsulated chlorinated salt and is at a higher pH than LDL. Both of these factors contribute to stability problems. Materials must be stable under the pH conditions while in the presence of chlorine that is liberated possibly during the shelf-life of the product and definitely during the use cycle, so as not to produce malodorous combinations. Usually the shelf-life conditions do not present too serious a problem except where moisture is produced and liberates the encapsulated chlorine. Testing for stability must be carried out in simulated 'as-marketed containers' in high humidity conditions as well as the usual accelerated conditions of 35–50°C. Upon evaluation, the perfumer is looking for degradation of the raw material odor or the formation of malodor.

In the case of the liquid ADD, the active ingredient is usually hypochlorite. The stability requirements are more stringent, as is the measurement of stability. Chlorine has the ability to attach itself to unsaturated compounds without necessarily changing the odor of the finished product. This problem necessitates the actual determination by titration of the amount of free chlorine that is lost by attacking the perfume. The raw material or resultant perfume not affected by the free chlorine after accelerating aging conditions is considered stable. Of course, even if it is stable, the other factor is: how does it smell? At times like these, and more so in the future as newer more aggressive active ingredients are developed, perfumery needs support from the analytical sector to determine stability not apparent on odor alone. Recent breakthroughs in bleaching active ingredients will require more sophisticated stability testing through chemical means to determine the effects of active ingredients and perfumes on each other. The nose will not always be enough.

Laundry detergents, fabric softeners, LDLs and ADDs are the most important products in the household products group. There are many more (too many to discuss in this chapter) which present specific challenges to the creative perfumery effort. These challenges and problems can be overcome if a systematic approach to raw material testing is undertaken with an understanding of the possible chemical reactivity that may occur between the perfume and the product. Successful perfuming of household products does not just happen; it is the result of hard, often tedious, work evaluating olfactorily for stability, and in some cases chemically testing each and every material used in the fragrance.

It should also be noted that in evaluating for stability there are many gray areas where products are neither completely stable (variations of odor profile) nor unstable. It becomes a judgment call as to whether or not the product should be used. Many of these uncertain materials, when used in combination with other materials, form chemical complexes which improve their stability, and they become a contributing part of the perfume. It is an experienced perfumer who can make this evaluation.

Soaps

Once the vanguard for the major soap manufacturers and perfume development, bar soaps have rested in a comfortable niche—providing a steady return with limited activity of new entries, save a few bar soap variations and line extensions already in place in the market. The 1970s saw a flurry of deodorant bar entries with powerful and diffusive fragrances containing up to 1·7% perfume—an unheard of concentration for mass market distribution. Since that time, perfume concentrations have dropped somewhat while perfume development activity in the deodorant soaps is limited to variations of the leading soaps.

For years, manufacturers have experimented with variations of the leading beauty bars as a way to attract more consumers and to obtain more shelf space.

Diffusion, discoloration and residual fragrance on the skin are the three major problems to be overcome in bar soap creativity. Diffusion cannot necessarily be overcome just by increasing the level of the perfume. Costs have to be considered along with the proper balancing of the perfume materials themselves. Aliphatic aldehydes play an important role in the proper diffusion of soap fragrances. Obviously, in aldehydic bouquets, they are prominent; however, they can play an important role in pushing the floral or non-floral notes out of the base. Woody chemicals, musks and nitriles also play important roles as diffusing notes in several of the deodorant bar fragrances. The musks also serve well as residual notes on skin.

Usually, the perfumer has many more experiments to make to achieve the proper balance in a soap bar or detergent bar as compared with perfuming detergent granules, because the surface area is significantly less. Development work should be carried out in white soap bases so that coloration or discoloration effects can be observed easily. Accelerated aging plays an important part in finding out which materials will cause discoloration.

Beauty bars continue to hold their following throughout the world, with several market leaders being given cosmetic-quality fragrance. This is the result of much work and many reformulations to achieve the necessary diffusion and quality of fragrance.

Deodorant bars have been more aggressive in the use of newer materials, bringing a style of boldness to that type of perfume. Strong, diffusive fragrances with a tendency to leave residual notes on the skin connoting freshness and deodorancy have been well received by the consumer.

Soap perfumery will always present some challenge to the perfumer. Soap bases vary considerably in the stock odors, altering the resultant perfume odor from base to base. This necessitates reformulations to accommodate the base odor.

ACKNOWLEDGMENT

The author acknowledges the following people for their contributions and encouragement in the preparation of this chapter: Theodore Barba, deLaire (New York); Donald Latici, Drom International (NJ, USA); Donald Stagg, Givaudan (Switzerland); Peter Finlay, Technical Insights (NJ, USA).

Chapter 13

The Impact of Market Research

J. STEPHAN JELLINEK

Dragoco Gerberding & Co., D-3450 Holzminden, Germany

1 THE CHANGING SCOPE OF THE PERFUMER'S AUTHORITY

In the past few decades, the scope of the perfumer's authority within his industry has undergone a change which is closely related to the impact of market research and to the way the perfumery profession has reacted to this impact.

Until about one generation ago, the perfumer was the acknowledged expert on all questions regarding perfume. He was an authority in a double sense: he knew what was the right fragrance to use in any given situation, and he knew how to make the fragrance. Today, he is considered to be an expert only in the creation of perfumes; the responsibilities regarding specification and selection have largely, and to an ever-increasing extent, passed to the marketing and the evaluation functions. These functions base their fragrance authority, quite legitimately in a demand-driven economy, upon their knowledge of consumer needs and preferences, a knowledge acquired largely by means of market research. The perfumer's role is being reduced to that of an executor of briefs.

The change in the perfumer's scope of authority has been a logical consequence of the evolution of the industry. The perfumer who knew what the fragrance should be like, and then went on to create it, worked in the finished products company. He bought his raw materials from essential oil and aroma chemical supply houses. This was the typical situation in the first few decades of our century. Starting in the 1920s, the creative function gradually began to shift to the industry which had originally been the raw material supplier and which continued to be called the 'essential oil industry' long after perfume compounds had become its main business. The perfumers in the finished products houses increasingly became the specifiers of what was needed and the evaluators of what had been submitted by others; this was the typical pattern of the 1940s and the 1950s. (Exceptions are some large soap and detergent manufacturers and a few alcoholic perfumery houses who maintained and continue to maintain an in-house creative capability.) As the financial consequences of product design decisions, including fragrance decisions, began to escalate, it became natural to supplement or replace the personal and intuitive decisions of the perfumer by the anonymous and quantifiable decision base provided by consumer research—a development which started in the late 1950s and continued to gain strength in the 1960s and 1970s.

Meanwhile, the perfume compound suppliers, now usually called the fragrance industry, grew and it became customary to have more than one perfumer work on important projects. This normally meant that more proposals were developed than could be submitted to the customer. A selection had to be made. Increasingly, this selection was based not on the individual, intuitive and perhaps not totally unbiased judgement of the chief perfumer but on the informed, impartial group decision of a new body, the Odour Evaluation Board.

In addition to impartiality, two arguments supported the establishment of these OEBs. The boards were not charged with creative tasks and could therefore take more time than the perfumer normally could, to evaluate perfume compositions under end-user conditions. In testing fabric softener perfumes, for example, they developed elaborate testing methods to determine the substantivity of fragrances and their lasting power on the dried fabric.

Also, it has often been said that the perfumer, because of his training and skill, cannot properly evaluate perfumes because he smells them differently from the consumer. He knows the individual perfume components, and cannot help recognizing them. In contrast to the consumer, he always smells analytically. The professional evaluators of the OEBs, on the other hand, take care not to become too knowledgeable about individual components and to maintain their ability to smell holistically, like the consumer.

In the early stages of these developments the creative perfumer, seeing his authority diminished, fought them. He fought especially the establishment of evaluation boards and their growing power. He has, however, long since accepted the new situation, and with it the thesis of the difference between goal-setting, evaluating and creative skills. This is a serious mistake.

2 CONSUMER PREFERENCES AND FRAGRANCE CREATION

I do not advocate the dissolution of odour evaluation boards. Because of their impartiality in comparing the creations of different perfumers, and especially because of the elaborate tests for which only they can take the necessary time, these bodies have an important role to play. But the thesis that the perfumer should concentrate on creation and should let others worry about the consumer is false: in order to be a successful creator, the perfumer must know as much as he possibly can about consumer needs and preferences.

For what is perfume creation, if not the step-by-step realisation, within the confines of all kinds of technical constraints, of an idea? What is it, if not a long series of trials and rejections until the best possible realisation of that idea is found? And what can guide the conception of the idea and the creative process if not a sense of 'rightness', of pleasingness which must be shared by the perfumer and the consumer if the perfume is to be successful? Each time the perfumer compares his last trial with his previous one and decides which one is better, he makes a decision which must be in line with the preferences of the consumers for whom the perfume or the perfumed product is intended. If it is not, his efforts move him away from success.

The process of developing a winning fragrance involves dozens or even hundreds of selection decisions. Only the last few are made by evaluation boards, marketing or market research. All of the others are made by the perfumer on the basis of that understanding of the consumer which he is not supposed to have.

It is true that the perfumer has been trained to smell analytically. But if he is to be successful, he must also always maintain his ability to smell each creation as a whole, the way the consumer does. He shares this creative schizophrenia, this ability alternately to become totally absorbed in his creation and to step away from it, with every other creative artist.

3 THE PERFUMER AND MARKET RESEARCH

Because understanding consumer preferences is so much part of commercially successful creation, it would seem natural for perfumers to take an interest in the new art of measuring and understanding these preferences, the art of market research. Yet few perfumers today take such an interest. Recently, an international seminar was held on 'Marketing Research for Flavours and Fragrances', the first of its kind ever. Among the more than 100 attending, there was one perfumer.

Although I believe that this lack of closeness between the community of perfumers and the market research profession has a great deal to do with the conceptual separation between creation on the one hand and briefing and evaluation on the other, there are, from the perfumer's point of view, a number of concrete obstacles and problems which have tended to cool his interest in market research. These include the specialised language of marketing research, the fact that consumer studies are usually conducted at the conclusion of projects, the inadequacy of the specific guidance obtained from most consumer studies, the fact that in consumer tests preference usually simply favours the familiar, and the related fact that consumer tests at best measure today's rather than tomorrow's preference. Let us examine these problems one by one.

4 THE LANGUAGE OF MARKET RESEARCH

As is typical of any professional community, market researchers have developed a specialised repertoire of techniques and a professional language. These are not so evident in qualitative and exploratory research, but the reports on quantitative studies abound with statistical analyses and expressions such as 'monadic testing' and 'multidimensional scaling', which tend to discourage the uninitiated. If the perfuming community and the market research profession are to arrive at a closer understanding, bridges will have to be built across this moat of unfamiliarity. It would be best if they were built starting from both sides.

Market researchers should present their findings, at perfumers' meetings and in perfumery journals, in a manner that is comprehensible also to the statistically non-initiated. The professional training of perfumers, on the other hand, should also include instruction in the basics of product testing.

5 THE TIMING OF CONSUMER TESTS

The central part of the job for most perfumers is the work on briefings. The perfumer considers market research to be useful to him (or her) to the extent that it can help him towards success in briefings. He would welcome consumer tests that would tell him, in the course of his work on a major briefing, whether he is on the right path, or which of several possible approaches is the most promising, or how he should modify his perfume-in-progress to optimise consumer acceptance.

Rare is the project where his need for this kind of information can be satisfied. Key problems are the expense and the time requirements of consumer testing. A good fragrance test must employ rather large numbers of consumers; a minimum of 100 per 'cell' is a typical recommendation. These must be consumers of the right kind; there is little point in testing, in France, the perfume for a product intended for the Scandinavian market; nor is it useful to test the fragrance for a baby product among unmarried and childless women. Usually, the perfumer would like to test several alternatives along with several products from the market which serve as references. This results in a need for a large number of 'cells' since each respondent cannot test too many products without fatigue or confusion. Last but not least, the test must be conducted under realistic exposure conditions. The perfume for a detergent can be properly evaluated only if that detergent is used in a washing machine and if the laundry is afterwards dried and ironed. To conduct such tests in a professional manner and to analyse their results normally requires tens of thousands of US dollars and takes at least four to six weeks of time. This kind of testing, at the perfume development stage, is possible only in the case of very major and long-term projects. In the majority of projects, consumer tests are conducted only at the end. At that point, it is too late for the perfumer to utilise the findings in his creative work on that project.

One logical remedy to this problem would lie in finding ways to make consumer tests less expensive and more rapid. Considerable progress has recently been made in the speeding up of data analysis and report production by the use of computers. In the logistics of the actual testing, the encounter between the consumer and the product, most measures taken towards streamlining the process also increase legitimate doubts about the validity of the results. Such steps may include a reduction of the geographical dispersion of the test subjects, a reduction in their number, or a reduction in the extent to which the respondent uses and involves himself with each sample, and an increase in the number of samples tested by each respondent. If too many compromises are made, the question arises whether it is worth conducting the test at all. How sure can one be that its results are more valid than decisions based on pure judgement?

There is a remedy to the problem that comes from a very different direction. While it is true that the findings generated in consumer tests at the conclusion of a product's development are of no use to the perfumer in his work on this specific product, they may be helpful in his work on future projects. We shall return to this observation later.

6 CONSUMERS CANNOT GIVE SPECIFIC GUIDANCE

The findings obtained in consumer tests are of two kinds: there are indications of the level of preference or of acceptance shown for the various samples tested; there are indications also of the reasons for the differences in preference levels and preference patterns. What the perfumer would most like to gain from market research is reasons for preferences and for deficiencies in acceptance, in terms which to him are operational. He would like to obtain guidance on how he should modify his formulation in order to improve its acceptance level, which components he should increase and on which he should cut down. Unfortunately, consumer tests never give him this kind of guidance.

One reason for this is that consumer language is too imprecise. If consumers find that a given perfume is 'too sweet', does this mean that the perfumer should cut back on the benzoe or the heliotropin, on the basilicum or the jasmine, on the musk or the methyl octine carbonate? Or should he perhaps increase the galbanum, the vetiver or the citrus complex? It could be any or all of these. And what should he do if the consumer says that a perfume smells 'a little sickly' or 'too soapy'?

The problem lies not only in this vagueness of language but also in the fact that pinpointing specific aspects of a complex fragrance requires a skill in analytical smelling which the consumer does not possess. The recent report on a test of beers in which one specific sample received the highest number of bad marks as being 'too bitter' yet came out top in terms of overall preference, is typical.[1] Typical also are the kind of test results where 28% of respondents saying 'too strong' are counterbalanced by 22%, saying 'not strong enough' and 50% saying 'just right'.

What are the possible remedies for this vexing problem? One possibility is to ask the consumer only the kind of questions which he can readily answer. These are questions which aim, not at having him analyse his overall experience with the perfume, but at describing it in terms of where he would place the fragrance in his everyday experience. In the case of an alcoholic perfume, for example: does he perceive it as a fragrance that would be worn by fashion models, by executive secretaries, or by young girls who wear no make-up? Is it the kind of fragrance one could wear all day long, or is it suitable primarily for an evening at the opera or a romantic tête-à-tête? In the case of a cream perfume, one could ask whether the cream smells more like an anti-acne cream, like a sun-protection cream or like a night cream for sensitive skin. Questions of this kind can readily be answered by consumers and sometimes they lead to findings with statistical significance. The problem to the perfumer is that they are not operational. If a perfume that is intended for a fashion-oriented target group smells too much like something for young girls without make-up, how should it be changed?

Within the marketing research community, a radical conclusion is finding increasing support: one should, it is said, not use the consumer at all for generating reasons for preferences but only to determine preferences and acceptance levels pure and simple.[2,3] The analysis, i.e. pinpointing and describing the specific odour notes that are discernible in the perfume, should be done by a group of sensory experts especially trained for the job. This approach is finding increasing application, especially in the food industry. The extent to which it can be useful also in fragrance development has yet to be established.

A related approach also uses the consumer to provide acceptance levels but dispenses with the expert panel. Development guidance is achieved by preparing samples in which the levels of several parameters are varied systematically and determining the consumer acceptance of each sample. The relationship between acceptance and composition is then established through mathematical analysis.[1] In view of the complexity of perfumes and the multiple interactions among the multitude of fragrance components, this approach hardly appears feasible for perfume development.

A different approach uses the consumer judgements, not in their raw form but as the material from which to construct, through complex mathematical operations, a spatial representation of the way in which consumers generally perceive a specific product area in terms of fragrance. An example of this kind of 'product mapping' is shown in Fig. 1.[4] Such mappings show the main criteria by which the consumer judges and distinguishes the various perfumes submitted and they show the degree to which each perfume is perceived to possess these criteria. In Fig. 1, the main criteria of judgement are 'for women' vs 'for family' and 'mild/gentle' vs 'aggressive/fresh'. Points representing concepts such as 'has a very natural smell' or 'an ideal soap' can be introduced into the mapping alongside the points representing actual samples and can shed additional light upon consumer perceptions. Various techniques can be used to introduce measures of preference into such mapping. In this way, directions for improvement of perfumes can be derived from mapping. Improving the acceptance for a fragrance then becomes a matter of 'moving it

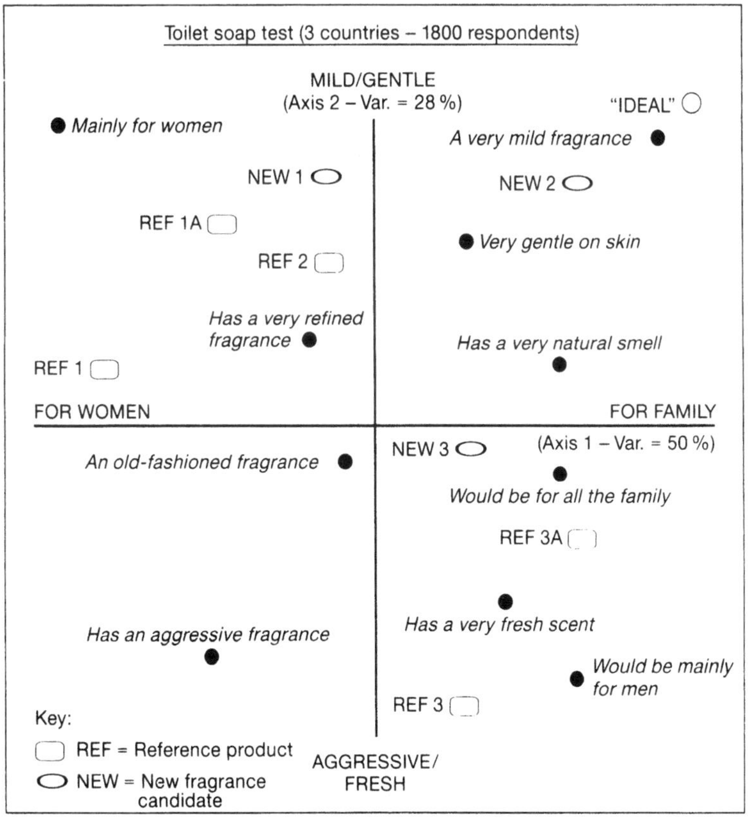

Fig. 1. Product images: toilet soap test (three countries; 1800 respondents).

from its current position in the mapping to a position associated with higher preference'. In Fig. 1 this would mean moving it closer to the 'ideal' point. The most elaborate mappings developed so far are associated with the Sandpiper system;[5–7] a database has been developed for this system covering consumer preference information across 80 different product fields in five major markets.

Through mapping, guidance can be extracted from consumer comments on perfumes, guidance which is far from perfect in perfumers' terms but better than anything based simply on compiling consumer responses.

7 SAFE FRAGRANCES ALWAYS WIN

There is an opinion, widespread among the perfumery profession, to the effect that decisions by consumer testing lead toward an avoidance of innovation and towards mediocrity. It is true that in the kind of test in which the respondent is given samples to evaluate without the advertising communication which normally accompanies major new product introductions and which prepares him to expect and accept some degree of novelty, he usually favours the samples that are close to products with which he is already familiar. It is true also that if the key score in the decision process is an average rating, it will usually favour products which have reasonably good acceptance and few rejections over alternatives which may be exceptionally well liked by some of the respondents but rejected by others (Table 1).[8]

This type of test and this kind of data analysis would be appropriate to products meant for a large market, in which fragrance does not play a central role and which are introduced into the market without much advertising support. If a new product does not fit this description, different market research approaches should be employed. They may include a selection of respondents designed to give appropriate weight to those who look for innovation; the presentation, in the test, of advertising, packaging, etc., which prepares the respondent to expect something new and different from the

Table 1
Fragrance and image—toilet soap test

Rating	*Product A*	*Product B*
5 Excellent	20%	50%
4 Very good	40%	—
3 Good	40%	—
2 Fair	—	—
1 Poor	—	50%
Average rating	3·8	3·0
Percentage of respondents preferring test product over brand used most often	10%	25%

customary; data interpretation which looks at the fit of fragrances to the concept underlying the new products; and an interpretation of the preference scores which puts more weight on the proportion of 'excellent' ratings and less weight on average ratings.

If test respondent selection, interview design and data interpretation in a product test are not appropriate to the specific product tested, this does not mean that market research in general is misleading but, rather, that this specific test has been poorly designed. It is certainly true that some of the fragrance selection tests conducted in the past have been poorly designed, but in recent years market research in general has become considerably more sophisticated.

A more active involvement of the perfumery community in marketing research could provide additional impetus towards the development and adoption of more differentiated, more specifically tailored approaches to fragrance testing.

8 TESTS MEASURE TODAY'S INSTEAD OF TOMORROW'S PREFERENCES

Considerable time often elapses between the day on which the design of a fragrance test is decided and the day on which the new product is introduced into the market. During this time interval, the marketing environment and consumer expectations may have changed. This possibility must be taken seriously, especially in product fields with close links to fashion and life-styles such as feminine fragrances and deodorant colognes. There is no sure way to overcome this difficulty, but one can take steps in the right direction. These include conducting the tests in areas which tend to be culturally ahead of the market at large; to select, within these markets, respondents who are trend leaders.

It is of the essence of new products in the luxury sector that they do not fall within the taste and preference pattern of the general public at any given time. They must lie outside of the customary and represent a kind of 'reaching out'. It takes subtlety to allow for this in market research, but it can be done.

A fine example of 'reaching out' market research is offered by the tests conducted prior to the US market introduction of *Poison*.[9] The test was conducted among 33 'women who were actively interested in trying new fragrances and were potential trendsetters'. The New York area was selected for the test 'because it was culturally receptive to European values and products and ahead of the Eastern half of the US'. The respondents were selected according to criteria of socio-demographics, life-style, clothing styles, tastes in fragrances and ability to express olfactory sensations. They could take part in the interview only after they had passed a final face-to-face screening for style and sense of smell. The interview included individual 'blind' scent testing, group sessions during which the imagery of the scents was explored and the name, the box and the bottle were unveiled, and use of the fragrances at home after the group sessions.

The analysis of the data looked for the following 'indicators of promising potential':

—passionate reactions: positive, ambivalent or negative;
—compatibility of themes associated with each of the scents on one hand, and with the concept on the other hand;
—convergence of images associated with each of the elements of the mix, with the scents and with the concept;
—common characteristics in terms of personality and expectations *vis-à-vis* fragrances of the respondents who liked or rejected the scents;
—appeal of the most promising scent to those among the respondents who were attracted to the concept;
—compatibility of the imagery associated with a scent with certain aspects of the fantasies expressed by those who liked that scent.

The responses were also analysed with a view to giving directions for the fine tuning of the four fragrance candidates included in the test.

The results revealed strong negative aspects of the concept, including 'violent face-value rejection of the name by most of the respondents' and 'rational-level reactions (which) were quite negative and even confused'. Of the three fragrances included in the first round of the test (two of these were actually the same fragrance at different concentrations), the one which was selected for further modification 'was forcefully rejected by many because of its harsh potency and its heavy sweetness. Its main assets were the intensity with which the respondents rejected or, more rarely, adopted it, and the convergence of the images associated with it'. The modified version of this fragrance, tested among the same respondents one month later, 'met with passionate reactions: positive, ambivalent (about half) or negative (the other half)'. This was the fragrance finally adopted after further enhancing the top note.

On the strength of these test results, obtained from 33 women, the Dior management had the courage not only to introduce the fragrance in the United States, but to stake an introductory advertising and promotional budget of $9 million on its success. This example shows how far market research can move away from straightforward routine preference tests.

Another case history of research tailored to the specific requirements of a project is offered by the tests conducted for the selection of fragrances for ten-year-old girls and boys launched by a Dutch manufacturer of children's clothes with retail outlets throughout the world.[10] The research was conducted exclusively among the young customers of the stores, it involved a test in the store, simulating the purchase act as well as a long-term product test in which the respondents kept detailed diaries on the occasions of using the fragrance and in which they reported things like the comments received from others. Mapping was applied to the data obtained. In this case the fragrances which received the highest number of positive reactions were selected.

9 THE USE OF MARKET RESEARCH BY THE PERFUMER

The picture which emerges from this review is one where market research can provide a great deal of insight into consumer response patterns to specific fragrances but where the possibilities of the perfumer using these insights in the course of concrete projects, to give him direction in his creative work, are limited. This leads us to conclude that the perfumer can and should study the findings generated by consumer studies, not so much to use them directly in the specific project on hand but in order to refine his general understanding of consumer responses to fragrances. To use an analogy: consumer tests can rarely provide the perfumer with the kind of answer that a foreign language phrase book gives when it tells you exactly what to say if you want the heat in your hotel room turned down. It can provide him, however, with a better understanding of the grammar of consumer responses and improve his skill in constructing his own phrases based on this general background.

Fragrance tests are usually conducted for specific confidential projects; their results are rarely published. Moreover, learning from product tests ideally involves smelling the fragrance tested at the same time as one reads about consumer responses to it. Because of this, only those published reports on consumer tests are really fully interpretable which refer to perfume materials and finished products generally available in the market-place. The body of such published reports is quite small.[11,12,13] Most learning opportunities, therefore, come from product tests conducted in the course of projects in which the perfumer's company is directly involved or which are conducted by his company. 'Post-mortem' review of test results, in which the samples tested and the consumer responses obtained are studied side by side, have great educational value also to perfumers not involved in the specific projects concerned.

The four types of data that are particularly valuable in this context are described in detail below.

Degree of Acceptance and Preference

Here, the perfumer can ask himself the following questions. Which are the 'safe' fragrances that do not elicit great enthusiasm but also very few rejections, and which are the 'love/hate' fragrances? Smelling these fragrances in the context in which they were presented to the consumer, can he (the perfumer) understand why they evoked these different response patterns?

Are there distinct differences between the first-impression reaction to a fragrance and the reaction after an in-home use test or some other form of prolonged exposure? If so, can the perfumer guess what may have caused these differences?

Do the samples which achieved high overall preference scores have anything in common that distinguishes them from the ones receiving low scores?

Naturally, in interpreting these findings, one must always take into consideration the specific group of people who participated in the test. The results reflect only *their* reactions and preferences, not those of people in general.

Spontaneous Free Associations

Where free associations are included in test reports, they can make interesting reading. Different people usually have quite different and sometimes opposite associations with a given fragrance. What is fresh and clean to one person may be sweet and cloying to another. Yet, looking at the total body of responses to each fragrance, patterns usually emerge. To make sense out of these patterns takes the combined skills of a detective, a psychologist and a poet, but it can be a stimulating and worthwhile exercise.

The Matching of Fragrances to Concepts

It is quite easy to get people to match fragrances with other things or concepts. Tests have been conducted in which respondents matched perfumes with different types of women[14] and with colours,[15] and a pilot test in which cream perfumes were matched with cream functions.[16] Beyond their immediate marketing uses in the project for which the test was run, such findings can also be educational in a general sense. They can confirm or sometimes modify the perfumer's own intuitions about what kind of fragrance goes well with young girls or with mature women, with pastel colours or strong primary colours, with sun protection or sensitive skin creams, etc. Again, one must always keep in mind that these associations may also vary from country to country or even among different population groups in the same country.

Mapping

Mappings developed from consumer data can also generate insights that help the perfumer in refining his understanding of specific markets. For example, the soap fragrance map presented in Fig. 1 shows that consumers' perception whether a soap is good especially for women or for the whole family is strongly affected by the fragrance. This, in fact, appears to be a major aspect of what the fragrance communicates; it accounted for 50% of the differences in the responses to the samples tested. Looking at the mapping even without knowing the reference and the new fragrances tested, the perfumer can see that fragrances that are considered 'very natural' or 'very fresh' tend to move a soap into the 'family soap' direction, while 'a very refined' fragrance leads to the soap being considered more suitable for women. The analysis is far more useful, of course, to the perfumer who can smell the soaps involved as he looks at the data. The odour of reference 1 is very much 'for women', that of reference 3a and new 2 is particularly good 'for family'.

The second perceptional dimension which is much affected by fragrance and accounts for 28% of the overall difference between the samples is the mild/gentle–aggressive/fresh dimension. This is interesting in itself, but again becomes far more illuminating if one can smell the differences between new 1 and new 2 which are highest on 'mild/gentle' and reference 3 which is highest on 'aggressive/fresh'.

The distances from the 'ideal' point (upper right) are interesting as well. Does the perfumer intuitively understand why new 1, new 2 and new 3 are closer to the ideal point than the corresponding reference samples? In this case, one ideal point only is given for the 1800 respondents from three countries. Segmentation studies in which respondents are grouped on the basis of differential response patterns often show different ideal points for different consumer segments.

Normally, the analyses of consumer data will confirm the perfumer's basic intuitions to a considerable extent. If it were not so, he could not be successful as a perfumer. Also, if it were not so, he would distrust the data and may indeed be right in that attitude. However, in confirming existing intuitions, the results may well refine, enrich and sometimes modify them; herein lies the value of the learning process.

The largest and most refined consumer tests are usually conducted not by the perfumers' companies but by their customers, at the conclusion of major projects. The customer may be reluctant to give the perfumer access to all of the data and especially to all of the fragrances involved in the test, if these include fragrances submitted by competitors to the perfumer's company. I can only hope that as the marketing and the perfuming community become more and more convinced of the value to the perfumer of learning from consumer responses, a liberal attitude towards information exchange will come to prevail. If the test data and test samples are made available to all of the perfumery houses competing in the project, the practice is not unfair and it can help to make everyone more knowledgeable.

10 SUMMARY AND CONCLUSION

During the past few decades, a separation has taken place between the specification-setting and evaluating functions and the creative function in perfumery. This separation has been a logical and natural development. However, it must not lead to an attitude on the perfumer's part that 'consumer responses are not his business'.

The amount of specific guidance that consumer research can give the perfumer in his work on the project in hand is limited today, and will probably always remain limited. However, the study of consumer test results along with the fragrances involved in the test can be a valuable adjunct to the perfumer's general professional experience.

REFERENCES

1. Crofts, P., The need for a systematic approach to product optimization and flavor development research. *Esomar Seminar on Research for Flavours and Fragrances, Lyon 1989*. Esomar, Amsterdam, 1989, pp. 81–9.
2. Köster, E. P., Recent developments in the study of perception: taste and smell. *Esomar Seminar on Research for Flavours and Fragrances, Lyon 1989*. Esomar, Amsterdam, 1989, pp. 1–16.
3. Wilton, Vivien S., Bridging the language gap between consumer and flavour/fragrance developer and marketer to improve product optimisation. *Esomar Seminar on Research for Flavours and Fragrances, Lyon* 1989. Esomar, Amsterdam, 1989, pp. 27–37.
4. Shalovsky, I. H., Research methods in the development of flavours and fragrances. *Esomar Seminar on Research for Flavours and Fragrances, Lyon 1989*. Esomar, Amsterdam, 1989, p. 105.
5. Bingham, J., A new approach to international fragrance research. In: *Esomar Seminar on International Marketing Research, Englefield Green, 1988*. Esomar, Amsterdam, 1988, pp. 75–93.
6. Cooke, Viki L., Stimulating creativity and communicating consumer opportunities to the perfumer. *Esomar Seminar on Research for Flavours and Fragrances, Lyon 1989*. Esomar, Amsterdam, 1989, pp. 55–69.
7. Cooke, Viki, International brands: on the scent of sweet success. *Cosmetics and Toiletries*, **104**(6) (1989) 62–5.
8. Jellinek, J. S., *The Use of Fragrance in Consumer Products*. Wiley, New York, 1975, p. 78.
9. Wolfe, O. H., High-touch research applied to fine fragrance development: the *Poison* case history. *Esomar Seminar on Research for Flavours and Fragrances, Lyon 1989*. Esomar, Amsterdam, 1989, pp. 121–33.
10. Goodman, D. & de Bock, H., Juliette in wonderland. *Esomar Seminar on Research for Flavours and Fragrances, Lyon 1989*. Esomar, Amsterdam, 1989, pp. 169–79.
11. Jellinek, J. S. & Köster, E. P., Perceived fragrance complexity and its relation to familiarity and pleasantness I. *J. Soc. Cosm. Chem.*, **30** (1979) 253–62; II. **34** (1983) 83–97.
12. Jellinek, J. S., du Bosque, Bettina Olias & Gschwind, Jürgen, Expert classification and consumer acceptance of perfumes. 15th IFFSCC International Congress, London 1988, Preprints (1988), Vol. 4, 71–99.
13. Jellinek, J. S., Perfumes and colors. *Dragoco Report*, **35**(1) (1988) 14–29.
14. Jellinek, J. S., Schlabritz, H. G. & Reichert, W., Wie mißt man Parfum-Charakter?, *Parfümerie und Kosmetik*, **58** (1977) 153–9.
15. Sponbergs, P. & Harx, A., Fragrances: a colorful connection. *Cosmetics and Toiletries*, **104**(6) (1989) 51–57.
16. Jellinek, J. S., The Skin Health Index as a guide to perfuming cosmetic emulsions. *Dragoco Report*, **30**(3) (1983) 55–65.

VI PRODUCTION OF PERFUMES

This part of the book is more technology- than science-oriented. It takes the reader from the chemistry of raw materials to compounding, it mentions aspects of quality control and toxicology, and it ends with an appraisal of the situation of the fragrance industry in a changing world.

It is further intended to sensitize the reader's view to the complexity of the fragance industry, and it will certainly illustrate the fact that survival in this business depends in the first place on professionalism in many fields.

Chapter 14

The Chemistry of Synthetic Raw Materials Production

JULIAN DORSKY

41, Lake Shore Drive, Rockaway, NJ 07866, USA

1 INTRODUCTION

Modern perfumers have an almost unlimited number of raw materials available to create fragrances. These raw materials are traditionally grouped into three classes: natural essential oils, synthetic chemicals and semi-synthetic chemicals derived by chemical processing from natural products. Another class of materials, a fourth group widely used by perfumers, consists of mixtures from the above three classes denominated as specialties or sub-compounds. Specialties are sometimes prepared 'in house' by perfumers for exclusive use, but are often purchased from commercial suppliers, in which case the exact composition may not be known to the user.

When setting out to create a fragrance, a perfumer can have before him an array of raw materials from the four groups numbering in the thousands.

Disregarding specialties and sub-compounds the number of materials available from the other three classes is about 3500. Most perfumers limit themselves to a fraction of this resource and some, in fact, pride themselves on the fact that their work palette comprises no more than 400. However, some of the 400 items can be specialties that in themselves are basic fragrances and contain many ingredients.

Several comprehensive lists of perfume raw materials are available. *Flavor and Fragrance Materials—1989*[1] is the most complete and up-to-date available. It comprises both natural and synthetic materials arranged alphabetically with much useful information such as synonyms, commercial suppliers, CAS numbers and references to government and industrial association classifications. This compilation of about 4000 materials includes 1754 flavor materials on the FEMA GRAS list (Flavor Extract Manufacturers Association Generally Recognized As Safe). Many of the GRAS materials (i.e. pyrazines, thiazoles)[2–4] are used in low concentrations in fragrances and are usually not found in other lists of fragrance raw materials.

Common Fragrance and Flavor Materials by Bauer & Garbe[5] covers about 500 of the most important synthetic chemicals and about 120 essential oils. It is up-to-date and contains reliable information on preparation, properties and use.

The *Guide to Fragrance Ingredients* from Haarmann & Reimer in 1984 lists about 200 of the most important chemicals and 150 essential oils.[6] A market study on essential oils and oleoresins published in 1986 under the auspices of the United Nations is a good source of information on the commercially most important essential oils and oleoresins.[7]

Perfume and Flavor Chemicals (*Aroma Chemicals*) by Steffen Arctander[8] is a comprehensive and authoritative list of chemical raw materials. Though very useful for the information it contains, this work was published in 1969 and does not include important raw materials introduced since then. An earlier work by Arctander devoted to natural products, *Perfume and Flavor Materials of Natural Origin,*[9] published in 1960, is also available.

There are several classical treatises on fragrance and flavor materials published 20 to 30 years ago.[10–12] They deserve mention because some useful information in these works is not included in recent publications.

2 PRODUCTION OF SYNTHETIC PERFUME MATERIALS

Organic chemists have carried on a romance with fragrant chemicals from the early days of organic chemistry. In the beginning the challenge was to isolate and identify fragrant chemicals from natural products. Next, effort was directed toward the synthesis and commercial production of nature-identical chemicals for the growing perfume industry. Finally, new chemicals were designed and synthesized expressly for their valuable odor properties. Progress in this field has been amply reported in scientific journals, patents and several books. *Perfumery and Flavoring Synthetics* by Paul Z. Bedoukian, first published in 1951 and now in its third edition,[13] has been the standard reference for the chemistry and technology of the principal synthetic chemicals used in the fragrance industry. Each year since 1945 an annual review by Bedoukian has covered the literature on perfumery and flavor materials.[14] Two other important treatises dealing with the chemistry of perfume chemicals have been published by Beets[15] and Theimer.[16] Billot & Wells[17] discuss natural and synthetic perfumery raw materials extensively from the perfumers' viewpoint, with little attention to chemistry.

The references cited above and many reviews in the literature usually treat the chemistry of each material individually or group the materials according to structure or functionality. In the presentation that follows the organization is different. Synthetic perfumery raw materials are grouped according to the industrial processes by which they are or have been manufactured. Examples of the important chemicals and intermediates produced by each process are discussed. Knowledge of commercial processes comes mainly from the author's personal experience in the production of perfume chemicals over many years. This knowledge was supplemented by reliable information from colleagues in the industry. Coverage of processes used in the industry and the examples of each process are by no means complete but it is believed that the most important, pound- and dollar-wise, have been covered.

Many of the examples cited use basic chemistry which has changed little since the processes were first commercialized. Significant improvements, however, have been made in technology, especially for those chemicals produced in large quantities.

Chemicals of the type discussed in the following pages are normally used in relatively high concentrations in perfumes. Combinations of these chemicals fill the drums and tank trucks of the industry. In number, however, they represent only a fraction of the thousands of materials used by perfumers. Most ingredients in perfume compositions are 'exotic' chemicals used in low concentration for their intense odor or special effects. Fenn,[18] in a recent review of perfume chemical usage, estimated that 'exotic' chemicals frequently comprise about 80% or more of the total number of ingredients in a fragrance.

We will now turn to 'unit reactions' which play the preponderant role in the industrial manufacture of synthetic odorants.

3 UNIT REACTIONS AS THE PRODUCTION BASIS

Oxidation

The most important industrial oxidizing agents are manganese dioxide, chromic acid and oxygen or air (Fig. 1). Both manganese dioxide and chromic acid require a strong acid, usually sulfuric acid. Ozone as an oxidizing agent is discussed elsewhere.

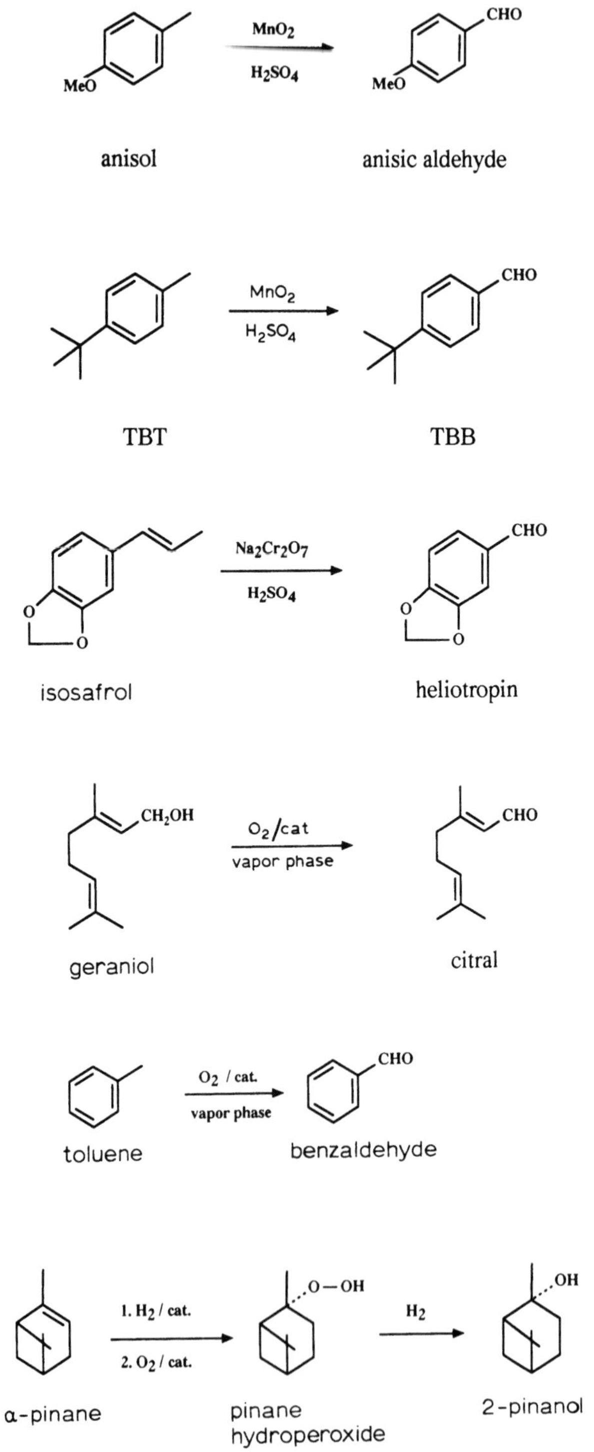

Fig. 1. Oxidation.

Manganese dioxide with sulfuric acid has been used extensively to oxidize a methyl group on the benzene ring to produce anisic aldehyde[19] and *p-tert.*-butylbenzaldehyde (TBB), the intermediate for Lilial®.[20] By-product manganous sulfate, which formally was indiscriminately discarded in the waste water, is now recovered and sold for soil enrichment or as an animal food supplement. Reduced magnesium can also be oxidized electrolytically and recycled. A process in which manganese(II) is oxidized in an electrolytic cell to the manganese(III) state for recycling is described in the literature.[21]

For many years chromic acid (sodium bichromate–sulfuric acid mixture) was widely used to produce heliotropin by oxidation of isosafrole.[22,23] To improve yields, sulfanilic acid was added and air was blown through the reaction mixture to remove by-product acetaldehyde. Chromic acid oxidation was also used industrially to produce anisic aldehyde from anethole in the days when cheap anethole was available from turpentine. In past years when the economics were favorable citral was produced in good yield from synthetic geraniol by chromic acid oxidation.

The basic technology for manganese dioxide and chromic acid oxidation was developed early in the 20th century and is still in use today. Over the years significant improvements have resulted from optimization of reaction conditions and utilization of modern process technology, but the basic processes are still the same. Since disposal of waste manganese and chromium are regulated in most parts of the world, these materials are either oxidized and recycled or recovered and disposed of in a safe manner.

Oxidation using gaseous oxygen or air is a common process used in the petrochemical industry to produce huge quantities of commodity chemicals. Air, and oxygen as well, are relatively cheap and the usual reaction conditions, catalytic vapor-phase reaction with short contact time, favor low-cost production. Vapor-phase air oxidation of toluene is the main source of benzaldehyde. Vanadium and molybdenum catalysts are reported to give high yields. The reaction is applicable to other alkylated toluenes as well. *p-tert.*-Butylbenzaldehyde has been made semicommercially in this way.

Vapor-phase oxidation of geraniol/nerol mixtures with oxygen has been an important source of citral for many years. Copper is a good catalyst for this reaction. Catalytic dehydrogenation of the alcohols over copper also yields citral but direct oxidation is superior on a commercial scale.[24]

Liquid-phase oxidation of *cis*-pinane to *cis*- and *trans*-pinanol catalyzed by the hydroperoxide is the second step in the commercial production of turpentine derivatives at SCM's 30 million pounds (13·6 million kilograms) per year Colonel's Island plant near Brunswick, Georgia, USA. The process starts with the catalytic hydrogenation of α-pinene to *cis*-pinane, which is oxidized to pinane hydroperoxide. Hydrogenation yields 2-pinanol. Linalool is produced by pyrolysis of 2-pinanol. Several commercial C_{10} terpene chemicals and their derivatives are made from linalool. At Colonel's Island *cis*-pinane is oxidized at about 95°C with oxygen and 2-pinane hydroperoxide as catalyst to a mixture of approximately 75% *cis*- and 25% *trans*-pinane hydroperoxide. There are early references in the literature[25–27] to this chemistry. SCM has developed the oxidation process to the point where it is run in a 30 000-gallon (136 000-liter) reactor.

Condensation Reactions

A number of important perfume chemicals are manufactured on a large scale by a variety of condensation reactions including aldol, Darzens and Schiff base types. Aldol condensations shall be considered in the broadest sense to include products formed by dehydration of the hydroxycarbonyl compounds formed by the condensation.

Five widely used products made via aldol condensation (Fig. 2) are pseudomethylionone (cyclized to methylionone).[28] α-amylcinnamic aldehyde,[29] α-hexylcinnamic aldehyde, alkyl-α-methyldihydrocinnamic aldehydes (Cyclamen aldehyde® and Lilial®[20]) and cinnamic aldehyde. The reactions are catalyzed by alkali under conditions which minimize self-condensation of the more reactive aldehyde, i.e. propionaldehyde in the preparation of the unsaturated intermediates for Cyclamen aldehyde® and Lilial®, and acetaldehyde for cinnamic aldehyde.

Before the 1960s, substantially all of the pseudoionones were produced via aldol condensation of citral with acetone or methyl ethyl ketone. When pseudomethylionone was prepared the reaction yielded a mixture of the

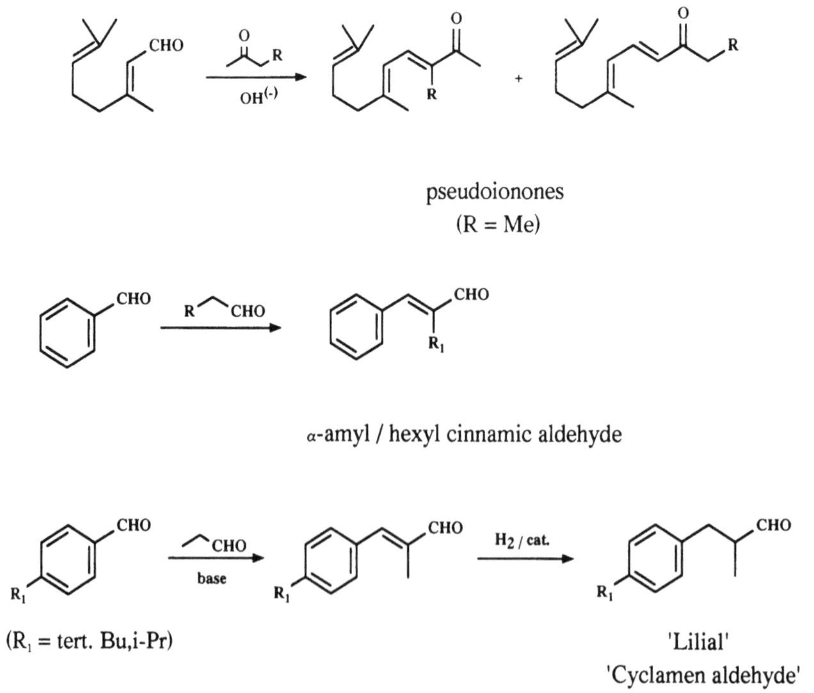

Fig. 2. Aldol condensation.

straight (normal) and branched (iso) chain products. Because isomethylionone was preferred for its odor quality, many proprietary processes were developed to increase the ratio of iso to normal. The most effective involved selection of special solvent mixtures and catalysts.[28]

A new route to pseudoionone and its methyl analogue based on acetone and acetylene via 6-methylhept-5-en-2-one and dehydrolinalool was commercialized by Hoffmann–La Roche.[30,31] The original chemistry yielded >95% iso which cyclized to the preferred branched-chain methylionone. However, the established market for this material was limited because of its high price, and change by perfumers to the pure iso form in low-priced fragrances was slow. It was therefore necessary to find conditions under which the new route would produce the popular mixture of iso and normal isomers, about 60:40.

The Perkin reaction used in the synthesis of coumarin from salicylic aldehyde and acetic anhydride involves the formation of an aldol followed by dehydration and lactonization.[32]

Figure 3 illustrates two examples of the Darzens reaction used to produce fragrance chemicals, methyl nonyl acetaldehyde[33] and ethyl methylphenylglycidate.[34] The Darzens reaction between an aldehyde or a ketone and an α-halo ester yields an α,β-epoxy ester (glycidic ester). On heating, these compounds decarboxylate to yield aldehydes. Ethyl methylphenylglycidate (Strawberry aldehyde) on decarboxylation forms hydratropic aldehyde and methyl nonyl acetaldehyde (Aldehyde C-12 MNA) is also a decarboxylation product. The intermediate glycidic ester has no value in perfumery.

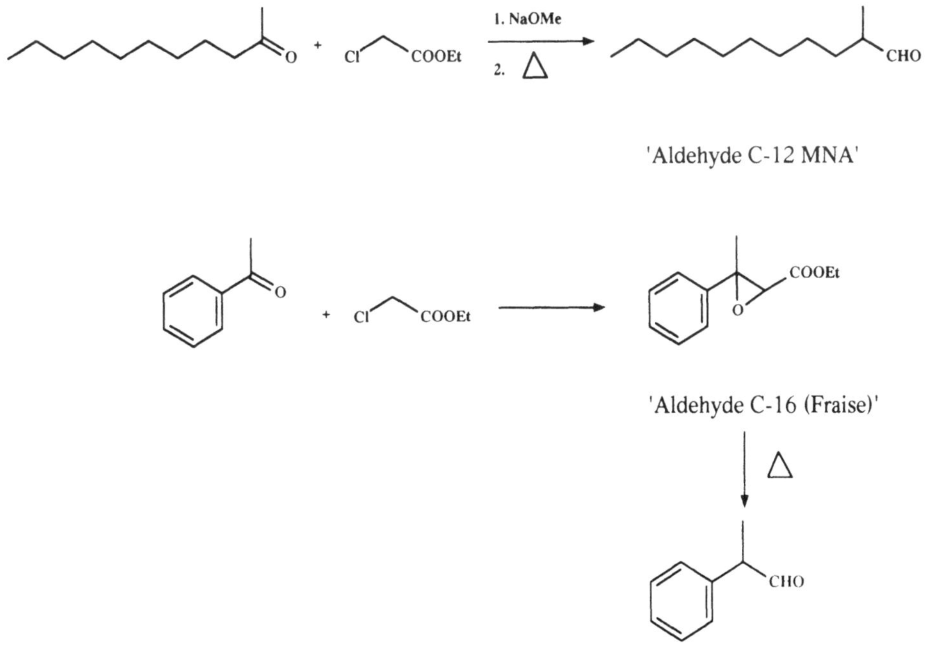

Fig. 3. Darzens reaction.

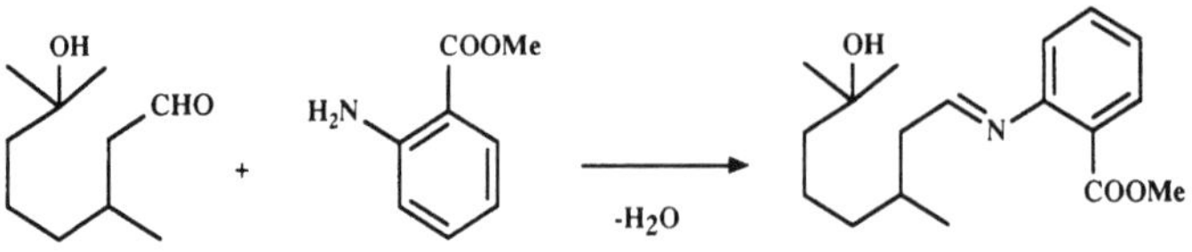

'Aurantiol'

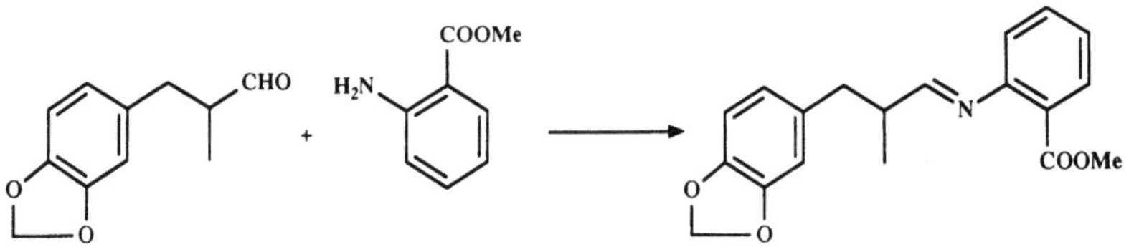

'Helioforte'

'Fleurantiol'

Fig. 4. Schiff bases.

Methyl nonyl acetaldehyde, a valued perfume ingredient, has recently been made commercially from cheap, petrochemical-derived undecanal and formaldehyde via the α-methylene derivative. Methyl octyl acetaldehyde has also been produced by this route.[35]

Frequently used Schiff bases are condensation products of aldehydes and methyl anthranilate. They are easily made by heating a stoichiometric mixture of an aldehyde and methyl anthranilate under vacuum in a short-column still to remove water formed in the condensation. The viscous yellow product is normally used without further purification. To avoid excessive discoloration the contents of the still are kept below 100°C. Figure 4 illustrates the formation of two widely used Schiff bases from hydroxycitronellal or α-methyl-β-3,4-methylenedioxyphenylpropionaldehyde, respectively. The latter became popular after its successful use in the best-selling perfume *Giorgio*.

Methyl dihydrojasmonate (Hedione®), an important component of synthetic jasmine and other floral fragrances, is industrially produced by a sequence of condensation reactions shown in Fig. 5. Aldol condensation of cyclopentanone and pentanal yields α-pentylidenecyclopentanone. Isomerization shifts the double bond to form 2-pentylcyclopent-2-en-1-one. Michael condensation of this intermediate with malonic ester affords methyl dihydrojasmonate.[36,37]

Hydrogenation

Liquid-phase catalytic hydrogenation is prominent among reactions used to produce fragrance chemicals on a large scale. Raney nickel catalyst is the most

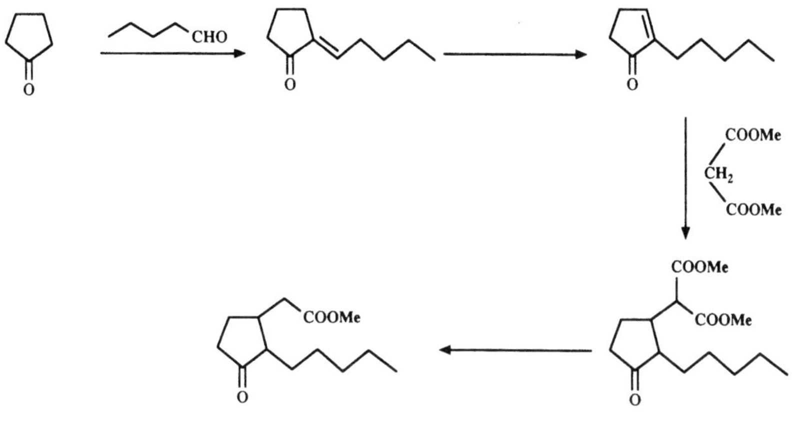

Fig. 5. Synthesis of methyl dihydrojasmonate.

widely used for reduction of carbon–carbon and carbon–oxygen unsaturation, including aromatic rings. Most hydrogenations are conducted under optimized conditions of temperature, pressure and solvent to achieve desired selectivity (Fig. 6). Hydrogenation of *p-tert.*-butylphenol with Raney nickel yields a mixture of *cis*- and *trans*-4-*tert.*-butylcyclohexanol rich in *trans* isomer.[38] A product containing mostly *cis* isomer, preferred for its finer odor, is obtained using palladium or rhodium catalyst.[39] The acetylated alcohol (Vertenex®) is used in large amounts in soaps and detergents.

Controlled hydrogenation of geraniol with Raney nickel saturates mainly the allylic double bond to yield citronellol, leaving the isopropenyl double bond unchanged. Similarly, citral can be selectively hydrogenated to citronellal. Current economics do not favor commercial use of this reaction but it was applied during World War II to produce citronellal when lemongrass oil (citral source) could be obtained but not citronella oil (citronellal source).

Raney nickel catalyst is used in the aforementioned Colonel's Island terpene chemistry to obtain high selectivity in the hydrogenation of α-pinene to *cis*-pinane.[40] Pinane hydroperoxide in this process is reduced to *cis*-pinanol with hydrogen and Raney nickel catalyst.

Under high temperature and pressure, Raney nickel can effect hydrogenolysis as well as hydrogenation. Thus, at 200°C and about 40 atm ($\simeq$ 4000 kPa) isocamphylguaiacol yields isocamphylcyclohexanol.[41]

Palladium is a versatile hydrogenation catalyst used to produce several fragrance chemicals on a large scale. The common form used industrially is 5% on carbon. Palladium is expensive; therefore catalyst is re-used several times if possible and then carefully recovered for reprocessing. The principal application for 5% palladium on carbon is selective hydrogenation of carbon–carbon double bonds in α,β-unsaturated carbonyl compounds, i.e. *p-tert.*-butyl-α-methylcinnamic aldehyde is selectively hydrogenated to the corresponding

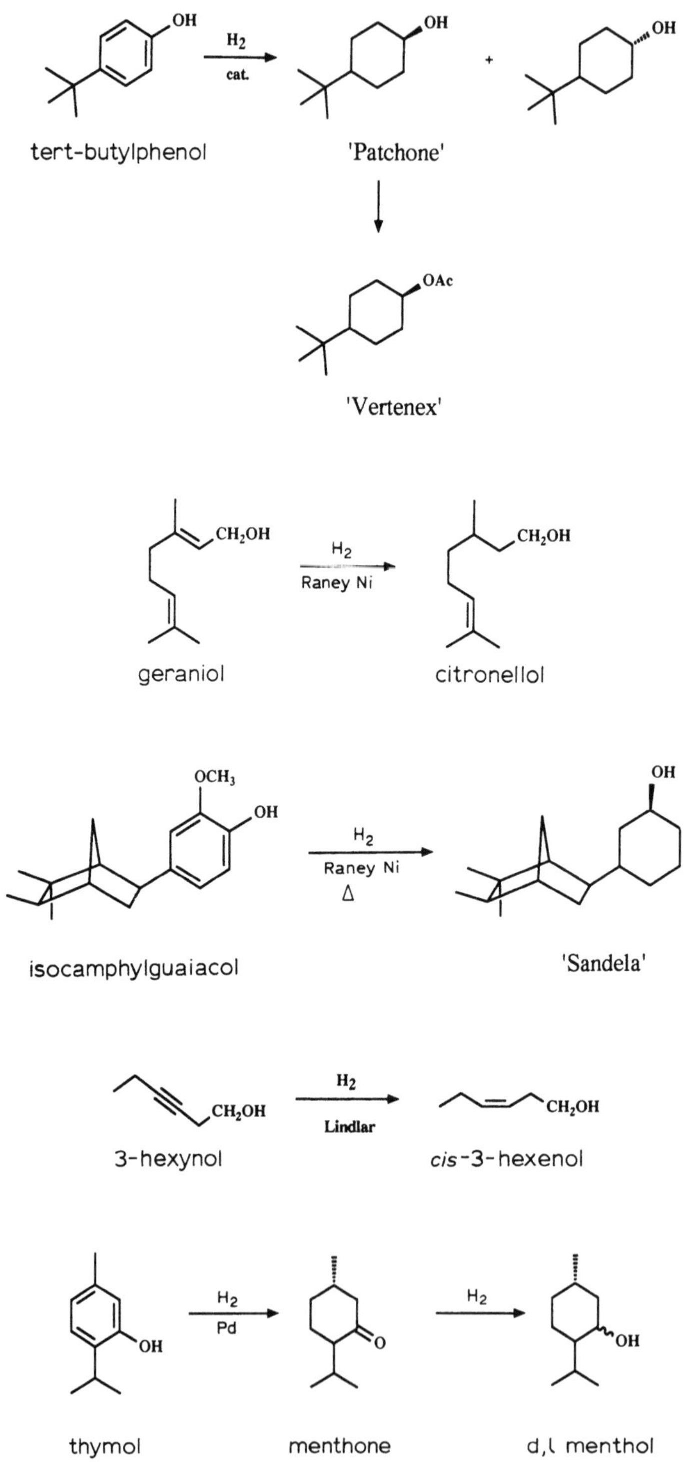

Fig. 6. Hydrogenation.

dihydrocinnamic aldehyde (Lilial®) with 5% palladium on carbon catalyst.[20] Hydrogenation proceeds rapidly without solvent at about 80°C and about 15 atm (≃1500 kPa). Under the right conditions hydrogen absorption will cease after the olefinic double bond has been saturated; 5% or less of the hydrogen is consumed in formation of the saturated alcohol by reduction of the carbonyl double bond. In contrast, with Raney nickel the hydrogenation is considerably less selective. The reaction must be terminated when most of the starting material has been consumed but by that time considerable alcohol has been formed.

Palladium modified with lead on calcium carbonate (Lindlar)[42] is useful for hydrogenating an acetylenic triple bond to a *cis* olefin. Hydrogenation of 3-hexynol with Lindlar catalyst yields *cis*-3-hexenol (leaf alcohol), an important perfume chemical used for its intense green note.[43]

An interesting application of palladium-catalyzed hydrogenation is the preparation of 2-substituted cyclohexanones from *ortho*-alkylated phenols. Thus, hydrogenation of thymol with 5% palladium on carbon yields racemic menthone.[44]

Commercial production of menthol from thymol by hydrogenation with copper chromite catalyst has been in use for many years. The hydrogenated product contains about 75% racemic menthol and 25% other isomers. Optically active (−)-menthol is obtained by various industrial procedures. These other isomers can be isomerized and racemized with copper chromite and other catalysts.[45,46] Raney nickel and Raney cobalt have also been used to produce menthol.

Copper chromite is an excellent dehydrogenation catalyst and is used to prepare aliphatic aldehydes from readily available alcohols, e.g. *n*-octanal from *n*-octanol.

Friedel–Crafts Reaction

Many perfume chemicals and intermediates are produced on a large scale by this reaction, which involves alkylation or acylation of an aromatic compound in the presence of a Lewis acid catalyst. Aluminum chloride is the most common catalyst but sulfuric acid, phosphoric acid, ferric chloride and boron trifluoride are also used. The alkylating agents may be olefins, alkyl halides, alcohols or epoxides. Aromatic or aliphatic acids, their halides, esters or anhydrides may be used as acylating agents.[47]

The alkylation of benzene with ethylene oxide using aluminum chloride catalyst is the main commercial route to β-phenylethyl alcohol, one of the most important of all perfume chemicals.[48,49]

In the production of Galaxolide®, one of the intermediates is prepared by alkylating 1,1,2,3,3-pentamethylindan with propylene oxide and aluminum chloride.[50]

Several tertiary butylated aromatic hydrocarbon intermediates for nitro musks are produced by alkylation with isobutylene and aluminum chloride or boron trifluoride. *tert.*-Butyl-*m*-xylene is the intermediate for Musk Xylene. The intermediate for Musk Ketone is prepared by acylation of *tert.*-butyl-*m*-xylene with acetic anhydride or acetyl chloride and aluminum chloride. Alkylation of *m*-cresyl methyl ether with isobutylene and aluminum chloride yields the intermediate for Musk Ambrette (but this product is practically no longer used because of its sensitizing properties). *tert.*-Butyltoluene (TBT), the starting material for Lilial®, is prepared by a facile alkylation of toluene. The butylated final products mentioned above are shown in Fig. 7.

There are four polycyclic musks offered commercially. All are acylated bicyclic aromatic hydrocarbons and are shown in Fig. 8.[51] Acetylation is usually accomplished with acetic anhydride or acetyl chloride and aluminum chloride catalyst at about 10°C in a solvent such as ethylene dichloride.

Methyl *β*-naphthyl ketone (Oranger crist®) is used in floral-type fragrances for its orange blossom odor. It is prepared by acylating naphthalene with acetic anhydride or acetyl chloride and aluminum chloride in a polar solvent. The product is a mixture of *α*- and *β*-naphthyl ketones, the proportions depending mainly on the solvent. Highest yields of *β*-ketone, preferred for its finer odor, are obtained in nitrobenzene.[52]

Cycloalkylation

The cycloalkylation reaction of Bruson & Kroeger[53] was used to prepare the hydrocarbon for one of the first commercial polycyclic musks, Versalide®, 7-acetyl-6-ethyl-1,1,4,4-tetramethyltetralin.[54] Alkylation of ethylbenzene with 2,5-dichloro-2,5-dimethylhexane and ferric chloride catalyst produced the

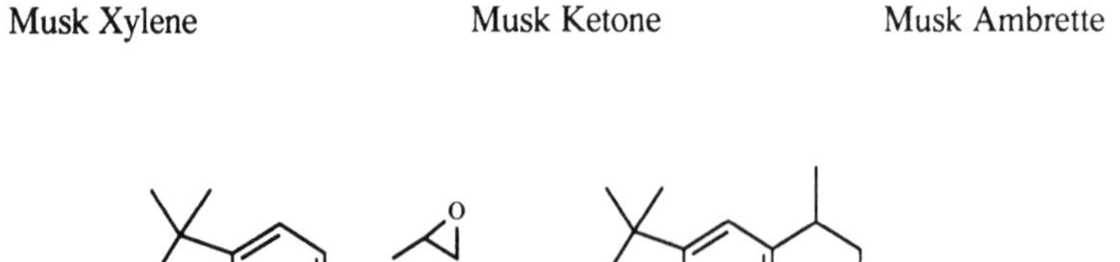

Musk Xylene   Musk Ketone   Musk Ambrette

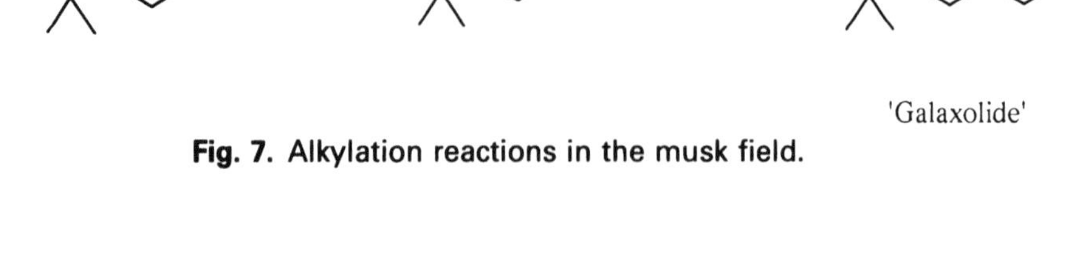

'Galaxolide'

Fig. 7. Alkylation reactions in the musk field.

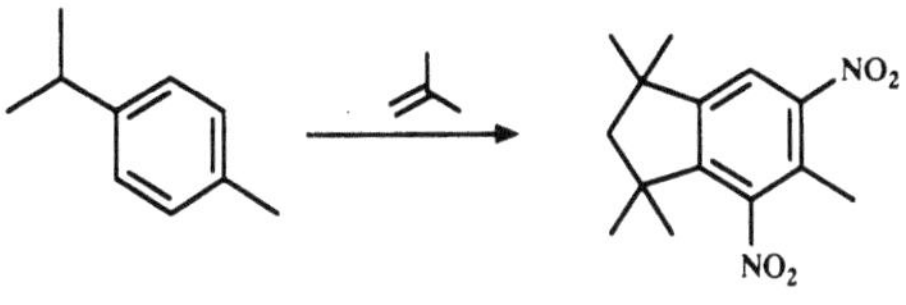

'Moskene'

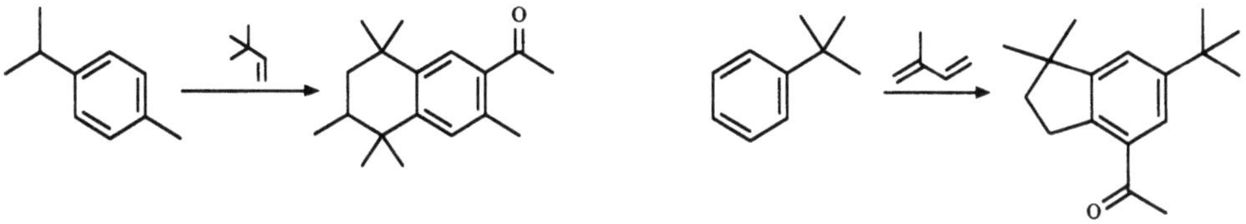

'Tonalide' (PFW)

'Fixolide' (GIV)

'Celestolide' (IFF)

'Crysolide' (GIV)

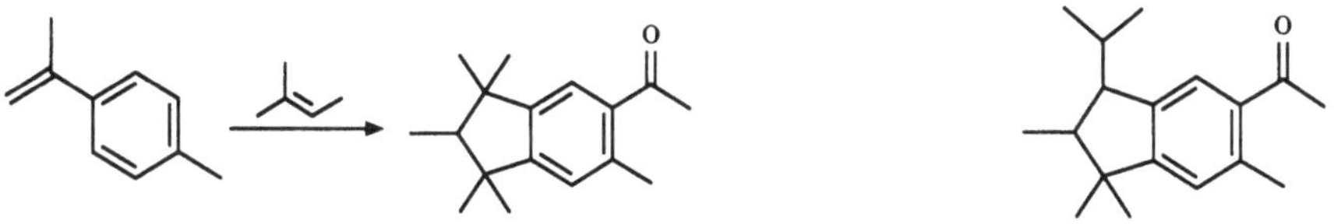

"Phantolide' (PFW)

'Traseolide' (NAAR)

Fig. 8. Cycloalkylation reactions (first step of musk production).

alkyltetralin which yielded Versalide® on acetylation. Versalide is no longer used by the fragrance industry because it produced neurotoxic effects in some animals.

The four commercial polycyclic musks shown in Fig. 8 are all derived from alkylindans or alkyltetralin, and are made by variations of the cycloalkylation.

Figure 8 also shows the reactions by which the hydrocarbon intermediates are formed. Formation of the Galaxolide and Celestolide precursors probably proceed by alkylation of the aromatic ring followed by cyclization. Sulfuric acid is the preferred catalyst for these reactions. Synthesis of the indan for the nitro musk Moskene® or the tetralin for Fixolide®/Tonalide® involves the formation of a *p*-cymene-derived carbonium ion by hydride transfer to the olefin. Saturated hydrocarbons are the by-products, neohexane in one reaction and isobutane in the other. Expensive neohexene, the hydride acceptor in the Fixolide-related hydrocarbon synthesis, adds considerably to the cost of Fixolide. The chemistry of the polycyclic musks and cycloalkylation are presented in detail by Wood.[51]

Isomerization/Rearrangement

Many perfume chemicals are produced in large quantities from α- and β-pinene, the main constituents of turpentine. The skeletal structures of all of these products differ from the parent terpenes and are formed by isomerization or rearrangement.[55] Only a few of the many possible examples will be cited. Treatment of α- or β-pinene with aqueous sulfuric acid forms terpin hydrate. Purification and dehydration yields perfume-grade α-terpineol. A crude and inexpensive mixture of terpineols sold as 'pine oil' is made by a one-step hydration of α-pinene with aqueous mineral acid. Isomerization of α-pinene over an acid catalyst in the absence of water forms camphene.[56] Addition of acetic acid to camphene produces isobornyl acetate. d,l-camphor is obtained by oxidation of borneol from saponification of isobornyl acetate.

Pyrolysis of α-pinene at 400–450°C yields dipentene; at a higher temperature allo-ocimenes are formed.[57] Alcohols from allo-ocimenols are useful in perfumes. The pyrolysis product of pinane, 3,7-dimethylocta-1,6-diene, is hydrated to a valued perfume chemical, dihydromyrcenol.[58]

Myrcene has been produced in large quantities by the pyrolysis of β-pinene.[59] Hydrochlorination and further reactions of the mixture of chlorides lead to linalool, geraniol/nerol, citral and other important terpenoids used in perfumery.[60] In the past a large portion of C_{10} terpene derivatives consumed by the perfume industry came from β-pinene via myrcene. Commercialization of the α-pinene route to these materials at Colonel's Island has reduced the importance of β-pinene and myrcene. For historical reasons it is interesting to note that at one time the demand for β-pinene was so great that it was actually produced commercially by isomerization of α-pinene.[61,62] The process is inefficient because the thermodynamic equilibrium mixture contains only about 5% of the higher-boiling β-isomer. The process was abandoned because recovery of β-pinene from the equilibriated mixture was too costly.

Other important isomerization reactions in the terpene series are: pinanol to linalool;[63] linalool to geraniol;[64] and dehydrolinalool to citral.[65–68] Some of these isomerization reactions are shown in Fig. 9.

Isomerization reactions are run on compounds other than terpenes. Prominent examples are safrole to isosafrole and eugenol to isoeugenol. In both cases an olefinic bond is shifted from the 1,2- to the 2,3-position. Isomerization can be effected by heating safrole or eugenol with 50% sodium hydroxide or potassium hydroxide[69] or catalytically with compounds of ruthenium[70] or of rhodium.[71]

Diels–Alder Reaction

Four commercially successful perfume chemicals made by Diels–Alder reactions involving addition of a dienophile to a diene are shown in Fig. 10. In each case two structures are possible depending on the mode of addition. The structures in Fig. 10 are the principal ones formed.

Three of the four materials are produced from myrcene. Lyral®, a mixture of 4- and 3-(4-hydroxy-4-methylpentyl)-3-cyclohexene-1-carboxaldehyde, was the first of the group introduced. It is prepared from myrcenol and acrolein with[72] or without[73] a catalyst. Myrac aldehyde® (IFF)/Myraldene®

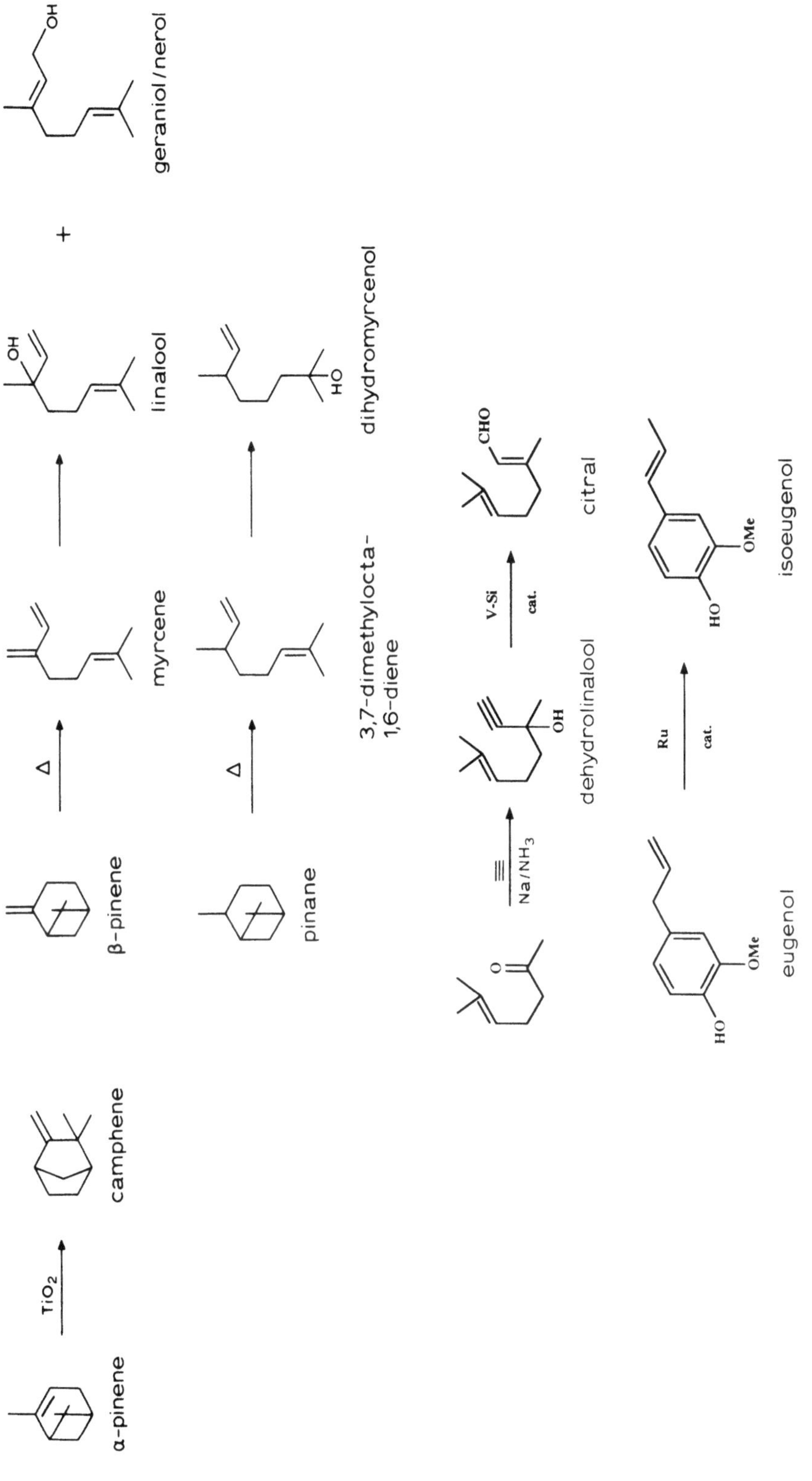

Fig. 9. Isomerization reactions.

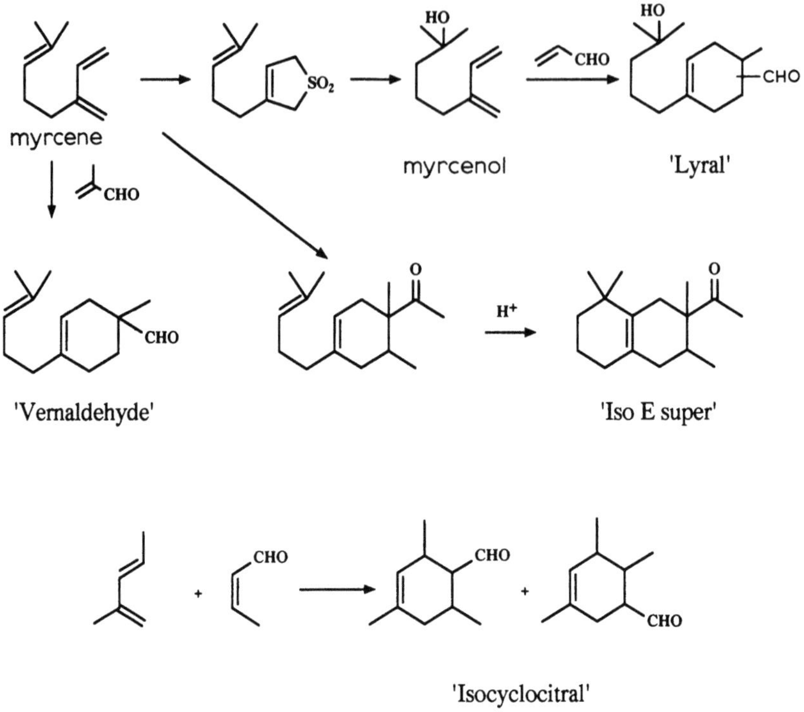

Fig. 10. Diels–Alder reactions.

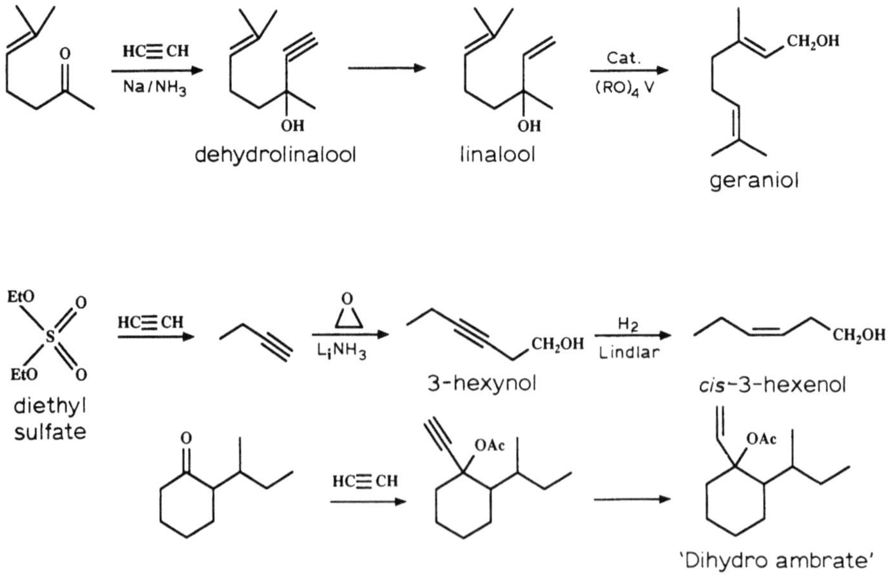

Fig. 11. Ethynylation reactions.

(Givaudan)[74] and Vernaldehyde®[75] are made from myrcene and acrolein and methacrolein respectively. Isocyclocitral[76] was the first commercial Diels–Alder reaction product introduced into perfumery. It is prepared from crotonaldehyde and 2-methyl-1,3-pentadiene. The diene is made by dehydration of the readily available diol, 2-methyl-2,4-pentanediol.

Isocyclemone (IFF), also known in the trade as Iso E Super®,[77] is the cyclized product of the Diels–Alder intermediate as shown in Fig. 10.

Ethynylation

The total synthesis of linalool, citral, ionones, nerolidol and finally vitamins E and A commercialized by Hoffmann–La Roche involves ethynylation with acetylene in several steps along the way.[78] The reaction of 6-methyl-hept-5-en-2-one with sodium acetylide in liquid ammonia to form dehydrolinalool shown in Fig. 11 is typical.

An ethynylation reaction is involved in the commercial synthesis of cis-3-hexen-1-ol, the leaf alcohol.[79] In the first step, diethyl sulfate is ethynylated with sodium acetylide in liquid ammonia. In the second step the resulting 1-butyne is added to ethylene oxide with lithium amide, also in liquid ammonia. The final step is a stereospecific hydrogenation of the triple bond to the *cis*-olefin with Lindlar palladium catalyst. Ambrate®—later replaced by Dihydroambrate—is made from secondary butylcyclohexanone via ethynylation of the ketone.[80]

Ozonolysis

Two important aldehydes widely used in perfumery, heliotropin and vanillin, are manufactured in Japan and China by ozonolysis of olefinic double bonds. The technology and amounts produced are closely guarded secrets. Isoeugenol is the starting material for vanillin[81] and isosafrole for heliotropin.[82] Ozonolysis requires a substantial investment in electrical equipment and consumes large amounts of electrical energy, which is expensive in much of the industrialized world. However, the reaction is clean and processing costs are low. Under the right circumstances ozonolysis apparently can be competitive with other methods of producing heliotropin and vanillin. Brassylic acid, the raw material for ethylene brassylate, was made in the past by ozonolysis of erucic acid,[83] but this process has been replaced by other, more economical, procedures.

Epoxidation

Epoxidation is more important for the preparation of intermediates than of perfume chemicals. Epoxidation of α-pinene with peracetic acid or air produces pinane epoxide which is rearranged to campholenic aldehyde. Aldol condensation with methyl ethyl ketone followed by hydrogenation results in a mixture of alcohols with a strong sandalwood odor (Sandalore®).[84] The

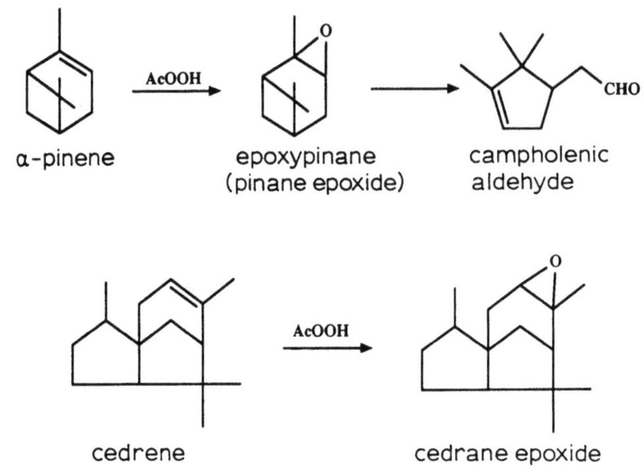

Fig. 12. Epoxidation products.

reaction of peracetic acid and cedrene produces an epoxide with a fine woody odor (Fig. 12).

Large amounts of β-phenylethyl alcohol have become available as a by-product from the co-production of propylene oxide and styrene via epoxidation with ethylbenzene hydroperoxide. The hydroperoxide is prepared by passing air into ethylbenzene in the presence of hydroperoxide as an initiator. The hydroperoxide is reduced to methyl phenyl carbinol and β-phenylethyl alcohol. Styrene is produced by dehydration of the alcohol mixture. Phenylethyl alcohol accumulates in the dehydration heavy fractions because it dehydrates at a slower rate than the secondary alcohol. Although the yield of phenylethyl alcohol is small, substantial amounts are produced because of the scale of the operation. Two large plants utilizing this process are operated by Air Reduction Corporation. The chemistry is summarized in Fig. 13.

It has to be mentioned that a more recent method of producing β-phenylethyl alcohol consists of the hydrogenation of styrene epoxide obtained from the aforementioned styrene.

Hydroformylation

The synthesis of aldehydes from olefins, carbon monoxide and hydrogen is known as hydroformylation or the 'Oxo-reaction'. After its discovery by Roelen in 1938, it was commercialized by the petrochemical industry for the large-scale production of aldehydes and alcohols. These chemicals were used mainly as intermediates for the manufacture of industrial products such as plasticizers, resins, solvents, etc. Some, however, were useful directly in perfumery or as intermediates. For industrial applications aldehydes are usually converted to alcohols, esters, amines and acids. In recent years

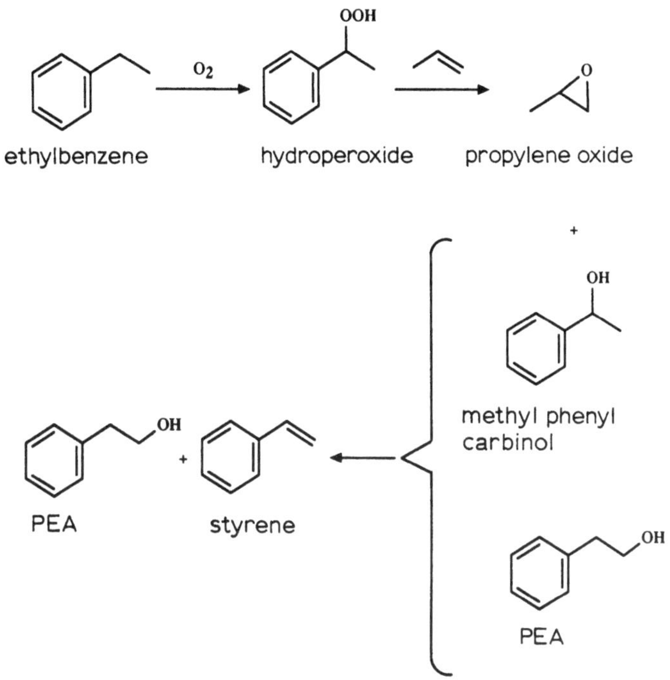

Fig. 13. Phenylethyl alcohol (PEA)—epoxidation by-product.

chemicals have been produced by hydroformylation specifically for use in perfumes (Fig. 14). Chalk[85] has made an informative and timely review of hydroformylation and its application to the synthesis of perfume chemicals. It will be mentioned again in Chapter 20, in which several examples of recent research work are given.

Grignard Reaction

The Grignard reaction is used to synthesize two tertiary alcohols for perfumery, dimethyl benzyl carbinol (used mainly as the acetate) and dimethyl phenethyl carbinol. These products have been made for many years using diethyl ether as the solvent in spite of the potential hazards of flammability and formation of explosive peroxides. Tetrahydrofuran, dibutyl ether and diamyl ether have also been used,[86] but these solvents present problems as well. In diethyl ether, dimethyl benzyl carbinol can be prepared from benzyl magnesium chloride and acetone. Similarly, dimethyl phenylethyl carbinol can be prepared from the phenethyl Grignard compound and acetone. Other routes to these materials are also available. For example, the addition of the methyl Grignard compound to benzalacetone also yields dimethyl phenethyl carbinol.

An alternative procedure for preparing these alcohols using homogeneous palladium catalysts was reported by Chalk.[87]

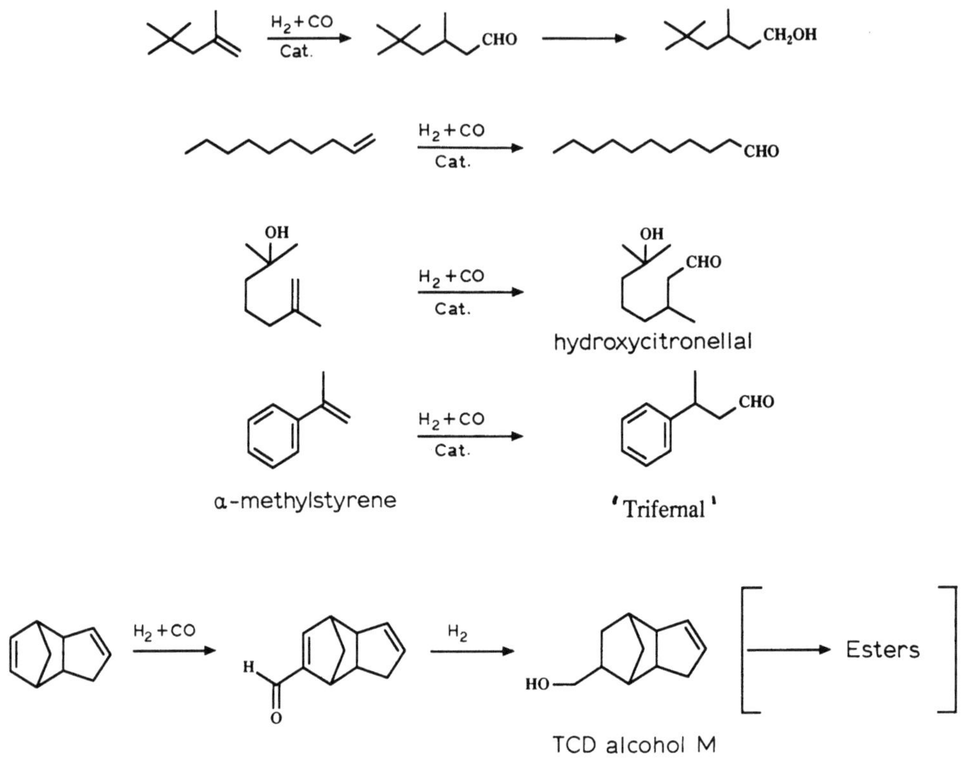

Fig. 14. Hydroformylation products.

FINAL REMARKS

The reactions discussed above are relatively simple and are widely used to produce bulk perfume chemicals because they are adaptable for plant-scale production. More complex reactions are often used to synthesize some of the 'exotic' perfume ingredients mentioned earlier in the introduction to this chapter. For these materials adaptability of a synthesis to plant-scale production is not too important. The practicality of a synthesis depends on whether the performance of the product warrants the effort and expense of preparing it. So-called 'exotic' chemicals are usually prepared in a special kilo lab or by the chemist who developed the synthesis. The number and diversity of reactions employed are limited only by the ingenuity of the chemist.

REFERENCES

1. Anon., *Flavor and Fragrance Materials—1989*. Allured Publishing Corp., Wheaton, IL, 1989.
2. Sabetay, S., *Soap, Parfum. Cosmet.*, **46** (1973) 83.
3. Riezebos, G., *Soap, Parfum. Cosmet.*, **45** (1972) 759.
4. Riezebos, G., *Parfums, Cosmet., Savons, Fr.*, **3**(2) (1973) 65.

5. Bauer, K. & Garbe, D., *Common Flavor and Fragrance Materials.* VCH Verlagsgesellschaft, Weinheim, 1985.
6. Hall, R., Klemme, D. & Neinhaus, J., *The H & R Book, Guide to Fragrance Materials.* Johnson, London, 1985.
7. Anon., *Essential Oils and Oleoresins.* International Trade Center UNCTAD/GATT, Geneva, 1986.
8. Arctander, S., *Perfume and Flavor Chemicals (Aroma Chemicals),* Vols I and II. Arctander Publishing, Elizabeth, NJ, 1969.
9. Arctander, S., *Perfume and Flavor Materials of Natural Origin.* Arctander Publishing, Elizabeth, NJ, 1960.
10. Guenther, E., *The Essential Oils,* Vols I–VI. Van Nostrand, Princeton, NJ, 1947–1952.
11. Treibs, W., *Die Aetherischen Ole,* Vols I–VII. Akademie Verlag, Berlin, 1956–1961.
12. Müller, A., *Internationaler Riechstoff-Kodex.* Alfred Hüthig-Verlag, Heidelberg, 1950.
13. Bedoukian, P. Z., *Perfumery and Flavoring Synthetics,* 3rd edn. Allured Publishing Corp., Wheaton, IL, 1986.
14. Bedoukian, P. Z., *Perfumery and Flavoring Materials,* Annual Review Articles 1945–1990 from the *Perfumer and Flavorist* (originally the *American Perfumer & Essential Oil Review*). Allured Publishing Corp., Wheaton, IL.
15. Beets, M. G. J., *Structure–Activity Relationships in Human Chemoreception.* Applied Science Publishers, London, 1978.
16. Theimer, E. T., *Fragrance Chemistry.* Academic Press, New York, 1982.
17. Billot, M. & Wells, F. V., *Perfumery Technology,* 2nd edn. Ellis Horwood, Chichester, 1981.
18. Fenn, R., *Perfumer & Flavorist,* **14**(2) (1989) 1.
19. Lewinsohn, *Perfumery & Essential Oil Record,* **14** (1923) 398.
20. Carpenter, M. S. & Easter, W. M., US Patent 2 875 131 (1959).
21. Becking, D. H., US Patent, 3 985 809 (1976).
22. Wagner, A., *Riechstoff Ind.,* **1** (1926) 65.
23. Fritzsche & Co., German Patent 207 702 (10 March 1909).
24. Polinski, L. M., Huang, J.-Der & Dorsky, J., Ger. Offenlegungschrift (Federal Republic of Germany), 2 338 291 (1973).
25. Fisher, G. S., Stinson, J. S. & Goldblatt, L. A., *J. Am. Chem. Soc.,* **75** (1953) 3675.
26. Fisher, G. S., Stinson, J. S., Moore, R. N. & Goldblatt, L. A., *Ind. Eng. Chem.,* **47** (1955). 1368.
27. Filliartre, C. & Lalande, R., *Bull. Soc. Chim. Fr.,* (1968) 4141.
28. Beets, M. G. J., British Patent 812 727 (1959).
29. Swiss Patent 127 159 (1927).
30. Kimmel, W. *et al., J. Org. Chem.,* **23** (1958) 153.
31. Saucy, G. & Marbet, R., *Chimia,* **14** (1960) 362.
32. Johnson, J. D., *Organic Reactions,* **1** (1942) 210.
33. Darzens, G., *Compt. Rend.,* **139** (1904) 1214 .
34. *Organic Synthesis,* Coll. Vol. III. John Wiley & Sons, Inc., New York (1955), p. 727.
35. Blumenthal, J. H., US Patent 3 463 818 (1969).
36. Mookherjee, B. D., US Patent 4 264 830 (1981).
37. Firmenich & Cie, British Patent 907 431 (1961).
38. Bollman, M. & Kroning, E., US Patent 2 582 743 (1952).
39. Sommerville, W. T. & Theimer, E. T., US Patent 2 927 127 (1960).
40. SCM, US Patent 4 018 842 (1976).
41. Dorsky, J. & Easter, W. M., US Patent 3 499 937 (1970).
42. Lindlar, H., *Helv. Chim. Acta,* **35** (1952) 446.
43. Bedoukian, P. Z., *J. Agr. Food Chem.,* **19** (1971) 1111.

44. Dankert, L. J. & Permoda, D. A., US Patent 3 124 614 (1964).
45. Booth, A. B., US Patent 2 843 636 (1958).
46. Yoshida, T., Komatsu, A. & Indo, M., *Agr. Biol. Chem.* (*Tokyo*), **29** (1965) 824.
47. Olah, G. A., *Friedel-Crafts and Related Reactions*, Vols 1–3. Interscience, New York, 1963–1965.
48. Schaarschmidt, A., Herman, L. & Szemco, B., *Ber.* **58B** (1925) 1914.
49. Carpenter, M. S., US Patent 2 013 710 (1935).
50. Theimer, E. T., US Patent 3 532 719 (1970).
51. Wood, T. F., Synthetic benzenoid musks. In *Fragrance Chemistry*, ed. E. T. Theimer. Academic Press, New York, 1982, p. 509.
52. Pfau, S. & Ofner, A., *Helv. Chim. Acta*, **9** (1926) 670.
53. Bruson, H. A., & Kroeger, J. W., *J. Am. Chem. Soc.*, **62** (1940) 36.
54. Carpenter, M. S., US Patent 2 897 237 (1959).
55. Naves, Y. R., *Russian Chem. Rev.*, **37**(10) (1968) 779.
56. Derfer, J. M., *Perfumer & Flavorist*, **3**(1) (1978) 45.
57. Hawkins, J. E. & Hunt, H. G., *J. Am. Chem. Soc.*, **73** (1951) 5379.
58. International Flavors & Fragrances (IFF), US Patent 3 487 118 (1969).
59. Goldblatt, L. A. & Palkin, S., *J. Am. Chem. Soc.*, **63** (1941) 3517.
60. Ansari, H. R., *Flavor Ind.*, **1** (1970) 252.
61. Derfer, J. M., US Patent 3 359 342 (1967).
62. Derfer, J. M., US Patent 3 360 581 (1967).
63. Ohloff, G. & Klein, E., *Tetrahedron*, **18** (1962) 37.
64. Kane, B. J., US Patent 4 254 291 (1981).
65. Charbordes, P., Kuntz, E. & Varagnat, J., *Tetrahedron*, **33** (1977) 1775.
66. Charbordes, P., German Patent 1 811 517 (1968).
67. Hindley, N., Pauling, H. & Andrews, D. A., German Patent 2 353 145 (1973).
68. Hindley, N., Pauling, H., Andrews, D. A. & Hindley, N. C., *Helv. Chim. Acta*, **59** (1976) 1233.
69. Nagai, J., *J. Chem. Soc. Japan/Ind. Chem. Sect.*, **29** (1926) 364.
70. Grandilhon, P., German Patent 1 936 727 (1969).
71. Chalk, A. J., German Patent 2 508 347 (1975).
72. Sanders, J. M. & Schreiber, W. L., US Patent 4 007 137 (1975).
73. Teegarden, R. W. & Steinbach, L., US Patent 2 947 780 (1960).
74. Dahill, R. T. & Kaiser, L., German Patent 2 321 472 (1973).
75. Kitchens, G. C., US Patent 2 842 598 (1958).
76. Naef & Cie, French Patent 672 025 (1929).
77. Hall, J. B. & Sanders, J. M., US Patent 3 929 677 (1975).
78. Ref. 13, pp. 270, 442.
79. Bedoukian, P. Z., *Am. Perfum*, **78** (1963) 31.
80. Nikawitz, E. J., Tavares, R. F. & Easter, W. M., US Patent 3 769 330 (1973).
81. Ref. 13, p. 409.
82. Ref. 13, p. 182.
83. Carlson, K. D., Sohnes, V. E., Perkins, R. B. & Huffman, E. L., *Ind. Eng. Chem. Prod. Res. Div.*, **16**(1) (1977) 95.
84. Naipawer, R. E. & Easter, W. M., US Patent 4 052 341 (1976).
85. Chalk, A. J. In *Flavors and Fragrances: A World Perspective*. Elsevier Science Publishers, Amsterdam, 1988, p. 867.
86. Ref. 16, p. 280.
87. Chalk, A. J. & Magennis, S. A., US Patent 4 070 374 (1978).

Chapter 15

Compounding

ALEXANDER BOECK
Henkel KGaA, D-4000 Düsseldorf, Germany
&
HANS-UDO FERGEN
Haarmann & Reimer GmbH, D-3450 Holzminden, Germany

1 INTRODUCTION

Perfumes and perfumed products are part and parcel of our everyday life. The demand worldwide for perfumes is enormous and constantly on the increase. At present over 300 000 tonnes of aroma chemicals and perfume oils valued at over 5000 million US dollars are used for cosmetics, washing and cleansing agents and other products. The perfume industry has become a major business.

While in earlier times perfumes were blended and their virtues extolled by priests, alchemists, apothecaries, perfumers or even charlatans, since the mid-19th century the production of perfumes has gradually become manufacturing in character and is now largely automatic and computer-controlled.

2 PERFUME OIL FORMULA

Perfume oils are homogeneous, clear, mostly yellow-tinted and often very complex mixtures of natural and synthetic fragrances. Currently about 20–30% of the demand for perfume oils in terms of quantity is covered by natural substances and 70–80% by synthetic products. In terms of numbers there are well over 3000 different, mostly liquid and non-crystalline, fragrances. Perfume oils generally consist of 20 to 200 different components.

Compared with earlier times it is far more expensive to mix perfume oils because of this increased variety of substances and the vast number of complex perfume oil formulas from the many products to be perfumed.

In all perfume houses the perfume oil formulas are among the best kept secrets and represent the know-how. They play a major role in the success of the companies. Therefore, access to the formulas is strictly controlled and numerous safety provisions are incorporated. Earlier formulas were kept in safes, but now, in the age of computers, formula safety is provided by card readers, passwords, user IDs and the like. The metering and mixing processes naturally represent a risk in the security system. However, careful selection of mixing staff and suitable coding of the individual fragrances and formulas minimize the risk. The expanding use of computer control in compounding also increases formula safety.

3 MANUFACTURE OF PERFUME OILS

The manufacture of perfume oils means metering of the individual components in accordance with the formula, followed by blending for homogenization.

In view of the wide variety of individual substances and perfume oil formulas, the latter should be devised in such a way as to provide a logical sequence of products to be metered one after the other.

The following method has proved successful:

—addition of products with low volatility in liquid form;
—dissolution, possibly under heat;
—cooling as required;
—addition of more volatile substances (aldehydes, esters, etc.);
—incorporation of crystalline substances and resins (in some cases it may be wise to add these substances at the beginning).

Metering Techniques

In the current state of the art there are basically two different types of metering, volumetric and gravimetric metering.

Volumetric metering devices work on the principle of displacement. The metering flow or quantity is not measured but produced by reproducible drive parameters. There is usually a relatively simple relationship between metering flow and drive parameters.

Some of the many volumetric metering devices include oval wheel meters and metering pumps for liquids, and screw-type metering devices and star feeder devices for solids.

In gravimetric processes the metering flow weight is constantly measured and indicated.

Since density fluctuations have an effect in all metering methods with mass measurement, gravimetric metering is the most accurate.

Volumetric appliances may fit the bill in a large number of cases, especially when no special demands are made. However, gravimetric methods have so far been preferred for metering perfume oil components.

In the production of perfume oils accurate metering of the individual, mostly high-quality, components is the decisive criterion for quality and hence a guarantee of constant product composition.

Weight is therefore the essential factor for quality assurance and materials handling.

Conventional Metering

The individual components are weighed by hand. In the past, mechanical scales (indicator scales) were used, which did not permit data input and recording. In recent years, however, electronic scales have taken over.

Depending on the procedure, the weighing system is fixed or mobile.

—The scales are installed and the individual components are brought from storage to the scales. Streamlining is provided in the form of rows of battery tanks for frequently used aroma chemicals with feed pipes to the weighing station (product to scales).

—The scales are mobile and are moved to the individual storage containers which are also suitably arranged in battery tank rows (scales to product).

To illustrate this, these two basic principles are depicted in Figs 1 and 2.

Automatic Metering on a Production Scale

The complexity leads automatically to thoughts of how the conventional metering process can be made easier.

A first step towards automation was to control the weighing system with punched cards. The aroma chemicals flowed from the storage containers to the central metering stations and metering was controlled in accordance with the given formula by punched cards on which the component data were printed in the form of punch-holes. This process has now been superseded by the rapid development of microprocessor technology.

The electronic scales thus form the heart of the system and, together with control mechanisms and computers, provide the basis for a wide variety of

Fig. 1. Conventional metering methods: product to scales.

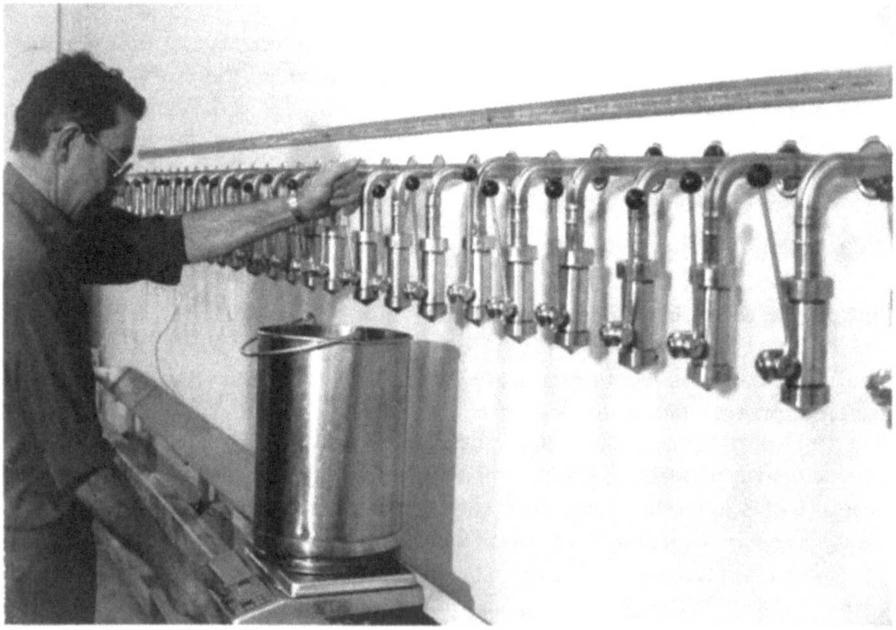

Fig. 2. Conventional metering method: scales to product.

tasks which go well beyond the simple recording of weight. The use of electronic weighing systems can be seen from two main viewpoints. On the one hand, weight and the respective operating data are to be determined and further processed. On the other, there is the weight-dependent control of the metering processes.

As a basic parameter, weight guarantees

—correct composition of the individual substances in terms of quantity;
—control and monitoring of material consumption;
—logical control of sequences.

Basic parameters for the design of an automatic metering station are the required throughput (size and number of batches) and the requisite weighing accuracy for the individual components. The time needed to make up a batch is known as the lead time, which is determined by the number and duration of weighing operations and the need for weighing accuracy.

Electronic control is the 'brain' of the installation. It controls and monitors the process sequence and organizes the logging.

Selection of automated equipment will depend by and large on data volume, the requisite memory capacity and the operating system.

While process control and interlock circuits in metering installations are mostly achieved with the aid of programmable logic controllers (PLC), process computers with and without external data memory are used for formula management, on-screen dialogue and extensive recording.

Depending on the production standard, there are quite different ideas on equipment set-up and operating systems for weighing devices with integrated or dislocated metering controls.

The main function groups of metering control are:

—nominal/actual comparison for coarse/fine metering per component with tolerance control and metering time monitoring;
—metering sequence control;
—keys for functions, data, formulas;
—data interfaces for peripheral equipment, e.g. electronic data processing, printer, remote indicators;
—control functions.

The demand for high accuracy levels calls not only for powerful computers but also for metering valves with appropriate regulating ranges, the lowest limit being drop metering. They must also be suitable for liquids with differing properties, from water-thin to highly viscous.

The installation set-up will naturally depend on the specific operational circumstances and requirements. In general, metering of liquid perfumes frequently used in large quantities is computer-controlled from the storage containers by way of an automatic, self-monitoring metering device. Crystalline substances and products used less often are added by hand at so-called hand weighing stations, although they are still controlled and monitored by the same computer system.

Through this 'composite' system, the entire formula is weighed under control and potential errors are practically excluded.

Since the weighing accuracy depends on the weighing range, there are basically two suitable methods of obtaining maximum accuracy:

—different weighing stations with varying weighing ranges (e.g. under 10 kg ± 1 g; 10–100 kg ± 10 g; 100–500 kg ± 50 g)
—only one weighing station at which only one component is weighed each time, which is removed on reaching a given weight by discharge to the mixing container.

This basic principle of computer-controlled weighing gives high mixing capacity and accuracy with extraordinarily little susceptibility to interference.

Figures 3–6 show examples of metering installations. The scales stand in the centre as the value indicator in the form of loadcells or electronic scales.

Figures 7–10 show metering systems found in practice.

Automatic Metering on a Laboratory Scale

There are basically two possibilities for computer-controlled metering of perfume oil components on a laboratory scale (sample quantities up to 500 g):

—comparable with production; and
—with a robot system.

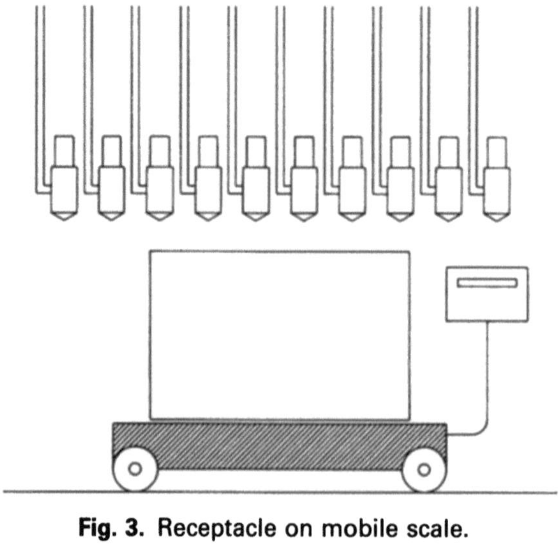

Fig. 3. Receptacle on mobile scale.

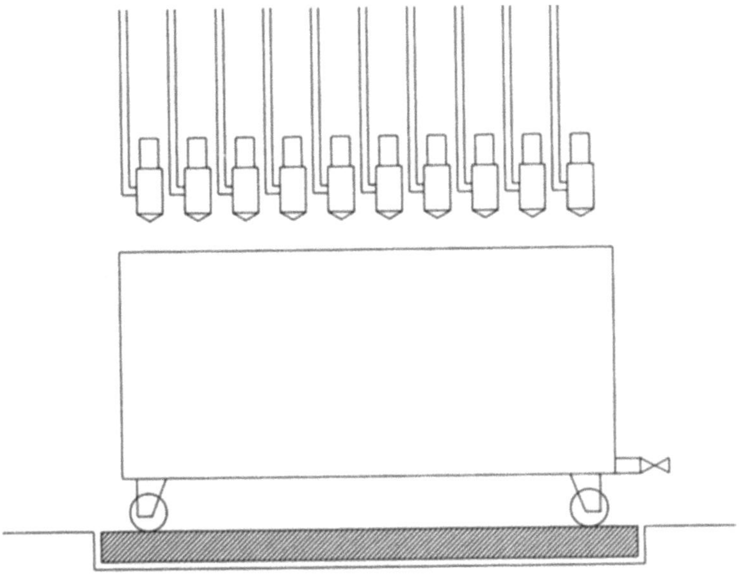

Fig. 4. Fixed scale with receptacle.

Metering Similar to Production Scale

The basic principle in this system is the same as the computer-controlled metering installation on a production scale, i.e. all metering and weighing processes are directed and controlled by a central processor. Once again there is automatic and manual weighing, although a higher degree of automation can be achieved in a laboratory mixing station. The liquid perfumes metered

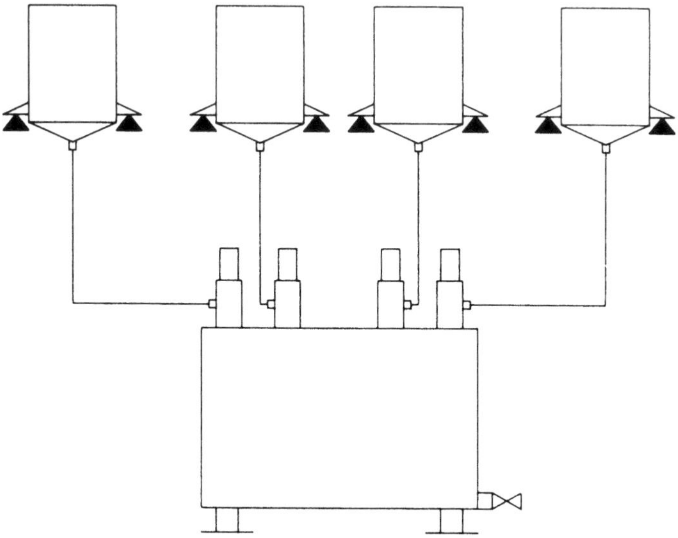

Fig. 5. Storage containers on loadcells.

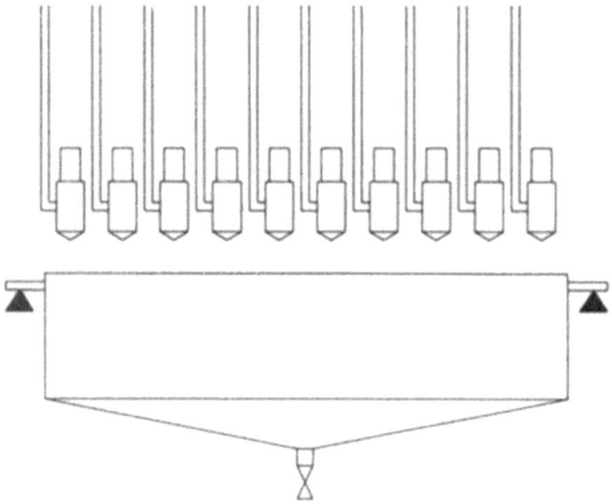

Fig. 6. Scale receptacle on loadcells.

Fig. 7. Multistage metering system.

Fig. 8. Valve arrangement.

Fig. 9. Metering system station.

Fig. 10. Storage receptacles, valves and product feed tubes of a metering station.

Fig. 11. Storage receptacles of a laboratory mixing station.

automatically are arranged in rectangles or circles on different levels of stainless steel receptacles of different sizes (e.g. 1–5 litres). Teflon or stainless steel pipes lead from all the receptacles to circular metering valve stations below.

To protect air-sensitive aroma chemicals and to control the metering rate, all storage containers are placed under slight nitrogen overpressure, which can be varied according to liquid flow capacity.

Under the metering rim, electronically controlled laboratory scales with weighing receptacles are operated; at the metering stations the perfume oils, depending on the formula in question, are weighed with extremely high accuracy.

Labels are affixed to the weighing receptacles which, in addition to the perfume oil number and the required quantity in grams, show a bar code which is read with the aid of a laser scanner.

By means of a rotating belt and a handling device for placing the weighing receptacles on the scales and putting them back on the belt after weighing, fully automatic 24-hour operating is possible.

Of course not all perfume oils, especially crystalline ones, can be metered fully automatically in the manner described above. The formula is completed in manual metering rooms linked to the same computer. The bar code is read with a scanner and the ingredients still missing appear on a terminal screen one after the other. If the last component has been added, a log is made with nominal and actual formula. Figs 11 and 12 show a laboratory mixing station.

Fig. 12. Valve stations and receptacles of a laboratory mixing station.

Robot System

The raw materials are arranged in circles or rectangles in a store consisting of several tiers. A fixed robot 'grabs' the raw materials required for the formula with a pipette affixed to the storage container, conveys it to the scales hooked up to the computer, dispenses the requisite quantity and returns the pipette to the storage container. It would also be feasible to use an expendable pipette for each metering process or for the whole storage container to be used by the robot for metering (e.g. tipping a bottle).

The robot has advantages such as

—no feed pipes from the storage container to the weighing station;
—automatic metering of many components, an extension also being relatively simple;
—relatively little space required by the entire installation;
—possible automatic metering also of crystalline substances ('by spoon');

but also disadvantages such as

—high breakdown tendency because of the high number of movable parts and metering steps;
—danger of dripping when conveying the product from the storage container to the scales and back;
—stability problems when air enters the storage container as the product is removed;
—high time outlay for the individual metering processes.

Mixing of Perfume Oils

Mixing follows metering to homogenize the formula components. Mixing perfume oils means blending and compounding two or more products. The components may be solid, liquid or pasty.

Mixing is a process which depends on various parameters, such as

—mechanical effect on the product;
—duration of mixing process;
—initial distribution of products;
—product properties.

These parameters can be affected by the

—choice of mixing principle (stirring, injection, recirculation pumps, gas turbulence or a combination of these methods);
—determination of the mixing time by fixed mixer throughput times and mixer parameters;
—metering of individual components;
—choice of atmospheric conditions (temperature, pressure, etc.).

In the original discontinuous mixing process each component is weighed and added individually to the mixer. Each component is now mixed to a varying degree and the blending process begins. Depending on the mixing principle, mixing time and product properties, this process will be faster or slower until it reaches optimum distribution for the respective combination.

The sequence of mixing processes can be shown graphically by the shifts in concentration of the components.

When the job of mixing consists of blending two components with identical specific properties, more or less ideal concentration ratios are obtained after a sufficiently long mixing time (see Fig. 13).

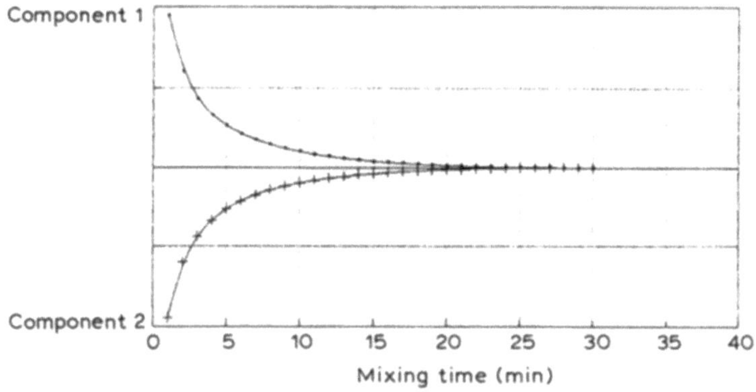

Fig. 13. Mixing (schematic illustration).

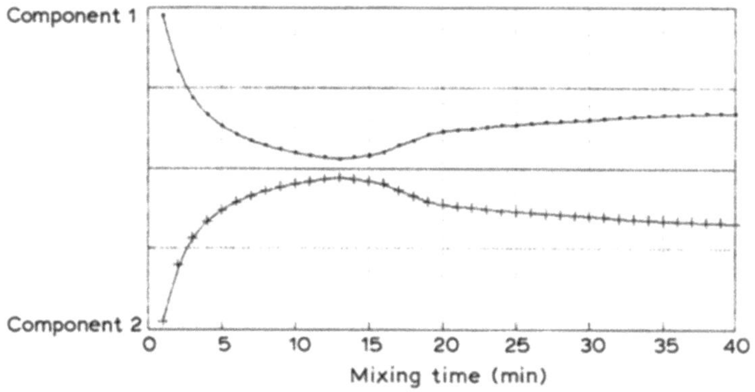

Fig. 14. Separation (schematic illustration).

A further characteristic feature is segregation on 'overmixing' or subsequent storage of product mix. This danger is especially acute with substances with very divergent specific gravities (see Fig. 14).

4 CONTROL BY BAR CODE

Computer control allows far better management of the entire perfume oil mixing process, as has already been mentioned.

The introduction of bar codes in perfume oil compounding is a further step towards guaranteed error-free weighing. Each fragrance or perfume oil is given a bar code, and identification is by laser scanner. In this way important steps in compounding can be controlled and human errors eliminated. Partial automation at least can also be achieved.

5 MACHINERY AND APPARATUS

It has been explained above that metering and mixing are the main operations in the manufacture of perfume oils. In view of the special tasks of the individual processes in production, certain requirements are made of machinery and apparatus.

Mixing Receptacles

Perfume oils are produced in different batch sizes. Therefore, different kinds of high-grade steel receptacle are needed, from the simple pot form to containers of 20 m^3 content.

Smaller receptacles are mounted on trolleys so as to be mobile. Larger appliances are fixed on work platforms. This arrangement allows additional devices for feeding and discharging liquids and protective gas as well as heating.

Figure 15 shows a diagram of a mixing apparatus.

Stirring and Mixing Systems

Rotating agitators, mostly fitted with individual drives, produce turbulent flow. The resultant turbulence mixes the components locally. At the same time the rotating flow displaces the entire quantity of liquid in the container. Centre and edge layers change places and compensate for locally produced mixtures.

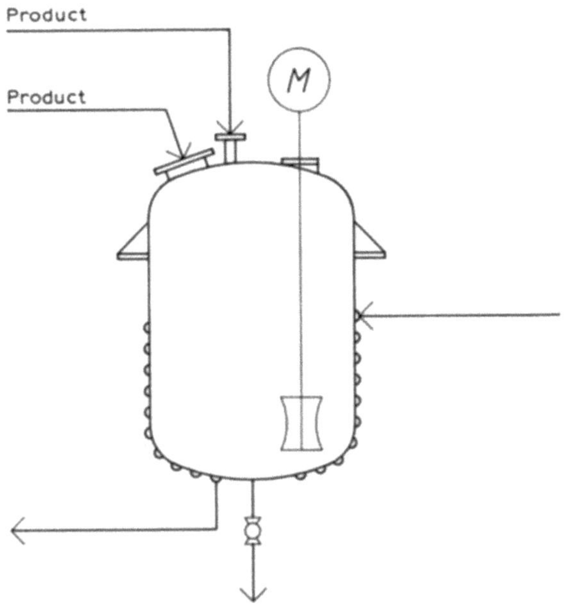

Fig. 15. Mixing tank (schematic).

The arrangement and shape of the rotating agitator parts are geared to the size and shape of the mixing container, the desired effect and the properties of the liquids.

A distinction must be made between the conventional slow mixers—such as blade and anchor mixers—and the more modern, fast mixers—such as propeller and turbine mixers.

These different types of mixer differ in their flow patterns. An overview of the prevalent flow in the mixer is given below:

Flow:	*Form*:
Tangential	Blade, anchor
Axial	Propeller
Radial	Turbine, disc

The trend is towards a departure from the slow and (in relation to the container size) large mixers. They are being replaced by fast propeller, turbine and disc mixers, which handle large quantities of liquid per unit time.

More recently, so-called rotor/stator systems, such as direct jet mixers, are being used for quicker and less abrasive mixing of perfume oils. These systems have considerable advantages since extremely strong suction and even stronger axial discharge are produced by special flow guide mechanisms. This results in an extremely rapid stirring of the medium without any air intake whatsoever (see Figs 16 and 17).

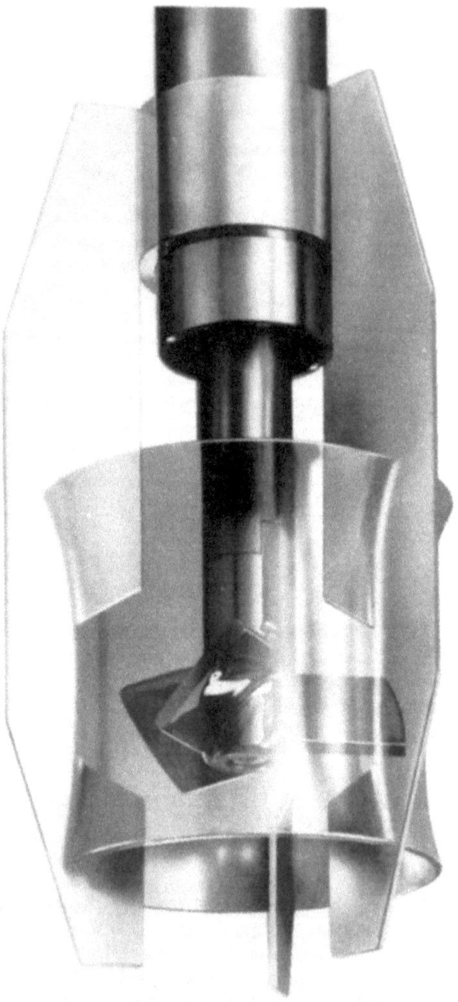

Fig. 16. Direct jet mixer-head.

Pumping Systems

To overcome differences in height and resistance to flow in pipes, fittings and apparatus pumps are used which convey the incompressible liquid from the region of low pressure, the suction chamber, to the region of high pressure, the pressure chamber.

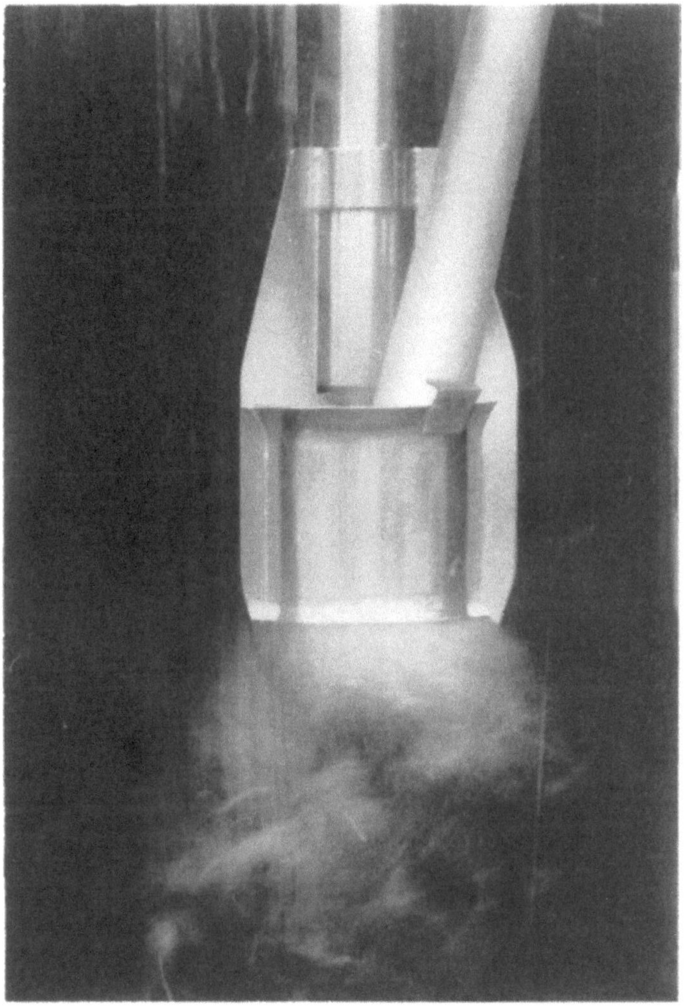

Fig. 17. Direct jet mixer-head, showing the homogeneous mixing of liquid and solid raw materials.

The mechanical conveyance of liquids is performed by different types of pumps:

—reciprocating pumps;
—diaphragm pumps;
—centrifugal pumps;
—rotary pumps.

Centrifugal pumps are often used in the perfume industry. These are compact in design, space-saving in operation and directly coupled with the

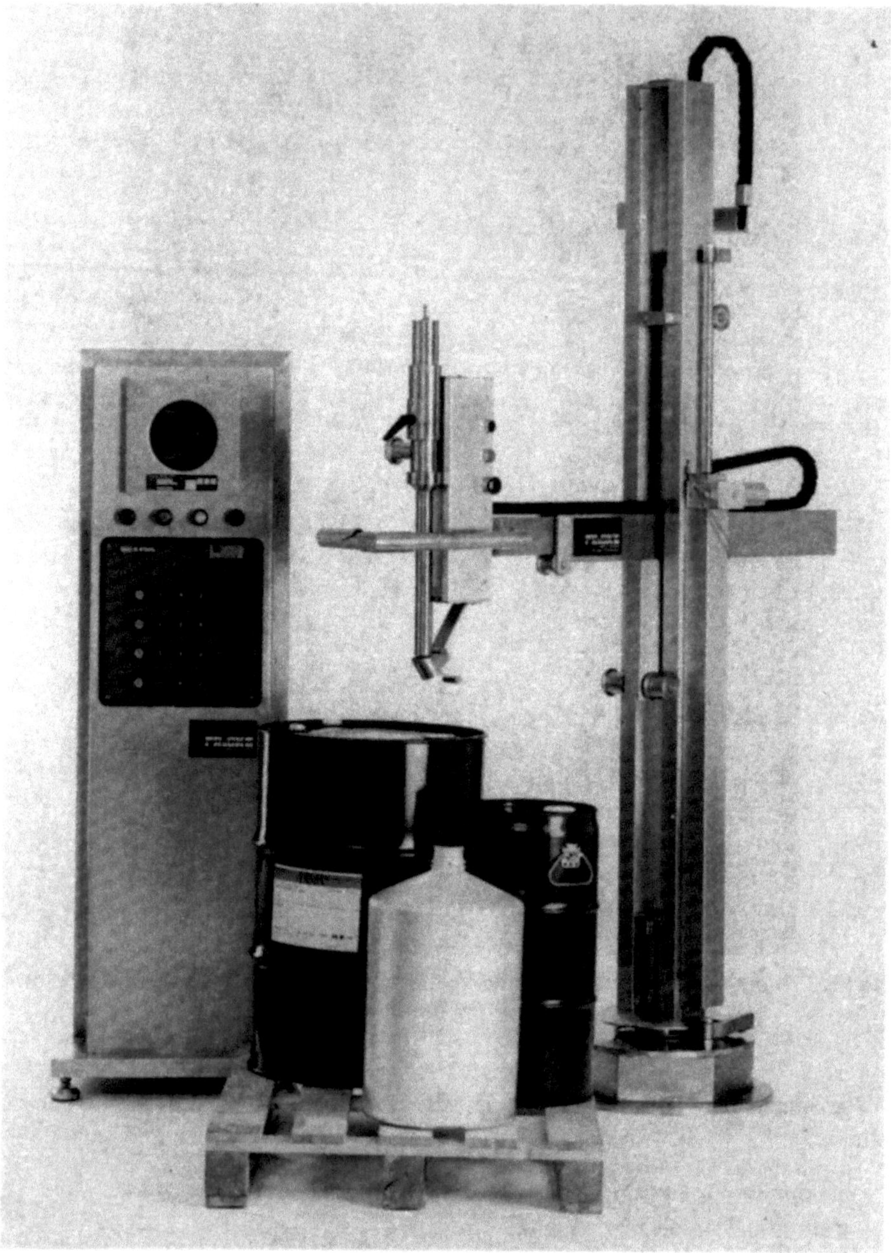

Fig. 18. Semi-automatic filling system.

drive motor. They are less prone to wear and cheap to operate. Centrifugal pumps convey large quantities at low and medium heights.

For special purposes, barrel-emptying pumps are used, which have the advantage of relatively small drives. Through the modular system a versatile combination of the individual motors with a pump-pipe arrangement to match the intended purpose is possible.

Filling Technique

At the end of the mixing process the perfume oils are filled into dispatch containers.

The properties of the product being decanted have a direct influence on the design of the filling installation, and especially on the design of the filling device, since liquids may be non-foaming (above-level filling), low foaming (part-below-level filling) or highly foaming (below-level filling).

The filling systems are controlled in different ways, these being electric, pneumatic and pneumatic–hydraulic.

Once again the scales play a central role, controlling coarse and fine filling. Filling installations can operate semi-automatically or fully automatically, depending on operating requirements. An example of a semi-automatic installation is shown in Fig. 18.

6 STORAGE OF FRAGRANCES AND PERFUME OILS

In general, storage is in aluminium cans or receptacles with internal protective coatings, in drums (25–200 kg) or in larger quantities in containers (mostly 800 kg) and storage tanks (up to 30 t) made of high-grade steel, which may be directly linked to the metering stations. Fragrances and perfume oils should be protected when stored, and not exposed to high temperatures.

Also suitable, but very expensive, are so-called high stacker shelves which take up little space and allow clearly arranged and computer-controlled storage of a large number of fragrances and perfume oils.

7 SAFETY ASPECTS

The head of a production plant is not only responsible for operational safety in the sense of safe functioning of the equipment and plant, he must also observe all legal regulations and provisions to protect the installations and their operating staff.

These requirements are found in the accident prevention regulations in force in nearly all countries, which can be divided into the following groups:

—basic legal provisions (e.g. Chemicals Law);
—industrial medicine, hygiene and physiology;
—substance properties: fire and explosion hazard, poisonous effect;
—working tools: equipment and plant needing licensing and monitoring (pressure vessels, steam boilers, means of transport, electrical plant);
—danger marking: premises, containers, pipes;
—respiratory protection;
—fire protection, fire-fighting;
—air protection measures (filter system, e.g. with activated charcoal; burning of exhaust air; biowashers);
—water protection measures (oil separators; retention basins).

8 FUTURE PROSPECTS OF COMPOUNDING

The conventional method ('indicator scales') will give way increasingly to the modern, considerably more accurate, easy-to-use, computer-controlled systems. The enormous variety of individual components, the large number of formulas and their complexity plus the high demands of exact compliance with formula provisions absolutely cry out for computer control.

Through automation, compounding can be carried out in a much calmer and cleaner environment (fewer 'barrel movements' and 'less spillage'). Working conditions will improve appreciably for operating personnel.

Automation will also provide a clearer picture of operation processes, improved control and, in the final analysis, better quality with greater readiness for delivery. In view of tougher and tougher competition and of continuing concentration in the industry, the highest possible quality standard is the prerequisite for success.

Chapter 16

The Toxicology and Safety of Fragrances

RICHARD A. FORD

Research Institute for Fragrance Materials, Inc., Englewoods Cliffs, NJ 07632, USA

1 INTRODUCTION

The consumer today hears more about toxicology and safety than even health professionals did 20 years ago. Yet it is the rare consumer that purchases a fine perfume, an aftershave, a cosmetic or any other of the hundreds of available fragranced products and questions whether it is safe to use. There are several reasons for this.

Firstly, the typical consumer thinks differently about products that go onto the skin than about those that are swallowed or inhaled. The latter are thought of as going into the body while substances applied to the skin are assumed incorrectly to remain outside the body. Secondly, when an adverse reaction to a fragranced product does take place, it is normally in the form of reddening of the skin, perhaps with some itching; effects that are not considered serious and can be stopped simply by no longer using the particular product that causes the problem. Thirdly, the fragrance industry has a remarkable record of safety due to the nature of ingredients that are used in fragrances, the small amounts that are used and the manner in which they are used. Most importantly, however, the fragrance industry's safety record is due to its program of testing, safety evaluation and self-regulation.

This chapter will discuss the industry program of testing and safety evaluation and will describe the principles and mechanisms which form the basis for the program. The types of tests conducted, the meaning of the varied results of these tests and how they form an integral part of the industry's self-regulatory program will also be described.

2 SELF-REGULATION IN THE FRAGRANCE INDUSTRY

The perception of safety by the consumer is reflected in the almost total absence of government regulations concerning the use of fragrances in consumer products. This is, of course, of great benefit to the industry. In no country is there any requirement for approval for either ingredients that go into fragrances or for the compounded fragrance itself. There are no requirements to test fragrance materials for safety for use in consumer products. There are no requirements to list the ingredients of fragrances on consumer products. Of course, there are other regulations that may require the testing or labeling of fragrance materials related to their transportation, manufacture, and/or potential environmental effects, but not in relationship to their safety in consumer products.

How can this be? There are several reasons but, perhaps, the most important is the fact that the industry, with great foresight, over 20 years ago set up a self-regulatory procedure that has yet to be seriously challenged by governments or consumers. This procedure involves two international organizations, the International Fragrance Association (IFRA) and the Research Institute for Fragrance Materials (RIFM).

IFRA is an association of national trade associations and represents over 100 fragrance manufacturers in 15 countries. Its headquarters are in Geneva, Switzerland. All member countries are represented on the IFRA Technical Advisory Committee (TAC), a committee that represents the scientific and technical expertise of the industry. The TAC reviews all available data (mostly supplied by RIFM) on fragrance ingredients and, when necessary, issues recommendations for the safe use of these materials. These recommendations, which include restrictions on use levels, minimum purity standards and even recommendations not to use some substances at all, are all published in the IFRA Code of Practice.[1] Specific examples of how guidelines are developed are given later.

All members of IFRA have agreed to adhere to the recommendations in the Code of Practice. The Code also includes other self-regulatory disciplines including guidelines for environmental protection. Copies of the Code are available at a minimal reproduction charge from IFRA and all companies engaged in the manufacture and use of fragrances should not only have a copy but adhere to the recommendations therein.

IFRA also represents the interests of the fragrance industry to national and international authorities, assuring them of knowledge not only of the guidelines in the Code of Practice, but of all other information necessary to demonstrate that the industry self-regulatory program is functional and adequate to assure that marketed fragrances present no significant risk to the consumer.

RIFM is the research organization of the industry and is headquartered in Englewood Cliffs, NJ, USA. All major manufacturers of fragrances around the world are members of RIFM. The critical objectives of RIFM are:

—to develop and disseminate scientific data to support the safe use of fragrance ingredients, and
—to provide a scientific basis for industry self-regulation worldwide.

RIFM has pursued these objectives in a number of ways. The largest effort has been to provide a basic package of toxicological data for all materials commonly used by the industry as fragrance ingredients. This effort has involved sponsoring over 7000 tests on approximately 1300 materials. For each material, at least the following types of data were required:

(1) acute oral toxicity—limit test;
(2) acute dermal toxicity—if oral toxicity is significant;
(3) skin irritation;
(4) skin sensitization; and
(5) phototoxicity—when the nature of the material is such that an effect is suspected.

In addition, where chemical structure and exposure data indicate the need, further studies such as 30- and 90-day subchronic studies and dermal absorption and metabolism data are collected.

All data from the RIFM-sponsored studies as well as from all other sources are reviewed and interpreted by an independent Expert Panel of dermatologists, toxicologists and pharmacologists with no connection to the fragrance or cosmetic industry.

Several things then happen to the data and the Panel's interpretations of them. They are transmitted through the RIFM/IFRA Joint Advisory Committee (JAC) to IFRA for consideration of possible guidelines. They are published, along with all other data obtained through literature searches and from member companies, in the form of substance monographs in the refereed journal *Food and Chemical Toxicology* (formerly *Food and Cosmetic Toxicology*).

In the event that the Expert Panel, after reviewing the data on a particular fragrance material, determines that there is a potential for harm under some reported conditions of use, the membership of RIFM as well as of IFRA is notified of this conclusion by an Advisory Letter. It should be emphasized that this Advisory Letter is not a recommendation to stop using the material in question. It is only to advise that under the conditions of the particular test that was conducted, a potential for harm was detected. All Advisory Letters are followed by an IFRA guideline that recommends the conditions under which the material can be used safely or, indeed, whether it should be used at all.

3 TOXICOLOGY AND SAFETY EVALUATION

There would be no disagreement with the statement that all fragrances should be safe. But what does that mean? It is common to use the word 'safe' in the manner of 'that action is safe' or 'this substance is safe', but it is incorrect. Safety is not an absolute, but a relative, term and it is a negative concept—it is the absence of harm. Safety cannot be measured, detected or proven. One can only estimate the probability of harm (risk) and, if the risk is low enough, reach a conclusion of safety. One person might say that jumping out of an airplane is safe if you are properly trained and have a good parachute. Another might conclude that it is safe to ride in an airplane but certainly not to jump out regardless of training and equipment. Of course, there are those that only feel safe when in solid contact with the ground. Who is correct? Obviously none of the situations is absolutely safe. There is some chance of harm regardless of whether you are a well-trained sky diver, an airplane passenger, or even if you are standing firmly on the ground. On the other hand, in each case a judgement has been made of the probability of harm (risk assessment) and a conclusion reached as to whether that risk is small enough to be acceptable. There has also been a judgement of relative risk. The person standing on the ground has concluded that the risk of a parachute not opening is greater than that of being hit by an object falling from the sky (say a falling parachutist).

Thus the elements of safety evaluation are:

(1) hazard determination—what harm can possibly occur?
(2) risk assessment—what is the probability that the harm will occur?
(3) comparison of relative risks—does decreasing one risk leave another that is even greater?
(4) safety determination—is the risk small enough to be acceptable?

It is relatively easy to determine hazard. Continuing the above analogy, hitting the ground too hard and being hit by a falling object are two obvious hazards. There might also be other more subtle hazards—getting struck by lightning while standing on the ground—but determining these is still usually easier than assessing the probability of their occurrence. This is especially so when it comes to assessing the risk from exposure to fragrance materials, where the hazards may be quite subtle and often can only be determined in animals, not humans.

The study of the potential hazards of substances is called toxicology. It is common to refer to certain substances as 'toxic' but this is as incorrect as calling them safe, albeit for different reasons. All substances have a potential for being toxic, i.e. they have an intrinsic capacity for causing harm. One of the most common measurements of toxic potential is the LD_{50}, the theoretical dose necessary to kill 50% of a population. The dose of table sugar needed to kill 50% of a group of rats weighing 300 g is about 10 g while only 1·0 g of table salt is needed and 0·002 g of cyanide. Therefore, cyanide is more toxic than salt which, in turn, is more toxic than sugar, but they all have toxic potential; it only takes different amounts to reach that potential. This basic principle was first stated in 1538 by Paracelsus, a Swiss physician who said 'Alle Dinge sind Gift, und nichts ist ohne Gift: allein die Dosis machts, dass ein Ding kein Gift sei'—which translated is, 'All things are poisons and nothing is without toxicity: Only the dose allows anything not to be poisonous'.

Toxicology, then, is the study of the toxic potential of substances and involves not only determining the hazard or adverse effect, but also the amount or dose of the material that causes the effect. At this point risk assessment might seem a straightforward procedure: simply determine the dose that causes no adverse effects, and that should be safe.

Unfortunately it is not that simple, for several reasons. First, if the only hazard was death as in the above examples, it would be easy to determine—an animal is either dead or not. But, of course, we are concerned about much more subtle effects. Does the substance cause irritation (which may be a very slight redness or a clear inflammation and soreness) after prolonged (how long?) exposure? Does it cause a change in the immune system sufficient (how much is sufficient?) to elicit an allergic response? Does exposure to the material over a lifetime (how long a lifetime and how much exposure?) cause illness or, even more subtly, increase the susceptibility to illness? Are there other effects too subtle to detect or measure? In studies of toxicity, we commonly speak of the NOAEL, or the 'no observed adverse effect level'. This emphasizes that at the particular dose level, no adverse effects were seen but there is no guarantee that some subtle effects occurred which were not observed. This problem is partially compensated for by testing higher doses that would cause a more pronounced and therefore more easily observed effect.

The second complication in risk assessment is biological variation; that is, no two animals, including (or perhaps especially) humans, are exactly alike.

Therefore, the amount of a substance that may cause no effect in the specific animal studied might cause some effect in another, even if the species, strain, sex, age and other controllable variables are the same. This is partially compensated for by studying not just one animal but several. The larger the number of animals studied, the greater the likelihood that normal biological variation can be compensated for, but it is no guarantee.

A third complication is that most forms of toxicity must be studied in animals, not in humans. It is obviously not ethical to test humans in an attempt to determine how much of a substance it takes to cause liver damage or some other serious effect. Unfortunately, there is no test animal that exactly mimics humans in their response to exposure to test substances, and this introduces another unknown variable. How do the observed effects or, for that matter, the dose that produces no effect in the test animal, relate to the dose/effect pattern in humans?

There are other complications to risk assessment but these are the most important. The toxicologist when studying a material tries to design his test to compensate as well as is possible and practical for these complications. One way, as mentioned, is to study groups of animals to help compensate for biological variability and to test at least one high dose to accentuate subtle adverse effects that otherwise might be missed at lower doses.

Another way to help compensate is by using a control group, i.e. another group of animals that is as identical as possible to those treated with the test material and is maintained under the same conditions. The test and control group can then be compared for differences. In this way disorders that occur normally in the test animal can be studied by finding out if more of the treated animals have the disorder than in the control group, or if it is more severe.

The problem of extrapolating results in test animals to risk in humans is the most difficult of the complications to compensate for. It is possible to study the mechanism of the effects in the animal and then to assess whether such an effect is likely to occur in humans when they are exposed to the material, i.e. to compare the animal under conditions of the test and the human under conditions in which he or she might be exposed to the substance, say from the use of a consumer product containing the material. This comparison includes: how much of the material gets into the body, say by dermal absorption, and at what rate; how the substance is distributed throughout the body; and how much and at what rate it is eliminated. This is called pharmacokinetics. It is also important to compare how the material is transformed while in the body-metabolism. While studies to obtain such data are often expensive, time-consuming and, in some cases, not even possible, they can provide invaluable information for use in risk assessment.

A short cut that has been used by toxicologists is the safety factor. This involves first studying the effects of the substance carefully in large enough groups of animals to compensate for variability in the test species, and at several doses to determine as carefully as possible the NOAEL. Then, by making the assumption that humans may be as much as 10 times more susceptible to the effect than the test animals, and that some humans might be 10 times as sensitive as others, the NOAEL in humans can be assumed, with good confidence, to be no less than 1/100 that in the animal. If the exposure from use of consumer products is less than that level, then the use would be concluded to be safe. Obviously this method has some uncertainty but it has served very well over the years. More sophisticated mathematical methods have been developed for risk extrapolation but they too have built-in uncertainty.

Very often the most difficult part of extrapolating from test conditions to the conditions of human exposure from use of consumer products is in determining just what that human exposure is.

4 EXPOSURE TO FRAGRANCE MATERIALS

No safety evaluation can be conducted without some estimate of exposure under the conditions of use. For fragrances this is a very difficult problem, for several reasons: the lack of knowledge of fragrance formulations; the complexity of the formulations; the lack of knowledge of what products the formulations are used in and at what concentrations; the variation of the amount of a fragranced product used in a single application; the variation of frequency of use of such products.

First let us address the route of exposure. At first thought many might guess that the principal route of exposure to fragrance materials is inhalation since, obviously, fragrances are meant to be inhaled. Certainly inhalation does occur. However, due to the exquisite sensitivity of the nose, only minuscule amounts need to be inhaled to accomplish the intended task. While there are scientific data to support this, it can be illustrated with a simple example. Let us consider a typical application of a fine perfume as 0·1 ml dabbed behind each ear. This is certainly enough to be detected (smelled) for a considerable period of time when in the near vicinity. Using a typical fragrance oil concentration in the perfume of 20% and a reasonably high concentration for any one ingredient of 5% in the fragrance oil, it can easily be calculated that the maximum amount of that ingredient evaporating into the air (ignoring any that is absorbed into the skin, comes off onto the clothes or is washed off) over the time period of detection, say one hour, is 2 μl. Assuming that the perfume can only be detected within 1 m of the person wearing it and there are no drafts or movement that may cause further spreading outside this small distance, this amount evaporates into a volume of approximately 8 m^3 and would, therefore, reach a maximum concentration of less than 40 parts in 10^9, a level well below the acceptable average concentrations for eight-hour inhalation of all but the most extremely toxic occupational chemicals.[2] If evaporation into a normal-sized room or an even larger space and normal movement and air circulation are taken into account, a much lower maximum concentration can be calculated. Obviously, inhalation is not a route of toxicological concern for fragrance ingredients because not only are air concentrations extremely low, but the toxicity is orders of magnitude lower than that of most industrial chemicals. On the other hand, inhalation (smelling) of a fragrance can have a significant effect on the body which can range from good feelings and pleasant memories to unpleasant effects depending on the relationship with past memories. This is no different from the other senses: seeing or hearing some things can have significant effects, both positive and negative, but these can hardly be called toxic effects.

The route of exposure that is most important for fragrance materials is via the skin. Before discussing how to measure or estimate exposure, it should be decided what units to use. In dermal exposure, percentage concentration is certainly the most commonly used, the easiest to determine and the easiest to work with. It is also, with the exception of irritation and sensitization reactions, the least relevant. Consider, for example, the amount of a substance applied to the skin that actually gets into the body. Obviously this depends not just on concentration but on the total amount available. It also depends on the area of the skin; that is, the more spread out the material, the faster it can be absorbed through the skin. As mentioned above, it is also desirable to consider how the material is distributed throughout the body and how it is eliminated. This is dependent, among other things, on the size of the animal. Therefore, for effects that might occur inside the body (systemic effects) the total dose is often compensated by the weight of the animal. Thus a dose might be expressed as weight of material divided by the weight of the animal, i.e. mg/kg. There are other units that can be used, but for the following discussion we will concentrate only on percentage concentration and what will be called systemic dose, i.e. mg of material/kg bodyweight.

It is fairly easy to determine the approximate percentage concentration of a perfume oil in a consumer product because this does not change very much within a given type of product: for instance, a hand soap typically contains 1–3% fragrance, a cologne 7–12% and a perfume 20–40%. It is far more difficult to determine the concentration of a particular ingredient in the product since it might be used in the fragrance formula at any level from several per cent down to only trace amounts, or even not at all. Of course, one can always take the highest concentration in a formula, and assume the formula is used at the highest concentration in the consumer product, but this will give an unrealistically exaggerated level in most cases. For example, an informal survey of the major fragrance manufacturers' use levels of phenylethyl alcohol in formulae intended for use in hydroalcoholic products revealed a maximum use level of 40%. If it is assumed that this formula is used at the maximum level of 40% in a perfume, the final concentration would be 16%. While there may be a product or products on the market containing 16% phenylethyl alcohol, this is certainly not typical and it would be a mistake to base a risk assessment only on this atypical use. On the other hand, when the reporting companies were asked for their ten highest use levels, again in formulae intended for use in hydroalcoholic products, the average of all these maximum levels was 5·5% which, when combined with a typical perfume concentration of 20% or even a high concentration of 40%, yields more realistic figures of 1–2% for the final concentration. But what is the justification for omitting the higher levels? Unless it is known that these high levels are seen only in formulae used at lower than the 20–40% level, there is none. At present there is no good way of dealing with this problem other than tracing each formula to its final use—a very difficult, if not impossible, task. Of course the job becomes even more difficult when expanded to all the other types of products that might contain the substance in question.

As mentioned, the percentage concentration is useful only in risk assessment for irritation and sensitization effects. For possible systemic effects, effects that occur in the body as opposed to on the skin, it is necessary to know or to estimate the total dose in terms of amount per bodyweight, mg/kg. This is even more complicated than determining a reasonable estimate of concentration. Even if the concentration in a particular product were known, it would be necessary to establish how much of that product was used, how often it was used, and, since we would be concerned about how much enters the body through the skin, the area over which the product was applied. This information would then need to be added to similar data from all other products that are used and contain the substance in question. This is clearly a formidable undertaking and at present only crude estimates of systemic exposure can be calculated. Once total exposure (dose) on the skin is estimated, it is then necessary to estimate how much might go through the skin and actually enter the body, as opposed to the amount that evaporates, comes off on the clothes or is washed off.

As mentioned, the amount of a material that enters the body through the skin (transdermally) is dependent on a number of factors: the amount applied to the skin; the vehicle in which it is applied; the concentration in the vehicle; the area of skin on which it is applied; the site on the body where it is applied; the amount that evaporates into the air or is washed off prior to being absorbed; whether the material is covered (occluded) or left open to the air. In addition, the amount that goes through the skin of a test animal is almost always different from that which goes through human skin.

Fortunately, there is an alternative to attempting to work through this complicated maze—conducting a simulated exposure study. This procedure involves exposing a small number of human volunteers (two or three are generally sufficient) to a simulated, but typical, use of the substance in question. Because of the small amounts of material involved, it is usually necessary to use material labeled with a very low level of ^{14}C. For example, when phenylethyl alcohol was studied,[3] the exposure was to 1 ml of a 1% alcoholic solution of phenylethyl alcohol. The solution was spread over a measured area of 100 cm^2 and allowed to remain on the skin for 6 h. During the six hours and for five days later all urine and feces were collected and analyzed. Likewise, periodic samples of the blood were collected to determine blood serum levels. Attempts were also made to analyze for the principle metabolite, phenylacetic acid, but the normal background levels of this metabolite were such that the small amount added as a result of absorption through the skin was not detectable. This study yielded a very accurate profile of the systemic exposure to phenylethyl alcohol and the metabolites that would result from a typical exposure to this material. These data were very valuable in conducting the risk assessment for phenylethyl alcohol (see below) and demonstrated safety factors substantially greater than 10 000.

This type of study is fairly expensive because it often involves making radio-labeled material and extensive studies in animals prior to the human study. However, it yields extremely valuable data for use in risk assessment.

5 POTENTIAL DERMAL HAZARDS

Fragrance materials are often found at the highest levels in products which are applied to the skin. It is, therefore, important to know the potential hazards to the skin that might occur. This section discusses the principle types of dermal reactions that have been caused by exposure to fragrance materials.

Irritation

The most common and generally least serious hazard for dermal contact is irritation. Irritation is a non-allergic (non-immunologic) inflammatory reaction that occurs at the site of exposure to the substance either on initial exposure (primary irritation) or after repeated exposure to the same site (cumulative irritation). Generally this effect appears within minutes to hours of initial contact and is dependent on concentration, solvent and time of contact. Shortly after the irritant is removed, the reaction will begin to diminish and it usually clears within a few days. Different sites of the body have differing susceptibility to irritants, with the eye generally being the most susceptible.

Practically all materials have some irritation potential; however, fragrance ingredients tend to be only mild to moderate irritants, if at all. This, combined with the fact that they are rarely used at more than a few per cent in consumer products, means that they can generally be considered as not irritating under the conditions of use. Many fragrance ingredients must be labeled as possible skin and eye irritants in the workplace, however, since they are encountered there in undiluted form.

There are several tests that are used to determine irritation potential. Most commonly the rabbit is used but other animals including humans have also been involved and recently there have been significant efforts to develop alternatives to animal tests—in-vitro tests. These alternatives are not yet reliable predictors of possible risk to humans, but may be useful as preliminary screens or in structure–activity studies.

The animal tests for irritancy potential usually involve applying 0·5 ml of the undiluted material to a defined area of one square inch (6·5 cm^2) on the clipped or shaved and sometimes abraded skin for periods ranging from 4 to 24 h. The application may remain uncovered (unoccluded) but is usually either wrapped so that some evaporation can occur (semi-occluded) or so that no evaporation occurs (occluded). After the material and any wrappings are removed, the site of application is examined within 0·5 h and 24 and 48 h later. The effects may include redness (erythema), swelling (edema) or both. In severe cases, corrosion can occur but, of the fragrance materials reported to RIFM to be in use, none has been so classified.

Eye irritancy potential is most commonly measured using the controversial Draize rabbit-eye test. However, because fragrance ingredients are found in consumer products only at very low levels, it is generally not necessary to conduct this test on these materials.

Sensitization

The second most common hazard from dermal exposure is skin sensitization or allergic contact dermatitis. This effect can be similar in appearance to irritation, but typically the reaction appears from 24 to 48 h after exposure and, unlike irritancy, often increases in intensity over 2–3 days after exposure to the material has ceased. Reactions can also persist for up to several weeks. Sensitization differs fundamentally from irritancy in that it involves the immune system, i.e. it is an allergic response. Thus, if there has been no previous exposure to induce the immune system, no reaction will occur on contact. However, once the allergy has been induced by repeated exposures, a reaction can occur at a concentration often considerably lower than that required for irritation. The reaction may also persist for several days or even weeks after exposure. Again these reactions are generally not very serious but a person who develops severe allergic contact dermatitis may have to avoid coming into contact with products that contain a significant percentage of the allergen. A considerable number of fragrance materials have been found to have sufficient sensitization potential to warrant a recommendation by IFRA either not to use the material at all or to limit the use to a certain concentration or purity (Table 1).

For a few materials, a phenomenon known as quenching[4] has been observed. These are materials that exhibit sensitization under normal conditions but in the presence of certain other materials, known as quenching agents, the sensitization seems to be suppressed. For example, citral, when tested by itself on humans, caused sensitization reactions at concentrations as low as 0·1%. However, when 20% *d*-limonene was added, the mixture produced no reactions at 5% (4% citral, 1% *d*-limonene). Based on this and other tests using other citrus terpenes, IFRA developed a recommendation to use citral in fragrances only when it contains 25% *d*-limonene, mixed citrus terpenes or α-pinene. This and other quenching recommendations are referred to in Table 1.

Table 1
Fragrance materials restricted by IFRA

Substance	*Recommendation*[a]	*Reason*
Acetylethyltetramethyltetralin (AETT)	Prohibited	Neurotoxicity
5-Acetyl-1,1,2,3,3,6-hexamethylindane	Restricted	Phototoxicity
Acetyl isovaleryl	Prohibited	Sensitization
Acetylated vetiver oil	Preparation	Sensitization by some samples
Alantroot oil (Elecampane oil)	Prohibited	Sensitization
Allyl esters	Restricted	Irritation
Allyl heptine carbonate	Restricted	Sensitization
Amylcyclopentenone (2-Pentyl-2-cyclopenten-1-one)	Restricted	Sensitization
Angelica root oil	Restricted	Phototoxicity
Anisylidene acetone (4-(*p*-Methoxyphenyl)-3-buten-2-one)	Prohibited	Sensitization
Benzylidene acetone (4-Phenyl-3-buten-2-one)	Prohibited	Sensitization
Bergamot oil	Restricted	Phototoxicity
Bitter orange oil expressed	Restricted	Phototoxicity
p-tert.-Butylphenol	Prohibited	Sensitization and depigmentation
Carvone oxide	Quenching[c]	Sensitization
Cassia oil	Restricted	Sensitization
Cinnamic alcohol	Restricted	Sensitization
Cinnamic aldehyde	Quenching[c]	Sensitization
Cinnamic aldehyde–methyl anthranilate Schiff base	Quenching[c]	Sensitization
Cinnamon bark oil, Ceylon	Restricted	Sensitization
Citral	Quenching[c]	Sensitization
Costus root oil, absolute and concrete from *Saussurea lappa* Clarke	Prohibited	Sensitization
Cumin oil	Restricted	Phototoxicity
Cyclamen alcohol (3-(4-Isopropylphenyl)-2-methylpropanol)	Prohibited[b]	Sensitization
Diethyl maleate	Prohibited	Sensitization
Dihydrocoumarin	Prohibited	Sensitization
2,4-Dihydroxy-3-methylbenzaldehyde	Prohibited	Sensitization
4,6-Dimethyl-8-*tert.*-butylcoumarin	Prohibited	Photosensitization
Dimethyl citraconate	Prohibited	Sensitization
Ethyl acrylate	Prohibited	Sensitization
Ethyl heptine carbonate	Restricted	Sensitization
Farnesol	Specifications	Sensitization by impurities
Fig leaf absolute	Prohibited	Phototoxicity and Sensitization
trans-2-Heptenal	Prohibited	Sensitization
Hexahydrocoumarin	Prohibited	Sensitization
trans-2-Hexenal	Restricted	Sensitization
trans-2-Hexenal diethyl acetal	Prohibited	Sensitization
trans-2-Hexenal dimethyl acetal	Prohibited	Sensitization
α-Hexylidene cyclopentanone	Restricted	Sensitization
Hydroabietyl alcohol	Prohibited	Sensitization
Hydroquinone monoethyl ether	Prohibited	Depigmentation
Hydroquinone monomethyl ether	Prohibited	Depigmentation
Hydroxycitronellal	Restricted	Sensitization

Table 1—*contd.*

Substance	*Recommendation*[a]	*Reason*
Isoeugenol	Restricted	Sensitization
6-Isopropyl-2-decalol	Prohibited	Sensitization
Lemon oil cold pressed	Restricted	Phototoxicity
Lime oil expressed	Restricted	Phototoxicity
Marigold oil and absolute (Tagetes oil and absolute)	Restricted	Phototoxicity
Menthadienyl-7-methyl formate	Restricted	Sensitization
7-Methoxycoumarin	Prohibited	Photosensitization and sensitization
α-Methylanisylidene acetone (1-(4-Methoxyphenyl-1-penten-3-one)	Prohibited	Sensitization
6-Methylcoumarin	Prohibited	Photosensitization
7-Methylcoumarin	Prohibited	Photosensitization
Methyl crotonate	Prohibited	Sensitization
4-Methyl-7-ethoxycoumarin	Prohibited	Photosensitization
6-Methyl-3,5-heptadienone	Restricted	Sensitization
Methyl heptine carbonate	Restricted	Sensitization
p-Methylhydrocinnamic aldehyde	Restricted	Sensitization
Methyl *N*-methylanthranilate (Dimethylanthranilate)	Restricted	Phototoxicity
3-Methyl-2(3)-nonene nitrile	Restricted	Sensitization
Methyl octine carbonate	Restricted	Sensitization
Musk ambrette	Prohibited[b]	Neurotoxicity and photosensitization
Nitrobenzene	Prohibited	Acute toxicity
Nootkatone	Specifications	Sensitization by impurities
Oakmoss absolute and resinoid (concrete)	Restricted	Sensitization
1-Octen-3-yl acetate (Amylvinyl-carbinyl acetate)	Restricted	Sensitization
Opoponax	Preparations	Sensitization by some samples
Oils from Pinacea family	Specifications	
Pentylidene cyclohexanone	Prohibited	Sensitization
Perilla aldehyde	Restricted	Sensitization
Peru balsam (exudation from *Myroxylon pereirae* (Royle) Klotzsh	Prohibited	Sensitization
Phenylacetaldehyde	Quenching[c]	Sensitization
Propylidene phthalide	Restricted	Sensitization
Pseudoionone (2,6-Dimethylundeca-2,6,8-trien-10-one)	Prohibited[b]	Sensitization
Pseudomethylionones	Prohibited[b]	Sensitization
Rue oil	Restricted	Phototoxicity
Safrole, isosafrole and dihydrosafrole	Prohibited[b]	Chronic toxicity
Savin oil	Specifications	Sensitization by some samples
Sclareol	Specifications	Sensitization by some samples
Styrax American and Asian	Preparation	Sensitization by some samples
Verbena absolute (*Lippia citriodora*)	Restricted	Sensitization
Verbena oil (*Lippia citriodora*)	Prohibited	Phototoxicity and sensitization

[a] For more information such as the reasons for the recommendations, the limits of restrictions, etc., see the IFRA Code of Practice.[1]
[b] Special exemptions exist.
[c] Quenching means the substance should only be used with a quenching agent.

Animal Tests for Sensitization

There are several tests for sensitization that are in common use with both animals and humans. The most common test animal for sensitization is the guinea pig, tests with which generally involve dosing 10–20 animals with the test substance several times over a period of 1–3 weeks (induction period). The animals are then left untreated for about two weeks (rest period) in order to allow time for the immune system to react. After this period, the animals are treated with a challenge dose of the test material and examined for reactions (erythema and edema) 24, 48 and 72 h later. Dosing may be unoccluded, semi-occluded, occluded or even by intradermal injection. It may also involve treating the animals with Freund's complete adjuvant (FCA), a mixture containing heat-killed bacteria, to induce the immune system.

Among the commonly used guinea-pig tests, the maximization test, which combines FCA, irritancy, intradermal injection and occluded patches, is generally considered the most sensitive. If a material does not cause sensitization in a guinea-pig maximization test, it is very unlikely to cause sensitization under conditions of use. On the other hand, a positive reaction in this test, or in any of the others, does not mean that the material will definitely be a problem under conditions of use. The Buehler and the open epicutaneous tests seem to parallel tests in humans better than others do.

Other animals have been used for sensitization tests but none seems to have a clear advantage over the guinea pig. There are no prospects for an in-vitro test for sensitization at present.

Human Tests for Sensitization

Extrapolation to humans is especially difficult when considering sensitization potential because of the extreme complexity of the immune system and the significant differences between humans and test animals. The best test species is the human.

There are two general types of human tests that are in use, although there are several variations of each. The human maximization test involves pretreatment with sodium lauryl sulfate (SLS), an irritant to enhance penetration, and five 48-hr occluded patches of the test material (usually in petrolatum), with a 24-h rest period between each patch, over a three-week period. This is followed by a two-week rest period and then a 48-h challenge patch with a pretreatment with SLS. The human volunteers are then examined for reactions on removal of the patches and at 24 and 48 h later. Because of the severity of this test, only 25–30 volunteers are used.

The human repeat insult patch test (HRIPT) is the most commonly used test with human volunteers. This method involves application of the test material in an appropriate solvent under either occlusion or semi-occlusion for 24 h every other day for three weeks. This is followed by a two-week rest period and then a 24-h challenge patch. Examination for reactions is at 24 and 48 h after patch removal. This test usually involves 100–200 volunteers.

While the human tests are more relevant than the animal tests, they, like the animal tests, must be looked at as only screening tests. Although they are designed to exaggerate human exposure, a negative reaction in these tests is no guarantee the tested material will not cause a reaction under conditions of use. On the other hand, a material that causes a positive reaction in these tests may be able to be used safely if the concentration is low enough or other factors such as vehicle, evaporation, quenching and individual susceptibility differ significantly from those in the test.

Diagnostic Patch Tests

Dermatologists routinely use this test to determine whether a patient is sensitized to a material. The test is essentially similar to the challenge patch in the HRIPT; the material, usually in petrolatum, is applied under an occluded patch for 48 h and the site examined for reaction 24, 48 and 72 h after removal of the patch. This test may involve a single material if it is suspected of being the cause of the sensitization, or it may involve a mixture as a screen. Two groups of dermatologists with special interest in the area of skin sensitization, the North American Contact Dermatitis Group (NACDG) and the International Contact Dermatitis Research Group (ICDRG), have developed a group of recommended fragrance materials and a fragrance mix for screening for possible fragrance allergy.

Photoreactions

Less common dermal reactions are those induced only by dermal exposure both to the substance and ultraviolet (UV) light. These reactions can be either irritant (photoirritant or phototoxic) or allergic (photosensitizing) in nature. Substances that do not absorb in the UV spectrum do not cause phototoxic effects. While these effects are generally not very serious, chronic phototoxic reactions can, in rare cases, lead to skin tumors or result in a condition known as persistent light reactivity. These conditions are very serious and can even be life-threatening.

Test methods for photoreactions are similar to those for irritancy and sensitization except that they involve UV irradiation in addition to the test material.

6 POTENTIAL SYSTEMIC HAZARDS

While it should be obvious that to cause an allergic reaction a substance must enter the body, that type of reaction is generally separated from other types of systemic toxicity (effects on the system) because it is manifested by a topical, i.e. skin, reaction. Nevertheless, it should be emphasized that all fragrance materials can enter the body through dermal absorption and therefore have the potential for systemic toxicity, be it on the immune system or otherwise.

To list and discuss all the forms of systemic toxicity is far beyond the scope of this chapter. However, there are certain systemic effects that fragrance materials have been suspected of having. This suspicion has invariably emanated from animal studies. None of these effects has ever been seen in humans as a result of exposure to fragrance materials. Why then should they be discussed? When an effect is seen in animal testing, it is necessary to conduct risk assessment as outlined above. Because of the many difficulties involved in such an assessment, including the substantial time and expense in obtaining the necessary data for it, a few materials have been dropped from use on the basis of animal tests alone, even though neither an effect in humans, nor even any significant risk, was demonstrated.

Acute Toxicity

Acute toxicity is an adverse effect that occurs after a very short exposure, often a single dose. The common method for expressing acute toxicity is the LD_{50}, i.e. the theoretical dose necessary to kill 50% of a population. This is determined by dosing test animals, usually the mouse or the rat, with different doses of test material up to a dose that causes some deaths or is considered so large that the material can be considered essentially non-toxic. This test has been reported for essentially all major fragrance ingredients; they are generally non-toxic, although a few can be classified as moderately toxic.

Chronic and Subchronic Toxicity

Tests that involve a significant fraction of the life of the test animal are called subchronic and those involving most of the life are called chronic. These tests often involve daily dosing and a wide variety of possible adverse effects are looked for. Specific examples of these tests are described below.

Neurotoxicity

First let us look at neurotoxicity, i.e. a toxic effect on the nervous system. In 1977, RIFM was provided with data from a study that had been recently completed and later published.[5] These data indicated that a musk fragrance ingredient, 6-acetyl-7-ethyl-1,1,4,4-tetramethyl-1,2,3,4-tetralin, or AETT as it was popularly known, when applied dermally to rats at a daily dose of 9 mg/kg for 26 weeks caused an intense blue coloring of the central nervous system accompanied by other evidence of damage to the nervous system. No effects were seen at the 3 mg/kg dose level and no effects were seen in monkeys at a dose as high as 50 mg/kg. While there was a clear NOAEL in rats, and no effects were seen in monkeys, the RIFM Expert Panel judged that there were insufficient data to demonstrate an acceptable risk for the reported use of this material in fragrances. RIFM, therefore, issued an Advisory Letter and transmitted the data along with the Panel's judgement to IFRA. IFRA immediately issued a recommendation that AETT should not be used in fragrances. Various government health officials were also notified, not only of the findings but also of the industry's action to stop all use of AETT.

This all occurred prior to the publication of the data and did not require notification in the *Federal Register,* public hearings, possible legal action, etc., all of which is common when the Food and Drug Administration takes action. Thus, action was taken much faster and at considerably less expense to the public. Of course, while the industry lost a valuable material, it also protected its valuable right of self-regulation.

Another material, musk ambrette, has also been demonstrated to cause neurotoxicity when applied dermally to rats. Here, however, RIFM undertook a program to develop the necessary data for an adequate risk assessment.

First an estimate of possible exposure to musk ambrette was made. Analyses of the use of musk ambrette in many different formulae that were used in a wide variety of consumer products combined with use patterns of these products led to an estimated possible high daily exposure of about 0·3 mg/kg. The effects in the rat were seen at a daily applied dose of 80 mg/kg to the skin and the NOAEL was concluded to be the next-lower dose, 40 mg/kg.[6] It is a simple procedure to divide these two figures to demonstrate a calculated safety factor of about 130. Normally this would be considered adequate to demonstrate reasonable safety but several difficulties arose on careful analysis of the data. First, there was one rat in the group of 30 that was treated with the 40 mg/kg dose and that developed some slight evidence of muscular weakness which could be attributed to neurotoxicity. While this could be for any number of reasons, and normally would have been ignored since no other rats at that dose had any effects, it was difficult to dismiss entirely. If the next-lower dose at which there were clearly no adverse effects, 10 mg/kg, is used for the safety factor calculation, a factor of only about 30 is obtained.

Further studies were then undertaken to determine how much musk ambrette can be absorbed through the skin. It was found that only 38% of the dose applied to the skin of the rats was actually absorbed. Thus, when 10 mg/kg was applied to the skin, only 3·8 mg/kg actually got into the rat. This can be compared with the results from a simulated exposure study conducted with human volunteers to determine how much might actually get into the body of a human under the conditions of a typical, but high, use. A single application of 2·2 mg applied as a 0·1% solution in ethanol to three subjects resulted in an average of about 2% being absorbed. Therefore, when the exposure to musk ambrette from use of such products is 0·3 mg/kg, the actual absorbed dose is about 0·006 mg/kg (2% × 0·3). The actual doses absorbed in humans and rats can then be used to calculate a safety factor: 3·8 mg/kg divided by 0·006 yields a safety factor of over 600.

In this manner, dermal absorption data were useful in demonstrating a much larger safety factor and perhaps could have saved the use of musk ambrette. Unfortunately, this material was also demonstrated to result in photosensitization under the conditions of actual use[7] and the industry again took action;

IFRA issued a recommendation to not use musk ambrette in any product which, under normal conditions of use, may come in contact with the skin.

Reproductive Toxicity

Another area where dermal absorption data have been useful in risk assessment is with a type of systemic toxicity known collectively as reproductive toxicity, i.e. the study of adverse effects on the process of reproduction that may occur from exposure to a substance. Such effects range from decreased fertility to birth defects (teratogenicity). Fortunately, only one fragrance ingredient has ever had any significant suspicion of causing reproductive effects and current data fully support the view that this suspicion is not justified.

One of the most commonly used fragrance ingredients is phenylethyl alcohol (PEA), a natural and characterizing ingredient of rose oil. It has been reported that PEA, when administered in large amounts by stomach intubation (gavage) to pregnant rats, causes adverse effects on the offspring.[8] Although this method of dosing is totally inappropriate for the evaluation and risk assessment of materials applied to the skin, a report such as this obviously raises questions which must be addressed. There are several problems, however, in interpreting the reported results in terms of possible risk to humans. Of course, all the standard difficulties in extrapolating results in rats to humans, as described above, exist. These difficulties are compounded here because of the great differences between the human and the rat in reproductive physiology, such as in the number of offspring (10–12 in the rat), gestation period (22 days for the rat), a much simpler and therefore more permeable uterine wall protecting the fetus of the rat, and several other factors that must be considered. Additionally, there was no clear dose response in the gavage study and a NOAEL was not determined. All of these problems were compounded by the lack of data on PEA when applied dermally.

Why then were the results from the gavage study of concern if there were so many problems? All of these problems only create unknowns that make risk assessment more difficult; they do not mean that the results are not relevant. It could only be concluded that the relevance to humans was unknown—it could have been that PEA was more of a risk to humans than to rats. On the other hand, since this material had been used in fragrances since antiquity, both as such and as a principal component of rose oil, it would be very surprising if it presented any significant risk. RIFM, therefore, undertook a series of studies to obtain the data necessary to conduct a proper risk evaluation. The data obtained from these studies[3] are far too vast for inclusion in this chapter but can be summarized as follows.

First it was demonstrated that dosing by the more appropriate route, i.e. dermally, even with very high doses, produced only slight effects that probably would be reversible and would not be described as adverse. It was also demonstrated that when the amount that goes through the skin was taken into account along with the much more rapid metabolism and excretion by humans, a safety factor of well over 10 000 could be demonstrated. The most important finding, however, was that in humans PEA is very rapidly and completely converted to phenylacetic acid, a normal component of the blood. Since the amount of phenylacetic acid that is formed in the blood as a result of the use of a fragrance containing even high levels of PEA, is but a small fraction of the amount that is normal in the blood, it can be concluded that there is no significant risk from such an exposure.

Target Organ Effects

Certainly neurotoxicity and reproductive toxicity can be described as target organ effects and, for that matter, so can irritation and sensitization. However, this phrase is commonly used to describe specific damage to organs such as the liver, kidney, spleen, etc.

There are several ways in which adverse effects to such organs can be detected. Typically, during a study which routinely involves daily dosing of test animals for periods ranging from two weeks up to a substantial portion of the life of the animal (two years or longer for rats and mice), periodic analysis of basic organ function is made along with daily physical examination of the animals. This may involve urinalysis, hematology or other such tests. At the end of the test period, the animals are usually killed for examination of the organs; sometimes, however, an additional period of time may be allowed in order to see if any effects are reversible, i.e. if the animal recovers. After the in-life portion of the study is terminated, the internal organs are examined individually for obvious signs of injury and many are weighed. The organs are then prepared for microscopic examination, the most sensitive detector of adverse effects.

Such studies may involve from 50 to 400 animals if rats or mice are used (by far the most common), with from 100 to 500 or more observations per animal. This results in so many data that it is necessary to use statistics to aid in analysis. It is common to hear that an observation has a reliability of greater than 95% or a P value of less than 0·05 (less than a 5% probability that the observation is not reliable). This means that the statistical analysis has indicated that the probability that the difference seen between test and control groups is in fact real, and not the result of a spurious event, is 95% or greater. It is important to note that this is not a guarantee that the effect is real, but only an indication of the probability that it is. It is in fact, an indication that the effect has a certain small probability of being wrong.

Of course, once all these data have been gathered and analyzed, it is still necessary to conduct the kind of risk evaluation that is described above with all of the inherent uncertainties. Indeed, as has been said, safety evaluation is a risky business.

Carcinogenicity

Tests for carcinogenicity, commonly called carcinogenicity bioassays, are chronic tests with a special emphasis on the target organ effect of tumor formation. There is considerable disagreement on a number of aspects of such bioassays. Typically, the mouse and rat are the animals of choice and they are administered the largest possible dose (maximum tolerated dose, MTD) daily throughout essentially all of their adult life. Some say this high dose is necessary to compensate for the fact that rats and mice live for only a small fraction of the life expectancy of humans. Others say such large doses completely overwhelm the capacity of the animal and therefore produce results of questionable relevance when this dose may be thousands or millions of times greater than that to which humans are exposed under conditions of use. It is the common practice to count both benign and malignant tumors and this is also a source of disagreement. There are those who say that benign tumors are indicative of the ability of the test material to increase cell growth and therefore should be considered as relevant, while others say that a benign tumor is of no consequence. There are those who say that if a single molecule of a substance can reach and alter the DNA of only a single cell (single-hit theory), that cell can reproduce and develop into a malignant tumor and therefore there is no safe dose (no NOAEL) for carcinogens. Others point out that the chance of such an encounter is vanishingly remote and even so there are DNA repair mechanisms and other defenses that can come into play. It is being more widely accepted that for susbstances that produce cancer by a so-called secondary mechanism, such as chronic irritation, a threshold or NOAEL exists. There are many other disagreements but all agree that extrapolation from the high-dose rodent bioassays to humans in a manner that results in risk determination with reasonable confidence is extremely difficult.

At the time of writing, no fragrance material has been the subject of an RIFM Expert Panel Advisory Letter based on potential carcinogenicity, nor has any material been restricted by IFRA because of carcinogenicity.

7 OCCUPATIONAL SAFETY

Given the large number of chemicals, both synthetic and natural, used in making fragrances, the industry has a remarkable record of occupational safety. A recent epidemiology study of workers in the fragrance and flavor industries showed no significant increases in the incidences of mortality from any cancer.[9] This is due almost exclusively to the low toxicity of the materials that are used.

This is not to say that this is an issue that can be ignored. The great majority of the toxicological data that have been gathered on fragrance materials has been oriented towards their use in consumer products where they are encountered at relatively low levels and in small amounts. It can, therefore, be argued that, since the undiluted forms of many ingredients are used in the workplace, the exposure to these materials should be kept to a minimum. IFRA has endorsed this position in their Code of Practice.[1]

By far the most common complaint among workers in the industry results from odors of materials which, when handled in concentrated form, can be quite unpleasant. This, however, can be a protection, since any significant exposure would likely be avoided and, if such an exposure accidentally occurs, it would normally be remedied quickly. IFRA has, in its Code of Practice,[1] presented some general industrial guidelines for health and safety in the workplace.

8 ENVIRONMENTAL CONSIDERATIONS

Concern for the environment has been one of the major issues of the 1980s and looks to be even greater in the 1990s. The fragrance industry has already felt a significant impact in this area due to legislation and regulation primarily aimed at larger industries. Development of new fragrance materials has slowed considerably, essentially because of the cost of registering these substances with the various governments.

Is concern about the environmental impact of fragrances justified? Certainly fragrances get into the environment primarily from evaporation and from being washed down the drain. There is no evidence, however, to indicate that fragrances have a significant impact on the environment. This is attributable to a number of factors including, as discussed before, the inherently safe nature of the materials used in compounding fragrances and the fact that the relatively simple types of chemical structures encountered among these materials justify an assumption of ready biodegradability in virtually all cases. Fragrances are produced in comparatively small volumes (relative to major industrial chemicals) and are used at low levels in the consumer products such as soaps and detergents that are produced in the largest volumes. Additionally, the wide range of types of products in which fragrances are used assures that they are widely dispersed in the environment and this, combined with their low toxicity, supports a conclusion of no significant impact even if biodegradability is not as facile as assumed.

There is a growing trend in the industry to develop fragrance for use in products labeled 'environmentally friendly', 'ozone friendly' or with other such phrases. This is a trend that can be very risky. First, use of these designations implies that other products not so labeled are more hazardous to the environment when there is no basis for that conclusion. Secondly, this practice insinuates that we have a clear understanding of the fate of these complex mixtures in the environment and ignores the fact that environmental toxicology is an even less exact science than is toxicology in general. Short of studying every fragrance formulation individually, assumptions must be made and the basis for these assumptions are no better than the assumptions outlined above based on simple chemical structures, low toxicity, small volume and wide dispersal in the environment.

9 NATURAL VERSUS SYNTHETIC

There is a growing trend towards the labeling of fragrances as 'natural' with the implication that these are safer. (They may be better and are certainly more expensive, but that is not the subject of this chapter.) The fallacy that 'natural' is 'safer' has been exploited in the food industry the world over, and is the principal foundation for the continuation of this belief among consumers.

The fact is that there is no link between the natural origin of fragrance materials and their safety. All available data support the conclusion that fragrances made with natural ingredients are no more safe than those made from synthetic ingredients. Many synthetic fragrance materials are safer than naturals because of better quality control in their production and a resulting greater purity. To imply the contrary by aggressive labeling and advertising claims has no basis in fact and should be avoided by the fragrance industry.

10 SUMMARY AND FINAL REMAKRS

Toxicology and safety evaluation are complex subjects and cannot be treated adequately in a short review such as this. It is hoped, however, that this cursory discussion gives some appreciation for the subjects and to how they impact on the industry. Just because the industry has had few problems attributable to the adverse effects of fragrances, is no reason to relax standards. Quite the contrary; if the industry is to continue to enjoy its image and is to remain self-regulating, it must be diligent in pursuing even higher standards of safety in the future through support of a scientifically based program of testing and safety evaluation, taking into account proper priorities. This includes careful adherence to the IFRA Code of Practice[1] and the recommendations therein.

REFERENCES

1. International Fragrance Association, Code of Practice. IFRA, Geneva, Switzerland.
2. American Conference of Governmental Industrial Hygienists, *Threshold Limit Values and Biological Exposure Indices for 1989–1990*. ACGIH, Cincinnati, OH, USA.
3. Ford, R.A., Metabolic and kinetic criteria for the assessment of reproductive hazard. In *Vth International Congress of Toxicology—Basic Science in Toxicology—Proceedings*, ed. G. N. Volands, J. Sims, F. M. Sullivan and P. Turner, Taylor and Francis, London, 1990, 59–68.

4. Opdyke, D. L., Inhibition of sensitization reactions induced by certain aldehydes. *Fd Cosmet. Toxicol.*, **14** (1976) 197–8.
5. Spencer, P. S., Sterman, A. B., Horoupian, D. & Bischoff, M., Neurotoxic changes in rats exposed to the fragrance compound acetyl ethyl tetramethyl tetralin. *Neurotoxicology,* **1** (1979) 221–37.
6. Spencer, P. S., Bischoff-Fenton, M. C., Moreno, O. M., Opdyke, D. L. & Ford, R. A., Neurotoxic properties of musk ambrette. *Toxic. Appl. Pharmac.*, **75** (1984) 571–5.
7. Cronin, E., Photosensitivity to musk ambrette. *Contact Dermatitis,* **11** (1984) 88–92.
8. Mankes, R., LeFevre, R., Bates, H. & Abraham, R., Effects of various exposure levels of 2-phenylethanol on the fetal development and survival in Long–Evans rats. *J. Toxicol. Environm. Hlth,* **12** (1983) 235–44.
9. Guberan, E. & Raymond, L., Mortality and cancer incidence in the perfumery and flavour industry of Geneva. *Br. J. Ind. Med.*, **42** (1985) 240–5.

Chapter 17

The Fragrance Industry in a Changing World

MILTON MANOWITZ & RICHARD E. NAIPAWER

Givaudan Corporation, Clifton, NJ 07015, USA

1 INTRODUCTION

Fragrances and their various applications permeate our daily lives. On any given day, the public unconsciously comes in contact with fragrances through various consumer products. Fragrances are used in soaps, laundry products, household cleaners, deodorants, hair care products, room air fresheners, personal care products and paper products, just to name a few.

The use of fragrances is not unique to modern society. It is well documented in the earliest written historical records that early civilizations used fragrances in very primitive but effective forms.

Modern perfumes, colognes and bath oils have evolved over thousands of years from the use of concentrated essential oils, plant exudates or tinctures of plant extracts by the Egyptians, Romans and Greeks.[1] The wealthy members of these early civilizations were able to afford these expensive products which were used to mask their body odors by application directly to the skin or by addition to their bath water. Today there are time-release fragrances, which are either micro-encapsulated or absorbed in a microsponge, and individualized fragrances which behave differently when applied to the body of each individual.

The modern room air freshener[2] has its foundation in the burning of whole plants and/or plant parts as incense offerings to the deity as part of religious ceremonies. Today the environment can be fragranced by using wicks, blotters, gels, aerosols or even very sophisticated electronic devices.

The early civilizations also used various essential oils as inhalants or balms for medicinal purposes, an application of fragrances which can be looked upon as the earliest form of aromatherapy.

In those early days fragrances were very simple, consisting of single essential oils or the extracts obtained from a single type of plant or plant part. In contrast, today's fragrances are complex mixtures of natural oils and extracts, and natural and synthetic chemicals. These mixtures are artistically balanced and are created to give optimal performance in the particular product application in which they are to be used.

Over several thousand years, the use of essential oils and plant exudates in various forms has evolved into a highly sophisticated industry which has come to be known as the fragrance industry.[3] As time has marched on, as civilizations have evolved and changed, as the socio-economic complexion of the world has changed, so has the fragrance industry. In modern times, this industry tries to keep pace with trends and changes, and sometimes even tries to set their pace and direction. The fragrance industry, for the purposes of this chapter, cannot be considered as a single self-contained industry comprising only compounders and marketers of fragrances. It must be viewed as a composite of several industries including manufacturers and suppliers of fragrance ingredients, creators and compounders of fragrances, companies that sell fragrances to the public and consumer product industries that incorporate fragrances into their final products.

This chapter will discuss various factors in the changing world that have affected the fragrance industry in the past few decades and are anticipated to continue to influence the shape of the industry in the near future. It will include the impact of safety and environmental regulations, economic factors, demographics, and certain recent public movements.

2 ESSENTIAL OILS, NATURALS AND SYNTHETICS

Since the dawn of the use of fragrance materials to adorn the body, essential oils, spices and extracts of various natural products were the exclusive or principal foundation of these odorants. However, the introduction of synthetic fragrance chemicals during the last century, both nature-identical and unknown in nature, has made a major impact on these compositions. At present, the synthetics comprise by far the preponderant ingredients found in fragrances and it appears this trend will continue to grow. Although essential oils have some distinct advantages and many perfumers consider them absolutely essential in their creation of new fragrances, they have presented some significant problems throughout the years. Quality can vary from year to year and from different geographical locations. Depending on weather conditions, the local political situation, world economics, etc., the supply and price can drastically change in a very short period of time. As a result of the industry's experience with this unreliability of quality, cost and availability, there has been a movement by some of the larger companies to compound and use their own imitations of the essential oils utilizing available analytical data and synthetic ingredients. This strategy has been made possible largely through scientific developments in the area of chemical analyses of the natural oils and newly devised techniques such as headspace analyses. It has also been significantly helped by the availability of new synthetic fragrance chemicals resulting from the research efforts by the fragrance industry's synthetic chemists.

It would appear that the quantities of natural essential oils relative to synthetics used by the fragrance industry will continue their slow decline in the future, being largely replaced by the reconstituted imitations and/or by individual synthetic ingredients.[4] The oils that will decrease the fastest will be the most expensive ones and those that have provided the poorest record from the standpoint of cost, quality, and availability. Synthetic ingredients, both nature-identical and artificial, will continue to grow. Some of the stock nature-identicals, such as eugenol and citral, will have two sources: some will be derived from natural products (i.e. clove oil and lemongrass oil, respectively), while most will be manufactured.

The development of new synthetics, which grew at a very rapid rate since 1950, has slowed considerably in the last decade and probably will continue to decline in the future. This decrease can be attributed to many factors, including the R&D costs in developing new ingredients compared with the economic returns. One part of these costs, that continue to grow rapidly, is the extensive toxicology studies that are required prior to placing a new product on the market. The targeted new ingredients for the future will have special properties in addition to their odor, such as substantivity, very new and unique odors, or extremely low price especially for the high-volume soap, detergent and household products markets.

During the past decade there has been a very strong demand in the processed food market for natural flavors, resulting from the consumer's association of 'natural' with 'healthy' and a growing suspicion of artificial flavors. However, no such trend has occurred yet in the fragrance field, although there are a few marketed specialty products, based mainly on essential oils, promoting their natural fragrance content. It would appear that if a strong 'natural' movement did develop in the future it would require overcoming the barrier of the substantially higher cost of the product and would present major problems to the perfumers because of the limitations in the number and amounts of natural fragrance ingredients available.

3 ACQUISITIONS AND MERGERS

The fragrance industry has often been described as a fragmented one, but it is now entangled in the era of acquisitions and mergers[5,6] that have been taking place with increasing frequency during the 1980s. Some of the takeovers or mergers that occurred were the following:

—Union Camp acquired BBA.
—Bell Flavors & Fragrances purchased Synfleur.
—Perry Brothers and Creations Aromatique were merged.
—PPF purchased Norda.
—Firmenich acquired Chem-Fleur.
—Beecham acquired Jovan.
—L'Oreal purchased Helena Rubinstein.
—P&G took over Richardson–Vicks.
—Kao acquired Jergens.
—Avon acquired Giorgio and Parfums Stern

However, it is the aggressive acquisition activity of Unilever, greatly accelerated in the late 1980s, that has startled the industry the most. They have been acquiring fragrance and fragrance product companies at a very rapid rate. Norda and Naarden were purchased and merged with PPF to form Quest. Chesebrough–Ponds and Calvin Klein Cosmetics and the Rimmel and Chicogo businesses were acquired. It is claimed that upon completion of the current (August 1989) acquisition of the Fabergé–Elizabeth Arden business, Unilever will control over 10% of the worldwide cosmetic, toiletries, and fragrance markets.[7]

The ultimate effects of these mini- and mega-mergers and acquisitions on the industry are yet to be determined. Nevertheless, it is generally believed that the character of the fragrance industry developed through previous decades will be significantly changed. No longer will it be as fragmented as in the past, consisting of many small and medium-sized companies and some large corporations. In fact, there are some predictions that the industry, especially the consumer product oriented companies, will be dominated by six or fewer giant international corporations.

The mergers and acquisitions will continue, although many of the resulting possible and probable disadvantages can be anticipated. The age of internationalism, environmental and regulatory problems, consumer protection pressures, rapid technical and scientific developments, emerging markets in Russia, China, and developing countries are a few of the factors that the industry will continue to encounter.[8] Meeting these problems is beyond the means of small and many of the medium-sized companies. Therefore, some further consolidation of the fragrance industry is necessary to provide the resources to contend with these factors in the changing world.[4]

4 COPYCATS

The fragrance industry, like so many other consumer product industries, is economically vulnerable to competition from cheaper imitations of their products that are placed on the market. This is especially true for the highly promoted expensive designer fragrances that are best-sellers with the public. Although some imitations, so-called copycat fragrances, were marketed before 1980, it has been only during the last decade that it has grown into a big business.[9] In part, this has been due to the high price of fine fragrances, which places them out of the reach of a large segment of the consumer public. However, another major contributor to the current increased popularity of the imitations has been the significant improvement of their odor resemblance to the originals. This has occurred through the use of new technological developments in analytical instrumentation. In previous years, determining the composition of a target fine fragrance for copying was a laborious and relatively inexact procedure depending to a large extent on an expert's nose. Today, through the use of gas chromatography, mass spectroscopy, nuclear magnetic resonance spectroscopy, high-pressure liquid chromatography and other sophisticated techniques, these analyses are quite rapid and precise, thereby facilitating the formulations of copycats.

The reaction of the fragrance industry to these blossoming imitation counterparts still appears to be a matter of opinion, at least officially. Some companies claim it has no impact on their sales since customers for the 'knock-offs', sold mainly in discount stores and drugstores, were not their customers. Other companies have been more concerned and some litigation has resulted, but without any apparent dramatic effects from the court decision. It has also been proposed that the copycats are actually beneficial to the future of the industry by introducing a taste for fine fragrances to the younger and less affluent consumer through the affordable 'knock-offs'. Presumably, this segment of society, upon achieving higher economic status, will purchase the more elegant, prestigious originals.

Speculating on the future of the copycat business and its ultimate impact on the fragrance industry can lead to various possibilities. For example, imitation fragrances could grow very little or even wither away owing to counter-measures (legal, technical, etc.) introduced by the manufacturers of originals. An example of a technical approach would be the formulation of fragrances using key, impact ingredients that have been patented, and are therefore unavailable to 'knock-off' producers, thus providing an extremely difficult obstacle to compounding a fragrance closely resembling the original. On the other hand, imitation fragrances could emerge as a viable industry of their own, providing significant competition to the manufacturers of originals. In that case, they could be absorbed by the producers of originals, who would formulate and market both the more expensive original and the cheaper 'knock-offs'. To a limited extent, this approach is in existence today, being used in the extension lines of some consumer products, mainly in the shampoo, laundry soap and detergent field. At present, imitation products are tolerated by most of the fragrance industry, but this acceptability could radically change in the future.

5 GEOPOLITICAL CHANGES

Historically, France was considered the center of the fragrance industry. During the 20th century, the industry spread slowly to various parts of Western Europe and to the United States, where viable fragrance manufacturing and marketing companies were established. In the past two decades, Japan also has become a more significant factor in the field and will continue to be a formidable competitor to the industry.[10,11] Japan, North America[12] and Western Europe[13] together probably represented about 80% of the market size for the worldwide fragrance industry during the 1980s.

The 21st century, however, could see a marked change in the size of the market and the geographical distribution of fragrance marketing and manufacturing companies. These changes would result from the dramatic geopolitical, sociological and economic transformations that have been taking place in the world, especially during the last decade. The vast population of China with its slow but steady improvement in living standards represents an attractive potential market to the industry. Companies from France, Japan, the United States and West Germany have already established production facilities in China, usually in partnership with a Chinese producer. China has been a long-time exporter of essential oils to the fragrance industry and also is now producing some of the synthetics, such as the nitro musks, in their factories.[14,15] The emergence of China during the 1980s struck a discordant note with the occurrence of the Tiananmen Square event in 1989. Continued development of the industry in China, to a large extent, will depend on the future suppressive or progressive actions of the government.

The planned unification of the European Economic Community in 1992 will have an international impact on the fragrance industry. The proposed elimination of all economic barriers among member countries should strengthen some fragrance companies and weaken others. The potential effect on tariff barriers between the EEC and non-member countries and the increase or decrease in policies of internationalism or isolationism could modify many of the industry's traditional methods of doing business.

One of the greatest geopolitical changes that will affect the industry is the sudden turn-around in the political, economic and social status of Eastern Europe that occurred in the late 1980s. The policies of *perestroika* and *glasnost* introduced by the USSR released a chain reaction of peaceful (to date) mass movements toward democratization and capitalism in countries such as Poland, Hungary, East Germany, Romania and Czechoslovakia. It appears that if this trend continues and the USSR and other East European countries are successful in overcoming their major financial problems, the fragrance industry would be presented with remarkable marketing opportunities to relieve the consumer product-starved appetites of these populations.

6 DEMOGRAPHICS

Since the termination of World War II, there has been a major expanding movement of various national, ethnic and racial groups from relatively poorer areas of the world to the USA and Western Europe. Many have moved for economic reasons, either searching for the 'pot of gold' or lured by the promises of employment from countries such as Germany, Switzerland and the USA. Others have migrated because of political or religious harassment or even persecution. Regardless of the reasons, these immigrant groups have often established relatively large, new subpopulations within these countries by retaining a large part of their culture, language, habits, etc. In the USA, the demographic 'melting pot', there are very large numbers of Blacks, Hispanics and Asians—to name just a few groups who are increasing numerically at a much more rapid rate than the white Caucasian population (Table 1).

Table 1
Estimated population projection in the USA (millions)[16]

Year	*White*	*Black*	*Hispanic*
1987	206	30	18
1990	210	31	19
2000	222	35	25

Marketing strategy to these groups by the perfume and fragranced product industries for the most part has not been directed toward their specific preferences and/or needs.[17] The same products that are developed for and sold to the native population are also sold to the Blacks, Hispanics and Asians. Exceptions are made, however, for some products that are retailed in Hispanic neighborhoods, where the advertising and labeling is done in Spanish. It is surprising, especially in the USA where these subpopulations numerically represent a large market, that more innovative approaches have not been examined. The Hispanic population alone in 1989 was estimated to be about 19 million people and is projected to grow to about 25 million by the year 2000. It is a moot question whether the currently established fragrance and fragrance product industries will address these statistics or whether new specialty fragrance companies will arise for this purpose.

7 AGE GROUPS

The fragrance and fragrance product industries have successfully developed and marketed certain products for specific age groups. Johnson & Johnson's record in the baby product line has been remarkable, with the baby powder odor immediately recognizable by a large segment of the population. The adolescent's problems and needs have also been addressed with products such as a line of light perfumes acceptable to parents as well as to teenagers, and with fragrances for use in functional formulations such as acne products. Industry is also aware of the fickle nature of teenagers who are always ready to try something new and novel and are very trend-oriented.[17] For the most part the adolescent market is one where the successes of a product are often measured in months rather than years. It is very lucrative and one that requires a continual influx of creativity and innovative ideas. The fine fragrance area, for the most part, has been geared to the 21- to about 45-year age group. This age classification is composed of people starting and developing their careers, families and strong buying habits. It is also the age category in which men are introduced to a variety of men's fragranced products and to the art of buying fine fragrance products for women.

On the other hand, the age range that has been most neglected by the industry has been the over 50-year-old category and especially those over 65 years of age. Many possible reasons for this have been suggested by representatives of the fragrance industry. They claim that previous generations of older people were not as concerned about their appearance and those that were concerned had firm buying habits established. There was also the belief that mature and aged individuals were more reluctant to spend money on fragrance products, especially expensive ones. Now there is a realization that some of these concepts about the mature and aged are currently incorrect and perhaps they were never valid. In addition, the significant upward shift in the age of the population is prompting the industry to examine further this potential market that has not been given thorough attention in the past. The US Bureau of the Census has estimated that the 50-and-over age group in 1987 represented between 25 and 26% of the total population in the USA. They have also estimated that this age group will grow to 28% of the population by the year 2000. Also, a significant segment of this group has accumulated sufficient available funds to permit regular expenditure on luxury fragrances and fragranced products. Although some inroads have been made, industry still requires the development of products and promotional approaches to this promising but essentially untapped market.

8 THE CHANGING STATUS OF WOMEN

During the past few decades there have been some far-reaching sociological changes in the status and aspirations of women[18] that have impacted on the work force throughout most industries. These changes have resulted in more than 60% of the women with small children in the USA now being fully employed. The women have decided for economic reasons, career purposes or just because of peer pressure not to stay at home as previous generations did. This movement has also strongly campaigned for treatment equal to that of men for most jobs in the workplace. Its greatest activity has been in the USA but Western Europe is also moving in this direction, and there has even been some stirring in Japan. The vast increase in the number of women now working instead of staying at home has created a large new target for the fragrance and fragranced products markets. This market will grow and will possibly be segmented according to work function and career.

Historically, men have completely dominated the operation of the fragrance and related industries. The business of creating, manufacturing and selling fragrances was almost exclusively a man's job. This is changing in parallel with the acceptance of the competence of women in other industries. The capabilities of women as perfumers and evaluators is a matter of record. There has also been a strong increase in the number of women in the fragrance sales force since 1980, and again their contributions have been notable.[4] Some have achieved promotion to upper middle-management jobs, but attainment of top management positions for women in the fragrance industry is still the exception. Ultimately, this should change and it will be interesting to observe in the future if this will significantly modify the policies, strategies and directions of an industry that had been exclusively directed by men for products predominantly used by women.

9 SAFETY, ENVIRONMENTAL AND REGULATORY ASPECTS

The worldwide movement during the last two decades of demanding and providing increasingly greater protection to the consumer, worker and environment has introduced new and very difficult problems and burdens to the fragrance industry, requiring vast changes in its philosophy and methods of operation. These burdens are expanding as new regulations and requirements are continually imposed by federal, state and local governmental agencies under pressure from the public, consumer and worker protection groups and the news media. The chemical industry in general, including the fragrance industries, has been the target of a large part of these new requirements. Among the federal agencies in the USA that have major responsibilities in this area are the Food & Drug Administration (FDA), the Environmental Protection Agency (EPA) and the Occupational Safety & Health Administration (OSHA). Although they are assumed to have exclusive areas of responsibility relevant to the regulatory processes, there can be areas of overlap. In addition, many states have passed their own regulatory legislation that may be even more stringent than, and sometimes conflicting with, the federal legislation. The increase in concern leading to the cascade of safety and environmental regulations which started in the USA is also rapidly developing in Western Europe, Japan, Canada and other countries.

The FDA, under authorization of the Food, Drug and Cosmetic Act,[19] is responsible for the safety of all cosmetics on the US market. This includes perfumes and colognes *per se,* and fragrances in cosmetics, toiletries and other products applied directly to the skin. It excludes fragrances in soaps, detergents, household products and various industrial applications. The fragrance industry and the FDA have worked cooperatively in solving problems arising from adverse toxicological properties of fragrance ingredients and no new major regulations have been imposed on the industry to date.

Under the authorization of several different Acts of the US Congress, the EPA has the responsibility to protect the public and the environment from possible exposure to toxic materials which may find their way into the air, water and/or soil. In the mid-1970s under the authority of the Toxic Substance Control Act (TSCA),[20] manufacturers, importers and exporters of chemicals and chemical intermediates must register each such substance with the EPA 90 days prior to the commencement of manufacturing, importing or exporting. The registration process requires the submission of acute toxicity data, analytical data and manufacturing process information to the EPA. Also, under TSCA Section 8(e), if one becomes aware of toxic effects of a chemical substance through animal testing or an epidemiological study, one is legally obliged to report this to the EPA within 15 days.

A similar registration law is in place in the EEC and is just going into effect in Australia and Canada (the Canadian Environmental Protection Act—CEPA).

Other laws under the jurisdiction of the EPA which affect both segments of the fragrance industry are the Comprehensive Environmental Response, Compensation and Liability Act of 1980 (CERCLA) and the Superfund Amendments and Reauthorization Act of 1986 (SARA Title III).[21]

In 1985, the OSHA Hazard Communication Standard[22] (29 CFR 1910), also known as the Federal Worker Right-to-Know Law, went into effect. This law requires that a Material Safety Data Sheet (MSDS) must be prepared and made readily available to workers who handle or may be exposed to a hazardous substance or mixture containing a hazardous substance. Implementation of this regulation has required the partial disclosure of formula information by the fragrance suppliers to their customers, the consumer product companies.

The Federal Worker Right-to-Know Law has precipitated several State Right-to-Know Laws in Massachusetts, Pennsylvania and New Jersey. These state laws have required more disclosure of fragrance formula compositions to state, local and health authorities. A similar worker law recently went into effect in Canada—Workplace Hazardous Materials Information System (WHMIS).[23] This law requires that the fragrance manufacturer must have on file and make available to government authorities on request the entire compositions of all fragrance formulations they manufacture.

Another piece of legislation which has affected the two segments of the fragrance industry is the recently enacted California Safe Drinking Water and Toxic Enforcement Act of 1986, which has become known as the notorious Proposition 65. This law requires that any consumer product containing an ingredient which is a potential carcinogen or reproductive toxin must be labeled accordingly. This law has required the consumer product industry to request this information from their raw material suppliers (i.e. the fragrance suppliers).

With more and more legislation, such as that discussed above, being enacted in the USA, Canada, Europe and other countries, it is becoming increasingly difficult for the fragrance suppliers to maintain trade secrecy, the bread and butter of this industry. These regulatory barriers and safety requirements can be both a deterrent and a stimulus to the industry's support of research and development projects for new and novel fragrance ingredients: a deterrent from the standpoint of the economic burdens and bureaucratic problems required to comply with the regulations, and a stimulus from the standpoint of developing unique fragrances that cannot be copied because of the patented new ingredients they contain.

10 ANIMAL WELFARE

A relatively new phenomenon, the animal welfare movement, has significantly affected the fragrance and fragranced product industries during the 1980s. It has been strongest in the United States but is growing in Western Europe and other countries. The original objective of the animal welfare proponents was a vast improvement in the living and experimental conditions for animals in research and toxicological testing facilities. Strong support for their platform was gained through illustrations of alleged sadistic experiments being conducted on animals in these laboratories. The well organized, politically potent animal rights activists in the USA have already achieved some of their goals through amendments to the Animal Welfare Act, and additional government regulations are being considered.[24] However, these groups now appear to be broadening their aspirations toward the complete banning of the use of animals for some testing and experimental purposes. Animal welfare has now become animal rights.[25]

The cosmetic, toiletry, and fragrance (CTF) industries have been one of their primary targets in this regard. Massive letter-writing campaigns, boycotts, demonstrations and picketing have been utilized by the militant 'rights' groups. The CTF industries, who are very sensitive to any adverse publicity, have responded in several ways. The Cosmetic, Toiletry and Fragrance Association (CTFA) has funded a US$3 million effort at Johns Hopkins University to develop in-vitro tests as alternatives to animal testing for determining the safety of their products. Several cosmetic companies have disbanded their animal testing facilities and stated they will no longer use animal testing in the research, development, and production of their products.[26] Presumably this will satisfy the animal rights activists and most CTF companies probably will follow suit.

The effect of the elimination of animal tests imposes serious handicaps on the CTF industries for proving the safety of their newly developed products. The number of available in-vitro tests is limited and the validity of existing ones still requires further corroboration. The use of human tests on new products without prior animal studies is an unacceptable procedure. As a consequence, the pace of new product research and development could be substantially reduced in the future.

11 OPPORTUNITIES, THREATS AND STRATEGIC PLANS

This chapter has selected and discussed some of the more significant factors in the rapidly changing world that will affect the fragrance industry. Factors such as fragrance trends[27,28] and contributions from research and development studies were omitted since they were included in other chapters. It should be mentioned, however, that the future of the recent trend toward fragrances that allegedly can affect an individual's mood—aromatherapy[29,30]—is still uncertain. Additional data are required to determine if this concept can be supported on a scientific basis, or if it is essentially a marketing tool.[31,32] A scientific breakthrough that would divulge the olfactory code for recognizing individual odorants should produce radical changes in the methodology of formulating new fragrances in the future.

The future of the fragrance industry at this time is replete with opportunities and threats. The current democratization process taking place in Eastern Europe should ultimately provide a large new market potential for the industry. China, with its huge population, represents another vast market possibility in time, but not until major changes are made in government policies. The migration and importation of labor from undeveloped countries to Western Europe and the United States could also be a forerunner of future markets.

On the other hand, the problems of complying with ever-increasing stringent and complex government regulations concerning safety, the environment, product labeling and 'right-to-know' have imposed difficult and expensive burdens on companies. It is anticipated that the number of regulations will multiply, requiring companies to expand their cadre of regulatory, legal and consulting personnel. The economic consequences of these burdens could ultimately lead to the demise of smaller businesses, leaving a few larger resource-rich companies to control the field. An offshoot of the regulatory cascade appears to be a growing trend to include some need for formula disclosure in the legislation, an action that strikes at a fundamental tenet of the fragrance industry. Formula disclosure, if carried to its pinnacle, would lead to generic fragrances with the exact composition of successful products on the market.

Advantageous situations and major problems will continue to surface in the rapidly changing world. More than ever, the fragrance industry needs to develop strategic plans, both contingent and long-range, to grasp the opportunities and solve the threatening problems that will emerge in the future.

REFERENCES

1. Guenther, E., *The Essential Oils,* Vol. I. Van Nostrand, New York, 1948, pp. 3–13.
2. Samuels, B., Air fresheners, Days of pine and roses, still? *Givaudanian* (Winter 1976) 3–6.
3. Dorland, W. E. & Rogers, J. A., Jr, *The Fragrance and Flavor Industry.* W. E. Dorland Co., Mendham, NJ, 1977.
4. Adamson, A. S., Consumer products perfumery in the 80s and 90s. *Soap/Cosmetics/Chemical Specialties,* **65** (June 1989) 26–8, 30, 73–4.
5. Unger, L., Worldwide merchant sales of flavors and fragrances, 1984–1990. *Perfumer & Flavorist,* **11** (Apr./May 1986) 63–72.
6. Unger, L., The worldwide flavor and fragrance industry, 1986–1990. *Perfumer & Flavorist,* **19** (Aug./Sept. 1988) 21–6.
7. Anon., Unilever Continues Buying Spree with $1·55 Bil. Fabergé/Arden Purchase, *FDC Reports (The Rose Sheet),* **10** (17 July 1989) 1–2.
8. Morton, J., New business acquisitions: R&D's role. *Cosmetics & Toiletries,* **103** (Oct. 1988) 71–7.
9. Fisher, A., Mark Laracy's obsession. Inc., **10** (May 1988) 70–82.
10. Tokuda, K., The fragrance market in Japan today. *Cosmetics & Toiletries,* **103** (June 1988) 31–6.
11. Nakajima, M. & Ikenouchi, Y., Perfumery in Japan. *Cosmetics & Toiletries,* **101** (June 1986) 51–67.
12. Grisanti, E. P., State of the fragrance industry in North and South America. *Cosmetics & Toiletries,* **103** (June 1988) 44–7.
13. Landers, G. J. B., Fragrance in the EEC. *Cosmetics & Toiletries,* **103** (June 1988) 37–43.
14. Daqing, Q., More foreign links for a rapidly growing industry. *Cosmetics & Toiletries,* **103** (Feb. 1988) 65–8.
15. Quince, M. L., Fragrance industry in China. *Cosmetics & Toiletries,* **103** (June 1988) 47–8.
16. Spencer, G., Projections of the population of the United States, by age, sex, and race: 1983 to 2080. In *Current Population Reports, Population Estimates and Projections,* Series P-25, No. 952. US Department of Commerce, Washington, DC, 1984.
17. Kintish, L., Fragrance industry considers unexplored markets. *Soap/Cosmetics/Chemical Specialties,* **65** (April 1989) 58–59 & 72.
18. Jessee, J., The emerging women. *Givaudanian,* No. 3 (1971) 3–6.
19. Walsh, G. P., Jr, *Federal Food, Drug and Cosmetic Act with Amendments.* US Government Printing Office, Washington, DC, 1981.
20. Toxic Substances Control Act. *Federal Register* (42 FR 13130), 9 March 1977, and 40 CFR Parts 702–775.
21. Hazardous Chemical Reporting; Emergency Planning and Community Right-to-Know Programs. *Federal Register* (52 FR 2836–2855), 27 January 1987.
22. Hazard Communication; Final Rule. *Federal Register* (48 FR 53280–53348), 25 November 1983.

23. Hazardous Products Act—Controlled Products Regulations, *Canada Gazette,* Part II, Vol. 122, No. 22, 1 January 1988, pp. 488–589.
24. Holden, C., Compromise in sight on animal regulations. *Science,* **245** (1989) 124–5.
25. Holden, C., A preemptive strike for animal research. *Science,* **244** (1989) 415–16.
26. Stinson, S., Revlon, others end tests of products on animals. *Chem. Eng. News,* **67** (10 July 1989) 6.
27. Donohue, J., Quest identifies fragrance trends. *Soap/Cosmetics/Chemical Specialties,* **65** (Mar. 1989) 36, 72.
28. Hall, N., European trends. *Cosmetics & Toiletries,* **103** (Sept. 1988) 97–101.
29. Van Toller, S., Odours, emotion, and psychophysiology. *Int. J. Cosmet. Sci.,* **10** (1988) 171–97.
30. Roebuck, A., Aromatherapy: fact or fiction. *Perfumer & Flavorist,* **13** (June/July 1988) 41–3.
31. Lin, T. J., Notes from the Orient. *Cosmetics & Toiletries,* **101** (Sept. 1986) 11–12.
32. Carsch, G., Aroma therapy. A status review and commentary. *Soap/Cosmetics/Chemical Specialties,* **63** (Aug. 1987) 36–42, 88.

VII TOPICS IN PERFUMERY RESEARCH

This book has so far dealt almost exclusively with fragrances, in spite of the facts that it is in general the fragrance and *flavour industry which is being considered and that, of course, flavours act to a great extent through their odour. Research is, therefore, normally active in both fields, and the molecules used by perfumers and flavourists are quite often the same. However, in accordance with the scope of this volume, the following chapters in essence neglect the flavour aspect.*

The first chapter in this section again discusses odour perception and the receptors involved, but the focus is now on research and future developments. The subsequent chapters deal with natural and synthetic raw materials. They should create at least an impression of recent research contributions and of the extraordinary elegance of some of the synthetic work in this field. Professionals and even consumers might thus develop a better understanding of the term 'nature-identity' in comparison with 'naturalness'.

The section is concluded by a short 'Outlook' intended to show the direction in which research will be driving the fragrance industry in the future.

Chapter 18

Receptors: Current Status and Future Directions

F. L. MARGOLIS

Department of Neuroscience, Roche Institute of Molecular Biology, Nutley, NJ 07110, USA

&

T. V. GETCHELL

Department of Physiology and Biophysics, College of Medicine, Chandler Medical Center, Lexington, Kentucky 40536, USA

1 INTRODUCTION

Molecules in the environment have no significance as signals unless they can be detected, interpreted and responded to. Information processing takes place at several sequential levels. These include the initial interaction between the stimulus and the cell membrane, the transduction of this event to the interior of the cell, and subsequent transmission of this information to processing centers and association areas in the central nervous system (CNS) for ultimate evaluation. This applies equally whether we are considering commercial fragrances, animal pheromones or cooking gas. From the point of view of the recipient, inability to detect or perceive a signal is functionally equivalent to its absence. There are numerous examples of this in nature where, for example, it can facilitate the separation of breeding populations since the selective inability to detect signal pheromones is functionally equivalent to the absence of the donor. In a chemosensory context more familiar to ourselves, this phenomenon is quite apparent to those individuals who are genetically unable to detect the bitter taste of phenylthiocarbamate. A dramatic demonstration in olfaction is the differential response of individuals to the presence of androstenone (boar taint). Thus, the perceived *umwelt* is primarily a function of the recipient and its detection and transduction mechanisms as compared with the source or donor.

Although the intimate molecular details of these events can be, and frequently are, specific, the general mechanisms involved manifest extensive homologies across species and sensory modalities. For chemoreception the critical differences will probably be a function of the nature of the match between the signal molecule itself and the initial detection event rather than in the mechanisms of the transduction and secondary response elements.

Insect pheromones convey messages to conspecifics but not to other unrelated species although the transduction events certainly include biochemical pathways utilized by other biological signal response systems. Even if humans could detect bombykol, a component of the silkworm moth mating pheromone, the biological significance of the signal would be modulated by the transduction and information processing components, ultimately leading to a very different interpretation of the significance of the signal. Thus, it will be useful in the context of this chapter to review briefly other signal detection and transduction systems.

In discussing the mechanisms that might apply to detection and transduction of chemosensory signals, it is useful to remember that these events are primarily mediated via neural pathways. One can envision the primary chemoreceptor events in olfaction and gustation as a problem in molecular pharmacology analogous to characterizing the mechanisms of post-synaptic responses to neurotransmitters or neuromodulators. However, in the chemosensory systems under consideration here the stimuli do not arise from adjacent cells or distant organs that contribute to the '*milieu interieur*', but instead derive from the external environment.

In this chapter we will address three primary questions:

(1) What detection and transduction mechanisms have already been described in chemosensory and non-chemosensory systems?
(2) What are possible models for olfactory reception?
(3) How can contemporary techniques of cell and molecular biology be used to address these issues?

This chapter will assume a molecular pharmacological orientation to the chemical senses with a major emphasis on vertebrate olfaction. Several recent sources of detailed information about chemoreception and signal transduction are available.[1–14]

2 RECEPTORS AND RECEPTOR CELL LOCATION

Olfactory receptor neuron cell bodies are located within the olfactory neuroepithelium in the posterior nasal vault. Centrally, the axons of these neurons project to the most rostral portion of the CNS, the olfactory bulb, where they synapse with two main populations of target neurons. One population consists of local circuit neurons (the juxtaglomerular neurons) that express multiple neuropeptides and neurotransmitters. The second population, the mitral cells, are the major output neurons.[15]

In the peripheral direction, the receptor cells send a dendritic process towards the surface of the olfactory neuroepithelium. It terminates in a knob-like process from which several long cilia project into the mucus that covers the surface of the epithelium.[9]

On teleological grounds alone, the cilia would be a logical candidate for the initial site of interaction of the stimulus with the cell. This view is consistent with ontogeny of ciliogenesis and odor responsivity[16–18] and is supported by calculations based on the kinetics of cellular response to ligand application, air/aqueous partition coefficients and stimulus diffusion constants across the thickness of the mucus.[9]

Biochemical evidence for the presence in the cilia of many molecular species associated with signal transduction processes are consistent with their involvement in detection and transduction. For example, olfactory cilia from several species express stimulus ligand binding activity,[19–26] contain high concentrations of adenylate cyclase[8,27–32] and G-proteins,[33–38] exhibit cyclic nucleotide gated channels,[39] and manifest odorant-activated cation-selective channels[39–42] and stimulus-ligand activated phosphatidylinositol–phosphate phosphodiesterase activity.[43] In addition, olfactory tissue-specific isoforms of G-proteins and adenylate cyclase have been identified and characterized.[30,36,44] Calmodulin, a protein involved in modulating many cellular transduction processes,[45] is also abundant in the olfactory receptor neurons.[28,46]

3 STIMULUS DYNAMICS

For air-breathing animals the volatile, hydrophobic nature of most odor stimuli would seem to present problems of delivery and removal since these ligands must enter the aqueous mucus phase that overlies the neuroepithelial sheet before they can interact with the presumed receptor sites present on the olfactory neurons. Although apparently counter-intuitive, many hydrophobic molecules that are real or potential olfactory ligands actually possess air–water partition coefficients that favor the aqueous phase by factors of 10–1000.[47–49] Thus, their concentrations in the mucus would exceed those in the air phase. This phenomenon could easily lead to an incorrect estimate of receptor sensitivity or response threshold if it was based solely on stimulus ligand concentration in the air phase. The hydrophobic nature of these molecules suggests that they might be easily trapped in lipophilic cell membranes and present a problem for removal. Clearly, efficient stimulus elimination is essential to prevent receptor desensitization.

Recently, soluble proteins present in mucus that are capable of binding many ligands have been isolated and cloned from nasal mucosa[50–54] and taste tissue.[55] It has been proposed that these proteins act as carriers to transport hydrophobic stimulus ligands from the environment through the mucus phase for delivery to olfactory[51,52] and gustatory receptor sites.[55] The amino acid sequence homology of these proteins to many members of the retinol binding-protein superfamily[50,52,53] is consistent with this hypothesis. Aphrodisin, a novel member of this protein superfamily has been isolated from hamster vaginal secretions and cloned.[56] The properties of this protein, and the behavioral responses to it, suggest that it may be a pheromonal protein acting directly on vomeronasal chemoreceptor neurons rather than as a ligand carrier. However, evidence to indicate that the presence of these proteins can alter the sensitivity of the receptor cells or that they interact with presumed receptor sites on the cilia of olfactory cells is still needed. In view of the need for efficient stimulus removal to prevent receptor desensitization, these binding proteins may participate in ligand removal. Their presence in the mucus on the surface of the neuroepithelium may contribute to stimulus removal by mechanical means.[9,47,49] Possibly they could function as a component of the mucus in the differential sorption of odor ligands along the nasal passages as postulated in the chromatographic model of olfactory reception.[57,58]

Conversion of these hydrophobic stimuli to more aqueous soluble molecules could also play a role in their removal. Thus, it is of interest that the olfactory tissue is very rich in P-450 mixed-function oxygenase activity[59–61] as well as glucuronylation activity (Lancet, D., pers. comm.). Both of these enzymes are involved in the intracellular conversion of initially hydrophobic molecules to metabolites that are much more hydrophylic.[62] The potential significance of this process is further indicated by the presence of olfactory tissue-specific isoforms of these enzymes.[63,64]

Thus, questions of stimulus access to and removal from the cellular sites of detection as well as the role of various components of the mucus in determining the dynamic aspects of this process remain a critical area of investigation.

4 SIGNAL TRANSDUCTION

Signal transduction necessitates mechanisms for amplification of the primary event by which the stimulus ligand interacts with a receptive cell. In eucaryotes these cascades can be divided into several biochemical categories.

The most direct signal transduction mechanism is the direct gating of an ion channel and the attendant passage of ions.[65–70] This category of receptors, for which the γ-aminobutyric acid, N-methyl-D-aspartic acid and nicotinic cholinergic receptors may be taken as examples, are usually multimeric protein complexes that span the cell membrane creating a pore through which charged ions flow to carry a transmembrane current. This pore is normally in a closed state so that few or no ions can gain passage. The binding of an appropriate ligand to a site on the membrane protein complex causes a transient conformational change of the protein, altering the dimensions of the pore, and directly gating the passage of ions. The ligand binding site is usually assumed to be located on the external surface of the cell membrane. However, in the α-adrenergic receptor it may be within the membrane and recent studies in the vertebrate visual and olfactory systems suggest that, at least for cyclic nucleotides, a binding site capable of regulating ion flux may be on the cytoplasmic face of the membrane.[10,39,71]

In contrast to the direct ligand-gated channels described above, many receptor systems are mediated by one or another type of second messenger system. These receptors consist of transmembrane proteins that are coupled to their respective amplification systems by intermediate guanine nucleotide-binding regulatory proteins (G-proteins). This group includes vertebrate rhodopsin, muscarinic, serotonergic and adrenergic receptors, and probably olfaction and taste. In these cases a cascade of chemical-enzymatic events results in the amplification of the initial stimulus-receptor event to transduce the initial signal into the appropriate cellular compartment. These systems include amplification via alterations in cAMP, cGMP, phosphoinositol metabolism, intracellular Ca^{2+} mobilization and protein phosphorylation.[1,10,33–35,39,43,72–83]

A third mechanism is exemplified by the insulin and EGF receptors, among others, in which the binding of the peptide ligand to a membrane receptor activates an associated tyrosine protein kinase to initiate a protein phosphorylation cascade.[84] Since the activity of many enzymes and their interactions with cellular regulatory proteins can be modulated by their state of phosphorylation, activation of this cascade can have dramatic effects on cellular metabolism, resulting in signal transduction in the recipient cell. The membrane-bound form of guanylate cyclase can also act as a signal detector/transducer. When a stimulus ligand binds to the appropriate extracellular site on this enzyme a signal (presumably resulting from a conformational change) is transmitted to the intracellular catalytic domain. In several systems this results in elevated cGMP as an intracellular second messenger.[76] In contrast, the soluble form of guanylate cyclase may play a role in invertebrate olfaction.[85]

Alternatively, the signal molecule itself (e.g. estrogen) can diffuse directly into the cell and bind to a cytoplasmic receptor protein to form a complex that is transported to the nucleus. Once in the nucleus this complex can interact at particular sites on the chromosomal DNA to regulate transcription of one or more specific genes. The nuclear receptor family includes various receptor proteins for steroids, retinoic acid and thyroid hormone. These receptors are multidomain proteins containing diverse structural elements that function as ligand binding domains, DNA binding domains or transcriptional regulation domains. These properties enable them to function as ligand-regulated transcription factors. Thus, binding of their respective ligands can serve either to activate or to inactivate expression of specific gene loci.[86,87]

The members of the first two classes of ligands capable of interacting with these receptors (steroids and retinoids) are terpenes in structure and biosynthetically derived from the five-carbon isoprene unit, as are many odor ligands. The ligand-binding domains of these nuclear receptors are postulated[88] to be descended from an ancient protein family, possibly related to the origins of olfactory receptor molecules. Potentially these domains could have been modified or incorporated into olfactory ligand receptor sites in the course of evolution.

Even in procaryotes, multicomponent signal transduction systems are known to play a critical role in responding to changes in the local environment. So-called 'two-component regulatory systems' have been described consisting of a membrane receptor (a transmembrane signaling device) and cytoplasmic regulatory proteins.[89–92] Environmental information gathered by the receptor is transferred to the cytoplasmic regulatory partner by phosphorylation. If the regulatory protein is a DNA binding protein then the modification results in an alteration (up or down regulation) in gene expression. In bacterial chemotaxis the output is altered flagellar rotation, rather than changes in gene expression.

Our assumption, as indicated above, is that the transduction modes utilized in chemosensory systems are analogous to those described for other biological signal detection systems. In contrast to the extensive range of studies characterizing receptors for various drugs, hormones, ion channels, and neurotransmitters and modulators, our knowledge regarding the molecular mechanisms associated with chemoreception is more limited but expanding rapidly.

Thus, evidence indicating a major role for adenylate cyclase in both olfaction and taste derives support from studies in several laboratories. These include guanine-nucleotide-dependent activation of the enzyme following exposure to stimuli.[27–29,31,32,35,38,43,73,80,83] The existence of a cyclic-nucleotide-gated channel in olfactory cilia[39] analogous to that seen in retinal membranes,[71] as well as the anatomical localization[33,37] and expression of olfactory neuron-specific isoforms of the cyclase[30] and of a G-protein all support this view.[36,44]

The ability of various concentrations of odor stimuli to activate adenylate cyclase in non-olfactory cells[93] and the reports of non-receptor-mediated chemosensory transduction[94,95] have encouraged additional studies of the role of Ca^{2+} and calmodulin in regulation of olfactory adenylate cyclase activity. These recent results indicate that olfactory adenylate cyclase is subject to Ca^{2+}/calmodulin regulation.[28] Based on this, and other data, a modified model of the role of adenylate cyclase in chemosensory transduction has been proposed.[28]

The very strong experimental support for the role of adenylate cyclase in chemoreception should not be permitted, at this early stage, to overshadow the possibility that other transduction cascades may also be utilized. Indeed, perhaps the enormous range of stimuli, chemosensory anatomy, species, habitats, etc., should alert us to this very possibility. From this point of view it is not at all surprising that reports have appeared implicating the involvement in chemosensory transduction of direct ligand-gated channels,[23,40–42,96–99] phosphoinositide hydrolysis[43] and Ca^{2+} translocation.[28,72] These observations illustrate the rapidity with which knowledge of the biochemical/molecular bases of chemosensory transduction has advanced in a very few years.

5 RECEPTOR MODELS

Clearly, there is growing evidence regarding the role of various transduction mechanisms in chemoreception, but the primary event itself is still largely unknown. By analogy to studies in other systems where more pharmacological data are available most, but not all,[94,95] researchers presume that the first event in the transduction sequence is a direct molecular interaction between the stimulus molecule and a membrane-associated protein or glycoprotein. Thus, one critical question that must be addressed is the nature of this class or group of molecules and the mechanisms by which they interact with the stimulus and transmit the information deriving from that interaction to the interior of the cell. What might be a reasonable model of such receptors and could it facilitate the design of experimental strategies for research? At the molecular level several possible strategies can be envisioned as biological solutions to the problem of identifying, categorizing and interpreting the information derived from specific stimuli in a complex mixture of stimuli, all of which compete for attention simultaneously.

One model assumes that the olfactory neuronal population is composed of a series of clones, each of which, in analogy to the immune system, synthesizes one kind of receptor molecule. Cellular response to a given stimulus molecule might then depend upon the selectivity properties of the specific receptor molecule expressed by an individual receptor neuron. Alternatively, individual neurons could each express a small number of different receptors so that each cell presents a mosaic of receptor molecules of diverse selectivity properties. In this model each cell is considered to express a limited subset of the total receptor molecule population expressed by the olfactory neuroepithelium.

Each of these alternative scenarios is based on a combinatorial model.[2,8,9] It is assumed in both cases that a given stimulus molecule could interact with multiple receptor cells. In the presence of a pure stimulus ligand, olfactory neurons derived from one or more clones would respond as a function of ligand–receptor interaction and evaluation at the combinatorial level would occur in the olfactory bulb of the CNS where the input from the individual cells is processed and evaluated. In the mosaic model a pure stimulus ligand could interact with multiple cells each expressing different populations of receptor molecules. Thus, the cellular response could be dependent on combinatorial interactions occurring between classes of receptor molecules at the level of the receptor cell itself, to be followed by further analysis in the olfactory bulb as in the clonal model. Thus, in both models, events in the olfactory bulb play a crucial role in analysis of the stimulus ligand.

The specificity of the individual receptor molecules in neurons descending from different clones could differ considerably. Thus, at one extreme the receptor molecule could be of highly restricted specificity and capable of responding only to a very limited range of molecular structures causing the cell to have an extremely limited response profile. At the opposite extreme the receptor molecule could be of significantly lower specificity and capable of responding to different concentrations of a wide range of molecular structures. The same range of olfactory neuron response profiles could be generated by the mosaic model in which cells differentially express defined populations of receptor molecules. An alternative model based on ligand interaction with the cell membrane in the absence of specific receptor molecules has also been proposed[94,95] but it is more difficult to envision how specificity would be achieved in this situation.

Although, as indicated earlier, one can envision a system with many different receptor molecules, each of relatively high specificity, as the means by which the specificity and sensitivity of the olfactory system is expressed, it is difficult to envision how this can be pre-established in anticipation of the arrival of a large number of unknown molecules.

To deal with the anticipated arrival of any of a very large number of unanticipated molecular structures, the immune system has resorted to a mechanism of extensive gene splicing and creation of an enormous potential array of clones expressing individual receptor molecules from which a small number of clones are selectively proliferated in response to challenge. However, the time constraints imposed on the immune and olfactory systems are very different, in that the olfactory system needs to respond optimally, essentially at the moment of ligand arrival, with no time available for clonal selection and expansion. Although stimulus ligand induction of olfactory receptor molecules may occur, it would presumably be related to much longer-term events than the initial stimulus response.

By their very nature, no ligand–protein interactions are absolutely specific.[100] This applies equally to antibodies, enzymes or receptors. Conceptual models for chemoreceptor molecules of highly restricted specificity can be drawn by analogy to other receptor systems, to highly specific enzyme–substrate interactions or to very selective antibodies. Although insect pheromone systems provide evidence for the existence of receptors with very high specificity,[2] an alternative solution to the problem of detection of a vast array of diverse, largely hydrophobic, molecules in the desired time frame and with variable thresholds could be achieved by the existence of a reasonably small population of receptor molecules with diverse and overlapping specificities distributed as a mosaic across the population of receptor cells.

A model for this concept was proposed a few years ago[101] based on the properties of the family of enzymes known as polysubstrate monooxygenases.[61,62] Although these enzymes have many of the structural properties proposed for olfactory receptor sites,[8,34] their enzymatic properties are equally significant in commending them as a model on which to base a conceptual framework. Thus, these integral membrane proteins are members of a superfamily with perhaps as many as 100 members. These enzymes act on a very wide range of small hydrophobic molecules of diverse structures, many of which are, or are related to, chemostimulatory ligands. The individual members of this family of enzymes tend to have broad overlapping substrate–ligand specificities whose affinities range from millimolar to micromolar with substrate turnover numbers that can vary by 1000-fold. These enzymes interact, and are associated with, a reductase that is largely cytoplasmic, in analogy to the interaction of various receptor systems with macromolecules involved in second messenger cascades.

Despite the wide tissue distribution of these enzyme activities[61,62] and their abundance in olfactory mucosa,[59,61,63,64] it is important to recall that they are being proposed as the basis of a hypothetical model, not as receptors *per se*. It is nevertheless conceivable that in the course of evolution, possibly by exon shuffling, substrate–ligand binding sites have been utilized for multiple purposes. These and related questions will only be answered when detailed structural data are available for chemoreceptor ligand-binding sites.

6 IDENTIFICATION AND CHARACTERIZATION OF RECEPTOR AND TRANSDUCTION MOLECULES

The state of our knowledge and the thrust of our discussion make it imperative to isolate and characterize receptor molecules, determine how they are distributed on receptor cells, evaluate what regulates their expression and characterize the nature of their interactions with cellular second messenger systems. The most direct approach to this problem is the isolation of the protein/glycoprotein receptor by the techniques of protein chemistry. Numerous laboratories have pursued this approach with some success[21,26] as recently reviewed[8,9,11,12] but, in general, detailed structural and functional characterization of the isolated molecule at the desired level of resolution is still unavailable. Part of the problem may relate to the low abundance of receptor and/or to the difficulty of obtaining sensitive, functional, in-vitro assays. These problems are not unique to taste and smell. However, in contrast to many pharmacologically well-characterized receptor systems,[1,102] the chemical senses suffer a lack of well-defined ligands that behave as classical agonist/antagonist pairs or exhibit distinct stereospecificity or act as potent high-affinity receptor toxins.

The advent of contemporary techniques of molecular and cellular biology and their highly successful application to similar problems in many other receptor systems argues that the time is ripe for their exploitation in the chemical senses. An indication that this revolution has already begun is evident in reports from several laboratories.[3,6,36,41,44,46,48,50,52,54,56,63,103–109] A brief summary of some of these approaches follows, with an indication of how they are/will be applied to the chemical senses.

Functional expression of mRNA in *Xenopus* oocytes is a powerful technique that has been used to identify, characterize and clone receptor, channel and transduction molecules in several receptor pathways.[48,66–70,105,110–116] Following the injection of exogenous mRNA, various cellular and membrane-associated components of diverse transduction pathways can be expressed by *Xenopus* oocytes. These macromolecules can interact functionally among themselves or with molecules endogenous to the oocyte. Monitoring the electrophysiological or biochemical responses of such oocytes to the application of an appropriate ligand can provide evidence for the presence of a specific active mRNA. Recent results from our laboratories demonstrate that this approach will be a powerful tool for chemosensory systems as well.[48,105] Thus, *Xenopus* oocytes injected with mRNA isolated from olfactory or taste tissue express receptors that respond to appropriate ligands with changes in transmembrane conductance. The functional properties of the chemosensory receptor sites expressed in this system reflect those described from studies performed *in vivo* in the organism from which the mRNA was isolated. These exciting observations open the way to cloning of molecular chemoreceptors. The power of this technique for chemosensory systems is that it provides a functional assay independently of any prior knowledge of the molecular nature of the transductive events.

If high-affinity, specific ligands are available that are themselves fluorescent or can be derivatized with fluorescent reporters, without loss of function, then cells that express specific receptor molecules can be identified and separated by the technique of fluorescent activated cell sorting (FACS). This approach, combined with that of expression cloning, has been used to analyze, identify and isolate genes for receptors.[117–119] Its applicability to cells expressing specific chemoreceptors awaits only the availability of appropriate ligands. The ligands used can be low-molecular-weight compounds such as chemoreceptor stimuli, or macromolecules such as monoclonal antibodies directed against specific receptors,[22,120] or even anti-idiotypic antibodies whose reactive centers mimic specific stimulus ligands. Such anti-idiotypic antibodies could also serve as affinity reagents for direct receptor isolation and characterization. The utility of affinity-driven ligands for identification and isolation of chemoreceptors and enzymes associated with ligand metabolism, especially those capable of being radiolabeled and photoactivated *in situ*, has already been demonstrated.[121–123]

In addition to the elegant technique of expression-cloning and analysis by FACS as described above, a wide range of molecular biological techniques are being applied to the study of the genes responsible for the detection and transduction events that characterize chemoreception. They will be alluded to briefly as an indication of the range of possibilities.

Many functional proteins have evolved into families of related molecules which exhibit homologies of structure and function but differences related to their specific roles in particular cells or metabolic pathways. Thus, many G-protein-linked receptor molecules share certain structural homologies,[77] as is also true for certain ligand-gated ion channels.[66] This has permitted the identification of previously unknown members of these families by screening cDNA or genomic libraries with probes derived from known members. This approach was used to characterize additional members of the G-protein family[44] leading to the identification of G_{olf},[36] a G-protein specifically expressed in the olfactory receptor neurons. An adenylate cyclase specifically or selectively expressed in the olfactory neurons was similarly identified.[30] By the same technique, novel tissue-specific isozymes of the family of P-450 mixed-function oxygenases[63] and UDP glucuronyl transferases were cloned from olfactory mucosa. The functions of these last two enzymes in chemoreception is unknown but may be involved in stimulus metabolism.

Although only beginning to be exploited, the existence of congenic strains of animals differing in chemosensory responsivity or sensitivity suggests that techniques of differential screening of cDNA or genomic libraries should facilitate isolation of the molecular species responsible for the specific chemosensory events.[124–126]

Additional approaches that can be or are being used to identify components in the detection–transduction sequence include reconstitution of isolated components into synthetic membranes to create model systems to analyze functional activity of partially purified fractions.[5,27,42] Real-time imaging of subcellular changes in Ca^{2+} ions has already been applied to the evaluation of initial events in taste response[73] and should be generally applicable to study and characterize response elements in chemoreception.

The functional role of these and other molecules can be evaluated by several in-vivo and in-vitro approaches based on recombinant DNA technology. Once identified, the responsible gene can be manipulated in a variety of ways to alter its expression including mutagenesis, deletion or inclusion in a chimeric molecule. It can be transfected into naive cells under the control of normal or novel regulatory elements that control subcellular localization, amount or timing of expression, permitting the study of function in a normal or altered cellular environment.[127] This same approach can be applied to the creation of transgenic and chimeric animals in which the genes coding for various components of the detection and transduction pathway can be programmed to be expressed at altered levels, in ectopic locations or varied temporally,[128] as already demonstrated with the gene for the olfactory marker protein (OMP).[104] Tissue-specific expression can be selectively blocked by expressing antisense RNA to prevent the normal mRNA from being translated.[110] In principle, the technique of homologous recombination can be used to eliminate selectively all or part of a gene, or even to replace it with a variant engineered by the investigator to determine the influence of these manipulations on the function in question.

7 CONCLUSIONS

In recent years dynamic changes have occurred in receptor research. The advent of protein microchemistry and microsequencing, coupled with the virtual explosion of recombinant DNA technology and the power of computers for modeling three-dimensional molecular structures, has totally altered the way in which receptor systems are studied. The initial beneficiaries of this revolution were the classical neurotransmitter, neuromodulator and hormone receptor systems. This exciting technological revolution is now beginning to be applied to research in chemoreceptor systems. Although we are just at the threshold of these events, the ferment is apparent. Approaches or questions that might have seemed fantasy a decade ago are almost commonplace today as a result of the advances in methodology. It has been the purpose of this chapter to try to illustrate where the impact is already evident and to indicate some of the questions that can now be experimentally addressed. Isolation, characterization and manipulation of the genes, mRNAs and proteins of chemoreceptor molecules and their associated second messenger transduction systems are clearly upon us. The ultimate impact of this information on human and animal medicine and hedonism can only be imagined.

REFERENCES

1. *Molecular Biology of Signal Transduction, Cold Spring Harbor Symp. Quantitat. Biol.*, **53** (1988) 1–1049.
2. Prestwich, G. D. & Blomquist, G. J., *Pheromone Biochemistry*. Academic Press, New York, 1987.
3. Brand, J. G., Teeter, J. H., Cagan, R. H. & Kare, M. R., Receptor events and transduction in taste and olfaction. In *Chemical Senses, Vol. 1*. Marcel Dekker, New York, 1989.
4. Finger, T. E. & Silver, W. L., *Neurobiology of Taste and Smell*. Plenum Press, New York, 1988.
5. Schild, D., ed., *Chemosensory Information Processing*. Springer-Verlag, Berlin, Heidelberg, New York, 1990.
6. Margolis, F. L. & Getchell, T. V., *Molecular Neurobiology of the Olfactory System*. Plenum Press, New York, 1988.
7. Roper, S. D., The cell biology of vertebrate taste receptors. *Ann. Rev. Neurosci.*, **12** (1989) 329–53.
8. Lancet, D., Vertebrate olfactory reception. *Ann. Rev. Neurosci.*, **9** (1986) 329–55.
9. Getchell, T. V., Margolis, F. L. & Getchell, M. L., Perireceptor and receptor events in vertebrate olfaction. *Progress in Neurobiol.*, **23** (1984) 317–45.
10. Hudspeth, A. J., MacLeish, P. R., Margolis, F. L. & Wiesel, T. N., Sensory transduction. *Discussion in Neurosciences*, **4** (1987) 1–164.
11. Anholt, R. R., Molecular physiology of olfaction. *Am. J. Physiol.*, **257** (1989) C1043–54.
12. Snyder, S. H., Sklar, P. B., Hwang, P. M. & Pevsner, J., Molecular mechanisms of olfaction. *TINS*, **12** (1989) 35–8.
13. Kinnamon, S.C., Taste transduction: a diversity of mechanisms. *TINS*, **11** (1988) 491–6.

14. Cagan, R. H., *Neural Mechanisms in Taste*. CRC Press, Boca Raton, FL, USA, 1989.
15. Shepherd, G. M., *The Synaptic Organization of the Brain*. Oxford University Press, New York, 1979.
16. Gesteland, R. C., Yancey, R. A. & Farbman, A. I., Development of olfactory receptor neuron selectivity in the rat fetus. *Neuroscience,* **7** (1982) 3127–36.
17. Menco, B. Ph. M. & Farbman, A. I., Genesis of cilia and microvilli of rat nasal epithelium during pre-natal development. *J. Cell Sci.,* **78** (1985) 311–36.
18. Brunjes, P. C. & Frazier, L. L., Maturation and plasticity in the olfactory system of vertebrates. *Brain Res. Rev.,* **11** (1986) 1–45.
19. Bryant, B. P., Harpaz, S. & Brand, J. G., Structure/activity relationships in the arginine taste pathways of the channel catfish. *Chem. Senses,* **14** (1989) 805–15.
20. Cagan, R. H., Recognition of taste stimuli at the initial binding interaction. In *Biochemistry of Taste and Olfaction,* ed. R. H. Cagan & M. R. Kare. Academic Press, New York, 1981, pp. 175–209.
21. Fesenko, E. E., Novoselov, V. I. & Brystrova, M. F., Properties of odour-binding glycoproteins from rat olfactory epithelium. *Biochim. Biophys. Acta,* **937** (1988) 369–78.
22. Goldstein, N. I. & Cagan, R. H., Biochemical studies of taste sensation: monoclonal antibody against L-alanine binding activity of catfish taste epithelium. *Proc. Natl Acad. Sci. USA,* **79** (1982) 7595–7.
23. Kalinoski, D. L., Bryant, B. P., Shaulsky, G., Brand, J. G. & Harpaz, S., Specific L-arginine taste receptors in the catfish, *Ictalurus punctatus*: biochemical and neurophysiological characterizations. *Brain Res.,* **488** (1989) 163–73.
24. Krueger, J. H. & Cagan, R. H., Binding of L-[^{3}H]alanine to a sedimentable fraction of catfish barbel epithelium. *J. Biol. Chem.,* **251** (1976) 88–97.
25. Bruch, R. C. & Rulli, R. D., Ligand binding specificity of a neutral L-amino acid olfactory receptor. *Comp. Biochem. Physiol.,* **91** (1988) 535–40.
26. Novoselov, V. I., Krapivinskaya, L. D. & Fesenko, E. E., Amino acid binding glycoproteins from the olfactory epithelium of skate (*Dasyatis pastinaca*). *Chem. Senses,* **13** (1988) 267–78.
27. Anholt, R. R. H., Functional reconstitution of the olfactory membrane: incorporation of the olfactory adenylate cyclase in liposomes. *Biochemistry,* **27** (1988) 6464–8.
28. Anholt, R. R. H. & Rivers, A. M., Olfactory transduction: crosstalk between second messenger systems. *Biochemistry,* **29** (1990) 4049–54.
29. Lowe, G., Nakamura, T. & Gold, G. H., Adenylate cyclase mediated olfactory transduction for a wide variety of odorants. *Proc. Natl Acad. Sci. USA,* **86** (1989) 5641–5.
30. Pfeuffer, E., Mollner, S., Lancet, D. & Pfeuffer, T., Olfactory adenylyl cyclase. Identification and purification of a novel enzyme form. *J. Biol. Chem.,* **264** (1989) 18803–7.
31. Shirley, S. G., Robinson, C. J., Dickinson, K., Aujla, R. & Dodd, G. H., Olfactory adenylate cyclase of the rat. Stimulation of odorants and inhibition by Ca^{2+}. *Biochem. J.,* **240** (1986) 605–6.
32. Sklar, B. B., Anholt, R. R. & Snyder, S. H., The odorant-sensitive adenylate cyclase of olfactory receptor cells. Differential stimulation by distinct classes of odorants. *J. Biol. Chem.,* **261** (1986) 15538–43.
33. Anholt, R. R. H., Mumby, S. M., Stoffers, D. A., Girard, P. R., Kuo, J. F. & Snyder, S. H., Transduction proteins of olfactory receptor cells: identification of guanine nucleotide binding proteins and protein kinase C. *Biochemistry,* **26** (1987) 788–95.
34. Lancet, D., Lazard, D., Heldman, J., Khen, M. & Nef, P., Molecular transduction in smell and taste. Ref. 1, pp. 343–8.

35. Bruch, R. C. & Kalinoski, D. L., Interaction of GTP-binding regulatory proteins with chemosensory receptors. *J. Biol. Chem.*, **262** (1987) 2401–4.
36. Jones, D. T. & Reed, R. R., G_{olf}: an olfactory neuron specific-G protein involved in odorant signal transduction. *Science*, **244** (1989) 790–5.
37. Mania-Farnell, B. & Farbman, A. I., Immunohistochemical localization of guanine nucleotide-binding proteins in rat olfactory epithelium during development. *Developmental Brain Res.*, **51** (1990) 103–12.
38. Pace, U. & Lancet, D., Olfactory GTP-binding protein: signal-transducing polypeptide of vertebrate chemosensory neurons. *Proc. Natl Acad. Sci. USA*, **83** (1986) 4947–51.
39. Nakamura, T. & Gold, G. H., A cyclic nucleotide-gated conductance in olfactory receptor cilia. *Nature*, **325** (1987) 442–4.
40. Frings, S. & Lindemann, B., Odorant response of isolated olfactory receptor cells is blocked by amiloride. *J. Membrane Biol.*, **105** (1988) 233–43.
41. Firestein, S. & Werblin, F., Odor-induced membrane currents in vertebrate-olfactory receptor neurons. *Science*, **244** (1989) 79–82.
42. Labarca, P., Simon, S. A. & Anholt, R. R., Activation by odorants of a multistate cation channel from olfactory cilia. *Proc. Natl Acad. Sci. USA*, **85** (1988) 944–7.
43. Huque, T. & Bruch, R. C., Odorant- and guanine nucleotide-stimulated phosphoinositide turnover in olfactory cilia. *Biochem. Biophys. Res. Commun.*, **137** (1986) 36–42.
44. Jones, D. T. & Reed, R. R., Molecular cloning of five GTP-binding protein cDNA species from rat olfactory neuroepithelium. *J. Biol. Chem.*, **262** (1987) 14241–9.
45. Rasmussen, C. D. & Means, A. R., Calmodulin, cell growth and gene expression. *TINS*, **12** (1989) 433–8.
46. Biffo, S., Goren, T., Khew-Goodall, Y. S., Miara, J. & Margolis, F. L., Expression of calmodulin mRNA in rat olfactory neuroepithelium. *Mol. Brain Res.*, 1991.
47. Getchell, T. V. & Getchell, M. L., Regulatory factors in the vertebrate olfactory mucosa. *Chem. Senses*, **15** (1990) 223–31.
48. Getchell, T. V., Grillo, M., Tate, S. S., Urade, R., Teeter, J. & Margolis, F. L., Expression of catfish amino acid taste receptors in *Xenopus* oocytes. *Neurochem. Res.*, **15** (1990) 449–56.
49. Getchell, M. L., Zielinski, B. & Getchell, T. V., Odorant and autonomic regulation of secretion in the olfactory mucosa. In *Molecular Neurobiology of the Olfactory System*, ed. F. L. Margolis & T. V. Getchell. Plenum Press, New York, 1988, pp. 71–93.
50. Lee, K. H., Wells, R. G. & Reed, R. R., Isolation of an olfactory cDNA: similarity to retinol-binding protein suggests a role in olfaction. *Science*, **235** (1987) 1053–6.
51. Pevsner, J., Hwang, P. M., Sklar, P. B., Venable, J. C. & Snyder, S. H., Odorant-binding protein and its mRNA are localized to lateral nasal gland implying a carrier function. *Proc. Natl Acad. Sci. USA*, **85** (1988) 2383–7.
52. Pevsner, J., Reed, R. R., Feinstein, P. G. & Snyder, S. H., Molecular cloning of odorant-binding protein: member of a ligand carrier family. *Science*, **241** (1988) 336–9.
53. Tirindelli, R., Keen, J., Cavaggioni, A., Eliopoulos, E. E. & Findlay, J. B. C., Complete amino acid sequence of pyrazine-binding protein from cow nasal mucosa. *Eur. J. Biochem.*, **185** (1989) 569–72.
54. Györgyi, T. K., Roby-Shemkovitz, A. J. & Lerner, M. R., Characterization and cDNA cloning of the pheromone-binding protein from the tobacco hornworm, *Manduca sexta*: a tissue-specific developmentally regulated protein. *Proc. Natl Acad. Sci. USA*, **85** (1988) 9851–5.

55. Schmale, H., Holtgreve-Grez, H. & Christiansen, H., Possible role for salivary gland protein in taste reception indicated by homology of lipophilic-ligand carrier proteins. *Nature,* **343** (1990) 366–9.
56. Henzel, W. J., Rodriguez, H., Singer, A. G., Stults, J. T., Macrides, F., Agosta, W. C. & Niall, H., The primary structure of aphrodisin. *J. Biol. Chem.,* **263** (1988) 16682–7.
57. Mozell, M. M., & Jagodowicz, M., Chromatographic separation of odorants by the nose: retention times measured across in vivo olfactory mucosa. *Science,* **181** (1973) 1247–9.
58. Nachbar, R. B. & Morton, T. H., A gas chromatographic (GLPC) model for the sense of smell. Variation of olfactory sensitivity with conditions of stimulation. *J. Theor. Biol.,* **89** (1981) 387–407.
59. Brittebo, E. B., The olfactory mucosa as a site of drug and steroid metabolism. *Acta Physiol. Scand.,* **124** (1985) Supplementum 542.
60. Reed, C. J., Lock, E. A. & De Matteis, F., Olfactory cytochrome P-450. *Biochem. J.,* **253** (1988) 569–76.
61. Dahl, A., The effect of cytochrome P450 dependent metabolism and other enzyme activities on olfaction. In *Molecular Neurobiology of the Olfactory System,* ed. F. L. Margolis & T. V. Getchell. Plenum Publishing Corp., 1988, pp. 51–70.
62. Nebert, D. W. & Gonzalez, F. J., P450 genes: structure, evolution, and regulation. *Ann. Rev. Biochem.,* **56** (1987) 945–93.
63. Nef, P., Heldman, J., Lazard, D., Margalit, T., Jaye, M., Hanukoglu, I. & Lancet, D., Olfactory-specific cytochrome P-450. cDNA cloning of a novel neuroepithelial enzyme possibly involved in chemoreception. *J. Biol. Chem.,* **264** (1989) 6780–5.
64. Ding, X. & Coon, M. J., Purification and characterization of two unique forms of cytochrome P-450 from rabbit nasal microsomes. *Biochemistry,* **27** (1988) 8330–7.
65. Nicoll, R. A., The coupling of neurotransmitter receptors to ion channels in the brain. *Science,* **241** (1988) 545–51.
66. Claudio, T., Recombinant DNA technology in the study of ion channels. *Trends Pharmacol. Sci.,* **7** (1986) 308–12.
67. Hollmann, M., O'Shea-Greenfield, A., Rogers, S. W. & Heinemann, S., Cloning by functional expression of a member of the glutamate receptor family. *Nature,* **342** (1989) 643–8.
68. Horikoshi, T., Asanuma, A., Yanagisawa, K., Anzai, K. & Goto, S., Taurine and β-alanine act on both GABA and glycine receptors in *Xenopus* oocyte injected with mouse brain messenger RNA. *Mol. Brain Res.* **4** (1988) 97–105.
69. Houamed, K. M., Bilbe, G., Smart, T. G., Constanti, A., Brown, D. A., Barnard, E. A. & Richard, B. M., Expression of functional GABA, glycine and glutamate receptors in *Xenopus* oocytes injected with rat brain mRNA. *Nature,* **310** (1984) 318–21.
70. Levitan, E. S., Blair, L. A. C., Dionne, V. E. & Barnard, E. A., Biophysical and pharmacological properties of cloned $GABA_A$ receptor subunits expressed in *Xenopus* oocytes. *Neuron,* **1** (1988) 773–81.
71. Kaupp, U. B., Niidome, T., Tanabe, T., Terada, S., Bonigk, W., Stuhmer, W., Cook, N. J., Kangawa, K., Matsuo, H., Hirose, T., Miyata, T. & Numa, S., Primary structure and functional expression from complementary DNA of the rod photoreceptor cyclic GMP-gated channel. *Nature,* **342** (1989) 762–6.
72. Akabas, M. H., Dodd, J. & Al-Awqati, Q., A bitter substance induces a rise in intracellular calcium in a subpopulation of rat taste cells. *Science,* **242** (1988) 1047–50.
73. Avenet, P., Hoffmann, F. & Lindeman, B., Transduction in taste receptor cells requires cAMP-dependent protein kinase. *Nature,* **331** (1988) 351–4.

74. Berridge, M. J. & Irvine, R. F., Inositol phosphates and cell signalling. *Nature,* **341** (1989) 197–205.
75. Freissmuth, M., Casey, P. J. & Gilman, A. G., G proteins control diverse pathways of transmembrane signaling. *FASEB J.*, **3** (1989) 2125–40.
76. Garbers, D. L., Guanylate cyclase, a cell surface receptor. *J. Biol. Chem.*, **264** (1989) 9103–96.
77. Lefkowitz, R. J., Benovic, J. L., Kobilka, B. & Caron, M. G., β-Adrenergic receptors and rhodopsin: shedding new light on an old subject. *TIPS* (1986) 444–8.
78. Persaud, K. C., Heck, G. L., DeSimone, S. K., Getchell, T. V. & DeSimone, J. A., Ion transport across the frog olfactory mucosa: the action of cyclic nucleotides on the basal and odorant-stimulated states. *Biochim. Biophys. Acta,* **944** (1988) 49–62.
79. Putney, J. W., Jr, Takemura, H., Hughes, A. R., Horstman, D. A. & Thastrup, O., How do inositol phosphates regulate calcium signaling? *FASEB J.*, **3** (1989) 1899–914.
80. Tonosaki, K. & Funakoshi, M., Cyclic nucleotides may mediate taste transduction. *Nature,* **331** (1988) 354–6.
81. Uhl, R., Wagner, R. & Ryba, N., Watching G proteins at work. *TINS* (1990) **13:** 64–70.
82. Connor, J. A., Fluorescence imaging applied to the measurement of Ca^{2+} in mammalian neurons. In *Calcium and Ion Channel Modulation,* ed. A. Grinnel, D. Armstrong & M. Jackson. Plenum Press, New York, 1988, pp. 395–406.
83. Striem, B. J., Pace, U., Zehavi, U., Naim, M. & Lancet, D., Sweet tastants stimulate adenylate cyclase coupled to GTP-binding protein in rat tongue membranes. *Biochem. J.*, **260** (1989) 121–6.
84. Hanks, S. K., Quinn, A. M. & Hunter, T. H., The protein kinase family: conserved features and deduced phylogeny of the catalytic domains. *Science,* **241** (1988) 42–52.
85. Ziegelberger, G., van den Berg, M. J., Kaissling, K.-E., Klumpp, S. & Schultz, J. E., Cyclic GMP levels and guanylate cyclase activity in pheromone-sensitive antennae of the silkmoths *Antheraea polyphemus* and *Bombyx mori. J. Neurosci.*, **10** (1990) 1217–25.
86. Evans, R. M., The steroid and thyroid hormone receptor super family. *Science,* **240** (1988) 889–95.
87. Green, S. & Chambon, P., Nuclear receptors enhance our understanding of transcription regulation. *Trends Genetics,* **4** (1988) 309–14.
88. Moore, D. M., Diversity and unity in the nuclear hormone receptors: a terpenoid receptor super family. *New Biol.*, **2,** (1990) 100–5.
89. Albright, L. M., Huala, E. & Ausubel, F. M., Procaryotic signal transduction mediated by sensors and regulators pair. *Genetics,* **23** (1989) 311–36.
90. Bourret, R. B., Hess, J. F., Borkovich, K. A., Pakula, A. A. & Simon, M. I., Protein phosphorylation in chemotaxis and two component regulatory systems of bacteria. *J. Bacteriol.*, **166** (1989) 706–12.
91. Stock, J. B., Ninfa, A. J. & Stock, A. M., Protein phosphorylation and the regulation of adaptive response in bacteria. *Microb. Rev.*, **53** (1989) 450–90.
92. Koshland, D. E., Jr, Chemotaxis as a model second-messenger system. *Biochemistry,* **27** (1988) 5829–34.
93. Lerner, M. R., Reagan, J., Gyorgyi, T. & Roby, A., Olfaction by melanophores: what does it mean? *Proc. Natl Acad. Sci. USA,* **85** (1988) 261–4.
94. Kashiwayanagi, M. & Kurihara, K., Evidence for non-receptor odor discrimination using neuroblastoma cells as a model for olfactory cells. *Brain Res.*, **359** (1985) 97–103.
95. Kashiwayanagi, M. & Kurihara, K., Neuroblastoma cell as model for olfactory

cell: mechanism of depolarization in response to various odorants. *Brain Res.*, **293** (1984) 251–8.
96. Kinnamon, S. C. & Roper, S. D., Evidence for a role of voltage-sensitive apical K^+ channels in sour and salt taste transduction. *Chem. Senses*, **13** (1988) 115–21.
97. Lindemann, B. & Avenet, P., Patch-clamp study of isolated taste receptor cells of frog. *J. Membrane Biol.*, **97** (1987) 223–40.
98. Maue, R. A. & Dionne, V. E., Patch-clamp studies of isolated mouse olfactory receptor neurons. *J. Gen. Physiol.*, **90** (1987) 95–125.
99. Teeter, J. H., Sugimoto, K. & Brand, J. G., Ionic currents in taste cells and reconstituted taste epithelial membranes. In *Chemical Senses, 1. Receptor Events and Transduction in Taste and Olfaction*, ed. J. G. Brand, J. H. Teeter, R. H. Cagan & M. R. Kare. Marcel Dekker, New York, 1989, pp. 151–70.
100. Brodersen, R., Honore, B., Pedersen, A. O. & Klotz, I. M., Binding constants for ligand–carrier complexes. *TIPS*, **9** (1988) 252–7.
101. Margolis, F. L., Olfactory receptor neurons: specific gene expression and a hypothetical model for stimulus receptors. *Disc. Neurosci.*, **4** (1987) 47–52.
102. Kenakin, T., Agonists, partial agonists, antagonists, inverse agonists and agonist/antagonists. *TIPS*, **8** (1987) 423–6.
103. Coon, H. G., Curcio, F., Sakaguchi, K., Brandi, M. L. & Swerdlow, R. D., Cell cultures of neuroblasts from rat olfactory epithelium that show odorant responses. *Proc. Natl Acad. Sci. USA*, **86** (1989) 1703–7.
104. Danciger, E., Mettling, C., Vidal, M., Morris, R. & Margolis, F. L., Olfactory marker protein gene: its structure and olfactory neuron-specific expression in transgenic mice. *Proc. Natl Acad. Sci. USA*, **86** (1989) 8565–9.
105. Getchell, T. V., Induction of odorant-evoked current transients in *Xenopus* oocytes injected with mRNA isolated from the olfactory mucosa of *Rana pipiens*. *Neurosci. Lett.*, **91** (1988) 217–21.
106. Margolis, F. L., Molecular cloning of olfactory-specific gene products. In *Neurobiology of the Olfactory System*, ed. F. L. Margolis & T. V. Getchell. Plenum Press, New York, 1988, pp. 237–65.
107. Pixley, S. K. & Pun, R. Y. K., Cultured rat olfactory neurons are excitable and respond to odors. *Developmental Brain Res.*, **53** (1990) 125–30.
108. Kinnamon, S. C., Cummings, T. A. & Roper, S. D., Isolation of single taste cells from lingual epithelium. *Chem. Senses*, **13** (1988) 355–66.
109. Rogers, K. E., Dasgupta, P., Gubler, U., Grillo, M., Khew-Goodall, Y. S. & Margolis, F. L., Molecular cloning and sequencing of a cDNA for rat OMP. *Proc. Natl Acad. Sci. USA*, **84** (1987) 1704–8.
110. Dascal, N., The use of *Xenopus* oocytes for the study of ion channels. *CRC Crit. Rev. Biochem.*, **22** (1987) 317–87.
111. Khorana, H. G., Knox, B. E., Hasi, E., Swanson, R. & Thompson, D. A., Expression of a bovine rhodopsin gene in *Xenopus* oocytes: demonstration of light-dependent ionic currents. *Proc. Natl Acad. Sci. USA*, **85** (1988) 7917–21.
112. Lotan, I., Volterra, A., Dash, P., Siegelbaum, S. A. & Goelet, P., Blockade of ion channel expression in *Xenopus* oocytes with complementary DNA probes to Na^+ and K^+ channel mRNAs. *Neuron*, **1** (1988) 963–71.
113. Masu, Y., Nakayama, K., Tamaki, H., Harada, Y., Kuno, M. & Nakanishi, S., cDNA cloning of bovine substance-K receptor through oocyte expression system. *Nature*, **329** (1987) 836–8.
114. Snutch, T. P., The use of *Xenopus* oocytes to probe synaptic communication. *Trends Neurosci.*, **11** (1988) 250–6.
115. Timpe, L. C., Jan, Y. N. & Jan, L. Y., Four cDNA clones from the *Shaker* locus of *Drosophila* induce kinetically distinct A-type potassium currents in *Xenopus* oocytes. *Neuron*, **1** (1988) 659–67.
116. Yoshii, K. & Kurihara, K., Inward rectifier produced by *Xenopus* oocytes

injected with mRNA extracted from carp olfactory epithelium. *Synapse*, **3** (1989) 234–8.
117. Rice, G. C., McCabe, S., Pennica, D., Woronicz, J., Borree, J. & Goeddel, D. V., Expression cloning by transient expression utilizing fluorescence activated cell sorting. *DNA Prot. Eng. Tech.*, **2** (1990) 10–14.
118. Margolskee, R. F., Kavathas, P. & Berg, P., Epstein–Barr virus shuttle vector for stable episomal replication of cDNA expression libraries in human cells. *Mol. Cell. Biol.*, **8** (1988) 2837–47.
119. Harada, N., Castle, B. E., Gorman, D. M., Itoh, N., Schreurs, J., Barrett, R. L., Howard, M. & Miyajima, A., Expression cloning of a cDNA encoding the murine interleukin 4 receptor based on ligand binding. *Proc. Natl Acad. Sci. USA*, **87** (1990) 857–61.
120. Price, S. & Willey, A., Effects of antibodies against odorant binding proteins on electrophysiological responses to odorants. *Biochim. Biophys. Acta*, **965** (1988) 127–9.
121. Delaleu, J. C. & Holley, A., Investigations of the discriminative properties of the frog's olfactory mucosa using a photoactivable odorant. *Neurosci. Lett.*, **37** (1983) 251–6.
122. Vogt, R. G., Prestwich, G. D. & Riddiford, L. M., Sex pheromone receptor proteins. *J. Biol. Chem.*, **263** (1988) 3952–9.
123. Shimazaki, K., Sato, M. & Nakao, M., Photoaffinity labeling of thaumatin-binding protein in monkey circumvallate papillae. *Biochim. Biophys. Acta*, **884** (1986) 291–8.
124. Whitney, G. & Harder, D. B., Single-locus control of sucrose octaacetate tasting among mice. *Behav. Genet.*, **16** (1986) 559–74.
125. Wysocki, C. J., Whitney, G. & Tucker, D., Specific anosmia in the laboratory mouse. *Behav. Genet.*, **7** (1977) 177–88.
126. Helfand, S. L. & Carlson, J. R., Isolation and characterization of an olfactory mutant in *Drosophila* with a chemically specific defect. *Proc. Natl Acad. Sci. USA*, **86** (1989) 2908–12.
127. Gorman, C., High efficiency gene transfer into mammalian cells. In *DNA Cloning II*, ed. D. M. Glover. IRL Press, Washington, DC, 1985, pp. 143–90.
128. Hanahan, D., Transgenic mice as probes into complex systems. *Science*, **246** (1989) 1265–75.

Chapter 19

Natural Products

MARTIN PETRZILKA & CHARLES EHRET
Roure, Centre de Recherches, F-06133 Grasse, France

1 INTRODUCTION

The use of natural products in perfumery originated in ancient times. At a time when perfumery was not an industry, but rather an art, i.e. during the ancient Greek, Roman and Arabic civilizations, extracts of plants or animals were used exclusively for the preparation of products intended for adornment, cleansing/conditioning of the skin or religious ceremonies. Only at the beginning of the 19th century was a real perfume industry built up in Europe, due to the development of distillation and extraction techniques for natural products. Towards the end of the same century this development ran in parallel with synthetic organic chemistry, resulting in the first classical compounds such as the ionones and hydroxycitronellal. Since then both of these activities have developed together.

Although in the past few years natural products appear to have regained the market, it should not be overlooked that during the last 20 years their use in perfumery has steadily decreased. Thus, the ratio of naturals/synthetics present in fine perfumes today averages around 45 : 55, whereas 20 years ago this ratio was nearer 70 : 30. Some of the more important reasons for this slower evolution of naturals are mentioned below.

—The substantial price increase of naturals has played a major part, especially of precious products such as orris, rose and tuberose. This was attributed to the continuous increase in labour costs. Consequently, the culture sites of most of the important aromatic plants have migrated from the West to the East, thus changing the olfactive qualities of these products.
—The cultivation of jasmine, which originally together with tuberose, orange flower and rose contributed to the reputation of the South of France and especially the area of Grasse, is today carried out in North Africa, mainly in Egypt, where it has already surpassed production in Italy and Spain. Moreover, the actual production of jasmine in Egypt is already competing with that in India, and will soon be challenging output in China.
—The rapid progress in instrumental analysis techniques has led to much closer reconstitutions of natural products.
—Despite the various attempts to extract exotic plants, only a few new high-impact natural products have been added to the existing palette of perfumers. This is in sharp contrast to the rapid evolution of synthetics.

Despite the above facts, the simultaneous use of natural and synthetic products has increased considerably during the past 30–40 years, and today, natural products appear to be more indispensable than ever before. Within fine perfumery, these masterpieces of Mother Nature contribute decisively, if properly combined with synthetic products, to the innovative creation of new perfumery conceptions.

2 RAW MATERIALS—STATE OF THE ART

In principle, the raw materials used in perfumery may be of either animal or plant origin. The former, although not very numerous, often play a crucial role in perfume compositions, since they provide them with substantivity, warmth and an erotic touch. Musk, castoreum, civet and amber constitute the four main representatives of this group. They are usually obtained by the ethanolic extraction of the glandular secretions of animals, or, in the case of amber, by extraction of powdered ambergris, which is a pathological excretion from the sperm whale. These odorants usually enter perfume compositions as tinctures (cold alcoholic extractions), infusions (hot alcoholic extractions) or resinoids. More and more, however, for reasons of animal protection, these products are replaced by synthetic substitutes, especially in the case of musk and amber.

In contrast to the animal kingdom, the flora provided by Mother Nature constitute an immense reservoir of fragrance raw materials. These may be isolated by the use of various extraction methods from either the whole plants or their specific parts such as flowers, seeds, leaves, stems, barks, roots or rhizomes.

For the sake of clarity the following terminology will be used throughout this chapter to distinguish between the various types of raw materials obtained by different extraction methods.

Essential Oils

Essential oils are obtained by the steam distillation of various plant raw materials. With the exception of blossoms and leaves, all other parts of plants usually have to be pretreated in order to increase the break-up of the fragrance-containing oil cells, and thereby the extraction yield. Thus, herbs and stems have to be cut, rhizomes and barks have to be ground, woods have to be grated and seeds have to be crushed.

Owing to the thermolability of certain constituents in some essential oils, steam distillation cannot be applied. For example, this is the case for all citrus oils (bergamot, bitter orange, lemon, lime, mandarin, orange) where a room-temperature procedure for pressing the peel is used instead (cold-pressed oils). Essential oils generally evaporate without leaving any residue.

Concretes

Concretes are obtained by the solvent extraction of different parts of plants. In addition to the volatile parts of essential oils, waxes and colouring matter are equally extracted by this process, furnishing the concretes as semi-solid wax-like masses. The solvent system used for extraction has to be chosen carefully with regard to polarity and boiling point. If the latter is too high, thermolabile compounds might be destroyed and loss of certain fragrance ingredients will occur during evaporation.

Because of their insoluble character and their substantive nature, concretes are often used in soap perfumery instead of the more expensive absolutes, obtained by further processing of concretes.

The old methods of *enfleurage* (extraction into cold fat) and *maceration* (extraction into warm fat) applied to certain delicate flowers (jasmine, orange flower, rose, tuberose) in the past, are not used nowadays, because they are very time-consuming and expensive.

Resinoids

As concretes, resinoids are obtained by solvent extraction. However, the vegetable raw materials subjected to extraction are usually resins, which are solid or semi-solid plant secretions or dry lichens growing on the bark of oaks and other trees. Further processing to remove co-extracted waxes and resins leads to the corresponding absolutes, sometimes also called the *résinoïdes lavés*.

Some typical examples are the resinoids of labdanum, olibanum, myrrh, opopanax and treemoss.

Absolutes

Absolutes consist of the alcohol-soluble components of the corresponding concretes or resinoids, and are generally obtained via the following four-step procedure:

—Homogenization of the concrete or resinoid in ethanol.
—Cooling this mixture to a temperature of *ca.* – 15°C.
—Cold filtration to remove precipitated waxes, resins and colouring matter.
—Concentration of the filtrate by distilling off the ethanol.

The resulting absolutes are high-quality fragrance materials, which are in general very expensive due to the low overall yields.

Although thousands of essential oils/absolutes are known, only 200–300 are actually involved in commercial transactions and only about 50 of these have commercial potential in perfumery. With respect to availability these may be split into three categories, as follows

Low-Quantity Products

This category includes the great classics of fine perfumery, i.e. extracts of flowers (jasmine, narcissus, rose, tuberose) and of dry or resinous plant secretions (myrrh, olibanum, opopanax). These extremely expensive fragrance materials, in general, are only available in relatively small quantities, i.e. from a few hundred kilograms up to about 15 metric tons per year (cf. Table 1).

Medium-Quantity Products

Most representatives of this category exhibit woody base notes (patchouli, sandalwood, vetiver) and/or notes of medium tenacity (bergamot, clary sage). Since they often contain a large number of complex mono- and/or sesquiterpenes, it is extremely difficult to replace them by reconstitutions. Their worldwide annual production amounts to from fifty up to several hundreds of metric tons per year (cf. Table 1).

High-Quantity Products

The essential oils in this category can be produced in high yields and therefore are available in large quantities (>1000 metric tons/year; cf. Table 1) and at low price. They often include one major constituent which is characteristic of the odour, e.g. citronellal in citronella oil, linalyl acetate in lavandin oil or

Table 1
Approximate Worldwide Annual Production of Main Natural Products

Main natural products of plant origin	*Type*	*Production, 1984 (metric tons)*	*Main natural products of plant origin*	*Type*	*Production, 1984 (metric tons)*	*Main natural products of plant origin*	*Type*	*Production, 1984 (metric tons)*
Low-quantity products			*Medium-quantity products*			*High-quantity products*		
Ambrette seed	B, E	0·5	Bergamot	E	115	Cedarwood (USA)	E	1640
Blackcurrant	C	0·9[b]	Clary sage	E	45	Cedarwood (China)	E	450
Cistus	E	2	Geranium	E	130	Citronella	E	2300
Chamomile	E	8	Guaiacwood	E	60	Clove bud	E	2000
Galbanum	E, R	12[c]	Lavender	E, C	200[c]	Cornmint	E	2100
Jasmine	C	11	Oakmoss	C	50[c]	Eucalyptus (cineole type)	E	1400
Myrrh	E, R	4[c]	Patchouli	E	500	Eucalyptus (citronellal type)	E	1700
Narcissus	C	0·9[a]	Sandalwood	E	70	Lavandin	C, E	900[c]
Neroli	E	1[c]	Treemoss	C, R	110[c]	Peppermint	E	2200
Olibanum	E, R	4	Vetiver	E	260	Spearmint	E	1400
Opopanax	E, R	3[c]	Ylang ylang	E	87			
Orange flower	C	1·5[c]						
Orris	B, R	3·5[c]						
Rose	C, E	15						
Tuberose	C	0·6[c]						
Violet leaf	C	0·2[a]						

Abbreviations: C, concrete; R, resinoid; E, essential oil; B, butter.
[a] Internal estimate for 1986.
[b] Internal estimate for 1987.
[c] Internal estimate for 1988.

cineole in eucalyptus oil. Moreover, due to the high availability of these inexpensive main constituents, the products of this class, in general, compete strongly with reconstitutions in very favourable quality/price ratios.

Table 1 follows this method of classification by listing the strategically important natural products utilized in contemporary perfumery, together with their approximate worldwide annual production. The latter information is rather difficult to elicit and originates from various sources.[1–3]

3 EVOLUTION OF RESEARCH

The complex compositions of natural fragrance materials such as essential oils and absolutes were not well elucidated until the 1950s because of the lack of analytical techniques. Only the main constituents were generally known, which usually represented less than 50% of the extracts.[4]

Research involving analysis dominated in the following 25 years, the main objectives being the detection, isolation and identification of the constituents of natural products. The progressive introduction of gas and liquid chromatography and the simultaneous development of spectroscopic methods have allowed a more efficient identification of trace components, which, as we shall see later, often contribute greatly to the odour pattern of a natural substance. Thus, in a period of 25 years, gas chromatography, for example, passing from simple packed columns to high-performance capillary columns, has developed in a most spectacular way, making possible as a result the separation of a few micrograms of a complex mixture into single components. Today, isolation of separate constituents is hardly necessary, if the GC is coupled to a mass spectrometer, provided the corresponding reference spectra are contained in a computer-linked library. This GC–MS technology has become a key method for the analysis of the volatile parts of essential oils and absolutes.

The rapid evolution of analytical methods has in turn initiated the synthesis of many specific nature-identicals. In general, these syntheses were carried out either for the purpose of structure confirmation or, more importantly, in order to make new fragrance ingredients available to the perfumers, serving either for the continuous refinement of reconstitutions or as building blocks in new creations. Among numerous examples,[5,6] only a few are cited below (Table 2) which directly emerged from the analytical study of natural products and today are used in significant quantities (for a more comprehensive discussion of this topic see Chapter 7).

During the last decade research in the field of natural products has diversified considerably. This evolution was not only driven by a desire to innovate, but was also strongly influenced by economic, environmental and toxicological considerations.

Table 2
Synthesis of Nature-Identicals

Nature-identical	*Natural source*
α-Damascone	Tea
β-Damascenone	Rose
Ethyl everninate	Oakmoss
Jasmone	Jasmine
Methyl jasmonate	Jasmine
Methyl β-orcinolcarboxylate	Oakmoss
Rose oxide	Rose
Safranal	Safran
1,3,5-Undecatrienes	Galbanum

4 CONSTITUENTS OF NATURAL PRODUCTS

Due to the revolutionary development of analytical methods our knowledge of fragrances of natural origin has evolved in a most spectacular way during the last 20 years. Thus, for most of the important essential oils, absolutes and resinoids, several hundreds of constituents are known today. These generally belong to a large number of chemical classes and, in addition, are often present in an optically active form, provided their structures contain asymmetric carbon atoms. They usually occur in concentrations from whole percentages down to the parts per million (ppm) level. To simplify matters the terminology below, which refers to three ranges of concentrations, will be used:

>1%	Main constituents
0·1–1%	Minor constituents
<0·1%	Trace components

The scent of a natural product is usually the result of a complex interplay of the various constituents listed above, thereby often creating synergisms which equally contribute decisively to the overall odour impression. The role of minor constituents, and even more that of trace components, might appear to be irrelevant at first sight, as they are present in extremely small quantities. However, if correlated with their odour values, it soon becomes clear that these compounds are equally important, if not crucial, for the reproduction of the odour pattern of a specific natural scent.

The significance of these odour values, which are dimensionless and indicate how many times the headspace concentration of a given compound exceeds its threshold value, and which therefore constitute a measure of intensity, are discussed extensively in Chapter 6.

For the reader's better comprehension, the various composition concepts chosen by nature and the main natural raw materials used extensively in perfumery have been divided into three main groups:

—*Group 1:* The typical odour patterns of the natural products belonging to this group (e.g. orris, sandalwood, vetiver, oakmoss, galbanum, lavandin) are essentially due to one or a few main constituents, which exhibit a high sensory impact. Minor constituents and trace components only play a role as balance compounds.

—*Group 2:* The crude character is given by the main constituents; however, the richness, the complexity, the originality and the special imprint is created by the minor constituents and trace components. Typical representatives of this group are rose, tuberose, jasmine, violet flowers and orange flowers.

—*Group 3:* None of the main constituents contributes decisively to the desired odour impression emitted by these natural products (e.g. mimosa, narcissus). Here, the olfactive character is the result of a very subtle equilibrium between the various minor and trace constituents created by Mother Nature.

Although hundreds of constituents are known for a majority of natural products, only the most characteristic ones (main constituents, minor constituents, trace components) are mentioned in the following review (cf. Tables 3, 4 and 5).

Constituents of Natural Products Belonging to Group 1

One of the main representatives of this category is orris absolute, obtained by the steam distillation of the rhizomes of *Iris pallida* or *Iris germanica,* cultivated in Italy and Morocco, respectively. Three years after planting, the rhizomes are plucked out, sometimes peeled and then dried. Finally, in order to liberate the irones **1–4**,[7] which constitute the olfactive principle of orris, the rhizomes have to be matured for about three years. As we shall see later, this long storage period may nowadays be replaced by a microbiological treatment lasting only a few days.

The primary steam distillate, called orris butter, still includes up to 70% of various acids, of which by far the most abundant is myristic acid. If subjected to an additional treatment, the acids are eliminated, leaving behind the precious orris absolute, containing numerous linear hydrocarbons (2·5%), methyl and ethyl esters of fatty acids (4%),[8] aldehydes, ketones, alcohols and pyrazines in minute amounts, besides the irones **1–4**[9] (>80%) (Scheme 1). An in-house study has resulted in the identification of about 60 of these trace components, which, on combination with irones, gave a good reconstitution of orris (Table 3).

Another example where a few main constituents determine practically the whole olfactive character of a natural product is sandalwood oil (*Santalum album* L.).[10] Here, the two sesquiterpenic alcohols α-santalol (**5**) and

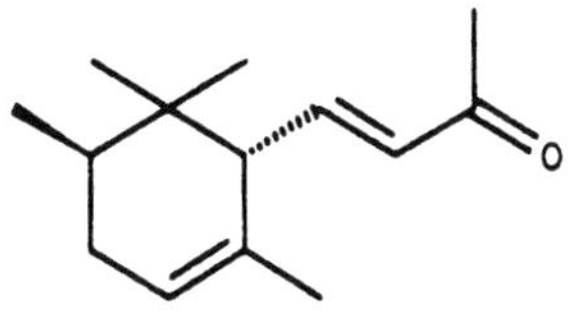

1
trans-α-Irone

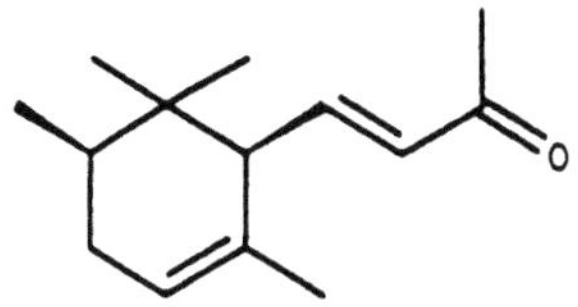

2
cis-α-Irone

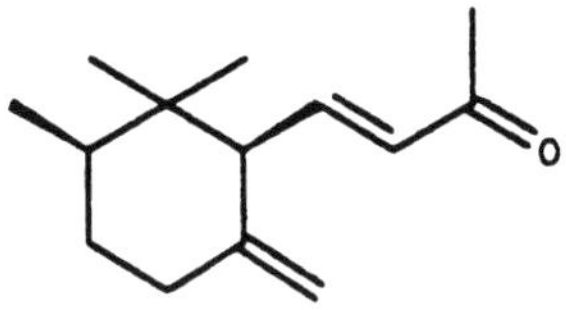

3
cis-γ-Irone

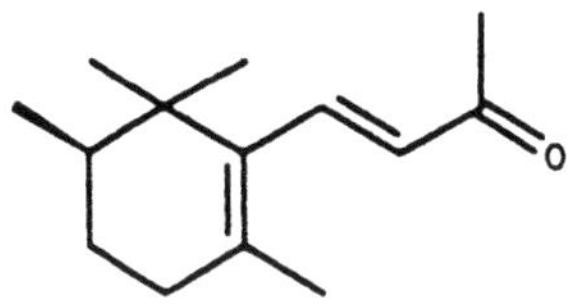

4
β-Irone

Scheme 1. Irones.

β-santalol (**6**) (Scheme 2), present at over 60%, play this crucial role. The other important isolates, given in Table 3, serve essentially as balance compounds.[11] The various phenols present in trace amounts (Table 3), for example, contribute to the smoky note of sandalwood oil.[11]

Like sandalwood oil, vetiver oil is mainly composed of sesquiterpenes or their derivatives.[12] In total about 50 constituents are known today, the most important of which are listed in Table 3. Although the main constituents identified are more numerous, together they represent only about 30% of the extract. Nevertheless they contribute decisively to the odour pattern of this essential oil.

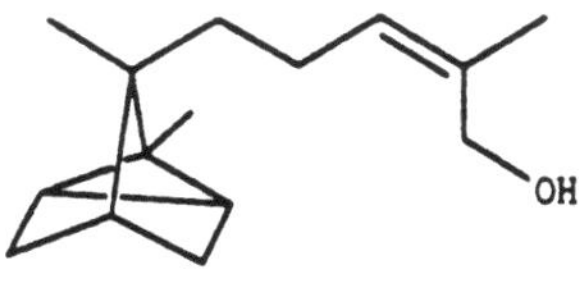

5

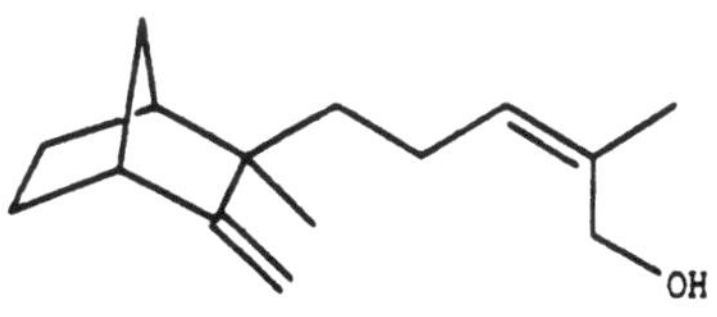

6

Scheme 2. Santalols.

In the case of the oakmoss absolute,[4] obtained by the extraction of the lichen *Evernia prunastri,* numerous minor constituents and trace components[13,14] serve to amplify the olfactive impression of a single key constituent (35%), i.e. methyl β-orcinolcarboxylate. Among these, various new monoethoxylated monoterpenes, **7–10** (Corbier, B., 1972, unpublished), may be cited (Scheme 3; Table 3).

Finally, the essential oils of galbanum (*Ferula galbaniflua*)[15,16] and lavandin (a hybrid of *Lavandula officinalis* and *Lavandula latifolia*)[17] constitute examples

Table 3
Constituents of Natural Products Belonging to Group 1

Main Constituents (>1%)	*(%)*	*Important minor constituents (0·1–1%)*	*Important trace components (<0·1%)*
Orris absolute			
cis-γ-Irone	42·0		*n*-Nonanal
cis-α-Irone	32·0		*n*-Decanal
trans-α-Irone	3·6		Linalool
β-Irone	1·4		Geraniol
Total	79·0		Benzyl alcohol
			Fatty acid esters
			Pyrazines
Sandalwood oil			
α-Santalol	46·0	α-Santalene	*o*- and *p*-Cresol
β-Santalol	20·0	β-Santalene	*p*-Vinylphenol
Total	66·0		Guaiacol
			4-Methylguaiacol
			6-Methoxy-4-allylguaiacol
			1-Furfurylpyrrole
Vetiver oil			
Khusimone	9·0	10-epi-γ-Eudesmol	α-Vetispirene
Khusimol	5·0	Elemol	Acoradiene
α-Vetivone	4·0	Junenol	Prezizaene
β-Vetivenene	4·0	α-Ylangene	α-Funebrene
β-Vetivone	3·0		Khusinol
Vetiselinenol	3·0		
Cyclocopacamphenol	2·0		
Zizaene	2·0		
Total	32·0		

Oakmoss absolute			
Methyl β-orcinolcarboxylate	35·0		1,8-Cineole
Orcinol monomethyl ether	10·0		Linalool
Ethyl everninate	3·0		Borneol
Total	48·0		Camphor
			8-Ethoxy-1-*p*-menthene[a]
			4-Ethoxy-1-*p*-menthene[a]
			Fenchyl methyl ether[a]
			Bornyl acetate[a]
Galbanum oil			
β-Pinene	50·0		Thiol esters[20]
Δ^3-Carene	20·0		Substituted pyrazines[21]
(3*E*, 5*Z*)-1,3,5-Undecatriene	1·0		13-Tetradecanolide[b]
(3*E*, 5*E*)-1,3,5-Undecatriene	0·5		14-Pentadecanolide[b]
Total	71·5		15-Hexadecanolide[b]
Lavandin oil			
Linalyl acetate	39·0	*cis*-β-Ocimene	Kobusone
Linalool	30·0	*trans*-β-Ocimene	β-Ionone
Camphor	7·0	Myrcene	4-Decanolide
1,8-Cineole	5·0	Camphene	Derivatives of linalyl acetate[22]
1-Terpinen-4-ol	2·0	α-Pinene	
Lavandulyl acetate	2·0	β-Pinene	
Caryophyllene	2·0	Caryophyllene oxide	
Borneol	2·0	Lavandulol	
Total	89·0		

[a] Unpublished results.
[b] Kaiser, R., Lamparsky, D., *Helv. Chim. Acta* **61** (1978) 2671–80

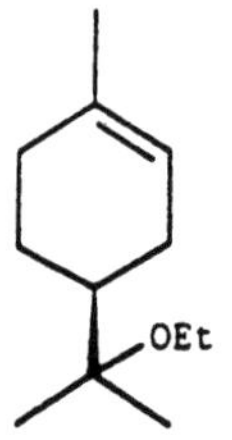

7

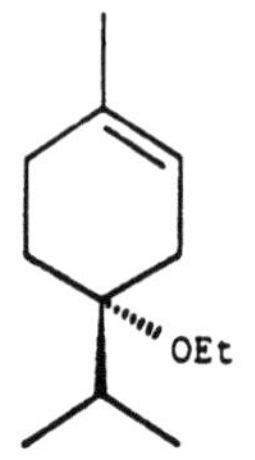

8

9

10

Scheme 3. Monoterpenes identified in oakmoss absolute.

where less than 10 main constituents, representing about 70% and 90%, respectively, of the total extract, are responsible for the general olfactive character of these oils. For accurate reconstitutions, however, numerous minor and trace constituents (Table 3) are equally necessary. For lavandin, which is typical, over 300 constituents are known.[18,19]

Constituents of Natural Products Belonging to Group 2

The absolutes and essential oils discussed in this section are mainly derived from precious flowers such as rose, jasmine, tuberose, violet and orange flowers. In contrast to Group 1, their olfactive character is not only based on a few main constituents, but also relies critically on the presence of minor and trace components. Thus, reconstitutions based on only the main constituents, usually representing 60–90% of the odoriferous part, give only a faint impression of the desired overall odour. In fact, it is the identification of these trace components which has led to the secret of these extracts, which are indispensable for fine perfumery.

Bulgarian rose oil (*Rosa damascena*) has been thoroughly investigated during the past years[23] and today almost 300 constituents are known. As shown by Ohloff & Demole[24] among these a number of minor and trace constituents (rose oxide, β-damascenone, β-damascone, 1-*p*-menthen-9-al, rose furan, β-ionone) play a crucial role in the odoriferous principle of this precious scent. Our own work has contributed to the addition of important trace components (Table 4) such as nerol oxide,[25] the diethyl acetals of *n*-pentanal and *n*-hexanal as well as the very interesting mixed acetals **11–13** (Ehret, C., 1971, unpublished) (Scheme 4).

Together with rose oil, the extracts of tuberose (*Polianthes tuberosa* L.) and

Table 4
Constituents of Natural Products Belonging to Group 2

Main constituents (>1%)	*(%)*	*Important minor constituents (0·1–1%)*	*Important trace components (<0·1%)*
Bulgarian Rose oil			
Citronellol	31·0	Phenylethyl acetate	Ethanal diethyl acetal
Geraniol	12·0	Phenylethyl tiglate	*n*-Pentanal diethyl acetal[a]
Nerol	8·0	Citronellyl acetate	*n*-Hexanal diethyl acetal[a]
Linalool	2·0	Caryophyllene	Ethanal ethyl citronellyl acetal[a]
Phenylethyl alcohol	2·0	Rose oxides	Ethanal ethyl neryl acetal[a]
Total	55·0		Ethanal ethyl geranyl acetal[a]
			β-Damascenone
			β-Damascone
			1-*p*-Menthen-9-al
			β-Ionone
			Rose furan
			Nerol oxide
Tuberose absolute			
Methyl palmitate	14·5	5-Decanolide	Massoialactone
Methyl benzoate	12·2	Jasmine lactone	Tuberolactone
Benzyl benzoate	10·4		5-Undecanolide
Ethyl oleate	5·3		5-Dodecanolide
α-Terpineol	4·0		5-Tetradecanolide
1,8-Cineole	3·8		4-Pentanolide
Methyl salicylate	3·6		4-Octanolide
Farnesol	3·5		4-Nonanolide
Methyl anthranilate	2·4		4-Decanolide[a]
Eugenol methyl ether	2·2		4-Undecanolide
Benzyl salicylate	2·1		4-Dodecanolide
Benzyl alcohol	1·8		Methyl dihydrocinnamate[a]
Palmitic acid	1·4		Hexyl methyl ether[a]
Oleic acid	1·3		Phenylacetonitrile[a]
Total	68·5		5-Dodecen-4-olide[a]
			1-Nonen-4-ol[a]
			α-Damascone[a]
			Anethole[a]
			6-Methyl-α-pyrone

(continued)

Table 4—*contd.*

Main constituents (>1%)	*(%)*	*Important minor constituents (0·1–1%)*	*Important trace components (<0·1%)*
***Tuberose absolute**—contd.*			
			p-Methoxyphenylacetonitrile[a]
			Epoxyeugenol[a]
			5*Z*-Octen-4-olide
			5*Z*-Decen-4-olide
			6*Z*-Nonen-4-olide
			6*Z*-Dodecen-4-olide
			6*Z*,9*Z*-Dodecadien-4-olide
			Tuberolide
Jasmine absolute			
Benzyl acetate	28·5	Methyl benzoate	Methyl anthranilate
Benzyl benzoate	17·0	α-Farnesene	δ-Jasmonic acid lactone[28]
Linalool	7·7	Nerolidol	Ethyl *cis*-(*Z*)-jasmonate
Phytol	7·4	Nerolidyl acetate	Ethyl *trans*-(*Z*) jasmonate
Isophytol	6·4	Jasmine lactone	4-Decanolide
Phytyl acetate	6·1	Methyl *trans*-(*Z*)-jasmonate	Methyl 4,5-dehydrojasmonate[b]
Indole	3·4	Methyl *N*-acetylanthranilate	
8-Geranyl linalool	3·2		
cis-Jasmone	2·5		
Eugenol	2·2		
Methyl linolenate	2·0		
Benzyl alcohol	1·9		
Methyl palmitate	1·4		
p-Cresol	1·0		
cis-3-Hexenyl benzoate	1·0		
Total	91·7		

Concrete of violet flowers (odoriferous part = 5%)			
α-Ionone	14·1	1,2,4-Trimethylbenzene[a]	Camphor[a]
Dihydro-β-ionone	11·4	*n*-Hexanal[a]	*cis*-3-Hexenyl formate[a]
2*E*,6*Z*-Nonadienal	7·2	*n*-Heptanal[a]	Hexyl isobutyrate[a]
1-Octen-3-ol	5·2	Ethyl amyl ketone[a]	α-Terpineol[a]
1,4-Dimethoxybenzene	4·5	2,5-Dimethyl-2-vinyl-4-hexenal	
cis-3-Hexen-1-ol	4·0	*cis*-Jasmone[a]	
Dihydro-α-ionone	3·9	β-Ionone[a]	
cis-3-Hexenyl acetate	2·3	1,8-Cineole[a]	
n-Nonanal	2·2	Benzyl alcohol	
Linalool	2·0	Butyl butyrate[a]	
2*E*-Nonen-1-al	1·7	Benzyl benzoate[a]	
2*E*,6*Z*-Nonadien-1-ol	1·5	Borneol[a]	
Total	60·0		
Absolute of orange flowers (odoriferous part = 70%)			
Linalool	30·0	Linalool oxide	*cis*-3-Hexenyl formate[a]
Linalyl acetate	7·5	Phenylethyl acetate	2-Aminobenzyl alcohol[a]
Farnesol	5·6	Phenylacetonitrile	Quinazoline[a]
Nerolidyl acetate	5·5	*cis*-Jasmone	γ-Vinyl-γ-methylbutenolide[a]
Methyl anthranilate	3·0	Methyl linoleate	6,7-Epoxylinalyl acetate[a]
Indole	2·0	Ethyl palmitate	
Ocimene	1·2	Methyl jasmonate	
α-Terpineol	1·2		
Limonene	1·0		
Total	57·0		

[a] Unpublished results.
[b] Kaiser, R., Lamparsky, D., Tetrahedron Lett. (1974) 3413–6.

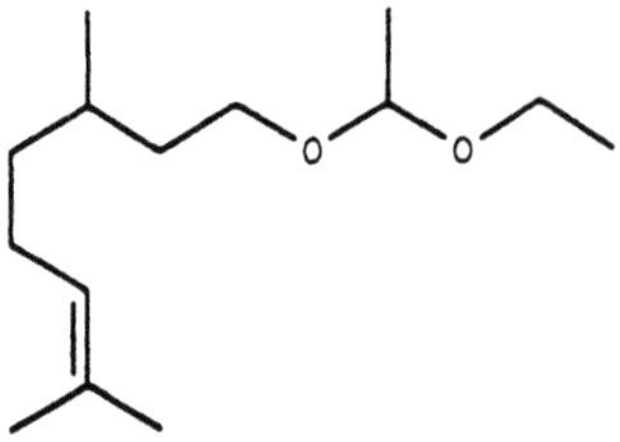

11

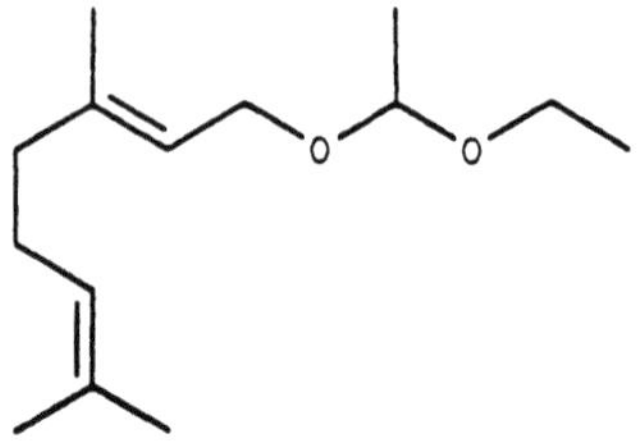

12

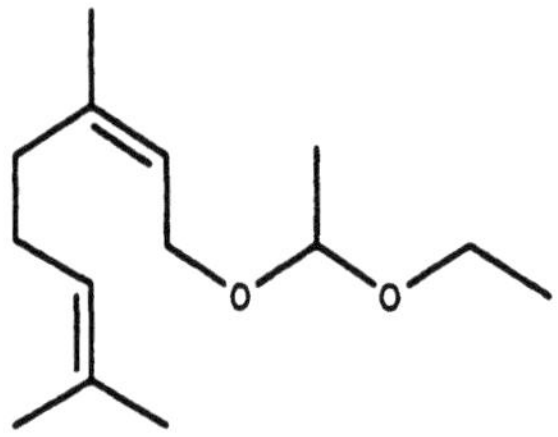

13

Scheme 4. Mixed acetals present in Bulgarian rose oil.

of jasmine (*Jasminum grandiflorum*) are the most important natural products in alcoholic perfumery. With the former, relatively few studies had been carried out. The main constituents (>1%) represent almost 70% of the extract, but as in the case of rose oil the olfactive richness is caused by the presence of various trace components. Recently, we have identified several new trace compounds (Maupetit, P., 1987, unpublished) which, together with the various known characteristic lactones described by Kaiser & Lamparsky[26] and by Maurer & Hauser,[27] appear to be essential for the reconstitution of this scent (Table 4).

In contrast, jasmine absolute has been the subject of numerous extensive studies.[28–30] Among the *ca.* 170 constituents quoted in the literature, 15 represent over 90% of the absolute. The most characteristic minor and trace components which contribute to the richness of the jasmine odour are listed in Table 4.

Until 1980, only 20 constituents of violet flowers (*Violata odorata*) were known.[4,31] Our systematic study of its concrete, the odoriferous part of which constitutes only 5%, has resulted in the identification of 80 new constituents. In Table 4, the main constituents (>1%) as well as a selection of the olfactively important minor and trace constituents are given (Ehret, C., 1975, unpublished). Today, due to its high cost price, the concrete of violet flowers is used less frequently.

The last representative in this category of natural products is the well-known[4] absolute of orange flowers, which is obtained by hexane extraction of the flowers of *Citrus aurantium* L. This absolute is characterized by the presence of various nitrogen-containing compounds,[32,33] such as indole (2%), methyl anthranilate (3%) and phenylacetonitrile (0·8%). A recent re-investigation (Ehret, C., 1987, unpublished) of this absolute allowed us to add some important new trace constituents to the list of the 150 compounds already known (cf. Table 4).

Constituents of Natural Products Belonging to Group 3

A typical example of this class of floral extracts is mimosa absolute (*Acacia dealbata/floribunda*). The olfactive part of this extract, which represents about 15% of the absolute, is characterized by a very rich woody floral note. Prior to 1985 very little information existed on its composition.[34] As a result of our studies, 130 constituents were identified,[35] a selection of typical representatives of which is given in Table 5. The few main constituents (>1%) identified, i.e. hydrocarbons and fatty acid esters, were found to have no olfactive impact at all. On the contrary, the constituents present in concentrations between 0·1 and 1%, i.e. mainly aldehydes (octanal, 2*E*-nonenal, 2*E*,6*Z*-nonadienal), esters and alcohols, already allowed the first attempt at a reconstitution. However, the complexity and richness of this scent is only created if, in addition, a number of trace components are integrated. Some of the more important ones are *β*-damascone (**14**), *cis*-jasmone (**15**) and various diethyl acetals (Scheme 5; Table 5).

14 **15**

Scheme 5. *β*-Damascone and *cis*-jasmone.

Another scent highly regarded for its intensity and complexity emerges from the absolute of narcissus (*Narcissus poeticus*). Initially, it exhibits a strong green, woody and earthy note, and only upon high dilution additional sweet floral, heavy and balsamic notes are discovered. Various research groups have studied the composition of extracts of different geographical and botanical origin,[36–38] thereby identifying about 200 constituents. Stimulated by this complex and attractive scent, we have carried out an extensive study of the absolute, focusing especially on olfactively interesting trace components. Thus, together with the 80 newly identified trace components, the total number of known constituents now amounts to about 280.[39]

The known main constituents (Table 5), which represent 90% of the volatile part of the absolute, are responsible for the spicy, sweet and balsamic notes. However, the most interesting aspects are created once again by the presence of a large number of minor and trace components:

Table 5
Constituents of Natural Products Belonging to Group 3

Main constituents (>1%)	*(%)*	*Important minor constituents (0·1–1%)*	*Important trace components (<0·1%)*
Mimosa absolute (odoriferous part = 15%)			
Heptadecene	45·0	Octanal	β-Damascone
Nonadecane	13·0	Decanal	*cis*-Jasmone
Ethyl palmitate	10·0	*cis*-3-Hexenyl acetate	Ethyl 2-ethylbutyrate
Ethyl linoleate	2·5	*n*-Hexanol	Butanal diethyl acetal
Phenylethyl alcohol	1·5	Benzaldehyde	Pentanal diethyl acetal
Nonanal	1·0	Linalool	Hexanal diethyl acetal
Isophytol	1·0	*n*-Heptanol	Heptanal diethyl acetal
Total	74·0	*n*-Octanol	Octanal diethyl acetal
		Ethyl benzoate	
		Anethole	
		Benzyl alcohol	
		Benzyl benzoate	
		2*E*-Nonenal	
		2*E*,6*Z*-Nonadienal	
		Anisaldehyde	
Narcissus absolute (odoriferous part = 20%)			
α-Terpineol	23·7	Methyl benzoate	2-Amylfuran
trans-Isoeugenol methyl ether	20·0	Ethyl benzoate	2*E*-Dodecenal
Benzyl benzoate	19·4	*cis*-3-Hexenyl benzoate	2*E*,6*Z*-Nonadienal
Coumarin	6·9	*cis*-3-Hexenyl acetate	2*E*-Nonenal
Benzyl alcohol	5·0	Benzyl valerate	1-*p*-Menthen-9-al
Δ^3-Carene	3·4	Phenylpropyl acetate	β-Ionone
Phenylethyl alcohol	2·2	Geraniol	1,2-Epoxy-β-ionone
Ethyl palmitate	2·2	Phenylethyl acetate	Carvacrol
Cinnamic alcohol	2·2	Isoeugenol	6,7-Epoxy-*cis*-β-ocimene
Phenylpropyl acetate	1·7	Eugenol	Thymol
1,8-Cineole	1·5	*p*-Cymen-8-ol	Methyl β-orcinolcarboxylate
Carophyllene	1·0	Methyl salicylate	Epoxyeugenol
Benzyl acetate	0·7	Phenylpropyl alcohol	Orcinol dimethyl ether
Total	89·9		Phenylacetonitrile
			Jasmine lactone

—β-ionone, epoxy-β-ionone, 2*E*,6*Z*-nonadienal and 2*E*-nonenal contribute to the violet note;
—various phenols add notes reminiscent of the orange flower and ylang-ylang;
—methyl β-orcinolcarboxylate, methyl everninate and orcinol dimethylether, i.e. constituents which are usually found in lichens, participate in the woody notes;
—phenylacetonitrile, jasmine lactone and 1-*p*-menthen-9-al finally introduce the floral shading.

5 BIOTECHNOLOGY APPLIED TO PERFUMERY

Biotechnology is a multidisciplinary science comprising chemistry, microbiology, cell biology, biochemistry, molecular biology, genetic engineering and process engineering. Essentially it brings into play the biological capacity of living cells (microorganisms, plant and animal cells) or of the enzymes derived therefrom (biocatalysts) to produce or to biotransform substances of generally organic origin. This technology has been applied in pharmaceutical, agricultural and food industries for a long time, for example in the production of cheese and wine via traditional fermentation.

The release of pleasant or unpleasant odours from cultures of microorganisms and especially fungi is a well-known phenomenon and has been the subject of numerous reviews.[40,41] Schindler & Schmid[41] demonstrated that microorganisms are capable of producing interesting volatile metabolites exhibiting food odours (fruity, vanilla, anisic), floral-type odours (narcissus, rose, jasmine) or woody–earthy-type odours. The structures of these compounds have been identified (mostly monoterpenes and aliphatic esters) and in general are well known to perfumers (Table 6). So far, however, their olfactive properties have been used mainly for taxonomical purposes, i.e. as an additional means of classification of the microorganisms themselves.[42–44]

In the field of perfumery, however, the interest in biotechnology arose much later, i.e. about ten years ago. This lack of readiness to engage in the new technology might have had various reasons:

—Raw materials of perfumes have traditionally originated from plants or animals without any known evidence for the intervention of microorganisms or enzymes.
—Previously, perfumery had been a luxury industry where raw material cost prices did not constitute a limitation.
—The continuous development of innovative synthetic products had allowed the completion of the palette of traditional natural products in a harmonious way.

Today biotechnology is becoming an attractive approach for perfumery,

Table 6
Odours and Fragrance Constituents from Microorganisms

Microorganism	*Odour*	*Constituents*
Bacillus subtilis	Soybean	Tetramethylpyrazine
Ceratocystis moniliformis	Fruity, banana, peach, pear, rose	3-Methylbutyl acetate, δ- and γ-decalactone, geraniol, citronellol, nerol, linalool, α-terpineol
Ceratocystis variospora	Fragrant, geranium	Citronellol, citronellyl acetate, geranial, neral, geraniol, linalool, geranyl acetate
Ceratocystis virescens	Rose, fruity	6-Methyl-5-hepten-2-ol acetate, citronellol, linalool, geraniol, geranyl acetate
Corynebacterium glutamicum	Soybean	Tetramethylpyrazine
Daedalea quercina	Apples	
Inocybe corydalina	Fruity, jasmine	Cinnamic acid methyl ester
Kluyveromyces lactis	Fruity, rose	Citronellol, linalool, geraniol
Lentinus cochleatus	Anisaldehyde	
Lenzites sepiaria	Slightly spicy	
Mycoacia uda	Fruity, grassy, almond	*p*-Methylacetophenone, *p*-tolyl-1-ethanol, *p*-tolylaldehyde
Penicillium decumbens	Pine, rose, apple, mushroom	Thujopsene, 3-octanone, 1-octen-3-ol, nerolidol, β-phenylethyl alcohol
Pholiota adiposa	Earthy	
Polyporus croceus	Narcissus	
Polyporus obtusus	Jasmine	
Poria xantha	Lemon	
Pseudomonas perolens	Musty, potato	2-Methoxy-3-isopropylpyrazine
Pseudomonas taetrolens	Musty, potato	2-Methoxy-3-isopropylpyrazine
Stereum murrayi	Vanilla	
Stereum rugosum	Fruity, banana	
Streptomyces odorifer	Earthy, camphor	*trans*-1,10-Dimethyl-*trans*-9-decalol, 2-*exo*-hydroxy-2-methylbornane
Trametes odorata	Honey, rose, fruity, anise	Methyl phenylacetate, geraniol, nerol, citronellol
Trametes suaveolens	Anisaldehyde	
Trichoderma viride	Coconut	6-Pentyl-2-pyrone

since it offers a number of important advantages:

—creation of new and more efficient plant species;
—independence of natural raw materials from market fluctuations (supply, quality, consistency);
—production of new natural products;
—stereospecific synthesis of new fragrance compounds, which would be inaccessible by other means.

In order to achieve these objectives, different techniques might be envisaged which are discussed in the following section.

Creation of New Plant Species

Contrary to the traditional methods of selection such as sexual reproduction which are used in other agricultural production (food, ornamental flowers) and which often require about ten years of research, the novel techniques introduced by biotechnology allow the rapid creation of new botanical varieties. These are adapted to a geographical site and/or are resistant to diseases and/or produce the desired products in higher yields and better qualities. Various possibilities present themselves to scientists.

—*Selection of novel varieties*:[45] The most interesting varieties with regard to yield and olfactive quality are selected from heterogeneous populations; this is followed by their multiplication using the technique of in-vitro micropropagation. By this method, thousands of absolutely identical plants may be produced.

—*Somaclonal variation*:[46] This method profits from the variations which occur *in vitro* by passing through the non-differentiated state. After regeneration of the whole plants, the best variants are then selected. This technique is rather laborious since, given its random character, the type of change exerted cannot be predicted *a priori*.

—*Protoplast fusion*:[46] This is the fusion of cells whose walls have been destroyed under the action of enzymes. It allows the crossover of sexually incompatible plants.

—*Mutagenesis*:[46] The phenomenon occurs if whole plants or cell cultures are subjected to mutagenic agents, e.g. chemical products or UV light.

—*Genetic engineering*:[46] The genetic code of a plant is modified by introducing foreign genes or parts of genes (cloning) known to induce specific qualities.

—*Encapsulation of somatic embryos*:[47] The germination of newly created species in the form of somatic (artificial) seeds is made possible.

Applying a variety of these techniques, the Research Institute of Tropical Agronomy on the island of Réunion is carrying out a complete programme with the goal of improving the species of *Geranium*, in order to resolve certain phytosanitary problems and to acquire a better knowledge of the essential oil production potential of this plant.[48]

Alternatively, Roberts *et al.*[49] have attempted to introduce *Geranium* into the state of Oregon (USA). The quality of the resulting essential oil is somewhere between those originating from China and Réunion. The same authors also described modified varieties of mint, the extracts of which are particularly rich in menthofuran (>75%).

Production of Fragrances by Microorganisms

Production of Metabolites

The idea of using microorganisms in the industrial production of essential oils is a relatively new one. Since 1970 Collins has systematically studied the composition of metabolites present in various fungi. In an excellent review[50] he discusses this concept with a view to eventual industrial production. The best yields achieved so far are in the range 2–3 g/litre of culture medium. Thus, by adsorbing the metabolites formed by *Ceratocystis variospora* on a resin (XAD 2), Schindler[51,52] obtained up to 2 g/litre of an essential oil which was analytically close to those of citronella and geranium. As shown by Lanza *et al.*[53] and Hubbal *et al.*,[54] the quality of the odours produced in this way depends critically on the nature of the culture medium and of the carbon and/or nitrogen sources used. Similarly, a mutant of *Corynebacterium* isolated by Demain *et al.*[55] is reported to produce 3 g of tetramethylpyrazine/litre in five days. Various other microorganisms are also able to produce pyrazines[56,57] (Table 6), compounds which are known to be olfactively important trace constituents of various essential oils.[6]

A number of important conclusions may be drawn from the actual state of research in this field:

(a) In order to assure the production of metabolites in a reproducible way, the culture media have to be defined precisely.
(b) The choice of the strain and type of culture medium (solid, liquid) has a significant influence on the yield of the metabolites.
(c) The production of fragrances via fermentation is only justified economically if the same compounds are not accessible by other means (synthesis, extraction from plants).

Biotransformations using Microorganisms

Any chemical reaction resulting from the treatment of a substrate (single compound, mixture of precursors, vegetable raw material) by living cells (microorganisms, cells of plant or animal origin) or by the enzymes derived therefrom may be defined as a biotransformation or bioconversion. If compared with the direct production of metabolites and/or with chemical synthesis, biotransformations exhibit the following advantages:

(a) Without the use of expensive reagents and multistep reaction schemes they allow for:

—the functionalization of non-activated carbon atoms,
—regio- and stereospecific syntheses,
—the resolution of racemates, and
—the release of fragrances from their precursors;

(b) the reactions may usually be carried out under mild conditions, e.g. at room temperature; and
(c) if the substrate is natural the resulting product is equally labelled natural.

Biotransformations of Terpenoids

A large number of communications, which have been reviewed,[58–61] bear witness to the strong activity in this field of research. The basic strategy behind all of these biotransformations is the valorization of easily accessible raw materials of either natural or synthetic origin. A few typical cases which have been of special interest for perfumery are quoted below.

(a) Patchoulol. Although patchoulol (**16**) is the main constituent of the essential oil of patchouli, it contributes very little to the olfactive character of this important natural product. On the other hand, Teisseire *et al.*[62] found that norpatchoulenol (**18**), present as a minor constituent (<0·5%) in the essential oil, constitutes one of the important vectors of this highly appreciated odour. Thus, the idea of biotransforming **16** to **18** or to another suitable intermediate was born and, after a huge screening of microorganisms,[63] finally was carried out by scientists at the Nippon Roche Research Centre.[64,65] Accordingly, treatment of **16** with fungi (*Pithomyces* sp.) resulted in the formation of the diol **17** in up to 45% yield, which could then be degraded to norpatchoulenol (**19**) with relative ease[63] (Scheme 6).

(b) Sclareol. Sclareol (**19**), an easily isolated main constituent (50–60%) of the concrete of clary sage, constitutes an ideal starting material for the synthesis of Ambrox® (**21**), since it already contains the four chiral centres of the latter with the correct relative and absolute configuration. Besides the many chemical approaches, biochemical ones have also been chosen to convert **19** to **21**. Thus, Farbood *et al.*[66] have succeeded in biodegrading the lateral side chain of **19** to the diol **20**, which in turn could easily be cyclized to **21** (Scheme 7).

(c) β-Ionone. Ionones represent characteristic constituents of the absolute of violet flowers, but they equally occur in trace amounts in many other essential oils. In view of an eventual flavouring of tobacco, various groups have studied their biotransformations and especially that of β-ionone (**22**), a compound which today is accessible in large quantities. Mikami *et al.*[67] were the first to describe such a tobacco base, obtained by biotransformation of β-ionone with *Aspergillus niger* JTS 191. The two main metabolites found in this case (Scheme 8) were the hydroxy-β-ionones **23** and **24**. The latter, a

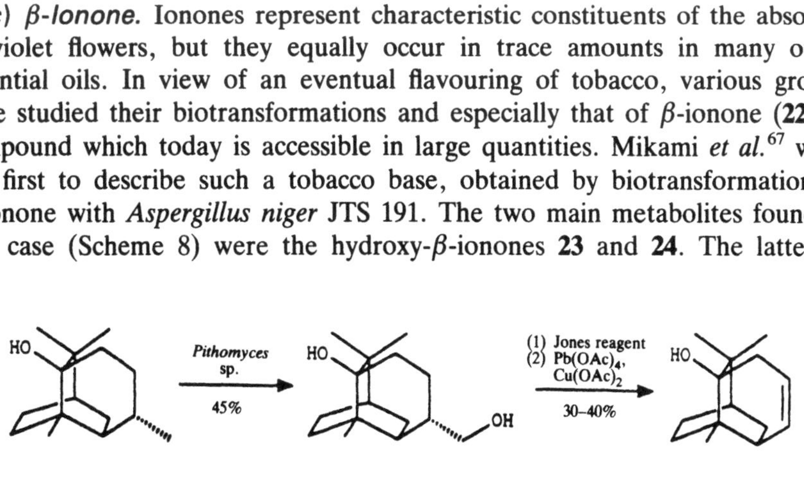

16 **17** **18**

Scheme 6. Biotransformation of patchoulol and synthesis of norpatchoulenol.

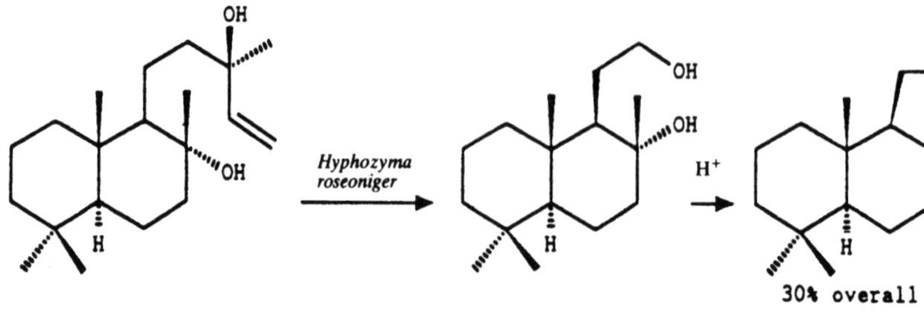

19 20 21

Scheme 7. Biotransformation of sclareol to Ambrox®.

product resulting from specific allylic oxidation, may be produced with even higher selectivity and in over 70% yield, if *Rhizopus oryzae* is used as the microorganism.[68]

A completely different course is taken by the reaction if β-ionone is treated with *Lasiodiplodia theobromae* ATCC 28570, as shown by Krasnobajew.[61] In fact, the products obtained under these conditions are the β-cyclohomogeraniol **25**, its hydroxylated analogues **26**, **27** and **28** as well as some minor products such as **29** and **30**[69] (Scheme 9). All of these compounds constitute Baeyer–Villiger-type degraded products, which lack two carbon atoms of the original side-chain of β-ionone. Similar results were obtained with α-ionone.[61]

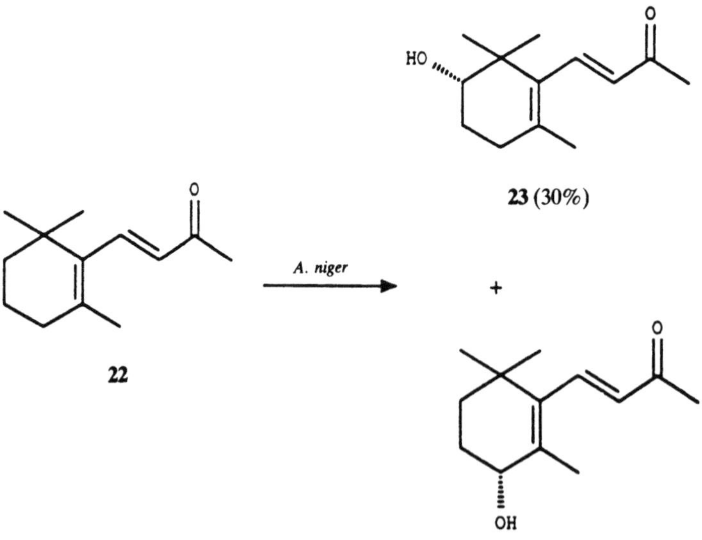

23 (30%)

22

24 (45%)

Scheme 8. Biotransformation of β-ionone to hydroxy-β-ionones.

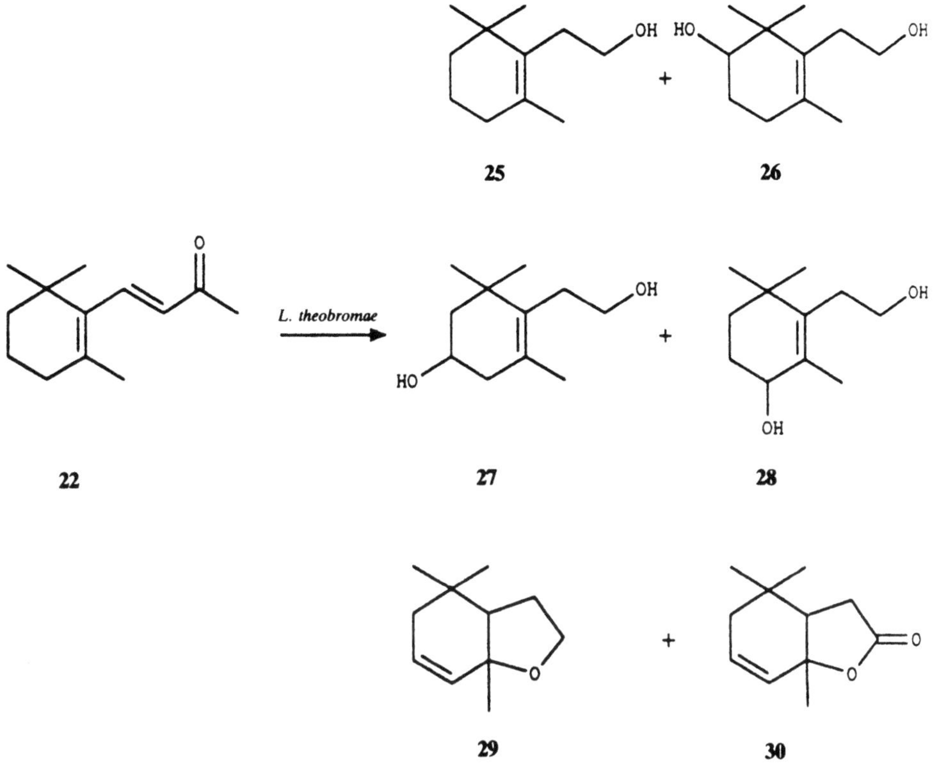

Scheme 9. Biodegradation of β-ionone by *L. theobromae* ATCC 28570.

(d) β-Damascone. This compound (**31**), first isolated from a Burley-type tobacco, is also found in various other plant extracts.[6,70] Its biotransformation with microorganisms of the species *Aspergillus*, *Botryosphaeria* and *Lasiodiplodia* leads to mixtures of mono- and di-hydroxylated β-damascones (e.g. **32–35** for *L. theobromae*), compounds which turn out to be very suitable for flavouring tobaccos[61] (Scheme 10).

Biotransformations of Plant Raw Materials

At the time of harvesting, certain fragrances of vegetable origin are only present in the form of their precursors (e.g. glycosides) and are only released after an additional period of stocking/maturation. Different chemical and biotechnological techniques may be applied in order to accelerate this maturation process and to increase the extraction yields as a consequence. Some typical examples are given below.

(a) Rosa damascena. In 1972 Georgiev *et al.*[71] showed that some constituents of the absolute of rose were present in the form of glycosides and could be released by acidic or enzymatic hydrolysis or by a microbiological treatment with microorganisms of the species *Aspergillus*, *Lycoperdon*, *Saccharomyces* or

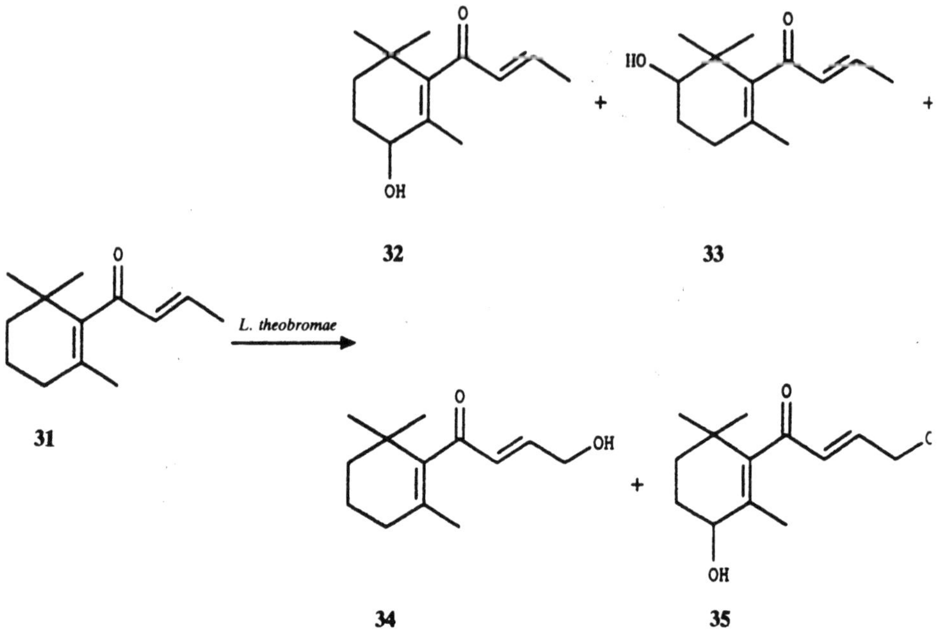

Scheme 10. Biotransformation of β-damascone.

Geotrichum, thereby increasing the total extraction yield by 10–15%. More recently, substantial increases in the contents of phenylethyl alcohol, linalool, citronellol and geraniol have been reported by Kupenov,[72] if the waste of the concrete of rose is treated with certain microorganisms (*Trichosporan* sp., *Penicillium* sp., *Candida* sp.) prior to their isolation via steam distillation.

(b) Jasminum grandiflorum. Similarly to the case of rose, certain constituents participating in the overall olfactive performance of jasmine exist in a glycosidic form in the freshly harvested flowers. According to Ambid,[73] who has studied their release using enzymatic hydrolysis, these represent about 20% of the concrete obtained by classical hexane extraction. When this technique was applied at an industrial scale, 100 kg of jasmine flowers gave 40 g of a mixture exhibiting a jasmine-like odour.

(c) Vanilla. Vanilla beans represent another example of a vegetable raw material which releases its olfactive principle (vanillin) only after a period of maturation and drying (3–6 months),[74] although not completely (*ca.* 2%). However, if the mature beans are pretreated with an enzyme of the pectinase or cellulase type before extraction, the isolated yield of vanillin may be increased by 10–20% (Ehret, C. unpublished). Alternatively, if the mature beans are subjected to a microbiological pretreatment, the resulting vanillin is of improved intensity and better organoleptic quality.[75]

(*d*) *Iris.* As mentioned earlier, freshly plucked rhizomes of *Iris pallida* or *Iris germanica* hardly contain any of the irones **1–4** (cf. Scheme 1), which constitute the principal characteristic odorants of the extract of orris.[76,77] These, over a period of maturation (three years), are presumably generated from homotriterpenoid precursors such as the cycloiridals **36** or **37**[78,79] (Scheme 11), the biosynthesis of which is proposed to pass through squalene cyclization.[78,80]

Thus, the formation of *cis*-α-irone (**2**) and *cis*-γ-irone (**3**) could be the result of an oxidative degradation of **36** and **37**, respectively.[79]

With an awareness of the economic interest that would be aroused by the reduction or the suppression of these long storage periods, several solutions have been proposed in this context. Thus, Buono *et al.*[81] showed that $KMnO_4$ oxidation of a lipophilic extract of fresh rhizomes allowed the isolation of irones, in a similar quantity and content to those obtained by classical maturation.

A different approach involving microbiological and/or enzymatic reactions has recently been found by Belcour *et al.* and further developed in our own laboratories.[82] It consists of treating fresh rhizomes with bacteria preferably belonging to the Enterobacteriacea or the Pseudomonacea families in the presence of a vegetable cell culture medium such as Murashige & Skoog or Gamborg medium. Some of the results achieved regarding isomer distribution and the yields obtained are summarized in Tables 7 and 8.

This biological treatment of rhizomes in fermentors may be carried out by using either bacteria or cell-free enzymatic fractions.

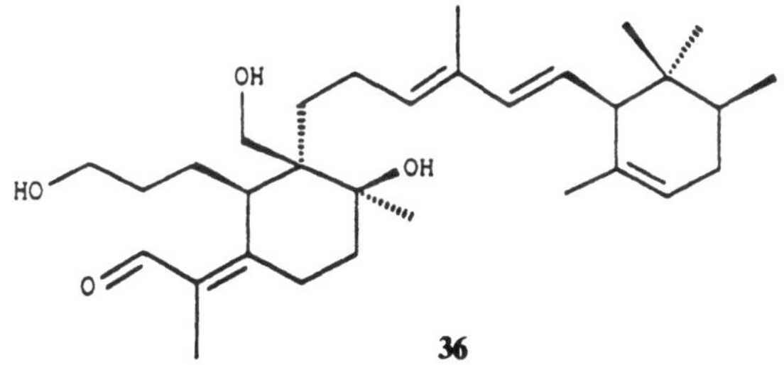

36

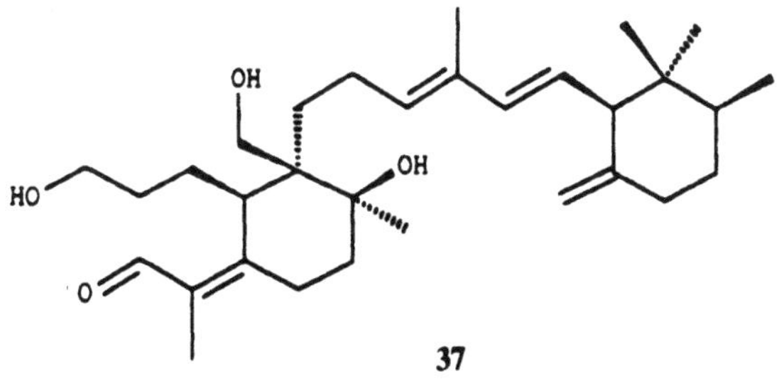

37

Scheme 11. Methyl-cycloiridals.

Table 7
Distribution of Irone Isomers

Irone isomer	*Isomer distribution resulting from classically matured rhizomes (%)*	*Isomer distribution resulting from microbiologically treated rhizomes (%)*
trans-α-Irone, **1**	6·5	4·1
cis-α-Irone, **2**	34·0	44·3
cis-γ-Irone, **3**	58·0	50·3
β-Irone, **4**	1·5	1·3

Table 8
Yields of Irones (Mixture of Isomers)

Age of rhizomes	*Yield from untreated rhizomes (mg irones/kg)*	*Yield from microbiologically treated rhizomes (mg irones/kg)*
Fresh	68	720
6 months	198	675
3 years	623	822

Production of Fragrances by Plant Cells

Production of Metabolites

With the development of plant cell culture techniques, initially applied successfully to pharmaceutical products,[83,84] the possibility of producing secondary metabolites of plants in vitro in view of the continuously rising prices of certain essential oils, was studied by various research groups. However, with the exception of a few isolated cases, little success has so far been achieved in the production of mono or sesquiterpenes via plant cell cultures. The results obtained in this field have been reviewed recently by Mulder-Krieger *et al.*[85] and it is not our aim to discuss them.

However, a number of problems have to be resolved before this technique may be used industrially:

—increase of the rate of growth;
—better knowledge of the biosynthetic pathways;
—development of culture media favouring the production of metabolites;
—improvement of the stability of cultures;
—development of techniques for the continuous isolation of metabolites.

In general, a re-differentiation of cells (culture of root structures or of microembryos) appears to be necessary in order to be able to produce significant quantities of metabolites.[85–87]

Two studies which are linked directly to perfumery are mentioned below.

(a) Geranium. The production of an essential oil of geranium using cultures of microembryos of *Pelargonium graveolens* was described by two Japanese research groups.[88–90] Whereas one of these[88,89] contained citronellol (38%) and isomenthone (13%) as the major components, the other one[90] was described to be rich in linalool/citronellol/geraniol (21·3%), citronellyl formate/citronellyl acetate/geranyl formate/geranyl acetate (26%) and menthone/isomenthone (8·8%).

(b) Chamomile. Cell cultures of *Matricaria chamomilla* accumulate only a few volatile compounds. However, an enhanced production was observed when a strain of chamomile, transformed by *Agrobacterium tumefaciens,* was cultivated in a two-phase system. Among the resulting metabolites, α-bisabolol, an important constituent of chamomile, was identified.[91]

Biotransformations using Plant Cells

Despite the fact that most of the plant cell cultures investigated so far seem unable to produce any terpenes *in vitro,* their potential with regard to biotransformations remains unaffected. The results achieved in this area have been widely reviewed.[85] We shall concentrate only on those regarding the valorization of certain monoterpenes.

(a) Tobacco cells. Cellular suspensions of *Nicotiana tabacum* are able to biotransform a large number of monoterpenes or their derivatives. Some of the observed reactions are illustrated in Schemes 12–14.

Depending on the starting material the transformations brought about by *Nicotiana tabacum* include regio- and/or enantioselective allylic oxidations of isolated double bonds[92,93] and diastereoselective reductions of α,β-unsaturated ketones.[94]

(b) Rose cells. Calli and cellular suspensions of *Rosa damascena* do not produce any metabolites *in vitro.* However, the enzymes known to be responsible for the conversion of mevalonic acid and isopentenyl pyrophosphate into geraniol could be extracted from these seemingly inactive calli.[95] They exhibit a strongly enhanced activity (100-fold) if compared with those resulting

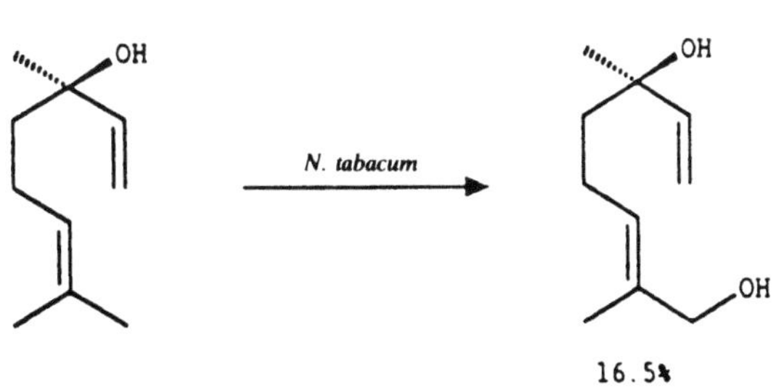

Scheme 12. Biotransformation of linalool.

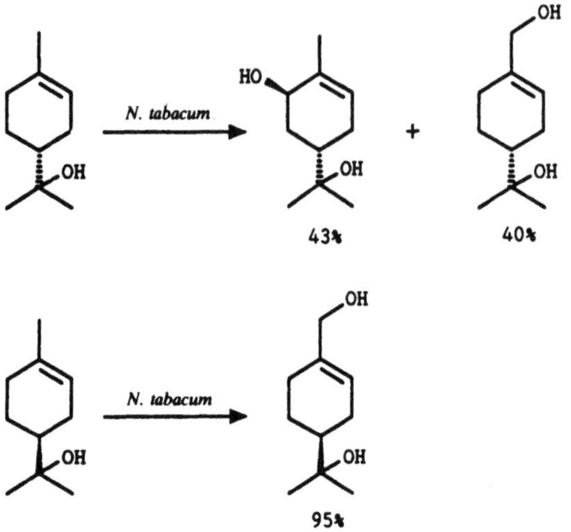

Scheme 13. Biotransformation of d- and l-α-terpineol.

from the extraction of the whole plant. On the other hand, these cellular suspensions could be shown to degrade monoterpenes oxidatively.[95] The cellular suspensions derived from the petals of *Rosa centifolia* also do not produce any metabolites; however, they are capable of biotransforming various monoterpenoid hydrocarbons, alcohols (linalool, geraniol, α-terpineol, citronellol) or acetates to compounds which occur naturally in the absolute of rose de Mai. The reactions involved in this process comprise oxidations, hydrolyses, isomerizations and cyclizations. The capability of these cellular suspensions to transform citronellyl acetate into rose oxide[96] is here of particular interest.

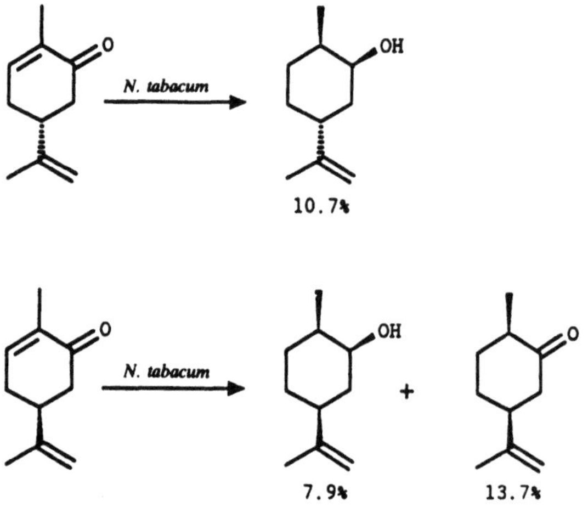

Scheme 14. Biotransformation of d- and l-carvone.

ACKNOWLEDGEMENT

We thank Dr A. Baydar for helpful comments and suggestions on the manuscript.

REFERENCES

1. Lawrence, B. M., *Perfumer & Flavorist,* **10** (1985), no. 5, 1–16.
2. Vigne, P., *Parfums, Cosmétiques, Arômes,* no. 78 (1987) 97–105.
3. Anon., *Essential Oils and Oleoresins* Intern. Trade Center UNCTAD/GATT Geneva, 1986.
4. Guenther, E., *The Essential Oils.* Van Nostrand, New York, 1952.
5. Teisseire, P., *L'Actualité Chimique* (Nov. 1977) 9–18.
6. Peyron, L., *Parfums, Cosmétiques, Arômes,* no. 33 (1980) 51–65.
7. Rautenstrauch, V. & Ohloff, G., *Helv. Chim. Acta,* **54** (1971) 1776–88.
8. Garnero, J., Joulain, D. & Buil, P., *Rivista Italiana EPPOS,* **10** (1978) 568–90.
9. Garnero, J. & Joulain, D., *Rivista Italiana EPPOS,* **13** (1981) 141–45.
10. Brunke, E. J. & Hammerschmidt, F. J., In Technical Data VIIIth, Int. Congress of Essential Oils, Oct. 1980, Fedarome, Grasse 1982, 221–9.
11. Demole, E., Demole, C. & Enggist P., *Helv. Chim. Acta,* **59** (1976) 737–47.
12. Garnero, J., *Rivista Italiana EPPOS,* **5** (1972) 315–36.
13. ter Heide, R., Provatoroff, N., Traas, P. C. & de Valois, P. J., *J. Agric. Food Chem.,* **23** (1975) 950–7.
14. Gavin, J., Nicollier, G. & Tabacchi, R., *Helv. Chim. Acta,* **61** (1978) 352–7.
15. Lawrence, B. M., *Perfumer & Flavorist,* **3** (1978), no. 4, 56–8.
16. Kaiser, R. & Lamparsky, D., *Helv. Chim. Acta,* **61** (1978) 2671–80.
17. Lamparsky, D., *Perfumer & Flavorist,* **11** (1986), no. 4, 7–20.
18. Kaiser, R. & Lamparsky, D., *Helv. Chim. Acta,* **66** (1983) 1835–42.
19. Kaiser, R. & Lamparsky, D., *Helv. Chim. Acta,* **66** (1983) 1843–9.
20. Burrell, J. W. K., Lucas, R. A., Michalkiewicz, D. M. & Riezebos, G., *Tetrahedron Lett,* **30** (1971) 2837–8.
21. Burrell, J. W. K., Lucas, R. A., Michalkiewicz, D. M. & Riezebos, G., *Chem. Ind.* (1970) 1409–10.
22. Mookherjee, B. D. & Trenkle, R. W., *J. Agric. Food Chem.,* **21** (1971) 298–307.
23. Kovats, E.sz., *J. Chromatogr.,* **406** (1987) 185–222.
24. Ohloff, G. & Demole, E., *J. Chromatogr.,* **406** (1987) 181–3.
25. Ehret, C. & Teisseire, P., *Recherches,* **19** (1974) 287–8.
26. Kaiser, R. & Lamparsky, D., *Tetrahedron Lett,* **20** (1976) 1659–60.
27. Maurer, B. & Hauser, A., *Helv. Chim. Acta,* **65** (1982) 462–76.
28. Kaiser, R. & Lamparsky, D., *Tetrahedron Lett,* **38** (1974) 3413–16.
29. Lawrence, B. M., *Perfumer & Flavorist,* **2** (1977), no. 6, 37–8.
30. Verzele, M., Maes, G., Vuye, A., Godefroot, M., Van Alboom, M., Vervisch, J. & Sandra, P., *J. Chromatogr.,* **205** (1981) 367–86.
31. Uhde, G. & Ohloff, G., *Helv. Chim. Acta,* **55** (1972) 2621–5.
32. Kaiser, R. & Lamparsky, D., In Technical Data VIIIth Int. Congress of Essential Oils, Oct. 1980. Fedarome, Grasse 1982, pp. 287--94.
33. Prager, M. J. & Miskiewicz, M. A., *J. Assoc. Off. Anal. Chem.,* **64** (1981) 131–8.
34. Peyron, L., *American Cosmet. Perfumery,* **87** (1972) 37–41.
35. Ehret, C., Paper presented at 33th meeting of Sepawa, Bad Dürkheim, FRG, 2–4 Oct. 1986.
36. Anonis, D. P., *Drug Cosmet. Ind.,* (July 1983) 34–38, 77–78.
37. Loo, A. & Richard, H. In *Flavors and Fragrances: a World Perspective,* ed. B. M.

Lawrence, B. D. Mookherjee & B. J. Willis, Proc. 10th Int. Congress of Essential Oils, Fragrances and Flavors, Washington, DC, USA. 16–20 Nov., 1986. Elsevier, Amsterdam, 1988, pp. 355–73.
38. Lamparsky, D., Paper presented at 6èmes Journées Internationales Huiles Essentielles, Digne-Les-Bains, France, 3–5 Sept. 1987.
39. Ehret, C., Maupetit, P., & Petrzilka, M., In Proceedings 11th Int. Congress of Essential Oils, Fragrances and Flavours, Vol. 5, Oxford & IBH Publishing Comp. PVT Ltd., New Delhi 1989, pp. 49–55.
40. Latrasse, A., Degorce-Dumas, J. R. & Leveau, J. Y., *Sciences des Aliments,* **5** (1985) 1–26.
41. Schindler, J. & Schmid, R. D., *Proc. Biochem.* (Sept./Oct. 1982) 2–8.
42. Badcock, E. C., *Trans. Br. Mycol. Soc.,* **23** (1939) 188–98.
43. Maga, J. A., *Chemical Senses and Flavors,* **2** (1976) 255–62.
44. Heim, R., *Champignons d'Europe.* Société Nouvelle des Éditions Boubée, Paris, 1984.
45. Klausner, A., *Bio/Technology,* **3** (1985) 534–8.
46. Dziezak, J. D., *Food Technology,* (Apr. 1986) 122–9.
47. Redenbaugh, K., Paasch, B. D., Nichol, J. W., Kossler, M. E., Viss, P. R. & Walker, K. A., *Bio/Technology,* **4** (1986) 797–801.
48. Demame, F., *Parfums, Cosmétiques, Arômes,* **62** (1985) 77–9.
49. Roberts, D., Lemberg, S. & Wiese, E., In IXth Int. Congress of Essential Oils (March, 1983), Technical Paper Book 1, Singapore 1983, pp. 30–9.
50. Collins, R. P., *Dev. Ind. Microb.,* **20** (1978) 239–45.
51. Schindler, J., *Ind. Eng. Chem. Prod. Res. Dev.,* **21** (1982) 537–9.
52. Schindler, J. & Bruns, K. (Henkel KGaA), German Patent Appln. 2840 143, (prior. 15 Sept. 1978).
53. Lanza, E., Hwee Ko, K. & Palmer, J. K., *J. Agric. Food Chem,* **24** (1976) 1247–50.
54. Hubbal, J. A. & Collins, R. P., *Mycologia,* **70** (1978) 117–29.
55. Demain, A., Jackson, M. & Trenner, N. R., *J. Bacteriol.,* **94** (1967) 323–6.
56. Kempler, G. M., *Adv. Appl. Microbiol.,* **29** (1983) 29–51.
57. McIver, R. C. & Reineccius, G. A. In *Synthesis of 2-methoxy-4-alkylpyrazines by Pseudomonas perolens,* ACS Symposium Series No. 317, American Chemical Society, Washington, DC, 1986, pp. 266–74.
58. Kieslich, K., *Microbial Transformations of Non-Steroid Cyclic Compounds.* Georg Thieme, Stuttgart, 1976.
59. Abraham, W. R., Hoffmann, H. M. R., Kieslich, K., Reng, G. & Stumpf, B., *Enzymes in Organic Synthesis.* CIBA Foundation Symposium, 1985, pp. 146–60.
60. Mikami, Y., *Biotechnol. & Genet. Rev.,* **6** (1988) 271–320.
61. Krasnobajew, V. In *Biotechnology,* ed. K. Kieslich. Verlag Chemie, Weinheim, 1984, pp. 97–125.
62. Teisseire, P., Maupetit, P. & Corbier, B., *Recherches,* **19** (1974) 8–35.
63. Teisseire, P., *Bull. Soc. Chim. France,* (1980) II 66–70.
64. Becher, E., Fujiwara, A., Fujiwara, M., Matzinger, P. K., Schuep, W., Shiomi, Y., Tazoe, M., Teisseire, P. J. & Ehret, C. (Hoffmann-La Roche), British Patent 1 586 860, appl. 20 Aug. 1976.
65. Suhara, Y., Itho, S., Ogawa, M., Yokose, K., Sawada, T., Sano, T., Ninomiya, R. & Maruyama, H. B., *Appl. Environ. Microbiol.,* **42** (1981) 187–91.
66. Farbood, M. I., Willis, B. J. & Christenson, P. A., (Fritzsche Dodge & Olcott), S. African Patent 85' 04 306, appl. 6 June 1985.
67. Mikami, Y., Watanabe, E., Fukunaga, Y. & Kisaki, T., *Agric. Biol. Chem.,* **42** (1978) 1075–7.
68. Yamazaki, Y., Fukunaga, Y., Hori, N. & Mikami, Y., (Japan Tobacco) Japanese Patent appl 63'030 445, appl. 24 July 1986.

69. Krasnobajew, V. & Helmlinger, D., *Helv. Chim. Acta,* **65** (1982) 1590–601.
70. Ohloff, G., *Perfumer & Flavorist,* **3** (Feb./March 1978).
71. Georgiev, E. V., Kupenov, L. G., Ganchev, G. P. & Kolovska, B. G., *Rivista Italiana EPPOS* (1972) 422–4.
72. Koupenov, L., *Nauchni Tr.-Vissh Inst. Khranit. Vkusova Prom.-st.,* **31** (1984) 185–92.
73. Ambid, C. (Sanofi), European Patent 309339, appl. 22 Sept. 1987.
74. Naves, Y. R., *Technologie et Chimie des Parfums Naturels.* Masson, Paris, 1974.
75. Hasegawa KK, Japanese Patent appl. 58' 58043-757, 8 Sept. 1981.
76. Rautenstrauch, V. & Ohloff, G., *Helv. Chim. Acta,* **54** (1971) 1776–88.
77. Rautenstrauch, V., Willhalm, B., Thommen, W. & Ohloff, G., *Helv. Chim. Acta,* **67** (1984) 325–30.
78. Jaenicke, L. & Marner, F.-J., The irones and their precursors. In *Progress in the Chemistry of Organic Natural Products,* Vol. 50. Springer-Verlag, Wien, 1986, pp. 1–25.
79. Krick, W., Marner, F.-J. & Jaenicke, L., *Z. Naturforsch,* **38c** (1983) 179–84.
80. Marner, F. J., Gladtke, D. & Jaenicke, L., *Helv. Chim. Acta,* **71** (1988) 1331–8.
81. Buono, G., Gil, G. & Triantaphylides, C., (Société Nationale Elf Aquitaine), French Patent 2 620 702 appl. 23 Sept. 1987.
82. Belcour, B., Courtois, D. & Ehret, C., (Roure), European Patent Appln 88810537 appl. 5 Aug. 1988.
83. Fowler, M. W., *Chem. Ind.* (1981) 229–33.
84. Deus, B. & Zenk, M. H., *Biotechnol. Bioeng,* **24** (1982) 1965–74.
85. Mulder-Krieger, T., Verpoorte, R., Baerheim Svendsen, A., & Scheffer, J. J. C., *Plant Cell, Tissue and Organ Culture,* **13** (1988) 85–154.
86. American Cyanamid Company, French Patent 72 17923, appl. 18 May 1972.
87. Bricout, J., Garcia-Rodriguez, M.-J. & Paupardin, C., *C.R. Acad. Sc. Paris,* **286D** (1978) 1585–8.
88. Takemoto, S., Katagi, H., Takahashi, E., Naito, K., Okuyama, G. & Inui, M., Paper no. 59 presented at 10th Int. Congress of Essential Oils, Fragrances and Flavors, Washington, DC, USA, 16–20 Nov. 1986.
89. Kanebo Ltd, Japanese Patent appl 59' 59 213 393, appl. 18 May 1983.
90. Yokoyama, M., Yomogida, K., Nakamura, S. & Yanagi, M. (Shiseido), Japanese Patent appl. 62' 62 54 796, appl. 4 Sept. 1985.
91. Bisson, W., Beiderbeck, R. & Reichling, J., *Planta Medica,* **47** (1983) 164–8.
92. Hirata, T., Aoki, T., Hirano, Y., Ito, T. & Suga, T., *Bull. Chem. Soc. Japan.* **54** (1981) 3527–9.
93. Suga, T., Hirata, T. & Lee, Y. S., *Chem. Lett.* (1982) 1595–8.
94. Hirata, T., Hamada, H., Aoki, T. & Suga, T., *Phytochemistry,* **21** (1982) 2209–18.
95. Banthorpe, D. V., Grev, T. J., Poots, I. C. & Fordham, W. D., *Phytochemistry,* **25** (1986) 2321–6.
96. Corbier, B. & Ehret, C. In *Flavors and Fragrances: A World Perspective,* ed. B. M. Lawrence, B. D. Mookherjee & B. J. Willis, Proc. 10th Int. Congress of Essential Oils, Fragrances and Flavors, Washington, DC, USA, 12–20 Nov. 1986. Elsevier, Amsterdam 1988, pp. 731–41.

Chapter 20

Synthetic Products

GEORG FRÁTER & DIETMAR LAMPARSKY

Givaudan Forschungsgesellschaft AG, CH-8600 Dübendorf, Switzerland

1 INTRODUCTION

When discussing highlights in synthetic chemistry with regard to usefulness in perfumery, we have to keep in mind that until the end of the last century, only natural ingredients were available for mixing fragrances. Since then, synthetic efforts to produce high-quality perfumes became more and more reliable. This is not the place to evoke all the interesting compounds created by the chemists' fantasy and their knowledge concerning the feasibility of transferring laboratory procedures to industrially applicable ones; but we would like to look at the development of new raw materials and at improvements realized in the synthesis of known fragrant molecules to produce them more economically.

Because of excellent reviews on many different topics in the fragrance field (e.g. Ref. 1) in the early 1980s, at first we tried to collect pertinent data only for the decade from 1980 to 1989. Furthermore, the choice of the material presented was guided—though not exclusively—by considerations of practicability.

A good survey of this domain will be achieved by consulting the patent literature. All companies dealing with the production of synthetic raw materials for the fragrance (and flavor) industry are concerned over the uniqueness of their new developments. Therefore they normally ask for patent protection of their inventive products and/or processes, so that they can offer their specialities as sole producer on the world market, at least for a certain period (normally 20 years).

It is well known to those working in the fragrance industry that it is getting more and more difficult to introduce a new compound to the perfumer's shelf. This has several reasons such as toxicological constraints, environmental considerations and, last but not least, the price structure, which is rarely in favor of the new chemical substance in comparison with well-established synthetics.

Today's prices are generally low and because production volumes of fragrance materials are relatively seldom on the large-scale side, very often there is no real price benefit from increasing volumes. Exceptions exist, however—Lilial®, Hedione®, Tonalide® or Fixolide®, linalool and linalyl acetate are just a few examples.

Cost figures play an important role, therefore, and consequently practical chemistry is highly in demand. It has to be simple, straightforward and/or imaginative. Thus, it should be patentable.

In view of the fact that a new product always causes considerable expense before its introduction to the market (research, development and toxicological testing), the filing of appropriate patent applications may give—at least to a certain extent—some kind of guarantee for a return of the invested money, if the commercialization is successful in due time.

Unfortunately, a too rigid regulation in practically all countries of interest to the fragrance industry has led to a strong decrease of innovations in the field of new synthetic odorants, especially in the last two to three years. This fact was well reflected recently by Fenn.[2] The author gives a list of 25 'most used' compounds, but we find only two 'new' aroma chemicals in this compilation. Looking more closely at their date of introduction to perfumery, they are also quite ripe in age: in the corresponding patents for Hedione® we find the year 1960[3] and for Iso E Super® the year 1973.[4] This means that successful launching and commercial development often take a long time in our industry, if large-scale production is the goal envisaged.

Better chances are given to so-called impact chemicals, i.e. chemicals with strong odor characteristics or low threshold values which favor their application in very small concentrations. That means that even costly new substances can be used successfully to give a typical *empreinte* (fingerprint) to a creation.

An example of this kind of new material is certainly the sulfur-containing monoterpene ketone 'Corps Cassis' (internal name, Givaudan)[5] which exists in two diastereomeric forms with different odor strengths.[6]

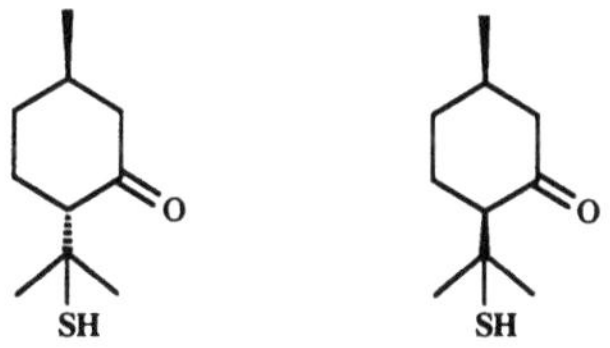

'Corps Cassis'
(ex D-(+)-pulegone)

Perfumers have used it in numerous perfumery creations, but only in a highly diluted version, to avoid the overdosage of this nature-identical compound which in the pure form has an odor that is even unpleasant, reminiscent, *inter alia,* of cats' urine. The substance was found in buchu leaf oil[7,8] where it represents the real 'impact compound', i.e. the most important odor-contributing substance.

This example also reflects very well one of the roots of fragrance research: intimate knowledge of the chemical composition of naturals often leads to the later development of new, purely synthetic odorants which make the perfumer independent of natural sources. This is highly valuable to the perfumer who will no longer be restricted by supply bottlenecks and/or price fluctuations due to bad harvesting conditions or pure speculation.

One of the best examples is surely today's availability of synthetic linalool and all the terpenoids derived from it. Only very few examples from the vast literature on terpene synthesis meet the prerequisites of a technologically feasible process, but the excellent review of P. Teisseire[9] describing the chemical development in the field of aliphatic terpenoids, is highly recommended for further information.

Returning to really new fragrant chemicals, a synthetic working chemist can extract more stimulus from the enthusiastic reviews of Boix Camps,[10] where one reads the following sentence: 'What have determined and still determine the creativity and the innovation in perfumery are the definite chemical aromatics'. And further: 'Effort is something that is absolutely essential: effort and a complete dedication, a "love" for the new developments that continue to arise from those ingenious structures that our most faithful collaborators, the chemists, are searching for. Today a perfume is, in part, their work and here I want to render them the homage of being the "wise men behind the scenes".' Moreover, the annual reviews by Bedoukian (e.g. Ref. 11) are—besides the patent literature already mentioned—another good source of information to penetrate the realm of fragrances. As a highly experienced chemist in this field, the author collected a huge amount of data every year covering the complete range of what was then currently relevant in perfumery.

On the following pages we will now focus on the most interesting developments achieved during the last decade. Information is arranged according to the most common and important odor classes. Thus, chemistry of musks, amber and sandalwood notes will be treated first, followed by a discussion of new material in the sections of generally woody, fruity, floral, green, herbaceous and spicy odor notes. In all these classes, continuous transitions like woody–fruity, fruity–floral, floral–green, green–herbaceous and so on, are the normal descriptors of the olfactory properties. It very seldom happens that a substance exhibits only one distinct odor type. Therefore, in most cases it is better to speak of chemicals having 'aspects' of the classes mentioned.

REFERENCES

1. Theimer, E. T., ed., *Fragrance Chemistry*. Academic Press, New York, 1982.
2. Fenn, R. S., Aroma chemical usage trends in modern perfumery. *Perfumer & Flavorist*, **14**(2) (1989) 1.
3. British Patent 907 431, priority CH 27 July 1960, to Firmenich.
4. German Patent 2 408 689, priority USA 27 Feb. 1973, to International Flavor & Fragrances (IFF).
5. Swiss Patent 529 829; US Patent 3 819 712, priority CH 10 Sept. 1969, to Givaudan.
6. Kaiser, R., Lamparsky, D. & Schudel, P., Analysis of buchu leaf oil. *J. Agric. Food Chem.*, **23** (1975) 943.
7. Lamparsky, D. & Schudel, P., *p*-Menthane-8-thiol-3-one, a new component of buchu leaf oil. *Tetrahedron Lett.*, (1971) 3323.
8. Sundt, E., Willhalm, B., Chappaz, R. & Ohloff, G., Das organoleptische Prinzip von Cassis Flavor im Buccublätterröl. *Helv. Chim. Acta*, **54** (1971) 1801.
9. Teisseire, P., Les grandes voies de synthèses des terpénoides aliphatiques. In *Progress in Terpene Chemistry*, ed. D. Joulain. Editions Frontières, Gif s/Yvette, 1986, 1–39.
10. Boix Camps, A., Perfumery technique in evolution. *Perfumer & Flavorist* (Part I) **10**(3) (1985) 1; (Part II) **10**(4) (1985) 15; Part (III) **11**(3) (1986) 9.
11. Bedoukian, P. Z., *Perfumer & Flavorist*, **14**(3) (1989) 1.

2 CHEMICALS WITH MUSKY AND AMBER ASPECTS

Musk Compounds

Musks are of central importance in perfumery, and so is their chemistry in the corresponding industry. Musk odorants can be divided into three main fields:

(a) The *macrocyclic musks*, which consist of the natural musks, like muscone (**1**), civetone (**2**) and ambrettolide (**3**), and the synthetic macrocyclics, e.g. ethylene brassylate (**4**) and 12-oxahexadecanolide (**5**) (Musc 174).

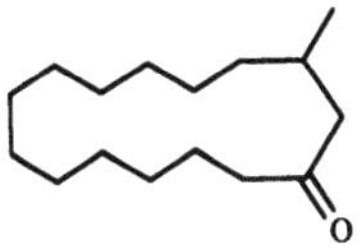

1

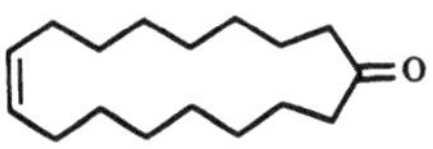

2

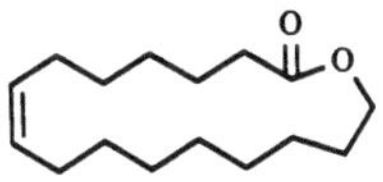

3

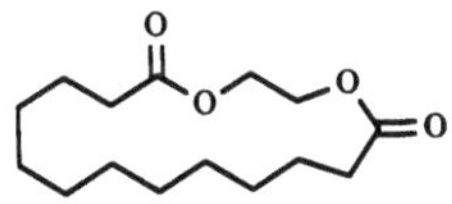

4

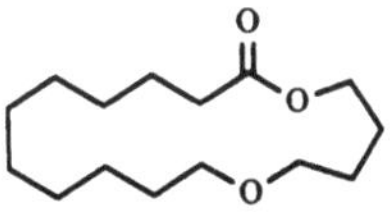

5

The natural musks have been reviewed by Mookherjee & Wilson,[1] whereas the synthetic macrocycles were described by Wood.[2] Further compilations concerning macrocyclics are in the Haarmann & Reimer booklet[3] and in the conference paper by Abe.[4]

(b) The *nitro musks,* which have a century's history,[5,6] still play a very important role as synthetic fragrance ingredients. The main representatives are musk xylene, musk ketone, musk ambrette, musk tibetene and moskene (see Ref. 7 for a review, and Chapter 14 for some remarks including the structural formulae of these compounds).

The great advantage of the nitro musks is their attractive price. For a long time the nitro musks were the largest-volume synthetic musks, but lately their output has decreased slightly, at about 5% annually since 1983. The reasons for the beginning of this decline are partly toxicological, partly practical (see Ref. 8), i.e. discoloration, but the main reason is the availability of very good non-nitro musks.

(c) *Polycyclic musks* or *aromatic non-nitro musks* today constitute the most important class. Their original discovery goes back some 40 years and was accidental. The target compound in a synthesis, compound **7**, turned out to be odorless, but the intermediate **6** had a 'fine, persistent musk odor'.[9,10]

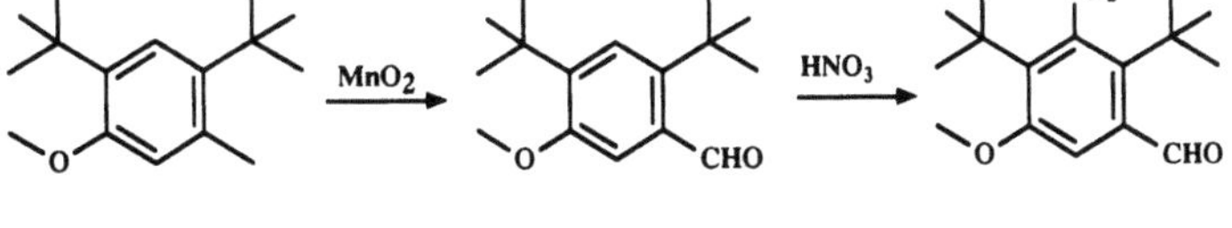

6 7

This group of very successful chemicals is well represented by Phantolide® (**8**), Celestolide® (**9**), Galaxolide® (**10**), Traseolide® (**11**) and Fixolide® or Tonalide® (**12**).

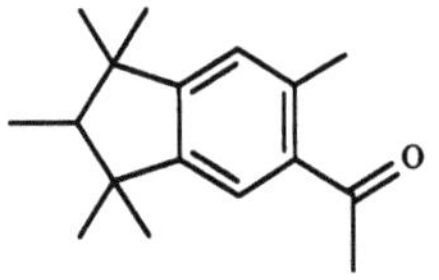

8

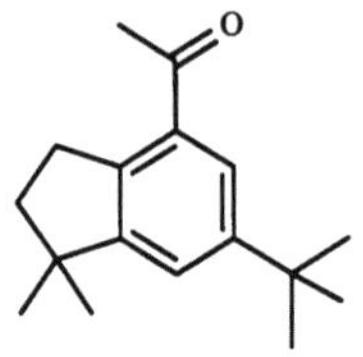

9

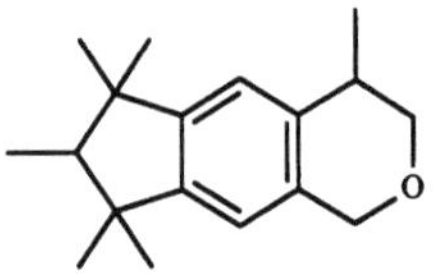

10

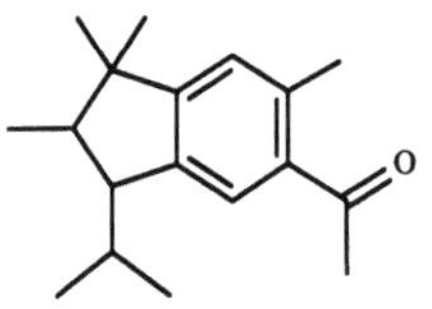

11

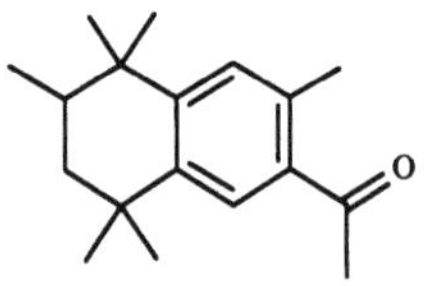

12

These chemicals are slightly more expensive than the nitro musks. Their advantage is high solubility, no discoloration and a good musk odor, which is nearer to the macrocyclic musks. For reviews see Refs 7 and 11.

In 1987 the worldwide production of musks amounted to about 7000 tons per year, with a market value of around $215 million. About 61% of the total tonnage were constituted by the polycyclic musks, 35% by the nitro musks and only the remaining 3–4% by the macrocyclic musks.[12]

After these remarks about economy, some recent developments in the syntheses of macrocyclic and polycyclic musks will now be discussed in more detail.

Macrocyclic Musks

In spite of the long history of this class of compounds, there is a continuing interest in their synthesis.

Aleuritic acid, a trihydroxyhexadecanoic acid of natural origin, served for a series of syntheses in this field, i.e. for the syntheses of **2**, **3** and other compounds.[13] Basically there are two approaches to the macrocyclic structures: either through ring enlargement of smaller rings or by means of ring closure. Most of the routes described here followed the former strategy.

A recent improvement in this direction was described by Fehr *et al.*[14] and used for the synthesis of Exaltone® (**13**) and muscone (**1**). The cyclopentenone **14** was converted into tosylhydrazone **15**, and treatment of the latter with N-bromosuccinimide in methanol/acetone at 0°C furnished the alkynone **16**.

This is a step forward compared with the previously known Eschenmoser fragmentation of epoxyketones,[15] which are often cumbersome to prepare.

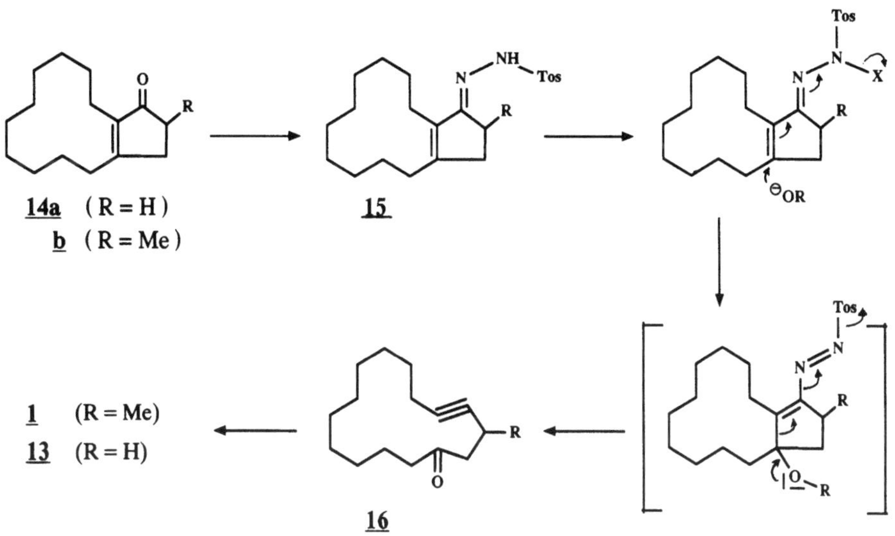

Bicyclopentenones already served as starting material for ring enlargements by three carbon atoms on the way to macrocyclic musks. A general method for their preparation is the Nazarov cyclization (for a review see Ref. 16).

The bicyclopentenone **19**, which is a precursor of muscone (**1**), has been prepared by the Nazarov cyclization from cyclododecanone (**17**) through the intermediate acetylene diol **18**.[17]

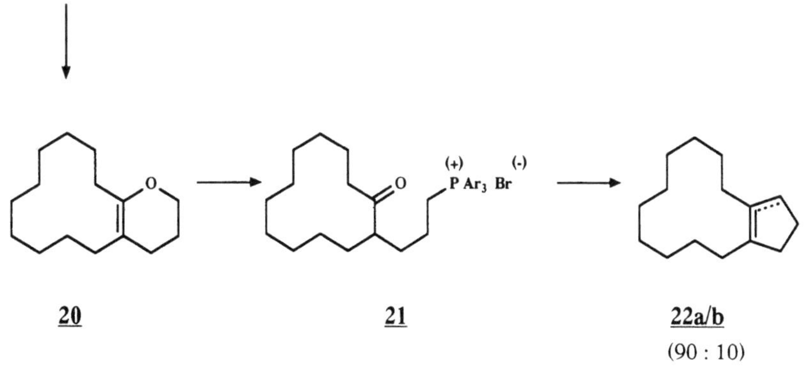

The bicyclo enol ether **20**, which is also easily available from **17**, can be converted into a mixture with a bicyclo[10.3.0] skeleton (**22a/b**) which itself is again the precursor of a three-carbon ring enlargement reaction in the synthesis of Exaltone® (**13**).[15] The reaction is either carried out in two steps, by first preparing the Wittig salt **21**,[18] or directly by heating **20** with triphenylphosphine, which obviously forms a ketophosphorane.[19]

Still another route to the bicyclo[10.3.0] hydrocarbons **22a/b** starts with the allylation of cyclododecanone (**17**) to **23a/b**.[20] Hydroformylation of **23a/b** furnishes **24a/b**, which can be cyclized to **25a/b**. The latter yields the decarbonylated product **22a/b** (R = H) or **26** (R = CH_3) by a metal-catalyzed pyrolysis.

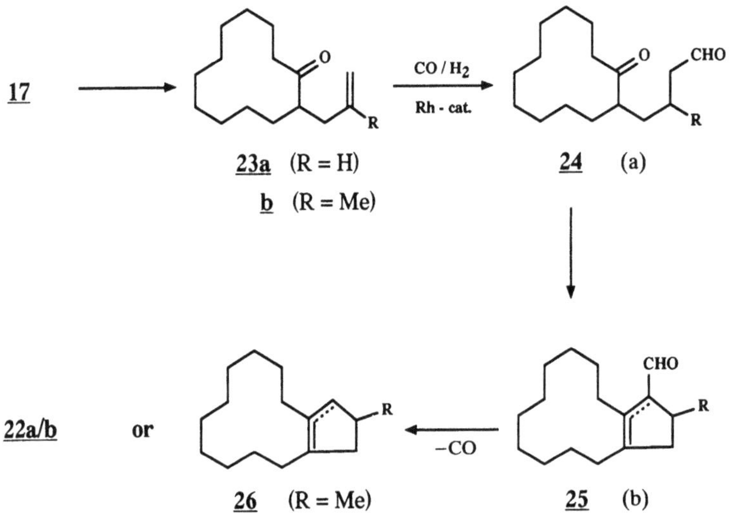

A sulfone route, which is analogous to the one described above with the aldehyde function (**17** to **25**), has also been described.[21,22] Thus the hydrocarbons **22** and **26** have been prepared from the sulfone **27**, which is a derivative of cyclododecanone (**17**). The ketosulfone was converted into the bicyclo sulfone **28**, from which sulfinic acid can be eliminated to yield the diene **29**. Hydrogenation of the latter gave **22** or **26**, respectively.

A clever variation on this theme starts once more from cyclododecanone (**17**) and from the allylated compounds **23a** and **23b**, but avoids the hydrocarbons **22** and **26**.[23,24] The allyl side-chain was converted to the primary bromo compound, which was oxidized to the lactone **30**. Intramolecular nucleophilic ring-opening furnished, among others, the hydroxyketone **31**, which was easily transformed into muscone (**1**) or Exaltone® (**13**).

A similar route through the sulfones **27** is also described. The key reaction is the ring enlargement of **32** to **33**.

An inventive approach to **14a** is the Pd(II)-catalyzed cyclization of **34**, which first forms the palladiun intermediate **35** (not isolated), which stabilizes itself to **36**.[25]

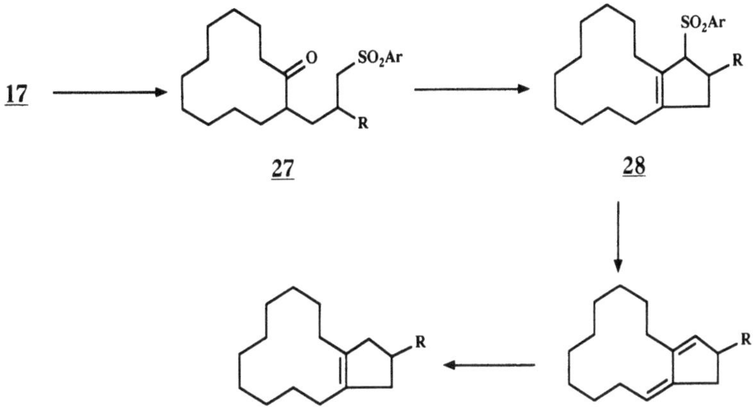

27 28 29

22a (R = H)
26a (R = Me)

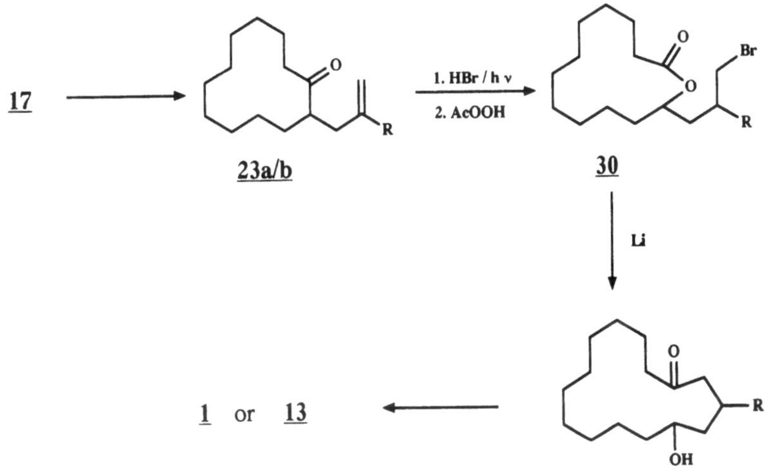

23a/b 30 31

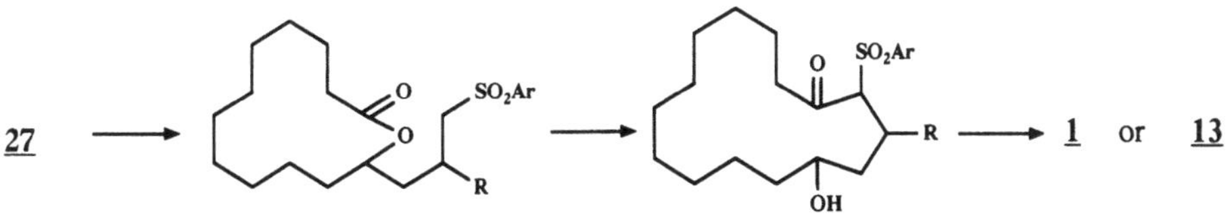

32 33

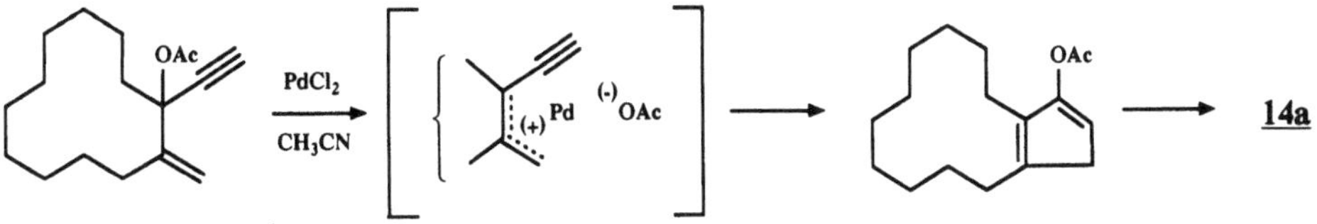

34 35 36

A new and ingenious way to muscone utilized an anion-accelerated [5,5] sigmatropic rearrangement, which corresponds to a ring enlargement by eight carbon atoms.[26] The first key step is the rearrangement of the alkynyl halohydrin **38** to the α-alkenyl alcohol **39**, which after oxidation has been converted to **40**. In the second key step, the ring enlargement was carried out with 50% yield to produce the *cis,trans* mixture of ketone **41**, which was finally hydrogenated to muscone (**1**).

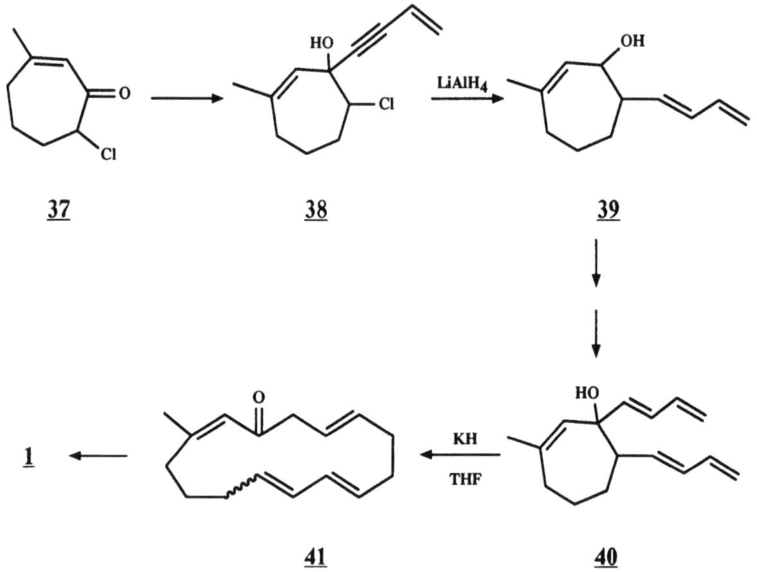

An interesting modern route to macrocycles or to their functionalized precursors is through olefin metathesis.[27] Thus, treatment of the ω,ω'-unsaturated ester **42** in high dilution ($c = 6 \times 10^{-3}$ mol/liter) furnishes dehydro-exaltolide (**43**) (*E,Z* mixture) in 65% yield.[28]

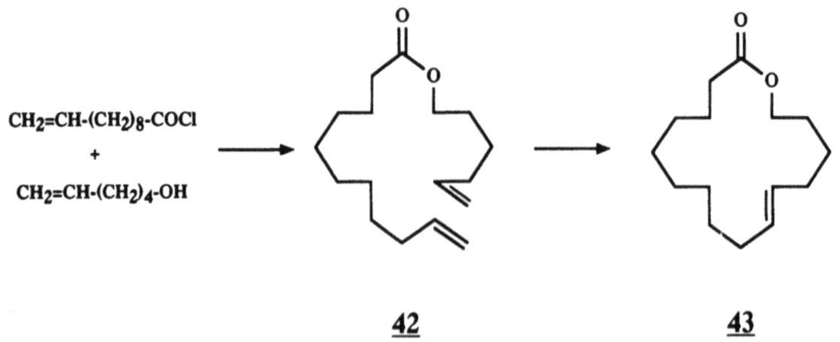

A synthesis of muscone (**1**) by means of olefin metathesis has also been described.[29] This is a new route to the well-known precursor of muscone, the C_{16}-diketone **47**. The synthesis starts with cycloheptene (**44**), which is metathetically transformed to 1,8-nonadiene (**45**), which again in a metathetical 'dimerization' furnishes 1,8,15-hexadecatriene (**46**). Oxidation of **46** gives the requisite diketone **47**, which upon aldol condensation and hydrogenation yields **1**.

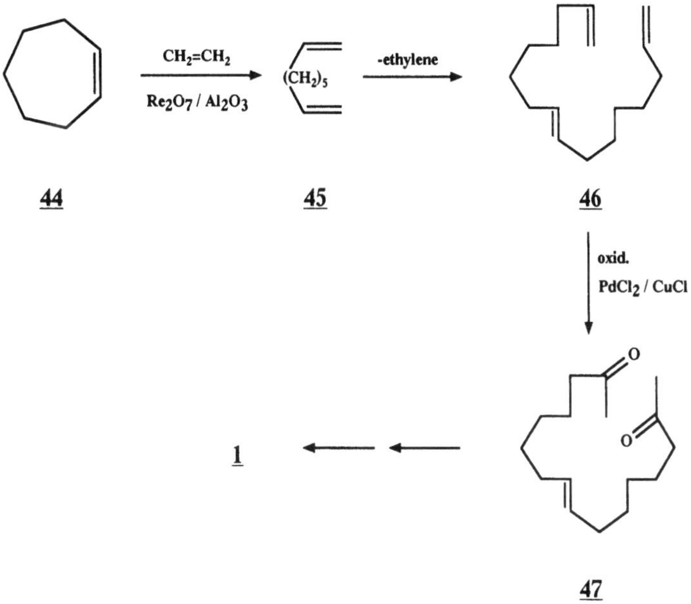

Recently the enzymatic oxidation of n-alkanes to α,ω-alkanedioic acids has been developed so successfully that these precursors have become interesting for a commercial process.[30] Oxidation of pentadecane, for example, by *Candida tropicalis* yielded α,ω-pentadecadioic acid (**48**) with a concentration of up to 100 g/liter. Acyloin condensation and reduction of the α-hydroxyketone **49** leads to cyclopentadecanone.

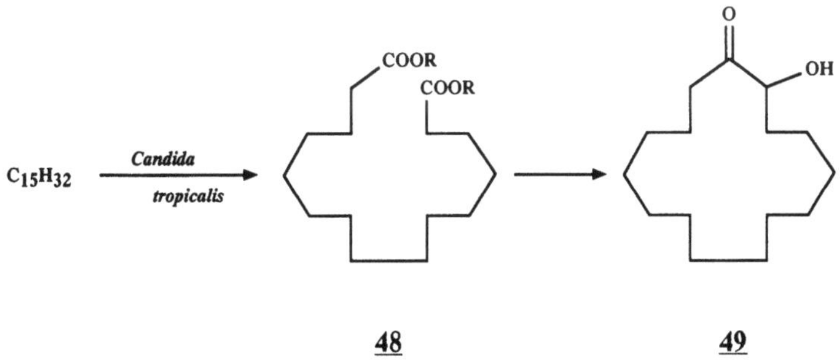

Another biotechnological route describes[31] the ω-oxidation of palmitic acid to a 7:1 mixture of ω-hydroxyhexadecanoic acid and 15-hydroxyhexadecanoic acid which is lactonized to hexadecanolide (**50**).

Dimethyl 1,15-pentadecadioate (**48**, R = CH_3) has also been prepared by a mixed Kolbe reaction; electrolysis of monomethyl adipate and monomethyl 1,11-undecadioate furnished **48**.[32]

$C_{15}H_{31}COOH \longrightarrow HOCH_2\text{-}(CH_2)_{14}\text{-}COOH + CH_3\text{-}CH(OH)\text{-}(CH_2)_{13}\text{-}COOH$

(7:1)

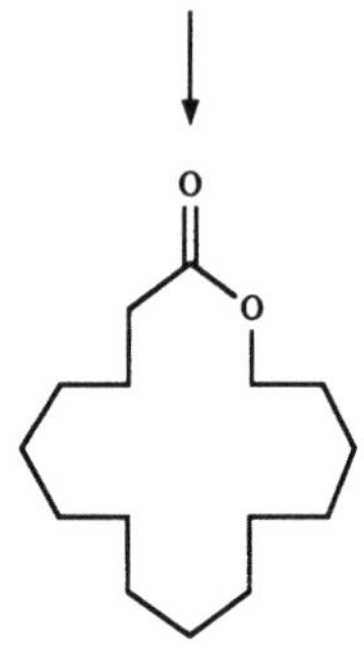

50

Polycyclic Musks

The above-mentioned successful method for the preparation of dehydro-exaltolide (**43**), i.e. the olefin metathesis, is also a key element in the synthesis of an aromatic bicyclo musk, Tonalide® or Fixolide® (**12**).[27] Di-isobutene is first isomerized to 2,4,4-trimethyl-2-pentene (**51**), which reacts with ethylene (WO_3–SiO_2 catalyst) to produce neohexene (**52**).[33] This reaction is carried out on the scale of *ca.* 1500 tons per year. Neohexene is used for the synthesis of hexamethyltetralin (**53**), which by acetylation gives compound **12**.

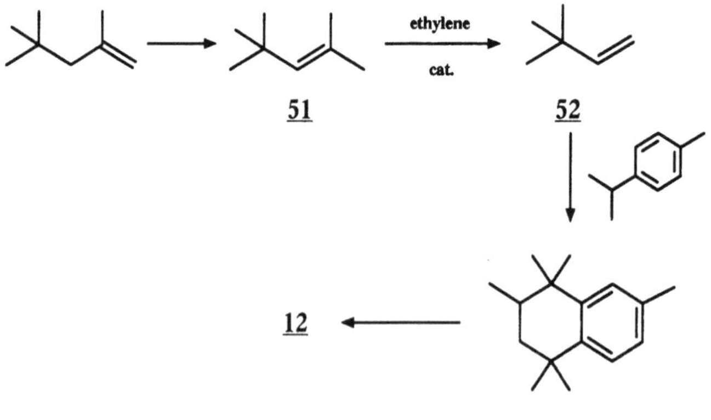

53

One of the more recent musks among the now already classical acetyltetralins and acetylindans is Traseolide® (**11**).[34] The tetramethylisopropylindan (**55**) is prepared from **54** and is finally acetylated to **11**.

The homologue **56** of the well known Galaxolide® (**10**) has also been described[35] as sweet, musky, caramel. The synthesis is analogous to that of Galaxolide®.

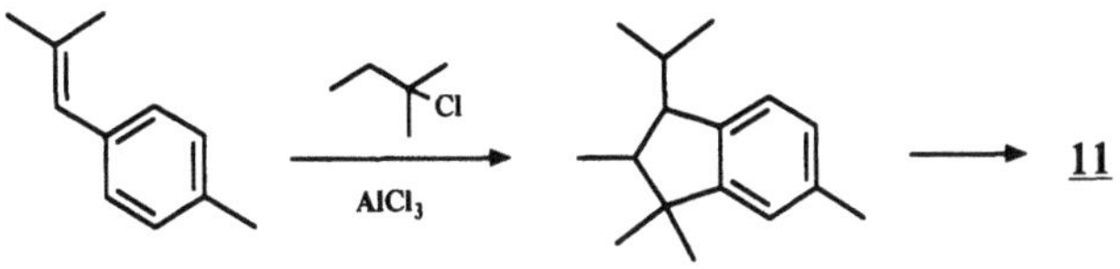

54 55

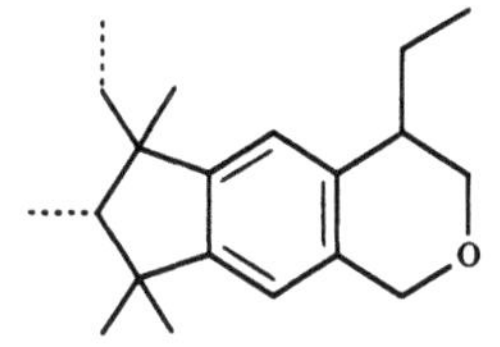

56

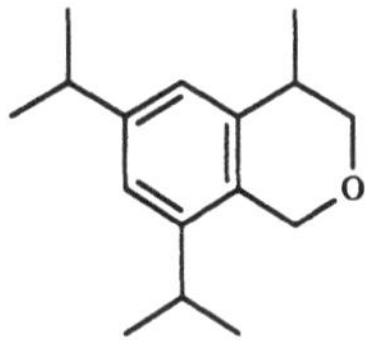

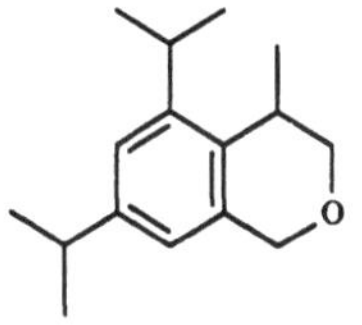

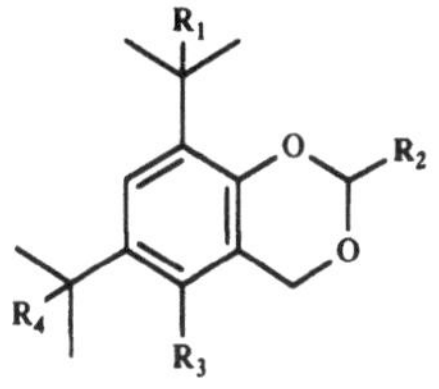

57a 57b 58

For the compound mixture **57a/b** which structurally resembles Galaxolide, a musk-like, woody and amber-like odor is claimed.[36]

Another invention is related to new musk-like aroma chemicals of type **58**, in which R_1, R_2, R_3 and R_4 stand for hydrogen or a lower alkyl group with the proviso that the molecule does not have more than 18 carbon atoms.[37] The strongest musky note is claimed for $R_1 = R_4 =$ methyl and $R_2 = R_3 =$ hydrogen.

Interestingly, the aldehyde **59** was patented only very recently,[38] and is described as a strong musk with a woody note. Its synthesis is most easily achieved through oxidation of the hydrocarbon **53**. A few months later, a patent application was filed for the new tetralin derivative **60** exhibiting a musk-like odor.[39]

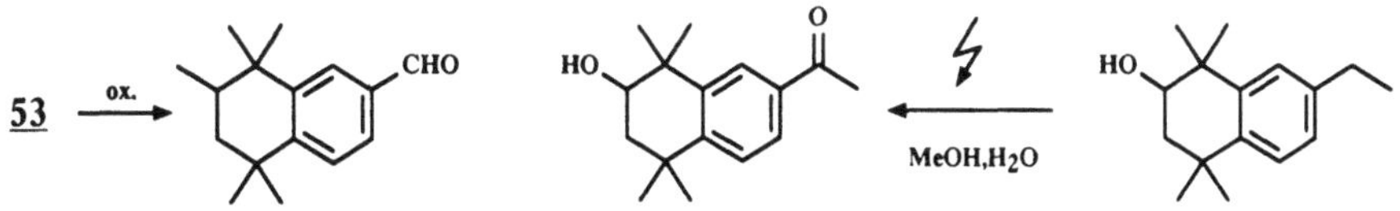

59 60

A completely new kind of structure in the field of musk chemicals has been introduced with the tricyclic oxirane **62**,[40] which exhibits a musky, slightly woody and fruity scent. Its preparation is performed by epoxidation of **61**, the

Birch reduction product of hexamethyltetralin (**53**). The same starting material, i.e. **61**, can serve for the synthesis of another new compound, the ketone **63**, with an odor profile of musky, slightly sandalwood, fruity and animal character.[41]

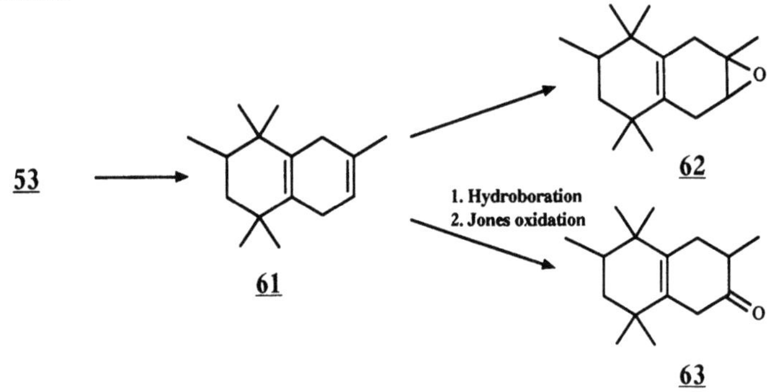

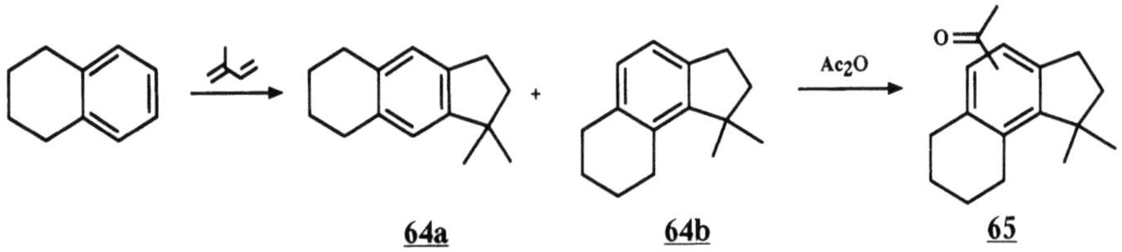

A novel tricyclic compound, **65**, has a noble musk fragrance, excellent in both quality and strength.[42] It can be manufactured at low cost from tetralin and isoprene via the hydrocarbon mixture **64a/b**.

Finally, Fehr *et al.*[43] describe, in a very recent and stimulating paper, a number of substances surprisingly omitted from the literature, but nevertheless of a certain interest in view of their musky character. The compound with the strongest musk odor claimed is the aldehyde (**69**), which was prepared by oxidation of the methyl group in **68**, the latter being the main cyclization product of an intermediate alcohol prepared from the ketone **67**.

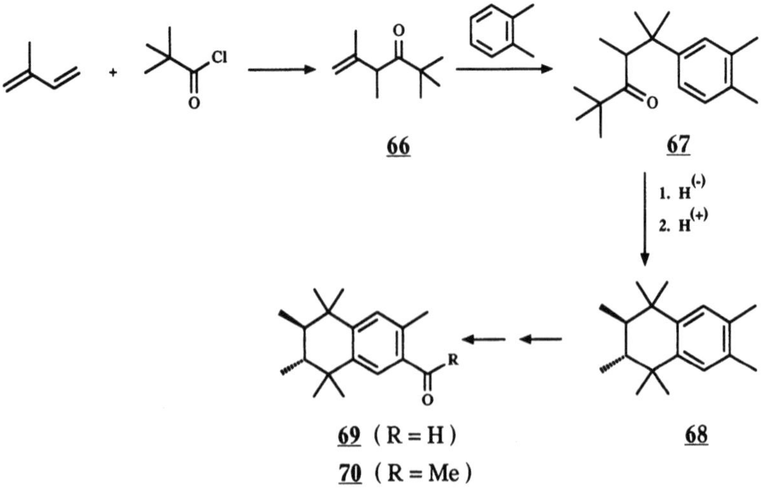

Amber Chemicals

As is well known for other scents also, the amber odor is a complex event. It comprises shades and tonalities of woody, musky and other directions, the differentiation of which is certainly a question of personal preference and capacity. However, an unequivocal opinion exists about the outstanding importance of amber chemicals in the creation of fine fragrances.

One of the classics of amber chemicals, (–)-Ambrox [(–)-**73**], is also the subject of a fascinating chapter of natural product chemistry. A comprehensive review[44] of this field appeared in 1982 and work on structure–activity relationship was also published by Firmenich chemists.[45–48]

A very valuable paper by Vlad[49] reviews the chemistry of derivatives of labdane diterpenoids up to 1981, and more recently a book has been written by the same group[50] (see also the short review by Cambie on ambergris substitutes[51]).

The chemicals (–)-ambrox and more recently (±)-ambrox are surely among the synthetically most complicated ingredients of the fragrance industry. This is also reflected in the price of (–)-ambrox of around $1000/kg.

Lately several new routes to (–)-ambrox and (±)-ambrox have been described. The interest in (±)-ambrox, which is normally accompanied by some of its isomers, is quite recent and motivated by the high price and limited availability of the natural starting material for the synthesis of (–)-Ambrox® (Firmenich), also called Ambroxane® (Henkel).

An attractive transformation of sclareol ((–)-**71**) into (–)-ambrox ((–)-**72**) by means of a copper-catalyzed decomposition of the hydroperoxide **72** has been published.[52,53] This transformation can be interpreted by the intermediacy of **74**.

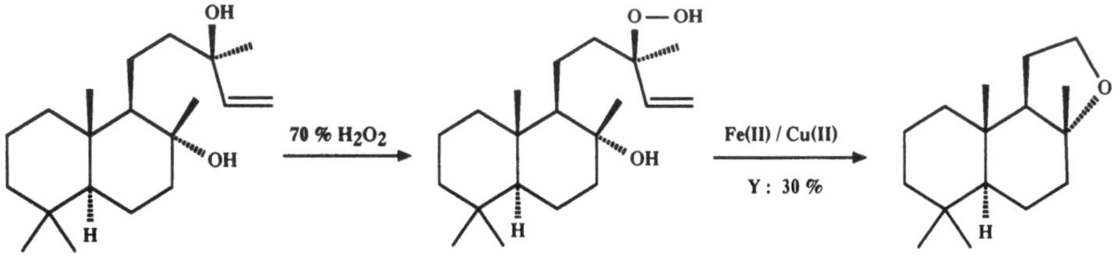

(-)- 71 72 (-)- 73

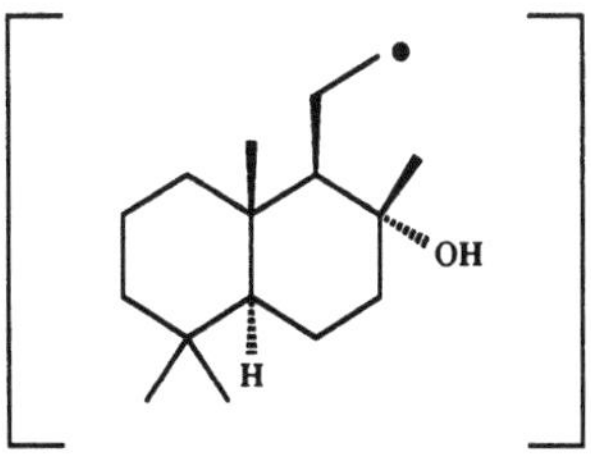

74

A related strategy is described in Refs 54 and 55 where, among others, sclareol monoacetate (**75**) has been treated with lead tetra-acetate and iodine to give **76**, which upon alkaline hydrolysis yielded (−)-**73** in 20% overall yield from **75**.

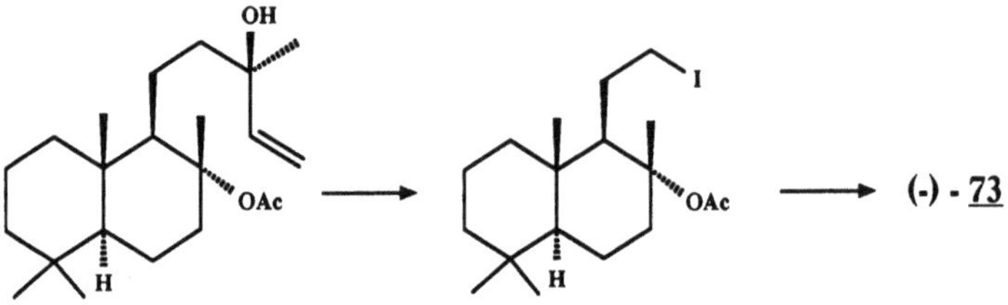

75 76

Ketone **77**, which is available from manool (**91**)[56] or from eperuic acid,[57] has been converted to a mixture of the aldehyde **78** and the acid **79**, which are claimed to be practical intermediates in the synthesis of (−)-ambrox [(−)-**73**].[58]

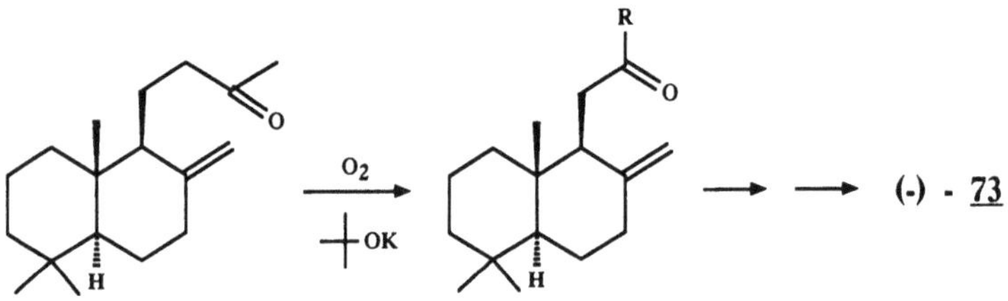

77

78 (R = H)

79 (R = OH)

Waegell and co-workers[59] describe a new variant of the transformation of sclareol [(−)-**71**] and 13-epi-sclareol to (−)-ambrox, the key step of which is the $Pd(OAc)_2$-catalyzed formation of the diene **80**.

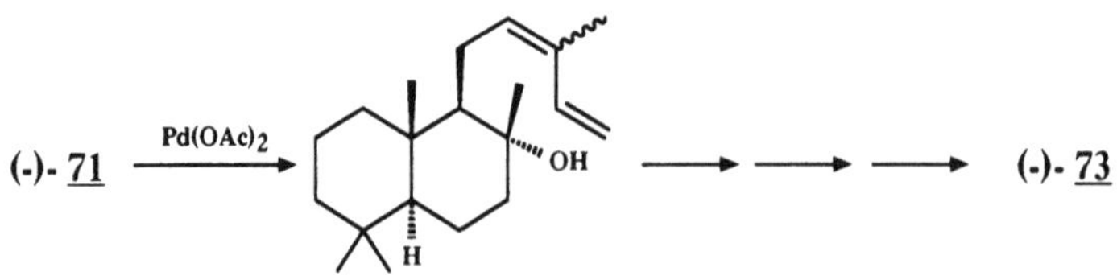

80

Cambie & Palmer[60] converted podocarpic acid into a series of nor- and bisnor-(−)-ambrox and nor- and bisnor-isoambrox (cf. **85**) and compared their odor. Differences in the intensity were observed, with some musky or minty notes beside the ambergris odor.

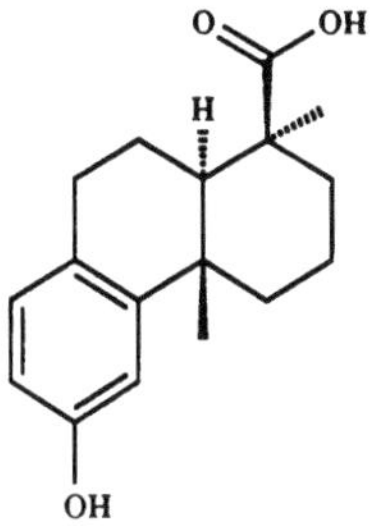

Podocarpic acid

In two other more academic syntheses both (–)-drimenol[61] and (–)-abietic acid[62] have been converted to (–)-ambrox [(–)-**73**].

A spectacular new synthesis of ambrox, of yet unknown configuration, consists of the enzymatic conversion of *E,E*-homofarnesol by the squalene–hopene cyclase from *Bacillus acidocaldarius*.[63]

Besides the above-described routes to (–)-**73** itself, a number of publications deal with the preparation of *rac*-sclareolide (**87**) and consequently of *rac*-ambrox ((±)-**73**) and of their respective diastereomers.

Starting with the easily available *β*-ketoester **81**, *rac*-ambrox (**73**) and *rac*-isoambrox (**85**) were prepared.[64–67] The ketoester **81** is transformed

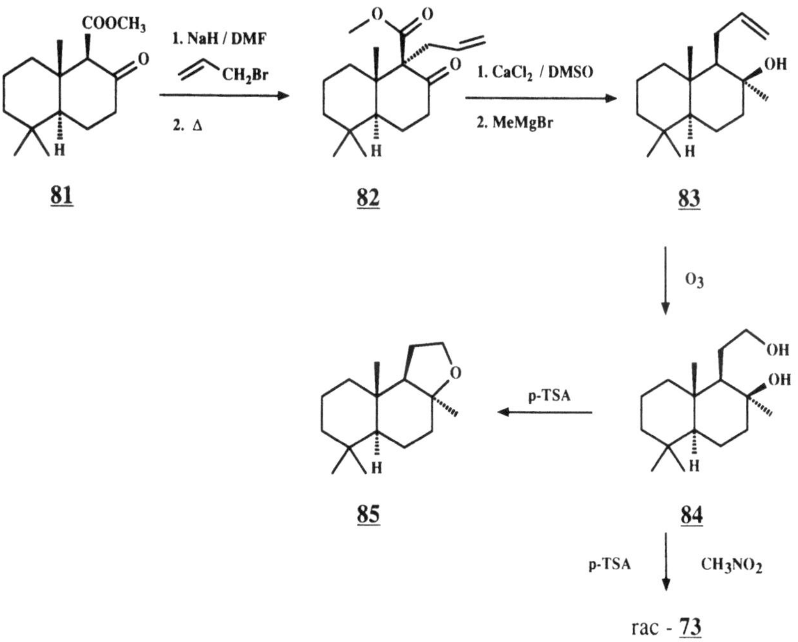

through *O*-allylation and rearranged into **82**, which is decarboxylated. The resulting ketone is treated with methylmagnesium halide to yield **83**, which is ozonized; the resulting hydroxyaldehyde is reduced to the diol **84**, which is cyclized by *p*-toluenesulfonic acid (TSA) to isoambrox (**85**) and by TosOH in CH_3NO_2 to *rac*-**73**. In this work it is stated that there is no difference in threshold value between (−)-**73** and (±)-**73** (see also Ref. 48).

The classical work of Lucius[68–70] has been revived in a few recent papers. Homofarnesenic acid (**86**) has been cyclized to sclareolide (**87**) and diastereomers thereof.[71–73] Monocyclohomofarnesenic acid (**88**) has been prepared either by Knoevenagel condensation from α-C_{14}-aldehyde[74] or by Wittig reaction from dihydro-β-ionone[75] and cyclized to **87** and isomers under different conditions.[76–79]

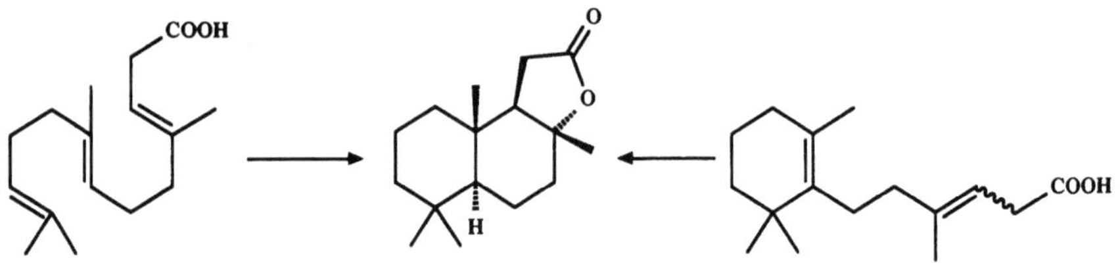

86 87 88

(+)-Ambreinolide (**89**) is an important lactone; on one hand it is an oxidation product of ambrein, and on the other it is the starting material for the very powerful amber compound **90**.[44] (For some interesting structure–activity findings concerning **90** and its diastereomers and homologs, see Ref. 80.)

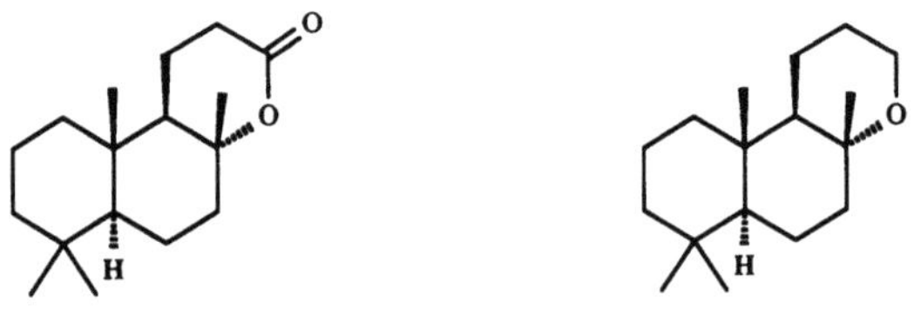

89 90

Ambreinolide (**89**) has been prepared repeatedly, but to our knowledge neither **89** nor the tetrahydropyranyl ether **90** is much used in perfumery.

Cambie & Palmer[81] transformed manool (**91**) in a multistep synthesis into **89**.

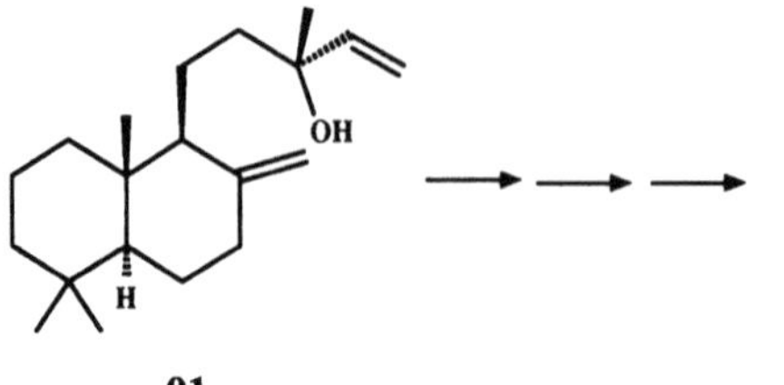

89

91

Compound (±)-**89** was also prepared from the ketone **92** by means of a high-pressure Diels–Alder reaction of the diene **93** and mesoxalate.[82]

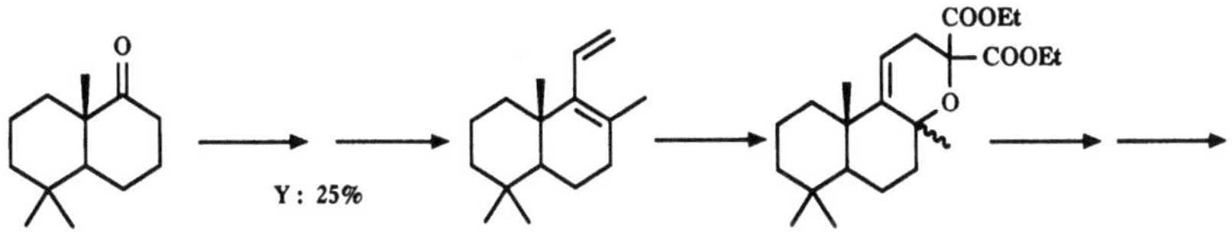

89

92 93 94

(65 : 35)

Abietic acid (**95**) served as a starting material for the synthesis of (+)-**89**,[83] as it did before for that of (−)-ambrox.[62]

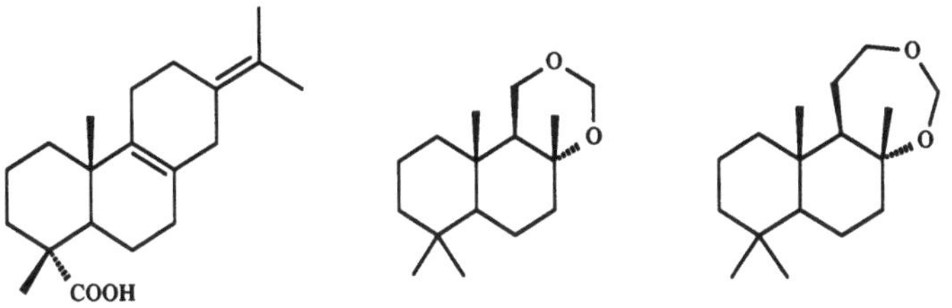

95 96 97

The acetals **96**[81] and **97**,[84] the oxa-analogs of the tetrahydropyran **89**, have been reported to be powerful amber compounds. Drimenol served once again as the starting material for the synthesis of **96**.[85]

We now turn to some other chemicals with distinct amber notes in their odor profiles. In the 1970s, a fragrant substance, Ambrate, stirred the interest of perfumers worldwide, but unfortunately—for toxicological reasons—it was withdrawn from the market and replaced by Dihydro Ambrate®, which constitutes a safe product (cf. Chapter 14, p. 414).

To just the same class of chemicals belongs Metambrate® (**98**) with an amber, woody (tobacco-like) character.[86] Metambrate® (**98**) is a mixture of isomers. The two main isomers are **98a** and **98b**, **98b** being more woody and ambergris-like, as well as about ten times stronger than the slightly camphoraceous and more earthy, powdery **98a**. It is possible to direct the synthesis in such a way that **98b** is preponderant in the mixture.

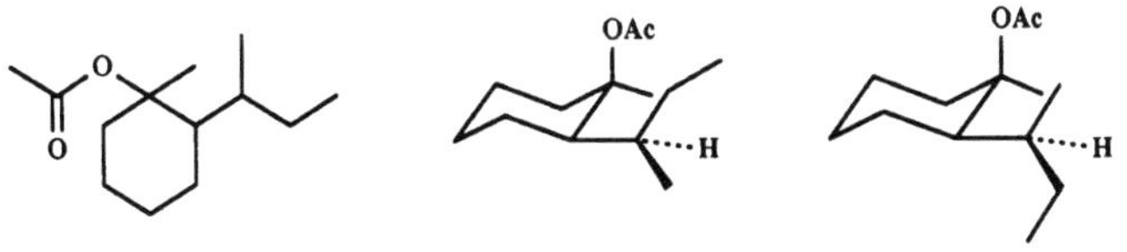

98 98a 98b

Oxyoctaline formate® (**100**) is even more on the woody side with a distinct amber note. Its synthesis from the so-called β-C_{14}-aldehyde, an intermediate in the technical synthesis of vitamin A,[87] is performed in two steps: selective hydrogenation to dihydro-β-C_{14}-aldehyde (**99**), followed by cyclization to **100**.[88]

99 100

α-Ambrinol (**103**) is an ingredient of ambergris.[89,90] Its odor and the structure–activity relationships of its isomers and analogs have been discussed at length by Ohloff.[44] There has been only relatively little activity towards its synthesis.

The first synthesis which seems practical starts from β-ionone (**101**) which, in low yield, produces the valuable intermediate **102** upon pyrolysis at 430°C. The latter can be hydrogenated to α-ambrinol (**103**).[91] The cyclization **101**→**102** is an intriguing step and could be interpreted as passing through the radical intermediate **102b** which could be formed again from **101b**, a (1,5)H shift-isomer of **101**.

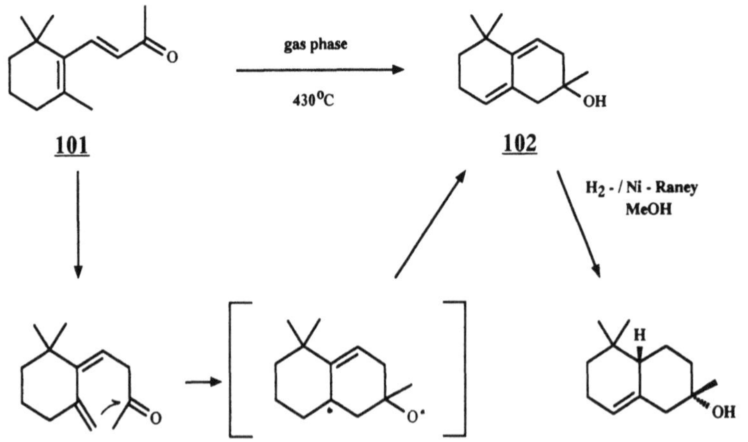

101b 102b 103

Another approach is the cyclization of dihydro-γ-ionone (**105**) which has been prepared from dihydro-α-ionone (**104**).[92,93]

Dihydro-γ-ionone (**105**) has also been prepared by a different route by Matsui and co-workers.[94] The key step in their synthesis is the introduction of the 'γ'-double bond by means of a Claisen rearrangement of an allylic alcohol

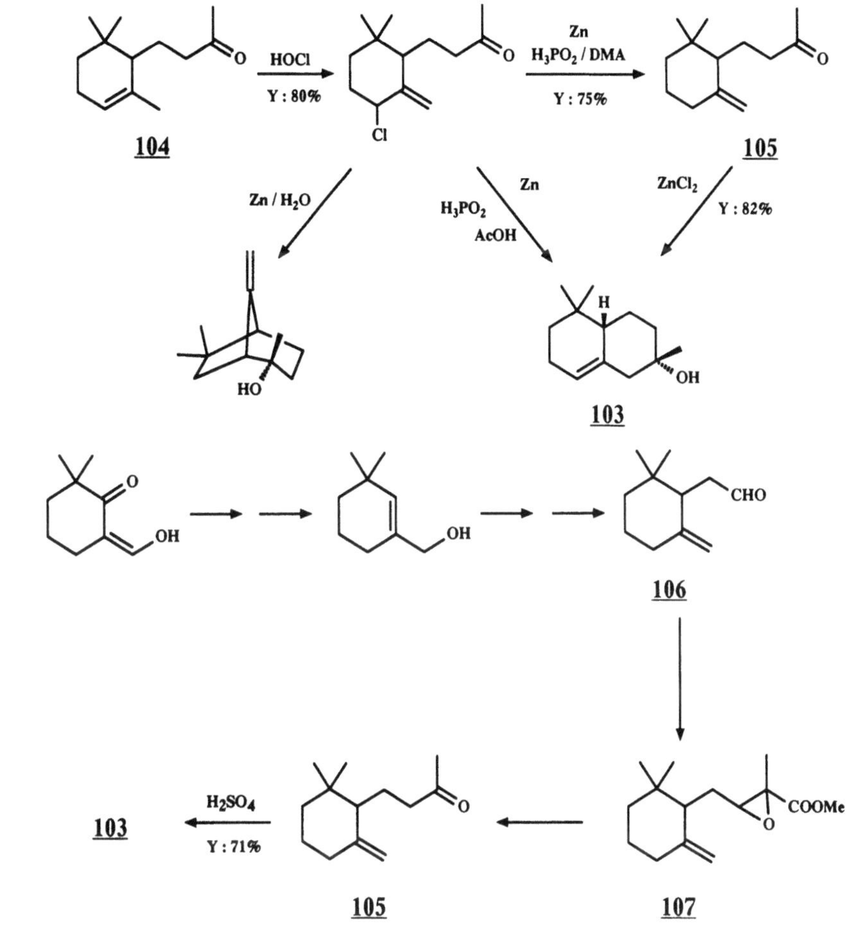

to the aldehyde **106**. Darzens reaction to a glycidic ester (**107**) and subsequent fragmentation furnished **105**, which was cyclized to α-ambrinol (**103**). Yet another route goes through the diene **108**, which after 1,4-functionalization to **109** and hydride reduction yielded α-ambrinol (**103**).[95]

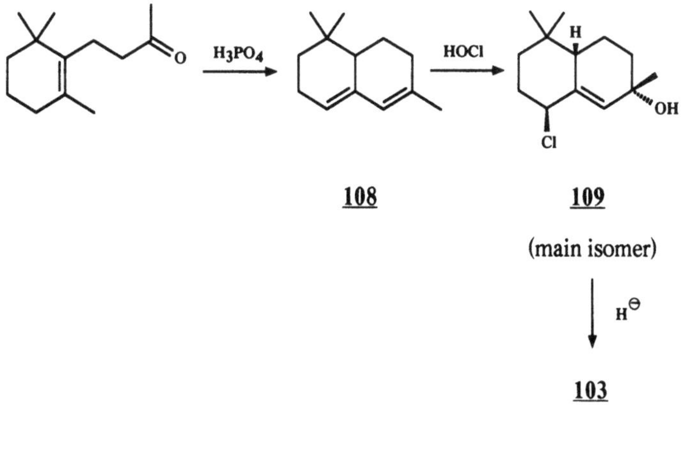

δ-Ambrinol was prepared by means of an intramolecular Diels–Alder reaction.[96] Grignard reaction of pentadienylmagnesium chloride on methylheptenone, with subsequent 1,3-rearrangement of the lithium salt of the resulting alcohol, yielded **110**, which underwent the intramolecular [4 + 2] addition furnishing four isomers. According to the authors, the isomers **111** and **112** display a strong ambergris odor and therefore represent just another interesting facet of substances smelling amber-like.

This wealth of compounds exhibiting the highly demanded amber odor characteristics is difficult to classify in a scheme of structure–activity relationships. Besides the so-called and well-known 'triaxial rule',[44] a more recent publication[97] describes another possible way to approach such problems and therefore underlines the on-going interest in this field.

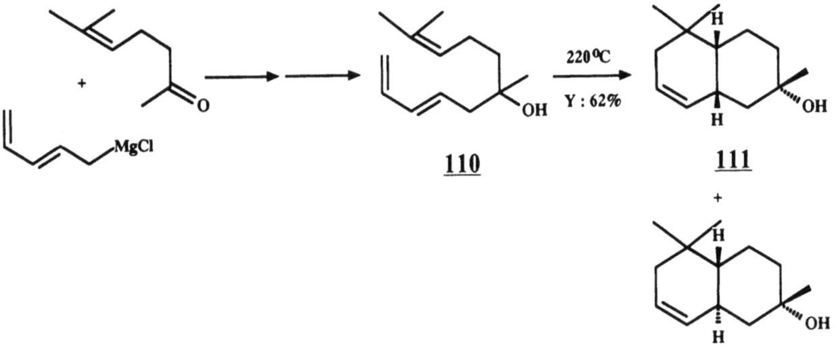

110 111 112

REFERENCES

Musk

1. Mookherjee, B. D. & Wilson, R. A. In *Fragrance Chemistry*, ed. E. T. Theimer. Academic Press, New York, 1982, p. 434.
2. Wood, T. F. In *Fragrance Chemistry*, ed. E. T. Theimer. Academic Press, New York, 1982, p. 495.
3. Croteau, R., ed., *Fragrance and Flavor Substances*, Proc. 2nd Int. Haarmann & Reimer Symposium, 1980 D & PS Verlag, 1980; Bauer, K. A. & Körber, A. K., p. 129; Körber, A. K. & Bauer, K. A., pp. 137, 155; Hagena, D. J. & Bauer, K. A., p. 145.
4. Abe, S. Paper presented at IXth Int. Congress of Essential Oils, Singapore, 1983, book 5, p. 34.
5. Bauer, A., *Ber. Dtsch Chem. Ges.*, **24** (1891) 2836.
6. German Patent 47 599 (priority Germany 3 July 1888).
7. Wood, T. F. In *Fragrance Chemistry*, ed. E. T. Theimer. Academic Press, New York, 1982, p. 509.
8. Bernard, T., Perineau, F., Bravo, R. & Delmas, M., *Parf. Cosm. Arômes*, **83** (Oct.–Nov. 1988) 65.
9. Carpenter, M. S. & Easter, W. M., Jr, US Patent 2 450 879, (priority US 12 Oct. 1948) to Givaudan.
10. Carpenter, M. S., Easter, W. M., Jr & Wood, T. F., *J. Org. Chem.*, **16** (1951) 591.
11. Wood, T. F., Chemistry of aromatic musks I–IX. *Givaudanian* (1968–1970).

12. Barbetta, L., Trowbridge, T. & Eldib, I. A., *Perf. & Flav.*, **13**, no. 3, (1988) 60.
13. Agarwal, S. C. & Majee, R. N., *Ind. Parf.*, **25**(3 + 4) (1981) 20.
14. Fehr, C., Ohloff, G. & Büchi, G., *Helv. Chim. Acta,* **62** (1979) 2655.
15. Eschenmoser, A., Felix, D. & Ohloff, G., *Helv. Chim. Acta,* **50** (1967) 708.
16. Santelli-Rouvier, C. & Santelli, M., *Synthesis* (1983) 429.
17. Baumann, M., Hoffmann, W. & Müller, N., *Tetrahedron Lett.* (1976) 3583.
18. US Patent 4 387 250 (priority CH 16 Jan. 1980) to Firmenich.
19. US Patent 4 328 383 (priority CH 18 Jan. 1980) to Firmenich.
20. German Patent 3 226 684 (priority CH 17 July 1981) to Firmenich.
21. US Patent 4 351 977 (priority CH 31 Mar. 1980) to Firmenich.
22. Fehr, C., *Helv. Chim. Acta,* **66** (1983) 2519.
23. US Patent 4 454 351 (priority CH 21 Apr. 1981) to Firmenich (see also German Patent 3 214 546).
24. Fehr, C., *Helv. Chim. Acta,* **66** (1983) 2512.
25. US Patent 4 608 193 (priority CH 6 Dec. 1982) to Firmenich.
26. Wender, P. A., Holt, D. A. & Sieburth, S. M., *J. Am. Chem. Soc.,* **105** (1983) 3348.
27. Warwel, S., Bachem, H., Deckers, B., Doering, N., Kätker, H. & Rose, E., *Seifen, Oele, Fette, Wachse,* **115** (1989) 538.
28. Villemin, D., *Tetrahedron Lett.,* **21** (1980) 1715.
29. Bachem, H., Ph.D. thesis, TH Aachen, 1984.
30. Okino, H., Taoka, A. & Nemura, N., *Developments in Food Science,* Vol. 18. *Flavors and Fragrances: A World Perspective.* Proceedings of the 10th International Congress of Essential Oils, Fragrances and Flavors, Washington, DC, U.S.A., 16–20 November 1986, ed. B. M. Lawrence, B. D. Mookherjee and B. J. Willis, Elsevier, Amsterdam 1988, pp. 753–60.
31. Leftcoat, R. & Willis, B. J., *Developments in Food Science,* Vol. 18, *Flavors and Fragrances: A World Perspective.* Proceedings of the 10th International Congress of Essential Oils, Fragrances and Flavors, Washington, DC, U.S.A., 16–20 November 1986, ed. B. M. Lawrence, B. D. Mookherjee and B. J. Willis, Elsevier, Amsterdam, 1988 pp. 743–51.
32. Japanese Patent 84 042 076 (priority JA 4 June 1981) to Asahi Chemical Ind.
33. Banks, R. L., Banasiak, D. S., Hudson, P. S. & Norell, J. R., *J. Mol. Catal.,* **15** (1982) 35.
34. US Patent 4 352 748 (priority NL 25 Feb. 1977) to Naarden International.
35. US Patent 4 608 193 (priority US 13 May 1985) to IFF.
36. Japanese Patent Appln 01 160 97 (priority JA 16 Dec. 1987) filed by KAO Corp.
37. US Patent 4 880 775 (priority US 1 Feb. 1987) to BASF-K & F.
38. European Patent Appln (priority CH 29 July 1987) filed by Givaudan.
39. European Patent Appln 339 521 (priority DE 27 Apr. 1988) filed by BASF.
40. US Patent 4 634 548 (priority CH 13 Jan. 1983) to Givaudan.
41. European Patent 195 975 (priority CH 25 Mar. 1985) to Givaudan.
42. US Patent 4 885 274 (priority JA 23 Nov. 1987) to Takasago.
43. Fehr, C., Galindo, J., Haubrichs, R. & Perret, R., *Helv. Chim. Acta,* **72** (1983) 1537.

Amber

44. Ohloff, G. In *Fragrance Chemistry,* ed. E. T. Theimer. Academic Press, New York, 1982, p. 535.
45. Winter, B., *Helv. Chim. Acta,* **72** (1989) 1278.
46. Vial, C., Thommen, W. & Näf, F., *Helv. Chim. Acta,* **72** (1989) 1390.
47. Ohloff, G. & Giersch, W., *Croatian Chem. Acta,* **58** (1985) 491.
48. Ohloff, G., Giersch, W., Pickenhagen, W., Furrer, A. & Frei, B., *Helv. Chim. Acta,* **68** (1982) 2022, and literature cited therein.

49. Vlad, P. F., *Russ. Chem. Rev.*, **51** (1982) 644.
50. Vlad, P. F. & Kolga, M. N., *Synthesis and Application of Odorous Compounds from Labdane Diterpenes*. Shtinka Publishers, Kishinew, USSR, 1988 (in Russian).
51. Cambie, R. C., *Chem. New Zealand* (1987) 31.
52. German Patent 3 610 063 (priority CH 26 Mar. 1985) to Firmenich.
53. Decorzant, R., Vial, C., Näf, F. & Whitesides, G. M., *Tetrahedron*, **43** (1987) 1871.
54. European Patent Appln 296 564 (priority US 23 June 1987) filed by BASF K & F. Corp., USA.
55. Christenson, P. A., *Tetrahedron*, **44** (1988) 1925.
56. Demole, E. & Wuest, H., *Helv. Chim. Acta*, **50** (1967) 1314.
57. Dey, A. K. & Wolf, H. R., *Helv. Chim. Acta*, **61** (1978) 1004.
58. US Patent 4 701 543 (priority CH 12 Aug. 1985) to Firmenich.
59. Coste-Manière, I. C., Zahra, J. P. & Waegell, B., *Tetrahedron Lett.*, **29** (1988) 1017.
60. Cambie, R. C. & Palmer, B. D., *Aust. J. Chem.*, **35** (1982) 601.
61. Gonzales-Sierra, M., Rúveda, E. A., López, J. T. & Cortés, M. J., *Heterocycles*, **26** (1987) 2801.
62. Koyama, H., Kaku, Y. & Ohno, M., *Tetrahedron Lett.*, **28** (1987) 2863.
63. Neumann, S. & Simon, H., *Biol. Chem. Hoppe-Seyler*, **367** (1986) 723.
64. US Patent 4 613 710 (priority US 8 Aug. 1984) to Firmenich.
65. US Patent 4 633 011 (priority US 10 Mar. 1986) to Firmenich.
66. US Patent 4 677 233 (priority US 23 June 1986) to Firmenich.
67. Büchi, G. & Wuest, H., *Helv. Chim. Acta*, **72** (1989) 996.
68. Lucius, G., *Ber. Dtsch Chem. Ges.*, **93** (1960) 2663.
69. Lucius, G., *Angew. Chem.*, **68** (1956) 247.
70. Lucius, G., *Arch. Pharm.*, **291** (1958) 57.
71. US Patent 4 503 240 (priority DE 28 Oct. 1982) to Consortium für Elektrochemische Industrie/Wacker.
72. Japanese Patent 86 035 193 (priority JA 9 Dec. 1983) to Japan Tobacco & Salt Pub. Corporation.
73. Emel'yanov, M. M., Lazanova, A. V., Moiseenkov, A. M., Smit, V. A. & Semenorskii, A. V., *Izv. Akad. Nauk. SSSR. Ser. Khim.* (1982) 2788.
74. Japanese Patent Appln 60 081 164 (priority JA 11 Oct. 1983) filed by Hasegawa.
75. Japanese Patent 84 014 475 (priority JA 6 Mar. 1981) to Japan Tobacco & Salt Pub. Corporation.
76. Saito, A., Matsushita, H., Tsujino, Y. & Kaneko, H., *Chem. Lett.* (1981) 757.
77. Saito, A., Matsushita, H. & Kaneko, H., *Chem. Lett.* (1983) 729; (1984) 591.
78. European Patent 165 458 (priority JA 14 May 1984) to Hasegawa.
79. Kawanobe, T., Kogami, K. & Matsui, M., *Agric. Biol. Chem.*, **50** (1986) 1475.
80. Ohloff, G., Vial, C., Demole, E., Enggist, P., Giersch, W., Jégon, E., Caruso, A. J., Polonsky, J. & Lederer, E., *Helv. Chim. Acta*, **69** (1986) 163.
81. Cambie, R. C. & Palmer, B. D., *Aust. J. Chem.*, **34** (1981) 1265.
82. Daniewski, W. M., Kubak, E. & Jurczak, J., *J. Org. Chem.*, **50** (1985) 3963.
83. Abad, A., Agulló, C., Arnó, M., Cuñat, A. C. & Zaragozá, J., *J. Org. Chem.*, **54** (1989) 5123.
84. Cambie, R. C., Joblin, K. N. & Preston, A. F., *Aust. J. Chem.*, **25** (1972) 1767.
85. Maturana, H., Sierra, J., López, J. & Cortés, M., *Synth. Comm.*, **14** (1984) 661.
86. US Patent 4 277 618 (priority CH 30 Oct. 1978), to Givaudan.
87. Isler, O., Huber, W., Ronco, A. & Kofler, M., *Helv. Chim. Acta*, **30** (1947) 1911.
88. Saucy, G., Ireland, R. E., Bordner, J. & Dickerson, R. E., *J. Org. Chem.*, **36** (1971) 1195.
89. Jegon, E., Polonsky, J., Lederer, E., Schulte-Elte, K., Egger, B. & Ohloff, G., *Nouv. J. Chim.*, **1** (1977) 529.

90. Mookherjee, B. D. & Patel, R. R. Paper presented at VIIth Int. Congress of Essential Oils, Kyoto, 1977.
91. Strickler, H., US Patent 4 163 866 (priority CH 28 July 1976) to Firmenich.
92. US Patent 4 341 908 (priority US 20 June 1980) to Fritzsche, Dodge & Olcott.
93. Christenson, P. A., Eilerman, R. G. & Willis, B. J. Paper presented at IXth Int. Congress of Essential Oils, Singapore, 1983, book 4, p. 101.
94. Kawanobe, T., Kogami, K., Hayashi, K. & Matsui, M. Paper presented at VIIIth Int. Congress of Essential Oils, Singapore, 1983, book 4, p. 182.
95. European Patent Appln 89 810 451.8 (priority CH 13 June 1989) to Givaudan.
96. Christenson, P. A., Willis, B. J., Wehrli, F. W. & Wehrli, S., *J. Org. Chem.*, **47** (1982) 4786.
97. Bersuker, I. B., Dimoglo, A. S., Gorbachov, M. Yu., Koltsa, M. N. & Vlad, P. F., *Nouv. J. Chim.*, **9** (1985) 211.

3 CHEMICALS WITH SANDALWOOD AND OTHER WOODY ASPECTS

Substances from Sandalwood Oil

The warm, sweet, woody fragrance of sandalwood oil is a precious tool in perfumery. Accordingly its chemistry has been studied in depth for a long time. The main components of the alcoholic part (~90% of the oil) are(+)-α-santalol (**1**) and (−)-β-santalol (**2**), whereas in the hydrocarbon part (5–10% of the oil) (+)-α-santalene (**3**), β-santalene (**4**) and epi-β-santalene (**5**) prevail. For the history of structure elucidation and synthesis of these compounds up to 1981 the reader can refer to excellent reviews.[1,2]

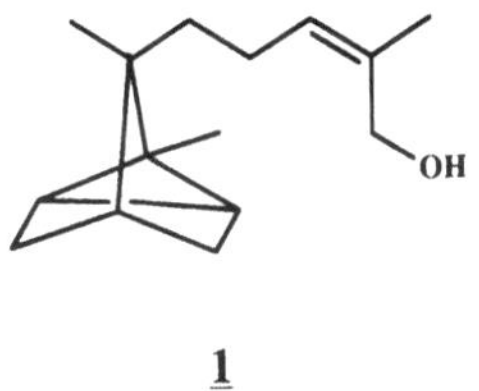

1

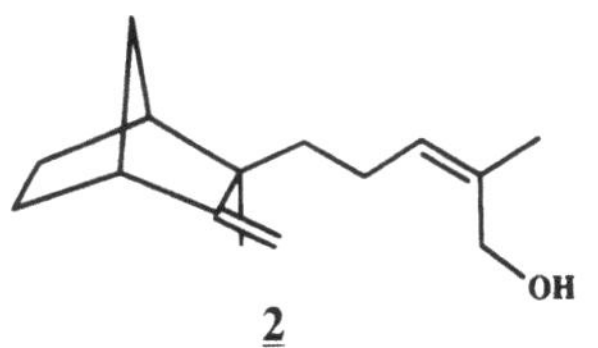

2

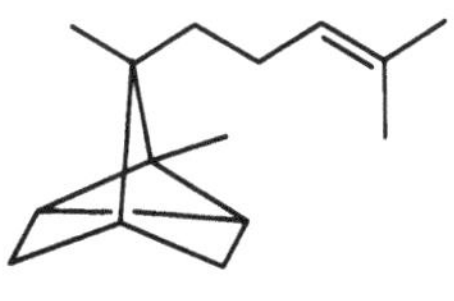

3

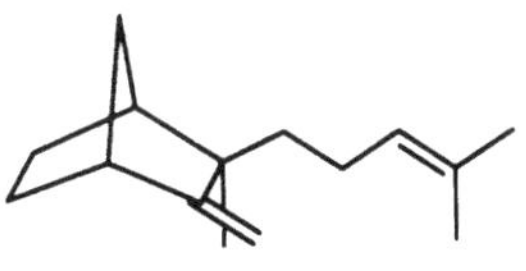

4

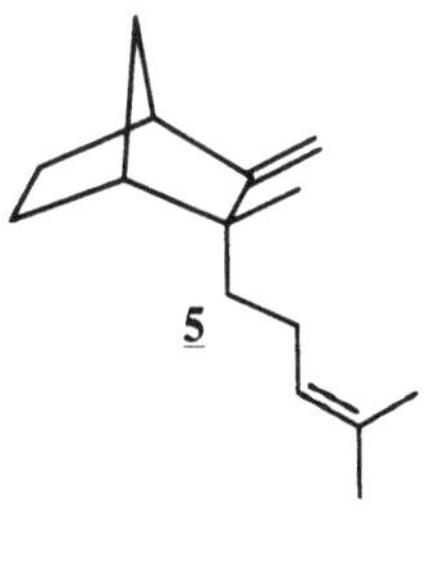

5

Relatively recently, (+)-epi-β-santalol (**6**)[3] and (*E*)-(−)-β-santalol (**7**)[4] have also been identified in sandalwood oil. Curiously the absolute configuration of the β-santalol series was only determined in 1982[5] through the transformation

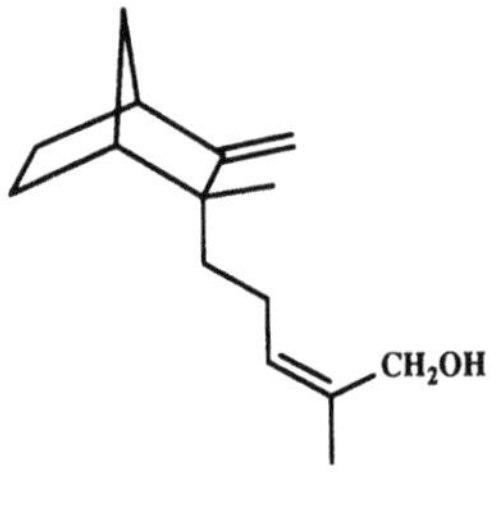

6

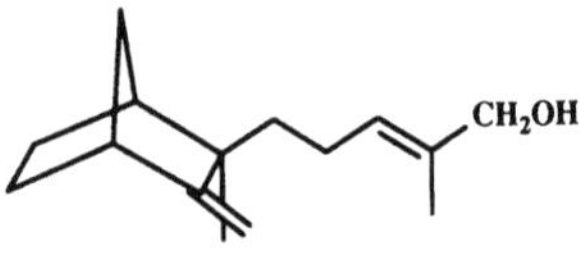

7

of (+)-α-santalyl acetate of known absolute configuration into (−)-β-santalyl acetate and (+)-epi-β-santalyl acetate.

In both α-santalol (**1**) and β-santalol (**2**) one of the synthetic problems is the stereoselective formation of a Z-allylic alcohol with a trisubstituted double bond. Besides the now well-known methodology based on electrophilic substitution of an intermediate β-oxido ylide in the Wittig reaction,[6] Still and co-workers[7] described a more direct, highly Z-selective Wittig reaction of α-alkoxyketones. With this method, α-santalol (**1**) has been synthesized from the Wittig ylide **8** in 85% yield and with 99% stereoisomer purity.

Another solution to this special problem has been found by Tamura *et al.* with the nucleophilic substitution of isoprene epoxide.[8,9] The lithium compound derived from the bromo compound **9** reacted in 58% yield to give α-santalol with a Z/E ratio of 88:12.

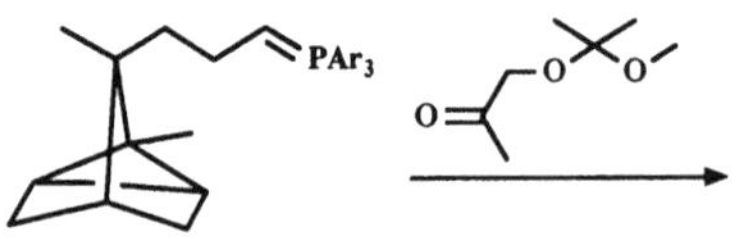

1

8

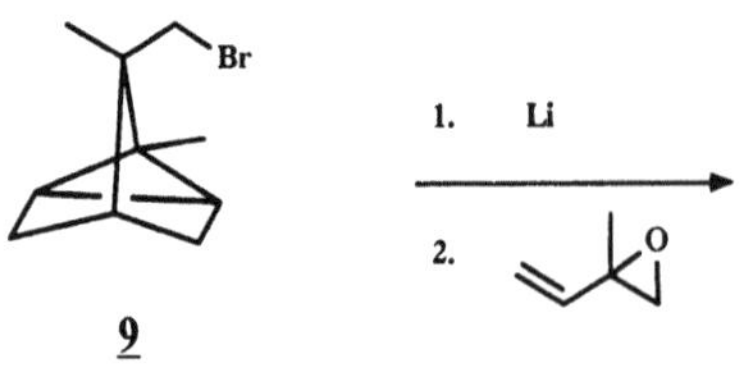

1 + (E)-isomer

9

A stereoselective synthesis of (±)-epi-β-santalene and (±)-epi-β-santalol was reported by Snowden *et al.*[10] The hetero-Diels–Alder addition product (**10**) of norbornene and acrolein was converted to the lactol **11**, which by means of a Wittig reaction and oxidation gave **12**. Alkylation and a methylene-Wittig reaction furnished (±)-**5**.

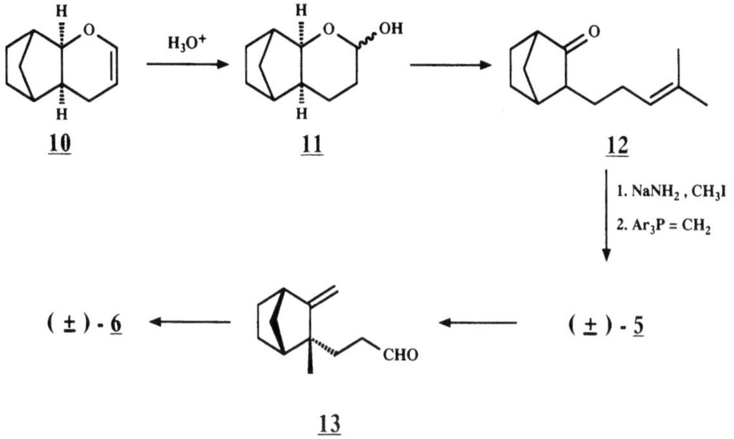

For the synthesis of epi-β-santalol the aldehyde **13** was prepared through selective oxidation of the trisubstituted double bond and transformed into (±)-**6** by the Corey–Wittig procedure.

The stereoselective approach of Sato *et al.*[11] to (±)-β-santalol [(±)-**2**] is based on the *exo*-alkylation of **14** with the completely preformed side-chain **15** to **16**. The sidechain **15** has been prepared from **17** in two steps.

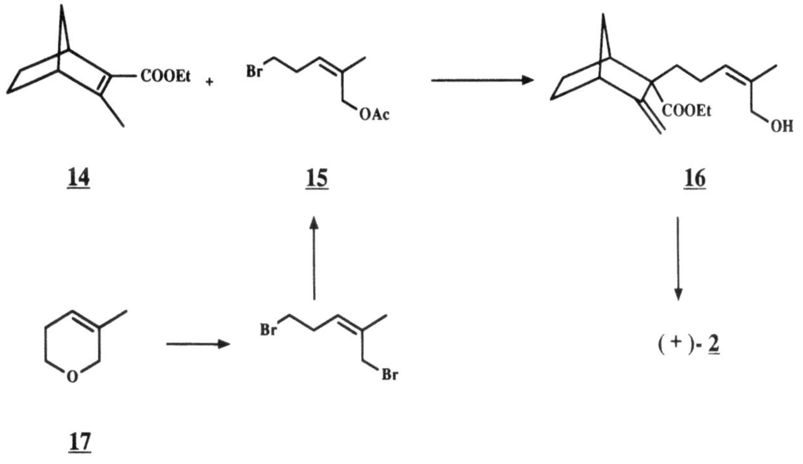

Another route to (±)-**2** was described by Solas & Wolinsky.[12] It starts with the alkylation of the easily available sultone **18** to **19**. Desulfurization was carried out in two steps through a phenylsulfone which was subsequently reduced with sodium amalgam. The product **20** was converted in several steps into (±)-**2**.

There have only been a few attempts to synthesize one of the sandal sesquiterpenes enantioselectively. In the following three examples the key step, i.e. the one that introduces the optical activity, is always a Diels–Alder addition.

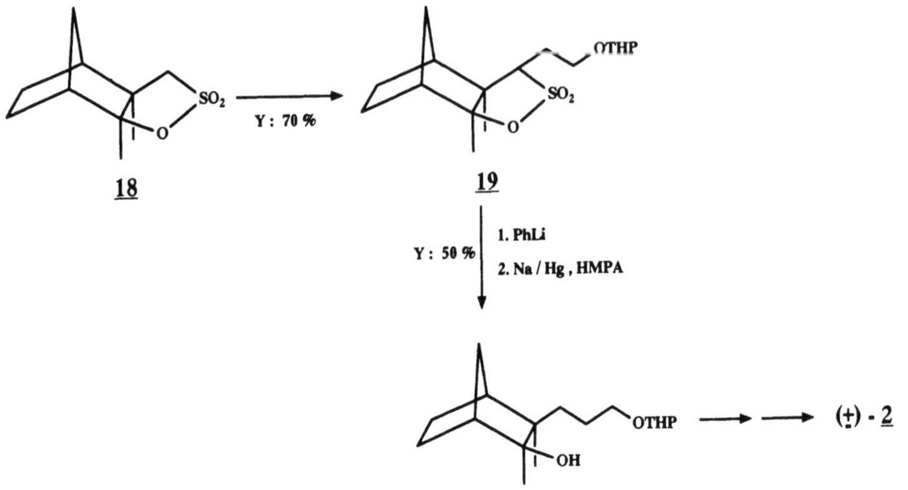

18 19 20

In the synthesis of (−)-*β*-santalene [(−)-**4**] of Oppolzer *et al.*[13,14] the allene ester **21** is added to cyclopentadiene to give the bicyclic compound **22** with very high *endo*-selectivity and 99% diastereo-*facial* differentiation. In four more steps **22** was converted into (−)-**4**.

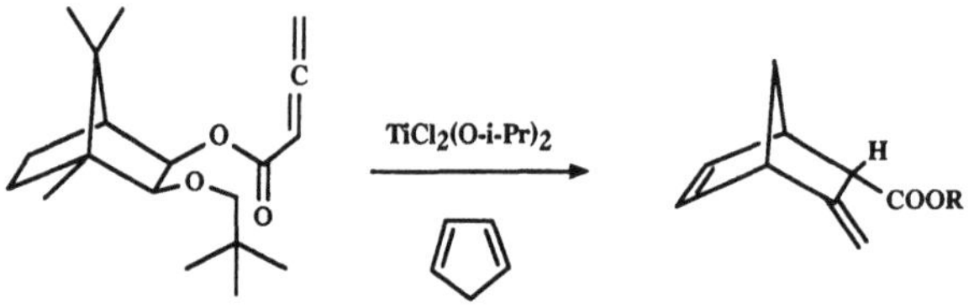

21 22

Koizumi and co-workers[15] describe the synthesis of (+)-epi-*β*-santalene [(+)-**5**]. Their dienophile is the optically active sulfoxide **23**, which was added to cyclopentadiene to yield the *endo*-isomer **24** as the main product (60%) with 94% diastereomeric excess. In three steps the ketoester **26** was prepared from **24**, which, after Wittig reaction, reduction and oxidation, yielded the aldehyde **27**. The latter was transformed into (+)-**5** in five more steps.

Helmchen *et al.*[16] succeeded in adding the chiral crotonate **28** to cyclopentadiene. The bicyclo ester **29** could be isolated in 72% yield and 97% diastereomeric excess. It has been converted to (−)-*β*-santalol [(−)-**2**] in several steps through the intermediates **30** and **31**. Helmchen's is the first and only synthesis of (−)-**2**. The aldehyde **31** also serves as an intermediate for the preparation of *β*-santalene.

In spite of all the ingenuous synthetic efforts described above, it is obvious that none of these syntheses is a practical one; i.e., technically they are mostly not feasible as well as too expensive.

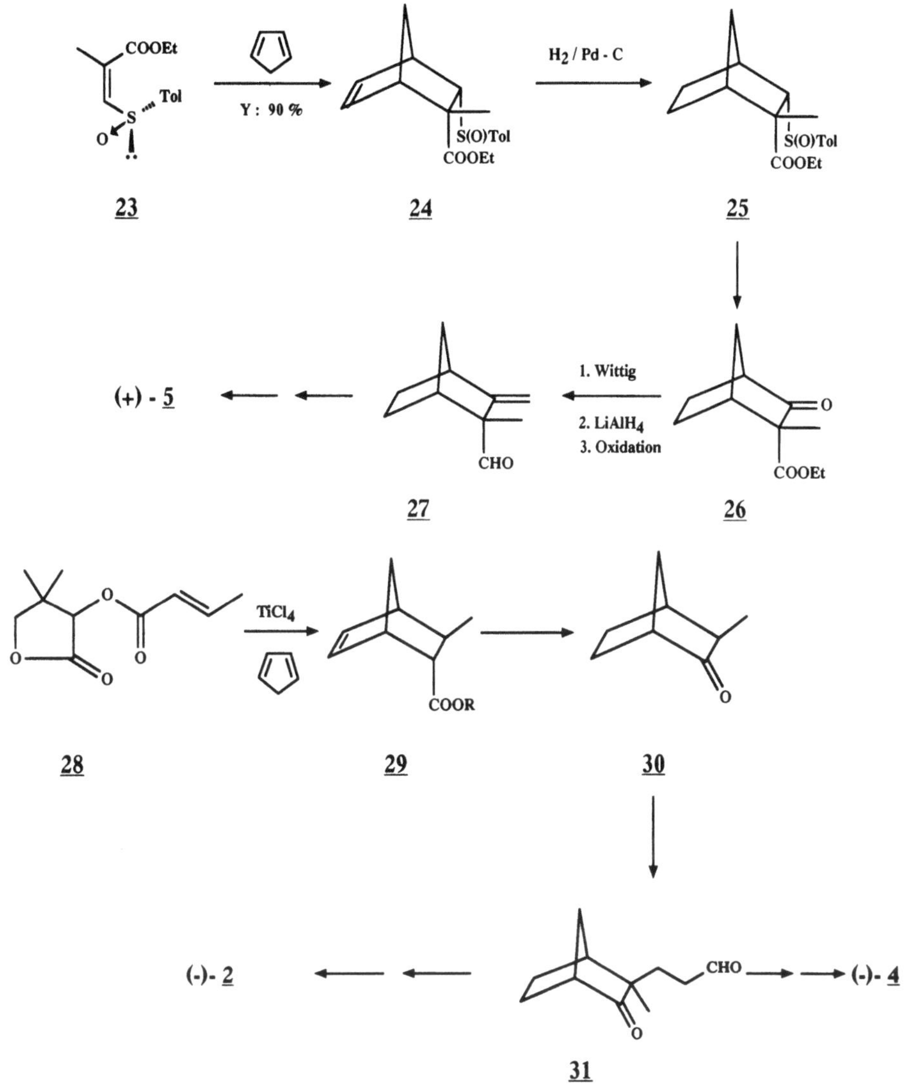

Synthetic Substitutes of Sandalwood Odor

World production of sandalwood oil is around 100 tons with a current price of about $180/kg.[17] With these limitations on the harvest and the often highly fluctuating price (it rose from $45 to $250/kg in 1974), chemists have been searching for cheaper synthetic substitutes for this important perfumery ingredient.

The first such compound was serendipitously found in Germany in the 1940s and became public knowledge after disclosure by the Allied Forces after World War II[18] (for a puristic definition of 'serendipity' see Ref. 19). It consists of the acid-catalyzed reaction of phenol or phenol derivatives like guaiacol with cheap terpene hydrocarbons like camphene. The odorless reaction products were then hydrogenated at high temperatures over Raney nickel; the complex product mixture had a strong sandalwood odor. Soon different products of this kind appeared on the market. The first was Sandela® by Givaudan around 1960, then 'Santalex-T® by Takasago, Sandel H&R® by Haarman & Reimer and Santalydol in the Soviet Union.

The interesting and far-from-trivial chemistry of the above mixtures is well reviewed in the literature.[1,20–22] It suffices to mention here that two minor ingredients in the complex mixture are responsible for the strong sandalwood odor, namely **32** and **33**[23] (see also Ref. 24).

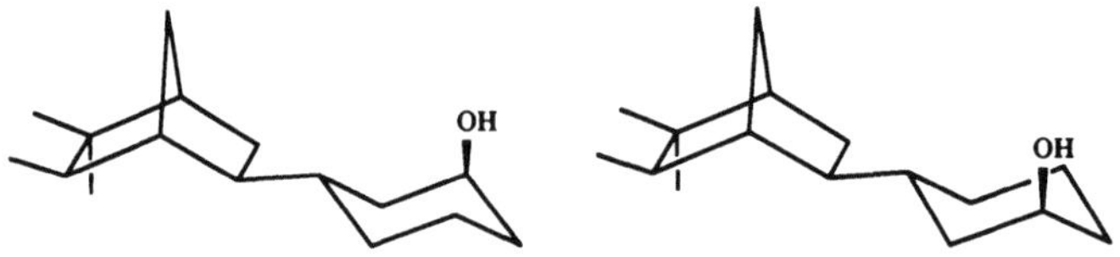

32 33

In addition to natural sandalwood oil and its ingredients on one hand, and the phenol-terpene hydrocarbon-derived sandalwood odorants on the other, a third group of chemicals with sandalwood scent has emerged recently. In fact, during the last 20 years we have witnessed tremendous development of this new class, all of which originate from campholenic aldehyde (**35**), an easily available rearrangement product of α-pinene epoxide (**34**) (for reviews see Refs 25 and 26).

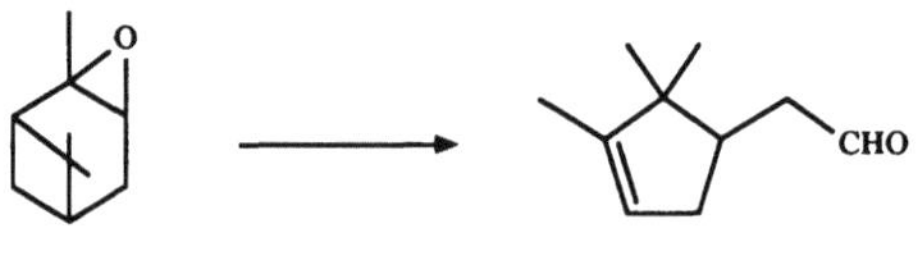

34 35

The great activity in this direction was triggered off by a patent from East Germany.[27] Successful chemicals of this series are Sandalore®, a mixture of **36** and **37**,[28] Sandacore (registered name, Kao Soap) (**38**), Bacdanol® (registered name, IFF) (**39**), Brahmanol® (**40**),[29] Ebanol® (**41**)[30] and compound **42**.[31]

Naturally, the importance of the sandalwood chemicals brought about consideration of structure–odor relationships. This usually becomes possible because, after the greater part of the work is done, a large collection of material is available.[1,25,26]

Notwithstanding, according to the description of a group of new chemicals (**43–46**),[32] of these only **45** displays a reportedly strong sandalwood scent, an observation which fits badly into the proposed rules (see Refs 1, 25, 26); that is, it suggests that they should be refined.

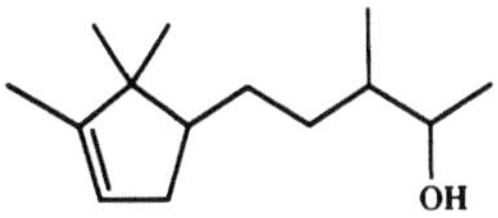

36

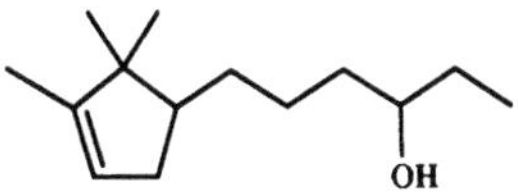

37

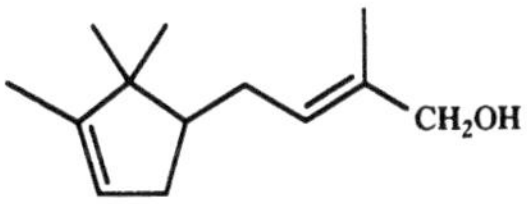

38

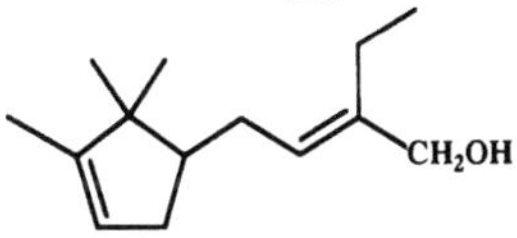

39

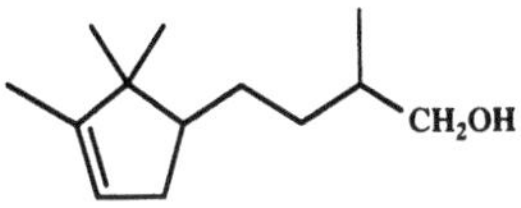

40

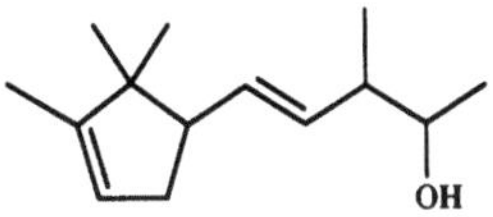

41

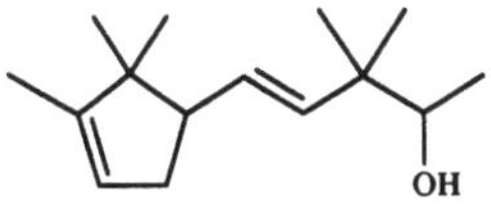

42

43

44

45

46

On the other hand, two recently presented chemicals with strong sandalwood odor, **47** and **48**,[17] seem to fit Naipawer's rule.[25]

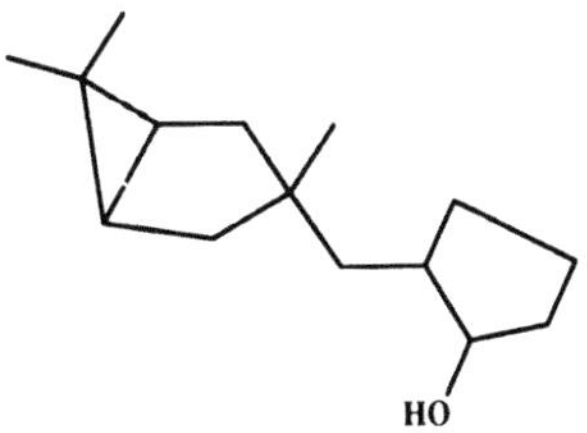

47

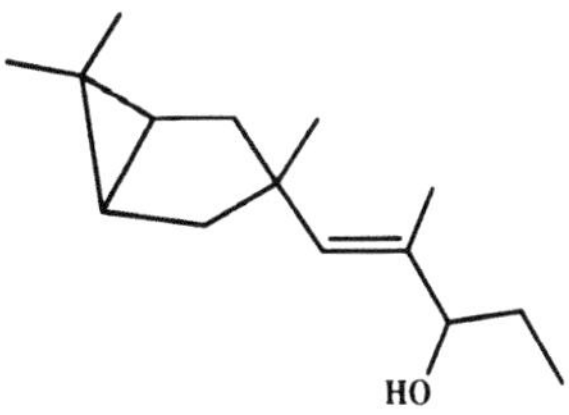

48

Woody Characters in General

The odor description 'woody' is again a not-unambiguous specification, as it depends on the individual usage. This wood-like resinous note is only rarely present in its pure form. In compounds it is mostly accompanied by ambery, musky, sandalwood- or cedarwood-like notes, or even by earthy, spicy-herbaceous shades.[33] The question of which structural requirements are necessary for a more woody, ambery or musky compound is by no means trivial, and is not yet solved. An attempt in this direction relying on basic structural types has been described,[34] although without great potential to make a possible prediction except in trivial cases.

A classical woody ingredient of today's perfumery is Iso-E-super® (**50**), which also displays a pleasant ambery note.[35] It is prepared from the Diels–Alder product (**49**) of myrcene and 3-methyl-3-pentene-2-one, by means of an acid-catalyzed cyclization.

49 50

A similar tricyclo acyl compound **53** with a warm, strong, woody odor has been synthesized from nopol (**51**), which after dehydration furnished the diene **52**, whose reaction with methyl vinyl ketone yielded **53**.[36,37]

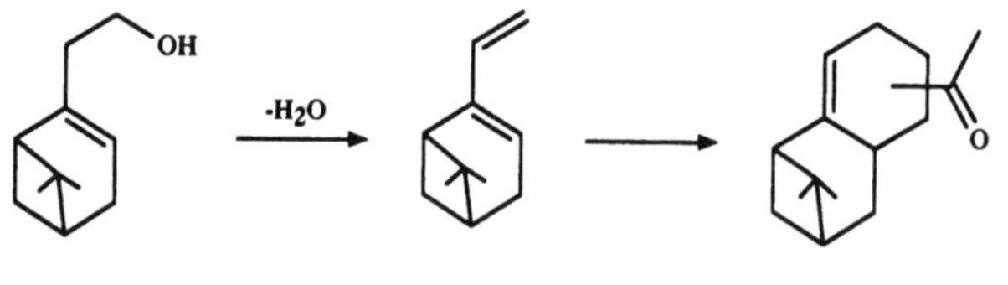

51 52 53

A variant of this chemistry is the synthesis of the acyl compounds **59, 60** and **61**.[38] The starting material is the mixture of the trimethylcyclopentanones **54a** and **54b**, which reacted with a vinyl Grignard reagent to give a mixture of **55a** and **55b**. Dehydration gave the dienes **56**, **57** and **58**, which upon thermal reaction with methyl vinyl ketone yielded the acyl compounds **59–61**, which were claimed to have a strong, warm, ambergris scent with woody, tobacco leaf, methylionone notes. Analogous reaction with acrolein instead of methyl vinyl ketone[39] furnished a mixture with a more woody, thujone-like, camphoraceous scent.

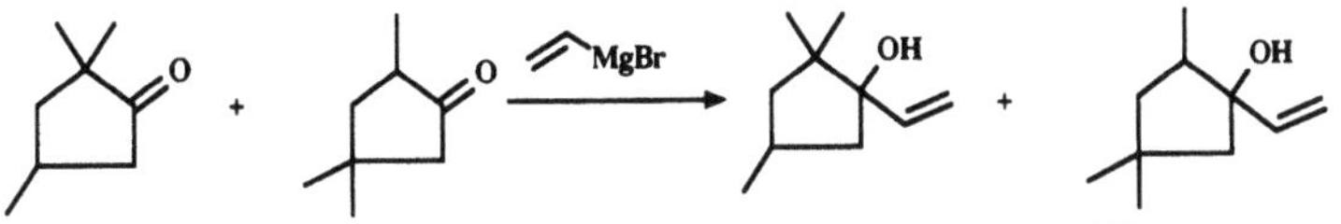

54a 54b 55a 55b

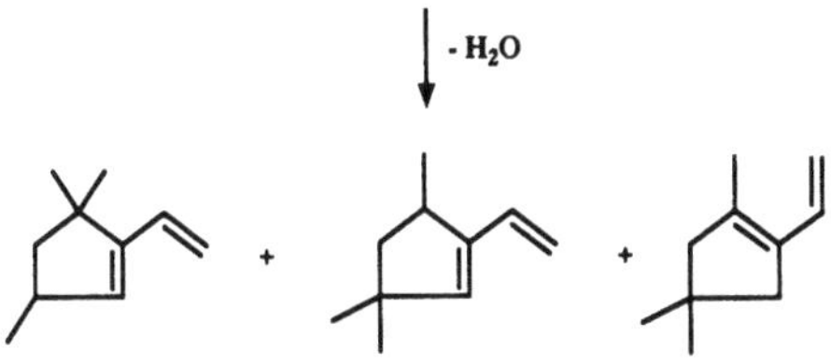

56 57 58

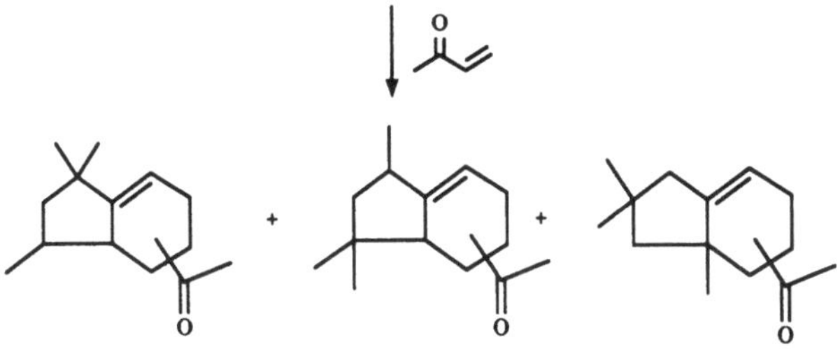

59 60 61

Still another tricyclo acyl compound has been described by the Naarden group.[40,41] A radical-initiated cyclization of caryophyllene (**62**) in the presence of acetaldehyde gave a mixture of four diastereomers (**63**) in 38% yield. This displays a pleasant woody odor with ambergris notes, whereas for the product mixture arising from the reaction with propanal a woody, somewhat spicy odor, is obtained.

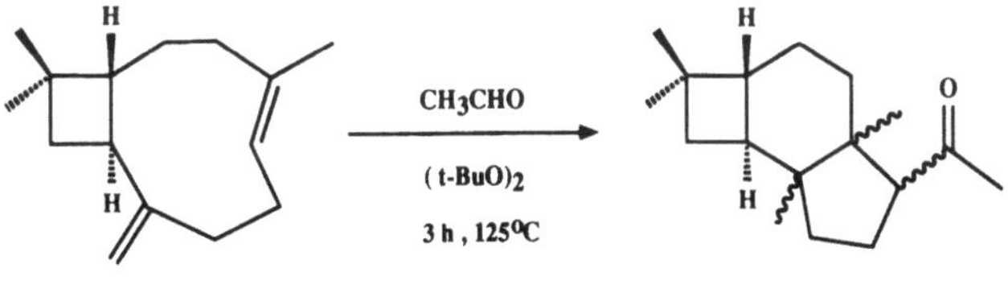

62 63

A strong woody note is attributed to **66**, which was prepared in two steps from the cheap hydrocarbon mixture **64**,[42] which is first cyclized to the tricyclic isomer **65** in 80% yield. Epoxidation then leads to **66**.

More on the woody–patchouli side is the odor of **70**.[43] This compound, with a tetracyclic structure resembling α-santalol, is a cyclization product of an intermediate in an epi-β-santalene synthesis.[44] Treatment of **67** with BF_3 etherate at 0°C yielded **70** in 62% yield; the mechanism of this skeletal rearrangement is depicted in **68** and **69**.

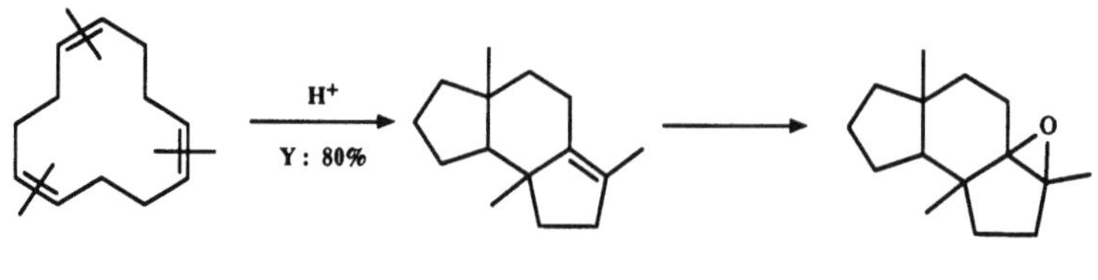

64 65 66

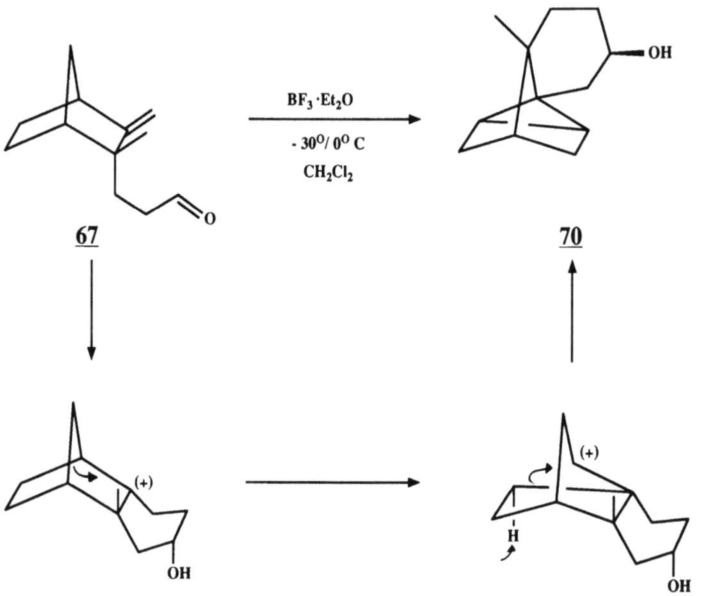

68 69

Speaking about the woody–patchouli character of the odor of **70**, it should be mentioned that 'patchouli' itself is a term meaning woody, earthy, camphoraceous and powdery all at the same time, well balanced together, as they are in patchouli oil. The best known and probably the most important constituent of patchouli oil is patchouli alcohol (**71**).[45] Ohloff and co-workers synthesized both enantiomers of **71**, and having the natural (−)-**71** and the unnatural (+)-**71** at hand, they could prove the importance of (−)-**71** for the odor characteristic of the oil on one hand, and its greater strength over its enantiomer on the other. Earlier, norpatchoulenol (**72**)[46–48] and the tetracyclic isomer **73**[49] have been claimed as the most important odorants in patchouli oil.

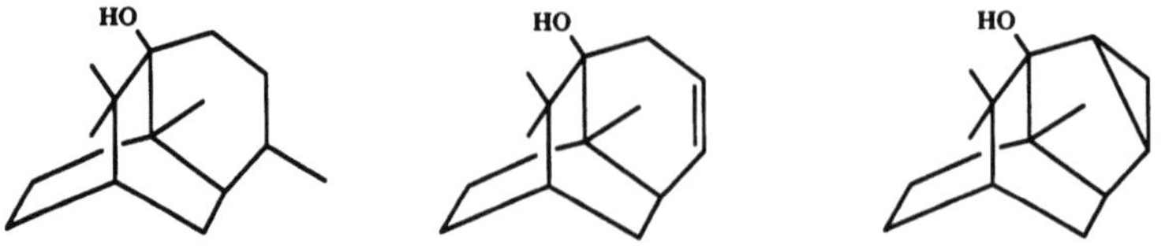

71 72 73

In a recent work it was attempted to translate the tricyclic structure of patchouli alcohol into highly substituted cyclohexanols and cyclohexenols.[50] Among the 45 derivatives described, the odor of **74** and **75** seemed to resemble most the odor of **71**, but without the well-balanced and powerful quality of the latter.

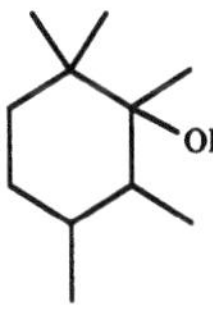

74

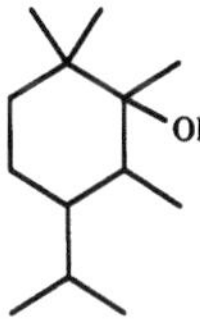

75

The Firmenich group looked again at an old woody fragrance material and found that the enantiomers of Polywood® [(−)-**76** and (+)-**76**] differ in their odor. The synthesis of the enantiomers was achieved by enantioselective hydrolysis of the chloroacetates by porcine pancreatic lipase.[51]

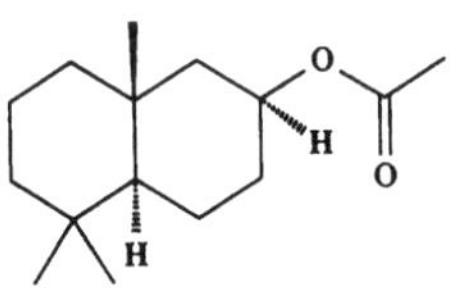

(-)- 76

Less rich, but still woody, dry, amber-like

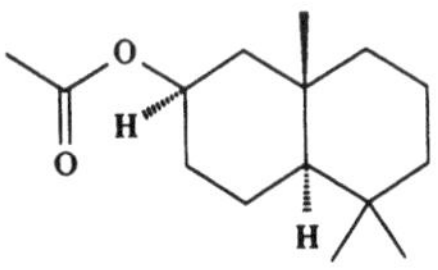

(+)- 76

very rich woody note

An interesting difference of odor and odor strength was revealed between the *trans,cis*-isomers **77** and **78**.[52–55] Though both compounds develop a woody and ambery scent, the *trans* isomer **77** shows a more animal and ambery character and is about 20 times more powerful than the commercial product Timberol®,[56] which is a mixture of about 90% **78** and about 10% **77**.

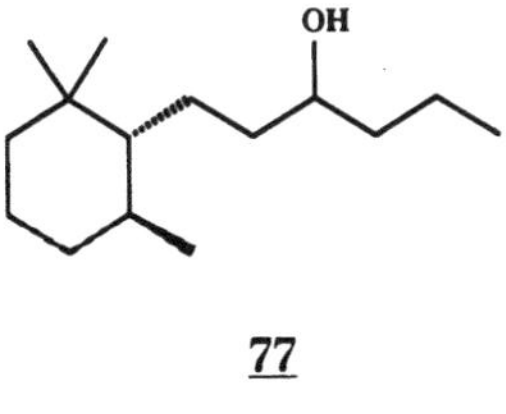

77

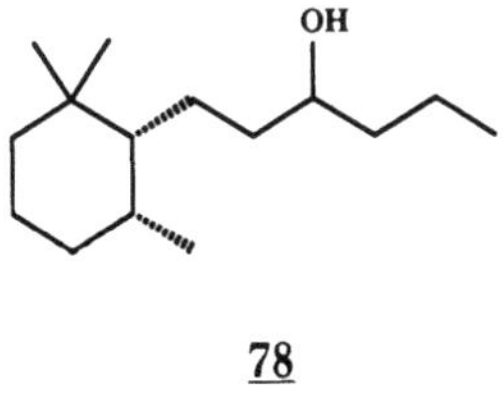

78

A rare *β*-diketone functional group in fragrance chemistry is present in Dione® (**80**). Dione is prepared in two steps from the acid **79** and has a woody, cedar-type, animal, powerful and diffusive odor.[57]

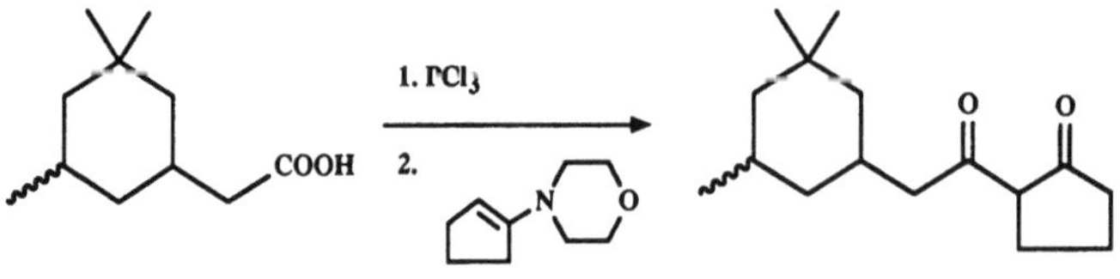

79 80

A very recent example in the field of woody chemicals is the ketal **81**,[58] with its woody, dry, ambery, Dione-like odor.

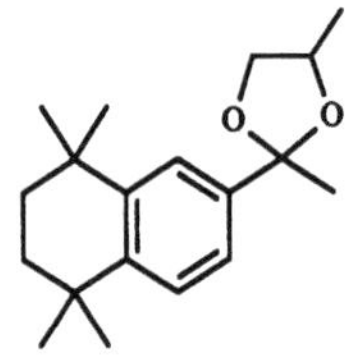

81

Rearrangement of the allylic alcohol **82** in formic acid with subsequent hydrolysis and oxidation gave rise to a very strong woody mixture mainly composed of three tricyclic ketones, **83**, **84** and **85**.[59] The rearrangement is thought to proceed through **86**, **87**, **88**, **89** and **90**.

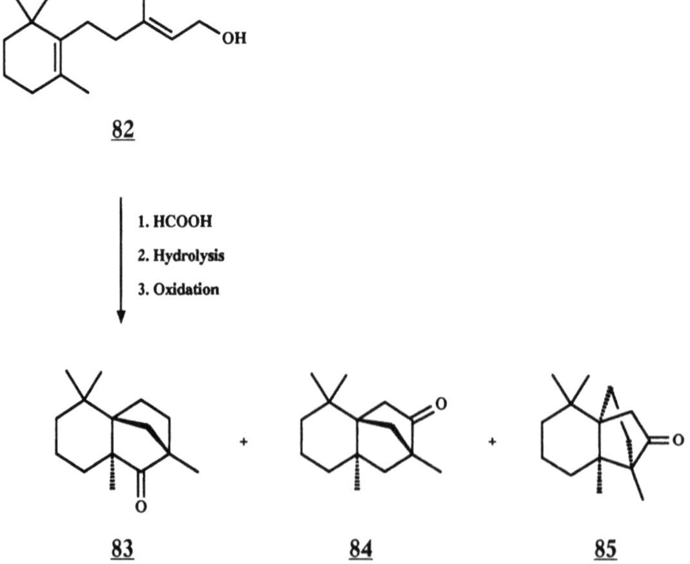

83 84 85

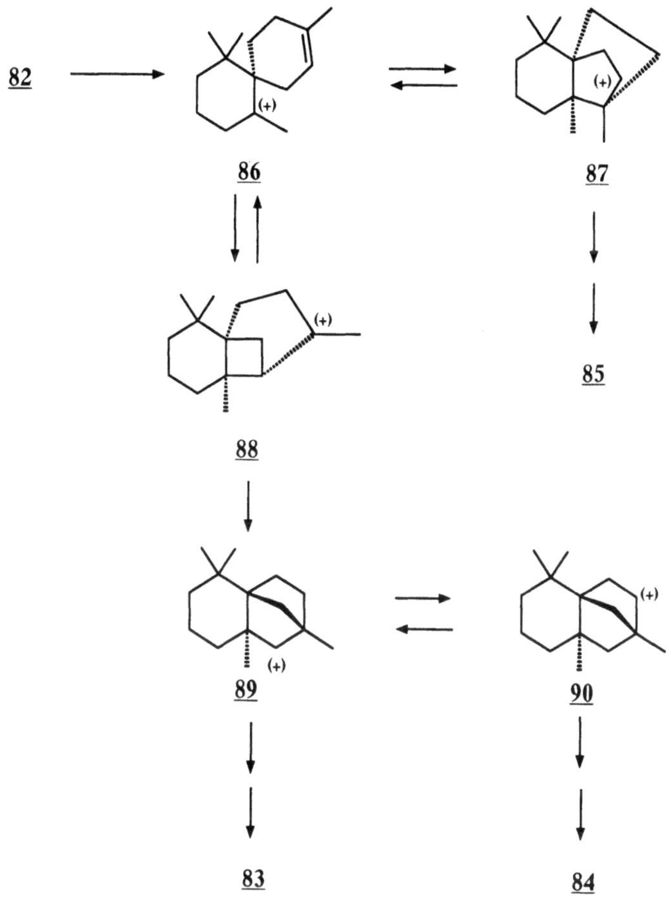

The tricyclic skeleton of **83** with a different substitution pattern has already been described. It was discovered independently in three different laboratories that (for example) aldehyde **91** undergoes a rearrangement to **92** in the presence of Lewis acids.[60–62]

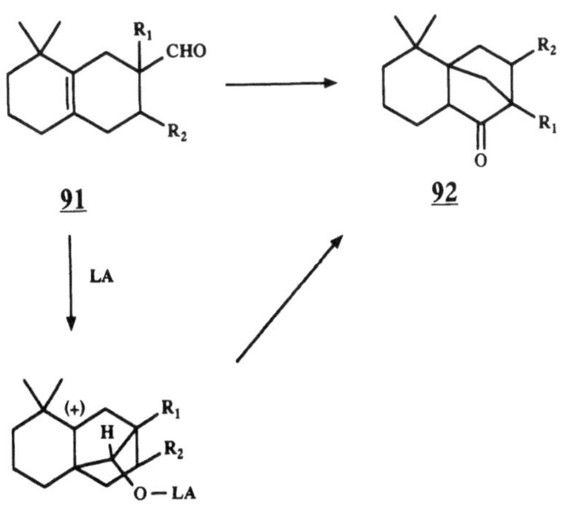

REFERENCES

Sandalwood

1. Brunke, E. J. & Klein, E. In *Fragrance Chemistry*, ed. E. T. Theimer, Academic Press, 1982, pp. 397–431.
2. Kumar Yadav, V. & Bisarya, S. C., *J. Sci. Ind. Res.*, **41** (1982) 650–5.
3. Brunke, E. J., Hammerschmidt, F. J. & Struwe, H., *Tetrahedron Lett.* (1980) 2405.
4. Brunke, E. J., *Dragoco Report*, **28** (1981) 91.
5. Brunke, E. J., Böhme, A. & Struwe, H., *Liebigs Ann. Chem.* (1982) 1105.
6. Corey, E. J. & Yamamoto, H., *J. Am. Chem. Soc.*, **92** (1970) 226.
7. Sreekumar, C., Darst, K. P. & Still, W. C., *J. Org. Chem.*, **45** (1980) 4620.
8. European Patent Appln 029 603 (priority JA, 22 Nov 1979) to Sumitomo.
9. Tamura, M. & Suzukamo, G., *Tetrahedron Lett.*, **21** (1981) 577.
10. Snowden, R. L., Sonnay, P. & Ohloff, G., *Helv. Chim. Acta*, **64** (1981) 25.
11. Sato, K., Myamoto, O., Inone, S. & Honda, K., *Chem. Lett.* (1981) 1183.
12. Solas, D. & Wolinsky, J., *J. Org. Chem.*, **48** (1983) 1988.
13. Oppolzer, W. & Chapuis, C., *Tetrahedron Lett.*, **24** (1983) 4665.
14. Oppolzer, W., Chapuis, C., Dupuis, D. & Guo, M., *Helv. Chim. Acta*, **68** (1985) 2100.
15. Arai, Y., Yamamoto, M. & Koizumi, T., *Bull. Chem. Soc. Jpn*, **61** (1988) 467.
16. Krotz, A. & Helmchen, G., Tetrahedron: Asymmetry **1** (8) (1990) 537.
17. Sukh Dev, Paper presented at 11th Int. Congress of Essential Oils, Fragrances and Flavours, New Delhi, 1989, Vol. 2, p. 1.
18. Byers, J. R., *Am. Perf. Essent. Oil Rev.*, **49** (1947) 483.
19. Kauffman, G. B., *Chem. Eng. News* **67** (49) (1989) 35 (book review).
20. Aulchenko, I. S. & Kheifits, L. A., *Am. Perf. Cosm.*, **85** (1970) 37.
21. Aulchenko, I. S., Cherkasova, T. P. & Kheifits, L. A., *IXth Int. Congress of Essential Oils, Singapore, 1983*, Vol. 5, p. 54.
22. Lange, G. K. & Bauer, K. A. In *Fragrances and Flavour Substances*, ed. R. Croteau. D & PS Verlag, D-3017 Pattensen, Germany, 1980.
23. Demole, E., *Helv. Chim. Acta*, **47** (1964) 319, 1766; **52** (1969) 2065.
24. Dahill, R. T., Dorsky, J. & Easter, W., *J. Org. Chem.*, **35** (1970) 251.
25. Naipawer, R. E., Purzicky, K. L., Shaffer, G. W. & Erickson, R. E. In *Essential Oils*, ed. B. D. Mookherjee & C. J. Mussinan. Allured Publishing Corp., Wheaton, IL, 1981.
26. Naipawer, R. E. In *Flavors and Fragrances: a World Perspective*. Proc. 10th Int. Cong. of Essential Oils, Fragrances and Flavors, Washington, 1986. Elsevier, Amsterdam, 1988, p. 805.
27. East German Patent 68 936 (priority 20 Sept. 1969).
28. US Patent 4 052 341 (priority US 4 Oct 1977) to Givaudan.
29. German Patent 2 827 957 (priority DE 26 June 1978) to Dragoco.
30. US Patent 4 696 766 (priority US 19 March 1986) to Givaudan.
31. US Patent 4 610 813 (priority CH 23 Mar. 1984) to Firmenich.
32. Witteveen, J. G. & van der Weerdt, A. J., *Rec. Trav. Chim. Pays-Bas*, **106** (1987) 29.

Woody in General

33. Brunke, E.-J., *Parf. Kosm.*, **64**(9) (1983) 470.
34. Thömel, F., *Chem. Zeitung*, **105**(9) (1981) 239.
35. US Patent 3 929 677 (priority US 27 Feb. 1973) assigned to IFF.

36. US Patent 4 322 599 (priority DE 3 Sept. 1979) assigned to Henkel.
37. US Patent 4 363 754 (priority DE 3 Sept. 1979) assigned to Henkel.
38. US Patent 4 302 363 (priority DE 25 June 1979) assigned to Henkel.
39. German Patent 3 212 326 (priority DE 2 Apr 1982) assigned to Henkel.
40. US Patent 4 453 014 (priority NL 16 Sept. 1981) assigned to Naarden International.
41. US Patent 4 594 183 (priority NL 16 Sept. 1981) assigned to Naarden International.
42. European Patent 172 376 (priority CH 26 July 1984) assigned to Firmenich.
43. European Patent 148 399 (priority CH 22 Dec. 1983) assigned to Firmenich.
44. Snowden, R. L., Sonnay, P. & Ohloff, G., *Helv. Chim. Acta,* **64** (1981) 25.
45. Näf, F., Decorzant, R., Giersch, W. & Ohloff, G., *Helv. Chim. Acta,* **64** (1981) 1387, and literature cited therein.
46. Teisseire, P., Maupetit, P. & Corbier, B., *Recherches (Paris),* **19** (1974) 8.
47. Teisseire, P., Maupetit, P., Corbier, B. & Rouillier, P., *Recherches (Paris),* **19** (1974) 36.
48. Oberhänsli, W. E. & Schönholzer, P., *Recherches (Paris),* **19** (1974) 62.
49. Mookherjee, B. D., Light, K. K. & Hill, I. D., Paper presented at 178th ACS National Meeting, Washington, DC, 10–14 Sept. 1979.
50. Weyerstahl, P., Splittgerber, H.-D., Walteich, J. & Wollny, T., *J. Ess. Oil. Res.,* **1** (1989) 1.
51. Gautier, A., Vial, C., Morel, C., Lander, M. & Näf, F., *Helv. Chim. Acta,* **70** (1987) 2039.
52. US Patent 4 711 875 (priority CH 9 Mar. 1983) assigned to Firmenich.
53. US Patent 4 623 750 (priority CH 9 Mar. 1983) assigned to Firmenich.
54. US Patent 4 626 381 (priority CH 11 Mar. 1983) assigned to Firmenich.
55. European Patent 121 828 (priority CH 12 Apr. 1983) assigned to Firmenich.
56. German Patent 2 807 584 (priority DE 22 Feb. 1978) assigned to Dragoco.
57. World Patent Appln 82 03 216 = European Patent 73 771 (priority CH 13 Mar. 1981) assigned to Givaudan.
58. European Patent Appln 379981 (priority CH 27 Jan. 1989) filed by Givaudan.
59. European Patent Appln 315 895 (priority CH 12 Nov. 1987) filed by Givaudan.
60. US Patent 4 250 338 (priority US 16 Nov. 1979) assigned to IFF.
61. European Patent 29 259 (priority GB 9 Oct. 1979) assigned to PFW.
62. Baldwin, J. E. & Lusch, M. J., *J. Org. Chem.,* **44** (1979) 1923.

4 CHEMICALS WITH FRUITY AND FLORAL ASPECTS

Fruity Odor Characteristics

With the identification of the sulfur-containing monoterpene **1** in buchu leaf oil,[1,2] a new era of interesting perfumery compounds—all exhibiting fruity notes—has been started. Although ketone **1** already has a tremendous odor strength, it has been beaten easily by the thia-analog (**2**) of α-terpineol which

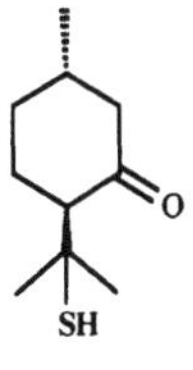

1

2

was mentioned as being 'apparently the most powerful flavor compound ever found in nature'.[3] This terpenic mercaptan was found in grapefruit juice, to which it really confers an extremely powerful impact with regard to the overall sensation of grapefruit. The compound was first synthesized in the following way:[4]

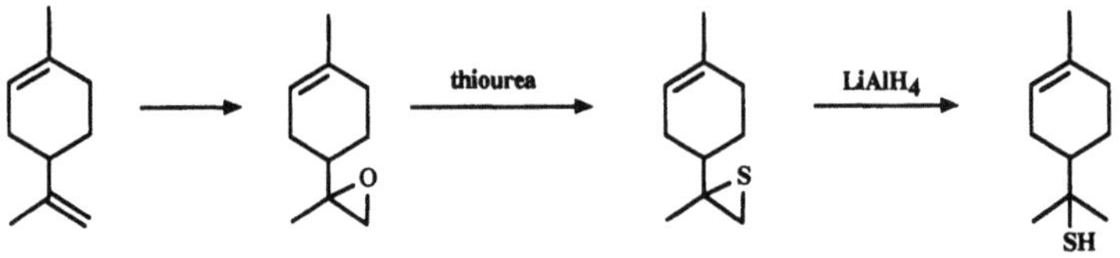

3

It is very typical of the fragrance and flavor industry that outstanding odorants immediately stir the attention of competitors. Thus, it is not surprising to come across a later but similar synthetic access to **2**, which claims to be more economical.[5] Furthermore, an attempt was made to establish an 'own process'-dependent substance claim in combination with α-terpineol.[6]

The addition of hydrogen sulfide to methyl (or ethyl) methacrylate in the presence of triethylamine affords the 3-mercapto-2-methylpropionic esters (**4**) with a fragrance of tropical fruits.[7] When added to fragrance compositions in minute amounts, both the esters mentioned enhance their natural characteristics, but higher esters are insufficient in perfuming.

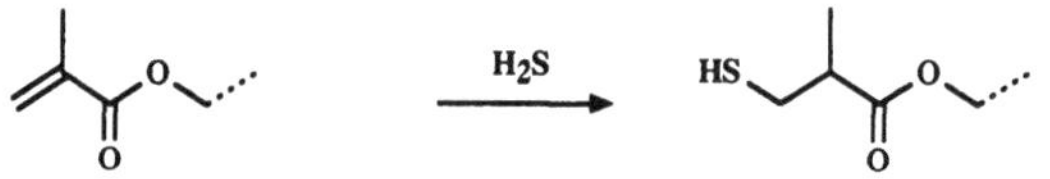

4

The range of sulfur compounds with terpene structures have extraordinary low threshold values. The sophisticated in-depth analysis of natural substrates now generates more and more such strong-smelling trace components. Whereas at the beginning of the 1960s, for instance, nootkatone (**5**) was thought to be the most important compound for the typical grapefruit flavor, in 1983 Demole *et al.*[8] reported on the identification of dehydronootkatone (**6**) in grapefruit juice. The odor of this sesquiterpene ketone is much stronger than that of nootkatone itself. The new compound can be prepared by dehydrogenation of nootkatone, which is accessible by a partial synthesis (allylic oxidation) from valencene, a sesquiterpene hydrocarbon occurring in the high-boiling fraction of orange oil.[9]

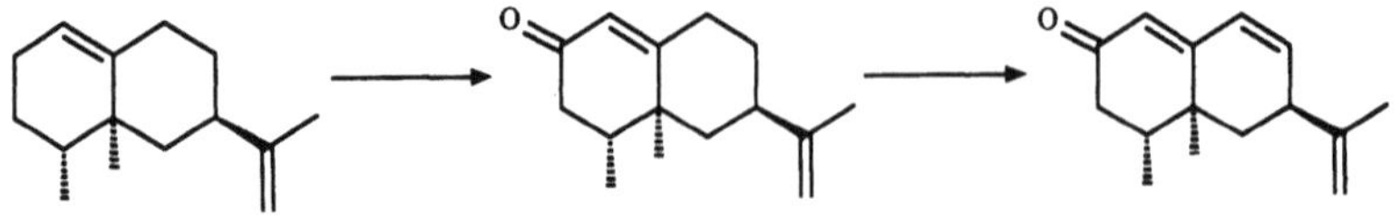

5 6

A purely synthetic bicyclic alcohol with 14 carbon atoms was patented in 1984. This chemical (**7**) has a strong odor of fruity character linked to the impressions of cassis and grapefruit, whereas similar compounds normally have more woody odor profiles.[10] The two-step reaction sequence uses first the

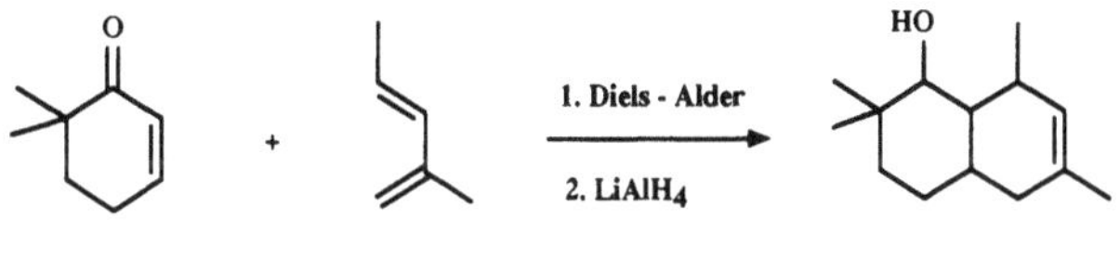

7

addition of a dienophile to 2-methyl-1,3-pentadiene (Diels–Alder reaction), followed by the reduction of the carbonyl group with lithium aluminum hydride. Both types of reaction are very useful in preparing fragrant molecules, but the hydride mentioned is normally too expensive for a large tonnage production scale. Generally, its replacement by sodium borohydride is tried whenever possible: thus the production costs are lowered to a considerable extent.

The absence of woody character in several keto-alcohols (**8**, **9**, **10**) with a cyclohexene moiety in their molecules is surprising.[11] They have sweet, fruity, raspberry-like and floral odor aspects and play a role as intermediates in the synthesis of 4-oxodamascone and 4-oxomegastigma-5,7,9-triene.

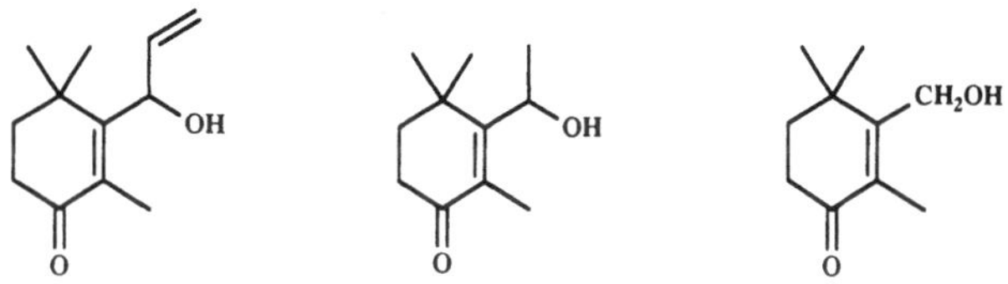

8 9 10

A typical raw-material basis used by the company BASF consists, *inter alia,* of isobutylene and phenol. Etherification followed by Friedel–Crafts reaction with acetyl chloride—also a widely used 'unit process' in the fragrance industry (see polycyclic musks)—delivers first the methyl ketone **11**. Its reduction and ring hydrogenation gives a *cis/trans* alcohol mixture, **12**, having woody odor characters which differ in nuance: the *cis* isomer is more sandalwood-like, whereas the *trans* isomer exhibits a cedarwood-like fragrance, according to a German patent.[12] Surprisingly, the acetate (**13**) of the mixture **12** has a fruity–herbaceous smell.

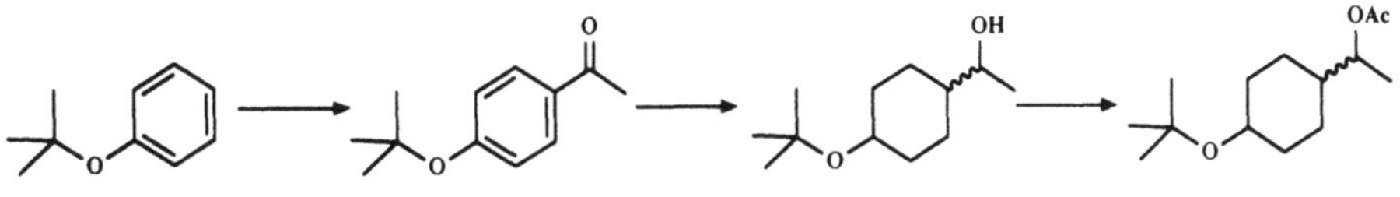

11 12 13

Fruity aspects, the characters of which also vary as a function of the stereochemistry of the substituents on the cyclohexene/cyclohexane ring, are described in a patent assigned to Naarden, now Quest International.[13] The α,β-unsaturated acid **14** is reacted with either penta-1,3-diene or (methyl)cyclopentadiene to give, after esterification, the ethyl ester mixtures **15** or **16**. In the case of **15**, the isomers were separated for exact odor

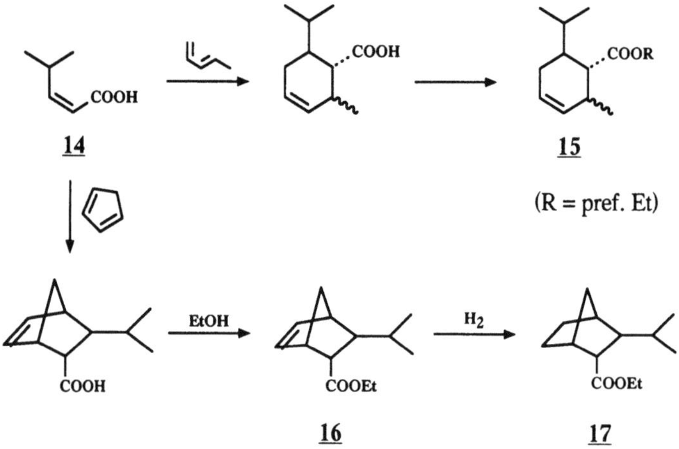

description: the *cis* isomer has a fruity (apple-like), rosy and woody fragrance, whereas the *trans* isomer is more flowery–fruity with spicy and balsamic effects.

The ester **16** also belongs to the fruity–spicy type and hydrogenation of the double bond leads to a fresh-fruity–green ester **17**, with woody and animal aspects.

The cassis, or blackcurrant, note inherent to the new sulfur-containing aroma chemicals has delighted the perfumers so much that they asked for similar odor notes, but without the unpleasant sulfury touch. Several companies have been at work and have presented their findings in corresponding patent applications. Two chemical classes are prominent: the ketoximes[14,15] and the spirocyclic ethers.[16,17]

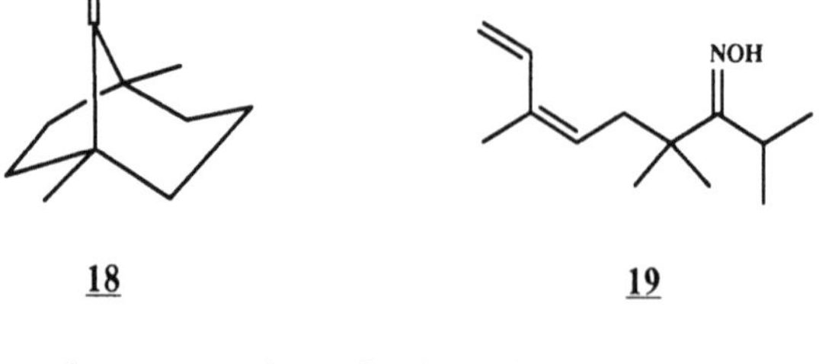

18 19

It may be that the preparation of oximes has been inspired by the successful introduction and use of the green-smelling Stemone® 20 years ago, but it is astonishing to have now, with Buccoxine® (**18**) (Dragoco) and 2,4,4,7-

tetramethyl-6,8-nonadien-3-one oxime (**19**) (Givaudan), two representatives with this exciting fruity–blackcurrant-like aspect on the perfumer's shelf.

The second group has the oxa-spiro[4.5]undecane skeleton resembling the well-known theaspiran **20** for which use claims were also granted to a fragrance house.[18,19] It was argued that compound **20** should be classed among the ten

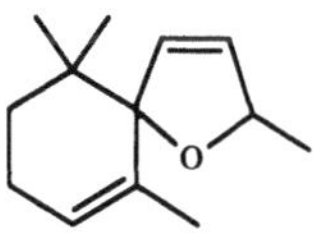

20

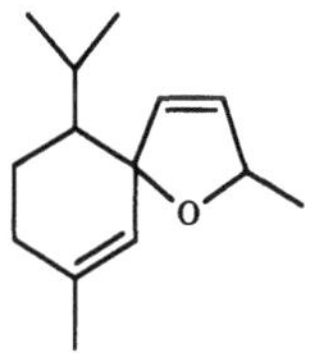

21

best chemicals on the market, but this statement may be a little too enthusiastic and perhaps reflects the personal experience of the reporter.[20]

The preparation of **21**[16] starts from piperitone, an α,β-unsaturated monoterpene ketone:

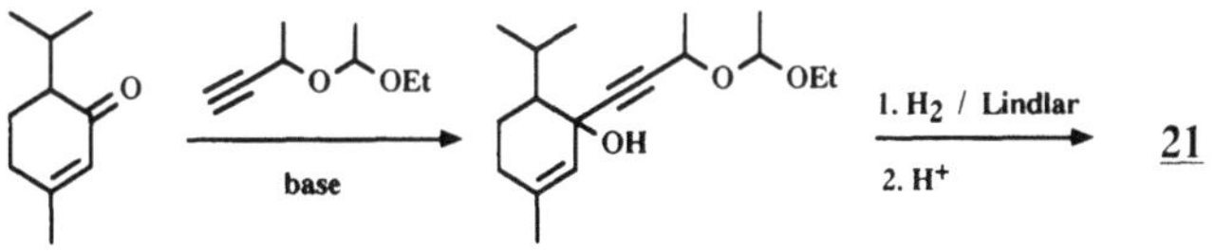

Instead of a base catalyst, propylmagnesium bromide can also function as promoter of the condensation.

In 1977 the mixture of cyclic spiro-ethers **22a/b**[17] was already described as having a fruity–green and woody fragrance similar to that of blackcurrant juice. A few years later, it was discovered that a similar ethyl-substituted compound **23** also exhibited very interesting cassis-like notes.[21]

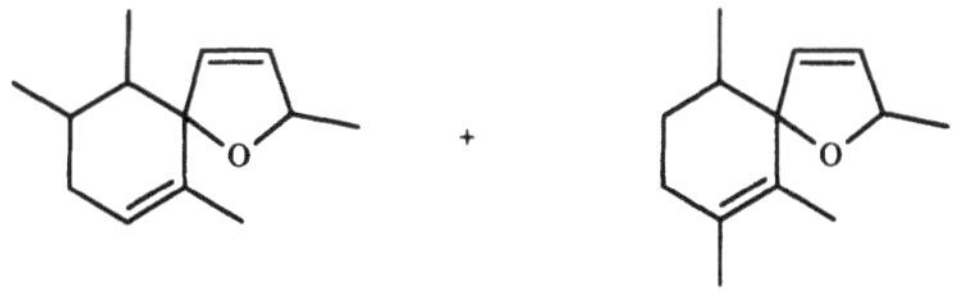

22a 22b

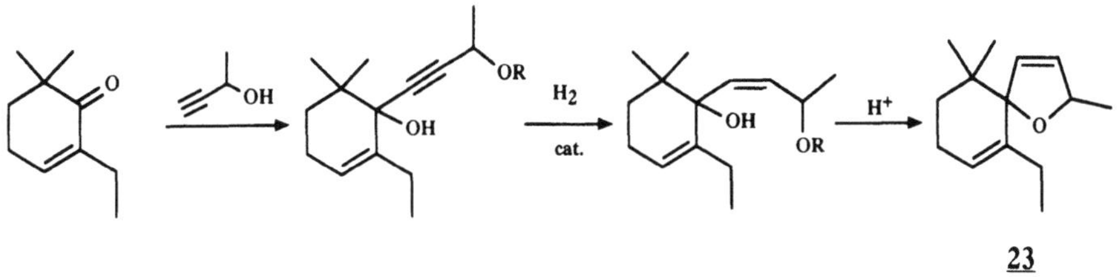

23

Regarding the substitution pattern in the structural formulae of **22b** and **23**, it turns out that another aroma chemical, **24a/b** (now commercially available; Givescone®), has exactly the same substitution pattern on the double bond of

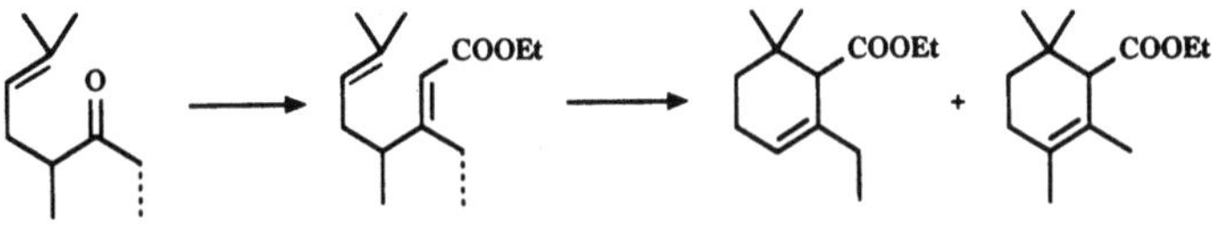

24a 24b

the cyclohexene ring.[22] The industrial production proceeds through condensation of a ketone with ethyl phosphonoacetate followed by cyclization. In the case of this mixture of esters, no cassis aspect could be observed in the fruity odor profile, which has spicy–woody and citrus–like nuances.

In connection with the study of processes leading to the damasc(en)ones (see page 580), it has been found that methyl α- and β-cyclogeranates (**25a/b**)—although known from the literature since 1959—not only represent appropriate starting materials, but are in themselves valuable fragrant chemicals. The α-isomer **25a**, in particular has very interesting fruity–floral, fresh and citrus-like odor facets.[23]

25a 25b

It seems worthwhile to mention here another bicyclic ketone, **27**, whose structural features have a remote resemblance to the famous damascones.[24] Its production starts from benzyl cyanide which is prenylated and cyclized to the intermediate **26** (Givrosia®), also a cheap fragrant chemical in its own right. Grignard reaction with ethyl chloride furnishes the ketone **27** with a fruity, rose- and tobacco-like fragrance. Its incorporation into perfume compositions adds shades of tea roses and plums and at the same time inparts volume, warmth and a certain sweetness.

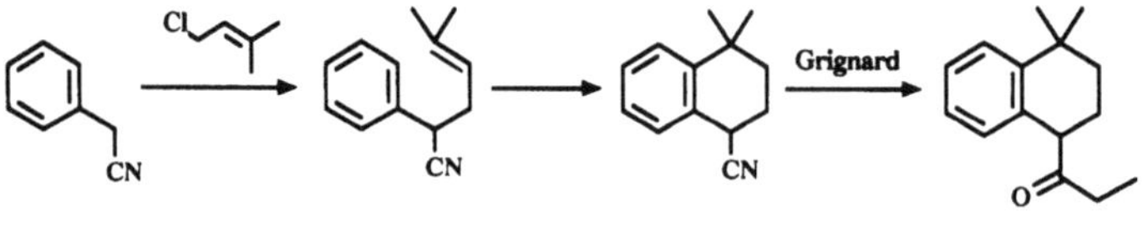

26 27

Similar alkylation reactions can be found in a series of new nitriles, **28–30**, with a strong fruity, lemon-like and often very tenacious smell; two of them have been commercialized by the German company Wacker-Chemie.[25]

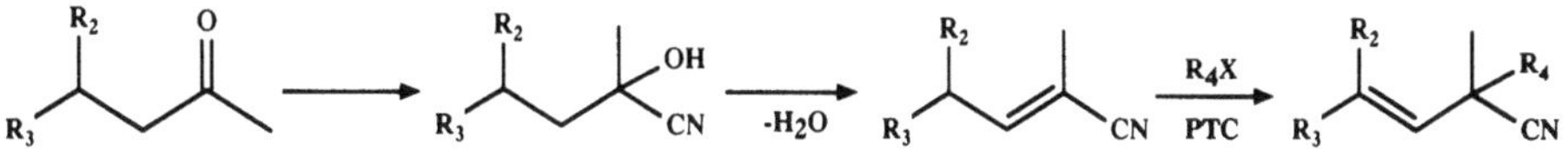

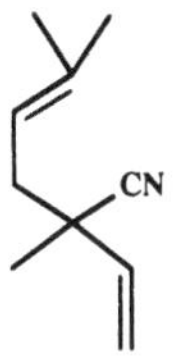

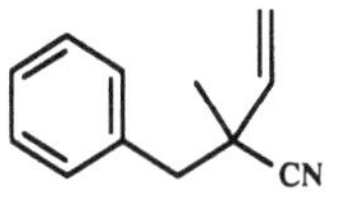

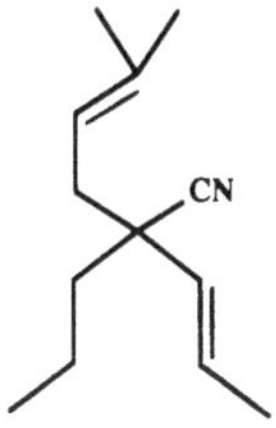

28 29 30

It follows from the reaction sequence that handling of HCN does not present real difficulties in industrial processes, as seen by the cyanohydrin intermediate formed from the ketone and hydrogen cyanide.

The similarity of odor characteristics for nitriles and aldehydes is often discussed, and it is therefore not astonishing to have such derivatives of the above-mentioned nitriles in a range of citrus-like smelling acetals, **31–33**.[26]

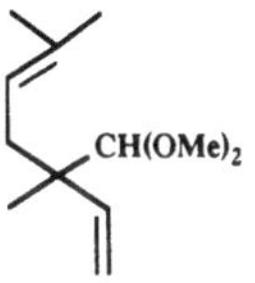

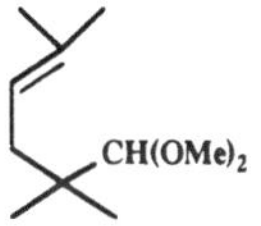

31 32 33

Aldehydes (**34**, **35**) with a green- or sweet-fruity, perilla-like aroma have been mentioned in a Japanese patent application.[27] They represent interesting intermediates in the preparation of alcohols (**36**/**37**) and ketones (**38**/**39**), which not only have citrus-like, fruity odors but also spicy–woody and even floral notes.[28]

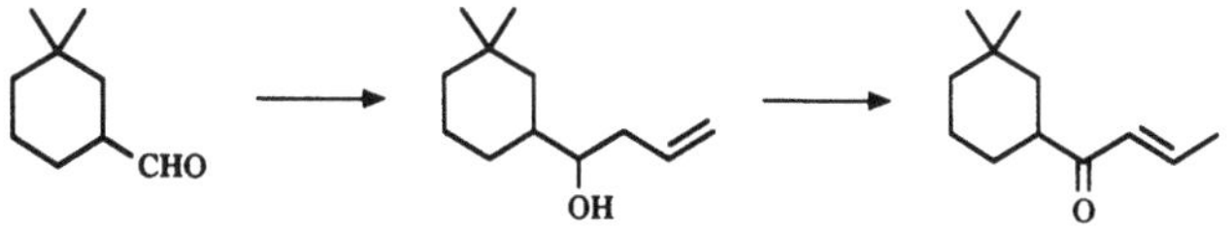

34 36 38

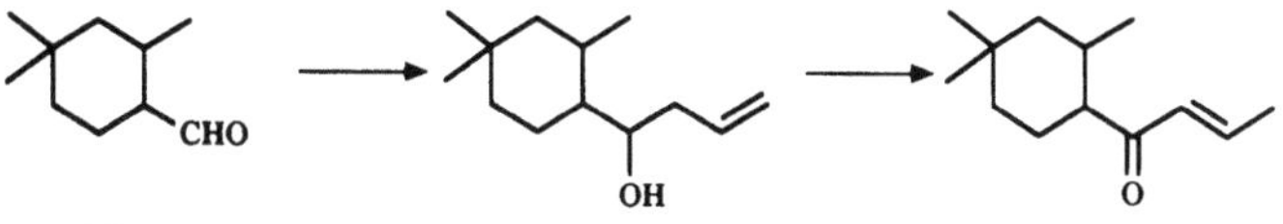

35 37 39

It is well known that many aliphatic esters have fragrances with a more or less pronounced fruity character. In an interesting lecture given at the SEPAWA Symposium in 1985, Sell[29] discussed the dependence of fruity notes on the molecular structure of such compounds.

In this context, two patents assigned to the German company Henkel are worth mentioning in view of the structural features of the compounds combined with the distinct fruity character. Thus, compound **40** (Melusate®), for instance, represents a fruity–green aspect, is inexpensive and has therefore a very good price/odor relationship.[30] The same trimethylhexanoic acid is the starting material for two amides (**41a/b**) with similar odor properties.[31]

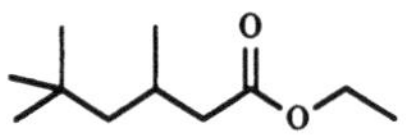

40

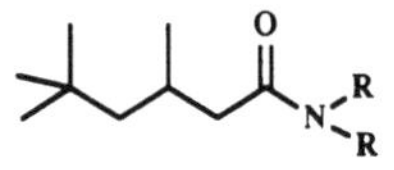

41a/b

(R = Me,Et)

A rather complicated reaction sequence leads to a saturated aliphatic propionate **42** with a fruity, pear-like fragrance, reminiscent of ambrette.[32]

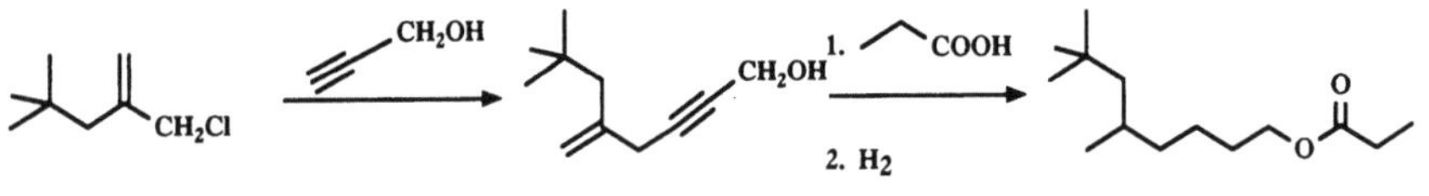

42

Fruity–flowery odor descriptions are found for some esters (**43a/b**) of an unsaturated aliphatic secondary alcohol.[33]

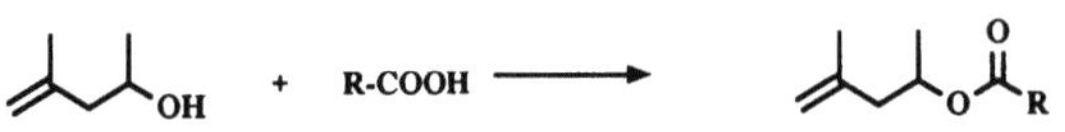

43a (R = Me)

43b (R = Sal)

The acetate **43a** of this series is more green–fruity and fugaceous, whereas the corresponding salicylate **43b** exhibits a more flowery–green odor.

Fruity–green–floral aspects are the main characteristics in another series of aliphatic esters (**44**), the parent acid being 2-methyl-2-ethylbutanoic acid.[34]

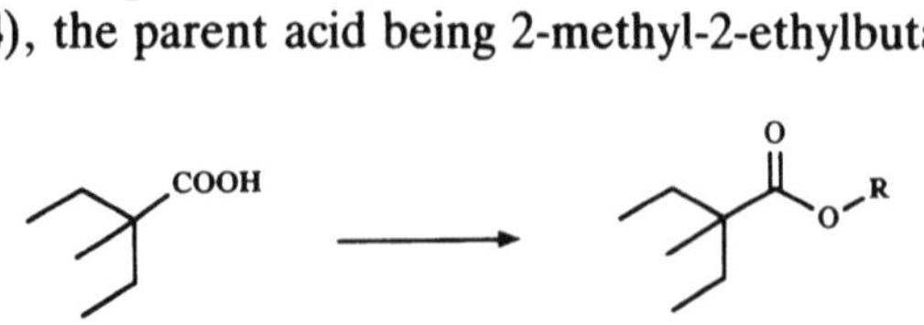

44

An interesting bifunctional compound **45** with fruity–floral, fresh and balsamic odor shadows has been prepared by using cheap ε-caprolactone as starting material.[35] Bifunctionality is also a structural key element in C_9 compounds prepared from hexanoic acid derivatives *via* a Grignard-type reaction.[36] 4-Oxononanal acetal (**46**), obtained in this way, as well as the parent aldehyde **47** (easily produced by acid hydrolysis of the acetal) both have fruity odor notes.

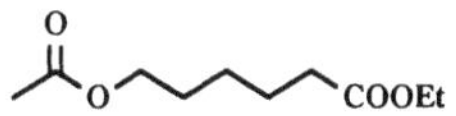

45

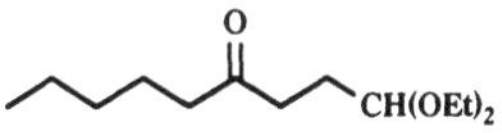

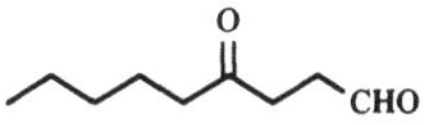

46 47

A recent German patent[37] describes cheap aliphatic ketones (**48**–**50**) and a pedigree is easily detectable within this series. Such considerations are often encountered in the research development work in the fragrance industry. The α,β-unsaturated ketone **49** is a known substance, but it has such interesting enhancer properties that patent protection was also sought.

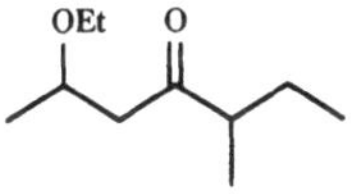

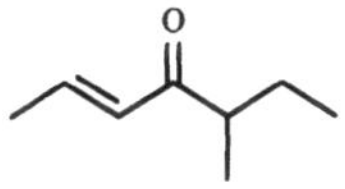

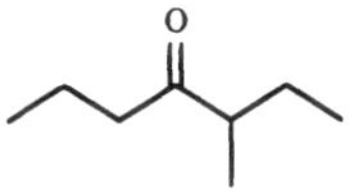

48 49 50

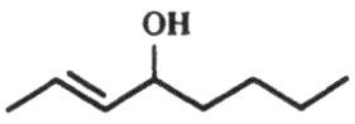

51

Finally, in its simple structure, the secondary alcohol **51** is reminiscent of the series of aliphatic compounds with fruity fragrances just mentioned. By incorporation of 2-octen-4-ol the impression of overripe fruit is imparted to flavor compositions according to the Patent Application filed about two years later by the same company,[38] thus underlining their interest in simple structures easy to prepare and evidently useful in specific applications in the flavor *and* fragrance fields.

In order to avoid difficulties of transition from fruity to floral themes, the present section ends with the mention of a new ketone, **52**, having a pleasant fruity–flowery smell.[39] Its preparation follows the pathway already shown previously for the esters **15**,[13] but now using 5-methyl-3-hexen-2-one instead of the carboxylic acid:

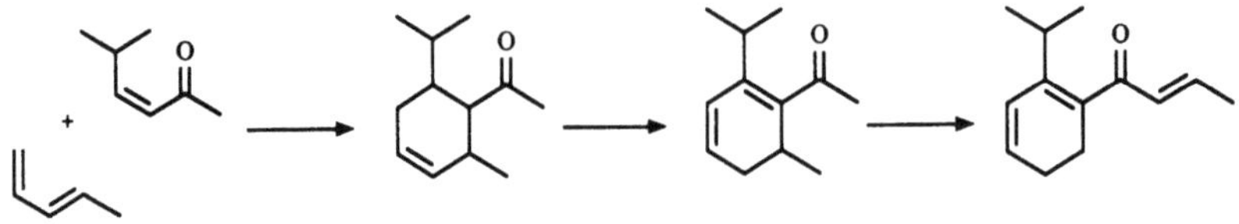

52

With structure **52** one can easily appreciate the obvious interest of the assignee in realizing isopropyl-substituted analogs of known fragrant molecules, i.e. damascenone in the present case.

Floral Odor Characteristics

The identification of the damascones **53a/b** and damascenones (**54**) in rose oil[40] and many other natural substrates has opened another new field of

53a/b 54

impressive research activities. It is quite clear that such a discovery has had some impact on the R&D efforts because the industrial production of these very fragrant materials needs outstanding capabilities to cope with the challenge always to be ahead with the most economic synthesis before competitors enter the same field.

Consequently, a great number of patent applications dealing with process research that aims at a promising route for access to the substances mentioned have been seen in the past 15 years. As outlined in the introduction to this chapter, we restrict ourselves to referring only to innovations published during the last decade. Thus, a direct route from esters **25** (or amides) was described in 1982:[41]

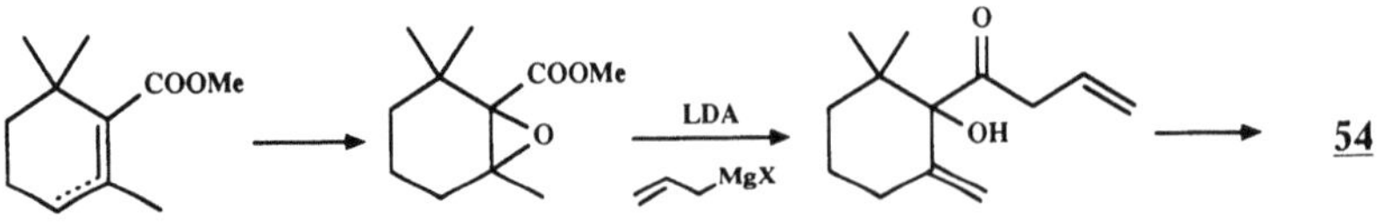

25a/b

It needs a minimum of 1 molar equivalent of a strong base like lithium diethylamide, which may be a disadvantage from the technical point of view.

Another three-step synthesis of damascenone (**54**) was elaborated from ketoisphorone (or 'oxophorone') via an allenic intermediate, **55**:[42]

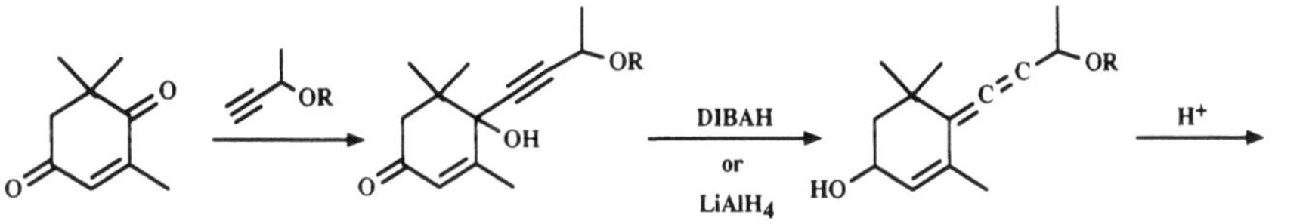

54

55

To make β-damascone more cheaply, a very interesting electrochemical route via an isoxazole derivative was patented in 1984.[43] It starts from β-ionone, which is first transformed into its ketoxime **56**. Anodic oxidation furnishes the heterocyclic intermediate **57**, thus transmitting the oxygen function from carbon atom number 3 to the desired carbon atom number 1 in the side chain. Cleavage of the heterocycle under acidic conditions finally gives β-damascone.

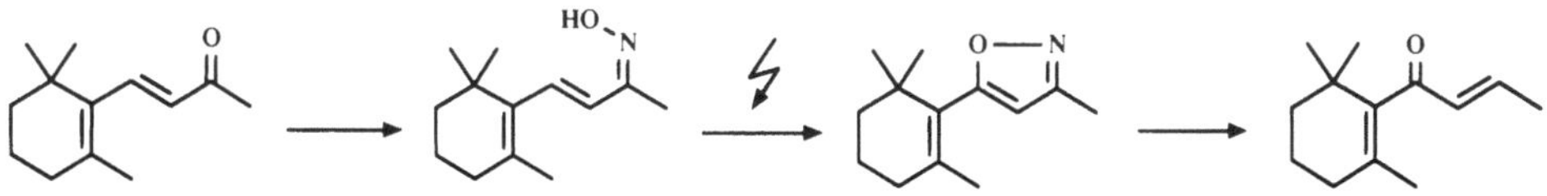

56 57 53b

Naef *et al.*[44] described a new route to α-damascone using a ketene (**58**) as key intermediate. Enantiomerically pure (*R*)(+)-**53a** and (*S*)(−)-**53a** are accessible by this efficient synthetic pathway using a highly enantioselective protonation of the ketone enolate prepared from **58**.[45]

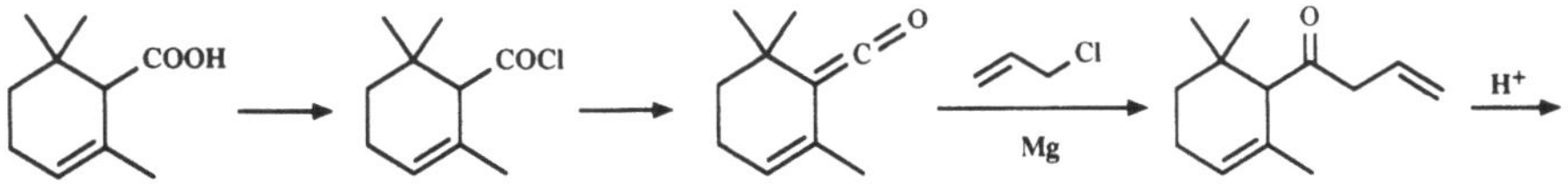

53a

58

Besides the α- and β-isomers, the other isomeric damascones also stirred a special interest in view of their refined odor characteristics. Thus, the preparation of γ-damascone (**59**) in a possibly economical way has been investigated using methyl β-cyclogeranate as starting material.[46]

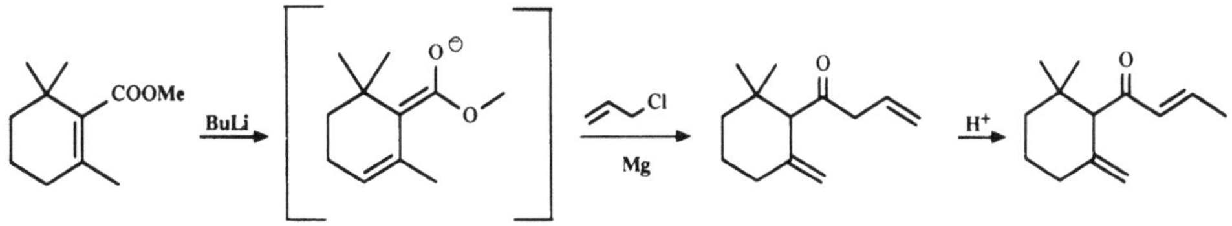

25b 59

Furthermore, it was revealed by the use of δ-damascone (**62**) in commercially available fragrance bases that even this isomer has some advantages in its own right. It can be produced from the ketoester **60** by first preparing the tertiary alcohol **61** in four steps.[47] An intermediate of this type was found to be useful throughout the complete damascone series because its treatment with strong bases followed by isomerization of the terminal double bond delivers the desired damascone isomers. In the case of δ-damascone (**62**), some ε-isomer is also present in the final product mixture.

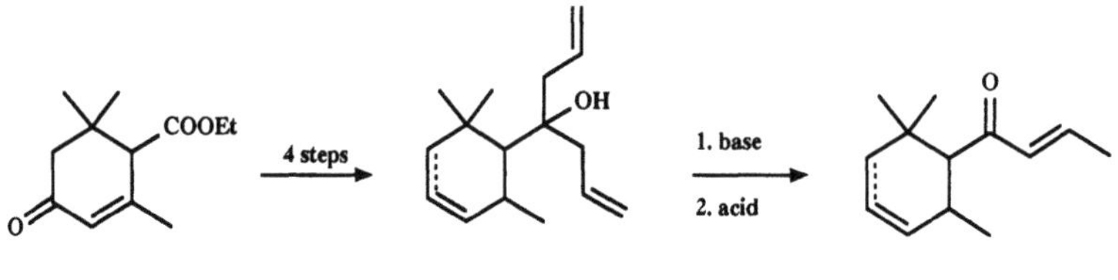

60 61 62

Finally, in the series of new fragrant damascone-derived chemicals, the side-chain epoxidized α-damascone **63** must be mentioned here. Its odor properties differ from the structurally prior known compound, α-damascone (**53a**), on account of its flowery–fruity note. The typical apple-like note of **53a** is completely absent; rather it develops a faint earthy, green and very natural fragrance without the camphoraceous character typical of α-damascone.[48]

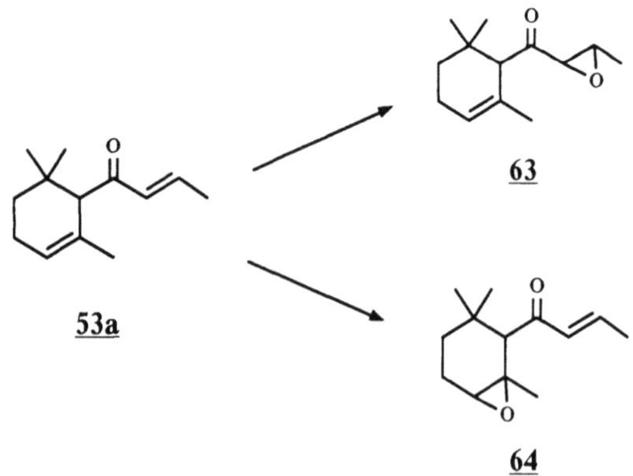

53a 63 64

It is furthermore worth mentioning that the isomeric epoxide **64** possesses a less noble character and is described as having a terpenic, camphoraceous and musty smell reminiscent of a winery.[49]

The two epoxides **63** and **64** are a very good example of the fact that a modification of the atomic arrangement in a given molecule often leads to an unexpected change in the olfactory properties, which is at the moment beyond our capacity to rationalize.

The versatility of the reaction **61**→**62** is furthermore shown by the sequence leading to the α-irone-like damascone homolog **65** as described in another patent assigned to the same company.[50]

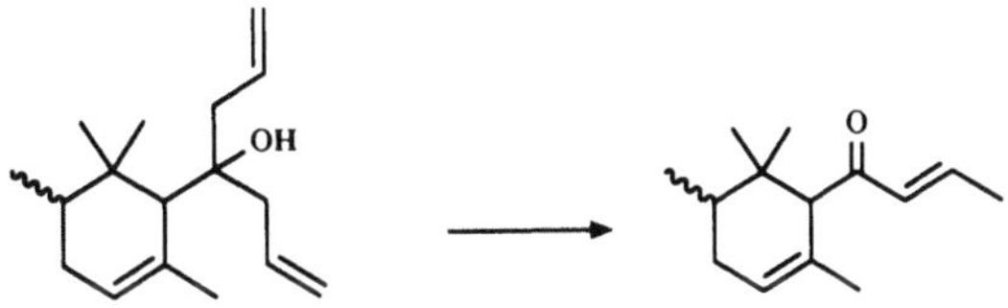

65

In the context of true irones and the aforementioned special interest in γ-isomers, i.e. the isomers having an exo-methylene group directly attached to the cyclohexane ring (valid for ionones, irones and damascones), research work in this field has also been reflected in some patent applications from a Japanese company.[51–53] For instance, an intermediate for racemic γ-irone has been prepared with a reduced number of steps, thus making the envisaged aroma chemical more economic. In the patent, only one step—the hydrogenation of 3,3,4-trimethylcyclohex-1-ene carbaldehyde (**66**) to the corresponding alcohol **67**—is claimed,[51] but the target compound is actually the γ-irone **68**.

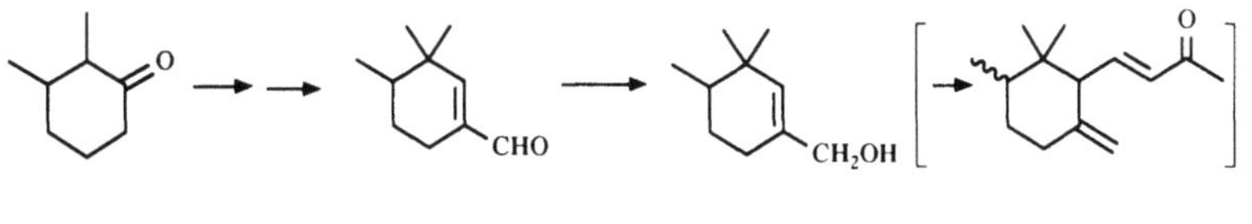

66 68

Another application deals with the transformation of an α-cyano-amide to the very interesting aldehydic intermediate **67**,[52] which is easily reacted with the Wittig salt of chloroacetone to yield cis-**68**.[53]

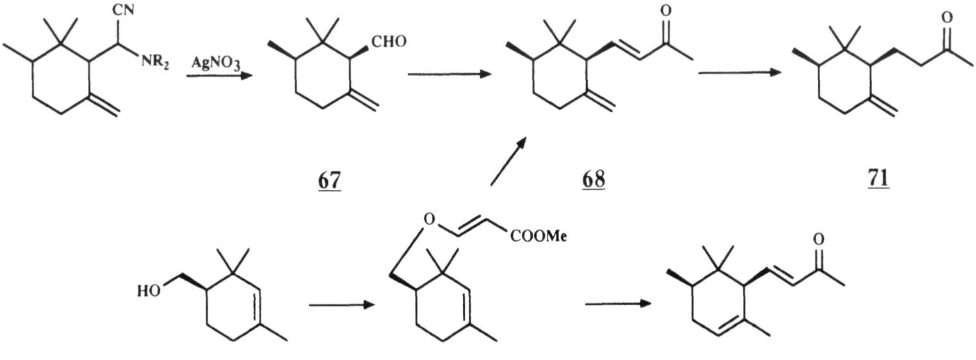

67 68 71

69 70

cis-γ-Irone (**68**) was also the target compound of a stereoselective synthesis from methyl 5-[2,2,4-trimethyl-3-cyclohexen-1-yl]-4-oxa-2(*E*)-pentenoate

(**69**),[54] which also constitutes the intermediate in a synthesis of racemic *trans*-α-irone (**70**).[55] The optically active (−)-(2*S*,6*S*)-*trans*-γ-irone in mixture with (+)-(2*S*,6*S*)-*trans*-α-irone has been prepared by the ene reaction of (−)-(*S*)-1,3,3-tetramethylcyclohex-1-ene with but-3-yn-2-one.[56]

A new manufacturing process for **67** uses an organic acid instead of silver nitrate.[57] It works also for the preparation of the intermediates in the γ-ionone and γ-damascone synthesis. Furthermore, selective hydrogenation of **68** to **71** elicited a certain interest in the application of this new fragrant chemical.[58]

When γ-cyclolavandulal (**72**) is reacted with allyl chloride and magnesium, we obtain a new secondary alcohol (**73**) with a substantially floral-like, fresh and mild aroma. Jones oxidation and isomerization of the terminal double bond with strong acids finally leads to γ-isodamascone (**74**).[59,60] Compound **72** is also the starting material for a Wittig reaction yielding the new γ-isoionone **75** with a floral, violet-like scent.[61]

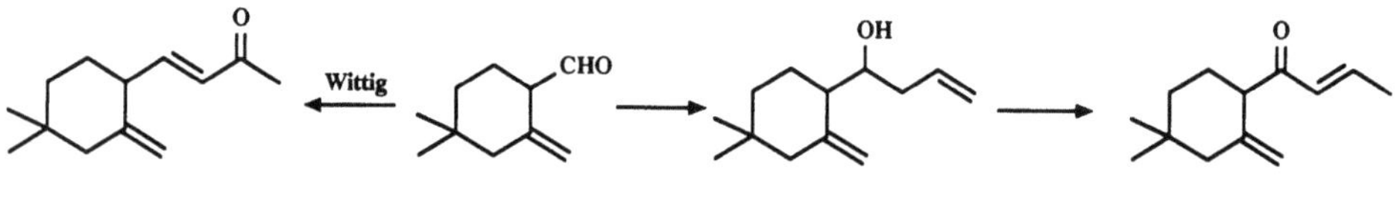

75 72 73 74

Looking at other odorants having a terpene skeleton and floral fragrance, we find in the recent patent literature three ethers which should be mentioned here (**76**–**78**).[62–64] A violet fragrance character is attributed to **76**, the

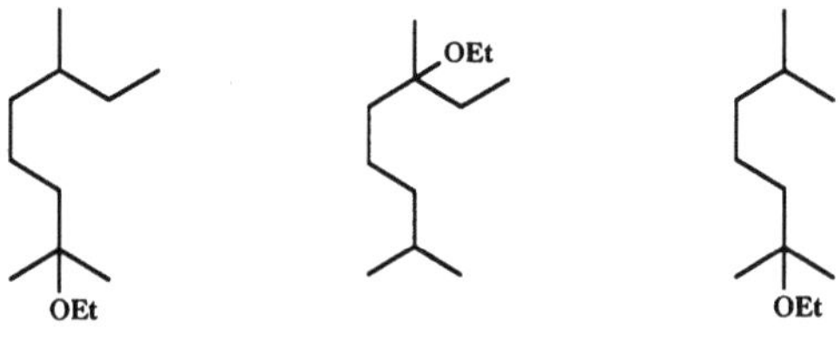

76	77	78
mild, fresh, clean green, floral (violet)	mild, fresh, green floral (freesia)	fresh, petals fruity – floral

manufacture of which is quite simple using first sodium hydride in paraffin to prepare the salt of 2,6-dimethyloctan-2-ol, followed by the reaction with diethyl sulfate, as usual in an ethyl ether synthesis.

Within the range of other important rose oil constituents, rose oxide (**80**) has surely found its place in the modern palette of synthetic, but nature-identical, aroma chemicals that perfumers throughout the world will use. With its identification in rose oil[65] and geranium oil,[66] a new topic of great interest in synthetic work was thus created in the mid-1960s.

The first syntheses, which will not be discussed here, always started from β-citronellol; but later, in 1981, the same starting material served for two synthetic pathways.[67,68] The first route passed via a sulfur-containing diol and the allylic alcohol **79** to **80**, whereas a Japanese company chose anodic methoxylation followed by catalytic desalkoxylation in the presence of a palladium or nickel complex and cyclization of the resulting 3,7-dimethyl-5,7-octadien-1-ol.

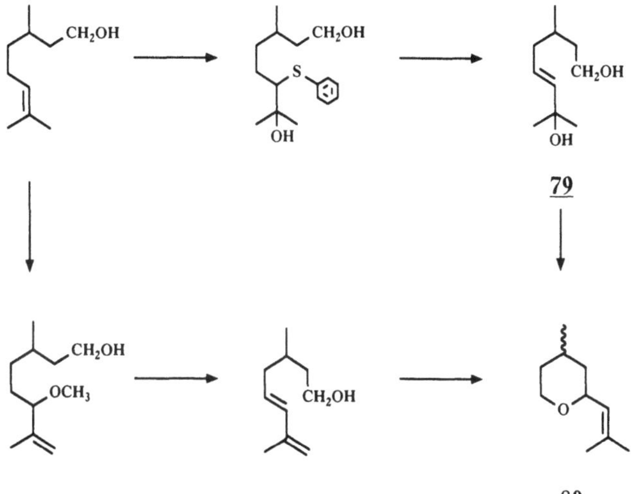

80

3-Methyl-δ-valerolactone is the starting material in a new scheme to synthesize **80**.[69] The Grignard reaction with methylallyl halide furnishes the tertiary alcohol **81** (cleavage of the lactone ring) and treatment with a strong base such as potassium *tert.*-butoxide gives the ketone **82**, which can be reduced to the corresponding diol and finally cyclized to **80**.

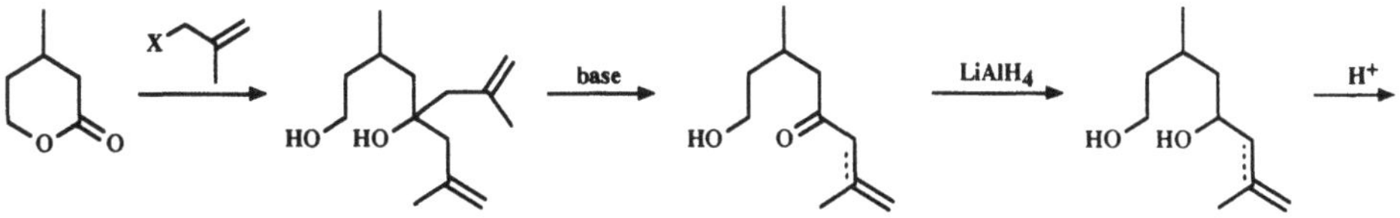

80

81 82

An interesting $C_5 + C_5$-concept was realized first by Tyman & Willis[70] since isoprenol and prenal became available on an industrial scale. The partial hydrogenation of **83** was conducted in the presence of hexachloroplatinic acid, giving 91% of the highly desirable *cis* isomer of **80**. More than 85% of the preferred *cis* isomer **80a** was also obtained by catalytic hydrogenation in the presence of platinum oxide and a cation exchanger.[71]

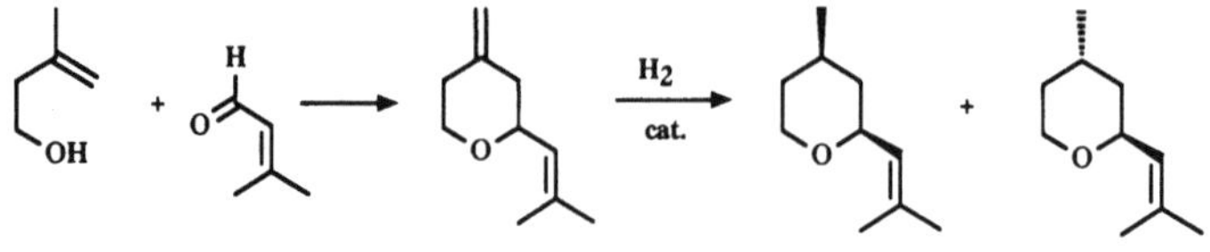

83 80a 80b

major minor

Excellent floral notes adhere to another tetrahydropyran derivative, **84**, which is commercially available as Florol®.[72] It could not be patented in view of earlier Russian publications.

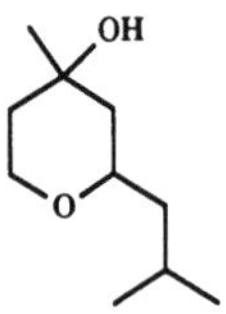

84

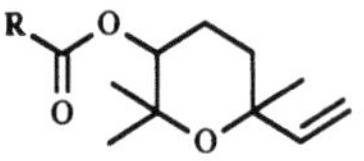

85

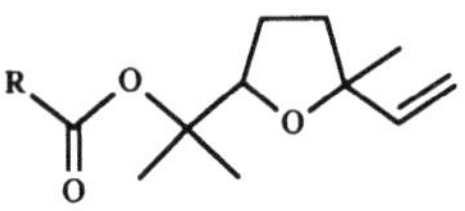

86

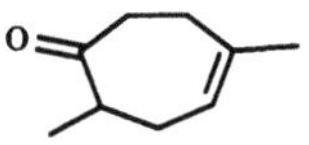

87

The well-known linalool oxides deliver esters having floral notes with long-lasting effects (e.g. **85**) or with shadows from camphoraceous to woody (e.g. **86**).[73,74] Karahanaenone (**87**) is another useful aromatic material. It can be prepared by a convenient one-step procedure from **86** (heating in the presence of an activated clay catalyst).[75]

A further derivative of the trisubstituted tetrahydrofuran ring system is the γ,δ-unsaturated aldehyde **88**.[76] This compound with a floral, lilac- or rose-like fragrance is produced by reacting the allylic primary alcohol with a lower-alkyl vinyl ether and heating the intermediate vinyl ether in an autoclave.

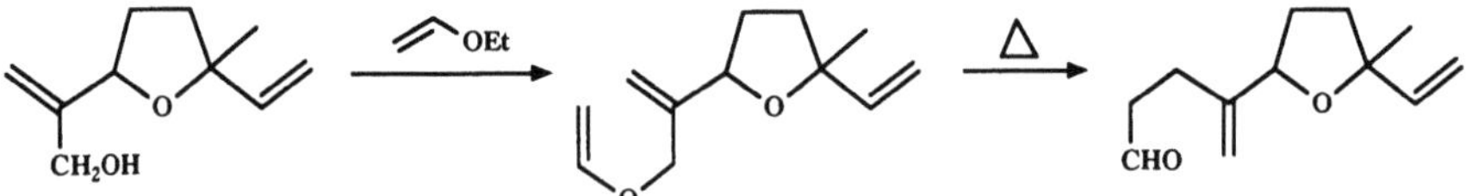

88

Floral–rosy notes are also claimed for a furan derivative with a carbonyl group in the side-chain (**89**), which can easily be prepared in a one-step reaction from the inexpensive raw material sylvane.[77]

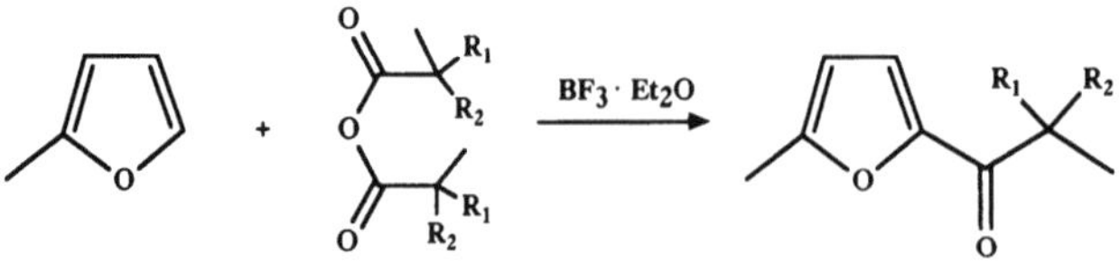

89

R1 = H,Me

R2 = Alkyl

An araliphatic ketone **90** is described as having an outstanding rose fragrance.[78] The use of propionic anhydride instead of propionyl chloride

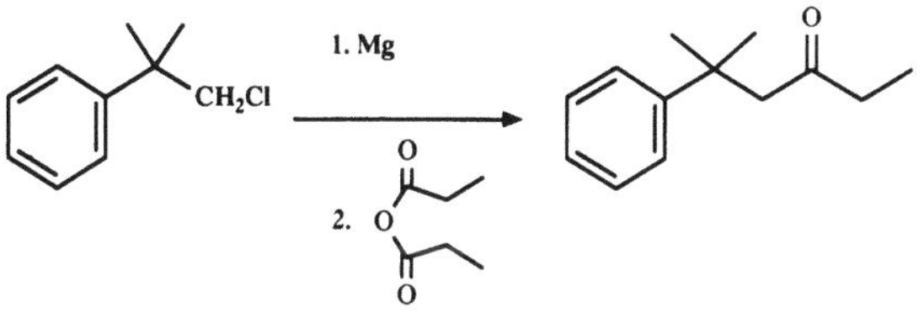

90

(which needs the assistance of cadmium chloride to perform well in the reaction) represents an ecological advantage in the industrial process. We remember here that waste disposal (e.g. of heavy metals) is a matter of first concern to the chemical industry today; we will see below another example of electrochemical routes to l-*tert*-butylbenzaldehyde illustrating the constant endeavour of the fragrance industry to cope with this problem.

Rosy fragrance aspects are furthermore attributed to new derivatives of 2-phenylallyl alcohol (**91**). These floral notes encompass a lilac and spicy shadow in the case of compound **92**, whereas the methylallyl ether **93** exhibits a greener, ozone-like smell with a fruity top note.[79]

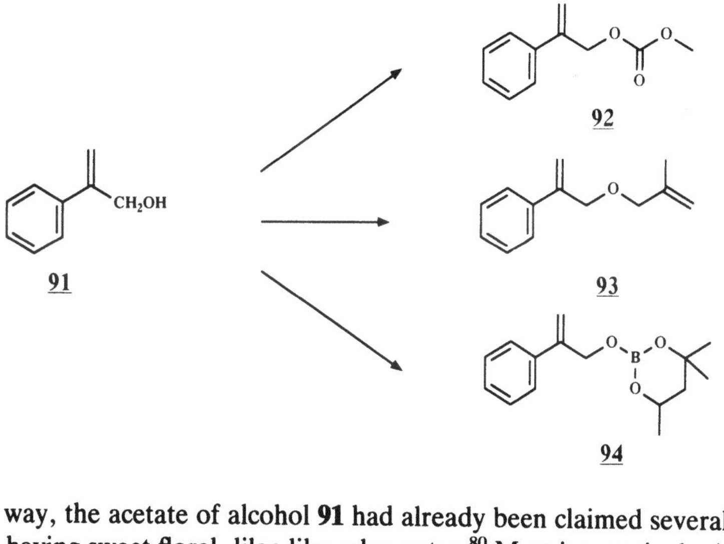

91 92 93 94

By the way, the acetate of alcohol **91** had already been claimed several years before as having sweet floral, lilac-like odor notes.[80] Most interestingly, in both the patents just mentioned we find the hint that even boronate esters (**94**) will have some perfumistic interest. The odor description of compound **94** is rose, cinnamon, carnation, spice, lilac.

The alcohol **91** itself, and processes for preparing the same and similar substances, are described in another patent assigned to International Flavors & Fragrances.[81] Isopropenylbenzene is halogenated and then reacted with an alkali salt of an organic acid, e.g. sodium acetate, to furnish first the ester

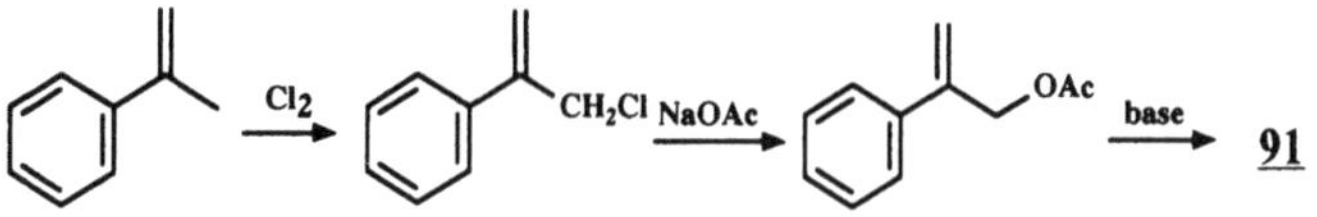

already mentioned which, on saponification, yields the corresponding alcohol with floral notes once again reminiscent of lilac, rose, hyacinth and peony.

It is furthermore reported that the well-known aldehyde **95** may be transformed to the homolog of compound **91**, namely the alcohol **96** with green–floral odor aspects, with the aid of aluminum isopropoxide, which represents a commonly used reagent in the fragrance industry.

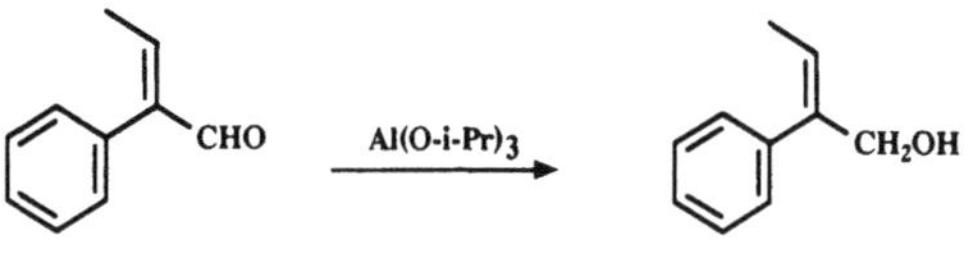

95 96

Isopropenylbenzene (mentioned above) is an industrially available, inexpensive chemical all over the world. This is clearly demonstrated by the interference of two patent applications of competitors residing in the United States[79,81] and West Germany.[82] They dealt with the acetate **98**, now commercially available as Bigaflor® (Henkel). Its production follows the pathway via an epoxide **97**, which at the same time is a useful intermediate for alkali-stable 1,3-dioxolane derivatives (**99**) with floral scents reminiscent of gardenia.[83]

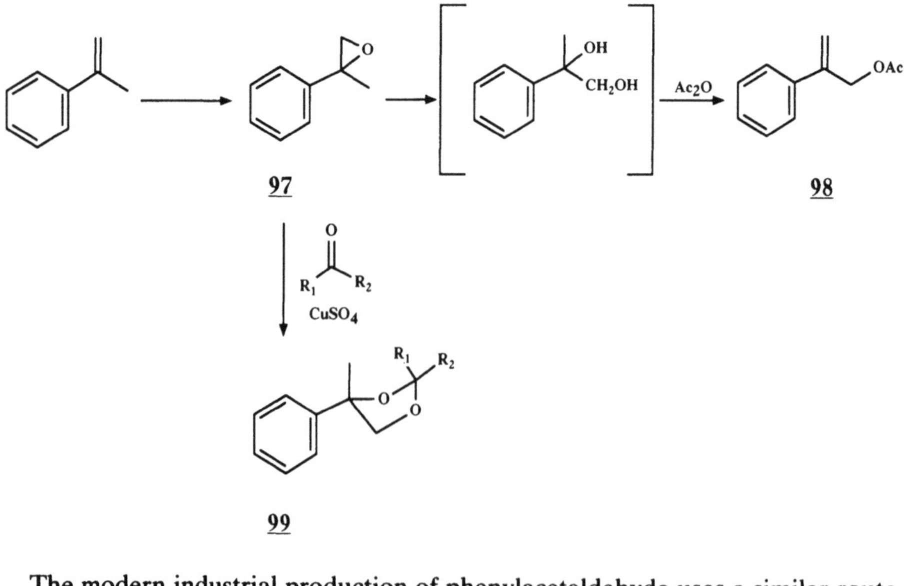

The modern industrial production of phenylacetaldehyde uses a similar route starting with the epoxidation of styrene[84] or the carbonylation of benzyl chloride.[85] Condensation with 2-methyl-1,3-dithiazine (*n*-butyl-lithium in equimolar amounts is needed) affords an α-hydroxy thioketal which after saponification leads to the hydroxyketone **100** and the corresponding diol. Both compounds have a long-lasting, hyacinth-like floral scent.[86]

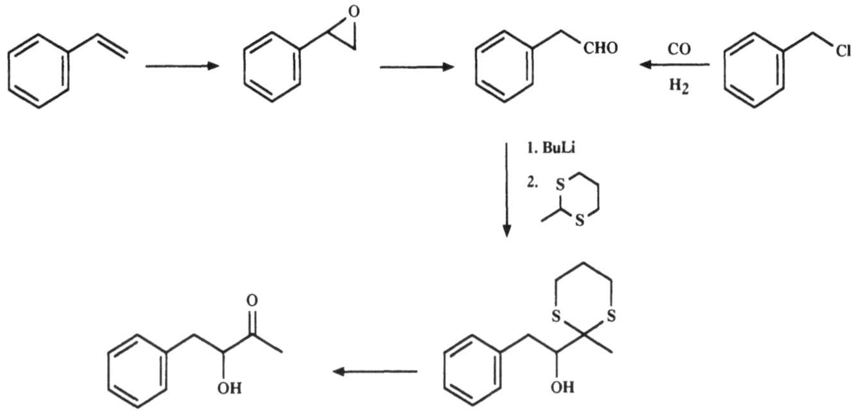

100

Amongst the rare aroma chemicals with a high annual production volume, Lilial® (**105**) is surely a compound that has ranked in one of the top places for more than 20 years. This fact has, of course, stirred the interest of many companies which have an adequate production line of basic chemicals like **101** and **104** at their disposal. Thus, we have seen interesting patent applications describing a great number of possible processes finally yielding the widely used aldehyde **105**, also known under the name of Lilestralis® (BBA) or Lysmeral® (BASF).

To make the intermediate *p*-tert.-butylbenzaldehyde (TBB; **104**) in the most economical way, special endeavours are made to use electrochemical processes. With lead as cathode and PbO_2 as anode in the presence of mineral acid and a higher alkanesulfonic acid,[87] with graphite anodes in the presence of

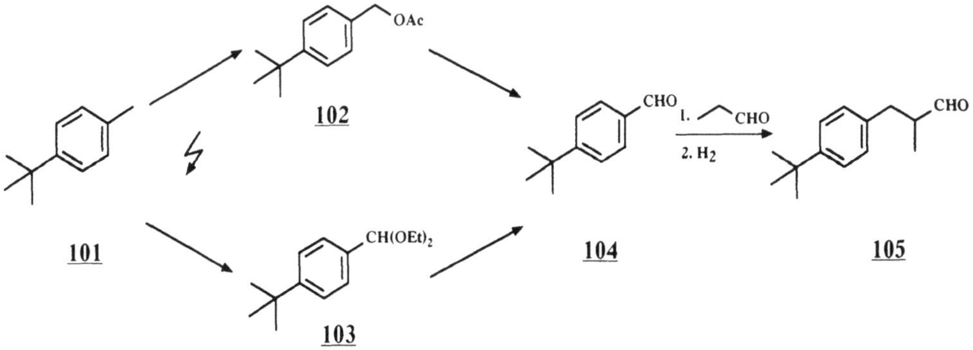

102 101 104 105 103

acetic acid (giving higher yields through the intermediate **102**),[88] or with a very elegant indirect oxidation using the manganese(III) ion as the oxidizing agent,[89] TBB has now become very cheaply available from toluene and isobutene. Another electrochemical oxidation of TBT (**101**) proceeds via the diethylacetal of TBB (**103**).[90]

The progress made by these variations of the electrochemical oxidation of TBT to TBB is really worth mentioning, because this process avoids the production of huge amounts of waste material in the form of manganese sulfate, which was the inevitable by-product of the former industrial production process using MnO_2 as the oxidizing agent.

To produce the desired fragrant chemical **105**, TBB has to be condensed with propionicaldehyde to an α,β-unsaturated intermediate, which after hydrogenation yields the product in question. This well-known sequence can be conducted in an economical way.[91] Nevertheless, further intensive studies were carried out to find other feasible access to 2-methyl-3-(*p*-tert.-butylphenyl)propaldehyde (**105**), *inter alia* the hydroformylation of the enol ether **106**[92] and the *tert.*-butylation of the enamine **107**,[93] or the alcohol **108**.[94]

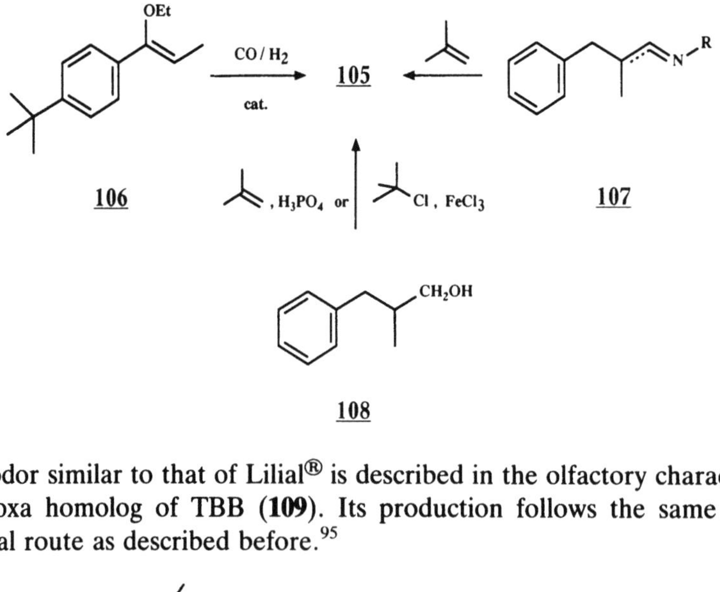

106 107

108

An odor similar to that of Lilial® is described in the olfactory characteristics of an oxa homolog of TBB (**109**). Its production follows the same electrochemical route as described before.[95]

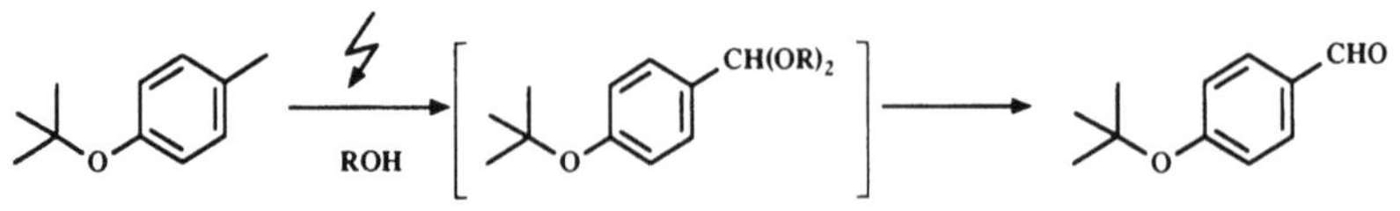

109

A more recent patent application deals with another *tert.*-butylated aromatic acetal analogously prepared. Whereas the dimethyl acetal **110a** has a floral–sweet fragrance, the diethyl acetal **110b** tends more in the direction of a green odorant.[96]

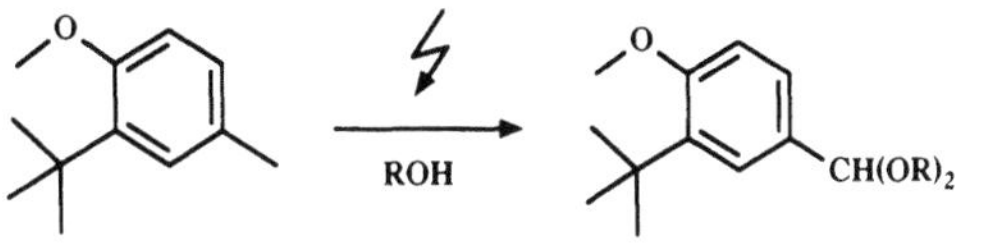

110a (R=Me)

110b (R=Et)

We already stated that it is very typical of industrial thinking to consider possible pedigrees for economic reasons. The aforementioned patents give a good impression of how best to use the raw material base of *p*-cresol and isobutene. The same holds for the experience gained of electrochemical oxidation as a promising unit operation.

The following series of patents from a Japanese company reflect the same facts with respect to the commercially useful heliotropin (or piperonal) **112**, which has already been discussed in Chapter 14 of this book (see p. 403). We note first an electrochemical oxidation of isosafrole glycol (**111**),[97] which—according to a patent application two years later—was followed by the transformation of the halide **113** to **112** via electrolysis in water.[98] This intermediate **113** is easily accessible by reacting methylene dioxybenzene with paraformaldehyde and hydrochloric acid. It serves not only for the production of heliotropin, but leads in a subsequent condensation reaction with isobutyraldehyde to a new fragrant material **114** with a floral smell.[99]

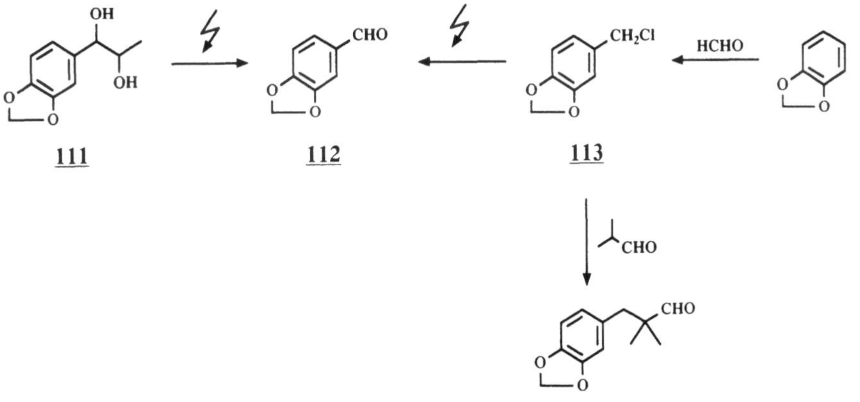

111 112 113 114

Such α,α-dialkylated substances often prove to be much more stable than the non- or only mono-alkylated analogs. We find a similar structural feature in the so-called 'Muguet-Alcohol' **115**,[100] which is produced from benzyl chloride through the intermediate aldehyde using phase transfer catalysis in the first step.

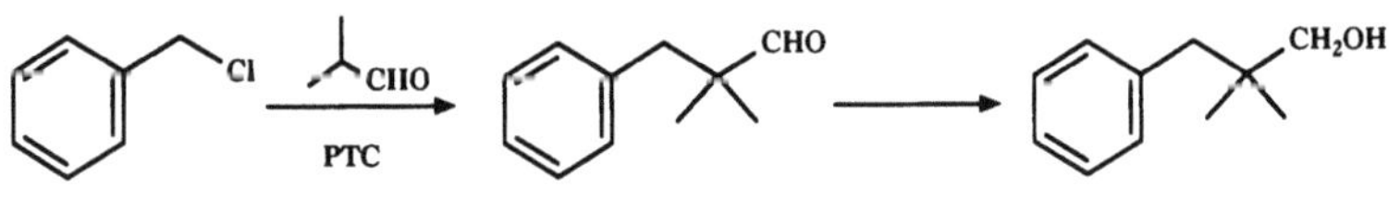

115

A group of secondary alcohols (**117**) with strong floral, fresh and fruity odor notes should be discussed next, as these substances derive from a former intermediate (**116**) of the Lilial® synthesis.[101] A vinyl-Grignard reaction leads to these interesting compounds, interesting because the molecular weight (R_3 and R_4 represent hydrogen or lower alkyl groups) may reach the borderline normally accepted for fragrant molecules.

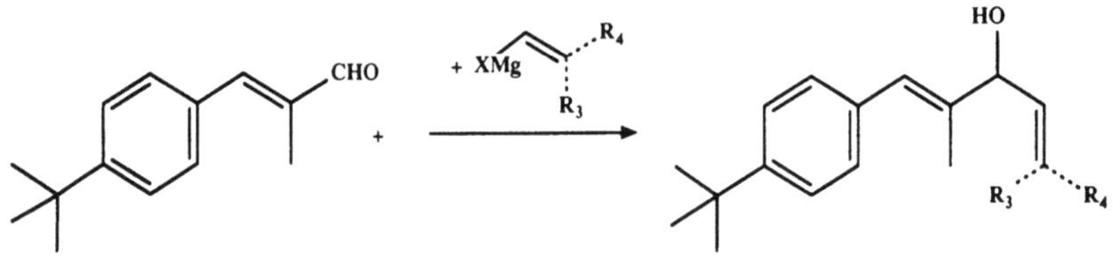

116 117

In the context of the structure of Lilial® (**105**), the isomeric aldehyde **118** must also be mentioned in view of its flowery odor characteristics.[102] In this compound, the α-carbon atom adjacent to the carbonyl group is not substituted.

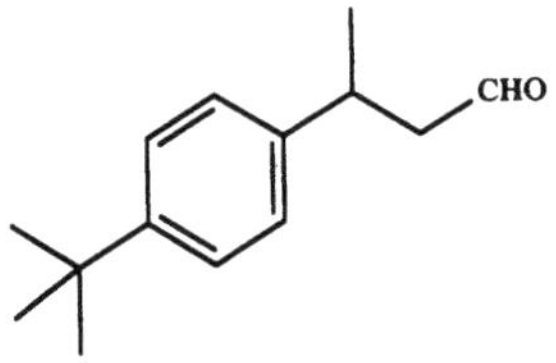

118

A very recent patent application describes another floral odorant (**120**), the synthesis of which is based on the industrially available *m*-di-isopropenyl benzene (**119**).[103] Here we make acquaintance with a very useful industrial procedure, now also applied by fragrance companies. The hydroformylation process needs only carbon monoxide and hydrogen together with a catalyst system (which is sometimes special).

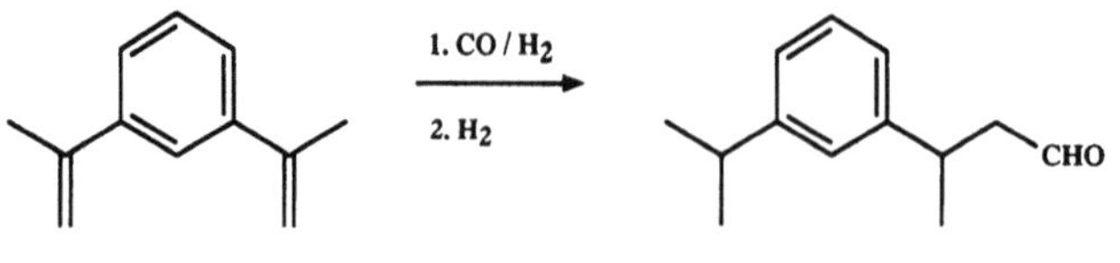

119 120

The *meta* substitution in the aromatic nucleus is also the subject of a patent application covering alcohols such as **121**, **122** or **123**.[104] As seen by the odor description beneath the formula number, they all have floral aspects and can easily be prepared from 3-methyl- or 3,5-dimethyl-substituted aromatics by industrially applied unit reactions such as condensation, alkylation and hydrogenation.

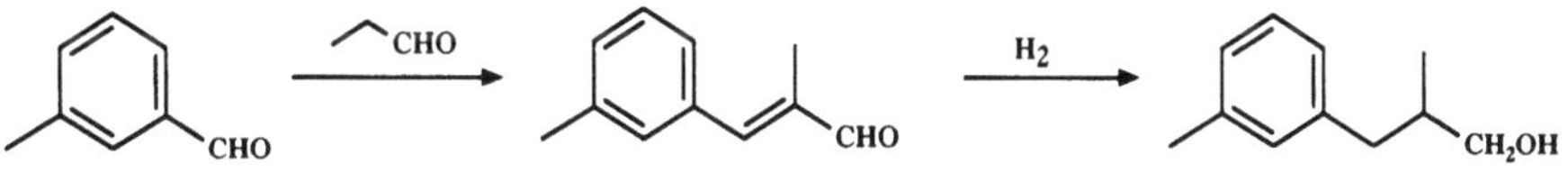

121

floral, aqueous, green

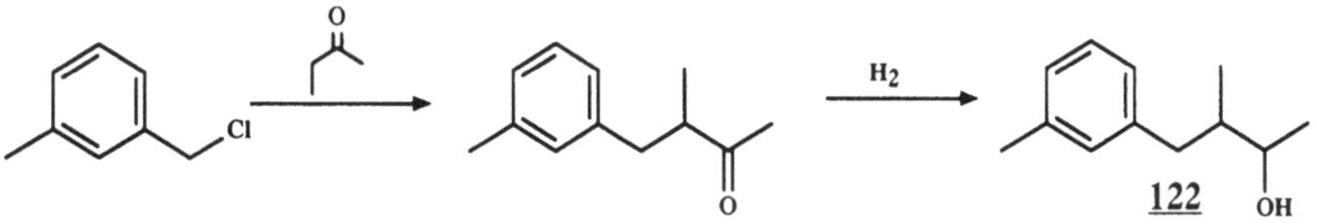

122

rosy, linden blossom, green

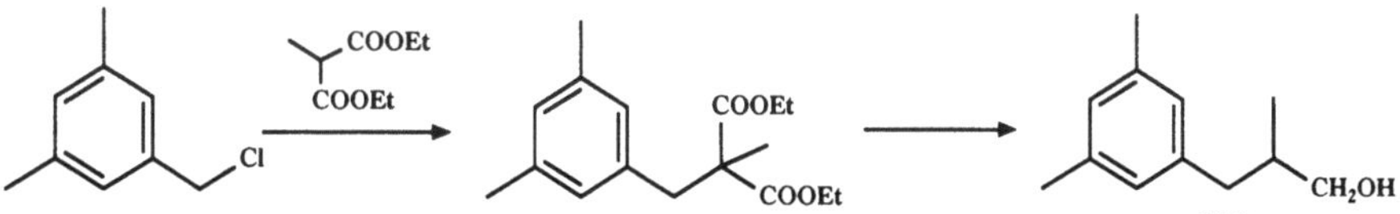

123

rosy, green

The *meta* substitution continues to be of interest, even in the field of saturated six-membered-ring derivatives. A *meta*-dialkylated cyclohexanone, **126**, for instance, offers a mild, soft, floral, violet-like fragrance with a woody touch, thus showing differences in kind in comparison with the well-known mono-*tert.*-butylated compounds (ketone, alcohol or acetate). The starting material is an inexpensive phenol (**124**), hydrogenation of which to the ring-saturated alcohol **125** is followed by dehydrogenation using copper chromite to yield the ketone **126**.[105]

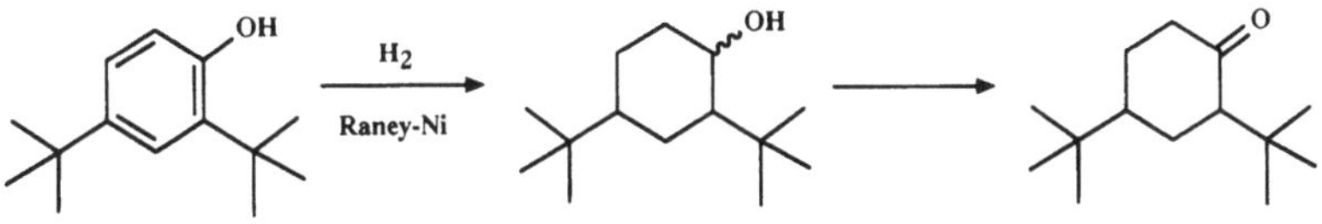

124 125 126

(cis/trans)

Alkyl-substituted *p*-carbalkoxycyclohexanones (**128**) show a similar substitution pattern. They have floral, jasmine-like, green and citrusy odor facets and are easily prepared by hydrogenation of an appropriate phenol and condensation with the aldehyde of desired chain length.[106]

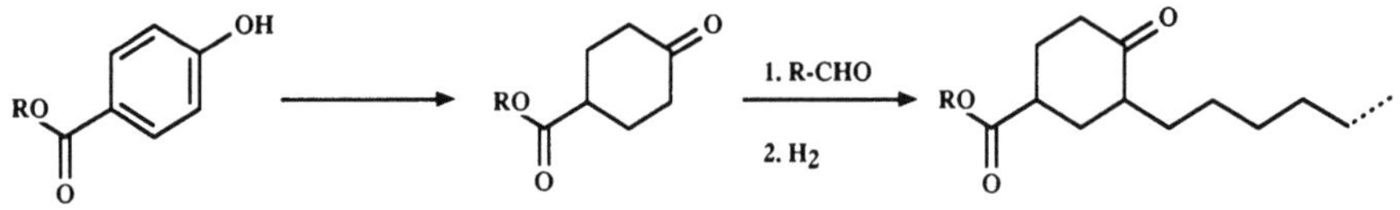

127 128

When cyclohexanone itself is aldolized with *n*-pentanal, the resulting 2-pentylidene cyclohexanone (**129**) may undergo an enol acetate formation (compound **130**) with acetic anhydride and an acid catalyst. Under the influence of heat, this intermediate rearranges to 2-pentyl-2-cyclohexenone (**131**) (primary goal of the experiments) and furthermore gives rise to the unexpected formation of diketone **132** with interesting odor properties (floral, jasmine-like, fatty and herbaceous).[107]

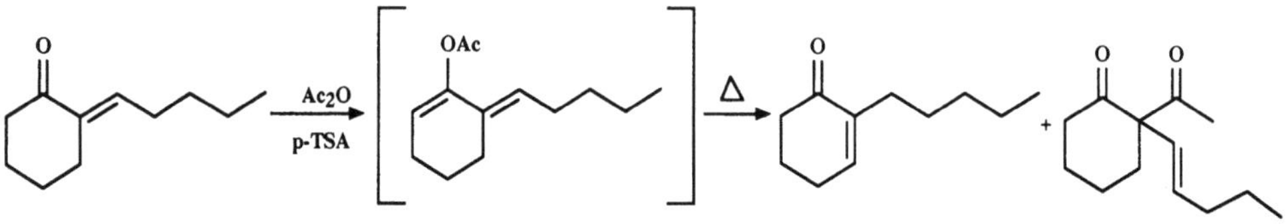

129 130 131 132

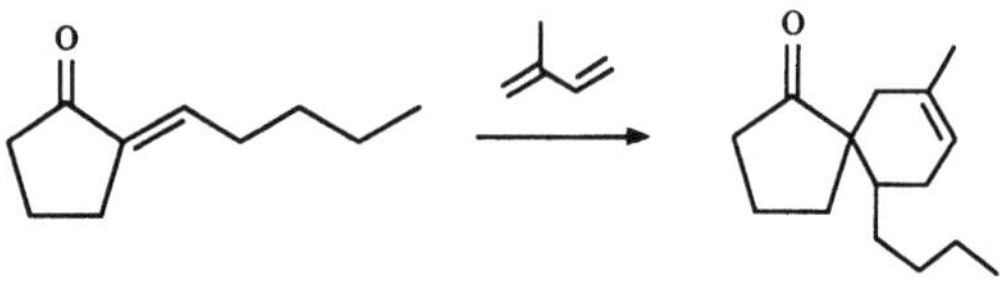

133

A similar starting material, 2-pentylidene cyclopentanone, reacts with isoprene to a bicyclic spiroketone **133** which also exhibits flowery, jasmine-like notes with more woody and spicy undertones.[108]

VEBA keton, a readily available starting material which has already been mentioned, can be aldolized with butyric aldehyde or pentanal. After acid-catalyzed isomerization, a jasmone-like smelling condensation product of type **134** is obtained.[109]

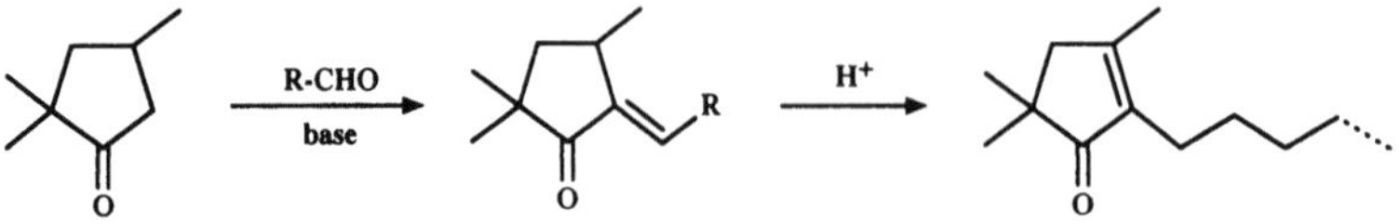

134

It would surely be an unpardonable omission to report on flowery, jasmine-like fragrances without mentioning any novelties in the field of jasmone itself and the jasmonate family.

Jasmone (**136a**) and dihydrojasmone (**136b**) are mentioned in a Swiss patent dealing with the transformation of compound **135** to **136**.[110]

135 136a/b

A series of inventive patents and patent applications provide a new process for the production not only of *cis*-jasmone (**136a**) and its dihydro derivative (**136b**), but also of important intermediates in the prostaglandin field. It has been well known since the beginning of the 1980s that the secondary alcohol **139** can be transformed to cyclopentenones of type **140**, the isomerization and reduction of which affords **136a** or **136b**.[111,112] The process was refined in the

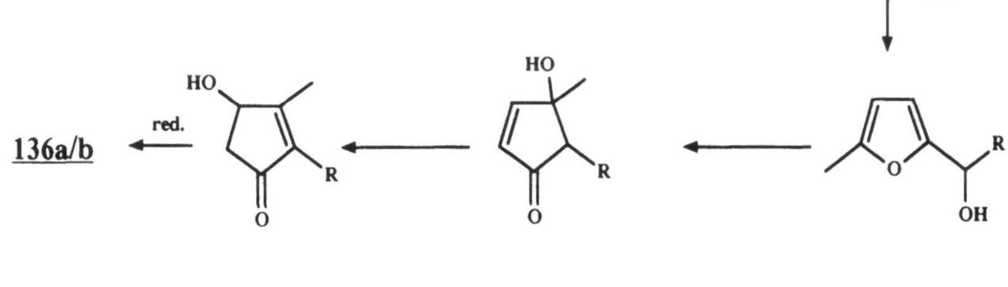

141 140 139

following years. Thus, the 5-halomethylfurfural (**137**) can be produced by acid decomposition of a saccharide in a water/organic solvent system in the presence of magnesium halide.[113] The catalytic reduction of **137** to **138** is described in another patent.[114] This route has some advantages because it avoids the use of toxic reagents like $POCl_3$ or phosgene, formerly used in the preparation of **138** from 5-methylfuran.

The production of the intermediate of type **139** is the subject of several improvements, especially in respect of the use of phase-transfer conditions in the coupling of **138** with propynyl or allyl chloride in the presence of zinc metal, an inorganic ammonium salt and an organic quaternary ammonium surfactant.[115–117]

The isomerization of compounds of type **140** proceeds very well when the acetate of **140** (with R = propynyl) is treated with sulfuric acid at 0°C. Compound **141** is obtained in 91% yield.[118]

Heating of 4-cyclopentenone derivatives in the presence of 1,8-diazabicyclo[5.6]undecene and treatment of the resulting intermediate with malonic acid esters yields the diester **142**, which is able to undergo further conversion to jasmonate(s).[119]

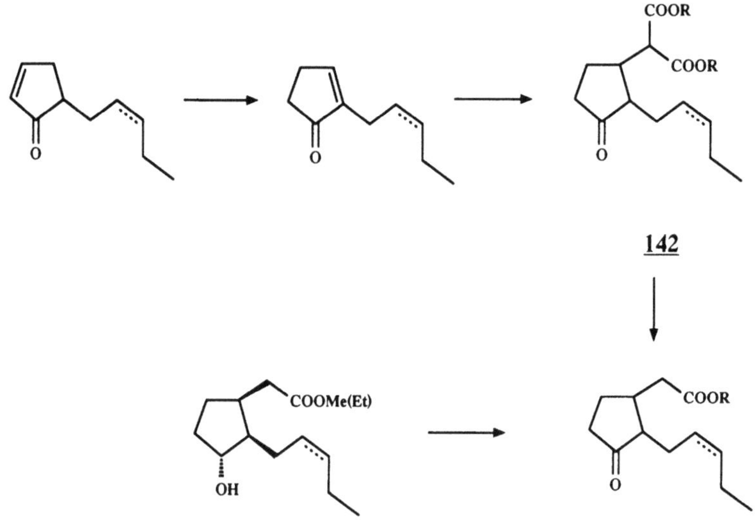

142

143

144a/b

In the field of jasmonate-type odorants, it should be noted that *cis*-2-alk(en)yl-3-alkoxycarbonylmethylcyclopentanol(s) (**143**) are new compounds which can easily be converted to the corresponding jasmonates (**144a**) and dihydrojasmonates (**144b**).[120,121] Finally, a new route for γ-jasmolactone (**145**) comprises the reaction of *cis*-4-heptenal with acrolein diethylacetal followed by reduction and oxidative cyclization.[122]

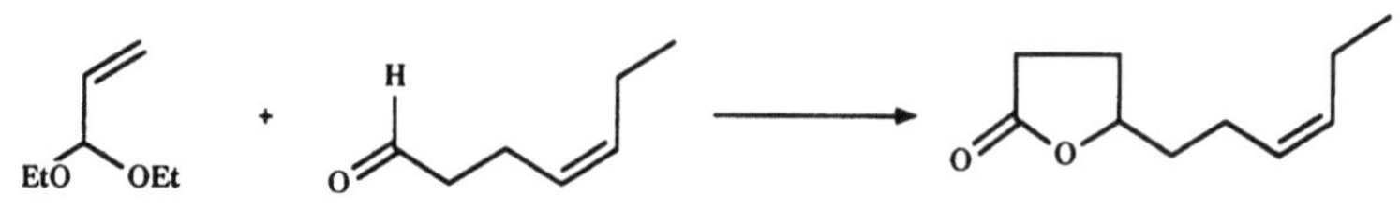

145

With respect to jasmine notes, it is also worthwhile to direct attention to a nitrile (**146**) patented in 1986,[123] because the floral fragrance is very tenacious and stable over a broad pH range.

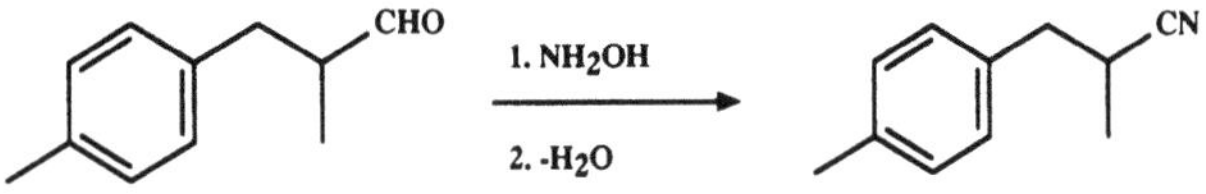

146

A systematic evaluation of odor and stability properties of new chemicals is of primordial interest to the fragrance industry. Thus, for instance, the search for better access to macrocyclic musks passed through the synthesis of a γ,δ-unsaturated aldehyde **147**, which proved its usefulness in its own right as a floral–fresh and very tenacious odorant.[124]

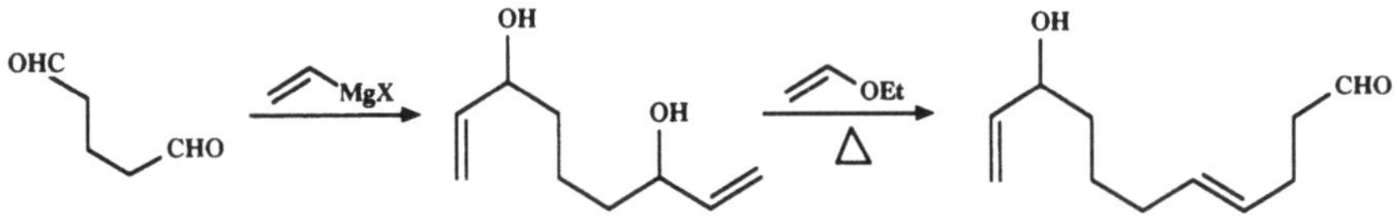

147

The γ,δ-unsaturation in long-chain aldehydes seems to be a key element with respect to diffusiveness and radiance, as shown some time earlier by the introduction of *trans*-4-decenal (**148**)[125] to the perfumer's shelf. This aldehyde is now commercially available.

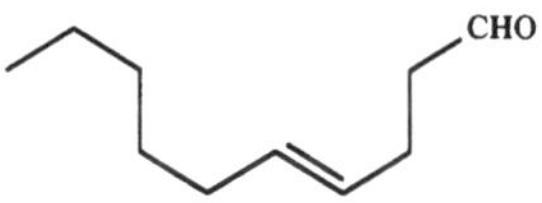

148

In the field of aliphatic compounds we have already encountered esters with a tertiary butyl group when discussing new odorants with fruity aspects. Two patent applications made in 1986 now reveal that fresh–floral smells can be attributed to aldehydes of the type **149**[126] and alcohols or their acetates (**150**) having similar structural features.[127]

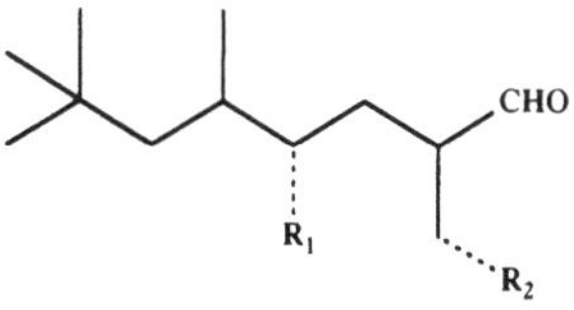

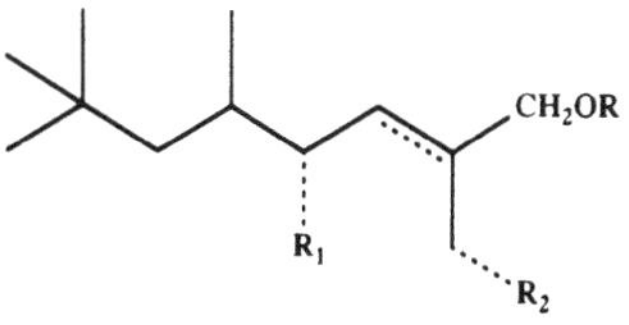

149 150

(R_1 , R_2 = H , Me) (R = H , Ac)

Pivalic acid esters (**151a–c**) are another interesting group of floral–fruity perfume chemicals.[128] They have at the same time lily-of-the-valley and rose aspects combined with pear- and bergamot-like shadows. The compound **151a** may have a methyl substituent in the 2-, 3- or 4-position of the aromatic ring.

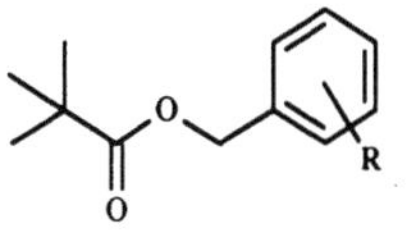

151a

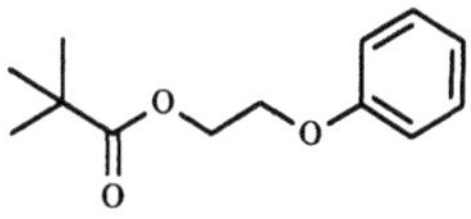

151b

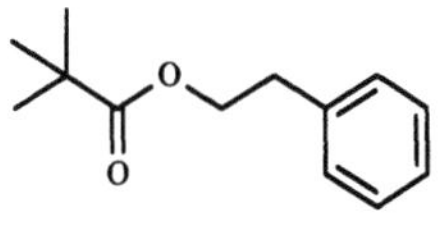

151c

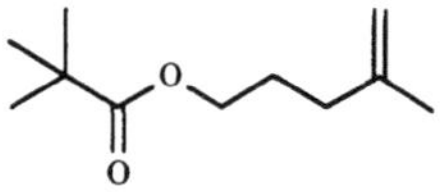

151d

Interestingly enough, another company has patented the completely aliphatic ester **151d** in view of its rose oxide-like odor combined with a so-called salicylate shadow.[129]

Esters of 2-methylpentanoic acid (**152**) are claimed in a further patent granted to the same company,[130] thus showing that even in the 1980s simple organic molecules are patentable subjects of interest to the fragrance industry. The exchange of the acid/alcohol part leads to 2-methylpentyl salicylate (**153**) with sweet–floral olfactory properties.[131]

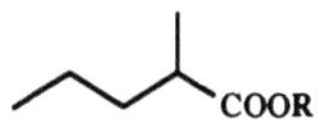

152
floral, fruity

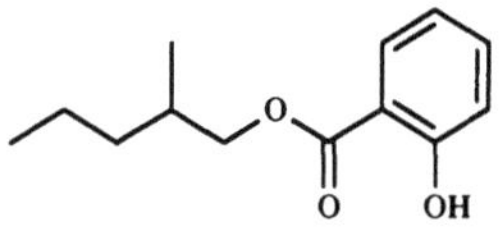

153
floral, sweet

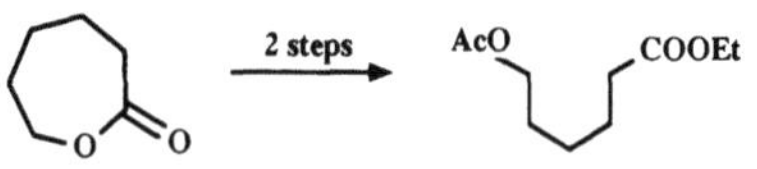

154

A very cheap bifunctional aliphatic ester is represented by formula **154**.[132] It will soon be available on the market under the name Fruitaflor® (Givaudan), which is already characteristic of the adjectives to be used in the odor description of this compound, which is produced from ε-caprolactone in a two-step reaction sequence: scission of the lactone ring with sodium hydrogen sulfate in the presence of ethanol and esterification of the resulting ω-hydroxyhexanoic acid ester with acetic anhydride.

With all these esters, it has been shown that the fragrance industry is always able to realize odorants with a reasonable price/odor value relationship, offering at least some advantages over the well-known fragrance compounds with similar odor characteristics. In this context, prenyl salicylate (**155**) should be mentioned in comparison with the known isoamyl salicylate, **156**. A modern esterification process was claimed for **155**, to which a powerful floral, sweet–herbaceous fragrance is attributed showing a better and more harmonious odor profile than **156**.[133]

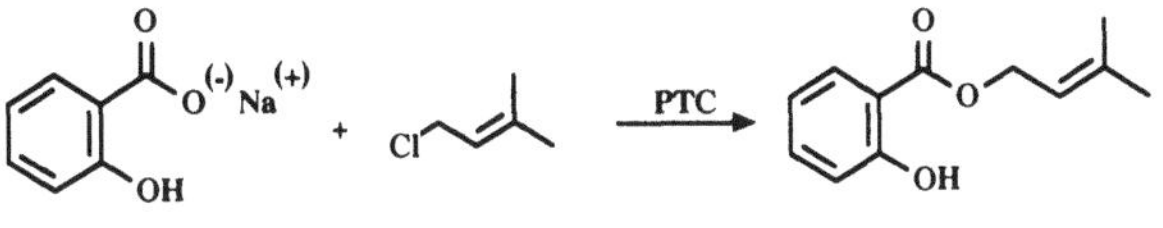

155

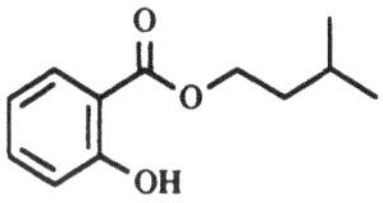

156

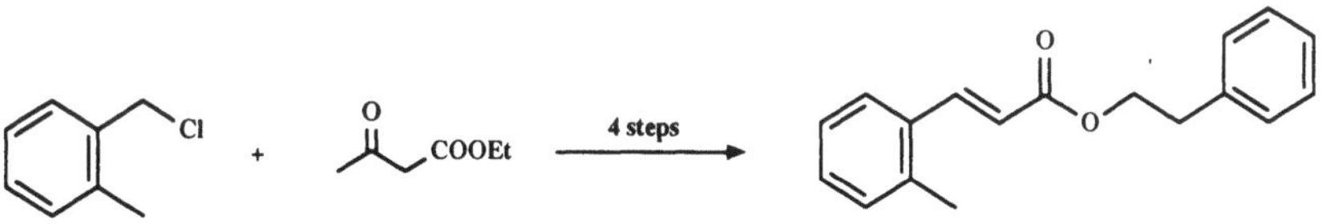

157

A similar small change in the molecular structure is observed in the case of the aromatic ester **157**, but the introduction of the methyl group in the *ortho* position of the cinnamic acid part of the molecule leads to a strong exaltation when this new chemical is used in flowery perfume compositions.[134]

We conclude the review of new materials exhibiting floral aspects in their odor profile by mentioning an interesting reaction sequence using the catalytic effect of the palladium chloride/triphenylphosphine system. Cheap raw materials thus deliver secondary (**158**) or primary alcohols (**159**) of interest to perfumery in view of their floral–green smell.[135]

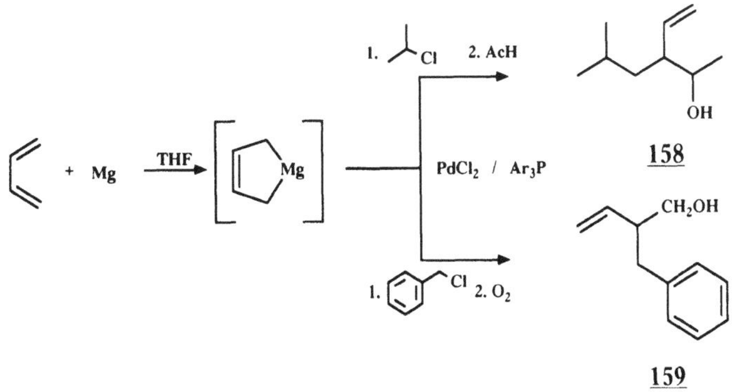

158

159

These compounds offer us a good link to the next group of new odorants to be discussed in the framework of this chapter. But we do not want to leave the flowery aspects without recommending a publication in which a perfumer tells us something about his point of view 'when speaking through flowers'.[137]

REFERENCES

Fruity

1. Lamparsky, D. & Schudel, P., *p*-Menthane-8-thiol-3-one, a new component of Buchu leaf oil. *Tetrahedron Lett.* (1971) 3323.
2. Sundt, E., Willhalm, B., Chappaz, R. & Ohloff, G., Das organische Prinzip von Cassis Flavor in Buccublätteröl. *Helv. Chim. Acta,* **54** (1971) 1801.
3. Demole, E., Enggist, P. & Ohloff, G., 1-*p*-Menthene-8-thiol: a powerful flavor impact constituent of grapefruit juice (*Citrus paradisi* MacFayden). *Helv. Chim. Acta,* **65** (1982) 1785.
4. European Patent 54 847 (priority CH 23 Dec. 1980) assigned to Firmenich.
5. US Patent 4 536 583 (priority US 1 Aug. 1983) assigned to IFF.
6. US Patent 4 584 128 (priority US 24 Aug. 1984) assigned to IFF.
7. Japanese Patent Appln 01 052 711 (priority JA 24 Aug. 1987) filed by Soda Koryo.
8. Demole, E. & Enggist, P., Further investigation of grapefruit juice flavor components: valencane- and eudesmane-type sesquiterpene ketones. *Helv. Chim. Acta,* **66** (1983) 1381.
9. European Patent 110 034 (priority CH 2 Nov. 1982) assigned to Firmenich.
10. European Patent 167 709 (priority CH 12 July 1984) assigned to Firmenich.
11. European Patent Appln 251 050 (priority DE 15 Mar. 1983) assigned to BASF (see also US Patent 4 550 211).
12. German Patent Appln 3 440 825 (priority 8 Nov. 1984) assigned to BASF (see also European Patent 183 970).
13. US Patent 4 843 061 (priority NL 13 June 1986) assigned to Naarden (see also European Patent 251 370).
14. European Patent 71 248 (priority DE 29 July 1981) assigned to Dragoco.
15. European Patent 45 861 (priority CH 8 Aug. 1980 and 15 June 1981) assigned to Givaudan (see also US Patent 4 426 321).
16. US Patent 4 537 702 (priority NL 13 May 1982) assigned to Naarden.
17. US Patent 4 179 448 (priority US 11 Oct. 1977) assigned to Firmenich.
18. US Patent 4 011 245 (priority CH 11 Mar. 1975) assigned to Givaudan.
19. US Patent 4 072 719 (priority CH 11 Mar. 1975) assigned to Givaudan.
20. Boix Camps, A., Perfumery techniques in evolution, Part III. *Perf. & Flavorist,* **11**(3), (1986) 9.
21. US Patent 4 465 618 (priority CH 8 Feb. 1980) assigned to Firmenich.
22. European Patent 21 100 (priority CH 13 June 1979) assigned to Givaudan (see also US Patent 4 375 001).
23. European Patent 56 109 (priority CH 13 Jan. 1981) assigned to Firmenich (see also US Patent 4 411 829).
24. European Patent 24 306 (priority CH 10 Aug. 1979) assigned to Givaudan (see also US Patent 4 406 828).
25. European Patent 135 719 (priority DE 5 Aug. 1983) assigned to Consortium Elektrochem. Industrie (Wacker).
26. US Patent 4 631 146 (priority DE 17 Nov. 1983) assigned to Consortium Elektrochem. Industrie (Wacker).

27. Japanese Patent Appln 61 158 943 (priority JA 29 Dec. 1984) filed by Hasegawa.
28. Japanese Patent Appln 61 158 958 (priority JA 29 Dec. 1984) filed by Hasegawa.
29. Sell, C. S., Abhängigkeit der Fruchtnoten von der molekularen Struktur, *Seifen-Oele-Fette-Wachse*, **112** (1986) 267.
30. US Patent 4 267 075 (priority DE 23 Dec. 1977) assigned to Henkel.
31. European Patent 39 032 (priority DE 28 Apr. 1980) assigned to Henkel.
32. US Patent 4 525 298 (priority CH 7 Dec. 1981) assigned to Firmenich (see also German Patent 3 245 047).
33. European Patent 34 334 (priority CH 13 Feb. 1980) assigned to Firmenich.
34. European Patent 134 613 (priority NL 17 Aug. 1983) assigned to Naarden.
35. European Patent 177 807 (priority CH 10 Oct. 1984) assigned to Givaudan.
36. Japanese Patent 87 020 962 (priority JA 2 Nov. 1979) assigned to Hasegawa.
37. European Patent 146 103 (priority DE 17 Dec. 1983) assigned to Haarmann & Reimer (see also US Patent 4 654 168).
38. European Patent 196 595 (priority DE 4 Apr. 1985) assigned to Haarmann & Reimer.
39. European Patent 231 556 (priority NL 23 Jan. 1986) assigned to Naarden (Quest Int.) (see also US Patent 4 760 050).

Floral

40. Demole, E., Enggist, P., Säuberli, U., Stoll, M. & Kovacs, E. sz, Structure et synthèse de la damascénone, constituant odorant de l'essence de rose bulgare (*Rosa damascena* Mill.). *Helv. Chim. Acta*, **53** (1970) 541.
41. European Patent 93 840 (priority CH 20 Apr. 1982) assigned to Firmenich.
42. European Patent 112 441 (priority CH 18 Nov. 1982) assigned to Firmenich.
43. European Patent 192 931 (priority CH 25 Jan. 1985) assigned to Firmenich (see also US Patent 4 670 109).
44. Naef, F. & Decorzant, R., Grignard and hydride addition to a ketene intermediate: a novel access to α-damascone and α-cyclocitral. *Tetrahedron*, **42** (1986) 3245; cf. European Patent 224 897 (priority CH 4 Dec. 1985) assigned to Firmenich.
45. Fehr, C. & Galindo, J., Synthesis of (*R*)-(+)- and (*S*)-(−)-α-damascone by tandem Grignard reaction–enantioselective protonation: evidence for the intermediacy of a chiral complex. *J. Am. Chem. Soc.*, **110** (1988) 6908.
46. European Patent 260 472 (priority CH 19 Sept. 1986) assigned to Firmenich.
47. US Patent 4 460 792 (priority CH 7 Sept. 1981) assigned to Firmenich (see also German Patent 3 233 175).
48. European Patent 181 475 (priority CH 15 Oct. 1984) assigned to Firmenich.
49. US Patent 3 931 326 (filed on 14 Nov. 1974 as a continuation-in-part application with priority CH 9 Nov. 1967) assigned to Firmenich.
50. European Patent 70 995 (priority CH 23 July 1981) assigned to Firmenich.
51. Japanese Patent Appln 61 243 034 (priority JA 19 Apr. 1985) filed by Hasegawa.
52. Japanese Patent Appln 61 243 035 (priority JA 19 Apr. 1985) filed by Hasegawa.
53. Japanese Patent Appln 61 243 041 (priority JA 19 Apr. 1985) filed by Hasegawa.
54. Nussbaumer, C. & Fráter, G., A stereoselective synthesis of (±)-*cis*-γ-irone. *Helv. Chim. Acta*, **71** (1988) 619.
55. Nussbaumer, C. & Fráter, G., A stereoselective synthesis of (±)-*cis*-α-irone. *J. Org. Chem.*, **52** (1987) 2096.
56. Helmlinger, D. & Fráter, G., Synthesis of (+)-(2*S*,6*S*)-*trans*-α-irone and of (−)-(2*S*,6*S*)-*trans*-γ-irone. *Helv. Chim. Acta*, **72** (1989) 1515.
57. Japanese Patent Appln 62 051 637 (priority JA 31 Aug. 1985) filed by Hasegawa.
58. Japanese Patent Appln 63 139 164 (priority JA 2 Dec. 1986) filed by Hasegawa.
59. Japanese Patent Appln 62 051 662 (priority JA 31 Aug. 1985) filed by Hasegawa.

60. Japanese Patent Appln 63 083 038 (priority JA 26 Sept. 1986) filed by Hasegawa.
61. Japanese Patent Appln 62 051 663 (priority JA 31 Aug. 1985) filed by Hasegawa.
62. Japanese Patent Appln 64 000 040 (priority JA 22 June 1987) filed by Takasago.
63. Japanese Patent Appln 64 000 041 (priority JA 22 June 1987) filed by Takasago.
64. Japanese Patent Appln 64 000 042 (priority JA 22 June 1987) filed by Takasago.
65. Seidel, C. F., Felix, D., Eschenmoser, A., Biemann, K., Palluy, E. & Stoll, M., Die Konstitution des Oxyds $C_{10}H_{18}O$ aus bulgarischem Rosenöl. *Helv. Chim. Acta,* **44** (1961) 598.
66. Naves, Y. R., Lamparsky, D. & Ochsner, P., Présence de tetrahydropyrannes dans l'huile essentielle de géranium. *Bull. Soc. Chim. Fr.* (1961) 645.
67. German Patent 3 240 740 (priority CH 4 Nov. 1981) assigned to Firmenich.
68. US Patent 4 340 544 (priority JA 31 July 1981) assigned to Sumitomo.
69. European Patent 71 708 (priority CH 24 July 1981) assigned to Firmenich (see also US Patent 4 438 283).
70. Tyman, J. P. H. & Willis, B. J., The reaction of 3-alkene-1-ols with aldehydes: synthesis of (±)-*cis*-2-(2-methyl-1-propenyl)-4-methyltetrahydropyran. *Tetrahedron Lett.* (1970) 4507.
71. European Patent 82 401 (priority DE 18 Dec. 1981) assigned to BASF.
72. *Florol,* Perfumer's Document issued by Firmenich.
73. Japanese Patent Appln 59 155 378 (priority JA 21 Feb. 1983) filed by Hasegawa.
74. Japanese Patent Appln 59 130 879 (priority JA 18 Jan. 1983) filed by Hasegawa.
75. Japanese Patent Appln 61 106 532 (priority JA 30 Oct. 1984) filed by Taiyo Koryo.
76. Japanese Patent Appln 59 122 481 (priority JA 28 Dec. 1982) filed by Taiyo Koryo.
77. German Patent 3 527 975 (priority DE 3 Aug. 1985) assigned to Haarmann & Reimer.
78. US Patent 4 450 296 (priority US 5 Feb. 1981) assigned to IFF.
79. US Patent 4 772 583 (priority US 18 Dec. 1987) assigned to IFF.
80. European Patent 35 183 (priority DE 27 Feb. 1980) assigned to Henkel.
81. US Patent 4 756 840 (priority US 27 Sept. 1980) assigned to IFF.
82. German Patent 3 007 232 (priority DE 27 Feb. 1980) assigned to Henkel.
83. European Patent 150 398 (priority DE 11 Jan. 1984) assigned to Henkel.
84. Japanese Patent 87 060 379 (priority JA 8 July 1980) assigned to Toyo Soda Mfg.
85. Japanese Patent 87 012 777 (priority JA 8 Feb. 1980) assigned to Sumitomo.
86. Japanese Patent Appln 63 083 018 (priority JA 29 Sept. 1986) filed by Hasegawa.
87. European Patent 29 995 (priority DE 1 Dec. 1979) assigned to BASF.
88. European Patent 72 914 (priority DE 19 Aug. 1981) assigned to BASF.
89. European Patent 131 001 (priority CH 29 Dec. 1982) assigned to Givaudan (see also US Patent 4 582 942).
90. European Patent 129 795 (priority DE 22 June 1983) assigned to BASF.
91. German Patent 3 105 446 (priority DE 14 Feb. 1981) assigned to BASF.
92. European Patent 52 775 (priority CH 21 Nov 1980) assigned to Givaudan (see also US Patent 4 435 585 and others).
93. European Patent 103 896 (priority CH 21 Sept. 1982) assigned to Givaudan (see also US Patent 4 568 772).
94. European Patent 45 571 (priority GB 9 July 1980) assigned to Bush, Boake, Allen (see also US Patent 4 486 607).
95. German Patent 3 214 658 (priority DE 21 Apr. 1982) assigned to BASF.
96. European Patent Appln 287 954 (priority DE 24 Apr. 1987) filed by BASF.
97. Japanese Patent 87 056 954 (priority JA 28 Sept. 1983) assigned by Takasago.
98. Japanese Patent Appln 62 005 977 (priority JA 3 July 1985) filed by Takasago.
99. Japanese Patent Appln 01 022 868 (priority JA 16 July 1987) filed by Takasago.
100. European Patent 76 493 (priority DE 2 Oct 1981) assigned to Dragoco (see also German Patent 3 139 358).

101. European Patent 185 872 (priority DE 26 Oct. 1984) assigned to BASF.
102. Belgian Patent 901 072 (priority US 2 July 1984) assigned to Chem-Fleur (now Firmenich).
103. US Patent 4 910 346 (priority US 10 Nov. 1988) assigned to Givaudan.
104. World Patent Appln 88 05 770 (priority DE 6 Feb. 1987) filed by Consortium Elektrochem. Industrie (Wacker).
105. European Patent Appln 296 798 (priority JA 22 June 1987) filed by Takasago.
106. US Patent 4 668 817 (priority US 21 Mar. 1986) assigned to IFF.
107. US Patent 4 788 337 (priority US 28 Jan. 1988) assigned to IFF.
108. US Patent 4 639 330 (priority US 21 Mar. 1986) assigned to IFF.
109. German Patent Appln 3 707 209 (priority DE 6 Mar. 1987) filed by Henkel.
110. Swiss Patent 629 467 (priority CH 30 Apr. 1982) assigned to Firmenich.
111. Japanese Patent 87 054 413 (priority JA 15 Dec. 1980) assigned to Sumitomo.
112. Japanese Patent 87 054 415 (priority JA 19 Aug. 1981) assigned to Sumitomo.
113. European Patent 79 206 (priority JA 6 Nov. 1981) assigned to Sumitomo.
114. European Patent 44 186 (priority JA 19 Mar. 1981) assigned to Sumitomo.
115. European Patent 247 589 (priorities JA 28 May, 28 July and 3 Dec. 1986) assigned to Sumitomo.
116. Japanese Patent Appln 63 035 569 (priority JA 30 July 1986) filed by Sumitomo.
117. Japanese Patent Appln 63 141 975 (priority JA 3 Dec. 1986) filed by Sumitomo.
118. European Patent Appln 133 251 (priority JA 22 July 1983) filed by Sumitomo.
119. Japanese Patent Appln 60 224 656 (priority JA 20 Apr. 1984) filed by Sumitomo.
120. Japanese Patent Appln 62 026 248 (priority JA 29 July 1985) filed by Hasegawa.
121. Japanese Patent Appln 63 083 049 (priority JA 29 Sept. 1986) filed by Hasegawa.
122. Japanese Patent Appln 63 119 480 (priority JA 5 Nov. 1986) filed by Toyotama Koryo.
123. European Patent 270 841 (priority DE 15 Nov. 1986) assigned to BASF (see also US Patent 4 791 098).
124. US Patent 4 271 048 (priority US 11 Aug. 1980) assigned to Firmenich (see also German Patent 3 131 617).
125. US Patent 3 920 752 (priority CH 26 Apr. 1969) assigned to Givaudan.
126. European Patent 252 378 (priority DE 3 Dec. 1986) assigned to BASF.
127. European Patent 269 999 (priority DE 3 Dec. 1986) assigned to BASF.
128. European Patent 56 500 (priority CH 16 Jan. 1981) assigned to Firmenich.
129. European Patent 175 115 (priority DE 13 Aug. 1984) assigned to Henkel.
130. European Patent 98 791 (priority DE 7 July 1982) assigned to Henkel.
131. World Patent Appln 85 03517 (priority DE 8 Feb. 1984) filed by Henkel (cf. European Patent 170 688 and others).
132. European Patent 177 807 (priority CH 10 Oct. 1984) assigned to Givaudan.
133. German Patent 3 341 607 (priority DE 17 Nov. 1983) assigned to Consortium Elektrochem. Industrie (Wacker).
134. European Patent Appln 341 707 (priority DE 13 May 1988) filed by Consortium Elektrochem. Industrie (Wacker).
135. US Patent 4 710 316 (priority DE 4 Sept. 1985) assigned to Consortium Elektrochem. Industrie (Wacker) (see also German Patent 3 531 609).
136. Kastner, D., Durch die Blume ge(sp)rochen—Blumennoten in der Parfümerie, *Parf. & Kosm.*, **66** (1985) 5.

5 CHEMICALS WITH GREEN–HERBACEOUS AND SPICY ASPECTS

The adjective 'green' stands for naturalness and freshness. Thus, 'green-smelling' substances should be very useful in such market segments where the consumer will be guided not only by product effectiveness, but also by a skillful combination of both. In 1981, Morris[1] clearly demonstrated the complex situation of 'green' fragrances and their composition. As a result, the olfactory term 'green' will be found very often in the description of odor *facets* of a definite fragrant chemical, whether it be otherwise fruity, floral or herbaceous. In practice, the type 'green' seems to be present throughout the complete spectrum of the odor palette and is therefore normally combined with another olfactory descriptor to define better the position of the chemical concerned.

Two years later, Lamparsky[2] looked at the 'green' notes from a chemical point of view. He presented a compilation of compounds arranged according to their functional groups and showed that research all over the world was very active in developing new, industrially feasible odorants which fit well into the 'green' olfactory concept.

Green–Herbaceous Odor Notes

In continuation of such a survey, we may first notice that the monoterpene hydrocarbon 3-carene (**1**), for instance, is a really interesting natural raw material for the synthesis of green-smelling compounds.[3]

The isocaranone (**2**) itself, as well as its oxime and the spiroketal **3**, exhibit green–herbaceous odor notes with very natural aspects. To a certain extent the same holds also for other substituted 1,3-dioxanes (**5a/b**), the synthesis of

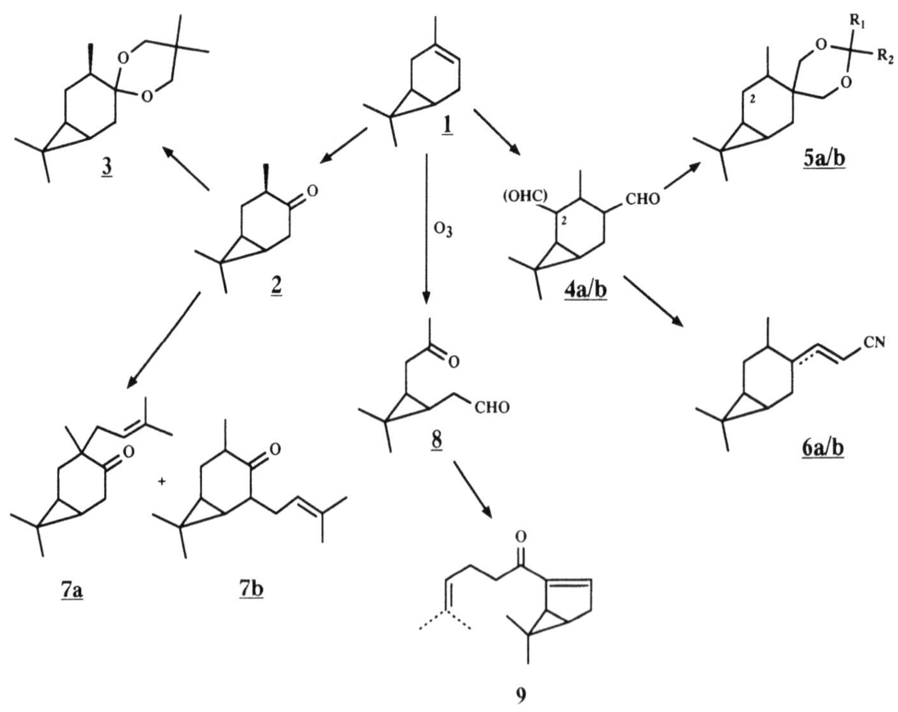

which is based on a hydroformylation and/or Prins reaction, respectively, followed by condensation of the intermediate diol with an appropriate carbonyl compound.[4] The reaction sequence delivers first a mixture of 2- and 4-caranecarboxaldehdye **4a/b** (30:70), as is also mentioned in a patent[5] dealing with new nitriles **6a/b** prepared from **1**. In this case, **4a/b** is condensed with cyanoacetic acid and then decarboxylated to a mixture of isomeric nitriles with a wide variety of odor effects (woody, floral, fruity, petitgrain, carrot, angelica root, musky). Surprisingly, also, the cyclic ketals of type **5a/b** have a distinct woody–amber odor recalling eucalyptus fruit in a very early green stage.

The mixture of ketones **7a/b** has a very diffusive green, fresh and natural odor reminiscent of privet flowers, pollen and hay. Prenylation of **2** gives this interesting fragrant material in one step with good yields.[6]

Another ketone (**9**) has a galbanum-like fragrance and can be prepared in several steps from **1**.[7] Most interestingly, a similar reaction sequence was used for the preparation of ketone **12**.[8] Ozonization of 1,5-dimethylcycloocta-1,5-diene, a dimerization product of isoprene, leads first to a ring scission followed by a new ring closure of the intermediate ketoaldehyde **10** to the cycloheptadienyl methyl ketone **11**, which reacts with an allyl halide to give ketone **12**.

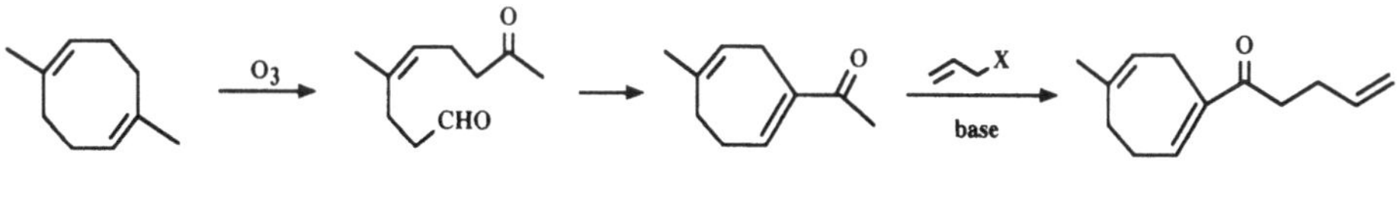

10 11 12

green, fresh, fruity
flowery

The ketone **13** surely has its roots in the compound **14**, identified in juniperberry oil.[9,10] The odor of **13** is described as green, flowery and fruity and the substance is prepared[11] in the way we have already encountered earlier (see Section 4, Refs 47, 50, 69).

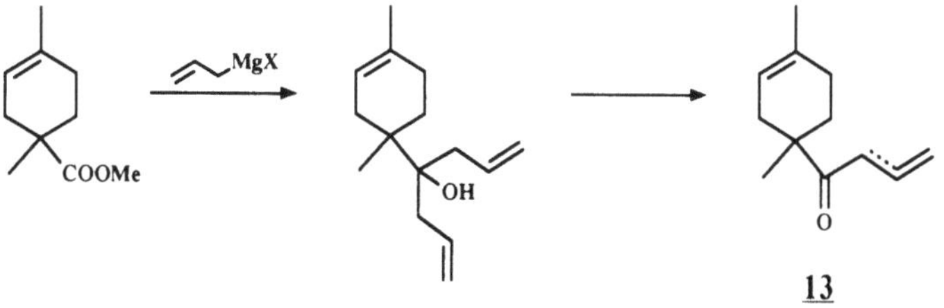

13

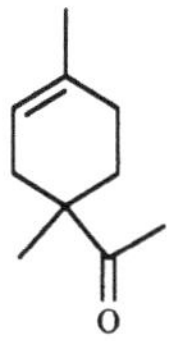

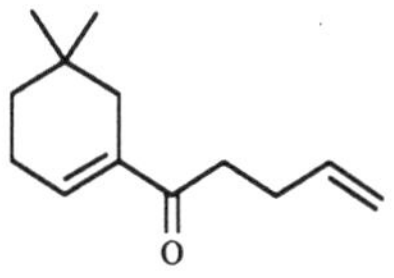

14 15

This is just the opportunity to also mention Neogal® (**15**, Firmenich), because this ketone represents an important ingredient in a great number of successfully launched 'green' bases.[12,13] Like the cyclic ketal **16** and the esters of type **17**, this compound is produced from dehydrolinalool, one of the most important intermediates in the fragrance industry.

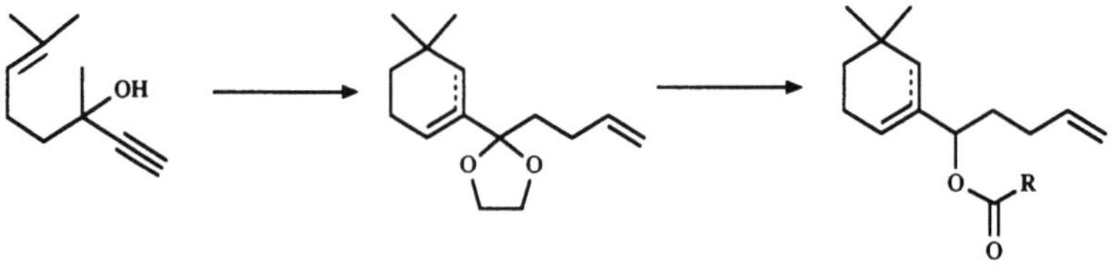

16 — green, herbaceous fruity

17 — green, herbal, aromatc galbanum-like

Green floral notes are also the main characteristics of a sulfur derivative (**18**) of dihydrolinalool which can easily be prepared by the addition of thioacetic acid to the monoterpene alcohol at slightly elevated temperature.[14]

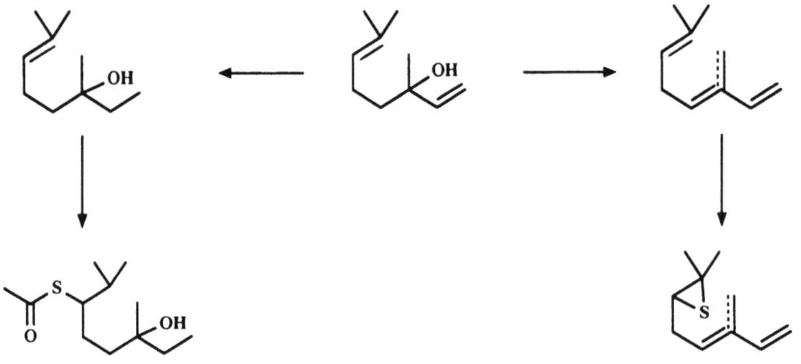

18 19

Myrcene or ocimene epoxides are converted to 6,7-epithio-dimethyloctadienes (**19**) when refluxed with sodium thiocyanate in isopropanol. These new compounds have a long-lasting stemmy green fragrance, with basil and petitgrain undertones.[15]

Galbanum-like, strong green fragrances are furthermore obtained from a cheap industrial raw material, 2,2,4-trimethylcyclopentanone. It is first converted into the diene **20**, which is capable of undergoing the Diels–Alder reaction with a vinyl ketone to deliver the bicyclic ketones **21** or **22**.[16]

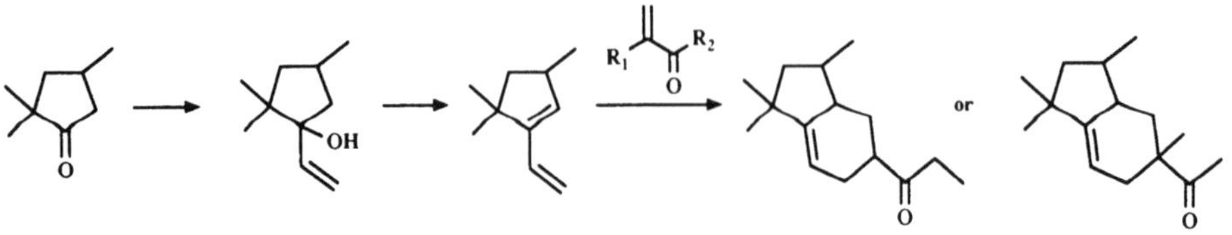

20 21 22

The Diels–Alder reaction is a very effective route to new fragrant substances; it has also proven its usefulness in a more recent patent application.[17] Here, the diene hydrocarbon **23** (obtained in two steps from isophorone) reacts with cheap acrolein as the philodienic component to yield a mixture of formyl trimethylbicyclo[2.2.2]oct-7-enes (**24a/b**) with a green–herbaceous, grassy and hay-like odor.

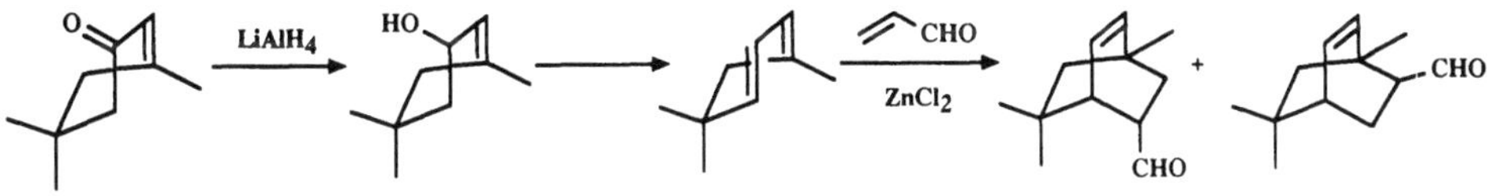

23 24a 24b

A bicyclic tertiary acetate **26** was described as an interesting odorant with a fresh, green and clary sage-like odor profile, providing a natural lift to compositions in which it is incorporated.[18] It can be manufactured from the tricyclic ketone **25** through Grignard reaction and acetylation of the resulting alcohol. Hydrogenation of the ring double bond diminishes the green aspects and a more woody, amber-like character is preponderant.

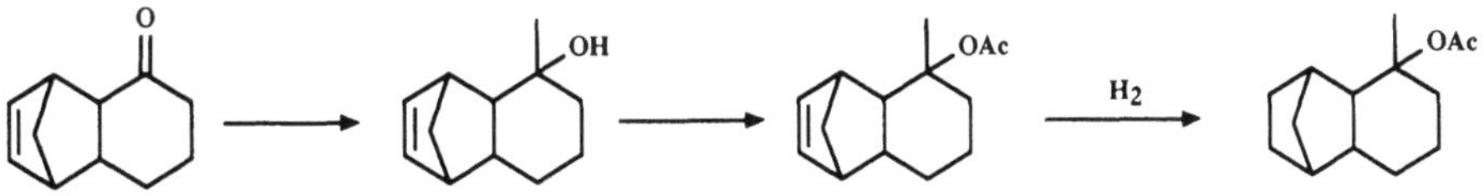

26

When discussing cycloaliphatic esters, it may be worthwhile to draw the attention to the allyl ester **27**. In comparison with the well-known Cyclogalbanate® (Dragoco), surprisingly this structurally very similar compound was—even in 1988—patentable over the known homolog in view of some unexpected properties with respect to the odor strength and substantivity. The olfactory description (green, galbanum-like, rosy, fruity) does not differ too much from the commercially available Cyclogalbanate.®

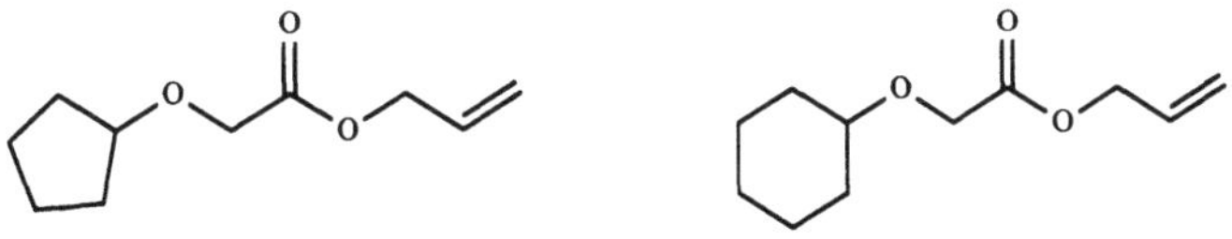

27 "Cyclogalbanate"

A green–fruity smell is also provided by the propionate **28**, whereas the isopropyl ether **29** shows up with a more herbaceous, petitgrain-like and bitter fragrance.[20] Both substances are available by simple addition reactions of propionic acid or isopropanol, respectively, to an industrially produced, inexpensive chemical.

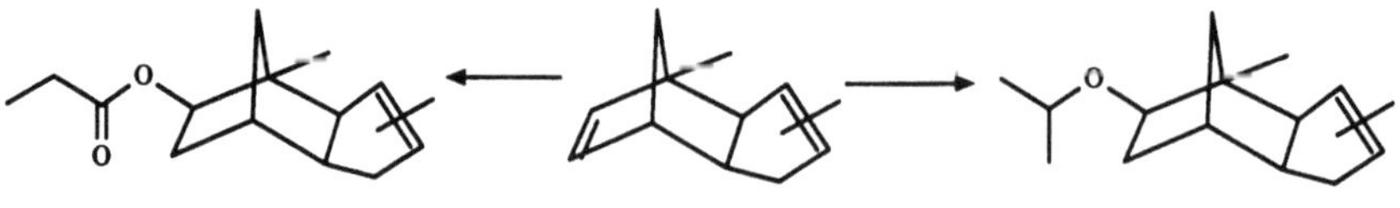

28 29

Another green–herbaceous, but also floral–fruity smelling group of new esters derives from the codimer of cyclopentadiene and butadiene. The one-step reaction with α,β-unsaturated acid yields tricyclic norbornyl esters of type **30**.[21] The addition of 1,3-butanediol furnishes an isomeric mixture of hydroxy ethers, **31a/b**. It exhibits a fruity–green–herbaceous odor with an alliaceous, parsley-like top note.[22]

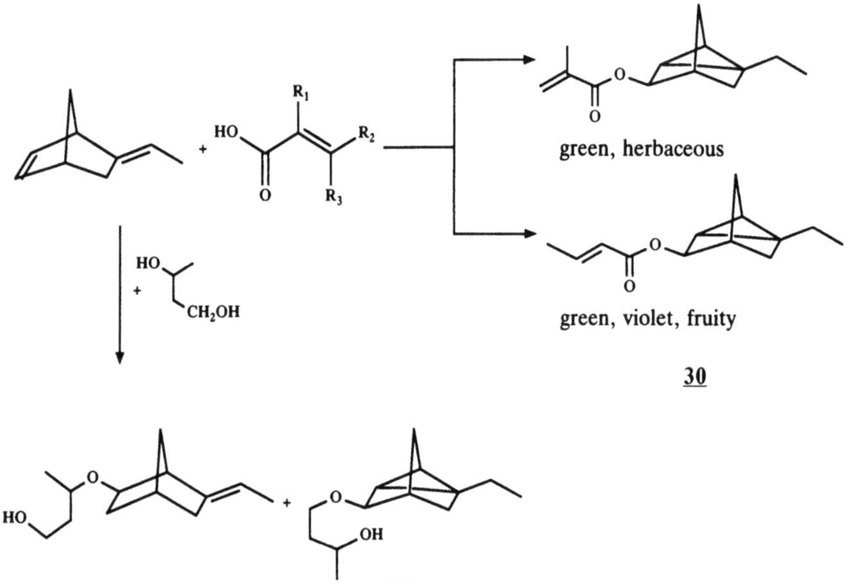

30

31a 31b

Especially useful in bleach perfuming are the ethers **32a–c**, produced simply and cheaply by addition of isopropanol to the aforementioned cyclopentadiene–butadiene codimer and hydrogenation.[23] A similar odorant mixture is commercially available under the name Isoproxen® (IFF).

32a 32b 32c

Alkyl ethers with an intact vinyl group can be hydroformylated to aldehydes of type **33**, which also have mainly green odor notes.[24]

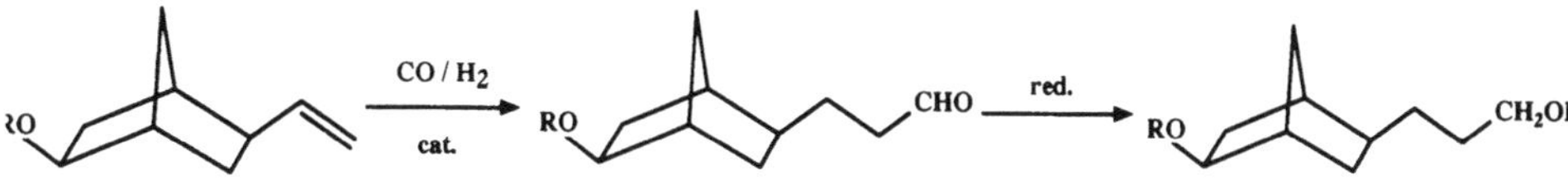

33

	33	
R = Me	green, fatty, woody	balsamic, floral
Et	green, fruity	woody

Norbornyl pyridine derivatives (**34a/b**) represent another interesting group of substances, because their odor qualities depend on the *ortho* or *para* position of the nitrogen in the heterocyclic nucleus.[25] In this case also, one of the starting materials belongs to the group of inexpensive, easily available industrial chemicals.

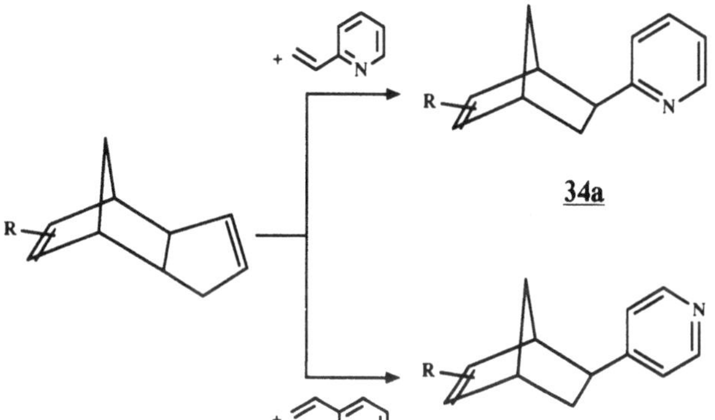

34a

R = H or Me
odor : green, herbaceous
petitgrain, amber

34b

R = H
odor : woody, amber, musk

R = Me
odor : woody, green, orris

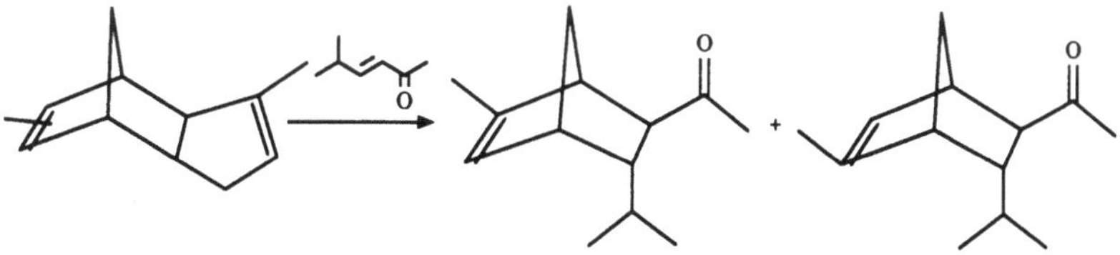

35a 35b

odor : herbaceous, spicy, green
fruity undertones

A similar transformation of dimethyl dicyclopentadiene has been described in a patent[26] which claims the use of exo and endo isomers of the unsaturated ketone **35a/b**. Hydrogenation of the ring double bond leads to a complete change of the olfactory character: we note a more woody–camphoraceous odor instead of the green odor note preponderant in **35**.

It is quite understandable that cheap, industrially available starting materials attract all companies working in the fragrance field: examples to demonstrate this fact are the patents dealing with the ester **36**[27] and the structurally interesting cyclic ether **37**,[28] commercially available under the name Rhuboflor® (Firmenich).

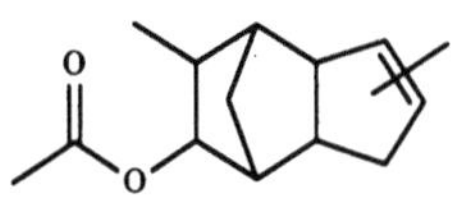

36
green, herbaceous
clay sage

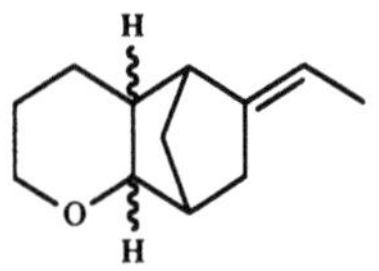

37
herbaceous, green
elderberry

The hydrocarbon **38** and the epoxide **39**[29] have odor descriptions very similar to those of the aforementioned compounds.

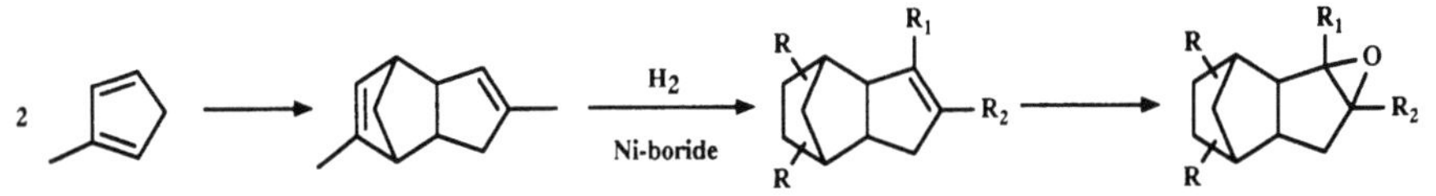

38
fresh, green, herbaceous

39
fresh, herbaceous

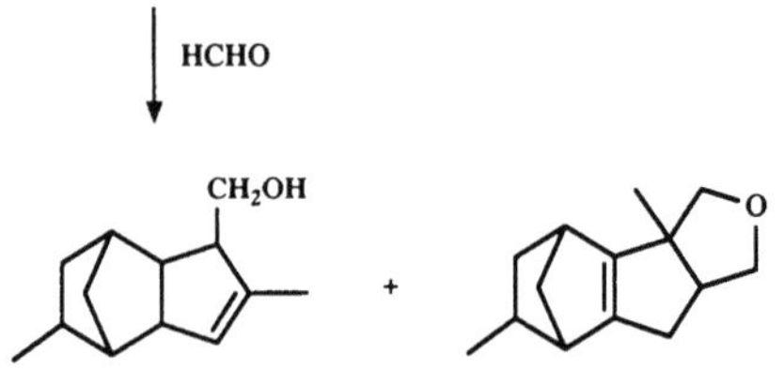

40

41

mixture: woody, fresh, spicy

The same raw material base is used in producing the mixture **40** + **41**.[30] The process for their preparation includes an interesting monohydrogenation in the presence of nickel boride and a single or double Prins reaction yielding mixtures of primary alcohols (**40**) and cyclic ethers of type **41** with odor notes that make them useful in perfumery.

A bicyclic spiroacetal (**43**)[31] has already been commercialized under the

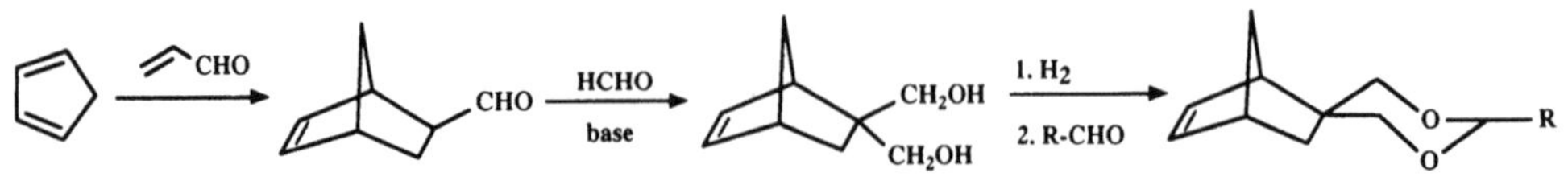

42

43

name Bonarox® (Henkel, R = methyl). Cyclopentadiene and acrolein react first to the bicyclic carbaldehyde **42**, the Prins reaction then furnishes the intermediate diol, which is hydrogenated and afterwards condensed with an appropriate aldehyde.

Once again, we see the usefulness of reaction types such as Diels–Alder, Prins, hydrogenation and acetalization, all giving high yields. Thus, they do not present any difficulty—even when combined in a four-step sequence—in delivering new odorants at reasonable costs.

Another good example of the combination of an industrially available raw material base, and well-known manageable unit operations, is given by the intensively fresh and green-smelling cyclic acetal **47**.[32] Acetylene, acetone, isobutene, carbon monoxide and hydrogen are the chemicals which form 2,5-dimethyl-3-hexyne-2,5-diol (**44**) and pivalic aldehyde (**45**), respectively.

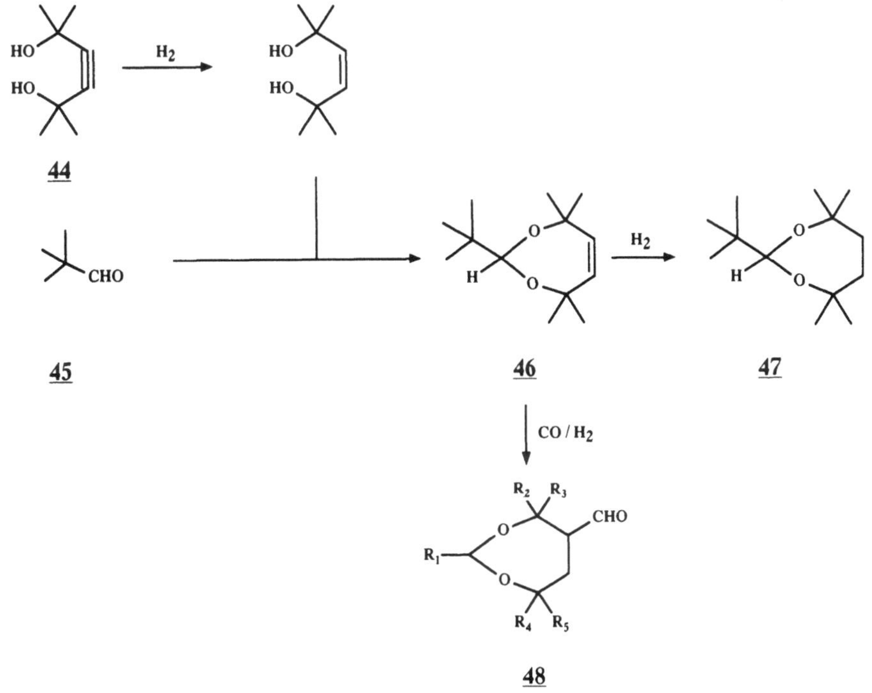

The partial hydrogenation of **44**, followed by condensation/cyclization with **45**, leads to a substituted 4,7–dihydro-1,3-dioxepin (**46**) which can be hydrogenated to **47** or hydroformylated to a compound of type **48**.[33] In the latter case, the use of hexanal as the carbonyl compound delivers, for instance, 2-pentyl-4,4,7,7-tetramethyl-5-formyl-1,3-dioxepane with a herbaceous–green fragrance, whereas other carbonyl compounds give rise to woody, fresh–aldehydic, floral or fruity notes. All the 4,7-dimethyl-substituted dioxepanes have more or less green and fresh smells and the same holds for the unsubstituted 5-formyl-1,3-dioxepanes.

The *tert.*-butyl group plays a predominant role in fragrant molecules originating from research at the German chemical giant BASF. Thus, for instance, 1-formyl-4-alkoxycyclohexanes (**49**) are interesting compounds.[34] Two pathways for their production are possible and a relationship to other, already mentioned, floral and musky fragrances is clearly visible.

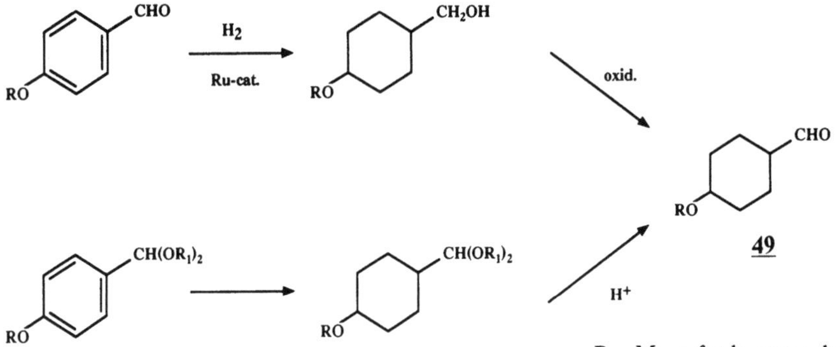

49

R = Me : fresh, green, herbaceous
t-Bu : green, floral

In the field of aromatic compounds, the green odor characteristic is quite rarely localized, but the more recent patent literature has nevertheless shown that here we also get some unexpected changes in odor with small changes in structure. The diethyl acetal (**50**), for instance, exhibits a strong green odor,[35] but the dimethyl acetal of the same aldehyde is a floral, sweet compound (see **110a** in Section 4), and ring hydrogenation of the parent aldehyde delivers an alcohol with musky character.[36]

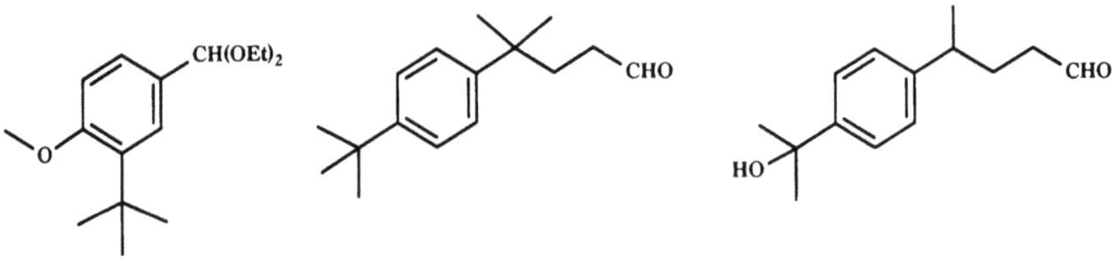

50 51 52

The odor description 'green, camphoraceous, herbaceous' is given to the *tert.*-butylated 4-methyl-4-phenylpentanal **51**.[37] A substance and a use claim are presented for the aldehyde **52** with a green, woody and balsamic smell.[38] A powerful and stable green olfactory note is ascribed to 2-methyl-4-phenylpentanol (**53**) and its acetate,[39] both showing fruity, floral and even vetiver-like or spicy facets in their complete odor profile.

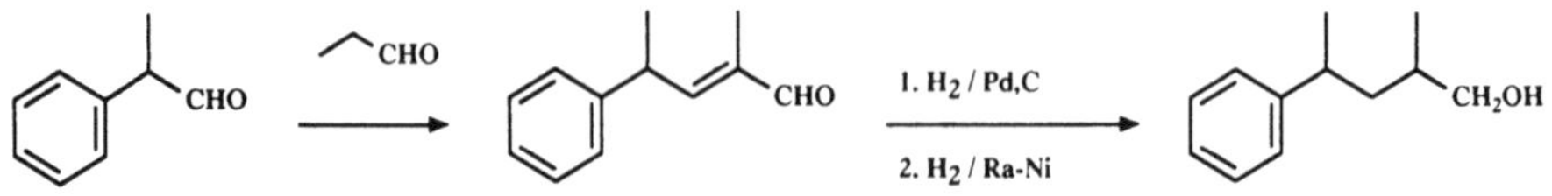

53

The next example in the 'green' series deals with the search for substitutes having exactly the same odor facets as a substance which had to be withdrawn from perfumers' collections due to allergenic effects on the skin. Such a toxic compound is the aldehyde **54**, with a fresh, green and citrus-like fragrance. The completely hydrogenated aldehyde **55a** seems a very good replacement—from the odor viewpoint—and can easily be produced by a liquid-phase dehydrogenation of the corresponding alcohol.[40] It was claimed that this process is much better than the one described by Martin & Phillips in 1978.[41] Thus, a known compound has shown attractive possibilities—not mentioned beforehand and therefore patentable in form-of-use claims—in the fragrancing of comsumer products.

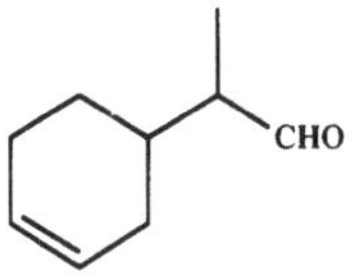

54

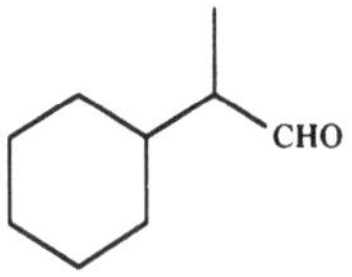

55a

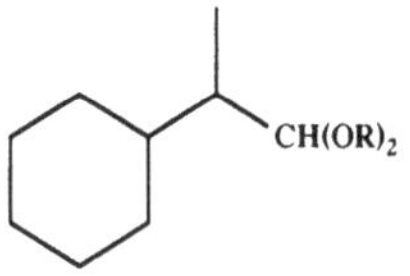

55b

Furthermore, acetals of the type **55b**, with R = methyl, ethyl or propyl, have been claimed together with the aldoxime or the Schiff base with methyl anthranilate as new substances. They may serve as green, floral and fruity-smelling ingredients in perfumery compositions.[42]

Switching now from cycloaliphatic to open-chain aliphatic aldehydes, we encounter first a hydroformylation product of an unsaturated ether, i.e. 3,6,8-trimethyl-5-oxa-non-7-enal **56**. It was claimed that various ether carboxaldehydes or carbinols of this type are useful new fragrance chemicals with a strong green, herbaceous and leafy character.[43]

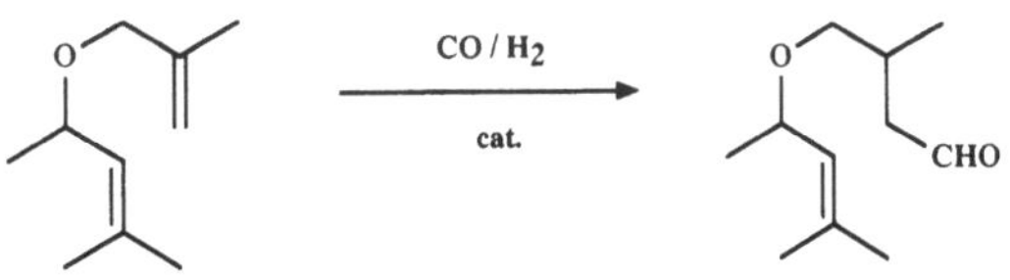

56

Intensively green and flowery is an unsaturated aldehyde (**57**) which bears a double bond in the γ,δ-position and shows a remote similarity to the oxa compound **56**.[44]

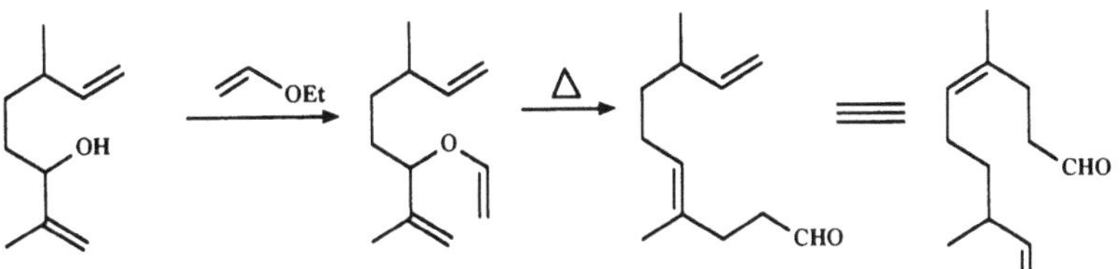

57

all-cis-4,7-Alkadienals of type **58** are also powerful green–fresh odorants with flowery, aldehydic or calamus and orris root-like undertones.[45] They can be

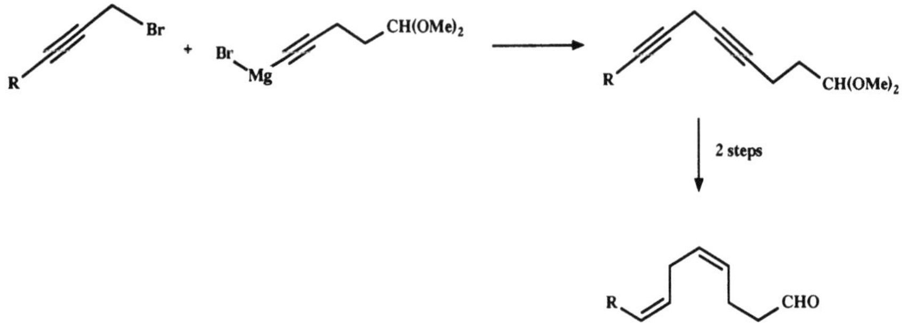

58

(R = alkyl or alkenyl C_2-C_5)

prepared by methods known *per se*: coupling of two acetylenic bromides, hydrogenation in the presence of Lindlar catalyst, and careful hydrolysis of an acetal by aqueous acetic acid.

Special emphasis has been given to those aldehydes having a third terminal double bond in the molecule, i.e. 4,7,9-decatrienal or 4,7,12-tridecatrienal, and it is worthwhile to mention that 4,7-decadienal has been identified in sweet flag oil (calamus oil) as a very characteristic component.[46]

In the field of essential oil constituents it is well known that 1,3,5-undecatrienes are important ingredients of galbanum, to which they confer the typical green character with a fatty–oily background.[47] Also known is the occurrence of 1,3(*E*),5(*Z*),8(*Z*)-undecatetraene (**59**) in algae.[48] It is therefore not astonishing that the *all-trans*-1,3,5,7-undecatetraene (**60**) is of a certain interest to the fragrance industry, due to its green, fresh and very natural aspects, without the fatty–oily undertone of the galbanum components mentioned above. Its fragrance is reminiscent of lemon peel and grapefruit and, when used as a flavoring substance, it provides an earthy–woody note similar to the American mandarin essence.[49]

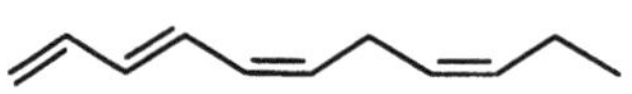

59

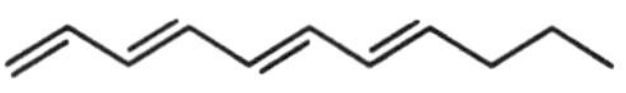

60

In this context, a functionalized butadiene telomer should also be mentioned. Two molecules of butadiene react with allyl alcohol in the presence of palladium(II) acetate and triphenylphosphine to yield the unsaturated aliphatic ether (**61**) with a strong green, lavender-like and slightly fruity smell.[50] This compound should enhance and improve floral–fruity and green bases.

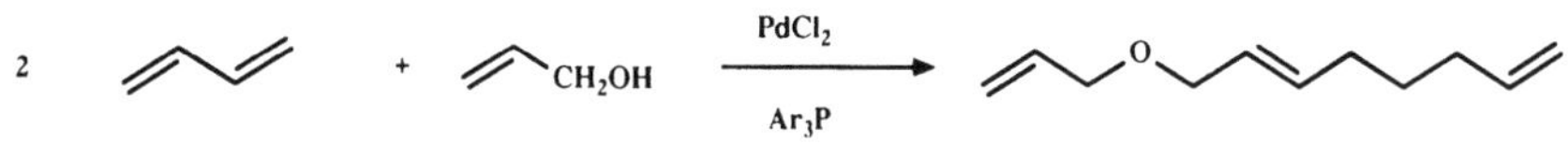

61

More recently, it was revealed that an intermediate (**62**) in the synthesis of the pheromone 7(*E*),9(*Z*)-dodecadienyl acetate provides even in small dosages a remarkable effect in perfume compositions, especially in the fresh, herbal types, where it modifies the top notes due to its volatility.[51]

62

strong, green, herbaceous

Another ω-unsaturated diol has been converted in the usual way to the new ether **63** with green and herbaceous odor characteristics reminiscent of galbanum.[52]

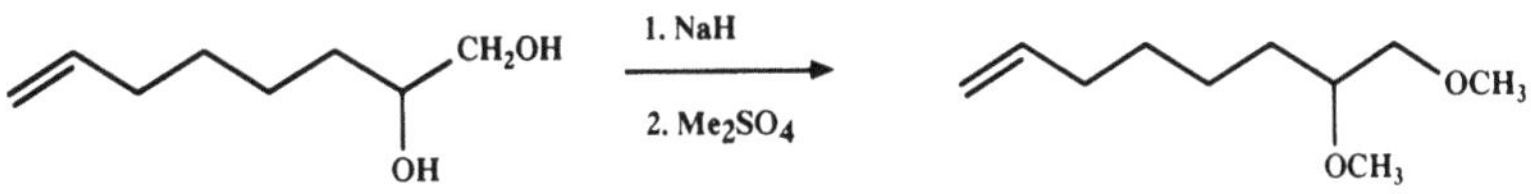

63

It is now interesting to see that the terminal double bond is not necessarily a prerequisite for a green-smelling aliphatic ether. Even an internal trisubstituted double bond in a molecule with 11 carbon atoms is sufficient to give useful odorant properties.[53] The dimerization of propanal furnishes 2-methyl-2-pentenal, the Grignard reaction of which with appropriate alk(en)yl magnesium halides leads to unsaturated alcohols of type **64** (green, floral, fruity and fatty odor notes) or, more specifically, to compound **65**,[54] commercially available under the name Undecavertol® (Givaudan). Its methyl ether **66**, for instance, is prepared in the same way as mentioned before and combines fresh, green, fruity and herbaceous odor aspects.[55]

Alkali-stable fragrances may also be produced by the reaction of Undecavertol® with vinyl ethyl ether. This way we get a mixed acetal of acetaldehyde (**67**) with a green–citrusy-like odor of bitter character.[55]

Even a fully saturated, branched primary alcohol may be a useful fragrant material, as shown by 2,4-dimethylheptanol (**68**), the end-product in a pedigree line of a German company that has a good technical knowledge of the alkylation reaction under phase-transfer conditions. Complete hydrogenation of the already commercialized unsaturated aldehyde Dimedienal® (Wacker-

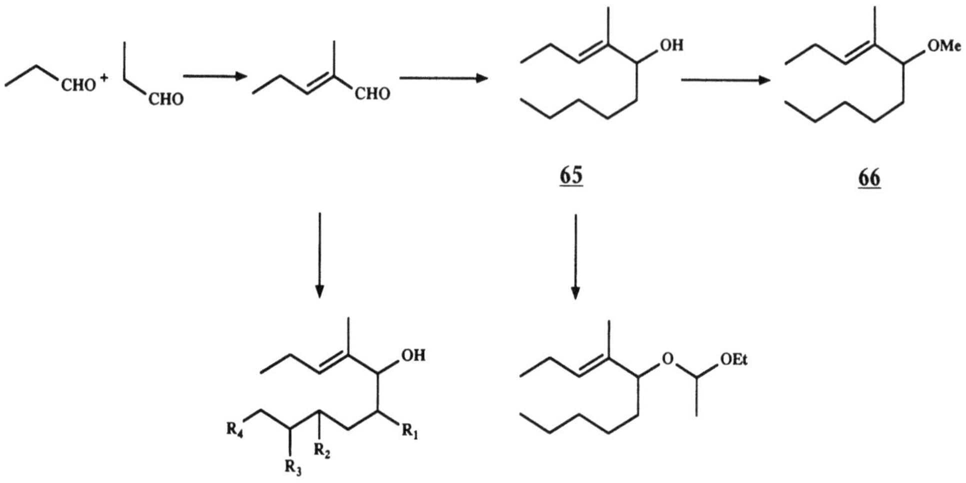

65 66

64 67

Chemie) delivers compound **68** with a herbaceous, leaf-like, fatty–woody fragrance.[56]

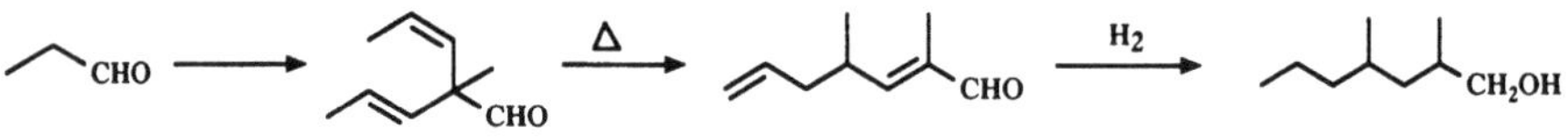

68

Finally, we also want to mention new odorants with green odor profiles in the field of aliphatic esters. In the first case,[57] a secondary alcohol with a substituted terminal double bond is esterified with lower fatty acids to give the type **69** (higher fatty acids as well as aromatic acids give esters with flowery–fruity fragrances!), and in the second case,[58] esters of unsaturated alcohols (including cinnamic alcohol) and 2-methylpentanoic acid (type **70**) are all claimed to have green, natural odor notes.

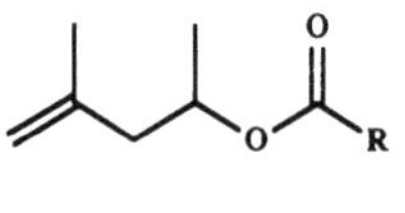

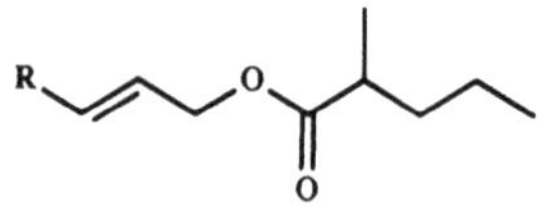

69 70

In this context it is interesting to refer to similar esters derived from the same acid which were discussed in Section 4 (cf. Ref. 130).

Herbaceous and Spicy Odor Notes

On the foregoing pages we have seen that the descriptor 'green' is often combined with the other olfactory impression 'herbaceous'; both simultaneously characterize the vegetal plant aspect from nature and, in the classification of fragrances (see Chapter 8), the green notes of raw materials are explained by additional adjectives like herbaceous and agrestic, too. Agrestic notes are encountered in numerous men's fragrances and we tend to connect their herbaceous character with natural essences like basil oil, bergamot, fennel, lavender, petitgrain, thyme and so on.

It is therefore not surprising to find similar words also in the odor description of new aroma chemicals like the nitriles **71a** and **71b** prepared from limonene or *p*-menth-1-ene, respectively, as the starting materials.[59]

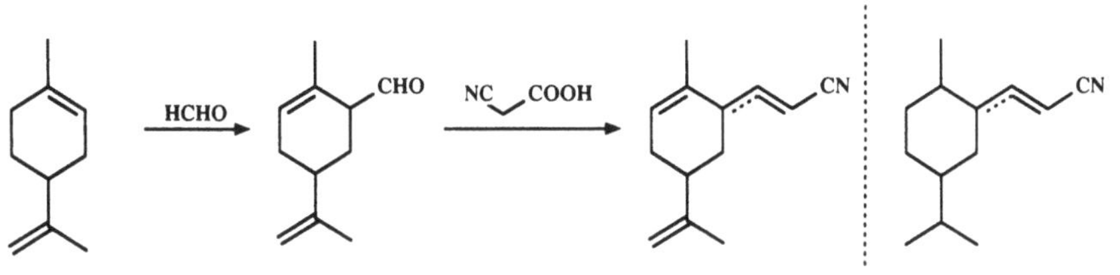

71a	71b
herbaceous, basil fennel	green, petitgrain woody

Another monoterpene hydrocarbon, β-pinene, is the basis for a new aldehyde, **72**, which develops bitter–herbaceous, aromatic and floral odor notes.[60] Its synthesis requires seven steps and includes a pyrolysis at 500°C in the last step:

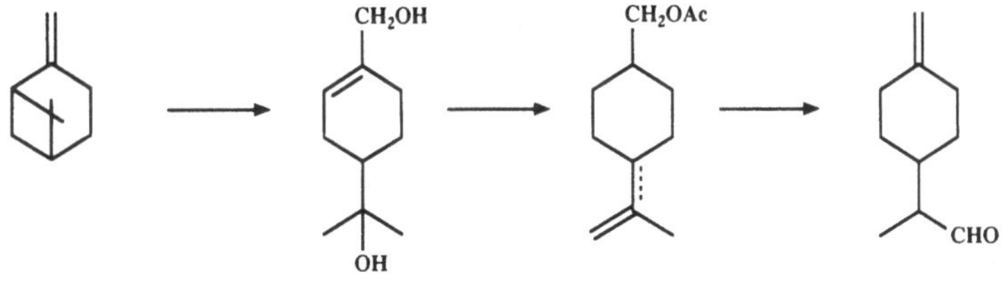

72

High temperatures (400–500°C) are also needed for the ring closure of the aldehyde **73** to the cyclopentane derivative **74**.[61] This new compound has a lavender character with camphoraceous, minty and floral shades. It also recalls some rosemary aspects and therefore fits very well into the agrestic line of fragrant materials.

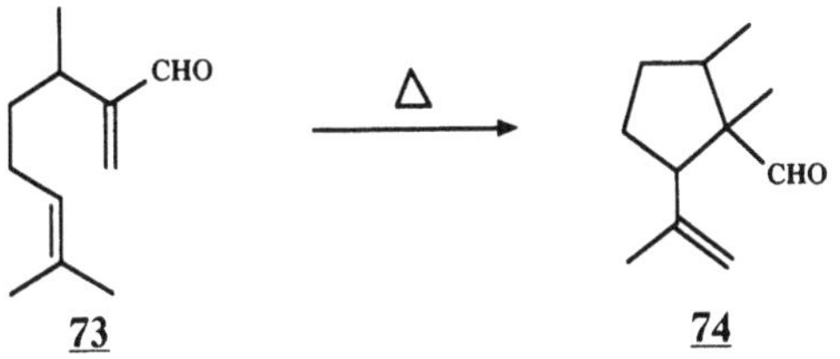

73 74

Lavender in general was the subject of a publication in 1986.[62] The author gave a very marketing-oriented review of the four main segments of perfumery, but emphasized also the chemical aspects that the lavender group presents. Here we would like to recall the natural occurrence of myrcene- and ocimene-derived ketones of the types **75** and **76**, which have been synthesized by means of a Diels–Alder reaction.[63]

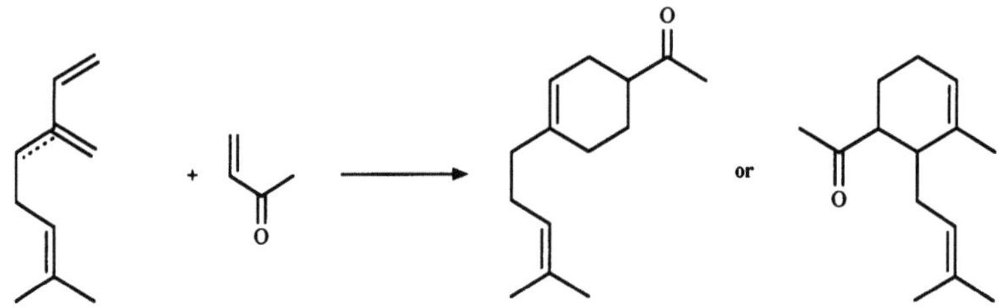

75 76

odor : green, herbaceous, woody

At a time it was feared that coumarin would be completely banned from the perfumer's shelf, an intensive search for substitutes could be noticed. The ketone **78** and the lactone **80** are representatives of these efforts.[64–66] Both substances are prepared from nopinone (**77**) (*ex β*-pinene): Robinson annulation of methyl vinyl ketone easily proceeds to compound **78**, which is commercially available under the name Tricyclone® (Firmenich), whereas the manufacture of **80** first needs the introduction of an auxiliary function as given in compound **79**.

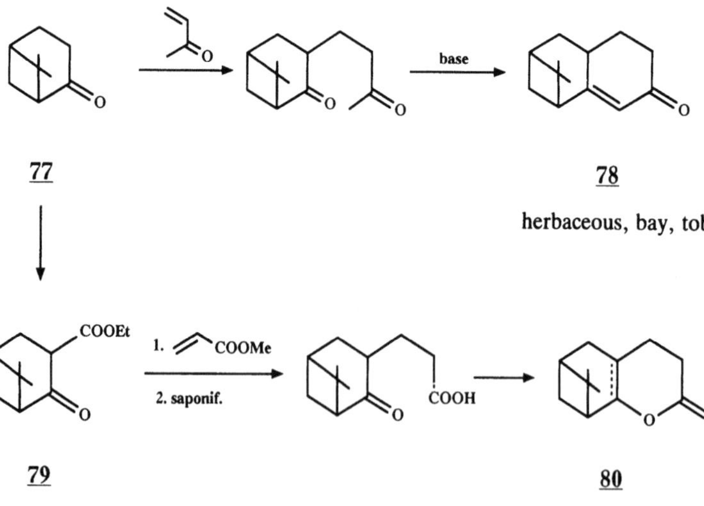

77 78

herbaceous, bay, tobacco

79 80

sweet, herbaceous, spicy
tobacco, tonka beans

A simple δ-lactone, **81**, is claimed to have coumarin and fenugreek notes, with celery, nutty and jasmine-like shadows at the same time.[67] An intensive coumarin and hay-like fragrance is ascribed to the γ-lactone **82a/b** and its fully hydrogenated derivative.[68]

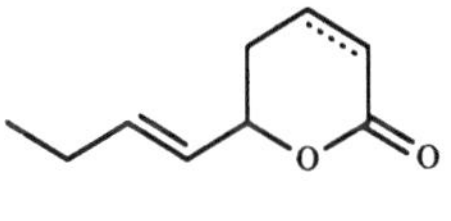

81

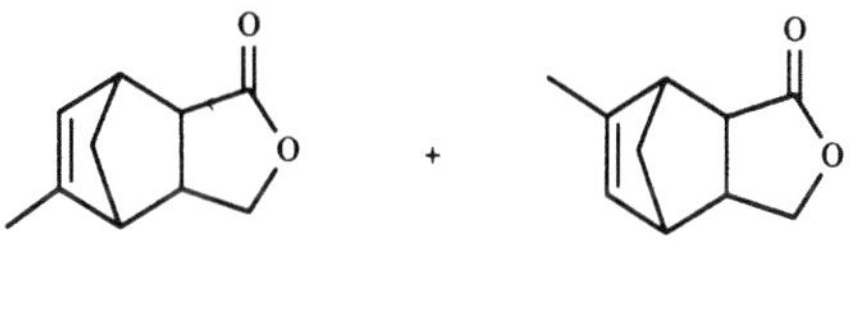

82a 82b

Very useful for the application in lavender-, chypre- and fougère-notes is the trimethyl-substituted δ-lactone **83a/b** (Trivalon®; Henkel) with a bitter and hay-like fragrance. The Baeyer–Villiger oxidation of the so-called VEBA ketone proceeds smoothly to the lactone and it is this simplicity of material and process That deserves attention.[69]

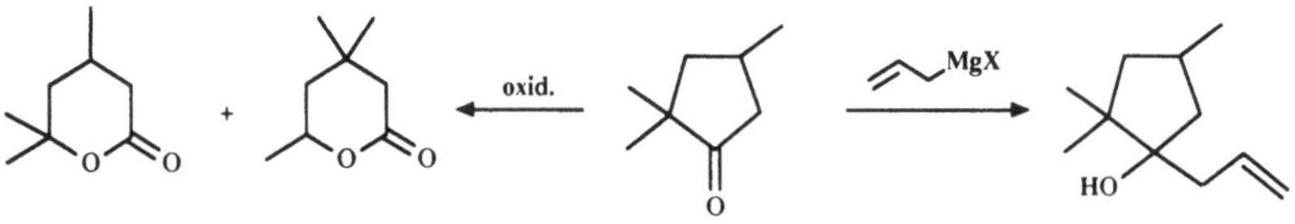

83a/b 84

We have encountered the inexpensive 2,2,4-trimethylcyclopentanone twice before (cf. pp. 565, 594) and now observe that a Grignard reaction with allyl chloride produces a further compound (**84**) of interest with respect to bleach-stable fragrances.[70] The tertiary alcohol **84** has a herbaceous, patchone-like odor, whereas other substituents introduced by the Grignard-reaction lead to more woody–earthy smells of the compounds prepared in this way.

Spirolactones of the type **87/89** represent another subject to be discussed within the framework of herbal–spicy odorants.[71] They are based on the monoterpene alcohol **85**, which reacts with ethylene oxide to the intermediate diol **86**. Cyclization of the corresponding hydroxyacid furnishes the δ-lactone **87** with a special effect in enhancing the spicy note of coumarin. Hydroformylation of **85** directly delivers the lactol **88**, which on oxidation with pyridinium dichromate yields the spiro-γ-lactone **89**.

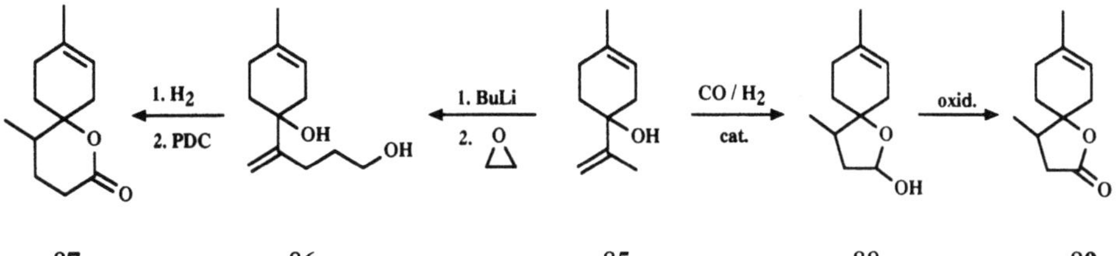

87 86 85 88 89

In the context of spirolactones it is interesting to compare the olfactory properties of a group of *p*-alkyl-substituted compounds (**90a–c**) with those of the unsubstituted parent γ-lactone, **91**. The former group[72] has in general a more woody–lactonic character and, when the substituent R is the methyl group, we note some coumarinic aspects in the complete odor profile. In contrast, compound **91** has a warm, spicy, lactonic fragrance,[73] which resembles tuberose and is therefore more flowery than those of the substituted homologs.

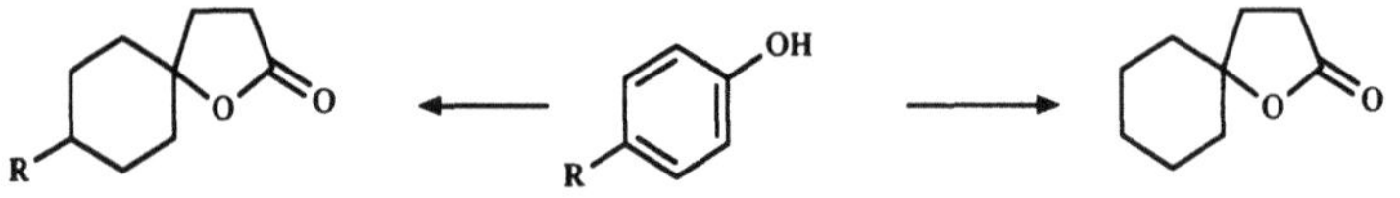

90a-c 91

(R = Me,Et,i-Pr)

A spiro-ether, **92**, commercialized under the name Thymoxane® (BBA), is the next example of a very simple molecule with an interesting smell (fresh, sweet, thyme, minty) and an easy manufacturing process.[74]

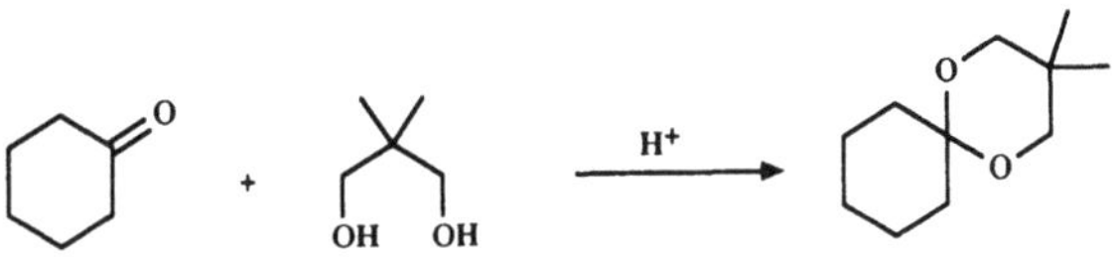

92

Simple chemical structures have always raised the interest of chemists working in the fragrance field. Thus, fully hydrogenated eugenol has been known *per se* since 1975, but it was noticed only in 1988 that the isomeric mixture **93** with the preponderant *all-cis*-isomer has a very interesting green–aromatic odor characteristic reminiscent of parsley, basil and other herbaceous notes.[75]

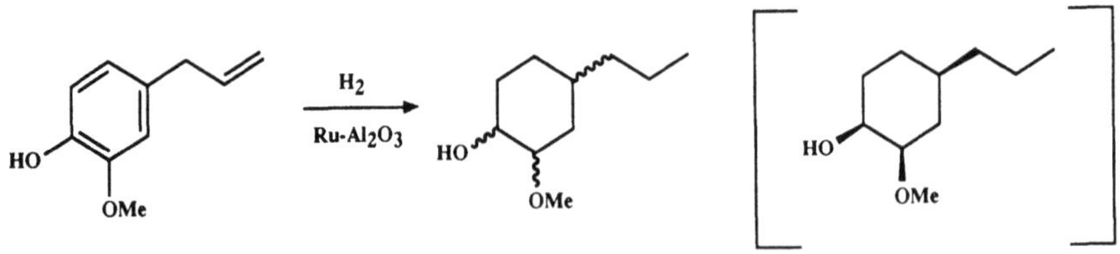

93

With eugenol we have now a good link to other aromatic compounds with spicy–vanilla-like odor profiles. First, two benzylic ethers (**94a,b**), derived from ethylvanillin, have already proved their usefulness in a great number of perfume compositions.[76] The methyl ether **94a** provides a very natural, diffusive and linear spicy odor, recalling white carnation flowers, whereas the isopropyl ether **94b** shows very smoky and leathery aspects in the basic spicy odor. As isoeugenol has been banned from collections of aroma chemicals, **94a**, especially, will surely find extensive use in safe substitutes for this kind of odor.

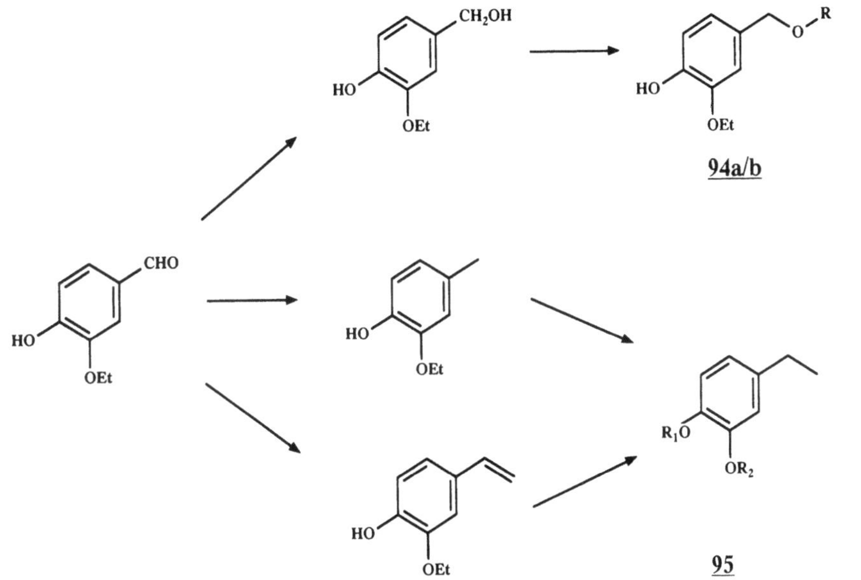

94a/b

95

(R_1,R_2 = H,Et,Pr)

A spicy, balsamic vanilla-like odor is predominant in the odor profile of phenolic ethers of type **95**.[77] They may be prepared according to methods described in the literature, i.e. Wittig reaction of ethylvanillin and hydrogenation of the vinyl group.

The bifunctional aromatic aldehydes of type **96** are described in a recent German patent application.[78] *p*-Hydroxybenzaldehyde is esterified with pivalic or isobutyric acid to yield the compounds **96** with a sweet, vanilla-like odor.

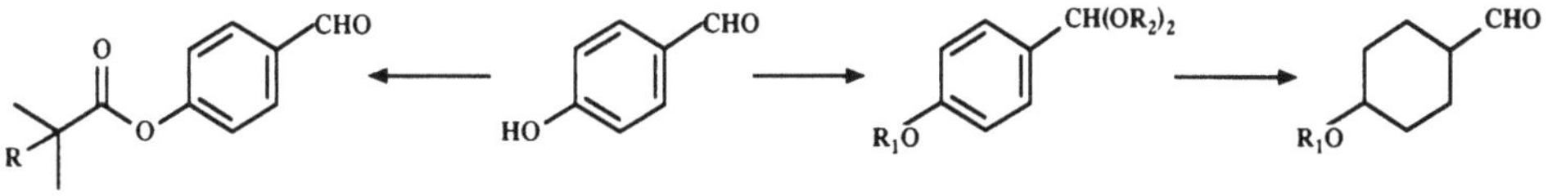

96

97

(R = H,Me)

(R_1 = Me,tert.Bu)

Using anisaldehyde or its *tert.*-butylated homolog as starting material, the ring hydrogenation furnishes the cyclohexane methanol derivative, which is de-hydrogenated to the ether-aldehydes **97** with fresh, green, herbaceous and floral aspects.[79]

We have already discussed the *tert.*-butylated aldehyde **51** as a green, camphoraceous odorant.[37] In the same patent, we find substance claims for compounds **98** and **99** and it is interesting to note that the substitution of the tertiary butyl group by a methyl group resulted in an odor change.

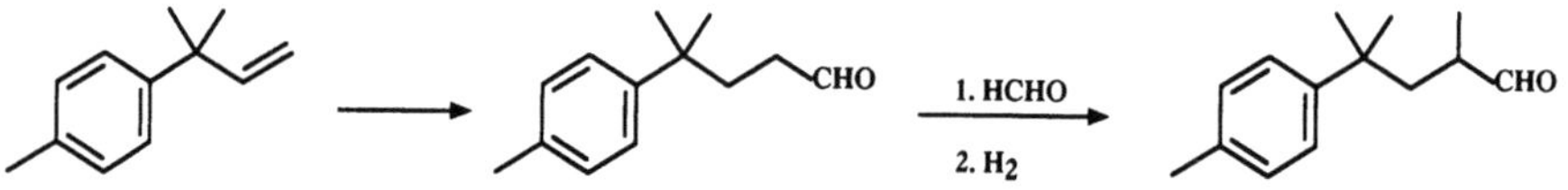

98 — spicy

99 — fresh leaves, herbal, woody

The intensive green character of 2,4-dimethylcyclohex-3-ene carboxaldehyde (**100**) is well known. The aldol condensation with propanal delivers an α,β-unsaturated aldehyde (**101**) for which spicy, cinnamon-like, as well as green and cumin-like odor properties are claimed.[80] Furthermore, we learned that isocyclocitral, with a third methyl substituent in the free *ortho* position to the aldehyde function, gives more floral smelling derivatives when reacted with propanal or butyric aldehyde.

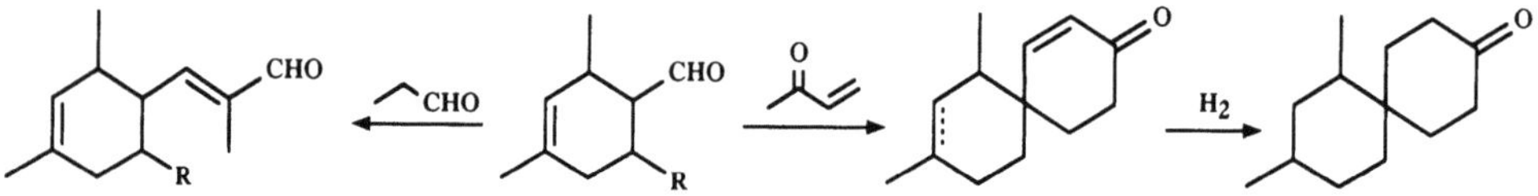

101 100 102

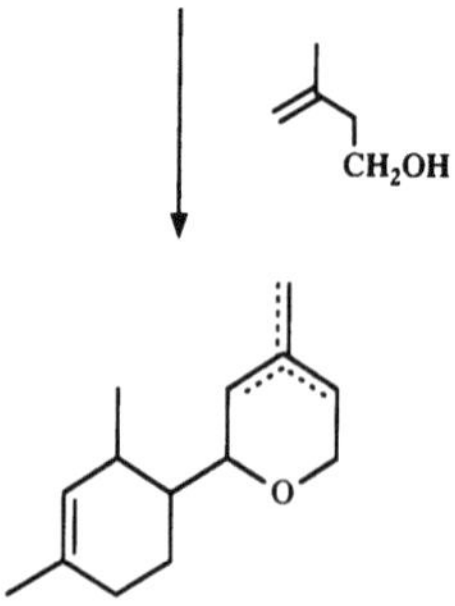

103

Reaction of **100** with methyl vinyl ketone followed by hydrogenation of the carbon–carbon double bond in the ring resulted in a spiroketone **102**. Natural fruity–spicy odors reminiscent of orris (iris) are mentioned for this compound in the corresponding patent.[81]

A skillful combination of the structural elements of Cyclal C® (**100**) and rose oxide has led to the cyclic ether **103** with a very natural, green, vegetable-like aroma.[82]

Esters and nitriles based on methylcyclopentadiene or its dimer should not be omitted from a discussion of herbaceous and spicy fragrant materials. The syntheses of compounds **104**[83] and **105**[84] are obvious; odor characteristic is described beneath the structure number. The odor profile of compound **103** is completely different from that of the parent alcohol, which predominantly has a woody, patchouli-like character.

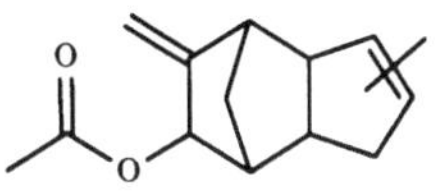

104

warm, aromatic, spicy
marigold, basil, sage

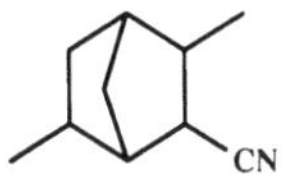

105

fresh, herbaceous
spicy

Campholenic aldehyde, an important intermediate in the manufacture of sandalwood chemicals (as mentioned earlier), can be transformed into the corresponding nitrile (**106**). In a very recent patent application,[85] both the α- and the γ-isomers were described as having spicy, fruity and woody odor notes with a certain sweetness (α-isomer) or freshness (γ-isomer).

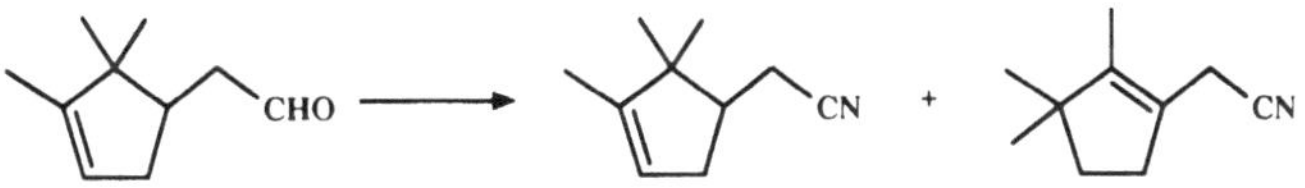

106

A remote structural similarity may be found for cyclopentenyl acetic acid (**107**), obtained from the well-known Cyclopidene® (Firmenich) by saponification.[86] The carboxylic acid is used as a perfumery ingredient to impart a sweet, honey-like scent to soaps and cosmetics.

107

In this review of green–herbaceous odorants, four patents dealing with aliphatic substances will terminate the compilation of relevant data extracted from the technical literature of the decade 1980–89. The recognition of a known intermediate, described by Wood *et al.*[87] for the preparation of musks, as a useful fragrance chemical *per se* is attributable to the research workers at Firmenich.[88] The ester **108** exhibits floral–fruity (pineapple) as well as

108

herbaceous, chamomile-like odor nuances, making it a valuable constituent in commercialized perfumery bases of that company. The same fragrance house has published analytical findings on the essential oil of *Artemisia herba-alba*,[89] which constitutes a rich source of so-called irregular monoterpenes. One of these components, the epoxy ketone **111**, is the subject of a patent.[90]

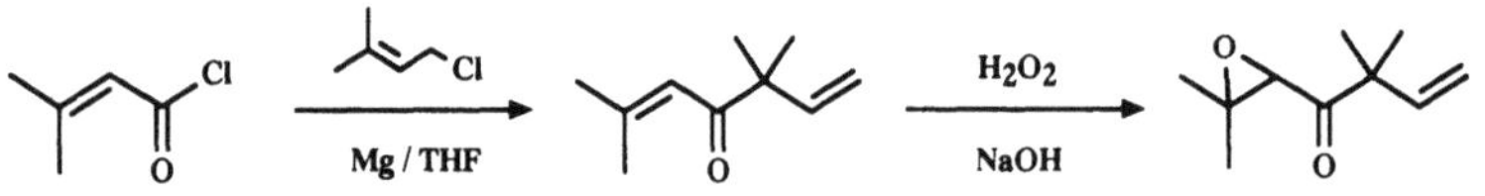

109 110 111

This herbaceous, aromatic, tobacco-, tea- and hay-like smelling substance can be prepared from the acid chloride **109** via coupling with prenyl chloride and subsequent epoxidation. The ketone **112**[91] structurally resembles the artemisia ketone (**110**). It is produced by an aldol condensation of 2-methyl-4-pentanone with isobutyraldehyde under phase-transfer conditions. The conjugated ketone **112** is especially preferred because of its fruity–chamomile and tagetes-like odor.

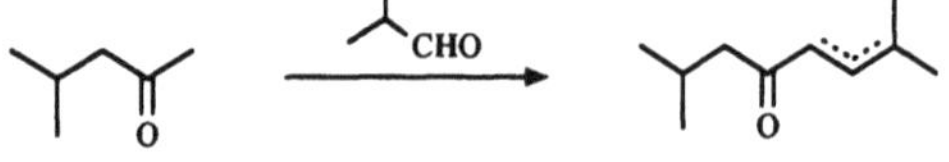

112

A more complicated chemical structure is the result of the conversion of alcohol **113** to the intermediate chloro compound **114**, which is the alkylating agent in the reaction with di-isopropyl ketone, finally leading to the unsaturated 2,4,4,7-tetramethylnona-6,8-diene-3-one (**115**), now commercialized under the name Labienone® (Givaudan). This ketone has a very natural agrestic fragrance of the lavender type, at the same time reminiscent of sage flowers.[92] Compound **115** is the parent ketone of the oxime (**19**) discussed in the fruity–floral section as an interesting odor substitute of the sulfurous blackcurrant note. This observation completes the circle traced by the main

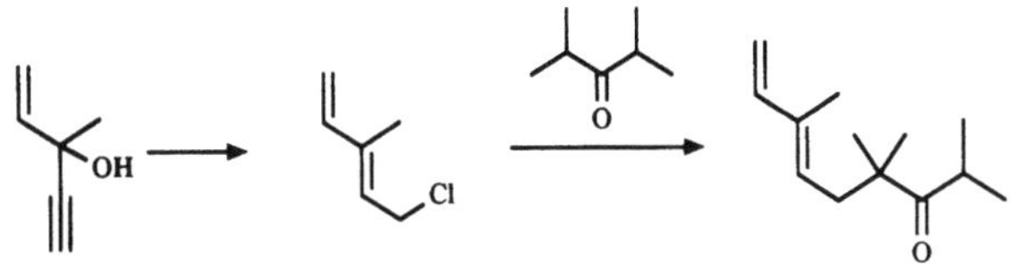

113 114 115

topics of this review, which has in general dealt with

(a) similar chemical structures leading to fragrances different in kind, and
(b) similar fragrance types provided by a great variety of chemical substances.

REFERENCES

Green

1. Morris, A. F., Perfumery notes: green notes. *Perf. & Flavorist,* **6**(1) (1981) 1.
2. Lamparsky, D., Grün-Noten aus der Sicht des Riechstoffchemikers, *Parf. & Kosm.,* **65** (1984) 5.
3. Lamparsky, D., Le Δ^3-carène: un hydrocarbure monoterpénique à l'origine de nouvelles matières odorantes synthétiques. In *Progress in Terpene Chemistry,* ed. D. Joulain. Editions Frontières, Gif s/Yvette, 1986, pp. 105–27.
4. US Patent 4 774 225 (priority CH 3 Nov. 1986) assigned to Roure.
5. US Patent 4 219 449 (priority GB 17 Feb. 1978) assigned to Polak's Frutal Works.
6. European Patent 11 877 (priority CH 5 Dec. 1978) assigned to Givaudan (see also US Patent 4 278 817 and others).
7. European Patent 52 115 (priority CH 22 May 1980) assigned to Firmenich.
8. US Patent 4 372 881 (priority CH 16 Oct. 1980) assigned to Firmenich (see also German Patent 3 140 666).
9. Thomas, A. F., 1,4-Dimethylcyclohex-3-enyl methyl ketone, a monoterpenoid with a novel skeleton. *Helv. Chim. Acta,* **56** (1973) 1800.
10. Kreiser, W., Haumesser, W. & Thomas, A. F., 1,4-Dimethylcyclohex-3-enyl methyl ketone. *Helv. Chim. Acta,* **57** (1974) 164.
11. US Patent 4 392 993 (priority CH 22 Aug. 1980) assigned to Firmenich (see also German Patent 2 418 335).
12. US Patent 4 147 672 (priority US 23 July 1976→CH 29 Jan. 1974) assigned to Firmenich.
13. US Patent 4 264 467 (division of US 4 147 672→CH 29 Jan. 1974) assigned to Firmenich.
14. US Patent 4 886 897 (priority US 18 Nov 1988) assigned to IFF.
15. US Patent 4 877 769 (priority US 28 Apr. 1989) assigned to IFF.
16. German Patent 3 212 326 (priority DE 2 Apr. 1982) assigned to Henkel.
17. European Patent Appln 340 326 (priority DE 5 May 1988) filed by Henkel.
18. US Patent 4 341 666 (priority CH 30 Mar. 1981) assigned to Firmenich.
19. US Patent 4 735 932 (priority US 22 May 1987) assigned to IFF.
20. US Patent 4 450 101 (priority US 20 Aug. 1981) assigned to IFF.
21. US Patent 4 435 316 (priority US 23 June 1983) assigned to IFF.

22. US Patent 4 601 835 (priority US 7 June 1985) assigned to IFF.
23. US Patent 4 772 411 (priority US 28 Jan. 1988) assigned to IFF.
24. US Patent 4 519 943 (priority US 26 Sept. 1983) assigned to IFF.
25. European Patent 154 844 (priority US 23 Feb. 1984) assigned to IFF (see also US Patent 4 539 143 and others).
26. US Patent 4 549 982 (priority 24 Aug. 1984) assigned to IFF.
27. German Patent 3 120 700 (priority CH 27 May 1980) assigned to Firmenich.
28. European Patent 66 684 (priority CH 21 May 1981) assigned to Firmenich.
29. US Patent 4 709 061 (priority DE 13 Nov. 1982) assigned to Dragoco (see also German Patent 3 242 042).
30. US Patent 4 683 083 (priority DE 13 May 1983) assigned to Dragoco (see also German Patent 3 317 476).
31. European Patent 43 507 (priority DE 3 July 1980) assigned to Henkel (see also US Patent 4 528 126).
32. European Patent 278 020 (priority DE 9 Feb. 1987) assigned to BASF (see also US Patent 4 707 302 and German Patent 3 603 661 with priority of 6 Feb. 1986).
33. US Patent 4 761 489 (priority DE 7 Nov. 1985) assigned to BASF (see also European Patent 221 445).
34. German Patent Appln 3 718 870 (priority DE 5 June 1987) filed by BASF.
35. European Patent Appln 287 954 (priority DE 24 Apr. 1987) filed by BASF.
36. European Patent Appln 279 954 (priority DE 19 Feb. 1987) filed by BASF.
37. European Patent Appln 291 849 (priority DE 19 May 1987) filed by BASF.
38. US Patent 4 491 537 (priority US 8 July 1983) assigned to IFF.
39. US Patent 4 458 699 (priority US 11 Feb. 1982) assigned to IFF.
40. European Patent Appln 326 959 (priority JA 28 Jan. 1988) filed by Kao Corp.
41. Martin, S. F. & Phillips, G. W., Homologation–alkylation of carbonyl compounds via regiospecifically generated metallo-enamines. *J. Org. Chem.*, **43** (1978) 3792.
42. European Patent Appln 358 091 (priority JA 1 Sept. 1988) filed by Kao Corp.
43. US Patent 4 619 780 (priority US 24 Aug. 1984) assigned to IFF.
44. German Patent 3 010 762 (priority GB 22 Mar. 1979) assigned to Bush, Boake, Allen Ltd.
45. European Patent 173 395 (priority NL 23 Aug. 1984) assigned to Naarden (see also US Patent 4 687 599).
46. van Lier, F. P., van der Linde, L. M. & van der Werdt, A. J. A., Isolation and synthesis of (*Z,Z*)-4,7-decadienal, the character impact compound in the oil of *Acorus calamus* L. In *Progress in Essential Oil Research*, ed. E.-J. Brunke. W. de Gruyter, Berlin, New York, 1986, pp. 215–25.
47. Lamparsky, D., Fingerprints in essential oil analysis: aim, results and correlations with sensory values. In *Capillary Gas Chromatography in Essential Oil Analysis*, ed. P. Sandra & C. Bicchi. Huethig-Verlag, Heidelberg, New York, 1987, pp. 155–213.
48. Marner, F. J., Boland, W. & Jaenicke, L., *Synthese isomerer 1,3,5,8-Undecatetraene. Ann.* (1982) 579.
49. European Patent 100 015 (priority CH 22 July 1982) assigned to Firmenich.
50. US Patent 4 515 711 (priority US 27 Sept. 1982) assigned to Givaudan (see also European Patent 104 602).
51. European Patent Appln 341 538 (priority DE 4 May 1988) filed by BASF.
52. US Patent 4 814 322 (priority DE 9 Apr 1987) assigned to Haarmann & Reimer (see also European Patent 285 980).
53. European Patent 86 945 (priority CH 27 Jan. 1982) assigned to Givaudan (see also US Patent 4 572 795).
54. European Patent 45 453 (priority CH 31 July 1980) assigned to Givaudan (see also US Patents 4 482 762 and 4 585 662).
55. European Patent 86 944 (priority CH 27 Jan. 1982) assigned to Givaudan.

56. German Patent 3 341 606 (priority DE 17 Nov. 1983) assigned to Consortium Elektrochem. Industrie (Wacker).
57. European Patent 34 334 (priority 13 Feb. 1980) assigned to Firmenich.
58. US Patent 4 613 457 (priority DE 25 Feb. 1983) assigned to Henkel (see also German Patent 3 306 560).

Herbaceous–Spicy

59. US Patents 4 235 805, 4 336 204 and 4 456 561 (priority GB 20 Mar. 1978) assigned to PFW (see also German Patent 2 910 579).
60. European Patent 92 031 (priority CH 15 Apr. 1982) assigned to Firmenich (see also US Patent 4 543 429 and others).
61. US Patent 4 474 992 (priority US 10 Mar. 1983) assigned to IFF.
62. Lamparsky, D., Perfumery notes: lavender. *Perf. & Flavorist,* **11**(4) (1986) 7.
63. Kaiser, R. & Lamparsky, D., New carbonyl compounds in the high-boiling fraction of lavender oil. *Helv. Chim. Acta,* **67** (1984) 1184 and earlier literature cited therein.
64. US Patent 4 307 122 (priority CH 13 Jan. 1978) assigned to Firmenich (see also German Patent 2 901 119).
65. German Patent 3 025 449 (priority CH 31 Aug. 1979) assigned to Firmenich.
66. Thomas, A. F., Thommen, W. & Becker, J., The configuration of tricyclo[7.1.1.0^{2,7}]undecanes and the stereochemistry of the reduction of Tricyclone. *Helv. Chim. Acta,* **64** (1981) 161.
67. European Patent 54 316 (priority CH 9 Dec. 1980) assigned to Firmenich.
68. European Patent 51 260 (priority DE 3 Nov. 1980) assigned to Henkel.
69. European Patent 4 978 (priority DE 26 Apr. 1978) assigned to Henkel.
70. European Patent 237 978 (priority DE 21 Mar. 1986) assigned to Henkel (see also US Patent 4 719 043).
71. US Patent 4 524 019 (priority CH 12 Aug. 1982) assigned to Firmenich (see also European Patent 100 935).
72. European Patent 105 157 (priority US 25 Aug. 1982) assigned to Givaudan (see also US Patent 4 519 944).
73. European Patent 200 890 (priority CH 19 Apr. 1985) assigned to Firmenich.
74. European Patent 151 521 (priority GB 1 Feb. 1984) assigned to Bush, Boake, Allen Ltd (see also US Patent 4 606 849).
75. European Patent Appln 334 005 (priority CH 25 Feb. 1988) filed by Firmenich.
76. European Patent 137 419 (priority CH 7 Oct. 1983) assigned to Givaudan.
77. European Patent 179 532 (priority NL 23 Oct. 1984) assigned to Naarden (see also US Patent 4 650 604).
78. German Patent Appln 3 718 201 (priority DE 29 May 1987) filed by BASF.
79. German Patent Appln 3 718 870 (priority DE 5 June 1987) filed by BASF.
80. US Patent 4 424 379 (priority US 16 Sept. 1981) assigned to IFF (see also US Patents 4 487 701 and 4 492 645).
81. European Patent 74 693 (priority NL 16 Sept. 1981) assigned to Naarden (Quest Int.).
82. US Patent 4 717 506 (priority JA 23 Jan 1985) assigned to Kao Corp.
83. US Patent 4 453 000 (priority CH 27 May 1981) assigned to Firmenich.
84. World Patent Appln 86 03 196 (priority DE 29 Nov. 1984) filed by Dragoco.
85. European Patent Appln 302 816 (priority 7 Aug. 1987) filed by Dragoco.
86. European Patent 115 278 (priority CH 26 Jan. 1983) assigned to Firmenich.
87. Wood, T. F., Easter, W. M., Jr, Carpenter, M. S. & Angiolini, J., Polycyclic musks I. Acyl- and dinitro-polyalkyltetralin derivatives. *J. Org. Chem.,* **28** (1963) 2248.

88. European Patent 178 532 (priority CH 18 Oct. 1984) assigned to Firmenich (see also US Patent 4 647 407).
89. Naef-Müller, R., Pickenhagen, W. & Willhalm, B., New irregular monoterpenes in *Artemisia vulgaris*. *Helv. Chim. Acta,* **64** (1981) 1424.
90. US Patent 4 447 627 (priority CH 11 June 1981) assigned to Firmenich (see also German Patent 3 221 876).
91. European Patent 141 569 (priority GB 25 Oct. 1983) assigned to BBA.
92. European Patent 45 861 (priority CH 8 Aug. 1980) assigned to Givaudan.

Outlook

PETER M. MÜLLER
GIVAUDAN Forschungsgesellschaft AG, CH-8600 Dübendorf, Switzerland

Research does not only drive developments, it is also driven by them—by political or economical changes, by evolution of management style or technology, by developments in consumer perceptions, and so on. All these factors are interdependent. Therefore, it is certainly difficult to make reliable predictions, and this outlook has to be regarded as sketchy or biased. Nevertheless, it is hopefully stimulating to read it.

ECOLOGY AND ASPECTS OF RAW MATERIAL PRODUCTION

It does not take a major prophetic talent to predict that mankind will more and more be concerned with ecological questions. Ecology even represents a 'new dimension' which economists try to take into account when they attempt to analyse, for example, the sum of values resulting from new investments in industrial production, or the total impact of decisions about specific production sites; governments have started to ask for environmental compatibility studies before approving the construction of new production facilities or the introduction of new products. However, ecology is a very complex topic—perhaps the most difficult field which science can approach today. It deals with 'the totality or patterns of relations between organisms and their environment'.[1]

But what has this to do with perfumery?

Well, perfumery with its aspects of art, science and technology is fully based on one of our senses, the sense of olfaction, and the senses are, of course, very generally guiding the 'relations between organisms and their environment'. Therefore it is not only possible, but even probable, that a science, namely that of the investigation of odorant chemicals from nature, which originally attempted nothing more than to achieve a better accessibility of precious raw materials, will contribute significantly to the understanding of ecologically relevant facts. This will become especially true if results from the study of odorants from plants (cf. Chapter 4) are merged with results from the study of semiochemicals (Chapter 3). Roman Kaiser touches briefly on this point in Chapter 7 of this book.

There is a second point of view which relates perfumery and ecology: the masking of odors. Clearly, this has resulted in the so-called 'functional perfumery' and has become an important aspect since, in the last century, the fragrance industry started to provide us with sufficient amounts of its products. Today, our lifestyle takes advantage of this development to quite an extent: we work indoors (often in small offices), we meet in crowded meeting rooms or convention halls, etc., and certainly we would not want to miss the perfuming of deodorants, for example, or the absence of malodors in many consumer products. This latter point has become so obvious that we have almost forgotten that the masking of disagreeable base odors is often the reason for perfuming. Thus, certain ecology-conscious consumer groups have started to prefer non-perfumed versions—often without realizing that the production of odorless bases (i.e. matrices) is in many cases ecologically really demanding and unfavorable if compared with simpler, but perfumed, products which cause much less of an ecological burden overall. A greater in-depth understanding of this kind of consideration is to be expected for the future, and this will also influence consumer attitudes.

The third ecology-related aspect to be discussed concerns synthetic raw materials. It can actually be dealt with quite shortly, as the total annual world production of aroma chemicals could in theory easily be loaded in one single supertanker.

Nevertheless, future research will focus on two related points: the increasing importance of environment-friendly syntheses of aroma chemicals, and the increasing attraction of naturals and of their biotechnological production.

There will be two principal biotechnological routes to improved yields of essential oils. The first will be controlled only indirectly by the fragrance industry. It concerns the strain improvement, by classical and non-classical methods, of plants yielding natural raw materials (essential oils, etc.). The second is related to the first, and is also based essentially on strain improvement. Here, however, the goal is finally to produce the aromatic oils in cell- or organ-cultures in fermenters rather than in the field (cf. Chapter 19). Both approaches will force the perfumer to learn to work with essential oils of a somewhat changed character in comparison with the products used traditionally.

OLFACTION AND PERCEPTION OF ODORS AND PERFUMES

Olfaction can be studied at many levels, and this book has obviously shown that none of them can be regarded as being fully understood. The easiest is probably the level of signals, i.e. the chemical composition of odors; but even here, not only is the study of natural scents (for example) instrumentally demanding, but it also leads to many questions as soon as time dependence or biological significance is considered (cf. Chapter 7). Nevertheless, it can be predicted that this field of investigation will develop in an evolutionary rather than a revolutionary sense.

This is less clear for the next level: the transport of the chemical signal to the receptor. Chapter 5 deals extensively with this point, and the reader is therefore referred to it. It will only be added here that the role of the so-called olfactory binding protein will probably be much better understood in the not-too-distant future.

The next level, that of ligand–receptor interaction, has the potential for fast progress in the near future—in spite of the fact that not even the number of receptor types is known and that there are still authors who have their doubts about the existence of olfactive receptors in the sense of classical membrane proteins (cf. Chapter 18). Thus, the situation is quite different from that in color vision research, where the correct number of receptor types was known long before its experimental confirmation at the molecular level.[2] The rapid progress expected will, of course, be based mainly on advances of molecular biology and cloning techniques (cf. Chapter 18), but the 'pharmacological approach to olfaction' will also have its impact, and this latter prediction will probably equally hold for the subsequent levels. They are represented by the generation of action potentials, neurotransmission and processing of the electrical signals arriving in the CNS (central nervous system).

The above-mentioned 'pharmacological approach to olfaction' concerns, in essence, techniques which have been developed for the investigation of therapeutics and which will be (or are) modified and applied to olfaction. Examples are as follows.

—*As a basic tool*: the measuring of dose/response curves of single (olfactively pure) odorants by use of olfactometric techniques or, perhaps, by use of 'secondary effects' such as 'brain waves' (cf. Chapter 6).

—*As a key step*: the indirect determination of the number of receptor types (or basic odor types) on the grounds of the comparison of dose/response curves, the observation of the development of tolerance (habituation) or cross-tolerance, and the exploitation of the individuality of panelists (correlating groups of individuals and their relative sensitivity for groups of odorants; this can be done by measuring odor perception threshold values in untrained subjects[3] or by monitoring descriptors and intensity estimates used by trained subjects).

—*As an investigation* oriented towards the understanding of the mechanism of signal transmission: the determination of the influence of specific drugs on odor perception threshold values or dose/response curves.

—*As an application-oriented dimension*: the investigation of real synergies of pairs of odorants (reliable sensory measurements at suprathreshold concentrations will probably lead to the breakthrough in this as-yet-contradictory field).

Of course, the individuality in odor perception is based on genetic dissimilarities. It has been discussed above in relation to 'counting receptors', but it has additional implications. One concerns the perfumer's work and affects marketing as well. It will be discussed here shortly by describing an approach which we have been pursuing for some time (Pesaro, M. & Decazes, J., Givaudan Forschungsgesellschaft AG, CH-8600 Dübendorf, unpublished). This approach consists in the independent description of given olfactively pure odorants by a number of perfumers who use a thesaurus of about 30 descriptive terms and a set of three intensity ratings. The results obtained with six perfumers and nine more-or-less arbitrarily selected substances (**1** to **9** in Fig. 1) are illustrated in Fig. 2. This latter combines in groups the five descriptors most frequently used by the perfumers for the description of the complete set of substances which are mentioned along with their main odor note below the respective bars. The height of the fractions adding up to the bars corresponds to the intensity rating (the higher the rating, the higher the fraction), and each type of shading corresponds to one specific perfumer.

The following conclusions should be mentioned:

—There are descriptors for which there are very obvious prototypes (e.g. **7** for 'earthy' or **2** and **3** for 'woody').

—The perfumers generally agree very well with regard to the main descriptors (prototypes), even though there are exceptions.

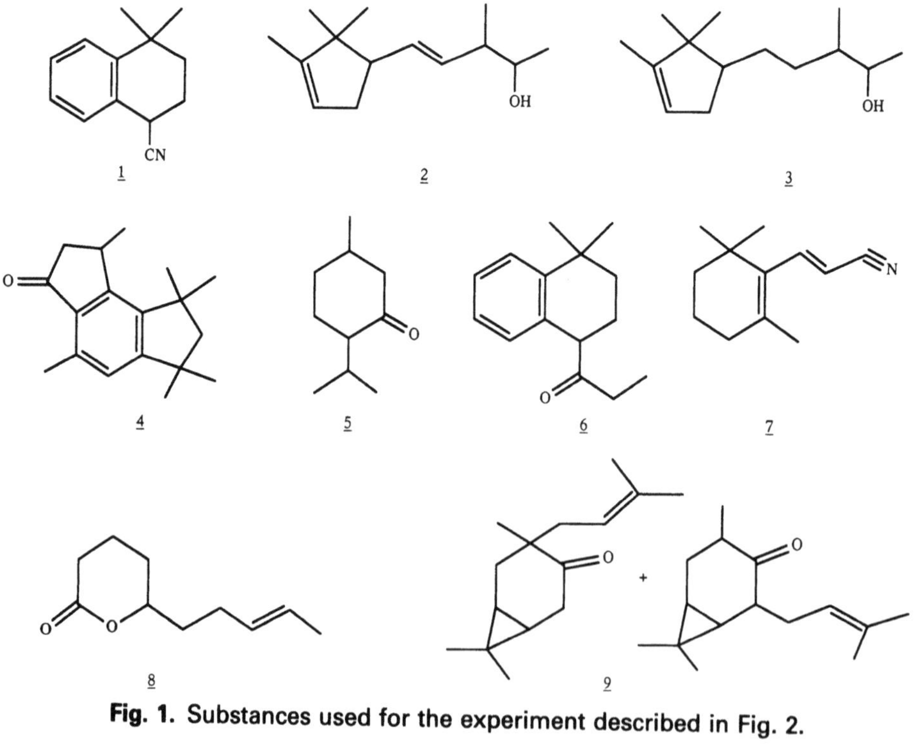

Fig. 1. Substances used for the experiment described in Fig. 2.

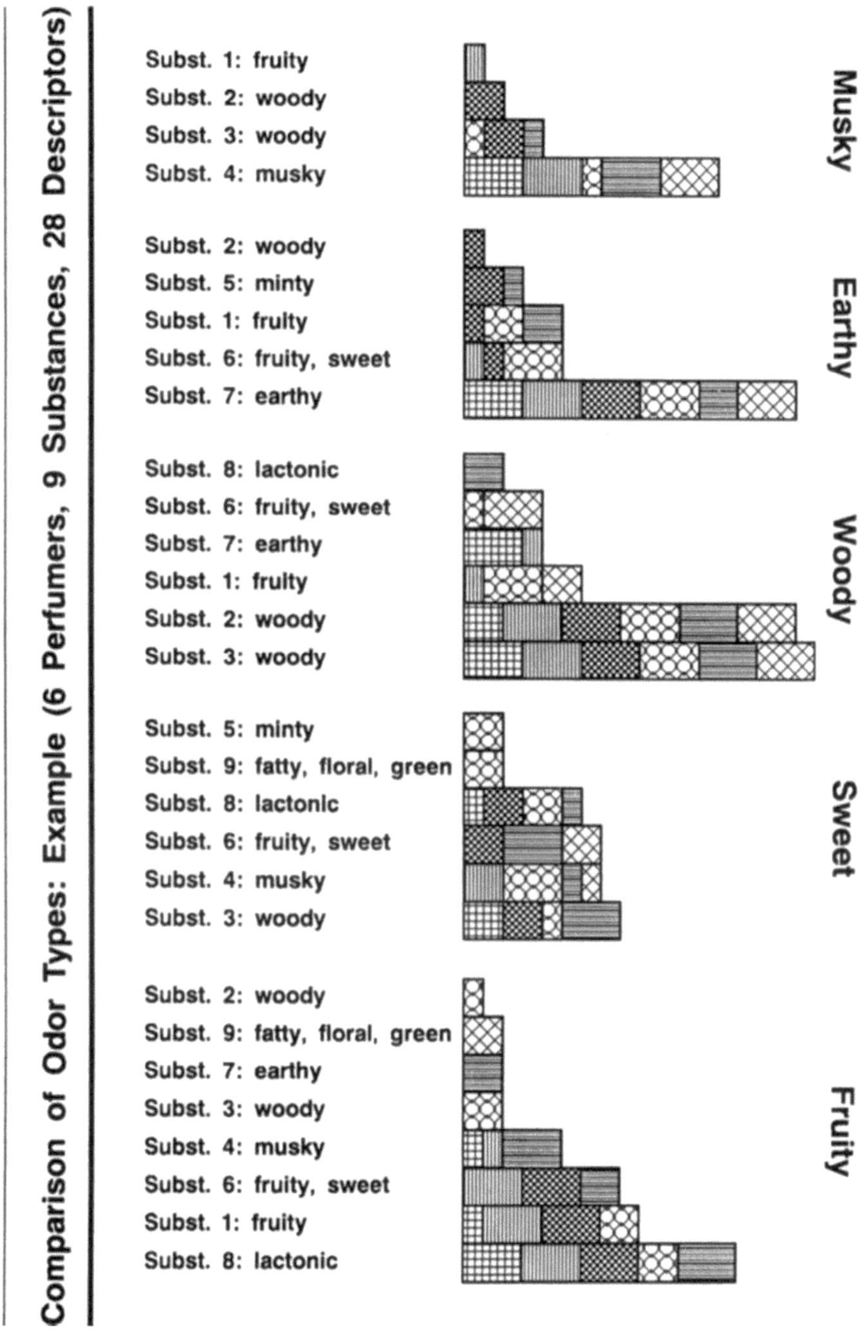

Fig. 2. Comparison of odor types; example with six perfumers, nine substances and 28 descriptors.

—There is quite a bit of disagreement with regard to 'secondary odor notes'. This might be very valuable in view of the determination of individual sensitivity differences.
—There are descriptors which are seemingly close like 'fruity' and 'lactonic', and descriptors like 'sweet' for which there are no prototypes (of course, results obtained with a larger number of substances would have to be presented in order to prove this statement unequivocally). These cases might actually represent odor descriptors belonging to a common receptor (the former), or lacking any specific receptor counterpart (the latter).

Obviously, it is very valuable to perform this kind of controlled (blind!) study. It not only contributes to scientific progress, but it also allows perfumers to characterize their own olfactive sensitivity very precisely in comparison with that of their colleagues or even with that of the average human being. Therefore, this simple approach will become more important in the future.

The step from the study of the perception of single odorants to that of mixtures significantly complicates the situation—in spite of the fact that even single odorants can have many facets. Therefore, this short outlook must remain incomplete and the reader is referred to Chapter 5 or, for example, to a recently published monograph on the perception of complex smells and tastes.[4] Only one aspect will be discussed here, namely the finding that receptors seemingly differ with regard to their selectivity. Insect pheromones are the obvious prototypes of odorants which, even as minor components of complex mixtures (or odors), are specifically perceived by a dedicated receptor. It is unclear whether evolution has retained up to the human level a remnant of this type of specific receptor, yielding upon activation a strongly accentuated pattern of neuronal responses which almost acts as an alarm. Of course, it would be highly interesting to have the answer to this question.

At the psychophysical level there is no difference between the perception of complex odors and perfumes. There is a new dimension only if one takes into account that perfumes are man-made creations which are meant to be experienced emotionally and/or intellectually. Considering this, one is immediately expanding the meaning of 'perception' and including, for example, an appreciation of the perfumes' compository principles. Today, these principles are mainly understood in terms of associations on the one hand and in terms of 'diffusivity', 'linearity', 'tenacity', 'richness' and 'accords' on the other. They will be supplemented or very much refined in the future when olfaction and its receptors are thoroughly understood. A dramatic development could then result, comparable with the step from classicism to impressionism in painting.

APPLICATION CONCEPTS

Application concepts are discussed in Part V of this book. Here, a few thoughts will be added and a few trends commented upon.

The fine fragrance segment will, of course, remain the segment of 'pure art' and always imply the largest degree of creative freedom. Consequently, application concepts refer in this context rather to the presentation and the distribution of perfumes than to new technical solutions involving them. New sub-groups of the population will be addressed and strongly contrasting product types will represent the cornerstones of the market. This will be reflected in the flasks, and one will find on the one hand little throw-away perfume containers which look like cigarette-lighters, and on the other hand expensive pieces of fine art which are hand-signed and produced as limited editions.

Furthermore, sophisticated two-compartment flasks might appear, containing carefully equilibrated subcompositions and allowing the user freely to adjust the perfume to his (or her) mood. Logically, extreme ease of handling is a prerequisite for success of this concept. Perfume-related technical studies might involve performance as a function of skin type or factors which minimize the development of tolerance (i.e. habituation).

'Mood fragrance' has become a buzz-word referring to fine fragrances as well as cosmetics and toiletries. Much has been published and a number of terms have been used or invented in this context: aromatherapy, osmotherapy, aromachology, etc. Folk medicine know-how has been consulted and sophisticated psychological and even pharmacological or 'in-vitro' experiments have been performed. Some were done in a methodologically clever way, and others are more than doubtful. Many have disregarded the problem that dosage under conditions of use is very hard to determine (cf. Chapter 16), and the point that 'smelling' is not 'inhaling' in the sense of a pharmacological administration. The list of open questions is endless, and the attraction of the field is accordingly high.

It cannot be the purpose of this short outlook to cope with the questions mentioned in the preceding paragraph. Thus, the reader will just be reminded of the strong associative power of odors and perfumes and of the strong 'feelings' and real physiological effects which can be triggered by surprising associations. Perfumers have known this all along, and the future will, hopefully, present us with more studies which are approaching this point in a subtle and scientifically sound way. Market research is not the least of the disciplines which can contribute to such a development.

A large number of new conceptual approaches can be envisaged for functional perfumery, which in a broad sense includes the segments of cosmetics and toiletries, soaps and detergents, and household and industrial products. Related research topics concern perfume release and distribution, affinity to support materials (matrices), microencapsulation and perfuming of plastics, odorant efficacy, chemical stability in aggressive media, etc. It will not be attempted here to predict where the major breakthroughs will occur: rather, it will be emphasized that the performance of perfumes in all the possible applications depends on their composition and on the characteristics of the perfumed product. A large number of empirical data have, thus, been collected in fragrance houses and made accessible by intelligently designed software accessing large data files. These files will continue to grow in the future, but more and more will it also become possible to select the relevant physicochemical and sensory parameters and to make predictions about the performance of mixtures on the grounds of measurements obtained with single components and on the grounds of their chemical structure.

REFERENCES

1. Merriam-Webster, A., Webster's Seventh New Collegiate Dictionary, G. & C. Merriam Company, Publishers, Springfield, Massachusetts, USA, 1971 (entry 2 under 'ecology', page 262).
2. Nathans, J., *Sci. American,* **260** (Feb. 1989) 28–35.
3. Müller, P. M., *Proc. 11th International Congress of Essential Oils, Fragrances and Flavors,* Vol. 2. Oxford and IBH Publishing Co. Pvt. Ltd, New Delhi, 1989, pp. 25–39.
4. Laing, D. G., Cain, W. S., McBride, R. L. & Ache, B. W., eds, *Perception of Complex Smells and Tastes.* Academic Press, Australia, Marrickville, NSW, 1989.

List of Contributors

Alexander Boeck joined the HENKEL Company, Düsseldorf, after finishing his studies of chemistry and his Ph.D. thesis in 1971. His career in this company started in the R&D department for household products and toiletries. He then spent two years in France and came back to attain his current position as Director and head of HENKEL's Fragrance centre.

Rodney Croteau earned his B.S. and Ph.D. at the University of Massachusetts, Amherst, and is currently Professor of Biochemistry at the Institute of Biological Chemistry, Washington State University. His research interests have centred on the biochemistry and function of natural products, with a particular focus on compounds of flavour and perfumery significance. The latter interest has led to a long-term association with the firm of HAARMANN & REIMER.

Julian Dorsky received his B.S. from Brooklyn College, M.S. from Pennsylvania State University and Ph.D from Purdue University. Before retiring he was corporate Vice President and Director of Research and Development for GIVAUDAN in New Jersey. During his long career in the fragrance supply industry he has been involved in research, development and production of many new and classical synthetic perfumery chemicals.

Charles Ehret received his advanced education in chemistry and his Ph.D. at the Laboratory of Chemistry of Natural Products of the University of Strasbourg, France, where he worked with Professor Ourisson. He spent postdoctoral years with Professor Kutney at the University of British Columbia in Vancouver and joined ROURE's research centre in Grasse in 1970. Now in the position of Research Group Manager for Natural Products and Biotechnology, he is deeply engaged in the investigation of natural products for perfumery.

Jean-Claude Ellena entered the field of perfumery in the Société CHIRIS in Grasse, France. He then went to Geneva where he received his professional education as the first student of the Givaudan Perfumery School. Eventually he became perfumer and a successful fine-fragrance senior perfumer and spent a year in the USA. In 1986 Jean-Claude Ellena took on the position as Director of Creation with GIVAUDAN in Paris.

Franz Etzweiler studied chemistry at the Technical College in Winterthur and supplemented his education at the Swiss Federal Institute of Technology in Lausanne. His professional career led him via a stay at the BATTELLE INSTITUTE in Geneva to GIVAUDAN in Dübendorf. There he took over responsibility for the engineering aspects of a large number of sensitive instruments which were developed in-house. Today, Franz Etzweiler is a Research Fellow in Norbert Neuner-Jehle's department. He heads the olfactometry and specialized headspace technology laboratory.

Charles Fehr earned his Ph.D. at the Swiss Federal Institute of Technology in Zürich and completed his scientific education with postdoctoral stages with Professor Oppolzer in Geneva and Professor Posner at the Johns Hopkins University in Baltimore. He joined FIRMENICH in 1978 where his interests as Senior Researcher are centred around novel synthetic methodologies and strategies and around studies of structure–activity relationships. In 1990 he won the Ruzicka award of the Swiss chemical society.

Hans-Udo Fergen started his career in the consumer goods industry in the field of manufacturing cosmetics. After having worked with BOEHRINGER Ingelheim, MERCK AG Darmstadt and BLENDAX Mainz, he joined HAARMANN & REIMER in Holzminden in 1975. Here, he became interested in rationalization, automation and computer-integrated manufacturing in fragrance compounding. This know-how and his chemical engineering degree led him to his current position as Director of Production for the Fragrance Division.

Richard A. Ford studied chemistry at the Oklahoma State and Wayne State Universities and received his Ph.D. in organic chemistry in 1967. He then did postdoctoral studies at Yale and spent several years as Assistant Professor at the Catholic University. His direct involvement in the field of flavours and fragrances started in 1969 when he became a Scientific Affairs Consultant of FEMA. Later he became a Food Chemical Consultant with Bernard Oser Associates and a Member of the Expert Panel of FEMA, and since 1982 he has been the President of the industry's Research Institute for Fragrance Materials (RIFM), which has a very significant responsibility regarding the safety aspects of perfumes. Richard A. Ford is a member of many professional societies and author of an impressive number of scientific publications.

Wittko Francke studied chemistry and earned his Ph.D. in 1973 with Professor K. Heyns at the University of Hamburg. His research interest in semiochemicals, insect pheromones and chemical signalling between plants and insects led to his incumbency in 1979 at the same university. Today, Wittko Francke is Professor of Organic Chemistry and author of a large number of scientific publications.

Georg Fráter obtained his Ph.D. working with Professor H. Schmid at the University of Zürich and spent postdoctoral years in Leiden, The Netherlands, and Edmonton, Canada. After more than ten years of agrochemical research with HOFFMANN–LA ROCHE, he joined GIVAUDAN in Switzerland and became head of its synthetic chemistry research. Georg Fráter has authored a large number of publications and contributed significantly to modern synthetic methodology.

John K. Funesti earned his B.S. in Chemistry at Fairleigh Dickinson University (1960), while working as a perfumer at Lever Brothers Company. Subsequently, he had Senior Managerial and Chief Perfumer positions with Givaudan and Lever Brothers Company and was Vice-President and Technical Director for Robertet Inc. He has had his own perfuming consulting business under the name of Y & F Creative Services. Today, he is Vice President, Perfumery of FLAVORSCENTS, Inc. He was President of the American Society of Perfumers twice (1974 and 1989), and he will be co-Chairman of the American delegation for the World Perfumery Congress 1991 in Palma de Mallorca.

Thomas V. Getchell, Ph.D., is a Professor of Physiology and Biophysics and Associate Dean for Research and Basic Sciences at the University of Kentucky College of Medicine. His studies have focused on the integrated cellular and molecular neurobiological activity of the olfactory sensory mucosa and olfactory receptor neurons. He is a recipient of the Senator Jacob K. Javits Neuroscience Investigator Award from the National Institutes of Health and was a TAKASAGO Lecturer in 1990.

Avery N. Gilbert studied psychology and biology at the Universities of California, Berkeley, and Pennsylvania, and he received his Ph.D. from the latter in 1982. He then spent several years at the MONELL CHEMICAL SENSES CENTER in Philadelphia, at first as postdoctoral Fellow and later as Assistant Member of the institute's faculty. Since 1988 he has been Director Olfactory Science at ROURE Inc. in the USA. Avery N. Gilbert is a member of the Scientific Affairs Committee of the Fragrance Materials Association and Chairperson of the Industry Advisory Committee of the Fragrance Research Fund Ltd. He has also received a respectable number of research grants and is member of several professional societies as well as author of many scientific papers. His co-authorship in the *National Geographic's* Smell Survey, the largest participatory experiment of its kind, deserves mention.

J. Stephan Jellinek obtained a practical foundation as a perfumer while receiving his education as an organic chemist (Ph.D., University of Wisconsin, 1955). He started his career as a research and creative perfumer and continued in marketing research and management positions with the GENERAL FOODS Corporation in the USA. Today he is International Marketing Manager at DRAGOCO. J. Stephan Jellinek is author of several books and of many publications in the area of perfumery.

Roman Kaiser studied chemistry at the Technical College in Winterthur, Switzerland. He completed his education by working as a Research Assistant in the Physicochemical Group of the same institute and joined GIVAUDAN's research department in 1968. For the last fifteen years, he has concentrated on the study of natural scents, especially those of flowers, by headspace technology. He became a pioneer of its application to perfumery. He is a Senior Research Fellow and a recipient of the Roche Research Prize.

Morley R. Kare was Professor of Physiology at the University of Pennsylvania and the Director of the MONELL CHEMICAL SENSES CENTER in Philadelphia, which is the world's most important institution focusing on basic research in olfaction and taste. Morley R. Kare has authored over two hundred publications and participated as editor or contributor to about fifty books. His main research interest was in the area of the comparative aspects of the chemical senses and in particular their function in nutrition. We deeply regret that Morley R. Kare died on July 30, 1990, just before the present book went into production.

Frank Karp holds a B.S. from the City College of New York and an M.S. from New York University. Currently a Senior Research Technologist at the Institute of Biological Chemistry, Washington State University, his specialty is enzymology, with a particular interest in natural products metabolism and cytochrome P-450-dependent reactions.

Dietmar Lamparsky obtained his Ph.D. from the Free University of Berlin in 1955. After a few years at the BATTELLE INSTITUTE in Frankfurt he joined GIVAUDAN in Switzerland in 1958 where he eventually became head of the Analytical Chemistry Department and one of the most productive researchers in the field of the chemistry of constituents of fragrant plants as well as of synthetic odorants. He is now retired, but still active as a consultant for the company.

Milton Manowitz received his Ph.D. in biochemistry from Georgetown University in 1954; the same year, he accepted the position as a Senior Research Chemist with

GIVAUDAN, New Jersey, USA. He was employed by GIVAUDAN for 35 years, attaining the position of Senior Vice-President R&D. His research activities and responsibilities have been closely involved with fragrance chemicals and processes, natural flavours, UV absorbers, industrial biocides, toxicology and government regulations. Since his retirement in 1989 he has been active as a consultant for the company.

Frank L. Margolis obtained his Ph.D. from Columbia University and has been working in olfaction for nearly 20 years. He has published a large number of papers and has edited, with Thomas V. Getchell, a book on *Molecular Neurobiology of the Olfactory System*. Currently he heads the Laboratory of Chemosensory Neurobiology at the ROCHE INSTITUTE FOR MOLECULAR BIOLOGY in Nutley, New Jersey, USA.

Peter M. Müller was awarded his Ph.D. in chemistry at the Swiss Federal Institute of Technology in Zürich. After two years at the Massachusetts Institute of Technology he joined HOFFMANN-LA ROCHE in Switzerland, where he worked for 10 years—initially in medicinal chemistry and later as head of an interdisciplinary research group. In 1985 he became the head of GIVAUDAN's world-wide research. One of his interests concerns the possibilities of applying theoretical concepts from pharmaceutical research to the field of olfaction.

Richard E. Naipawer obtained his Ph.D. in organic chemistry from Rutgers University in New Jersey, USA. Today, in a Director of Research position, he looks back on more than 20 years of research with GIVAUDAN in Clifton, New Jersey. The development of the 'sandalwood rule' illustrates his early interest in structure–odour relationships. He is a Fellow of the American Institute of Chemists and a Life Member of Phi Lambda Upsilon.

Norbert Neuner-Jehle took his Ph.D. at the University of Göttingen in 1966. He spent two further years in Göttingen and in Darmstadt (Technische Hochschüle), and he built up the mass spectrometry laboratory of this second university. He joined GIVAUDAN in Switzerland in 1968, and today he heads its Analytical and Physicochemical Research Department. Analytical and sensory studies of odours are at the centre of Norbert Neuner-Jehle's research interests and, together with Franz Etzweiler, he has developed a series of highly sensitive measuring techniques.

Günther Ohloff was awarded his Ph.D. by the Technische Hochschule in Dresden. He started in industrial R&D with SCHIMMEL & Co., Miltitz near Leipzig, and continued with DRAGOCO in Holzminden. After a second phase in academia (Tutorial Fellow at the Max Planck Institute in Mülheim/Ruhr) he joined FIRMENICH in Geneva in 1963 and eventually became its Corporate Vice-President Scientific Research. The awards he received (Ruzicka Award of the Swiss Federal Institute of Technology; the Ernest Guenther Award of the American Chemical Society; the Hackforth–Jones Award for outstanding work in perfumery by PPL, Ashford, Kent; the Otto-Wallach-Plakette of the German Chemical Society); and the R. H. Wright Award in olfactory research; the Dr *honoris causa* from the University of Geneva and the honorary duties he took on such as the Presidency of the Swiss Chemical Society as well as a number of visiting professorships, illustrate the outstanding nature of Günther Ohloff's scientific career. Now he has retired, but still works hard to continue his long list of publications.

Martin Petrzilka studied chemistry, and earned his Ph.D. at the Swiss Federal Institute of Technology in 1973. He then spent a year as Research Fellow in Professor Corey's group at Harvard University and several years as *chargé de recherche* at the University of Geneva. In 1979 he joined HOFFMANN–LA ROCHE's central research department in Basel where he first focused his interest on synthetic methodology and research on liquid crystals, and where he then progressed to more managerial positions enlarging the scope of his scientific work. Since 1987 Martin Petrzilka has headed ROURE's research centre in Grasse, France.

Edmond Roudnitska, born in 1905 in Nice, France, has played a decisive part in our century's history of perfumery, and he looks back today on many decades of success, culminating in the creation of an outstanding series of really fine fragrance classics. Starting with the FRAGRANCE FOUNDATION in 1926, he worked essentially as a freelance perfumer. In 1946 he founded ART & PARFUM at Cabris near Grasse. Edmond Roudnitska dedicated much of his energy to the study of general aesthetic principles and to their meaning for perfumery, and his first publications in this field date back as far as 1938.

Benno Streschnak started his professional career with HAARMANN & REIMER, Holzminden, after having received his Ph.D. degree in chemistry in 1955. He was working in research and later in perfumery and its application. In 1961 he left H & R and joined HENKEL KG a. A., becoming responsible for their perfumery, which he had to build up. He eventually attained the position of Department Director and retired in 1987, but he is still active as a consultant—recently, for the AZUR FRAGRANCES Company.

Maurice Thiboud started as a trainee in creative perfumery at BRUNO COURT SA in Grasse, France. His career continued as Perfumer and Senior Perfumer positions at LASERSON & SABETAY in Paris and MASCHMEIJER in Amsterdam. He joined GIVAUDAN in Geneva in 1958 and stayed there until 1989, attaining the position of Group Vice-President and Group Chief Perfumer. Now 'retired', he is still active part-time and enjoying the position as Head of the Givaudan Perfumery Training Centre.

Beat Winter studied chemistry at the Swiss Federal Institute of Technology in Zürich and earned his Ph.D. from the University of Geneva, where he worked in Professor Schaffner's group until 1976. He then spent postdoctoral years at Ohio State University and at the University of California in Berkeley. In 1979 Beat Winter joined FIRMENICH in Geneva, where his main research interests are focused on the design and synthesis of new odorant substances and on structure–activity relationships.

Index

MIX
Papier aus verantwortungsvollen Quellen
Paper from responsible sources
FSC® C105338

If you have any concerns about our products,
you can contact us on
ProductSafety@springernature.com

In case Publisher is established outside the EU,
the EU authorized representative is:
Springer Nature Customer Service Center GmbH
Europaplatz 3, 69115 Heidelberg, Germany

Printed by Libri Plureos GmbH
in Hamburg, Germany